Tharwat F. Tadros

Applied Surfactants

Further Titles of Interest

L. L. Schramm

Emulsions, Foams, and Suspensions

Fundamentals and Applications

2005

ISBN 3-527-30743-5

E. Smulders

Laundry Detergents

2002

ISBN 3-527-30520-3

H. M. Smith (Ed.)

High Performance Pigments

2002

ISBN 3-527-30204-2

W. Herbst, K. Hunger

Industrial Organic Pigments

Production, Properties, Applications

Third, Completely Revised Edition

2004

ISBN 3-527-30576-9

G. Buxbaum, G. Pfaff (Eds.)

Industrial Inorganic Pigments

Third, Completely Revised and Extended Edition

2005

ISBN 3-527-30363-4

K. Hunger (Ed.)

Industrial Dyes

Chemistry, Properties, Applications,

2003

ISBN 3-527-30426-6

Tharwat F. Tadros

Applied Surfactants

Principles and Applications

WILEY-VCH
WILEY-VCH Verlag GmbH & Co. KGaA

Author

Prof. Dr. Tharwat F. Tadros
89 Nash Grove Lane
Wokingham
Berkshire RG40 4HE
United Kingdom

1st Edition 2005
1st Reprint 2006

Library of Congress Card No.: Applied for

British Library Cataloguing-in-Publication Data: A catalogue record for this book is available from the British Library

Bibliographic information published by Die Deutsche Bibliothek
Die Deutsche Bibliothek lists this publication in the Deutsche Nationalbibliografie; detailed bibliographic data is available in the Internet at 〈http://dnb.ddb.de〉

Printed in the Federal Republic of Germany
Printed on acid-free paper

Composition Asco Typesetters, Hong Kong
Printing Strauss GmbH, Mörlenbach
Bookbinding Litges & Dopf Buchbinderei GmbH, Heppenheim

ISBN-13: 978-3-527-30629-9
ISBN-10: 3-527-30629-3

Dedicated to our Grandchildren
Nadia, Dominic, Theodore and Bruno

Contents

Applied Surfactants: Principles and Applications. Tharwat F. Tadros

ISBN: 3-527-30629-3

Preface

Surfactants find applications in almost every chemical industry, such as in detergents, paints, dyestuffs, paper coatings, inks, plastics and fibers, personal care and cosmetics, agrochemicals, pharmaceuticals, food processing, etc. In addition, they play a vital role in the oil industry, e.g. in enhanced and tertiary oil recovery, oil slick dispersion for environmental protection, among others. This book has been written with the aim of explaining the role of surfactants in these industrial applications. However, in order to enable the chemist to choose the right molecule for a specific application, it is essential to understand the basic phenomena involved in any application. Thus, the basic principles involved in preparation and stabilization of the various disperse systems used – namely emulsions, suspensions, microemulsions, nano-emulsions and foams – need to be addressed in the various chapters concerned with these systems. Furthermore, it is essential to give a brief description and classification of the various surfactants used (Chapter 1). The physical chemistry of surfactant solutions and their unusual behavior is described in Chapter 2. Particular attention was given to surfactant mixtures, which are commonly used in formulations. Chapter 3 gives a brief description of the phase behavior of surfactant solutions plus a description of the various liquid crystalline phases formed. The adsorption of surfactants at the air/liquid and liquid/liquid interface is described in Chapter 4, with a brief look at the experimental techniques that can be applied to measure the surface and interfacial tension. The adsorption of surfactants on solid surfaces is given in Chapter 5, with special attention given to the adsorption of polymeric surfactants, which are currently used for the enhanced stabilization of emulsions and suspensions. The use of surfactants for preparation and stabilization of emulsions is described in Chapter 6, paying particular attention to the role of surfactants in the preparation of emulsions and the mechanisms of their stabilization. The methods that can be applied for surfactant selection are also included, as is a comprehensive section on the rheology of emulsions. Chapter 7 describes the role of surfactants in preparation of suspensions and their stabilization, together with the methods that can be applied to control the physical stability of suspensions. A section has been devoted to the rheology of suspensions with a brief description of the techniques that can be applied to study their flow characteristics. Chapter 8 describes the role of surfactants in foam formation and its stability. Chapter 9 deals with the role of surfactants in formation and stabilization of nano-emulsions – the latter having recently been

Applied Surfactants: Principles and Applications. Tharwat F. Tadros

ISBN: 3-527-30629-3

applied in personal care and cosmetics as well as in health care. The origin of the near thermodynamic stability of these systems is adequately described. Chapter 10 deals with the subject of microemulsions, the mechanism of their formation and thermodynamic stability, while Chapter 11 deals with the topic of the role of surfactants in wetting, spreading and adhesion. The surface forces involved in adhesion between surfaces as well as between particles and surfaces are discussed in a quantitative manner.

Chapters 12 to 15 deal with some specific applications of surfactants in the following industries: personal care and cosmetics, pharmaceuticals, agrochemicals and the food industry. These chapters have been written to illustrate the applications of surfactants, but in some cases the basic phenomena involved are briefly described with reference to the more fundamental chapters. This applied part of the book demonstrates that an understanding of the basic principles should enable the formulation scientist to arrive at the optimum composition using a rational approach. It should also accelerate the development of the formulation and in some cases enable a prediction of the long-term physical stability.

In writing this book, I was aware that there are already excellent texts on surfactants on the market, some of which address the fundamental principles, while others are of a more applied nature. My objective was to simplify the fundamental principles and illustrate their use in arriving at the right target. Clearly the fundamental principles given here are by no means comprehensive and I provide several references for further understanding. The applied side of the book is also not comprehensive, since several other industries were not described, e.g. paints, paper coatings, inks, ceramics, etc. Describing the application of surfactants in these industries would have made the text too long.

I must emphasize that the references given are not up to date, since I did not go into much detail on recent theories concerning surfactants. Again an inclusion of these recent principles would have made the book too long and, in my opinion, the references and analysis given are adequate for the purpose of the book. Although the text was essentially written for industrial scientists, I believe it could also be useful for teaching undergraduate and postgraduate students dealing with the topic. It could also be of use to research chemists in academia and industry who are carrying out investigations in the field of surfactants.

Berkshire, January 2005 *Tharwat Tadros*

1
Introduction

Surface active agents (usually referred to as surfactants) are amphipathic molecules that consist of a non-polar hydrophobic portion, usually a straight or branched hydrocarbon or fluorocarbon chain containing 8–18 carbon atoms, which is attached to a polar or ionic portion (hydrophilic). The hydrophilic portion can, therefore, be nonionic, ionic or zwitterionic, and accompanied by counter ions in the last two cases. The hydrocarbon chain interacts weakly with the water molecules in an aqueous environment, whereas the polar or ionic head group interacts strongly with water molecules via dipole or ion–dipole interactions. It is this strong interaction with the water molecules that renders the surfactant soluble in water. However, the cooperative action of dispersion and hydrogen bonding between the water molecules tends to squeeze the hydrocarbon chain out of the water and hence these chains are referred to as hydrophobic. As we will see later, the balance between hydrophobic and hydrophilic parts of the molecule gives these systems their special properties, e.g. accumulation at various interfaces and association in solution (to form micelles).

The driving force for surfactant adsorption is the lowering of the free energy of the phase boundary. As we will see in later chapters, the interfacial free energy per unit area is the amount of work required to expand the interface. This interfacial free energy, referred to as surface or interfacial tension, γ, is given in mJ m^{-2} or mN m^{-1}. Adsorption of surfactant molecules at the interface lowers γ, and the higher the surfactant adsorption (i.e. the denser the layer) the larger the reduction in γ. The degree of surfactant adsorption at the interface depends on surfactant structure and the nature of the two phases that meet the interface [1, 2].

As noted, surface active agents also aggregate in solution forming micelles. The driving force for micelle formation (or micellization) is the reduction of contact between the hydrocarbon chain and water, thereby reducing the free energy of the system (see Chapter 2). In the micelle, the surfactant hydrophobic groups are directed towards the interior of the aggregate and the polar head groups are directed towards the solvent. These micelles are in dynamic equilibrium and the rate of exchange between a surfactant molecule and the micelle may vary by orders of magnitude, depending on the structure of the surfactant molecule.

Surfactants find application in almost every chemical industry, including detergents, paints, dyestuffs, cosmetics, pharmaceuticals, agrochemicals, fibres, plastics.

Applied Surfactants: Principles and Applications. Tharwat F. Tadros

ISBN: 3-527-30629-3

Moreover, surfactants play a major role in the oil industry, for example in enhanced and tertiary oil recovery. They are also occasionally used for environmental protection, e.g. in oil slick dispersants. Therefore, a fundamental understanding of the physical chemistry of surface active agents, their unusual properties and their phase behaviour is essential for most industrial chemists. In addition, an understanding of the basic phenomena involved in the application of surfactants, such as in the preparation of emulsions and suspensions and their subsequent stabilization, in microemulsions, in wetting spreading and adhesion, etc., is of vital importance in arriving at the right composition and control of the system involved [1, 2]. This is particularly the case with many formulations in the chemical industry.

Commercially produced surfactants are not pure chemicals, and within each chemical type there can be tremendous variation. This is understandable since surfactants are prepared from various feedstocks, namely petrochemicals, natural vegetable oils and natural animal fats. Notably, in every case the hydrophobic group exists as a mixture of chains of different lengths. The same applies to the polar head group, for example with poly(ethylene oxide) (the major component of nonionic surfactants), which consists of a distribution of ethylene oxide units. Hence, products that may be given the same generic name could vary a great deal in their properties, and the formulation chemist should bear this in mind when choosing a surfactant from a particular manufacturer. It is advisable to obtain as much information as possible from the manufacturer about the properties of the surfactant chosen, such as its suitability for the job, its batch to batch variation, toxicity, etc. The manufacturer usually has more information on the surfactant than that printed in the data sheet, and in most cases such information is given on request.

1.1 General Classification of Surface Active Agents

A simple classification of surfactants based on the nature of the hydrophilic group is commonly used. Three main classes may be distinguished, namely anionic, cationic and amphoteric. A useful technical reference is McCutcheon [3], which is produced annually to update the list of available surfactants. van Os et al. have listed the physicochemical properties of selected anionic, cationic and nonionic surfactants [4]. Another useful text is the *Handbook of Surfactants* by Porter [5]. In addition, a fourth class of surfactants, usually referred to as polymeric surfactants, has long been used for the preparation of emulsions and suspensions and their stabilization.

1.2 Anionic Surfactants

These are the most widely used class of surfactants in industrial applications [6, 7] due to their relatively low cost of manufacture and they are used in practically every type of detergent. For optimum detergency the hydrophobic chain is a linear alkyl group with a chain length in the region of 12–16 carbon atoms. Linear chains are preferred since they are more effective and more degradable than branched ones. The most commonly used hydrophilic groups are carboxylates, sulphates, sulphonates and phosphates. A general formula may be ascribed to anionic surfactants as follows:

- Carboxylates: $C_nH_{2n+1}COO^-X$
- Sulphates: $C_nH_{2n+1}OSO_3^-X$
- Sulphonates: $C_nH_{2n+1}SO_3^-X$
- Phosphates: $C_nH_{2n+1}OPO(OH)O^-X$

with $n = 8$–16 atoms and the counter ion X is usually Na^+.

Several other anionic surfactants are commercially available such as sulphosuccinates, isethionates and taurates and these are sometimes used for special applications. These anionic classes and some of their applications are briefly described below.

1.2.1
Carboxylates

These are perhaps the earliest known surfactants since they constitute the earliest soaps, e.g. sodium or potassium stearate, $C_{17}H_{35}COONa$, sodium myristate, $C_{14}H_{29}COONa$. The alkyl group may contain unsaturated portions, e.g. sodium oleate, which contains one double bond in the C_{17} alkyl chain. Most commercial soaps are a mixture of fatty acids obtained from tallow, coconut oil, palm oil, etc. The main attraction of these simple soaps is their low cost, their ready biodegradability and low toxicity. Their main disadvantages are their ready precipitation in water containing bivalent ions such as Ca^{2+} and Mg^{2+}. To avoid such precipitation in hard water, the carboxylates are modified by introducing some hydrophilic chains, e.g. ethoxy carboxylates with the general structure $RO(CH_2CH_2O)_nCH_2COO^-$, ester carboxylates containing hydroxyl or multi COOH groups, sarcosinates which contain an amide group with the general structure $RCON(R')COO^-$.

The addition of the ethoxylated groups increases water solubility and enhances chemical stability (no hydrolysis). The modified ether carboxylates are also more compatible both with electrolytes and with other nonionic, amphoteric and sometimes even cationic surfactants. The ester carboxylates are very soluble in water, but undergo hydrolysis. Sarcosinates are not very soluble in acid or neutral solutions but are quite soluble in alkaline media. They are compatible with other anionics, nonionics and cationics. Phosphate esters have very interesting properties being intermediate between ethoxylated nonionics and sulphated derivatives. They have good compatibility with inorganic builders and they can be good emulsifiers. A specific salt of a fatty acid is lithium 12-hydroxystearic acid, which forms the major constituent of greases.

1.2.2 Sulphates

These are the largest and most important class of synthetic surfactants, which were produced by reaction of an alcohol with sulphuric acid, i.e. they are esters of sulphuric acid. In practice, sulphuric acid is seldom used and chlorosulphonic or sulphur dioxide/air mixtures are the most common methods of sulphating the alcohol. However, due to their chemical instability (hydrolysing to the alcohol, particularly in acid solutions), they are now overtaken by the chemically stable sulphonates.

The properties of sulphate surfactants depend on the nature of the alkyl chain and the sulphate group. The alkali metal salts show good solubility in water, but tend to be affected by the presence of electrolytes. The most common sulphate surfactant is sodium dodecyl sulphate (abbreviated as SDS and sometimes referred to as sodium lauryl sulphate), which is extensively used both for fundamental studies as well as in many industrial applications. At room temperature ($\sim$25 °C) this surfactant is quite soluble and 30% aqueous solutions are fairly fluid (low viscosity). However, below 25 °C, the surfactant may separate out as a soft paste as the temperature falls below its Krafft point (the temperature above which the surfactant shows a rapid increase in solubility with further increase of temperature). The latter depends on the distribution of chain lengths in the alkyl chain – the wider the distribution the lower the Krafft temperature. Thus, by controlling this distribution one may achieve a Krafft temperature of $\sim$10 °C. As the surfactant concentration is increased to 30–40% (depending on the distribution of chain length in the alkyl group), the viscosity of the solution increases very rapidly and may produce a gel. The critical micelle concentration (c.m.c.) of SDS (the concentration above which the properties of the solution show abrupt changes) is 8×10^{-3} mol dm^{-3} (0.24%).

As with the carboxylates, the sulphate surfactants are also chemically modified to change their properties. The most common modification is to introduce some ethylene oxide units in the chain, usually referred to as alcohol ether sulphates, e.g. sodium dodecyl 3-mole ether sulphate, which is essentially dodecyl alcohol reacted with 3 moles EO then sulphated and neutralised by NaOH. The presence of PEO confers improved solubility than for straight alcohol sulphates. In addition, the surfactant becomes more compatible with electrolytes in aqueous solution. Ether sulphates are also more chemically stable than the alcohol sulphates. The c.m.c. of the ether sulphates is also lower than the corresponding surfactant without EO units.

1.2.3 Sulphonates

With sulphonates, the sulphur atom is directly attached to the carbon atom of the alkyl group, giving the molecule stability against hydrolysis, when compared with the sulphates (whereby the sulphur atom is indirectly linked to the carbon of the hydrophobe via an oxygen atom). Alkyl aryl sulphonates are the most common type of these surfactants (e.g. sodium alkyl benzene sulphonate) and these are usually prepared by reaction of sulphuric acid with alkyl aryl hydrocarbons, e.g. dodecyl benzene. A special class of sulphonate surfactants is the naphthalene and alkyl naphthalene sulphonates, which are commonly used as dispersants.

As with the sulphates, some chemical modification is used by introducing ethylene oxide units, e.g. sodium nonyl phenol 2-mole ethoxylate ethane sulphonate, $C_9H_{19}C_6H_4(OCH_2CH_2)_2SO_3^-Na^+$.

Paraffin sulphonates are produced by sulpho-oxidation of normal linear paraffins with sulphur dioxide and oxygen and catalyzed with ultraviolet or gamma radiation. The resulting alkane sulphonic acid is neutralized with NaOH. These surfactants have excellent water solubility and biodegradability. They are also compatible with many aqueous ions.

Linear alkyl benzene sulphonates (LABS) are manufactured from alkyl benzene, and the alkyl chain length can vary from C_8 to C_{15}; their properties are mainly influenced by the average molecular weight and the spread of carbon number of the alkyl side chain. The c.m.c. of sodium dodecyl benzene sulphonate is 5×10^{-3} mol dm^{-3} (0.18%). The main disadvantages of LABS are their effect on the skin and hence they cannot be used in personal care formulations.

Another class of sulphonates is the α-olefin sulphonates, which are prepared by reacting linear α-olefin with sulphur trioxide, typically yielding a mixture of alkene sulphonates (60–70%), 3- and 4-hydroxyalkane sulphonates (~30%) and some disulphonates and other species. The two main α-olefin fractions used as starting material are C_{12}–C_{16} and C_{16}–C_{18}.

A special class of sulphonates is the sulphosuccinates, which are esters of sulphosuccinic acid (**1.1**).

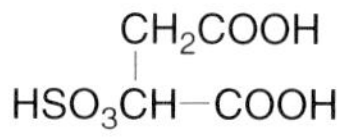

1.1

Both mono and diesters are produced. A widely used diester in many formulations is sodium di(2-ethylhexyl)sulphosuccinate (sold commercially under the trade name Aerosol OT). The diesters are soluble both in water and in many organic solvents. They are particularly useful for preparation of water-in-oil (W/O) microemulsions (Chapter 10).

1.2.4 Phosphate-containing Anionic Surfactants

Both alkyl phosphates and alkyl ether phosphates are made by treating the fatty alcohol or alcohol ethoxylates with a phosphorylating agent, usually phosphorous pentoxide, P_4O_{10}. The reaction yields a mixture of mono- and di-esters of phosphoric acid. The ratio of the two esters is determined by the ratio of the reactants and the amount of water present in the reaction mixture. The physicochemical properties of the alkyl phosphate surfactants depend on the ratio of the esters. Phosphate surfactants are used in the metal working industry due to their anti-corrosive properties.

1.3
Cationic Surfactants

The most common cationic surfactants are the quaternary ammonium compounds [8, 9] with the general formula $R'R''R'''R''''N^+X^-$, where X^- is usually chloride ion and R represents alkyl groups. A common class of cationics is the alkyl trimethyl ammonium chloride, where R contains 8–18 C atoms, e.g. dodecyl trimethyl ammonium chloride, $C_{12}H_{25}(CH_3)_3NCl$. Another widely used cationic surfactant class is that containing two long-chain alkyl groups, i.e. dialkyl dimethyl ammonium chloride, with the alkyl groups having a chain length of 8–18 C atoms. These dialkyl surfactants are less soluble in water than the monoalkyl quaternary compounds, but they are commonly used in detergents as fabric softeners. A widely used cationic surfactant is alkyl dimethyl benzyl ammonium chloride (sometimes referred to as benzalkonium chloride and widely used as bactericide) (**1.2**).

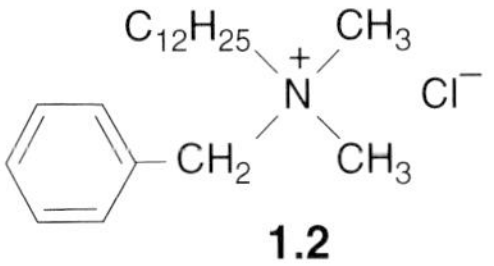

1.2

Imidazolines can also form quaternaries, the most common product being the ditallow derivative quaternized with dimethyl sulphate (**1.3**).

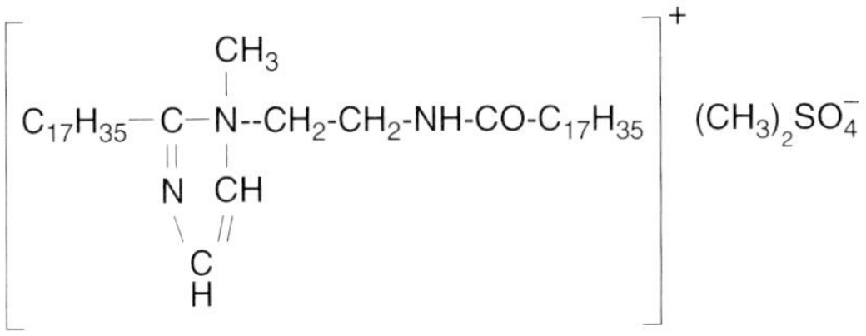

1.3

Cationic surfactants can also be modified by incorporating poly(ethylene oxide) chains, e.g. dodecyl methyl poly(ethylene oxide) ammonium chloride (**1.4**).

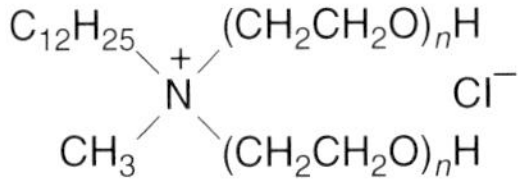

1.4

Cationic surfactants are generally water soluble when there is only one long alkyl group. They are generally compatible with most inorganic ions and hard water, but they are incompatible with metasilicates and highly condensed phosphates. They are also incompatible with protein-like materials. Cationics are generally stable to pH changes, both acid and alkaline. They are incompatible with most anionic surfactants, but they are compatible with nonionics. These cationic surfactants are insoluble in hydrocarbon oils. In contrast, cationics with two or more long alkyl chains are soluble in hydrocarbon solvents, but they become only dispersible in water (sometimes forming bilayer vesicle type structures). They are generally chemically stable and can tolerate electrolytes. The c.m.c. of cationic surfactants is close to that of anionics with the same alkyl chain length.

The prime use of cationic surfactants is their tendency to adsorb at negatively charged surfaces, e.g. anticorrosive agents for steel, flotation collectors for mineral ores, dispersants for inorganic pigments, antistatic agents for plastics, other antistatic agents and fabric softeners, hair conditioners, anticaking agent for fertilizers and as bactericides.

1.4
Amphoteric (Zwitterionic) Surfactants

These are surfactants containing both cationic and anionic groups [10]. The most common amphoterics are the N-alkyl betaines, which are derivatives of trimethyl glycine $(CH_3)_3NCH_2COOH$ (described as betaine). An example of betaine surfactant is lauryl amido propyl dimethyl betaine $C_{12}H_{25}CON(CH_3)_2CH_2COOH$. These alkyl betaines are sometimes described as alkyl dimethyl glycinates.

The main characteristic of amphoteric surfactants is their dependence on the pH of the solution in which they are dissolved. In acid pH solutions, the molecule acquires a positive charge and behaves like a cationic surfactant, whereas in alkaline pH solutions they become negatively charged and behave like an anionic one. A specific pH can be defined at which both ionic groups show equal ionization (the isoelectric point of the molecule) (described by Scheme 1.1).

$$N^+\cdots COOH \rightleftharpoons N^+\cdots COO^- \rightleftharpoons NH\cdots COO^-$$

acid pH <3 — isoelectric — pH >6 alkaline

Scheme 1.1

Amphoteric surfactants are sometimes referred to as zwitterionic molecules. They are soluble in water, but the solubility shows a minimum at the isoelectric point. Amphoterics show excellent compatibility with other surfactants, forming mixed micelles. They are chemically stable both in acids and alkalis. The surface activity of amphoterics varies widely and depends on the distance between the charged groups, showing maximum activity at the isoelectric point.

Another class of amphoterics is the N-alkyl amino propionates having the structure R-$NHCH_2CH_2COOH$. The NH group can react with another acid molecule (e.g. acrylic) to form an amino dipropionate R-$N(CH_2CH_2COOH)_2$. Alkyl imidazoline-based products can also be produced by reacting alkyl imidozoline with a chloro acid. However, the imidazoline ring breaks down during the formation of the amphoteric.

The change in charge with pH of amphoteric surfactants affects their properties, such as wetting, detergency, foaming, etc. At the isoelectric point (i.e.p.), the properties of amphoterics resemble those of non-ionics very closely. Below and above the i.e.p. the properties shift towards those of cationic and anionic surfactants, respectively. Zwitterionic surfactants have excellent dermatological properties. They also exhibit low eye irritation and are frequently used in shampoos and other personal care products (cosmetics).

1.5 Nonionic Surfactants

The most common nonionic surfactants are those based on ethylene oxide, referred to as ethoxylated surfactants [11–13]. Several classes can be distinguished: alcohol ethoxylates, alkyl phenol ethoxylates, fatty acid ethoxylates, monoalkaolamide ethoxylates, sorbitan ester ethoxylates, fatty amine ethoxylates and ethylene oxide–propylene oxide copolymers (sometimes referred to as polymeric surfactants).

Another important class of nonionics is the multihydroxy products such as glycol esters, glycerol (and polyglycerol) esters, glucosides (and polyglucosides) and sucrose esters. Amine oxides and sulphinyl surfactants represent nonionics with a small head group.

1.5.1 Alcohol Ethoxylates

These are generally produced by ethoxylation of a fatty chain alcohol such as dodecanol. Several generic names are given to this class of surfactants, such as ethoxylated fatty alcohols, alkyl polyoxyethylene glycol, monoalkyl poly(ethylene oxide) glycol ethers, etc. A typical example is dodecyl hexaoxyethylene glycol monoether with the chemical formula $C_{12}H_{25}(OCH_2CH_2O)_6OH$ (sometimes abbreviated as $C_{12}E_6$). In practice, the starting alcohol will have a distribution of alkyl chain lengths and the resulting ethoxylate will have a distribution of ethylene oxide chain lengths. Thus the numbers listed in the literature refer to average numbers.

The c.m.c. of nonionic surfactants is about two orders of magnitude lower than the corresponding anionics with the same alkyl chain length. The solubility of the alcohol ethoxylates depends both on the alkyl chain length and the number of ethylene oxide units in the molecule. Molecules with an average alkyl chain length of 12 C atoms and containing more than 5 EO units are usually soluble in water at room temperature. However, as the temperature of the solution is gradually raised the solution becomes cloudy (due to dehydration of the PEO chain) and the temperature at which this occurs is referred to as the cloud point (C.P.) of the surfactant. At a given alkyl chain length, C.P. increases with increasing EO chain of the molecule. C.P. changes with changing concentration of the surfactant solution and the trade literature usually quotes the C.P. of a 1% solution. The C.P. is also affected by the presence of electrolyte in the aqueous solution. Most electrolytes lower the C.P. of a nonionic surfactant solution. Nonionics tend to have maximum surface activity near to the cloud point. The C.P of most nonionics increases markedly on the addition of small quantities of anionic surfactants. The surface tension of alcohol ethoxylate solutions decreases with a decrease in the EO units of the chain. The viscosity of a nonionic surfactant solution increases gradually with an increase in its concentration, but at a critical concentration (which depends on the alkyl and EO chain length) the viscosity increases rapidly and, ultimately, a gel-like structure appears owing to the formation of an hexagonal type liquid crystalline structure. In many cases, the viscosity reaches a maximum, after which it decreases due to the formation of other structures (e.g. lamellar phases) (see Chapter 3).

1.5.2 Alkyl Phenol Ethoxylates

These are prepared by reaction of ethylene oxide with the appropriate alkyl phenol. The most common such surfactants are those based on nonyl phenol. These surfactants are cheap to produce, but suffer from biodegradability and potential toxicity (the by-product of degradation is nonyl phenol, which has considerable toxicity). Despite these problems, nonyl phenol ethoxylates are still used in many industrial properties, owing to their advantageous properties, such as their solubility both in aqueous and non-aqueous media, good emulsification and dispersion properties, etc.

1.5.3 Fatty Acid Ethoxylates

These are produced by reaction of ethylene oxide with a fatty acid or a polyglycol and have the general formula $RCOO\text{-}(CH_2CH_2O)_nH$. When a polyglycol is used, a mixture of mono- and di-esters ($RCOO\text{-}(CH_2CH_2O)_n\text{-}OCOR$) is produced. These surfactants are generally soluble in water provided there are enough EO units and the alkyl chain length of the acid is not too long. The mono-esters are much more soluble in water than the di-esters. In the latter case, a longer EO chain is required to render the molecule soluble. The surfactants are compatible with aqueous ions, provided there is not much unreacted acid. However, these surfactants undergo hydrolysis in highly alkaline solutions.

1.5.4
Sorbitan Esters and Their Ethoxylated Derivatives (Spans and Tweens)

Fatty acid esters of sorbitan (generally referred to as Spans, an Atlas commercial trade name) and their ethoxylated derivatives (generally referred to as Tweens) are perhaps one of the most commonly used nonionics. They were first commercialised by Atlas in the USA, which has since been purchased by ICI. The sorbitan esters are produced by reacting sorbitol with a fatty acid at a high temperature (> 200 °C). The sorbitol dehydrates to 1,4-sorbitan and then esterification takes place. If one mole of fatty acid is reacted with one mole of sorbitol, one obtains a mono-ester (some di-ester is also produced as a by-product). Thus, sorbitan mono-ester has the general formula shown in structure **1.5**.

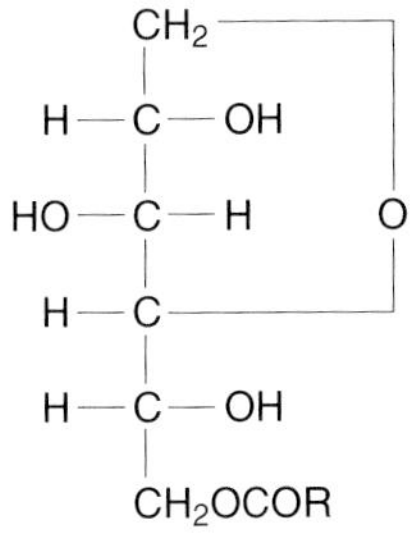

1.5

The free OH groups in the molecule can be esterified, producing di- and tri-esters. Several products are available depending on the nature of the alkyl group of the acid and whether the product is a mono-, di- or tri-ester. Some examples are given below:

- Sorbitan monolaurate – Span 20
- Sorbitan monopalmitate – Span 40
- Sorbitan monostearate – Span 60
- Sorbitan mono-oleate – Span 80
- Sorbitan tristearate – Span 65
- Sorbitan trioleate – Span 85

Ethoxylated derivatives of Spans (Tweens) are produced by the reaction of ethylene oxide on any hydroxyl group remaining on the sorbitan ester group. Alternatively, the sorbitol is first ethoxylated and then esterified. However, the final product has different surfactant properties to the Tweens. Some examples of Tween surfactants are given below.

- Polyoxyethylene (20) sorbitan monolaurate – Tween 20
- Polyoxyethylene (20) sorbitan monopalmitate – Tween 40

- Polyoxyethylene (20) sorbitan monostearate – Tween 60
- Polyoxyethylene (20) sorbitan mono-oleate – Tween 80
- Polyoxyethylene (20) sorbitan tristearate – Tween 65
- Polyoxyethylene (20) sorbitan tri-oleate – Tween 85

The sorbitan esters are insoluble in water, but soluble in most organic solvents (low HLB number surfactants). The ethoxylated products are generally soluble in water and have relatively high HLB numbers. One of the main advantages of the sorbitan esters and their ethoxylated derivatives is their approval as food additives. They are also widely used in cosmetics and some pharmaceutical preparations.

1.5.5 Ethoxylated Fats and Oils

Several natural fats and oils have been ethoxylated, e.g. linolin (wool fat) and caster oil ethoxylates. These products are useful for pharmaceutical products, e.g. as solubilizers.

1.5.6 Amine Ethoxylates

These are prepared by addition of ethylene oxide to primary or secondary fatty amines. With primary amines both hydrogen atoms on the amine group react with ethylene oxide and, therefore, the resulting surfactant has the structure **1.6**.

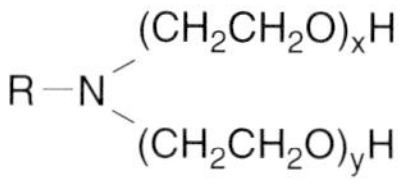

1.6

The above surfactants acquire a cationic character if there are few EO units and if the pH is low. However, at high EO levels and neutral pH they behave very similarly to nonionics. At low EO content, the surfactants are not soluble in water, but become soluble in an acid solution. At high pH, the amine ethoxylates are water soluble provided the alkyl chain length of the compound is not long (usually a C_{12} chain is adequate for reasonable solubility at sufficient EO content).

1.5.7 Ethylene Oxide–Propylene Oxide Co-polymers (EO/PO)

As mentioned above, these may be regarded as polymeric surfactants. These surfactants are sold under various trade names, namely Pluronics (Wyandotte), Synperonic PE (ICI), Poloxamers, etc. Two types may be distinguished: those prepared by reaction of poly(oxypropylene glycol) (difunctional) with EO or mixed EO/PO, giving block copolymers (**1.7**).

$$HO(CH_2CH_2O)_n\text{-}(CH_2\underset{\displaystyle CH_3}{\underset{|}{C}}HO)_m\text{-}(CH_2CH_2)_nOH \quad \text{abbreviated} \quad (EO)_n(PO)_m(EO)_n$$

1.7

Various molecules are available, where *n* and *m* are varied systematically.

The second type of EO/PO copolymers are prepared by reaction of poly(ethylene glycol) (difunctional) with PO or mixed EO/PO. These will have the structure $(PO)_n(EO)_m(PO)_n$ and are referred to as reverse Pluronics.

Trifunctional products (**1.8**) are also available where the starting material is glycerol.

$CH_2\text{-}(PO)_m(EO)_n$
|
$CH\ \text{-}(PO)_n(EO)_n$
|
$CH_2\text{-}(PO)_m(EO)_n$

1.8

Tetrafunctional products (**1.9** and **1.10**) are available where the starting material is ethylene diamine.

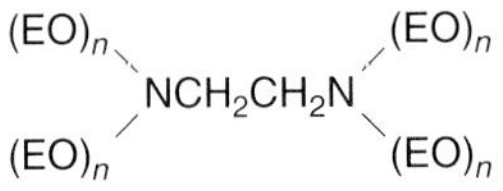

1.9

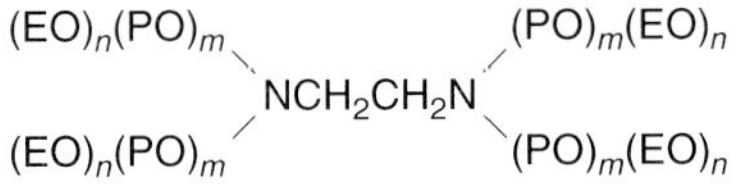

1.10

1.5.8
Surfactants Derived from Mono- and Polysaccharides

Several surfactants have been synthesized starting from mono- or oligosaccharides by reaction with the multifunctional hydroxyl groups. The technical problem is one of joining a hydrophobic group to the multihydroxyl structure. Several surfactants have been made, e.g. esterification of sucrose with fatty acids or fatty glycerides to produce sucrose esters (**1.11**).

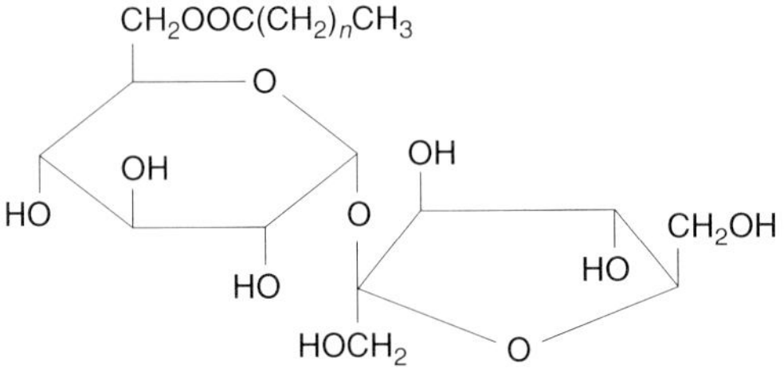

1.11

The most interesting sugar surfactants are the alkyl polyglucosides (APG) (**1.12**).

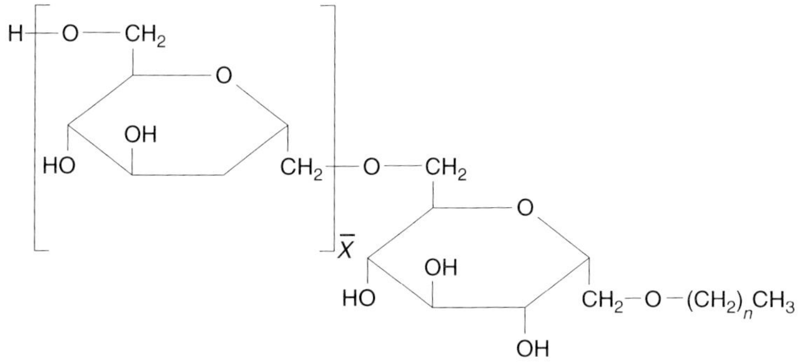

1.12

These are produced by reaction of a fatty alcohol directly with glucose. The basic raw materials are glucose and fatty alcohols (which may be derived from vegetable oils) and hence these surfactants are sometimes referred to as "environmentally friendly". A product with $n = 2$ has two glucose residues with four OH groups on each molecule (i.e. a total of 8 OH groups). The chemistry is more complex and commercial products are mixtures with $n = 1.1$–3. The properties of APG surfactants depend upon the alkyl chain length and the average degree of polymerisation. APG surfactants have good solubility in water and high cloud points (> 100 °C). They are stable in neutral and alkaline solutions but are unstable in strong acid solutions. APG surfactants can tolerate high electrolyte concentrations and are compatible with most types of surfactants.

1.6 Speciality Surfactants – Fluorocarbon and Silicone Surfactants

These surfactants can lower the surface tension of water to below 20 mN m^{-1} (most surfactants described above lower the surface tension of water to values above 20 mN m^{-1}, typically in the region of 25–27 mN m^{-1}). Fluorocarbon and silicone surfactants are sometimes referred to as superwetters as they cause enhanced wetting and spreading of their aqueous solution. However, they are much more expensive than conventional surfactants and are only applied for specific applications whereby the low surface tension is a desirable property.

Fluorocarbon surfactants have been prepared with various structures, consisting of perfluoroalkyl chains and anionic, cationic, amphoteric and poly(ethylene oxide) polar groups. These surfactants have good thermal and chemical stability and they are excellent wetting agents for low energy surfaces.

Silicone surfactants, sometimes referred to as organosilicones, are those with a poly(dimethyl siloxane) backbone. They are prepared by incorporation of a water-soluble or hydrophilic group into a siloxane backbone. The latter can also be modified by incorporation of a paraffinic hydrophobic chain at the end or along the polysiloxane back bone. The most common hydrophilic groups are EO/PO and the structures produced are rather complex and most manufacturers of silicone surfactants do not reveal the exact structure. The mechanism by which these molecules lower the surface tension of water to low values is far from well understood. The surfactants are widely applied as spreading agents on many hydrophobic surfaces.

Incorporating organophilic groups into the backbone of the poly(dimethyl siloxane) backbone can give products that exhibit surface active properties in organic solvents.

1.7
Polymeric Surfactants

There has been considerable recent interest in polymeric surfactants due to their wide application as stabilizers for suspensions and emulsions. Various polymeric surfactants have been introduced and they are marketed under special trade names (such as Hypermers of ICI). One may consider the block EO/PO molecules (Pluronics) as polymeric surfactants, but these generally do not have high molecular weights and they seldom produce speciality properties. Silicone surfactants may also be considered as polymerics. However, the recent development of speciality polymeric surfactants of the graft type ("comb" structures) have enabled one to obtain specific applications in dispersions. An example is the graft copolymer of a poly(methyl methacrylate) backbone with several PEO side chains (sold under the trade name Hypermer CG6 by ICI), which has excellent dispersing and stabilizing properties for concentrated dispersions of hydrophobic particles in water. Using such a dispersant, one can obtain highly stable concentrated suspensions. These surfactants have been modified in several ways to produce molecules that are suitable as emulsifiers, dispersants in extreme conditions such as high or low pH, high electrolyte concentrations, temperatures etc. Other polymeric surfactants that are suitable for dispersing dyes and pigments in non-aqueous media have also been prepared, whereby the side chains were made oil soluble, such as polyhydroxystearic acid.

Another important class of polymeric surfactants that are used for demulsification is those based on alkoxylated alkyl phenol formaldehyde condensates, with the general structure **1.13**.

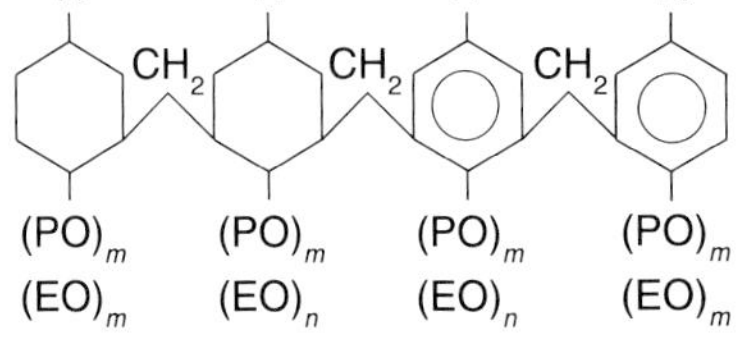

1.13

Several other complex polymerics are manufactured for application in the oil industry, e.g. polyalkylene glycol modified polyester with fatty acid hydrophobes, polyesters, made by polymerization of polyhydroxy stearic acid, etc.

1.8 Toxicological and Environmental Aspects of Surfactants

1.8.1 Dermatological Aspects

A large fraction of dermatological problems in normal working life can be related to exposure of unprotected skin to surfactant solutions [2]. Several formulations contain significant amount of surfactants, e.g. cutting fluids, rolling oil emulsions, some household cleaning formulations and some personal care products. Skin irritation of various degrees of seriousness is common, and in some cases allergic reactions may also appear. The physiological aspects of surfactants on the skin have been investigated by various dermatological laboratories, starting with the surface of the skin and progressing via the horny layer and its barrier function to the deeper layer of the basal cells. Surfactant classes that are generally known to be mild to the skin include polyol surfactants (alkyl polyglucosides), zwitterionic surfactants (betaines, amidobetaines and isethionates) and many polymeric surfactants. Alcohol ethoxylates are relatively mild, but not as mild as the polyol based non-ionics (the alkyl polyglucosides). In addition, alcohol ethoxylates may undergo oxidation to give by-products (hyperoxides and aldehydes) that are skin irritants. These classes are commonly used in personal care and cosmetic formulations.

For a homologous series of surfactants there is usually a maximum in skin irritation at a specific alkyl chain length; maximum irritation usually occurs at a C_{12} chain length. This reflects the maximum in surface activity at this chain length and the reduction in the c.m.c. Anionic surfactants are generally greater skin irritants than non-ionics. For example, sodium dodecyl sulphate, which is commonly used in tooth paste, has a relatively high skin toxicity. In contrast, the ether sulphates are milder and are recommended for use in hand dishwashing formulations. Sometimes, addition of a mild surfactant (such as alkyl polyglucoside) can greatly improve the dermatological properties. Some amphoteric surfactants such as betaines can also reduce the skin irritation of anionic surfactants.

1.8.2 Aquatic Toxicity

Aquatic toxicity is usually measured on fish, daphnia and algae. The toxicity index is expressed as LC_{50} (for fish) or EC_{50} (for daphnia and algea), where LC and EC stand for lethal and effective concentration, respectively. Values below 1 mg l^{-1}

after 96 h testing on fish and algae and 48 h on daphnia are considered toxic. Environmentally benign surfactants should, preferably, be above 10 mg l^{-1}.

1.8.3 Biodegradability

Biodegradation is carried out by bacteria in nature. By enzymatic reactions, a surfactant molecule is ultimately converted into carbon dioxide, water and oxides of the other elements. If the surfactant does not undergo natural biodegradation then it is stable and persists in the environment. For surfactants the rate of biodegradation varies from 1–2 h for fatty acids, 1–2 days for linear alkyl benzene sulphonates, and several months for branched alkyl benzene sulphonates. The rate of biodegradation depends on the surfactant concentration, pH and temperature. The temperature effect is particularly important, since the rate can vary by as much a factor of five between summer and winter in Northern Europe.

Two criteria are important when testing for biodegradation: (1) Primary degradation that results in loss of surface activity. (2) Ultimate biodegradation, i.e. conversion into carbon dioxide, which can be measured using closed bottle tests.

The rate of biodegradation also depends on the surfactant structure. For example, the surfactant must be water soluble. Lipophilic amphiphiles such as fluorocarbon surfactants may accumulate in the lipid compartments of the organism and break down very slowly. The initial degradation may also lead to intermediates with much lower water solubility and these degrade very slowly. An example of this is the alkyl phenol ethoxylates, which degrade by oxidative cleavage from the hydroxyl end of the polyoxyethylene chain. This leads to a compound with much smaller EO groups that is very lipophilic and degrades very slowly.

A third important factor in biodegradation is the presence of cleavable bonds in the alkyl chain, which depend on branching. Extensive branching of the alkyl chain tends to reduce the rate of biodegradation. This is probably due to steric hindrance preventing close approach of the surfactant molecule into the active site of the enzyme.

References

1 F. Tadros (ed.): *The Surfactants*, Academic Press, London, 1984.

2 K. Holmberg, B. Jonsson, B. Kronberg, B. Lindman: *Surfactants and Polymers in Solution*, 2nd edition, John Wiley & Sons, Chichester, 2003.

3 McCutcheon: *Detergents and Emulsifiers*, Allied Publishing Co, New Jersey, published annually.

4 N. M. Os van, J. R. Haak, L. A. M. Rupert: *Physico-chemical Properties of Selected Anionic, Cationic and Nonionic Surfactants*, Elsevier, Amsterdam, 1993.

5 M. R. Porter, *Handbook of Surfactants*, Blackie, London, 1994.

6 W. M. Linfield, W. M. Linfield (ed.): *Anionic Surfactants*, Marcel Dekker, New York, 1967.

7 E. H. Lucasssen-Reynders: *Anionic Surfactants – Physical Chemistry of Surfactant Action*, Marcel Dekker, New York, 1981.

8 E. Jungerman: *Cationic Surfactants*, Marcel Dekker, New York, 1970.
9 N. Rubingh, P. M. Holland (ed.): *Cationic Surfactants – Physical Chemistry*, Marcel Dekker, New York, 1991.
10 B. R. Buestein, C. L. Hiliton: *Amphoteric Surfactants*, Marcel Dekker, New York, 1982.
11 M. J. Schick (ed.): *Nonionic Surfactants*, Marcel Dekker, New York, 1966.
12 M. J. Schick (ed.): *Nonionic Surfactants: Physical Chemistry*, Marcel Dekker, New York, 1987.
13 N. Schonfeldt: *Surface Active Ethylene Oxide Adducts*, Pergamon Press, Oxford, 1970.

2 Physical Chemistry of Surfactant Solutions

2.1 Properties of Solutions of Surface Active Agents

The physical properties of surface active agents differ from those of smaller or non-amphipathic molecules in one major aspect, namely the abrupt changes in their properties above a critical concentration [1]. Figure 2.1 illustrates with plots of several physical properties (osmotic pressure, turbidity, solubilisation, magnetic resonance, surface tension, equivalent conductivity and self-diffusion) as a function of concentration for an ionic surfactant [1].

At low concentrations, most properties are similar to those of a simple electrolyte. One notable exception is the surface tension, which decreases rapidly with increasing surfactant concentration. However, all the properties (interfacial and bulk)

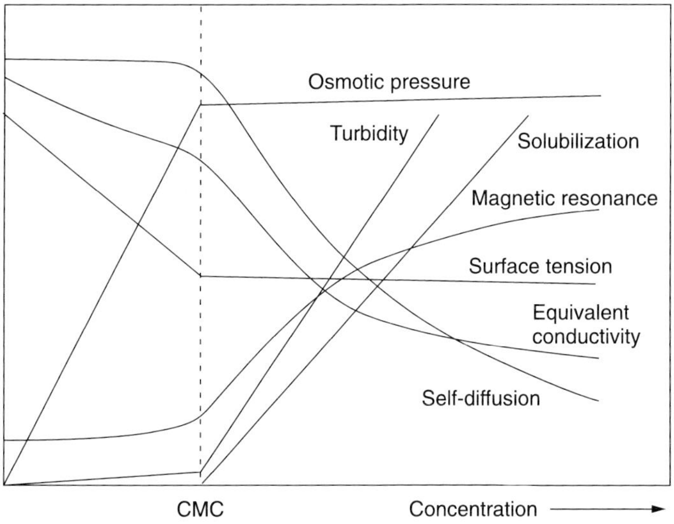

Fig. 2.1. Changes in the concentration dependence of a wide range of physico-chemical changes around the critical micelle concentration (c.m.c.) (after Lindman et al. [1]).

Applied Surfactants: Principles and Applications. Tharwat F. Tadros

ISBN: 3-527-30629-3

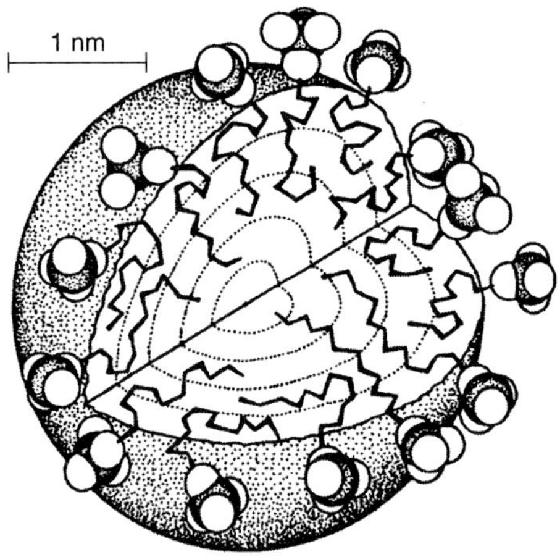

Fig. 2.2. Illustration of a spherical micelle for dodecyl sulphate [2].

show an abrupt change at a particular concentration, which is consistent with the fact that at and above this concentration, surface active ions or molecules in solution associate to form larger units. These associated units are called micelles (self-assembled structures) and the first formed aggregates are generally approximately spherical. A schematic representation of a spherical micelle is given in Figure 2.2.

The concentration at which this association phenomenon occurs is known as the critical micelle concentration (c.m.c.). Each surfactant molecules has a characteristic c.m.c. at a given temperature and electrolyte concentration. The most common technique for measuring the c.m.c. is by determining the surface tension, γ, which shows break at the c.m.c., after which it remains virtually constant with further increases in concentration. However, other techniques such as self-diffusion measurements, NMR and fluorescence spectroscopy can be applied. Mukerjee and Mysels compiled various c.m.c.s in 1971 [3]; while clearly not up-to-date this is an extremely valuable reference. As an illustration, Table 2.1 gives the c.m.c. of several surface active agents to show some of the general trends [2]. Within any class of surface active agent, the c.m.c. decreases with increasing chain length of the hydrophobic portion (alkyl group). As a general rule, the c.m.c. decreases by a factor of 2 for ionics (without added salt) and by a factor of 3 for nonionics on adding one methylene group to the alkyl chain. With nonionic surfactants, increasing the length of the hydrophilic group, poly(ethylene oxide), causes an increase in c.m.c.

In general, nonionic surfactants have lower c.m.c.s than their corresponding ionic surfactants of the same alkyl chain length. Incorporation of a phenyl group in the alkyl group increases its hydrophobicity to a much smaller extent than increasing its chain length with the same number of carbon atoms. The valency of the counter ion in ionic surfactants has a significant effect on the c.m.c. For example, increasing the valency of the counter ion from 1 to 2 reduces the c.m.c. by roughly a factor of 4.

Tab. 2.1. C.m.c. values of some surface active agents.

Surface active agent	*C.m.c. (mol dm^{-3})*
(A) Anionic	
Sodium octyl-l-sulphate	1.30×10^{-1}
Sodium decyl-l-sulphate	3.32×10^{-2}
Sodium dodecyl-l-sulphate	8.39×10^{-3}
Sodium tetradecyl-l-sulphate	2.05×10^{-3}
(B) Cationic	
Octyl trimethyl ammonium bromide	1.30×10^{-1}
Decetryl trimethyl ammonium bromide	6.46×10^{-2}
Dodecyl trimethyl ammonium bromide	1.56×10^{-2}
Hexacetyltrimethyl ammonium bromide	9.20×10^{-4}
(C) Nonionic	
Octyl hexaoxyethylene glycol monoether C_8E_6	9.80×10^{-3}
Decyl hexaoxyethylene glycol monoether $C_{10}E_6$	9.00×10^{-4}
Decyl nonaoxyethylene glycol monoether $C_{10}E_9$	1.30×10^{-3}
Dodecyl hexaoxyethylene glycol monoether $C_{12}E_6$	8.70×10^{-5}
Octylphenyl hexaoxyethylene glycol monoether C_8E_6	2.05×10^{-4}

The c.m.c. is, to a first approximation, independent of temperature. This is illustrated in Figure 2.3, which shows that the c.m.c. of SDS varies (by ca. 10–20%) non-monotonically over a wide temperature range. The shallow minimum around 25 °C can be compared with a similar minimum in the solubility of hydrocarbon in water [4]. However, nonionic surfactants of the ethoxylate type show a monotonic decrease [4] of c.m.c. with increasing temperature, as is illustrated in Figure 2.3 for $C_{10}E_5$.

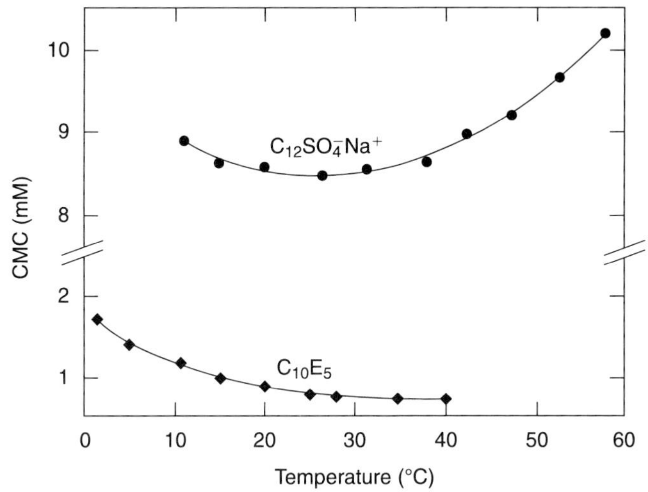

Fig. 2.3. Temperature dependence of the c.m.c. of SDS and $C_{10}E_5$ [4].

The effect of addition of cosolutes, e.g. electrolytes and non-electrolytes, on the c.m.c. can be very striking. For example, the addition of a 1:1 electrolyte to a solution of anionic surfactant dramatically lowers the c.m.c., by up to an order of magnitude. The effect is moderate for short-chain surfactants, but is much larger for long-chain ones. At high electrolyte concentrations, the reduction in c.m.c. with increasing number of carbon atoms in the alkyl chain is much stronger than without added electrolyte. This rate of decrease at high electrolyte concentrations is comparable to that of nonionics. The effect of added electrolyte also depends on the valency of the added counter ions. In contrast, for nonionics, addition of electrolytes causes only a small variation in the c.m.c.

Non-electrolytes such as alcohols can also cause a decrease in the c.m.c. Figure 2.4 illustrates this for several alcohols [5], added to an anionic surfactant, namely potassium dodecanoate. The alcohols are less polar than water and are distributed between the bulk solution and the micelles. The more preference they have for the micelles, the more they stabilize them. A longer alkyl chain leads to a less favourable location in water and more favourable location in the micelles.

The presence of micelles can account for many of the unusual properties of solutions of surface active agents. For example, it can account for the near constant surface tension above the c.m.c. (Figure 2.1). It also accounts for the reduction in molar conductance of the surface active agent solution above the c.m.c., which is consistent with the reduction in mobility of the micelles as a result of the counter ion association. The presence of micelles also accounts for the rapid rise in light scattering or turbidity above the c.m.c.

The presence of micelles was originally proposed by McBain [6] who suggested that below the c.m.c. most of the surfactant molecules are unassociated, whereas in the isotropic solutions immediately above the c.m.c., micelles and surfactant ions (molecules) are thought to co-exist, the concentration of the latter changing very slightly as more surfactant is dissolved. However, the self-association of an amphiphile occurs in a stepwise manner with one monomer added to the aggregate at a time. For a long-chain amphiphile, the association is strongly cooperative up to a certain micelle size where counteracting factors became increasingly important. Typically, micelles are closely spherical over a rather wide concentration range above the c.m.c. Indeed, Adam [7] and Hartley [8] first suggested that micelles are spherical and have the following properties: (1) the association unit is spherical with a radius approximately equal to the length of the hydrocarbon chain; (2) the micelle contains about 50–100 monomeric units – the aggregation number generally increases with increasing alkyl chain length; (3) with ionic surfactants, most counter ions are bound to the micelle surface, thus significantly reducing the mobility from the value expected from a micelle with non-counterion bonding; (4) micellization occurs over a narrow concentration range due to the high association number of surfactant micelles; (5) the interior of the surfactant micelle has essentially the properties as a liquid hydrocarbon. This is confirmed by the high mobility of the alkyl chains and the ability of the micelles to solubilize many water-insoluble organic molecules, e.g. dyes and agrochemicals.

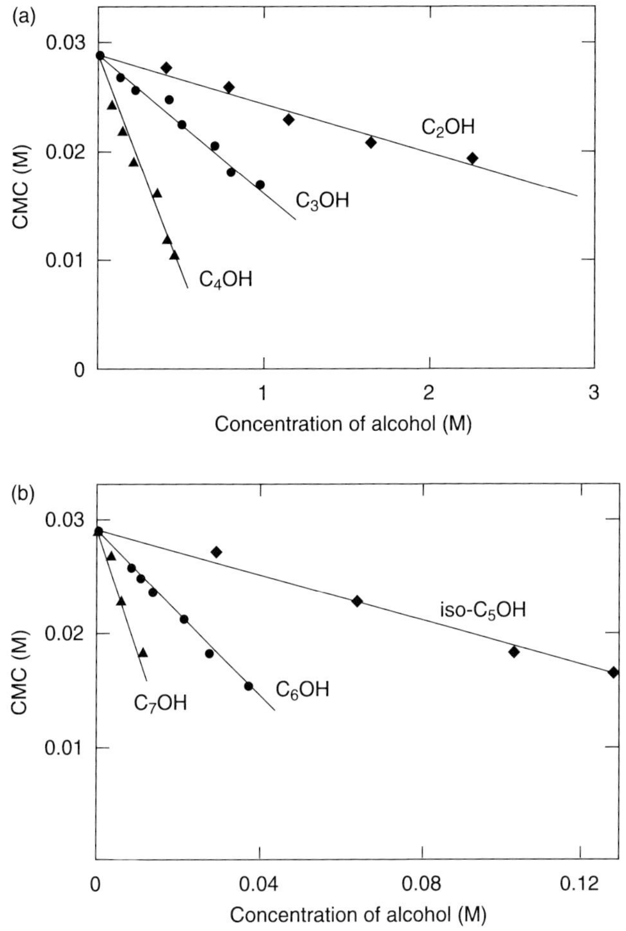

Fig. 2.4. Effect of alcohols on the c.m.c. of potassium dodecanoate [5] association. The presence of micelles also accounts for the rapid increase in light scattering or turbidity above the c.m.c.

To a first approximation, micelles can, over a wide concentration range above the c.m.c., be viewed as microscopic liquid hydrocarbon droplets covered with polar head groups, which interact strongly with water molecules. The radius of the micelle core constituted of the alkyl chains appears to be close to the extended length of the alkyl chain, i.e. in the range 1.5030 nm. As we will see later, the driving force for micelle formation is the elimination of the contact between the alkyl chains and water. The larger a spherical micelle, the more efficient this is, since the volume-to-area ratio increases. Notably, not all surfactant molecules in the micelles are extended. Only one molecule needs to be extended to satisfy the criterion that the radius of the micelle core is close to the extended length of the alkyl chain. Most surfactant molecules are in a disordered state. In other words, the interior of the micelle is close to that of the corresponding alkane in a neat liquid oil. This explains the large solubilization capacity of the micelle towards a broad range of non-polar and weakly polar substances.

At the surface of the micelle, associated counter ions (in the region of 50–80% of the surfactant ions) are present. However, simple inorganic counter ions are very loosely associated with the micelle. The counter ions are very mobile (see below) and there is no specific complex formed with a definite counter-ion–head group distance. In other words, the counter ions are associated by long-range electrostatic interactions.

A useful concept for characterizing micelle geometry is the critical packing parameter, CPP (discussed in Chapter 6). The aggregation number N is the ratio between the micellar core volume, V_{mic}, and the volume of one chain, v,

$$N = \frac{V_{mic}}{v} = \frac{(4/3)\pi R_{mic}^3}{v} \tag{2.1}$$

where R_{mic} is the radius of the micelle.

The aggregation number, N, is also equal to the ratio of the area of a micelle, A_{mic}, to the cross sectional area, a, of one surfactant molecule (Eq. 2.2).

$$N = \frac{A_{mic}}{a} = \frac{4\pi R_{mic}^2}{a} \tag{2.2}$$

Combining Eqs. (2.1) and (2.2),

$$\frac{v}{R_{mic}a} = \frac{1}{3} \tag{2.3}$$

Since R_{mic} cannot exceed the extended length of a surfactant alkyl chain, l_{max},

$$l_{max} = 1.5 + 1.265 n_c \tag{2.4}$$

This means that, for a spherical micelle,

$$\frac{v}{l_{max}a} \preceq \frac{1}{3} \tag{2.5}$$

The ratio $v/(l_{max}a)$ is denoted as the critical packing parameter (CPP).

Although, the spherical micelle model accounts for many physical properties of solutions of surfactants, several phenomena remain unexplained, without considering other shapes. For example, McBain [9] suggested the presence of two types of micelles, spherical and lamellar, to account for the drop in molar conductance of surfactant solutions. The lamellar micelles are neutral and hence they account for

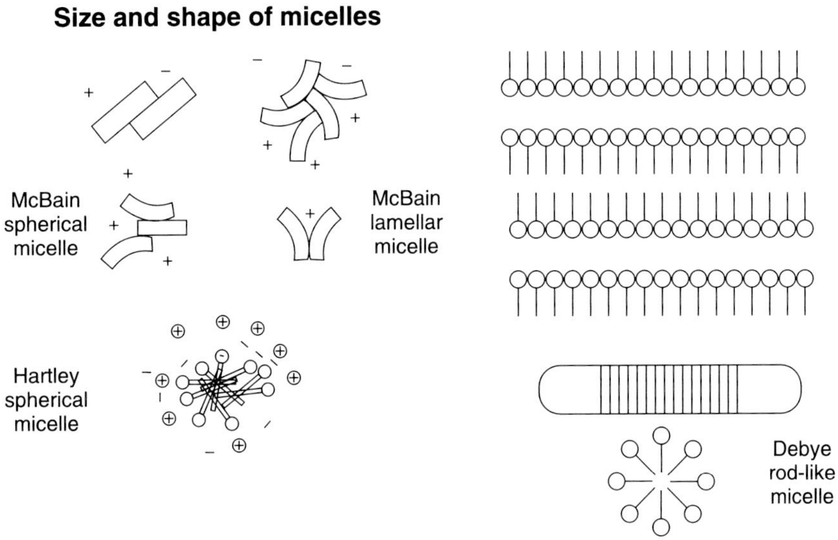

Fig. 2.5. Various shapes of micelles (according to McBain [6, 9], Hartley [8] and Debye [11]).

the reduction in the conductance. Later, Harkins et al. [10] used McBain's model of lamellar micelles to interpret their X-ray results in soap solutions. Moreover, many modern techniques such as light scattering and neutron scattering indicate that in many systems the micelles are not spherical. For example, Debye and Anacker [11] proposed a cylindrical micelle to explain light scattering results on (hexadecyltrimethyl)ammonium bromide in water. Evidence for disc-shaped micelles has also been obtained under certain conditions. A schematic representation of the spherical, lamellar and rod-shaped micelles, suggested by McBain, Hartley and Debye, is given in Figure 2.5.

2.2 Solubility–Temperature Relationship for Surfactants

This will be dealt with in detail in Chapter 3, and so only the main trends are summarized here. Many ionic surfactants show dramatic temperature-dependent solubility. The solubility may be very low at low temperatures and then increases by orders of magnitude in a relatively narrow temperature range. This phenomenon is generally denoted as the Krafft phenomenon, with the temperature for the onset of increasing solubility being known as the Krafft temperature. The latter may vary dramatically with subtle changes in the surfactant's chemical structure. In general, the Krafft temperature increases rapidly as the alkyl chain length of the surfactant increases. It also depends on the head group and counter ion. Addition of electrolytes increases the Krafft temperature.

2.3 Thermodynamics of Micellization

As mentioned above, the process of micellization is one of the most important characteristics of surfactant solution and hence it is essential to understand its mechanism (the driving force for micelle formation). This requires analysis of the dynamics of the process (i.e. the kinetic aspects) as well as the equilibrium aspects whereby the laws of thermodynamics may be applied to obtain the free energy, enthalpy and entropy of micellization. Below a brief description of both aspects will be given and this will be followed by a picture of the driving force for micelle formation.

2.3.1 Kinetic Aspects

Micellization is a dynamic phenomenon in which n monomeric surfactant molecules associate to form a micelle S_n, i.e.,

$$nS \rightleftharpoons S_n \qquad (2.6)$$

Hartley [8] envisaged a dynamic equilibrium whereby surface active agent molecules are constantly leaving and entering, from solution, the micelles. The same applies to the counter ions with ionic surfactants, which can exchange between the micelle surface and bulk solution.

Experimental investigations using fast kinetic methods such as stopped-flow, temperature and pressure jumps, and ultrasonic relaxation measurements have shown that there are two relaxation processes for micellar equilibrium [12–18], characterized by relaxation times τ_1 and τ_2. The first, τ_1, is of the order of 10^{-7} s (10^{-8} to 10^{-3} s) and represents the life-time of a surface active molecule in a micelle, i.e. it represents the association and dissociation rate for a single molecule entering and leaving the micelle, which may be represented by Eq. (2.7).

$$S + S_{n-1} \underset{k^-}{\overset{k^+}{\rightleftharpoons}} S_n \qquad (2.7)$$

where k^+ and k^- represent the association and dissociation rate respectively for a single molecule entering or leaving the micelle.

The slower relaxation time τ_2 corresponds to a relatively slow process, namely the micellization–dissolution, represented by Eq. (2.6); τ_2 is of the order of milliseconds (10^{-3}–1 s) and hence can be conveniently measured by stopped-flow methods. The fast relaxation time τ_1 can be measured using various techniques depending on its range. For example, τ_1 in the range of 10^{-8}–10^{-7} s is accessible to ultrasonic absorption methods, in the range of 10^{-5}–10^{-3} s it can be measured by pressure jump methods. τ_1 depends on surfactant concentration, chain length and temperature, and increases with increasing chain length of surfactants, i.e. the residence time increases with increasing chain length.

The above discussion emphasizes the dynamic nature of micelles and it is important to realise that these molecules are in continuous motion and that there is a constant interchange between micelles and solution. The dynamic nature also applies to the counter ions which exchange rapidly with life times in the range 10^{-9}–10^{-8} s. Furthermore, the counter ions appear to be laterally mobile and not to be associated with (single) specific groups on the micelle surfaces [2].

2.3.2
Equilibrium Aspects: Thermodynamics of Micellization

Two general approaches have been employed to tackle micelle formation. The first and simplest approach treats micelles as a single phase, and is referred to as the phase separation model. Here, micelle formation is considered as a phase separation phenomenon and the c.m.c. is then the saturation concentration of the amphiphile in the monomeric state whereas the micelles constitute the separated pseudophase. Above the c.m.c., a phase equilibrium exists with a constant activity of the surfactant in the micellar phase. The Krafft point is viewed as the temperature at which solid hydrated surfactant, micelles and a solution saturated with undissociated surfactant molecules are in equilibrium at a given pressure.

In the second approach, micelles and single surfactant molecules or ions are considered to be in association–dissociation equilibrium. In its simplest form, a single equilibrium constant is used to treat the process represented by Eq. (2.1). The c.m.c. is merely a concentration range above which any added surfactant appears in solution in a micellar form. Since the solubility of the associated surfactant is much greater than that of the monomeric surfactant, the solubility of the surfactant as a whole will not increase markedly with temperature until it reaches the c.m.c. region. Thus, in the mass action approach, the Krafft point represents the temperature at which the surfactant solubility equals the c.m.c.

2.3.3
Phase Separation Model

Consider an anionic surfactant, in which n surfactant anions, S^-, and n counter ions M^+ associate to form a micelle, i.e.,

$$nS^- + nM^+ \rightleftharpoons S_n \tag{2.8}$$

The micelle is simply a charged aggregate of surfactant ions plus an equivalent number of counter ions in the surrounding atmosphere and is treated as a separate phase.

The chemical potential of the surfactant in the micellar state is assumed to be constant, at any given temperature, and this may be adopted as the standard chemical potential, μ_m°, by analogy to a pure liquid or a pure solid. Considering the equilibrium between micelles and monomer, then,

$$\mu_m^\circ = \mu_1^\circ + RT \ln a \tag{2.9}$$

where μ_1 is the standard chemical potential of the surfactant monomer and a_1 is its activity, which is equal to $f_1 x_1$, where f_1 is the activity coefficient and x_1 the mole fraction. Therefore, the standard free energy of micellization per mol of monomer, ΔG°_m, is given by,

$$\Delta G^\circ_m = \mu^\circ_m - \mu^\circ_1 = RT \ln a_1 \simeq RT \ln x_1 \tag{2.10}$$

where f_1 is taken as unity (a reasonable value in very dilute solution). The c.m.c. may be identified with x_1 so that

$$\Delta G^\circ_m = RT \ln[\text{c.m.c.}] \tag{2.11}$$

In Eq. (2.10), the c.m.c. is expressed as a mole fraction, which is equal to $C/(55.5 + C)$, where C is the concentration of surfactant in mole dm^{-3}, i.e.,

$$\Delta G^\circ_m = RT \ln C - RT \ln(55.5 + C) \tag{2.12}$$

ΔG° should be calculated using the c.m.c. expressed as a mole fraction as indicated by Eq. (2.12). However, most quoted c.m.c.s are given in mole dm^{-3} and, in many cases, ΔG°_s have been quoted when the c.m.c. was simply expressed in mol dm^{-3}. Strictly speaking, this is incorrect, since ΔG° should be based on x_1 rather than on C. ΔG° obtained when the c.m.c. is expressed in mol dm^{-3} is substantially different from that found with c.m.c. expressed in mole fraction. For example, for dodecyl hexaoxyethylene glycol the quoted c.m.c. is 8.7×10^{-5} mol dm^{-3} at 25 °C. Therefore,

$$\Delta G^\circ = RT \ln \left[\frac{8.7 \times 10^{-5}}{55.5 + 8.7 \times 10^{-5}} \right] = -33.1 \text{ kJ mol}^{-1} \tag{2.13}$$

when the mole fraction scale is used. However,

$$\Delta G^\circ = RT \ln 8.7 \times 10^{-5} = -23.2 \text{ kJ mol}^{-1} \tag{2.14}$$

when the molarity scale is used.

The phase separation model has been questioned for two main reasons. Firstly, according to this model a clear discontinuity in the physical property of a surfactant solution, such as surface tension, turbidity, etc. should be observed at the c.m.c. This is not always found experimentally and the c.m.c. is not a sharp break point. Secondly, if two phases actually exist at the c.m.c., then equating the chemical potential of the surfactant molecule in the two phases would imply that the activity of the surfactant in the aqueous phase would be constant above the c.m.c. If this was the case, the surface tension of a surfactant solution should remain constant above the c.m.c. However, careful measurements have shown that the surface tension of a surfactant solution decreases slowly above the c.m.c., particularly when using purified surfactants.

2.3.4 Mass Action Model

This model assumes a dissociation–association equilibrium between surfactant monomers and micelles – thus an equilibrium constant can be calculated. For a nonionic surfactant, where charge effects are absent, this equilibrium is simply represented by Eq. (2.1), which assumes a single equilibrium. In this case, the equilibrium constant K_m is given by Eq. (2.15).

$$K_\mathrm{m} = \frac{[\mathrm{S}_n]}{[\mathrm{S}]^n} \tag{2.15}$$

The standard free energy per monomer is then given by

$$-\Delta G^\circ_\mathrm{m} = -\frac{\Delta G}{n} = \frac{RT}{n} \ln K_\mathrm{m} = \frac{RT}{n} \ln[\mathrm{S}_n] - RT \ln[\mathrm{S}] \tag{2.16}$$

For many micellar systems, $n > 50$ and, therefore, the first term on the right-hand side of Eq. (2.16) may be neglected, resulting in Eq. (2.17) for $\Delta G^\circ_\mathrm{m}$,

$$\Delta G^\circ_\mathrm{m} = RT \ln[\mathrm{S}] = RT \ln[\mathrm{c.m.c.}] \tag{2.17}$$

which is identical to the equation derived using the phase-separation model.

The mass action model allows a simple extension to be made to the case of ionic surfactants, in which micelles attract a substantial proportion of counter ions, into an attached layer. For a micelle made of n-surfactant ions, (where $n - p$) charges are associated with counter ions, i.e. having a net charge of p units and degree of dissociation p/n, the following equilibrium may be established (for an anionic surfactant with Na^+ counter ions),

$$n\mathrm{S}^- + (n-p)\mathrm{Na}^+ \rightleftharpoons \mathrm{S}_n^{p^-} \tag{2.18}$$

$$K_\mathrm{m} = \frac{[\mathrm{S}_n^{p^-}]}{[\mathrm{S}^-]^n[\mathrm{Na}^+]^{(n-p)}} \tag{2.19}$$

Phillips has given a convenient solution for relating ΔG_m to [c.m.c.] [17], arriving at Eq. (2.20),

$$\Delta G^\circ_\mathrm{m} = [2 - (p/n)]RT \ln[\mathrm{c.m.c.}] \tag{2.20}$$

For many ionic surfactants, the degree of dissociation (p/n) is ~ 0.2 so that,

$$\Delta G^\circ_\mathrm{m} = 1.8RT \ln[\mathrm{c.m.c.}] \tag{2.21}$$

Comparison with Eq. (2.17) clearly shows that, for similar ΔG_m, the [c.m.c.] is about two orders of magnitude higher for ionic surfactants than with nonionic surfactant of the same alkyl chain length (Table 2.1).

In the presence of excess added electrolyte, with mole fraction x, the free energy of micellization is given by the expression,

$$\Delta G^{\circ}_{\mathrm{m}} = RT \ln[\text{c.m.c.}] + [1 - (p/n)] \ln x \tag{2.22}$$

Eq. (2.22) shows that as x increases the [c.m.c.] decreases.

It is clear from Eq. (2.20) that as $p \to 0$, i.e. when most charges are associated with counter ions,

$$\Delta G^{\circ}_{\mathrm{m}} = 2RT \ln[\text{c.m.c.}] \tag{2.23}$$

whereas when $p \sim n$, i.e. the counter ions are bound to micelles,

$$\Delta G^{\circ}_{\mathrm{m}} = RT \ln[\text{c.m.c.}] \tag{2.24}$$

which is the same equation as for nonionic surfactants.

Although the mass action approach could account for a number of experimental results, such as the small change in properties around the c.m.c., it has not escaped criticism. For example, the assumption that surfactants exist in solution in only two forms, namely single ions and micelles of uniform size, is debatable. Analysis of various experimental results has shown that micelles have a size distribution that is narrow and concentration dependent. Thus, the assumption of a single aggregation number is an oversimplification and, in reality, there is a micellar size distribution. This can be analyzed using the multiple equilibrium model, which can be best formulated as a stepwise aggregation [2],

$$\mathrm{S}_1 + \mathrm{S}_1 \rightleftharpoons \mathrm{S}_2 \tag{2.25}$$

$$\mathrm{S}_2 + \mathrm{S}_1 \rightleftharpoons \mathrm{S}_3 \tag{2.26}$$

$$\mathrm{S}_{n-1} + \mathrm{S}_1 \rightleftharpoons \mathrm{S}_n \tag{2.27}$$

As noted in particular in the analysis of kinetic data [10–15], there are aggregates over a wide range of aggregation numbers, from dimers and well beyond the most stable micelles. However, for surfactants with not too high a c.m.c., the size distribution curve has a very deep minimum, the least stable aggregates being present in concentrations many orders of magnitude below those of the most abundant micelles. For surfactants with predominantly spherical micelles, the polydispersity is low and there is then a particularly preferred micellar size.

2.3.5 Enthalpy and Entropy of Micellization

The enthalpy of micellization can be calculated from the variation of c.m.c. with temperature. This follows from

$$-\Delta H^\circ = RT^2 \frac{\mathrm{d}\,\ln[\mathrm{c.m.c.}]}{\mathrm{d}T} \tag{2.28}$$

The entropy of micellization can then be calculated from the relationship between ΔG° and ΔH°, i.e.,

$$\Delta G^\circ = \Delta H^\circ - T\Delta S^\circ \tag{2.29}$$

Therefore ΔH° may be calculated from surface tension $-\log C$ plots at various temperatures. Unfortunately, the errors in locating the c.m.c. (which in many cases is not a sharp point) lead to a large error in ΔH°. A more accurate and direct method of obtaining ΔH° is microcalorimetry. As an illustration, the thermodynamic parameters $\Delta G^\circ, \Delta H^\circ$, and $T\Delta S^\circ$ for octylhexaoxyethylene glycol monoether (C_8E_6) are given in Table 2.2.

Table 2.2 shows that ΔG° is large and negative. However, ΔH° is positive, indicating that the process is endothermic. In addition, $T\Delta S^\circ$ is large and positive, implying that in the micellization process there is a net increase in entropy. As we will see in the next section, this positive enthalpy and entropy points to a different driving force for micellization from that encountered in many aggregation processes.

The influence of alkyl chain length of the surfactant on the free energy, enthalpy and entropy of micellization has been demonstrated by Rosen [19] who listed these parameters as a function of alkyl chain length for sulphoxide surfactants. The results given in Table 2.3 show that the standard free energy of micellization

Tab. 2.2. Thermodynamic quantities for micellization of octylhexaoxyethylene glycol monoether.

Temperature (°C)	*ΔG° (kJ mol^{-1})*	*ΔH° (kJ mol^{-1}) (from c.m.c.)*	*ΔH° (kJ mol^{-1}) (from calorimetry)*	*$T\Delta S^\circ$ (kJ mol^{-1})*
25	−21.3 + 2.1	8.0 + 4.2	20.1 + 0.8	41.8 + 1.0
40	−23.4 + 2.1		14.6 + 0.8	38.0 + 1.0

Tab. 2.3. Change of thermodynamic parameters of micellization of alkyl sulphoxide with increasing chain length of the alkyl group.

Surfactant	*ΔG° (kJ mol^{-1})*	*ΔH° (kJ mol^{-1})*	*$T\Delta S^\circ$ (kJ mol^{-1})*
$C_6H_{13}S(CH_3)O$	−12.0	10.6	22.6
$C_7H_{15}S(CH_3)O$	−15.9	9.2	25.1
$C_8H_{17}S(CH_3)O$	−18.8	7.8	26.4
$C_9H_{19}S(CH_3)O$	−22.0	7.1	29.1
$C_{10}H_{21}S(CH_3)O$	−25.5	5.4	30.9
$C_{11}H_{23}S(CH_3)O$	−28.7	3.0	31.7

becomes increasingly negative as the chain length increases. This is to be expected since the c.m.c. decreases with increasing alkyl chain length. However, ΔH° becomes less positive and $T\Delta S$ becomes more positive with increasing surfactant chain length. Thus, the large negative free energy of micellization is made up of a small positive enthalpy (which decreases slightly with increasing chain surfactant length) and a large positive entropy term $T\Delta S^\circ$, which becomes more positive as the chain is lengthened. The next section shows that these results can be accounted for in terms of the hydrophobic effect, which will be described in detail.

2.3.6
Driving Force for Micelle Formation

Until recently, the formation of micelles was regarded primarily as an interfacial energy process, analogous to the process of coalescence of oil droplets in an aqueous medium. If this was the case, micelle formation would be a highly exothermic process, as the interfacial free energy has a large enthalpy component. As mentioned above, experimental results have clearly shown that micelle formation involves only a small enthalpy change and, indeed, is often endothermic. The negative free energy of micellization is the result of a large, positive entropy. This led to the conclusion that micelle formation must be predominantly an entropy driven process. Two main sources of entropy have been suggested. The first is related to the so called "hydrophobic effect", which was first established from a consideration of the free energy, enthalpy and entropy of transfer of hydrocarbon from water to a liquid hydrocarbon. Table 2.4 lists some results, along with the heat capacity change ΔC_p on transfer from water to a hydrocarbon, as well as $C_p^{\circ,\text{gas}}$, i.e. the heat capacity in the gas phase [2]. The table shows that the principal contribution to ΔG° is the large positive ΔS°, which increases with increasing hydrocarbon chain length, whereas ΔH° is positive, or small and negative. To account for this large positive entropy of transfer several authors [19, 20] have suggested that the

Tab. 2.4. Thermodynamic parameters for transfer of hydrocarbons from water to liquid hydrocarbon at 25 °C.

Hydrocarbon	*ΔG° (kJ mol^{-1})*	*ΔH° (kJ mol^{-1})*	*ΔS° (kJ mol^{-1} K^{-1})*	*ΔC_p° (kJ mol^{-1} K^{-1})*	*$\Delta C_p^{\circ,gas}$ (kJ mol^{-1} K^{-1})*
C_2H_6	−16.4	10.5	88.2	–	–
C_3H_8	−20.4	7.1	92.4	–	–
C_4H_{10}	−24.8	3.4	96.6	−273	−143
C_5H_{12}	−28.8	2.1	105.0	−403	−172
C_6H_{14}	−32.5	0	109.2	−441	−197
C_6H_6	−19.3	−2.1	58.8	227	−134
$C_6H_5CH_3$	−22.7	−1.7	71.4	−265	−155
$C_6H_5C_2H_5$	−26.0	−2.0	79.8	−319	−185
$C_6H_5C_3H_8$	−29.0	−2.3	88.2	−395	–

water molecules around a hydrocarbon chain are ordered, forming "clusters" or "icebergs". On transfer of an alkane from water to a liquid hydrocarbon, these clusters are broken, thus releasing water molecules that now have a higher entropy. This accounts for the large entropy of transfer of an alkane from water to a hydrocarbon medium. This effect is also reflected in the much higher heat capacity change on transfer, ΔC_p°, when compared with the heat capacity in the gas phase, C_p°.

The above effect is also operative on transfer of surfactant monomer to a micelle, during the micellization process. Surfactant monomers will also contain "structured" water around their hydrocarbon chain. On transfer of such monomers to a micelle, these water molecules are released and they have a higher entropy.

The second source of entropy increase on micellization may arise from the increase in flexibility of the hydrocarbon chains on their transfer from an aqueous to a hydrocarbon medium [21, 22]. The orientations and bendings of an organic chain are probably more restricted in an aqueous phase than in an organic phase.

Notably, with ionic and zwitterionic surfactants, an additional entropy contribution, associated with the ionic head groups, must be considered. Upon partial neutralization of the ionic charge by the counter ions when aggregation occurs, water molecules are released. This will be associated with an entropy increase that should be added to the entropy increase resulting from the above hydrophobic effect. However, the relative contribution of the two effects is difficult to assess quantitatively.

2.3.7
Micellization in Other Polar Solvents

In strongly polar solvents, such as formamide and ethylene glycol, micelles are formed with qualitatively the same features as in water. Figure 2.6 illustrates this, showing the surface tension results for cetylytrimethylammonium bromide (CTAB) in formamide and ethylene glycol [23]. The c.m.c. is higher in formamide than in water (100 mM compared with 1 mM in water). The micelles are also smaller and the aggregation number is lower.

The results in polar solvents, other than water, indicate that self-assembly is much less cooperative. Consequently, the degree of counter-ion binding is also lower.

2.3.8
Micellization in Non-Polar Solvents

For simple amphiphilic compounds, the association is of low co-operativity in non-polar solvents and leads only to smaller and polydisperse aggregates. The aggregation number is in the range 1–5. However, introduction of even quite small amounts of water can induce a co-operative self-assembly, leading to inverse micelles.

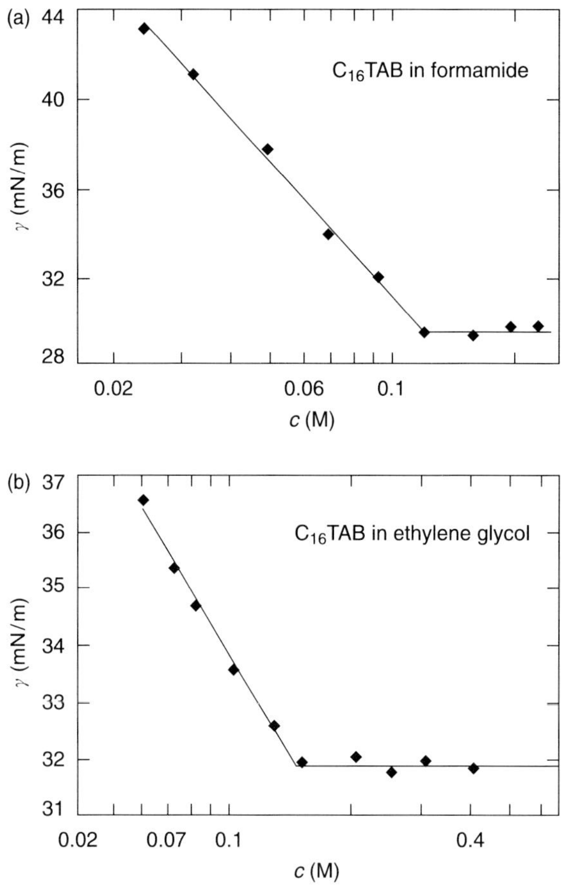

Fig. 2.6. Surface tension –log *C* results for (a) CTAB in formamide and (b) ethylene glycol [23].

2.4
Micellization in Surfactant Mixtures (Mixed Micelles)

Most industrial applications use more than one surfactant molecule in the formulation. It is, therefore, necessary to predict the type of possible interactions and whether this leads to synergistic effects. Two general cases may be considered: Surfactant molecules with no net interaction (with similar head groups) and systems with net interaction. These are discussed separately below [2].

2.4.1
Surfactant Mixtures with no Net Interaction

This is the case when mixing two surfactants with the same head group but with different chain lengths. In analogy with the hydrophilic–lipophilic balance (HLB) for surfactant mixtures (Chapter 6), one can also assume the c.m.c. of a surfactant mixture (with no net interaction) to be an average of the two c.m.c.s of the single components [2],

$$\text{c.m.c.} = x_1\text{c.m.c.}_1 + x_2\text{c.m.c.}_2 \tag{2.30}$$

where x_1 and x_2 are the mole fractions of the respective surfactants in the system. However, the mole fractions should not be those in the whole system, but those inside the micelle. This means that Eq. (2.30) should be modified,

$$\text{c.m.c.} = x_1^{\mathrm{m}}\text{c.m.c.}_1 + x_2^{\mathrm{m}}\text{c.m.c.}_2 \tag{2.31}$$

The superscript m indicates that the values are inside the micelle. If x_1 and x_2 are the solution composition, then,

$$\frac{1}{\text{c.m.c.}} = \frac{x_1}{\text{c.m.c.}_1} + \frac{x_2}{\text{c.m.c.}_2} \tag{2.32}$$

The molar composition of the mixed micelle is given by

$$x_1^{\mathrm{m}} = \frac{x_1\text{c.m.c.}_2}{x_1\text{c.m.c.}_2 + x_2\text{c.m.c.}_1} \tag{2.33}$$

Figure 2.7 shows the calculated c.m.c. and the micelle composition as a function of solution composition using Eqs. (2.32) and (2.33) for three cases where c.m.c.$_2$/c.m.c.$_1$ = 1, 0.1 and 0.01. As can be seen, the c.m.c. and micellar composition change dramatically with solution composition when the c.m.c.s of the two surfactants vary considerably, i.e. when the ratio of c.m.c.s is far from 1. This fact is used when preparing microemulsions (see Chapter 10) where the addition of a medium-chain alcohol (like pentanol or hexanol) changes the properties considerably. If component 2 is much more surface active, i.e. c.m.c.$_2$/c.m.c.$_1$ $\ll$ 1, and it is present in low concentrations (x_2 is of the order of 0.01), then from Eq. (2.33) $x_1^{\mathrm{m}} \sim x_2^{\mathrm{m}} \sim 0.5$, i.e. at the c.m.c. of the systems the micelles are composed of up to 50% component 2. This illustrates the role of contaminants in surface activity, e.g. dodecyl alcohol in sodium dodecyl sulphate (SDS).

Figure 2.8 shows the c.m.c. as a function of molar composition of the solution and in the micelles for a mixture of SDS and nonylphenol with 10 moles ethylene oxide (NP-E$_{10}$). If the molar composition of the micelles is used as the x-axis, the c.m.c. is more or less the arithmetic mean of the c.m.c.s of the two surfactants. If, however, the molar composition in the solution is used as the x-axis (which at the c.m.c. is equal to the total molar concentration), then the c.m.c. of the mixture shows a dramatic decrease at low fractions of NP-E$_{10}$. This decrease is due to the preferential absorption of NP-E$_{10}$ in the micelle. This higher absorption occurs because NP-E$_{10}$ surfactant has a higher hydrophobicity than SDS.

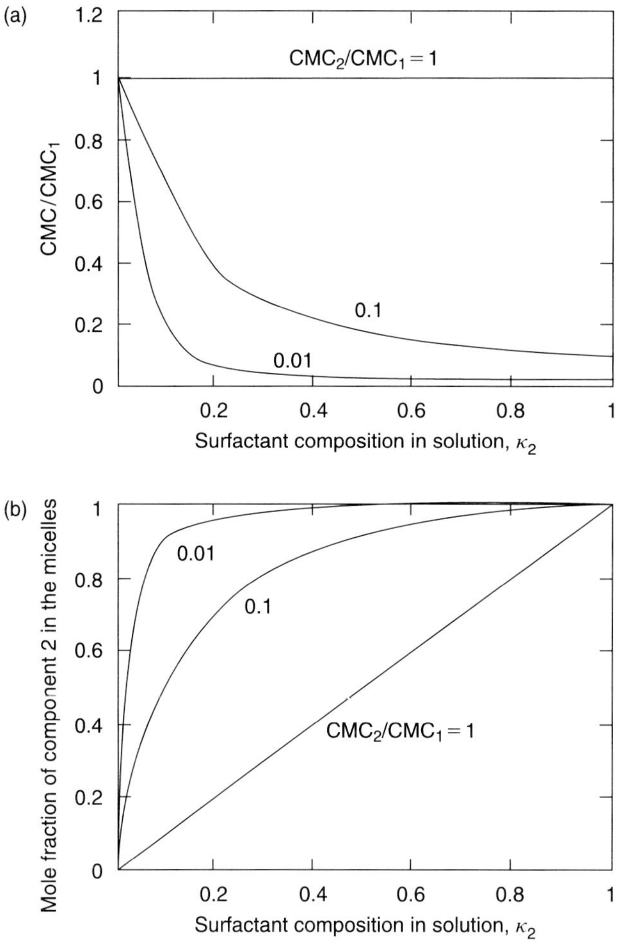

Fig. 2.7. Calculated c.m.c. (a) and micellar composition (b) as a function of solution composition for three ratios of c.m.c.s.

2.4.2
Surfactant Mixtures with a Net Interaction

With many industrial formulations, surfactants of different kinds are mixed together, for example anionics and nonionics. Nonionic surfactant molecules shield the repulsion between the negative head groups in the micelle and hence there will be a net interaction between the two types of molecules. Another example is the case when anionic and cationic surfactants are mixed, whereby very strong interaction occurs between the oppositely charged surfactant molecules. To account for this interaction, Eq. (2.31) has to be modified by introducing activity coefficients of the surfactants, f_1^m and f_2^m in the micelle (Eq. 2.34).

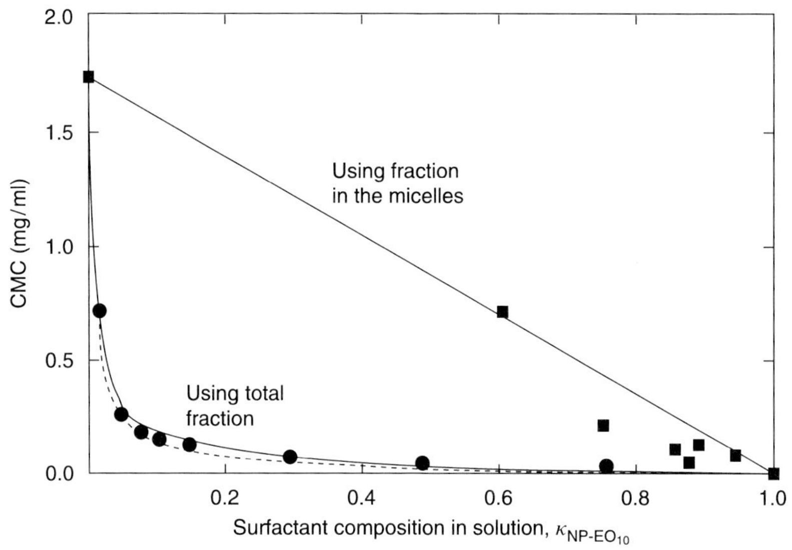

Fig. 2.8. C.m.c. as a function of surfactant composition, x_1, or micellar surfactant composition, x_1^m for the system SDS + NP-E_{10}.

$$\text{c.m.c.} = x_1^m f_1^m \text{c.m.c.}_1 + x_2^m f_2^m \text{c.m.c.}_2 \tag{2.34}$$

An expression for the activity coefficients can be obtained using the regular solutions theory,

$$\ln f_1^m = (x_2^m)^2 \beta \tag{2.35}$$

$$\ln f_2^m = (x_2^m)^2 \beta \tag{2.36}$$

where β is an interaction parameter between the surfactant molecules in the micelle. A positive β means that there is a net repulsion between the surfactant molecules in the micelle, whereas a negative β means a net attraction.

The c.m.c. of the surfactant mixture and the composition x_1 are given by Eqs. (2.37) and (2.38).

$$\frac{1}{\text{c.m.c.}} = \frac{x_1}{f_1^m \text{c.m.c.}_1} + \frac{x_2}{f_2^m \text{c.m.c.}_2} \tag{2.37}$$

$$x_1^m = \frac{x_1 f_2^m \text{c.m.c.}_2}{x_1 f_2^m \text{c.m.c.}_2 + x_2 f_2^m \text{c.m.c.}_1} \tag{2.38}$$

Figure 2.9 shows the effect of increasing β on the c.m.c. and micellar composition for two surfactants with a c.m.c. ratio of 0.1. As β becomes more negative, the c.m.c. of the mixture decreases. Values of β in the region of -2 are typical for

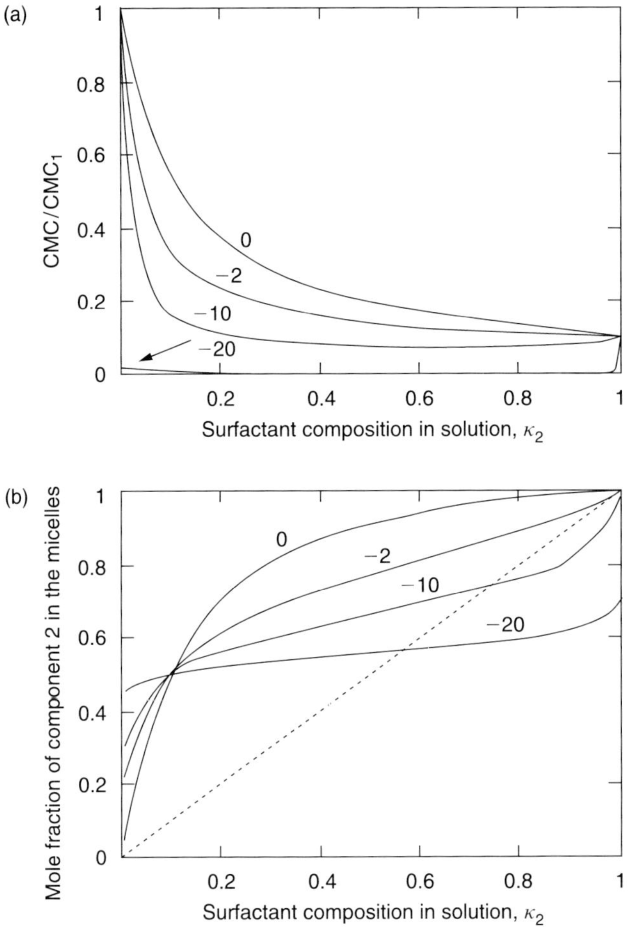

Fig. 2.9. C.m.c. (a) and micellar composition (b) for various β for a system with a c.m.c.$_2$/c.m.c.$_1$ ratio of 0.1.

anionic/nonionic mixtures, whereas values in the region of −10 to −20 are typical of anionic/cationic mixtures. With increasingly negative β, the mixed micelles tend towards a mixing ratio of 50:50, reflecting the mutual electrostatic attraction between the surfactant molecules.

Both the predicted c.m.c. and micellar composition depend on the ratio of the c.m.c.s as well as on β. When the c.m.c.s of the single surfactants are similar, the predicted c.m.c. is very sensitive to small variations in β. Conversely, when the ratio of the c.m.c.s is large, the predicted value of the mixed c.m.c. and the micellar composition are insensitive to variations of β. For mixtures of nonionic and ionic surfactants, β decreases with increasing electrolyte concentration. This is due to the screening of the electrostatic repulsion on the addition of electrolyte. With some surfactant mixtures, β decreases with rising temperature, i.e. the net attraction decreases with increasing temperature.

2.5 Surfactant–Polymer Interaction

Mixtures of surfactants and polymers are very common in many industrial formulations. With many suspension and emulsion systems stabilized with surfactants, polymers are added for several reasons, e.g. as suspending agents ("thickeners") to prevent sedimentation or creaming of these systems. In many other systems, such as in personal care and cosmetics, water-soluble polymers are added to enhance the function of the system, e.g. in shampoos, hair sprays, lotions and creams. The interaction between surfactants and water-soluble polymers furnishes synergistic effects, e.g. enhancing the surface activity, stabilizing foams and emulsions, etc. It is, therefore important to study systematically the interaction between surfactants and water-soluble polymers.

One of the earliest studies of surfactant/polymer interaction came from surface tension measurements. Figure 2.10 shows some typical results for the effect of addition of poly(vinylpyrrolidone) (PVP) on the $\gamma - \log C$ curves of SDS [23].

In a system of fixed polymer concentration and varying surfactant concentrations, two critical concentrations appear, denoted T_1 and T_2. T_1 represents the concentration at which interaction between the surfactant and polymer first occurs. This is sometimes termed the critical aggregation concentration (CAC), i.e. the onset of association of surfactant to the polymer. Because of this there is no further

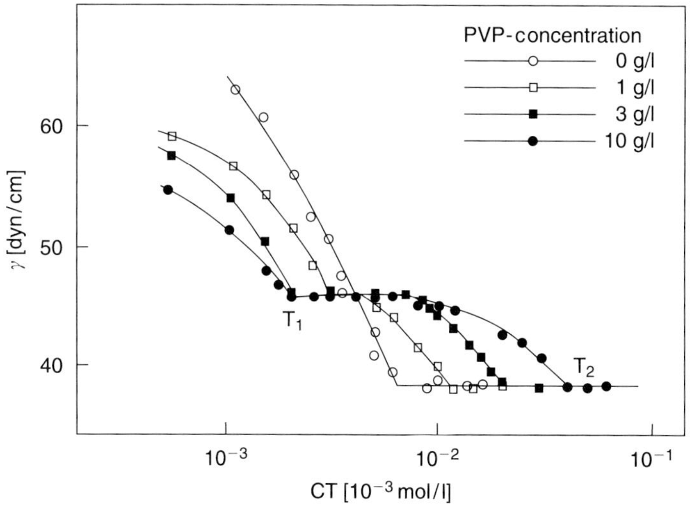

Fig. 2.10. γ versus $\log C$ curves for SDS solutions in the presence of different concentrations of PVP.

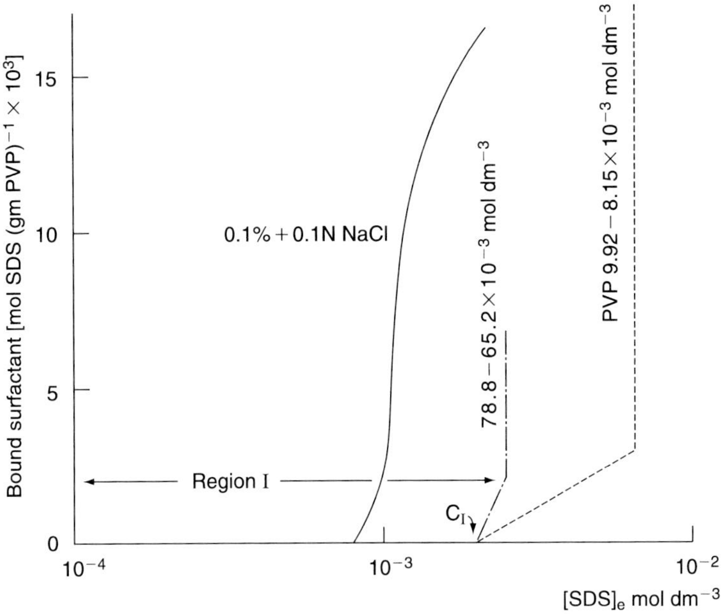

Fig. 2.11. Binding isotherms of a surfactant to a water-soluble polymer.

increase in surface activity and thus no lowering of surface tension. T_2 represents the concentration at which the polymer becomes saturated with surfactant. Since T_1 is generally lower than the c.m.c. of the surfactant in the absence of polymer, "adsorption" or "aggregation of SDS on or with the polymer is more favourable than normal micellization. As the polymer is saturated with surfactant (i.e. beyond T_2) the surfactant monomer concentration and the activity starts to increase again and there is a lowering of γ until the monomer concentration reaches the c.m.c., after which γ remains virtually constant and normal surfactant micelles begin to form.

The above picture is confirmed if the association of surfactant is directly monitored (e.g. by using surfactant selective electrodes, by equilibrium dialysis or by some spectroscopic technique). Binding isotherms are illustrated in Figure 2.11. At low surfactant concentration there is no significant interaction (binding). At the CAC, a strongly co-operative binding is indicated and at higher concentrations a plateau is reached. Further increases in surfactant concentration produces "free" surfactant molecules until the surfactant activity or concentration joins the curve obtained in the absence of polymer. The binding isotherms of Figure 2.11 show a strong analogy with micelle formation and the interpretation of these isotherms in terms of a depression of the c.m.c.

Several conclusions could be drawn from the experimental binding isotherms of mixed surfactant/polymer solutions: (1) The CAC/c.m.c. depends only weakly on polymer concentration over wide ranges. (2) CAC/c.m.c. is, to a good approxima-

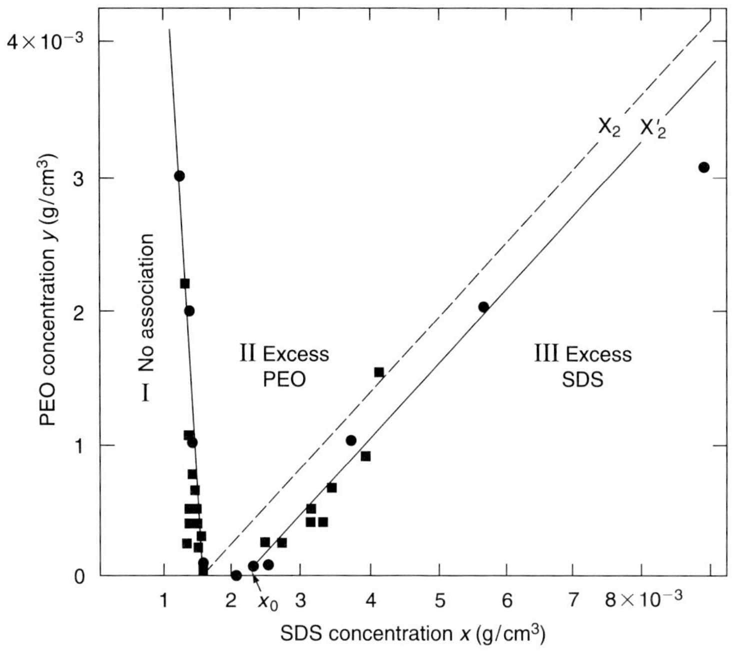

Fig. 2.12. Association between surfactant and homopolymer in different concentration domains [24].

tion, independent of polymer molecular weight down to low values. For very low molecular weight the interaction is weakened. (3) The plateau binding increases linearly with polymer concentration. (4) Anionic surfactants show a marked interaction with most homopolymers (e.g. PEO and PVP) while cationic surfactants show a weaker but still significant interaction. Nonionic and zwitterionic surfactants only rarely show a distinct interaction with homopolymers.

Figure 2.12 gives a schematic representation of the association between surfactants and polymers for a wide range of concentration of both components [24]. At low surfactant concentration (region I) there is no significant association at any polymer concentration. Above the CAC (region II), association increases up to a surfactant concentration, increasing linearly with increasing polymer concentration. In region III, association is saturated and the surfactant monomer concentration increases till region IV is reached, where there is co-existence of surfactant aggregates at the polymer chains and free micelles.

2.5.1
Factors Influencing the Association Between Surfactant and Polymer

Several factors influence the interaction between surfactant and polymer: (1) Temperature; increasing temperature generally increases the CAC, i.e. the interaction becomes less favourable. (2) Addition of electrolyte; this generally decreases the CAC, i.e. it increases the binding. (3) Surfactant chain length; an increase in the alkyl chain length decreases the CAC, i.e. it increases association. A plot of log(CAC) versus the number of carbon atoms, n, is linear (similar to the log[c.m.c.] – n relationship obtained for surfactants alone). (4) Surfactant structure; alkyl benzene sulphonates are similar to SDS, but introduction of EO groups in the chain weakens the interaction. (5) Surfactant classes; weaker interaction is generally observed with cationics than with anionics. However, the interaction can be promoted by using a strongly interacting counter ion for the cationic (e.g. CNS^-). Interaction between ethoxylated surfactants and nonionic polymers is weak. The interaction is stronger with alkyl phenol ethoxylates. (6) Polymer molecular weight; a minimum molecular weight of ~4000 for PEO and PVP is required for "complete" interaction. (7) Amount of polymer; the CAC seems to be insensitive to (or lowers slightly) with increasing polymer concentration. T_2 increases linearly with increasing polymer concentration. (8) Polymer structure and hydrophobicity; several uncharged polymer, such as PEO, PVP and poly(vinyl alcohol) (PVOH), interact with charged surfactants. Many other uncharged polymers interact weakly with charged surfactants, e.g. hydroxyethyl cellulose (HEC), dextran and polyacrylamide (PAAm). The following orders of increased interaction have been listed for (1) anionic surfactants: PVOH $<$ PEO $<$ MEC (methyl cellulose) $<$ PVAc [partially hydrolyzed poly(vinyl acetate)] $<$ PPO $\sim$ PVP, and for (2) cationic surfactants: PVP $<$ PEO $<$ PVOH $<$ MEC $<$ PVAc $<$ PPO. The position of PVP can be explained by the slight positive charge on the chain, which causes repulsion with cations and attraction with anionics.

2.5.2
Interaction Models

NMR data has shown that every "bound" surfactant molecule experiences the same environment, i.e. the surfactant molecules might be bound in micelle-like clusters, but with smaller size. Assuming that each polymer molecule consists of several "effective segments" of mass M_s (minimum molecular weight for interaction to occur), then each segment will bind a cluster of n surfactant anions, D^-, and the binding equilibrium may be represented by,

$$P + nD^- \rightleftharpoons PD_n^{n-} \tag{2.39}$$

and the equilibrium constant is given by,

$$K = \frac{[PD_n^{n-}]}{[P][D^-]^n} \tag{2.40}$$

K is obtained from the half-saturation condition,

$$K = [D^-]_{1/2}^n \tag{2.41}$$

By varying n and using the experimental binding isotherms one obtains $M_s = 1830$ and $n = 15$. The free energy of binding is given by,

$$\Delta G^\circ = -RT \ln K^{1/n} \tag{2.42}$$

ΔG° was found to be -5.07 kcal mol^{-1}, which is close to that for surfactants.

Najaragan [25] has introduced a comprehensive thermodynamic treatment of surfactant/polymer interaction. The aqueous solution of surfactant and polymer was assumed to contain both free micelles and "micelles" bound to the polymer molecule. The total surfactant concentration, X_t, is partitioned into single dispersed surfactant, X_1, surfactant in free micelles, X_f, and surfactant bound as aggregates, X_b,

$$X_t = X_1 + g_f (K_f X_1)^{g_f} + g_b n X_p \left[\frac{(K_b X_1)^{g_b}}{1 + (K_b X_1)^{g_b}} \right] \tag{2.43}$$

g_f is the average aggregation number of free micelles, K_f is the intrinsic equilibrium constant for formation of free micelles, n is the number of binding sites for surfactant aggregates of average size g_b, K_b is the intrinsic equilibrium constant for binding surfactant on the polymer and X_p is the total polymer concentration (mass concentration is nX_p).

Polymer–micelle complexation may affect the conformation of the polymer, but is assumed not affect K_b and g_b. The relative magnitudes of K_b, K_f and g_b determine whether complexation with the polymer occurs as well as the critical surfactant concentration exhibited by the system. If $k_f > k_b$ and $g_b = g_f$ then free micelles occur in preference to complexation. If $K_f < K_b$ and $g_b = g_f$, then micelles bound to polymer occur first. If $K_f < K_b$, but $g_b \ll g_f$, then the free micelles can occur prior to saturation of the polymer. A first critical surfactant concentration (CAC) occurs close to $X_1 = K_b^{-1}$. A second critical concentration occurs near $X_1 = K_f^{-1}$. Depending on the magnitude of nX_p, one may observe only one critical concentration over a finite range of surfactant concentrations.

Figure 2.13 shows the relationship between X_1 and X_t for different polymer concentrations (SDS/PEO system), using $K_b = 319$, $K_f = 120$, $g_b = 51$ and $g_f = 54$.

In the region from 0 to A, the surfactant molecules remain singly dispersed. In the region from A to B, polymer-bound micelles occur; X_1 increases very little in this region (large size of polymer bound micelles). If g_b is small (say 10), then X_1 should increase more significantly in this region. If nX_p is small, the region AB is confined to a narrow surfactant concentration range. If nX_p is very large, the saturation point B may not be reached. At B the polymer is saturated with surfactant. In the region AC, increases in X_t are accompanied by an increase in X_1. At C the formation of free micelles becomes possible; CD denotes the surfactant concentration range over which any further addition of surfactant results in the formation of free micelles. The point C depends on the polymer mass concentration (nX_p).

The above theoretical predictions were verified by the results of Guilyani and Wolfram [26] using specific ion electrodes. This is illustrated in Figure 2.14.

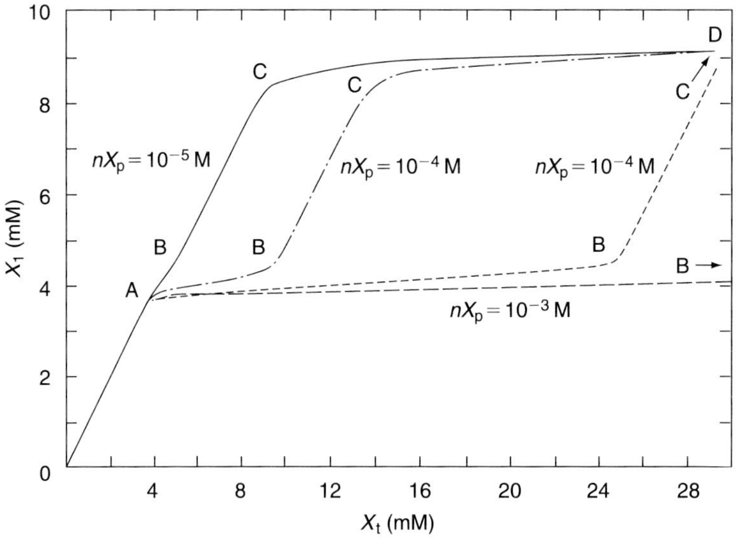

Fig. 2.13. Variation of X_1 with X_t for the SDS/PEO system.

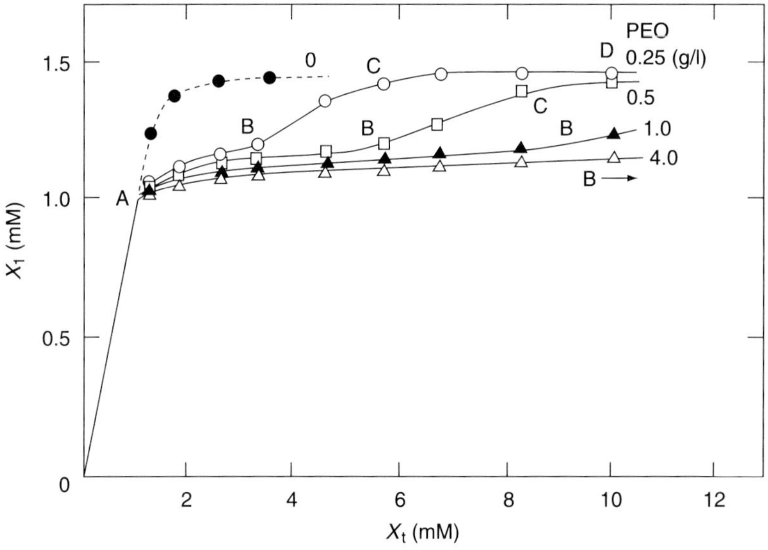

Fig. 2.14. Experimentally measured values of X_1 versus X_t for SDS/PEO system.

2.5.3
Driving Force for Surfactant–Polymer Interaction

The driving force for polymer–surfactant interaction is the same as that for the process of micellization (see above). As with micelles the main driving force is the reduction of hydrocarbon/water contact area of the alkyl chain of the dissolved surfactant. A delicate balance between several forces is responsible for the surfactant/polymer association. For example, aggregation is resisted by the crowding of the ionic head groups at the surface of the micelle. Packing constraints also resist association. Molecules that screen the repulsion between the head groups, e.g. electrolytes and alcohol, promote association. A polymer molecule with hydrophobic and hydrophilic segments (which is also flexible) can enhance association by ion–dipole association between the dipole of the hydrophilic groups and the ionic head groups of the surfactant. In addition, contact between the hydrophobic segments of the polymer and the exposed hydrocarbon areas of the micelles can enhance association. With SDS/PEO and SDS/PVP, the association complexes are approximately three monomer units per molecule of aggregated surfactant.

2.5.4
Structure of Surfactant–Polymer Complexes

Generally, there are two alternative pictures of mixed surfactant/polymer solutions, one describing the interaction in terms of a strongly co-operative association or binding of the surfactant to the polymer chain and one in terms of a micellization of surfactant on or in the vicinity of the polymer chain. For polymers with hydrophobic groups the binding approach is preferred, whereas for hydrophilic homopolymers the micelle formation picture is more likely. The latter picture has been suggested by Cabane [27], who proposed a structure in which the aggregated SDS is surrounded by macromolecules in a loopy configuration. A schematic picture of this structure, sometimes referred to as "pearl-necklace model", is given in Figure 2.15.

The consequences of the above model are: (1) More favourable free energy of association (CAC < c.m.c.) and increased ionic dissociation of the aggregates. (2) An altered environment of the CH_2 groups of the surfactant near the head group. The micelle sizes are similar with polymer present and without, and the aggregation numbers are typically similar or slightly lower than those of micelles forming in the absence of a polymer. In the presence of a polymer, the surfactant chemical potential is lowered with respect to the situation without polymer [28].

2.5.5
Surfactant–Hydrophobically Modified Polymer Interaction

Water-soluble polymers are modified by grafting a low amount of hydrophobic groups (of the order of 1% of the monomers reacted in a typical molecule), result-

Fig. 2.15. Schematic representation of the topology of surfactant/polymer complexes (according to Cabane [27]).

ing in the formation of "associative structures". These molecules are referred to as associative thickeners and are used as rheology modifiers in many industrial applications, e.g. paints and personal care products. An added surfactant will interact strongly with the hydrophobic groups of the polymer, leading to a strengthened association between the surfactant molecules and the polymer chain. Figure 2.16 gives a schematic picture for the interaction between SDS and hydrophobically modified hydroxyethyl cellulose (HM-HEC), showing the interaction at various surfactant concentrations [1].

Initially the surfactant monomers interact with the hydrophobic groups of the HM polymer, and at some surfactant concentration (CAC) the micelles can cross-link the polymer chains. At higher surfactant concentrations, the micelles, which are now abundant, will no longer be shared between the polymer chains, i.e. the cross-links are broken. These effects are reflected in the variation of viscosity with surfactant concentration for HM polymer (Figure 2.17). The viscosity of the polymer rises with increasing surfactant concentration, reaching a maximum at an optimum concentration (maximum cross-links) and then decreases with further increase of surfactant concentration. For the unmodified polymer, the changes in viscosity are relatively small.

2.5.6
Interaction Between Surfactants and Polymers with Opposite Charge (Surfactant–Polyelectrolyte Interaction)

The case of surfactant polymer pairs in which the polymer is a polyion and the surfactant is also ionic, but of opposite charge, is of special importance in many

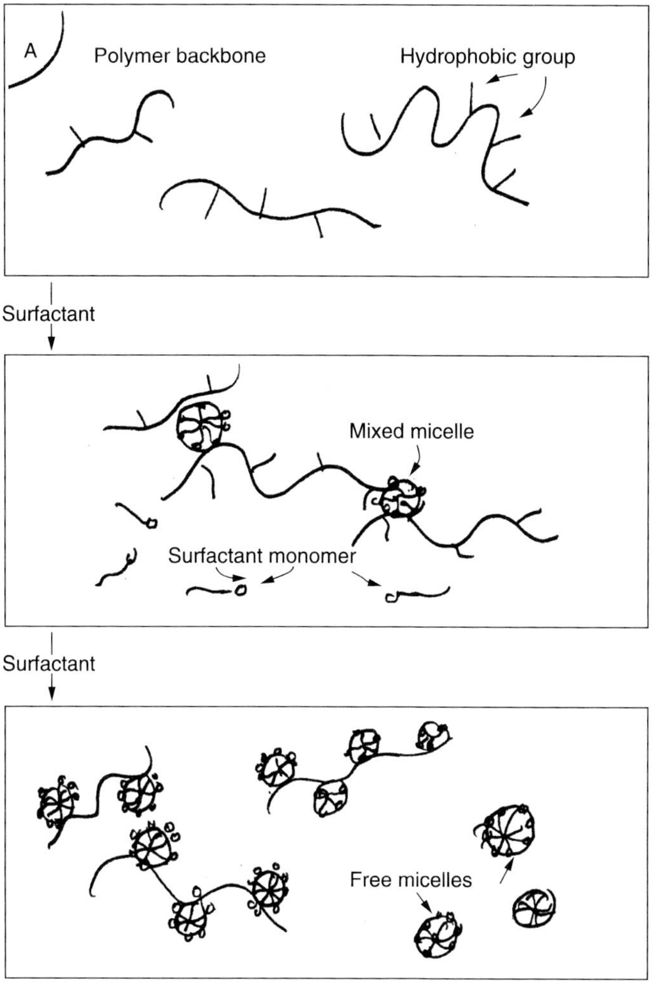

Fig. 2.16. Schematic representation of the interaction between surfactant and HM polymer.

cosmetic formulations, e.g. as hair conditioners. This is illustrated by the interaction between SDS and cationically modified cellulosic polymer (Polymer JR, Union Carbide) (Figure 2.18) using surface tension (γ) measurements [29]. The $\gamma - \log C$ curves for SDS in the presence and absence of the polyelectrolyte are shown in Figure 2.18, which also shows the appearance of the solutions. At low surfactant concentration, there is a synergistic lowering of surface tension, i.e. the surfactant–polyelectrolyte complex is more surface active. The low surface tension is also present in the precipitation zone. At high surfactant concentrations, γ approaches that of the polymer-free surfactant in the micellar region. These trends are schematically illustrated in Figure 2.19.

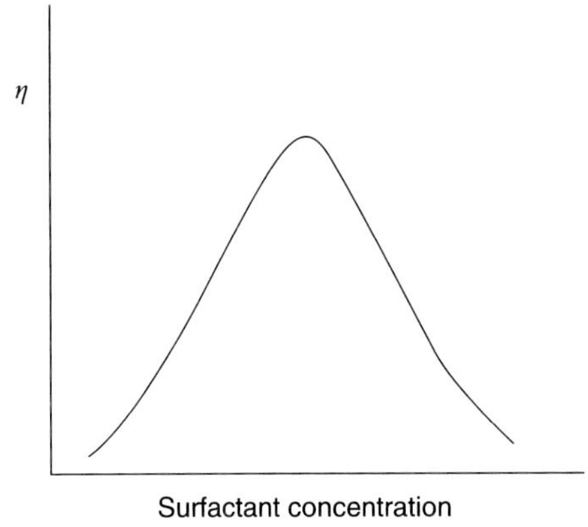

Fig. 2.17. Viscosity–surfactant concentration relationship for HM-modified polymer solutions.

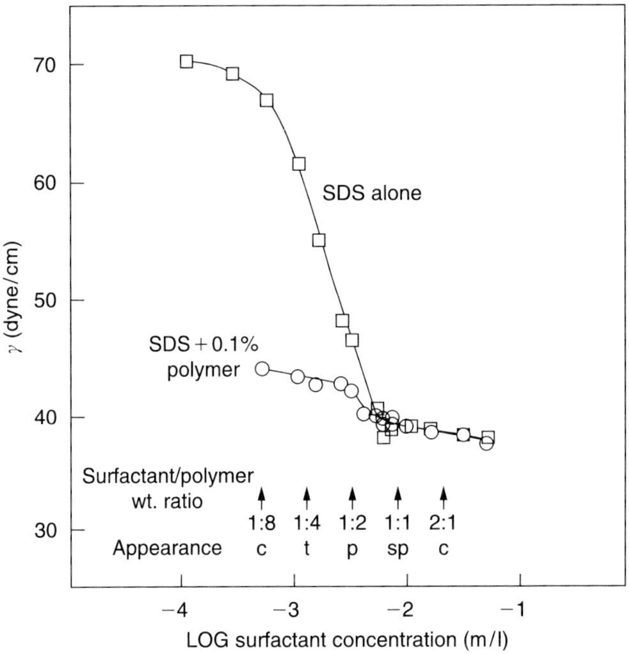

Fig. 2.18. γ – log C curves of SDS with and without addition of polymer (0.1% JR 400) – c (clear), t (turbid), p (precipitate), sp (slight precipitate).

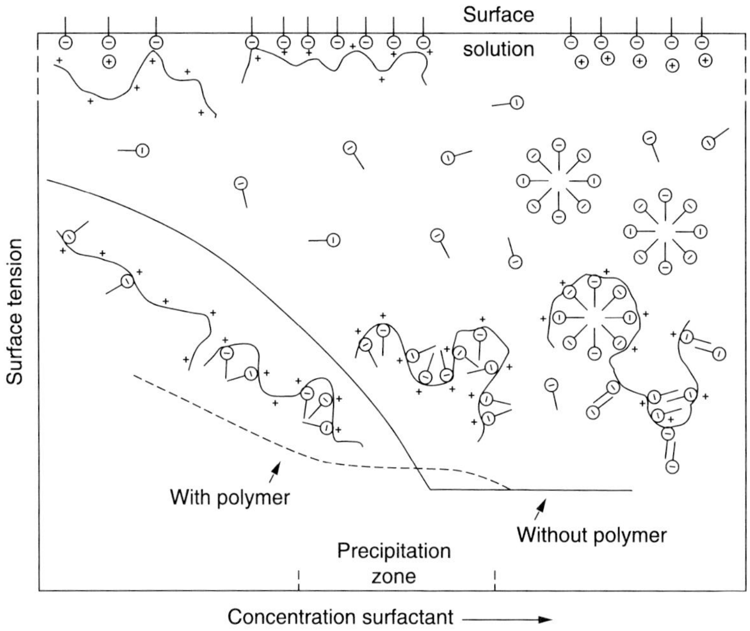

Fig. 2.19. Schematic representation of surfactant–polyelectrolyte interaction.

The surfactant–polyelectrolyte interaction has several consequences on application in hair conditioners. The most important effect is the high foaming power of the complex. Maximum foaming occurs in the region of highest precipitation, i.e. maximum hydrophobization of the polymer. Probably, the precipitate can stabilize the foam. Direct determination of the amount of surfactant bound to the polyelectrolyte chains revealed several interesting features. Binding occurs at very low surfactant concentration (1/20th of the c.m.c.). The degree of binding β reached 0.5 ($\beta = 1$ corresponds to a bound DS^- ion for each ammonium group). β versus SDS concentration curves were identical for polymeric homologues with a degree of cationic substitution (CS) > 0.23. Precipitation occurred when $\beta = 1$.

The binding of cationic surfactants to anionic polyelectrolytes also shows some interesting features. The binding affinity depends on the nature of the polyanion. Addition of electrolytes increases the steepness of binding, but the binding occurs at higher surfactant concentration as the electrolyte concentration is increased. Increasing the alkyl chain length of the surfactant increases binding, a process that is similar to micellization.

Viscometric measurements have revealed a rapid increase in the relative viscosity at a critical surfactant concentration. However, the behaviour depends on the type of polyelectrolyte used. As an illustration, Figure 2.20 shows the viscosity–SDS concentration curves for two types of cationic polyelectrolyte: JR-400 (cationically modified cellulosic) and Reten (an acrylamide/(β-methylacryloxytrimethyl)ammonium chloride copolymer, ex Hercules).

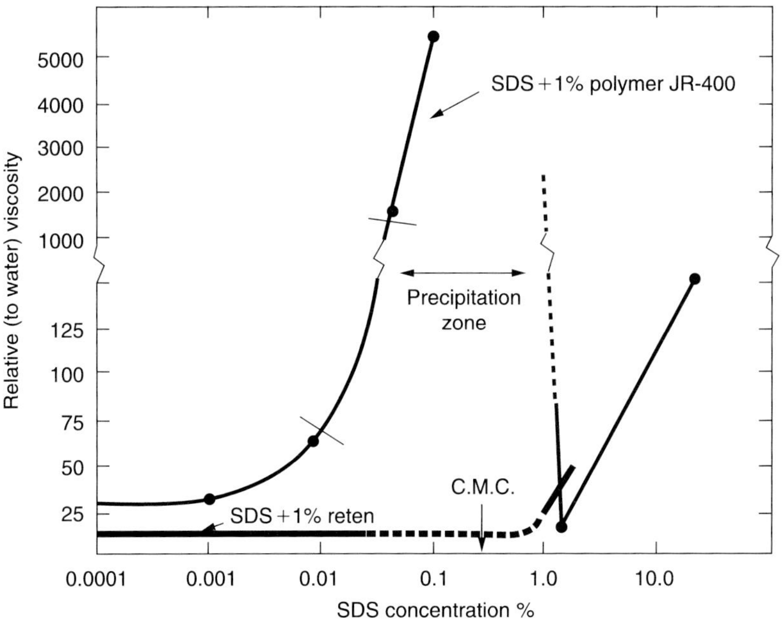

Fig. 2.20. Relative viscosity of 1% JR-400 and 1% Reten as a function of SDS concentration.

The difference between the two polyelectrolytes is striking and suggests little change in the conformation of Reten on addition of SDS, but strong intermolecular association between polymer JR-400 and SDS.

References

1 (a) B. Lindman: *Surfactants*. T. F. Tadros (ed.): Academic Press, London, 2003, pp. 1984. (b) K. Holmberg, B. Jonsson, B. Kronberg, B. Lindman: *Surfactants and Polymers in Aqueous Solution*, 2nd edition, John Wiley & Sons, USA, 2003.

2 J. Istraelachvili: *Intermolecular and Surface Forces, with Special Applications to Colloidal and Biological Systems*, Academic Press, London, 1985, pp. 251.

3 P. Mukerjee, K J Mysels: *Critical Micelle Concentrations of Aqueous Surfactant Systems*, National Bureau of Standards Publication, Washington, 1971.

4 P. H. Elworthy, A. T. Florence, C. B. Macfarlane: *Solubilization by Surface Active Agents*, Chapman & Hall, London, 1968.

5 K. Shinoda, T. Nagakawa, B. I. Tamamushi, T. Isemura: *Colloidal Surfactants, Some Physicochemical Properties*, Academic Press, London, 1963.

6 J. W. McBain, *Trans. Faraday Soc.*, **1913**, *9*, 99.

7 N. K. Adam, *J. Phys. Chem.*, **1925**, *29*, 87.

8 G. S. Hartley: *Aqueous Solutions of Paraffin Chain Salts*, Hermann and Cie, Paris, 1936.

9 J. W. McBain: *Colloid Science*, Heath, Boston, 1950.

10 W. D. Harkins, W. D. Mattoon, M. L. Corrin, *J. Am. Chem. Soc.*, **1946**, *68*, 220; *J. Colloid Sci.*, **1946**, *1*, 105.

11 P. Debye, E. W. Anaker, *J. Phys. Colloid Chem.*, **1951**, *55*, 644.

12 E. A. G. Anainsson, S. N. Wall, *J. Phys. Chem.*, **1975**, *78*, 1024; **1975**, *79*, 857.

13 E. A. G. Aniansson, S. N. Wall, M. Almagren, H. Hoffmann, W. Ulbricht, R. Zana, J. Lang, C. Tondre, *J. Phys. Chem.*, **1976**, *80*, 905.

14 J. Rassing, P. J. Sams, E. Wyn-Jones, *J. Chem. Soc., Faraday Trans. II*, **1974**, *70*, 1247.

15 M. J. Jaycock, R. H. Ottewill, *Fourth Int. Congr. Surf. Activity*, **1964**, *2*, 545.

16 T. Okub, H. Kitano, T. Ishiwatari, N. Isem, *Proc. R. Soc.*, **1979**, *81*, A36.

17 J. N. Phillips, *Trans. Faraday Soc.*, **1955**, *51*, 561.

18 M. Kahlweit, M. Teubner, *Adv. Colloid Interface Sci.*, **1980**, *13*, 1.

19 M. L. Rosen: *Surfactants and Interfacial Phenomena*, Wiley-Interscience, New York, 1978.

20 C. Tanford: *The Hydrophobic Effect*, 2nd edition, Wiley, New York, 1980.

21 G. Stainsby, A. E. Alexander, *Trans. Faraday Soc.*, **1950**, *46*, 587.

22 R. H. Arnow, L. Witten, *J. Phys. Chem.*, **1960**, *64*, 1643.

23 M. M. Breuer, I. D. Robb, *Chem. Ind.*, **1972**, 530.

24 B. Cabane, R. Duplessix, *J. Phys. (Paris)*, **1982**, *43*, 1529.

25 R. Nagarajan, *Colloids Surf.*, **1985**, *13*, 1.

26 T. Gilyani, E. Wolfram, *Colloids Surf.*, **1981**, *3*, 181.

27 B. Cabane, *J. Phys. Chem.*, **1977**, *81*, 1639.

28 D. F. Evans, H. Winnerstrom: *The Colloidal Domain. Where Physics, Chemistry, Biology and Technology Meet*, John Wiley and Sons VCH, New York, 1994, p. 312.

29 E. D. Goddard, *Colloids Surf.*, **1986**, *19*, 301.

3
Phase Behavior of Surfactant Systems

In dilute solutions surfactants tend to aggregate to form micelles with aggregation numbers in the region of 50–100. These micelles are in most cases spherical units, producing an isotropic solution (L_1 phase) with low viscosity. However, these micelles may grow, forming cylindrical micelles that are anisotropic and show features of structures on a macroscopic scale, e.g. flow birefringence. Even in this case, the solution appears as a single phase. However, at much higher surfactant concentrations, a series of mesomorphic phases, referred to as liquid crystalline phases appear whose structure depends on the surfactant nature and concentration. In general one can distinguish between three types of behaviour for a surfactant or polar lipid as the concentration is increased [1]. (1) Surfactants with high solubility in water, whereby the physicochemical properties such as viscosity and light scattering vary smoothly from the critical micelle concentration (c.m.c.) region up to saturation. In this case, the micelles remain small and are, in general, spherical. (2) Surfactants with high water solubility but as the concentration increases they show dramatic changes in their physicochemical properties such as viscosity and flow birefringence. In this case, there are marked changes in the self-assembly, i.e. formation of liquid crystalline structures. (3) Surfactants with low water solubility that show phase separation at low concentrations, e.g. separation of solid hydrated phase.

For surfactants with short-chain alkyl groups, C_8 or C_{10}, the solution shows a gradual variation in properties with no phase separation. Figure 3.1 illustrates this, showing the variation of relative viscosity with micelle volume fraction for spherical micelles for a system of $C_{12}E_5$ with an equal weight of solubilized decane [2]. The viscosity varies smoothly and approximately as predicted for a dispersion of spherical micelles.

For longer chain surfactants, e.g. C_{14}, the viscosity starts to increase rapidly at a critical concentration. Figure 3.2 illustrates this through the variation of zero shear viscosity with surfactant concentration for $C_{16}E_6$ [1], showing a rapid increase in viscosity above 0.1 wt%. Here, the surfactant micelles grow – at first to short prolates or cylinders and then to long cylindrical or thread-like micelles [1] (Figure 3.3). In some cases very long thread-like micelles, varying from 10 nm to several hundred nms, are produced.

Applied Surfactants: Principles and Applications. Tharwat F. Tadros

ISBN: 3-527-30629-3

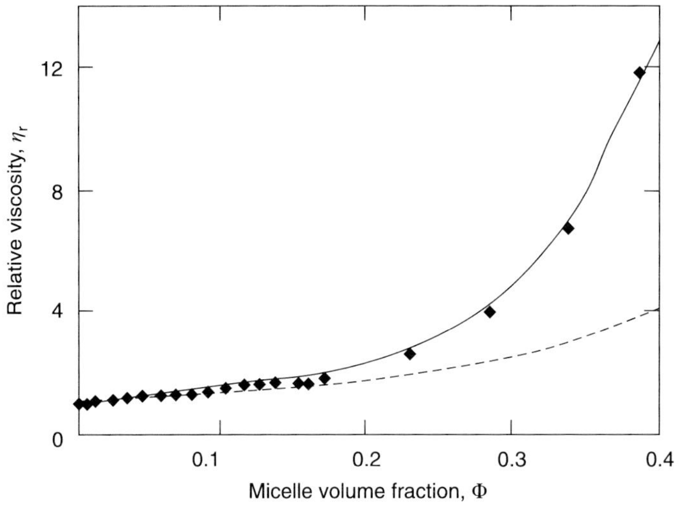

Fig. 3.1. Relative viscosity as a function of micelle volume fraction for solutions of spherical micelles [2] of $C_{12}E_5$ with equal weight of solubilized decane. Dashed and solid curves give theoretical predictions for two models of spherical micelles [2].

Micellar growth is common with many ionic surfactants and is strongly influenced by temperature and the addition of electrolyte. Figure 3.4 shows this in a plot of the aggregation number of sodium dodecyl sulphate (SDS) versus NaCl concentration at two temperatures [3]. Micellar growth clearly increases with increasing NaCl concentration and decreasing temperature.

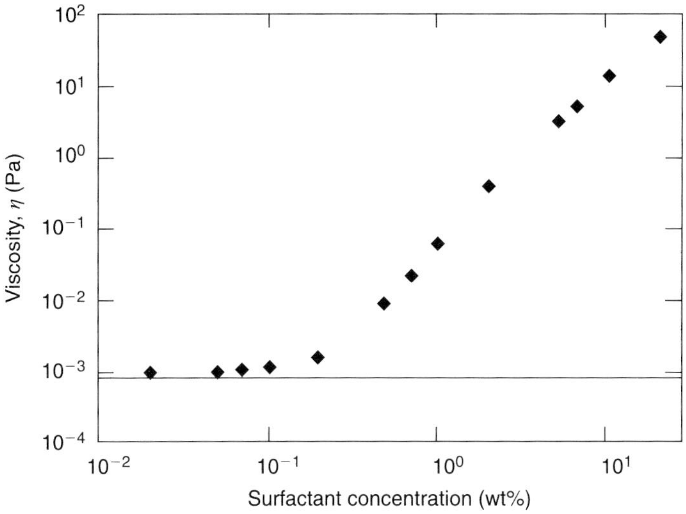

Fig. 3.2. Variation of zero shear viscosity with concentration for $C_{16}E_6$.

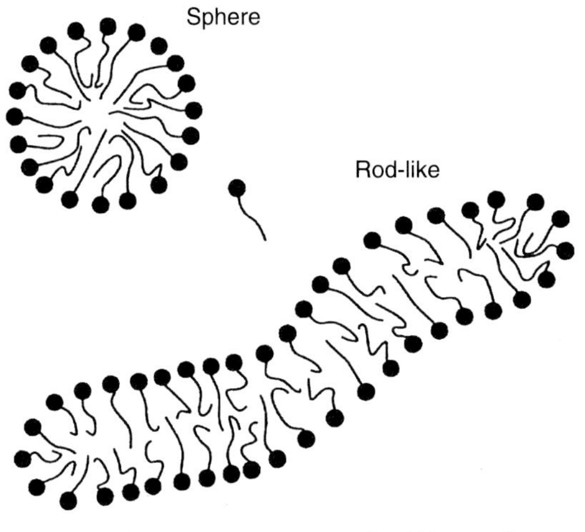

Fig. 3.3. Schematic representation of rod-like micelles [1].

Micellar growth also increases with increasing alkyl chain length and also with increasing surfactant concentration. The nature of the counter ion also affects micellar growth. For example, with (hexadecyltrimethyl)ammonium bromide there is a major micellar growth, whereas with SDS the micellar growth is insignificant with Li^+ or Na^+ but dramatic with K^+ or Cs^+. Non-polar solubilizates such as alkanes (located in the hydrocarbon core of the micelle) prohibit micellar growth, whereas alcohols or aromatic compounds (located in the outer part of the micelle) tend to induce micellar growth.

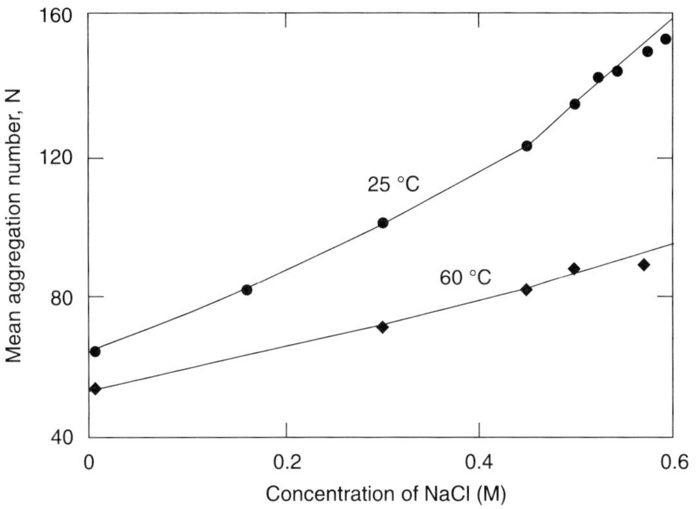

Fig. 3.4. Aggregation number of SDS as a function of NaCl concentration at two temperatures.

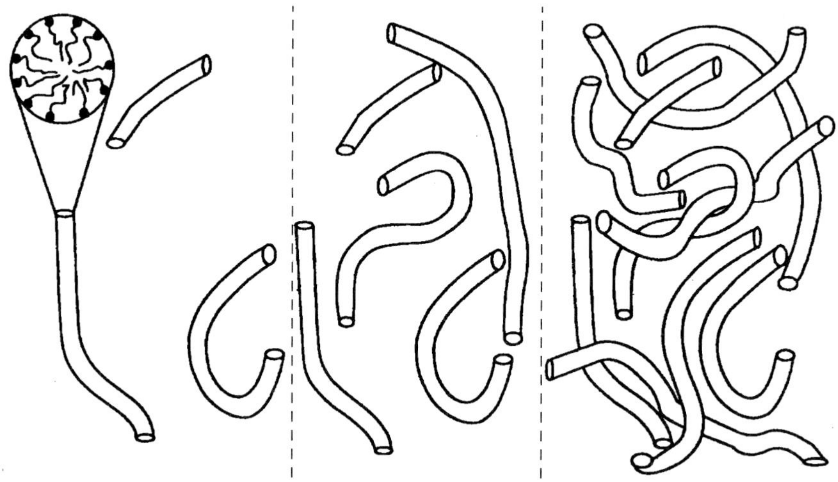

Fig. 3.5. Schematic representation of the overlap of thread-like micelles as the volume fraction is increased.

With nonionic surfactants of the ethoxylate type, micellar growth with increasing concentration is more marked the shorter the EO chain. With 4 to 6 EO units there is a dramatic growth, whereas with 8 or more EO units there is negligible growth. These nonionic surfactants show much more pronounced growth at higher temperatures, i.e. opposite to the case of ionic surfactants.

One of the main features of the above-mentioned long thread-like micelles is their behaviour when the concentration or volume fraction ϕ of the units is gradually increased. This is schematically shown in Figure 3.5.

In dilute solutions, where the micelles do not overlap, they behave as independent entities. Above a critical volume fraction, ϕ^* (the so-called semi-dilute region), the micelles begin to overlap and above this volume fraction the micelles are entangled and there is a transient network that is characterized by a correlation length [1]. This behaviour is similar to that observed with polymer solutions, and the viscosity of long linear micelles can be analyzed in terms of the motion of micelles, i.e. using the reptation model of polymer systems. In this case, the micelles creep like a "snake" through tubes in a porous structure given by the other micelles. The zero shear viscosity depends on the micelle size or its aggregation number N and volume fraction ϕ according to Eq. (3.1), which clearly shows that the viscosity increases strongly with both increasing micellar size and volume fraction.

$$\eta = \text{constant } N^3\phi^{3.75} \tag{3.1}$$

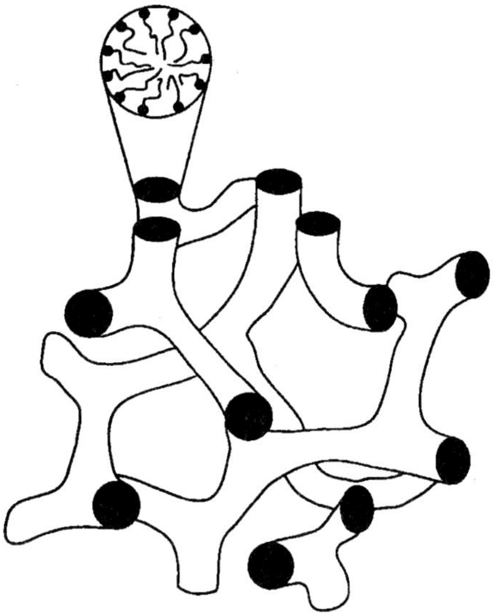

Fig. 3.6. Schematic representation of branched micelles (bicontinuous structures).

The linear growth of micelles is the dominant type, but disc-like or plate-like structures can also form, albeit over a narrow range of conditions. Linear growth can also lead to branched structures, which at a high enough concentration may lead to interconnected structures (Figure 3.6) that are referred to as "bicontinuous", since the solutions are not only continuous in the solvent but also in the surfactant [1].

3.1
Solubility–Temperature Relationship for Ionic Surfactants

With ionic surfactants, the solubility first increases gradually with rising temperature, and then, above a certain temperature, there is a very sudden increase of solubility with further increase in temperature [4, 5]. Figure 3.7 illustrates this with the results for sodium decyl sulphonate in water. The same figure also shows the variation of c.m.c. with temperature [4, 5]. The solubility of the surfactant clearly increases rapidly above 22 °C. The c.m.c. increases gradually with increasing temperature.

At a particular temperature, the solubility becomes equal to the c.m.c., i.e. the solubility curve intersects the c.m.c. and this temperature is referred to as the Krafft temperature of the surfactant, which for sodium decyl sulphate is 22 °C. At the Krafft temperature an equilibrium exists between solid hydrated surfactant, micelles and monomers (i.e. the Krafft point is a "triple-point"). Since the Krafft boundary represents the region below which crystals separate, the energy of the

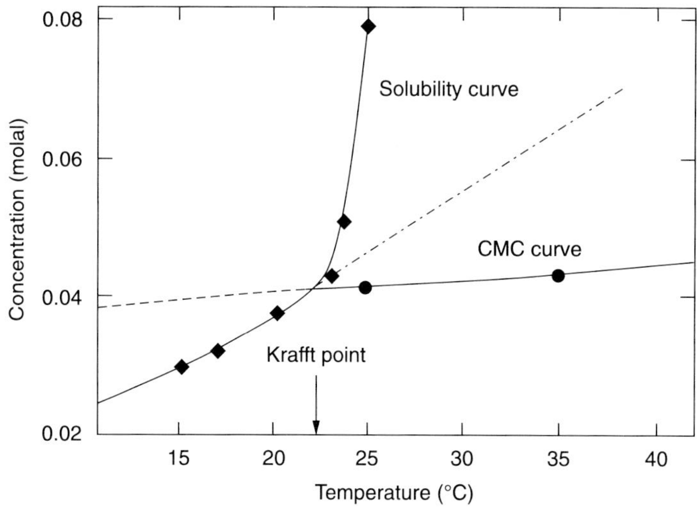

Fig. 3.7. Solubility and c.m.c. versus temperature for sodium decyl sulphonate in water. (♦) solubility, (•) c.m.c.

crystal lattice is the most important parameter controlling the Krafft temperature. Surfactants with ionic head groups, or compact highly polar head groups and long straight alkyl chains, will have high Krafft temperatures. The Krafft temperature increases with alkyl chain length of the surfactant molecule. It can be reduced by introducing branching in the alkyl chains. Using alkyl groups with a wide distribution of alkyl chain length can also reduce the Krafft temperature (see Chapter 1). The Krafft temperature is very important in the application of surfactants. As mentioned above, the solubility of a surfactant increases significantly above the Krafft temperature and, hence, most industrial applications require surfactants with a low Krafft temperature.

3.2 Surfactant Self-Assembly

Surfactant micelles and bilayers are the building blocks of most self-assembly structures. One can divide the phase structures into two main groups [1]: (1) those that are built of limited or discrete self-assemblies, which may be characterized roughly as spherical, prolate or cylindrical. (2) Infinite or unlimited self-assemblies whereby the aggregates are connected over macroscopic distances in one, two or three dimensions. The hexagonal phase (see below) is an example of one-dimensional continuity, the lamellar phase of two-dimensional continuity, whereas the bicontinuous cubic phase and the sponge phase (see later) are examples of three-dimensional continuity. Figure 3.8 illustrates these two types schematically.

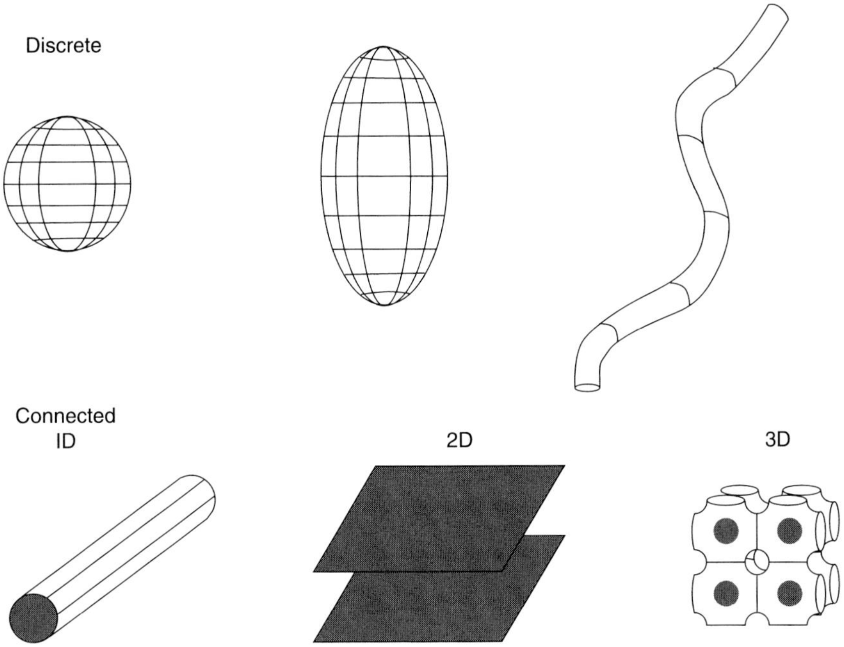

Fig. 3.8. Schematic representation of self-assembly structures [1].

3.3 Structure of Liquid Crystalline Phases

The above-mentioned unlimited self-assembly structures in 1D, 2D or 3D are referred to as liquid crystalline structures. They behave as fluids and are usually highly viscous. At the same time, X-ray studies of these phases yield a small number of relatively sharp lines that resemble those produced by crystals [6]. Since they are fluids they are less ordered than crystals, but because of the X-ray lines and their high viscosity it is also apparent that they are more ordered than ordinary liquids. Thus, the term liquid crystalline phase is very appropriate for describing these self-assembled structures. A brief description of the various liquid crystalline structures that can be produced with surfactants is given below, and Table 3.1 shows the most commonly used notation to describe these systems.

3.3.1 Hexagonal Phase

This phase is built up of (infinitely) long cylindrical micelles arranged in an hexagonal pattern, with each micelle surrounded by six other micelles (Figure 3.9). The radius of the circular cross-section (which may be somewhat deformed) is again close to the surfactant molecule length [7].

Tab. 3.1. Notation of the most common liquid crystalline structures.

Phase structure	*Abbreviation*	*Notation*
Micellar	mic	L_1, S
Reversed micellar	rev mic	L_2, S
Hexagonal	hex	H_1, E, M_1, middle
Reversed hexagonal	rev hex	H_2, F, M_2
Cubic (normal micellar)	cub_m	I_1, S_{1c}
Cubic (reversed micelle)	cub_m	I_2
Cubic (normal bicontinuous)	cub_b	I_1, V_1
Cubic (reversed bicontinuous)	cub_b	I_2, V_2
Lamellar	lam	L_α, D, G, neat
Gel	gel	L_β
Sponge phase (reversed)	spo	L_3 (normal), L_4

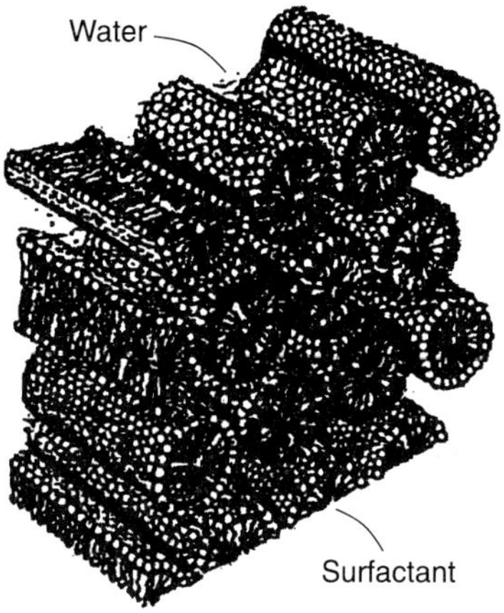

Fig. 3.9. Schematic representation of the hexagonal phase [7].

3.3.2
Micellar Cubic Phase

This phase is built up of regular packing of small micelles, which have properties similar to those of small micelles in the solution phase. However, the micelles are short prolates (axial ratio 1–2) rather than spheres since this allows a better packing (Figure 3.10) [8]. The micellar cubic phase is highly viscous.

3.3.3
Lamellar Phase

This phase is built of layers of surfactant molecules alternating with water layers (Figure 3.11) [7]. The thickness of the bilayers is somewhat lower than twice the surfactant molecule length. The thickness of the water layer can vary over wide

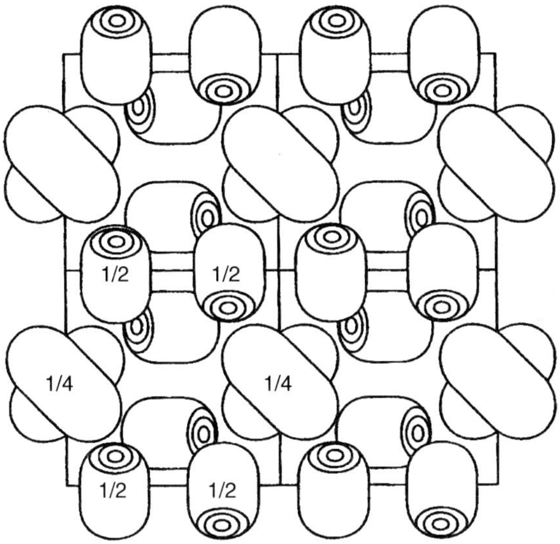

Fig. 3.10. Representation of the micellar cubic phase [8].

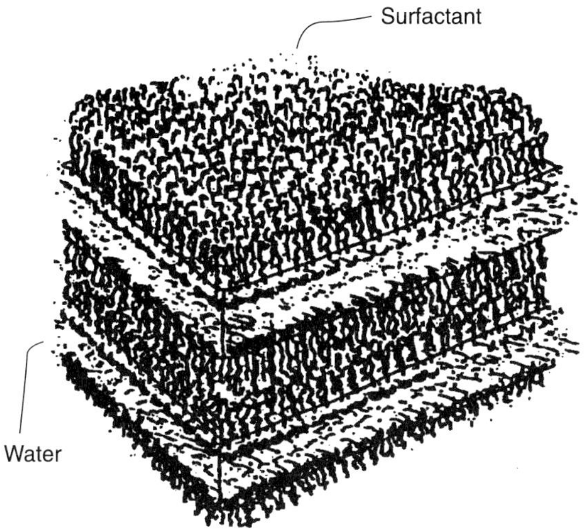

Fig. 3.11. Representation of the lamellar phase [7].

ranges, depending on the nature of the surfactant. The surfactant bilayer can range from being stiff and planar to very flexible and undulating.

3.3.4
Bicontinuous Cubic Phases

These phases can be several different structures, where the surfactant molecules form aggregates that penetrate space, forming a porous connected structure in

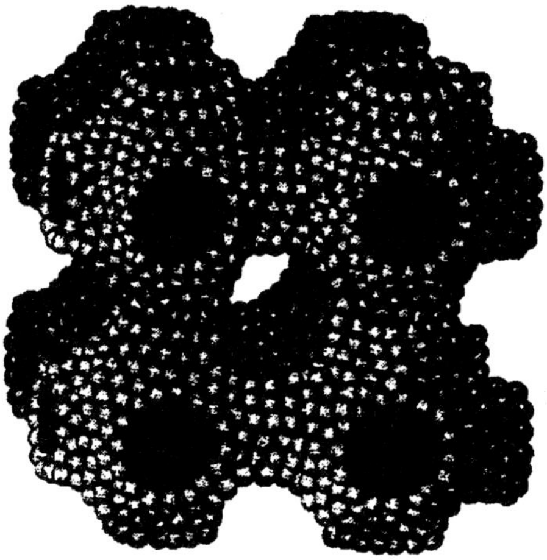

Fig. 3.12. Bicontinuous structure with the surfactant molecules aggregated into connected films characterized by two curvatures of opposite sign [9].

three dimensions. They can be considered as structures produced by connecting rod-like micelles (branched micelles) (Figure 3.6) or bilayer structures (Figure 3.12) [9].

3.3.5
Reversed Structures

Except for the lamellar phase, which is symmetrical around the middle of the bilayer, the different structures have a reversed counter part in which the polar and non-polar parts have changed roles. For example, a hexagonal phase is built up of hexagonally packed water cylinders surrounded by the polar head groups of the surfactant molecules and a continuum of the hydrophobic parts. Similarly, reversed (micellar-type) cubic phases and reversed micelles consist of globular water cores surrounded by surfactant molecules. The radii of the water cores are typically in the range 2–10 nm.

3.4
Experimental Studies of the Phase Behaviour of Surfactants

One of the earliest (and qualitative) techniques for identifying different phases is the use of polarizing microscopy. This is based on the scattering of normal and polarized light, which differs for isotropic (such as the cubic phase) and anisotropic (such as the hexagonal and lamellar phases) structures. Isotropic phases are clear and transparent, while anisotropic liquid crystalline phases scatter light and appear more or less cloudy. Using polarized light and viewing the samples through cross

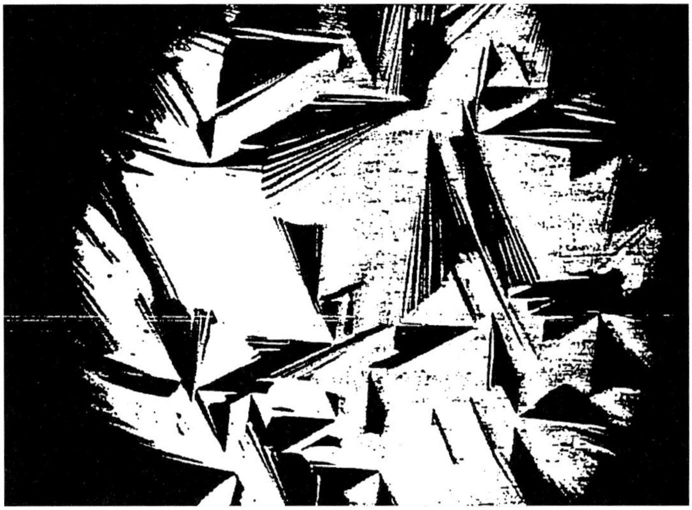

(a)

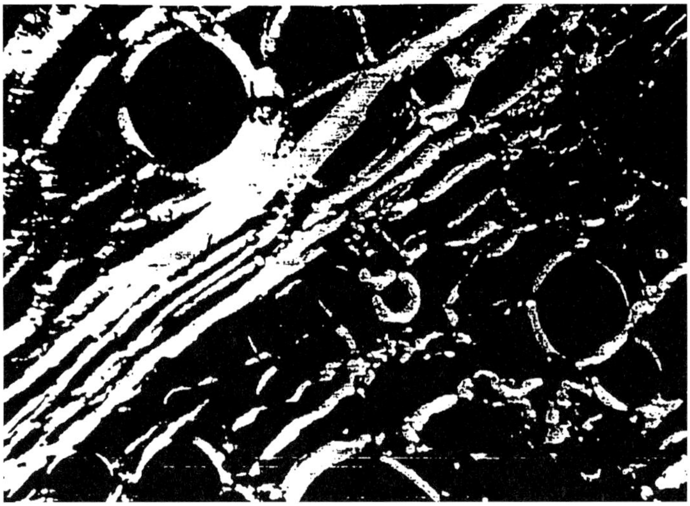

(b)

Fig. 3.13. Texture of the hexagonal (a) and lamellar phase (b) obtained using polarizing microscopy.

polarizers gives a black picture for isotropic phases, whereas anisotropic ones give bright images. The patterns in a polarization microscope are distinctly different for different anisotropic phases and can therefore be used to identify the phases, e.g. to distinguish between hexagonal and lamellar phases [10]. Figure 3.13 shows a typical optical micrograph for the hexagonal and lamellar phases (obtained using

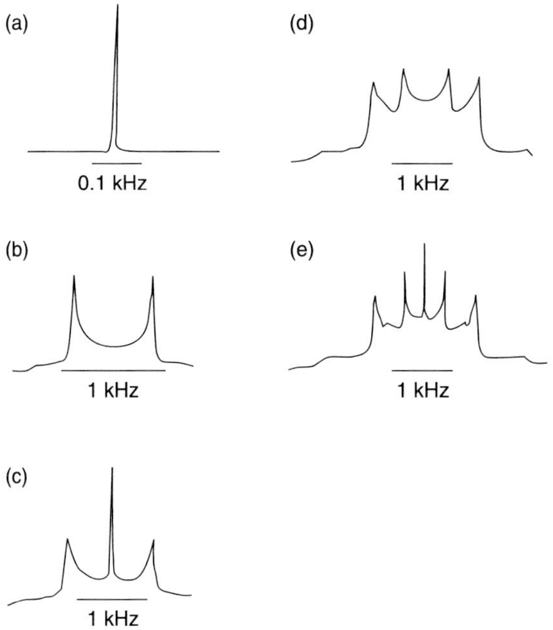

Fig. 3.14. ^{2}H NMR spectra of surfactants in heavy water (D_2O) [11].

polarizing microscopy). The hexagonal phase shows a "fan-like" appearance, whereas the lamellar phase shows "oily streaks" and "Maltese crosses".

Another qualitative method is to measure the viscosity as a function of surfactant concentration. The cubic phase is very viscous and often quite stiff, appearing as a clear "gel"; the hexagonal phase is less viscous and the lamellar phase is much less viscous. However, viscosity measurements do not allow an unambiguous determination of the phases in the sample.

The most qualitative techniques for identification of the various liquid crystalline phases are based on diffraction studies, either light, X-ray or neutron. Liquid crystalline structures have a repetitive arrangement of aggregates and observation of a diffraction pattern can give evidence of long-range order and so distinguish between alternative structures.

NMR spectroscopy is also very useful in identifying different phases – one observes the quadrupole splittings in deuterium NMR [11] (e.g. Figure 3.14).

For isotropic phases such as micelles, cubic and sponge phases one observes a narrow singlet (Figure 3.14a). For a single anisotropic phase, such as hexagonal or lamellar structures, a doublet is obtained (Figure 3.14b). The magnitude of the "splitting" depends on the type of liquid crystalline phase, which is twice as much for the lamellar phase than for the hexagonal phase. For one isotropic and one anisotropic phase, one obtains one singlet and one doublet (Figure 3.14c). For two anisotropic phases (lamellar and hexagonal) one observes two doublets (Figure

3.14d). In a three-phase region with two anisotropic phases and one isotropic phase, one observes two doublets and one singlet (Figure 3.14e).

Normal and reversed phases are easily distinguished using conductivity measurements. For normal phases, which are "water rich", the conductivity is high. In contrast, for reversed phases, which are "water poor", the conductivity is much lower (by several orders of magnitude).

3.5
Phase Diagrams of Ionic Surfactants

Figure 3.15 shows the phase diagram of the sodium dodecyl sulphate (SDS)–water system [6]. SDS has a relatively high Krafft temperature and the phase behaviour is expressed as a temperature (y-axis)–composition (wt%) relationship. Above the Krafft temperature, a large micellar region is obtained that extends to ~40 wt% SDS. This is followed by the hexagonal phase. A mixture of liquid crystalline phases is observed over a narrow concentration range, after which the lamellar phase appears, followed by a solid phase at much higher SDS concentration. Due to the high Krafft temperature, different solid phases play a much more important role at ambient temperature.

Figure 3.16 shows the phase diagram of cetyltrimethylammonium chloride–water system [12]. This surfactant has a high Krafft temperature, and so solid phases play only a minor role. The isotropic phase exists at room temperature up

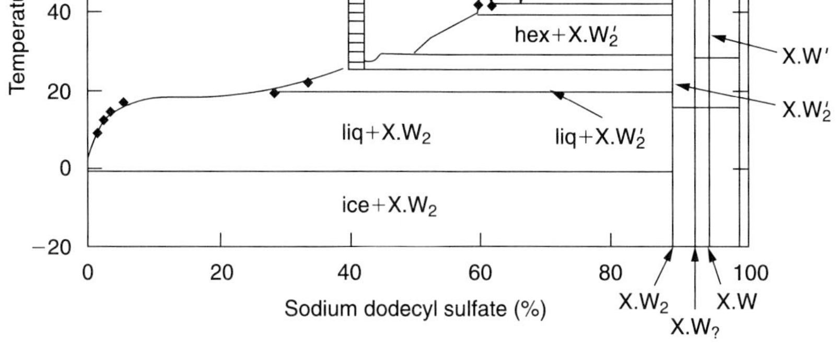

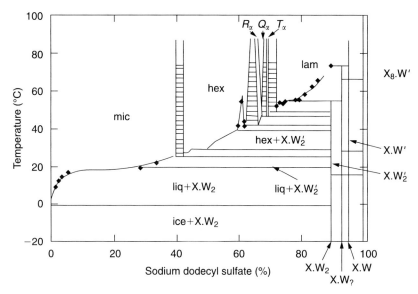

Fig. 3.15. Phase diagram of SDS–water system [6].

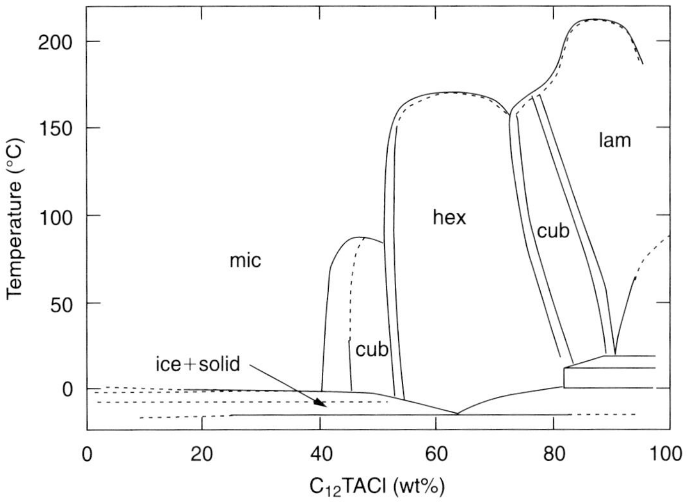

Fig. 3.16. Phase diagram of CTACl–water system [12].

to high concentrations (~40 wt%). The next phase is a cubic phase built up of discrete globular micelles. Between the two phases, there is a two-phase region where the two phases coexist. Owing to the impossibility of packing globular micelles at high volume fraction, the micelles deform and become elongated to furnish an hexagonal phase.

After the hexagonal phase there is transformation into another cubic phase of the bicontinuous type. Then, we find the lamellar phase and, finally, solid hydrated surfactant.

The phase behaviour of double chain surfactants such as sodium bis(2-ethylhexyl)sulphosuccinate is very different (Figure 3.17). The most important feature of the phase diagram is the large area of the lamellar phase, which extends over a wide concentration range. This lamellar phase is followed by a bicontinuous cubic phase and then a reversed hexagonal phase.

3.6
Phase Diagrams of Nonionic Surfactants

The phase behaviour of surfactants is best illustrated using nonionic surfactants of the poly(ethylene oxide) type. Figure 3.18 illustrates this with the phase diagram for the binary system, dodecyl hexaoxyethylene glycol monoether–water [14].

This phase diagram shows the various phases formed when the surfactant concentration and temperature is changed. Let us first consider a dilute nonionic surfactant solution, say 1%; this solution is isotropic (denoted by I) at low temperatures, but on increasing the temperature, a critical point is reached above which the solution becomes turbid. This critical temperature is defined as the cloud point

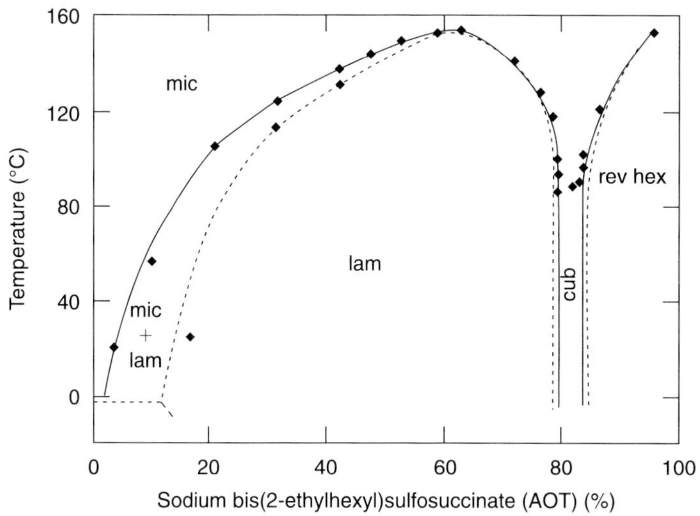

Fig. 3.17. Binary phase diagram of Aerosol OT–water system [13].

of the surfactant at this particular concentration. On further heating the solution above its cloud point, it separates into two liquid layers (defined by 2L), one rich in water and one rich in surfactant. Thus, the line that separates the 2L from the isotropic solution I may be defined as the cloud point curve. Clearly, the phase separation that first decreases with increasing surfactant concentration (in the dilute region) reaches a minimum (which may be defined as the lower consolute temperature, LCT) and then increases. The point x is characteristic of a nonionic surfactant–water mixture and hence may be defined as the cloud point of that particular solution. In most trade literature of surfactants, a cloud point is defined at a particular surfactant concentration (usually 1%). Figure 3.18 shows that the cloud point clearly depends on the surfactant concentration, which needs to be specified to have any meaning. It is sometimes qualitatively stated that the solubility of nonionic surfactants decreases with increasing temperature using cloudiness or phase separation as the solubility limit. Strictly speaking, this is an incorrect statement since that depends on which side of the minimum in the consolute boundary one is (Figure 3.18).

The two-phase region is sometimes referred to as the "miscibility gap". In an homologous series of polyoxyethylene surfactants, increasing the ethylene oxide (EO) chain length causes an increase in the LCT and decrease in the concentration range over which the two-phase region extends. Conversely, increasing the alkyl chain length at a given EO units lowers the LCT and widens out the concentration range for two phases.

The origin of cloudiness with nonionic surfactants has been the subject of considerable debate. Using light scattering, Corkill et al. [15] suggested that cloudiness is associated with a rapid increase in the micellar aggregation number, with

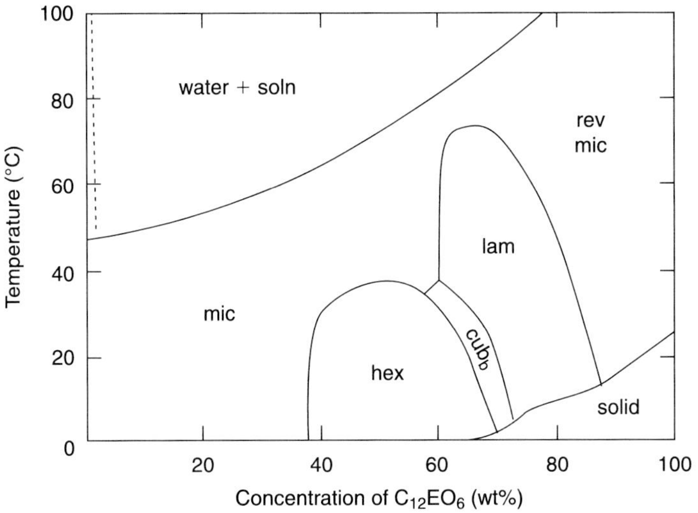

Fig. 3.18. Phase diagram for dodecyl hexaoxyethylene glycol monoether–water mixture.

the formation of long cylinders. However, using neutron scattering, Magid [16] showed only a modest (if any) increase in micellar size, but the intermicellar interaction increases markedly as the phase boundary is approached. The nature of the attractive interaction between micelles whose external surface consists of PEO chains was considered to be due a strong entropy dominance [17, 18]. It was suggested that the water molecules hydrogen bonded to the PEO chains are more structured (lower enthalpy and entropy) than those in the bulk. When the hydration layers of two approaching PEO chains overlap, the partial expulsion of the water molecules in the contact zone causes an increase in enthalpy and entropy of the system. At the LCT, the entropy gain exceeds the repulsive enthalpy contribution and the loss in entropy due to increased concentration, thus phase separation occurs. Confirmation of this hypothesis came from direct force measurement between smooth mica surfaces containing an adsorbed layer of $C_{12}E_5$ [19]. At low temperatures (below the C.P. curve), the force is repulsive but it becomes attractive above the LCT of the free surfactant.

The phase diagram of Figure 3.18 shows some characteristic regions at high surfactant concentrations, namely the M and N region. The M-phase is the region of the hexagonal or middle phase, which consists of cylindrical units that are hexagonally close-packed. In this region, the viscosity of the surfactant solution is extremely high and the system appears like a transparent gel. It shows characteristic textures under polarized light and hence the middle phase may be identified by optical microscopy by referring to published pictures [20]. The N-phase is the lamellar or neat phase, consisting of sheets of molecules in a bimolecular packing with head groups exposed to the water layers in between them. This is less viscous than the M-phase and shows different textures under polarized light [20]. Several other

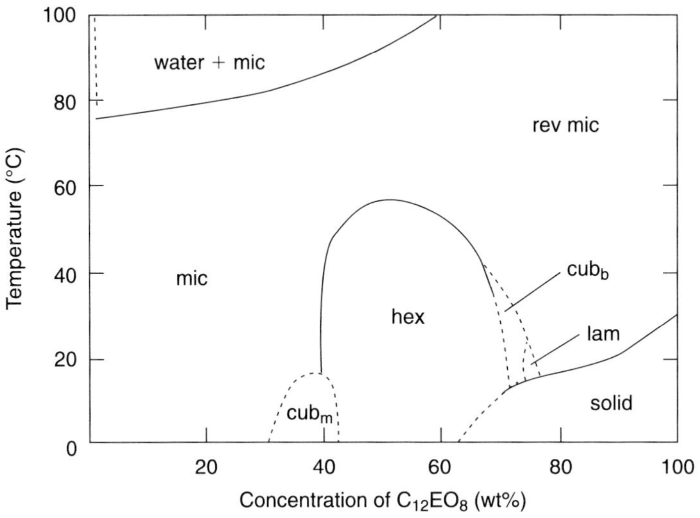

Fig. 3.19. Phase diagram of the $C_{12}E_8$–water system.

liquid crystalline phases may be identified with other nonionic surfactant systems, such as the cubic viscous isotropic phase. Corkill and Goodman have illustrated the above-mentioned three phases [21]. However, several other liquid crystalline phases (mesophases) have been identified, as shown in Figure 3.19 for the phase diagram of $C_{12}E_8$–water system, using different symbols to those used in Figure 3.18 to distinguish the various phases. Table 3.1 summarises the major lyotropic liquid crystalline phases found in surfactant/water systems.

Several ideas have been put forward to explain the driving force for formation of the different liquid crystalline phases. One of the simplest methods for predicting the shape of an aggregated structure is based on the critical packing parameter concept (P) introduced by Israelachvili and his co-workers [22, 23]. This concept will be discussed in detail in the chapter on emulsions (selection of emulsifiers). Basically, P is the ratio between the cross sectional area of the alkyl chain (that is given by v/l_c, where v is the volume of the hydrocarbon chain and l_c is the maximum length to which the alkyl chain can extend) and the optimum head group area a_0, i.e.,

$$P = \frac{V}{a_0 l_c} \tag{3.2}$$

Spherical micelles require P to be less than $\frac{1}{3}$, cylindrical micelles require $\frac{1}{3} < P < \frac{1}{2}$, whereas lamellar micelles require $P \sim 1$.

Using the above concept, one may predict the shape of a micelle in a dilute solution. For a nonionic surfactant such as $C_{12}E_6$, the preferred shape will be a spherical micelle. As the volume fraction of the surfactant is increased, repulsion between the micelles tends to space them out, forming first a cubic array of spherical units. As the volume fraction of the surfactant is increased further, the free energy of the system can be minimized by changing to a packing geometry of packed cylindrical units. The energy required to change the surface curvature from spherical to cylindrical is more than offset by changing to a geometry whereby the average separation between the micellar surfaces is greater. By a similar argument, one may rationalize the formation of lamellar phases, by relieving the "strain" of increasing the volume fraction even further. This simple argument explains the sequence of phases formed in practice (hexagonal → lamellar).

Lyotropic liquid crystalline phases show flow properties and degrees of molecular ordering that are intermediate between liquids and crystalline solids. Rheological studies have shown the liquid crystalline phases to be viscoelastic [24]. Even in the most viscous mesophases, the cubic phase, the X-ray diffraction studies showed broad wide-angle peak that is characteristic of a spacing of 0.45 nm, indicating that the hydrocarbon chains are in a liquid state [25, 26]. More evidence was obtained using NMR measurements [27]. These results indicate the "liquid-like" nature of the hydrocarbon chains in the liquid crystalline structures. The state and mobility of the water molecules in the liquid crystalline structures have been deduced from calorimetric measurements [28]. In some lamellar systems, three different types of water could be distinguished: unbound water similar to that of bulk water, ultrathin water layers close to the surfactant head groups, which melt at lower temperatures than bulk water, and a third water structure that exists at low temperatures (at −10 to −20 °C).

As mentioned above, the hexagonal phases H_1 and H_2 show a characteristic fan-like texture when observed under polarizing microscopy (see Fig. 3.13). Low-angle X-ray diffraction produces a series of spots that can be indexed on the basis of a two-dimensional hexagonal lattice [21]. The patterns are consistent with the structure consisting of an hexagonally packed array of cylindrical aggregates similar to cylindrical micelles (Figure 3.1). The diameter of the cylinders is usually about 10–30% less than the length of two surfactant chains and the alkyl chains in the cylinders are in a liquid-like state [29]. The hexagonal phases are very viscous even though they contain 30–60% water [24]. This is attributed to the hexagonal structure, which allows the cylinders to move freely only along their length. Owing to their high viscosity, hexagonal structures should be avoided in formulating emulsions (e.g. in the food industry). The reversed hexagonal phase H_2 consists also of cylindrical aggregates, with the head groups and the water cores inside the aggregates and the alkyl chains pointing outwards. The water core has a diameter in the region of 1–2 nm, but since all space between adjacent cylinders has to be filled with alkyl chains, the range of separation of the cylinders is much smaller than that in the normal H_1 hexagonal phase.

The lamellar liquid crystalline phase consists of several bilayers of surfactant molecules and shows a mosaic-type texture (with Maltese crosses) when viewed under polarizing light. Low-angle X-ray diffraction shows spacing characteristic of a lamellar structure and the repeat unit is the back-to-back bilayer of the surfactant molecules with their alkyl groups in contact. The phase is built up from these flexible bilayers which are arranged parallel to each other [25]. The surfactant bilayer thickness is also 10–30% less than two surfactant chains, while the thickness of the water layers separating the head groups varies depending on composition. Lamellar phases often extend down to ~50% surfactant. Below this, the stable phase changes to an hexagonal phase or an isotropic micellar solution.

Cubic phases occur in various parts of the phase diagram and they most likely have different structures. The cubic phases are optically isotropic and hence they show no texture under polarizing microscopy. Even with low-angle X-ray diffraction, they show poor quality patterns, making it difficult to accurately determine the spacing between the aggregates. Early investigations suggested closed globular aggregates that are arranged in a cubic close packed array (face centered or body centered, but later it was suggested that the building units are not spherical but consist of short rods or ellipsoids). The cubic phases are more viscous than the hexagonal phase.

Notably, the concentration and temperature domains on which these lyotropic mesomorphic liquid crystalline phases are formed vary widely for different surfactants. Major changes also occur on addition of electrolytes or another organic phase such as a long-chain alcohol.

References

1 K. Holmberg, B. Jonsson, B. Kronberg, B. Lindman: *Surfactants and Polymers in Aqueous Solution*, John Wiley & Sons, Chichester, 2003.

2 M. S. Leaver, U. Olsson, *Langmuir*, **1994**, *10*, 3449.

3 N. J. Turro, A. Yeketa, *J. Am. Chem. Soc.*, **1978**, *100*, 5951.

4 F. Krafft: *Ber. Dtsch. Chem. Gessel*, **1899**, *32*, 1596.

5 K. Shinoda: *Principles of Solution and Solubility*, Marcel Dekker, New York, 1974.

6 R. G. Laughlin: *The Aqueous Phase Behaviour of Surfactants*, Academic Press, London, 1994.

7 K. Fontell, *Mol. Cryst. Liq. Cryst.*, **1981**, *63*, 59.

8 K. Fontell, C. Fox, E. Hanson, *Mol. Cryst. Liq. Cryst.*, **1985**, *1*, 9.

9 D. F. Evans, H. Wennerstrom: *The Colloid Domain. Where Physics, Chemistry and Biology Meet*, John Wiley & Sons, VCH, New York, 1994.

10 F. B. Rosevaar, *J. Soc. Cosmet. Chem.*, **1968**, *19*, 581.

11 A. Khan, K. Fontell, G. Lindblom, B. Lindman, *J. Phys. Chem.*, **1982**, *86*, 4266.

12 R. R. Balmbra, J. S. Clunie, J. F. Goodman, *Nature*, **1969**, *222*, 1159.

13 J. Rogers, P. A. Winsor, *J. Colloid Interface Sci.*, **1969**, *30*, 247.

14 J. S. Clunie, J. F. Goodman, P. C. Symons, *Trans. Faraday Soc.*, **1969**, *65*, 287.

15 J. M. Corkill, J. F. Goodman, T. Walker, *Trans. Faraday Soc.*, **1967**, *63*, 759.

16 L. J. Magid: Structure and Dynamics by Small Angle Neutron Scattering, in *Nonionic Surfactants, Physical Chemistry.* M. Schick (ed.), Marcel Dekker, New York, 1987.

17 R. Kjellander, *J. Chem. Soc., Faraday Trans. II*, **1982**, *78*, 2025.

18 R. Kjellander, *J. Chem. Soc., Faraday Trans. II*, **1984**, *80*, 1323.

19 P. M. Claesson, R. Kjellander, P. Stenius, H. K. Christenson, *J. Chem. Soc., Faraday Trans. I*, **1986**, *82*, 2735.

20 J. M. Corkill, J. F. Goodman, *Adv. Colloid Interface Sci.*, **1969**, *2*, 297.

21 J. N. Israelachvili: *Intermolecular and Surface Forces*, Academic Press, London, 1985.

22 J. N. Israelachvili, D. J. Mitchell, B. W. Ninham, *J. Chem. Soc., Faraday Trans. I*, **1976**, *72*, 1525.

23 G. T. Dimitrova, T. F. Tadros, P. F. Luckham, *Langmuir*, **1995**, *11*, 1101.

24 V. Luzatti, H. Mustacchi, A. Skoulios, *Discussions Faraday Soc.*, **1958**, *25*, 43.

25 K. Fontell, *Colloid Polym. Sci.*, **1990**, *268*, 264.

26 K. D. Lawson, T. L. Flautt, *Mol. Crystals*, **1966**, *1*, 241.

27 N. Cassilas, J. E. Puigh, R. Olayo, T. J. Hart, E. I. Franses, *Langmuir*, **1989**, *5*, 384.

28 G. J. T. Tiddy, *Phys. Rep.*, **1980**, *57*, 1.

29 K. Fontell, K. K. Fox, E. Hansson, *Mol. Liq. Cryst. Lett.*, **1985**, *1*, 9.

4
Adsorption of Surfactants at the Air/Liquid and Liquid/Liquid Interfaces

4.1
Introduction

As mentioned in the general introduction, surfactants play a major role in the formulation of most chemical products. In the first place they are used to stabilize emulsions and microemulsions. Secondly, surfactants are added in emulsifiable concentrates for their spontaneous dispersion on dilution.

In the above-mentioned phenomenon, the surfactant needs to accumulate at the interface, a process that is generally described as adsorption. The simplest interface is that of the air/liquid and, in this case, the surfactant will adsorb with the hydrophilic group pointing towards the polar liquid (water), leaving the hydrocarbon chain pointing towards the air. This lowers the surface tension γ. Typically, surfactants show a gradual reduction in γ, till the c.m.c. is reached, above which the surface tension remains virtually constant. Hydrocarbon surfactants of the ionic, nonionic or zwitteronic ionic type lower the surface tension to limiting values, reaching 30–40 mN m^{-1}, depending on the nature of the surfactant. Lower values may be achieved using fluorocarbon surfactants, typically of the order of 20 mN m^{-1}. It is, therefore, essential to understand the adsorption and conformation of surfactants at the air/liquid interface.

With emulsifiable concentrates, emulsions and microemulsion, the surfactant adsorbs at the oil/water interface, with the hydrophilic head group immersed in the aqueous phase, leaving the hydrocarbon chain in the oil phase. Again, the mechanism of stabilization of emulsions and microemulsions depends on the adsorption and orientation of the surfactant molecules at the liquid/liquid interface. As we will see, macromolecular surfactants (polymers) are nowadays used to stabilize emulsions and hence it is essential to understand their adsorption at the interface. Suffice to say that, at this stage, surfactant adsorption is relatively simpler than polymer adsorption. This is because surfactants consist of a small number of units and they are mostly reversibly adsorbed, allowing one to apply some thermodynamic treatments. In this case, it is possible to describe the adsorption in terms of various interaction parameters such as chain/surface, chain solvent and surface solvent. Moreover, the configuration of the surfactant molecule can be simply described in terms of these possible interactions. In contrast, polymer adsorption is fairly complicated. In addition to the usual adsorption considerations described

Applied Surfactants: Principles and Applications. Tharwat F. Tadros

ISBN: 3-527-30629-3

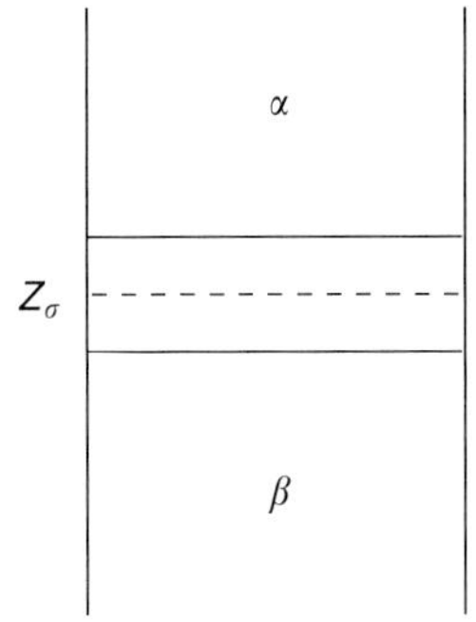

Fig. 4.1. Gibbs convention for an interface.

above, one of the principle problems to be resolved is the configuration of the polymer molecule on the surface. This can acquired various possible ways, depending on the number of segments and chain flexibility.

4.2 Adsorption of Surfactants

Before describing surfactant adsorption at the air/liquid (A/L) and liquid/liquid (liquid (L/L) interface it is essential to define the interface. The surface of a liquid is the boundary between two bulk phases, namely liquid and air (or the liquid vapour). Similarly, an interface between two immiscible liquids (oil and water) may be defined providing a dividing line is introduced since the interfacial region is not a layer that is one molecule thick, but usually has a thickness δ with properties that differ from the two bulk phases α and β [1]. However, Gibbs [2] introduced the concept of a mathematical dividing plane Z_{-} in the interfacial region (Figure 4.1).

In this model the two bulk phases α and β are assumed to have uniform thermodynamic properties up to Z. This picture applies for both the air/liquid and liquid/liquid interface (with A/L interfaces, one of the phases is air saturated with the vapour of the liquid).

Using the Gibbs model, it is possible to obtain a definition of the surface or interfacial tension γ, starting from the Gibbs–Deuhem Eq. (4.1), i.e.,

$$\mathrm{d}G^\sigma = -S^\sigma\,\mathrm{d}T + A\,\mathrm{d}\gamma + \sum n_\mathrm{i}\,\mathrm{d}\mu_\mathrm{i} \tag{4.1}$$

where G^σ is the surface free energy, S^σ is the entropy, A is the area of the interface, n_i is the number of moles of component i with chemical potential μ_i at the interface. At constant temperature and composition of the interface (i.e. in absence of any adsorption),

$$\gamma = \left(\frac{\partial G^\sigma}{\partial A}\right)_{T, n_i} \tag{4.2}$$

Obviously, from Eq. (4.2), for a stable interface γ should be positive, i.e. the free energy should increase if the area of the interface increases, otherwise the interface will become convoluted, increasing the interfacial area, until the liquid evaporates (for A/L case) or the two "immiscible" phases dissolve in each other (for the L/L case).

Eq. (4.2) shows clearly that surface or interfacial tension, i.e. the force per unit length tangentially to the surface measured in units of mN m^{-1}, is dimensionally equivalent to an energy per unit area measured in mJ m^{-2}. Thus, it has been stated that the excess surface free energy is identical to the surface tension, but this is true only for a single component system, i.e. a pure liquid (where the total adsorption is zero).

There are generally two approaches for treating surfactant adsorption at the A/L and L/L interfaces. The first approach, adopted by Gibbs, treats adsorption as an equilibrium phenomenon whereby the second law of thermodynamics may be applied using surface quantities. The second approach, referred to as the equation of state approach, treats the surfactant film as a two-dimensional layer with a surface pressure π that may be related the surface excess Γ (amount of surfactant adsorbed per unit area). These two approaches are summarized below.

4.2.1 Gibbs Adsorption Isotherm

Gibbs [2] derived a thermodynamic relationship between the surface or interfacial tension γ and the surface excess Γ (adsorption per unit area). The starting point of this equation is the Gibbs–Deuhem equation, Eq. (4.1). At constant temperature, but in the presence of adsorption, Eq. (4.1) reduces to Eq. (4.3).

$$d\gamma = -\sum \frac{n_i^\sigma}{A} d\mu_i = -\sum \Gamma_i \, d\mu_i \tag{4.3}$$

where $\Gamma_i = n_i^\sigma / A$ is the number of moles of component i adsorbed per unit area.

Eq. (4.2) is the general form for the Gibbs adsorption isotherm. The simplest case of this isotherm is a system of two components in which the solute (2) is the surface active component, i.e. it is adsorbed at the surface of the solvent (1). For such a case, Eq. (4.3) may be written as,

$$-d\gamma = \Gamma_1^\sigma \, d\mu_1 + \Gamma_2^\sigma \, d\mu_2 \tag{4.4}$$

and if the Gibbs dividing surface is used, $\Gamma_1 = 0$ and,

$$-d\gamma = \Gamma_{1,2}^\sigma \, d\mu_2 \tag{4.5}$$

where $\Gamma^{\sigma}_{2,1}$ is the relative adsorption of (2) with respect to (1). Since,

$$\mu_2 = \mu_2^{\circ} + RT \ln a_2^{\mathrm{L}} \tag{4.6}$$

or,

$$\mathrm{d}\mu_2 = RT \,\mathrm{d} \ln a_2^{\mathrm{L}} \tag{4.7}$$

then,

$$-\mathrm{d}\gamma = \Gamma^{\sigma}_{2,1} RT \,\mathrm{d} \ln a_2^{L} \tag{4.8}$$

or

$$\Gamma^{\sigma}_{2,1} = -\frac{1}{RT}\left(\frac{\mathrm{d}\gamma}{\mathrm{d}\ln a_2^{\mathrm{L}}}\right) \tag{4.9}$$

where a_2^{L} is the activity of the surfactant in bulk solution that is equal to $C_2 f_2$ or $x_2 f_2$, where C_2 is the concentration of the surfactant in moles dm^{-3} and x_2 is its mole fraction.

Equation (4.9) allows one to obtain the surface excess (abbreviated as Γ_2) from the variation of surface or interfacial tension with surfactant concentration. Note that $a_2 \sim C_2$ since in dilute solutions $f_2 \sim 1$. This approximation is valid since most surfactants have low c.m.c. (usually less than 10^{-3} mol dm^{-3}) but adsorption is complete at or just below the c.m.c.

The surface excess Γ_2 can be calculated from the linear portion of the $\gamma - \log C_2$ curves before the c.m.c. Such $\gamma - \log C$ curves are illustrated in Figure 4.2 for the air/water and o/w interfaces; $[C_{SAA}]$ denotes the concentration of surface active

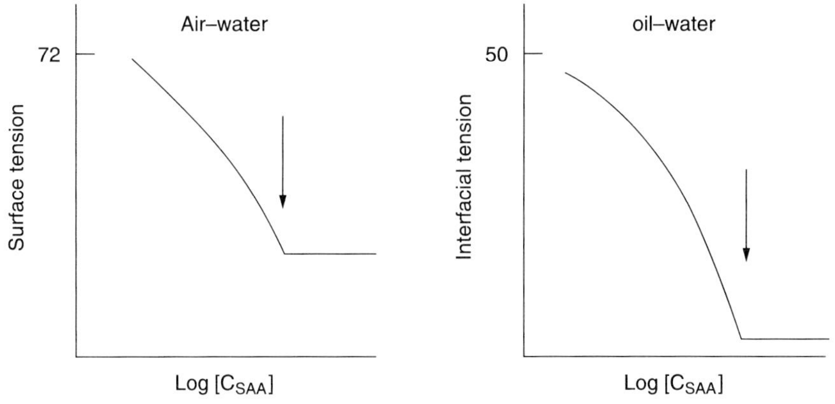

Fig. 4.2. Variation of surface and interfacial tension with log$[C_{SAA}]$ at the air–water and at the oil–water interface.

agent in bulk solution. For the A/W interface, γ decreases from the value for water (72 mN m^{-1} at 20 °C) to about 25–30 mN m^{-1} near the c.m.c. This is clearly schematic since the actual values depend on the surfactant nature. For the o/w case, γ decreases from about 50 mN m^{-1} (for a pure hydrocarbon–water interface) to ~1–5 mN m^{-1} near the c.m.c. (again depending on the nature of the surfactant).

As mentioned above, Γ_2 can be calculated from the slope of the linear position of the curves shown in Figure 4.2 just before the c.m.c. is reached. From Γ_2, the area per surfactant ion or molecule can be calculated since,

$$\text{Area/molecule} = \frac{1}{\Gamma_2 N_{av}} \tag{4.10}$$

where N_{av} is the Avogadro's constant. Determining the area per surfactant molecule is very useful since it gives information on surfactant orientation at the interface. For example, for ionic surfactants such as sodium dodecyl sulphate, the area per surfactant is determined by the area occupied by the alkyl chain and head group if these molecules lie flat at the interface, whereas for vertical orientation the area per surfactant ion is determined by that occupied by the charged head group, which at low electrolyte concentrations will be in the region of 0.40 nm^2. Such an area is larger than the geometrical area occupied by a sulphate group, as a result of the lateral repulsion between the head groups. On addition of electrolytes, this lateral repulsion is reduced and the area/surfactant ion for vertical orientation will be <0.4 nm^2 (in some cases reaching 0.2 nm^2). Conversely, if the molecules lie flat at the interface the area per surfactant ion will be considerably higher than 0.4 nm^2.

Another important point can be made from the $\gamma - \log C$ curves. At the concentration just before the break point, one has the condition of constant slope, indicating that saturation adsorption has been reached. Just above the break point,

$$\left(\frac{\partial \gamma}{\partial \ln a_2}\right)_{p,T} = 0 \tag{4.11}$$

indicating the constancy of γ with $\log C$ above the c.m.c. Integration of Eq. (4.11) gives,

$$\gamma = \text{constant} \times \ln a_2 \tag{4.12}$$

Since γ is constant in this region, then a_2 must remain constant. This means that addition of surfactant molecules above the c.m.c. must result in association to form units (micellar) with low activity.

As mentioned before, the hydrophilic head group may be unionized, e.g. alcohols or poly(ethylene oxide) alkane or alkyl phenol compounds, weakly ionized such as carboxylic acids or strongly ionized such as sulphates, sulphonates and quaternary ammonium salts. Adsorption of these different surfactants at the air/water and oil/water interface depends on the nature of the head group. With nonionic surfactants, repulsion between the head groups is small and these surfactants are usually strongly adsorbed at the surface of water from very dilute solutions. As already noted, nonionic surfactants have much lower c.m.c.s than ionic surfactants with the same alkyl chain length. Typically, the c.m.c. is in the region of 10^{-5}–10^{-4} mol dm^{-3}. Such nonionic surfactants form closely packed adsorbed layers at concentrations lower than their c.m.c.s. The activity coefficient of such surfactants is close to unity and is only slightly affected by addition of moderate amounts of electrolytes (or change in the pH of the solution). Thus, nonionic surfactant adsorption is the simplest case since the solutions can be represented by a two-component system and the adsorption can be accurately calculated using Eq. (4.9).

With ionic surfactants, however, the adsorption process is relatively complicated since one has to consider the repulsion between the head groups and the effect of presence of any indifferent electrolyte. Moreover, the Gibbs adsorption equation has to be solved, taking into account the surfactant ions, the counter ion and any indifferent electrolyte ions present. For a strong surfactant electrolyte such as an Na^+R^-

$$\Gamma_2 = \frac{1}{2RT} \frac{\partial \gamma}{\partial \ln a \pm} \tag{4.13}$$

The factor of 2 in Eq. (4.13) arises because both surfactant and counter ion must be adsorbed to maintain neutrally, and $d\gamma/d \ln a\pm$ is twice as large as for an unionized surfactant.

If a non-adsorbed electrolyte, such as NaCl, is present in large excess then any increase in concentration of Na^+R^- produces a negligible increase in Na^+ ion concentration and, therefore, $d\mu_{Na}$ becomes negligible. Moreover, $d\mu_{Cl}$ is also negligible, so that the Gibbs adsorption equation reduces to,

$$\Gamma_2 = -\frac{1}{RT}\left(\frac{\partial \gamma}{\partial \ln C_{NaR}}\right) \tag{4.14}$$

i.e. it becomes identical to that for a nonionic surfactant.

The above discussion clearly illustrates that in calculating Γ_2 from the $\gamma - \log C$ curve one has to consider the nature of the surfactant and the composition of the medium. For nonionic surfactants the Gibbs adsorption Eq. (4.9) can be directly used. For ionic surfactant, in absence of electrolytes the right hand side of the Eq. (4.9) should be divided by 2 to account for surfactant dissociation. This factor disappears in the presence of a high concentration of an indifferent electrolyte.

4.2.2
Equation of State Approach

In this approach, one relates the surface pressure π with the surface excess Γ_2. The surface pressure is defined by Eq. (4.15),

$$\pi = \gamma_0 - \gamma \tag{4.15}$$

where γ_0 is the surface or interfacial tension before adsorption and γ that after adsorption.

For an ideal surface film, behaving as a two-dimensional gas the surface pressure π is related to the surface excess Γ_2 by the equation,

$$\pi A = n_2 RT \tag{4.16}$$

or

$$\pi = (n_2/A)RT = \Gamma_2 RT \tag{4.17}$$

Differentiating Eq. (4.15) at constant temperature,

$$\mathrm{d}\pi = RT\,\mathrm{d}\Gamma_2 \tag{4.18}$$

Using the Gibbs equation,

$$\mathrm{d}\pi = -\mathrm{d}\gamma = \Gamma_2 RT\,\mathrm{d}\ln a_2 \simeq \Gamma_2 RT\,\mathrm{d}\ln C_2 \tag{4.19}$$

Combining Eqs. (4.18) and (4.19),

$$\mathrm{d}\ln \Gamma_2 = \mathrm{d}\ln C_2 \tag{4.20}$$

or

$$\Gamma_2 = KC_2^{\alpha} \tag{4.21}$$

Eq. (4.21) is referred to as the Henry's law isotherm, which predicts a linear relationship between Γ_2 and C_2.

Clearly, Eqs. (4.15) and (4.18) are based on an idealized model in which the lateral interaction between the molecules has not been considered. Moreover, in this model the molecules are considered to be dimensionless. This model can only be applied at very low surface coverages where the surfactant molecules are so far apart that lateral interaction may be neglected. Moreover, under these conditions the total area occupied by the surfactant molecules is relatively small compared with the total interfacial area.

At significant surface coverages, the above equations have to be modified to take into account both lateral interaction between the molecules as well as the area occupied by them. Lateral interaction may reduce π if there is attraction between the chains (e.g. with most nonionic surfactant) or it may increase π as a result of repulsion between the head groups in the case of ionic surfactants.

Various equation of state have been proposed, taking into account the above two effects, to fit the $\pi - A$ data. The two-dimensional van der Waals equation of state is probably the most convenient for fitting these adsorption isotherms, i.e.,

$$\left(\pi + \frac{(n_2)^2\alpha}{A_2}\right)(A - n_2 A_2^{\circ}) = n_2 RT \tag{4.22}$$

where A_2° is the excluded area or co-area of type 2 molecule in the interface and α is a parameter that allows for lateral interaction.

Eq. (4.19) leads to the following theoretical adsorption isotherm, using the Gibbs's equation,

$$C_2^\alpha = K_1\left(\frac{\theta}{1-\theta}\right)\exp\left(\frac{\theta}{1-\theta}-\frac{2\alpha\theta}{a_2^\circ RT}\right) \tag{4.23}$$

where θ is the surface coverage ($\theta = \Gamma_2/\Gamma_{2,\max}$), K_1 is constant that is related to the free energy of adsorption of surfactant molecules at the interface $[K_1 \propto \exp(-\Delta G_{\text{ads}}/kT)]$ and a_2° is the area/molecule.

For a charged surfactant layer, Eq. (4.20) has to be modified to take into account the electrical contribution from the ionic head groups, i.e.,

$$C_2^\alpha = K_1\left(\frac{\theta}{1-\theta}\right)\exp\left(\frac{\theta}{1-\theta}\right)\exp\left(\frac{e\Psi_0}{kT}\right) \tag{4.24}$$

where Ψ_0 is the surface potential. Eq. (4.24) shows how the electrical potential energy (Ψ_0/kT) of adsorbed surfactant ions affects the surface excess. Assuming that the bulk concentration remains constant, Ψ_0 increase as θ increases. This means that $[\theta/(1-\theta)]\exp[\theta/(1-\theta)]$ increases less rapidly with C_2, i.e. adsorption is inhibited as a result of ionization.

4.3
Interfacial Tension Measurements

These methods may be classified into two categories: those in which the properties of the meniscus is measured at equilibrium, e.g., pendent drop or sessile drop profile and Wilhelmy plate methods, and those where the measurement is made under non-equilibrium or quasi-equilibrium conditions such as the drop volume (weight) or the de Nouy ring method. The latter methods are faster, although they suffer from premature rupture and expansion of the interface, causing adsorption depletion. They are also unsuitable for measuring the interfacial tension in the presence of macromolecules, since in this case equilibrium may require hours or even days. For measurement of low interfacial tensions (< 0.1 mN m^{-1}) the spinning drop technique is applied. Below, a brief description of these techniques is given.

4.3.1
Wilhelmy Plate Method

Here [3] a thin plate made from glass (e.g., a microscope cover slide) or platinum foil is either detached from the interface (non-equilibrium condition) or it weight is measured statically using an accurate microbalance. In the detachment method, the total force F is given by the weight of the plate W and the interfacial tension force,

$$F = W + \gamma p \tag{4.25}$$

where p is the "contact length" of the plate with the liquid, i.e., the plate perimeter. Provided the contact angle of the liquid is zero, no correction is required for Eq. (4.25). Thus, the Wilhelmy plate method can be applied in the same manner as the du Nouy's technique described below.

The static technique may be applied to follow the interfacial tension as a function of time (to follow the kinetics of adsorption) till equilibrium is reached. In this case, the plate is suspended from one arm of a microbalance and allowed to penetrate the upper liquid layer (usually the oil) until it touches the interface, or alternatively the whole vessel containing the two liquid layers is raised until the interface touches the plate. The increase in weight ΔW is given by the following equation,

$$\Delta W = \gamma p \cos \theta \tag{4.26}$$

where θ is the contact angle. If the plate is completely wetted by the lower liquid as it penetrates, $\theta = 0$ and γ may be calculated directly from ΔW. Care should always be taken that the plate is completely wetted by the aqueous solution. For that purpose, a roughened platinum or glass plate is used to ensure a zero contact angle. However, if the oil is denser than water, a hydrophobic plate is used so that when the plate penetrates through the upper aqueous layer and touches the interface it is completely wetted by the oil phase.

4.3.2
Pendent Drop Method

If a drop of oil is allowed to hang from the end of a capillary that is immersed in the aqueous phase it will adopt an equilibrium profile (Figure 4.3) that is a unique

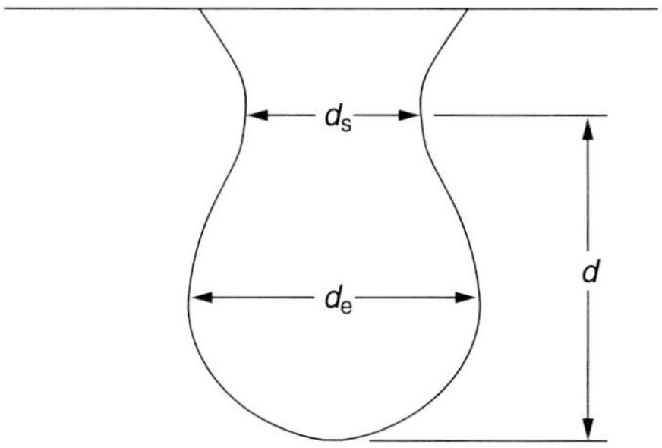

Fig. 4.3. Schematic representation of the profile of a pendent drop.

function of the tube radius, the interfacial tension, its density and the gravitational field.

The interfacial tension is given by Eq. (4.4),

$$\gamma = \frac{\Delta\rho g d_e^2}{H} \tag{4.27}$$

where $\Delta\rho$ is the density difference between the two phases, d_e is the equatorial diameter of the drop (Figure 4.3) and H is a function of d_s/d_e, where d_s is the diameter measured at a distance d from the bottom of the drop (Figure 4.3). The relationship between H and the experimental values of d_s/d_e has been obtained empirically [4] using pendent drops of water. Accurate values of H have been obtained by Niederhauser and Bartell [5].

4.3.3 Du Nouy's Ring Method

Basically one measures the force required to detach a ring or loop of wire from the liquid/liquid interface [6]. As a first approximation, the detachment force is taken to be equal to the interfacial tension γ multiplied by the perimeter of the ring, i.e.,

$$F = W + 4\pi R\gamma \tag{4.28}$$

where W is the weight of the ring. Harkins and Jordan [7] introduced a correction factor f (that is a function of meniscus volume V and radius r of the wire) for more accurate calculation of γ from F, i.e.,

$$f = \frac{\gamma}{\gamma_{\text{ideal}}} = f\left(\frac{R^3}{V}, \frac{R}{r}\right) \tag{4.29}$$

Values of the correction factor f were tabulated by Harkins and Jordan [7]. A theoretical account of f was given by Freud and Freud [8].

When using the du Nouy method to obtain γ the ring must be kept horizontal during the measurement. Moreover, the ring should be free from contaminant, which is usually achieved by using a platinum ring that is flamed before use.

4.3.4 Drop Volume (Weight) Method

Here one determines the volume V (or weight W) of a drop of liquid (immersed in the second, less dense liquid) which becomes detached from a vertically mounted capillary tip having a circular cross section of radius r. The ideal drop weight W_{ideal} is given by the expression,

$$W_{\text{ideal}} = 2\pi r\gamma \tag{4.30}$$

In practice, a weight W is obtained that is less than W_{ideal} because a portion of the drop remains attached to the tube tip. Thus, Eq. (4.30) should include a correction factor ϕ, which is a function of the tube radius r and some linear dimension of the drop, i.e., $V^{1/3}$ (Eq. 4.31).

$$W = 2\pi r\gamma\phi\left(\frac{r}{V^{1/3}}\right) \tag{4.31}$$

Values of $(r/V^{1/3})$ have been tabulated by Harkins and Brown [9]. Lando and Oakley [10] used a quadratic equation to fit the correction function to $(r/V^{1/3})$. A better fit has been provided by Wilkinson and Kidwell [11].

4.3.5
Spinning Drop Method

This method is particularly useful for measuring very low interfacial tensions ($< 10^{-1}$ mN m^{-1}), which are especially important in applications such as spontaneous emulsification and the formation of microemulsions. Such low interfacial tensions may also be reached with emulsions, particularly when mixed surfactant films are used. A drop of the less dense liquid A is suspended in a tube containing the second liquid B. On rotating the whole mass (Figure 4.4) the drop of the liquid moves to the centre. With increasing speed of revolution, the drop elongates as the centrifugal force opposes the interfacial tension force that tends to maintain the spherical shape, i.e., that having minimum surface area.

An equilibrium shape is reached at any given speed of rotation. At moderate speeds of rotation, the drop approximates to a prolate ellipsoid, whereas at very high revolutions, the drop approximates to an elongated cylinder. This is schematically shown in Figure 4.4.

When the shape of the drop approximates a cylinder, the interfacial tension is given by Eq. (4.32) [12],

$$\gamma = \frac{\omega^2 \Delta\rho r_0^4}{4} \tag{4.32}$$

where ω is the speed of rotation, $\Delta\rho$ is the density difference between the two liquids A and B and r_0 is the radius of the elongated cylinder. Eq. (4.32) is valid when the elongated cylinder is much longer than r_0.

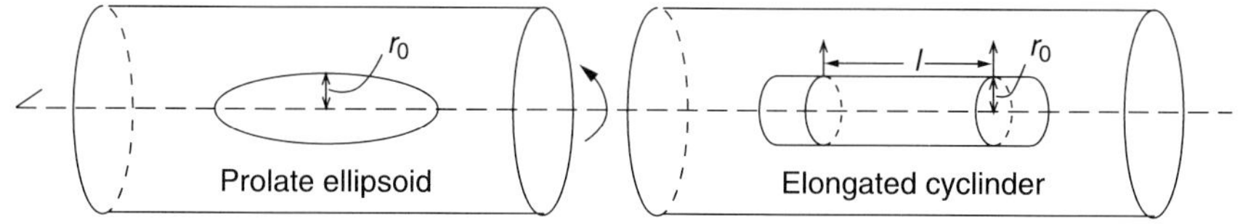

Fig. 4.4. Schematic representation of a spinning drop: (a) prolate ellipsoid, (b) elongated cylinder.

References

1 E. A. Guggenheim: *Thermodynamics*, North-Holland, Amsterdam, 1967, 45.
2 J. W. Gibbs: *Collected Works*, Longman, Harlow, 1928, 219, Vol. 1.
3 L. Wilhelmy, *Ann. Phys.*, **1863**, *119*, 177.
4 F. Bashforth, J. C. Adams: *An Attempt to Test the Theories of Capillary Action*, University Press, Cambridge, 1883.
5 D. O. Nierderhauser, F. E. Bartell: *Report of Progress, Fundamental Research on Occurence of Petroleum*, American Petroleum Institute, Lord Baltimore Press, Baltimore, 1950, 114.
6 P. L. Du Nouy, *J. Gen. Physiol.*, **1919**, *1*, 521.
7 W. D. Harkins, H. F. Jordan, *J. Am. Chem. Soc.*, **1930**, *52*, 1715.
8 B. B. Freud, H. Z. Freud, *J. Am. Chem. Soc.*, **1930**, *52*, 1772.
9 W. D. Harkins, F. E. Brown, *J. Am. Chem. Soc.*, **1919**, *41*, 499.
10 J. L. Lando, H. T. Oakley, *J. Colloid Interface Sci.*, **1967**, *25*, 526.
11 M. C. Wilkinson, R. L. Kidwell, *J. Colloid Interface Sci.*, **1971**, *35*, 114.
12 B. Vonnegut, *New Sci. Intrum.*, **1942**, *13*, 6.

5
Adsorption of Surfactants and Polymeric Surfactants at the Solid/Liquid Interface

5.1
Introduction

The use of surfactants (ionic, nonionic and zwitterionic) and polymers to control the stability behaviour of suspensions is of considerable technological importance. Surfactants and polymers are used in the formulation of dyestuffs, paints, paper coatings, agrochemicals, pharmaceuticals, ceramics, printing inks, etc. They are a particularly robust form of stabilisation, which is useful at high disperse volume fractions and high electrolyte concentrations, as well as under extreme conditions of high temperature, pressure and flow. In particular, surfactants and polymers are essential for stabilising suspensions in non-aqueous media, where electrostatic stabilisation is less successful.

The key to understanding how surfactants and polymers (to be referred to as polymeric surfactants) function as stabilisers is to know their adsorption and conformation at the solid/liquid interface. This is the objective of the present chapter, which is a survey of the general trends observed and some of the theoretical treatments.

Since surfactant and polymer adsorption processes are significantly different, the two subjects will be treated differently – surfactant adsorption is relatively simpler than polymer adsorption. This is because surfactants consist of a small number of units and they mostly are reversibly adsorbed, allowing one to apply thermodynamic treatments. In this case, it is possible to describe the adsorption in terms of the various interaction parameters, namely chain–surface, chain–solvent and surface–solvent. Moreover, the conformation of the surfactant molecules at the interface can be deduced from these simple interactions parameters. In contrast, polymer adsorption is fairly complicated. In addition to the usual adsorption considerations described above, one of the principle problems to be resolved is the conformation of the polymer molecule at the surface. This can be acquired in various ways depending on the number of segments and chain flexibility. This requires the application of statistical thermodynamic methods.

Applied Surfactants: Principles and Applications. Tharwat F. Tadros

ISBN: 3-527-30629-3

5.2 Surfactant Adsorption

As noted above, surfactant adsorption may be described in terms of simple interaction parameters. However, in some cases these interaction parameters may involve ill-defined forces, such as hydrophobic bonding, solvation forces and chemisorption. In addition, the adsorption of ionic surfactants involves electrostatic forces, particularly with polar surfaces containing ionogenic groups. Thus, the adsorption of ionic and nonionic surfactants will be treated separately. Surfaces (substrates) can be also hydrophobic or hydrophilic and these may be treated separately.

5.2.1 Adsorption of Ionic Surfactants on Hydrophobic Surfaces

The adsorption of ionic surfactants on hydrophobic surfaces such as carbon black, polymer surfaces and ceramics (silicon carbide or silicon nitride) is governed by hydrophobic interaction between the alkyl chain of the surfactant and the hydrophobic surface. Here, electrostatic interaction will play a relatively smaller role. However, if the surfactant head group is of the same sign of charge as that on the substrate surface, electrostatic repulsion may oppose adsorption. In contrast, if the head groups are of opposite sign to the surface, adsorption may be enhanced. Since adsorption depends on the magnitude of the hydrophobic bonding free energy, the amount of surfactant adsorbed increases directly with increasing alkyl chain length in accordance with Traube's rule.

The adsorption of ionic surfactants on hydrophobic surfaces may be represented by the Stern–Langmuir isotherm [1]. Consider a substrate containing N_s sites (mol m^{-2}) on which Γ moles m^{-2} of surfactant ions are adsorbed. The surface coverage θ is (Γ/N_s) and the fraction of uncovered surface is $(1-\theta)$.

The rate of adsorption is proportional to the surfactant concentration expressed in mole fraction $(C/55.5)$ and the fraction of free surface $(1-\theta)$, i.e.

$$\text{Rate of adsorption} = k_{ads}\left(\frac{C}{55.5}\right)(1-\theta) \tag{5.1}$$

where k_{ads} is the rate constant for adsorption.

The rate of desorption is proportional to the fraction of surface covered θ,

$$\text{Rate of desorption} = k_{des}\theta \tag{5.2}$$

At equilibrium, the rate of adsorption is equal to the rate of desorption and the ratio of (k_{ads}/k_{des}) is the equilibrium constant K, i.e.,

$$\frac{\theta}{(1-\theta)} = \frac{C}{55.5}K \tag{5.3}$$

The equilibrium constant K is related to the standard free energy of adsorption by,

$$-\Delta G^{\circ}_{\text{ads}} = RT \ln K \tag{5.4}$$

R is the gas constant and T is the absolute temperature. Eq. (5.4) can be written in the form,

$$K = \exp\left(-\frac{\Delta G^{\circ}_{\text{ads}}}{RT}\right) \tag{5.5}$$

Combining Eqs. (5.3) and (5.5),

$$\frac{\theta}{1-\theta} = \frac{C}{55.5} \exp\left(-\frac{\Delta G^{\circ}_{\text{ads}}}{RT}\right) \tag{5.6}$$

Eq. (5.6) applies only at low surface coverage ($\theta < 0.1$) where lateral interaction between the surfactant ions can be neglected.

At high surface coverage ($\theta > 0.1$) one should take the lateral interaction between the chains into account, by introducing a constant A, e.g. using the Frumkin–Fowler–Guggenheim Eq. (5.1),

$$\frac{\theta}{(1-\theta)} \exp(A\theta) = \frac{C}{55.5} \exp\left(-\frac{\Delta G^{\circ}_{\text{ads}}}{RT}\right) \tag{5.7}$$

Various authors [2, 3] have used the Stern–Langmuir equation in a simple form to describe the adsorption of surfactant ions on mineral surfaces,

$$\Gamma = 2rC \exp\left(-\frac{\Delta G^{\circ}_{\text{ads}}}{RT}\right) \tag{5.8}$$

Various contributions to the adsorption free energy may be envisaged. To a first approximation, these contributions may be considered to be additive. In the first instance, ΔG_{ads} may be taken to consist of two main contributions, i.e.

$$\Delta G_{\text{ads}} = \Delta G_{\text{elec}} + \Delta G_{\text{spec}} \tag{5.9}$$

where ΔG_{elec} accounts for any electrical interactions and ΔG_{spec} is a specific adsorption term that contains all contributions to the adsorption free energy that depend on the "specific" (non-electrical) nature of the system [4]. Several authors have subdivided ΔG_{spec} into supposedly separate independent interactions [4, 5], e.g.

$$\Delta G_{\text{spec}} = \Delta G_{\text{cc}} + \Delta G_{\text{cs}} + \Delta G_{\text{hs}} + \cdots \tag{5.10}$$

where ΔG_{cc} is a term that accounts for the cohesive chain–chain interaction between the hydrophobic moieties of the adsorbed ions, ΔG_{cs} is the term for chain–

substrate interaction, whereas ΔG_{hs} is a term for the head group–substrate interaction. Several other contributions to ΔG_{spec} may be envisaged e.g. ion–dipole, ion-induced dipole or dipole-induced dipole interactions.

Since there is no rigorous theory that can predict adsorption isotherms, the most suitable method to investigate adsorption of surfactants is to determine the adsorption isotherm experimentally. Measurement of surfactant adsorption is fairly straightforward. A known mass m (g) of the particles (substrate) with known specific surface area A_s ($m^2\ g^{-1}$) is equilibrated at constant temperature with surfactant solution with an initial concentration C_1. The suspension is stirred for sufficient time to reach equilibrium. The particles are then removed from the suspension by centrifugation and the equilibrium concentration C_2 is determined using a suitable analytical method. The amount of adsorption Γ (mole m^{-2}) is calculated as follows,

$$\Gamma = \frac{(C_1 - C_2)}{mA_s} \tag{5.11}$$

The adsorption isotherm is represented by plotting Γ versus C_2. A range of surfactant concentrations should be used to cover the whole adsorption process, i.e. from the initial values low to the plateau values. To obtain accurate results, the solid should have a high surface area (usually $>1\ m^2$).

Several examples may be quoted from the literature to illustrate the adsorption of surfactant ions on solid surfaces. For a model hydrophobic surface, carbon black has been chosen [6, 7]. Figure 5.1 shows typical results for the adsorption of sodium dodecyl sulphate (SDS) on two carbon black surfaces, namely Spheron 6 (untreated) and Graphon (graphitised), which also describe the effect of surface treatment. Adsorption of SDS on untreated Spheron 6 tends to show a maximum that is removed on washing. This suggests the removal of impurities from the carbon black that become extractable at high surfactant concentration. The plateau adsorption is $\sim 2 \times 10^{-6}$ mol m^{-2} (~ 2 μmol m^{-2}). This plateau value is reached at ~ 8 mmol dm^{-3} SDS, i.e. close to the c.m.c. of the surfactant in the bulk solution. The area per surfactant ion in this case is $\sim 0.7\ nm^2$. Graphitisation (Graphon) removes the hydrophilic ionisable groups (e.g. –C=O or –COOH), producing a more hydrophobic surface. The same occurs by heating Spheron 6 to 2700 °C. This leads to a different adsorption isotherm (Figure 5.1), showing a step (inflection point) at a surfactant concentration in the region of ~ 6 mmol dm^{-3}. The first plateau is at ~ 2.3 μmol m^{-2} whereas the second plateau (which occurs at the c.m.c. of the surfactant) is ~ 4 μmol m^{-2}. In this case, the surfactant ions probably adopt different orientations at the first and second plateaus. In the first plateau region, a "flatter" orientation (alkyl chains adsorbing parallel to the surface) is obtained whereas at the second plateau a vertical orientation is more favourable, with the polar head groups directed towards the solution phase. Addition of electrolyte (10^{-1} mol dm^{-3} NaCl) enhances the surfactant adsorption, due to a reduction in lateral repulsion between the sulphate head groups.

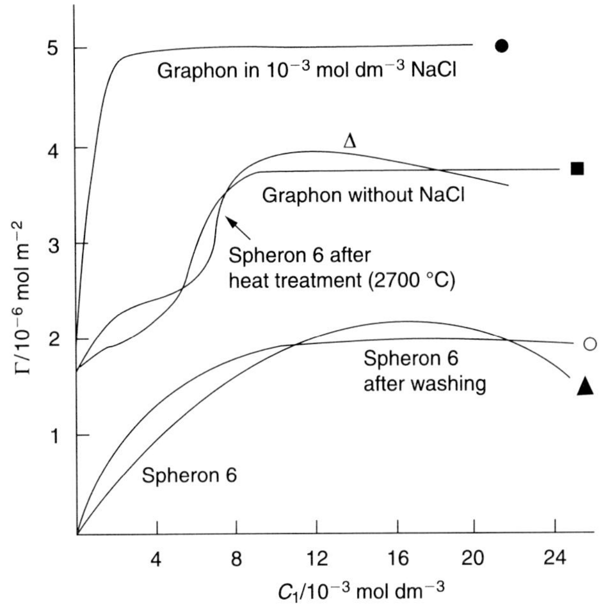

Fig. 5.1. Adsorption isotherms for sodium dodecyl sulphate (SDS) on carbon substrates. Graphon in 10^{-1} mol dm^{-3} NaCl (●), and without added electrolyte (△); Spheron 6 (▲) and after washing (○) and after heat treatment at 2700 °C (■).

The adsorption of ionic surfactants on hydrophobic polar surfaces resembles that for carbon black [8, 9]. For example, Saleeb and Kitchener [8] found a similar limiting area for cetyltrimethylammonium bromide on Graphon and polystyrene (~0.4 nm^2). As with carbon black, the area per molecule depends on the nature and amount of added electrolyte. This can be accounted for in terms of reduction of head group repulsion and/or counter ion binging.

Surfactant adsorption close to the c.m.c. may appear Langmuirian, although this does not automatically imply a simple orientation. For example, rearrangement from horizontal to vertical orientation or electrostatic interaction and counter ion binding may be masked by simple adsorption isotherms. Therefore, adsorption isotherms must be combined with other techniques such as microcalorimetry and various spectroscopic methods to obtain a full picture of surfactant adsorption.

5.2.2
Adsorption of Ionic Surfactants on Polar Surfaces

The adsorption of ionic surfactants on polar surfaces that contain ionisable groups may show characteristic features due to additional interaction between the head group and substrate and/or possible chain–chain interaction. This is best illustrated by the results of adsorption of sodium dodecyl sulphonate (SDSe) on alu-

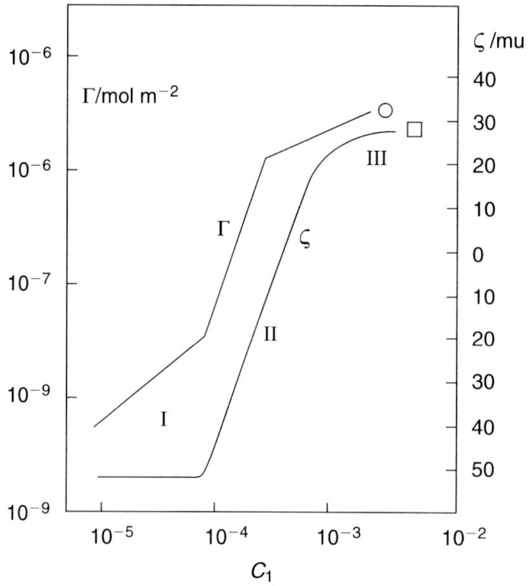

Fig. 5.2. Adsorption isotherm for sodium dodecyl sulphonate (SDSe) on alumina (○) and the corresponding ζ-potential of alumina particles (□) as a function of the equilibrium surfactant concentration; pH 7.2 and ionic strength 2×10^{-3} mol dm^{-3}.

mina at pH 7.2 obtained by Fuerstenau [10] (Figure 5.2). At this pH, the alumina is positively charged (the isoelectric point of alumina is at pH ~ 9) and the counter ions are Cl^- from the added supporting electrolyte. In Figure 5.2, the saturation adsorption Γ_1 is plotted versus equilibrium surfactant concentration C_1 on logarithmic scales. The figure also shows the results of zeta potential (ζ) measurements (which are a measure of the magnitude sign of charge on the surface). Both the adsorption and zeta potential results show three distinct regions. The first region, showing a gradual increase of adsorption with increasing concentration, with virtually no change in the zeta potential, corresponds to an ion-exchange process [11]. In other words, the surfactant ions simply exchange with the counter ions (Cl^-) of the supporting electrolyte in the electrical double layer. At a critical surfactant concentration, the desorption increases dramatically with further increase in surfactant concentration (region II). Here, the positive zeta potential gradually decreases to zero (charge neutralisation) after which a negative value is obtained, which increases rapidly with increasing surfactant concentration. The rapid increase in region II was explained in terms of "hemi-micelle formation, which was originally postulated by Gaudin and Fuerestenau [12]. In other words, at a critical surfactant concentration (denoted the c.m.c. of "hemi-micelle formation" or, better, as the critical aggregation concentration CAC) the hydrophobic moieties of the adsorbed surfactant chains are "squeezed out" from the aqueous solution by forming two-dimensional aggregates on the adsorbent surface. This is analogous to the process of micellisation in bulk solution. However, the CAC is lower than the c.m.c., indicating that the substrate promotes surfactant aggregation. At a certain surfactant concentration in the hemi-micellisation process, the isoelectric point is exceeded and, thereafter, adsorption is hindered by the electrostatic repulsion between the hemi-micelles and, hence, the slope of the adsorption isotherm is reduced (region III).

5.2.3
Adsorption of Nonionic Surfactants

Several types of nonionic surfactants exist, depending on the nature of the polar (hydrophilic) group. The most common type is that based on a polyoxyethylene glycol group, i.e. $(CH_2CH_2O)_nOH$ (where n can vary from as little as 2 to as high as 100 or more units), linked either to an alkyl (C_xH_{2x+1}) or alkyl phenyl (C_xH_{2x+1}–C_6H_4–) group. These surfactants may be abbreviated as C_xE_n or $C_x\phi E_n$ (where x refers to the number of C atoms in the alkyl chain, ϕ denotes C_6H_4, and E denotes ethylene oxide). These ethoxylated surfactants are characterised by a relatively large head group compared to the alkyl chain (when $n > 4$). However, there are nonionic surfactants with small head group such as amine oxides ($-N \rightarrow O$) head group, phosphate oxide ($-P \rightarrow O$) or sulphinyl-alkanol ($-SO-(CH_2)_n-OH$) [13]. Most adsorption isotherms in the literature are based on the ethoxylated-type surfactants.

The adsorption isotherm of nonionic surfactants are in many cases Langmuirian, like those of most other highly surface active solutes adsorbing from dilute solutions, and adsorption is generally reversible. However, several other adsorption types are produced [13], which are illustrated in Figure 5.3. The steps in the isotherm may be explained in terms of the various adsorbate–adsorbate, adsorbate–adsorbent and adsorbate–solvent interactions. These orientations are schematically illustrated in Figure 5.4. In the first stage of adsorption (denoted by I), surfactant–surfactant interaction is negligible (low coverage) and adsorption occurs mainly by van der Waals interaction. On a hydrophobic surface, the interaction is dominated

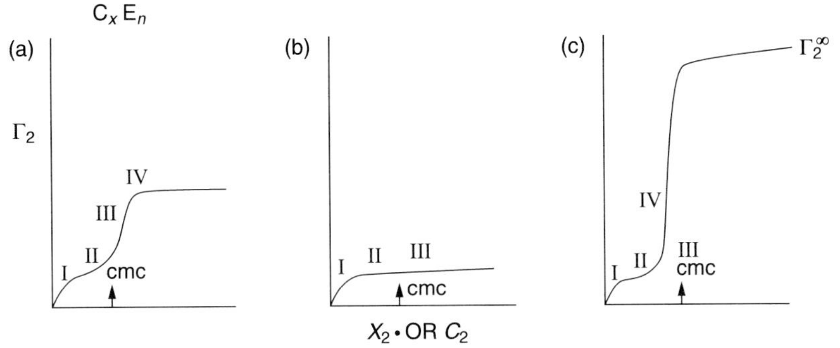

Fig. 5.3. Adsorption isotherms, corresponding to the three adsorption sequences shown in Figure 5.4 (I–IV), indicating the different orientation; the c.m.c. is indicated by an arrow.

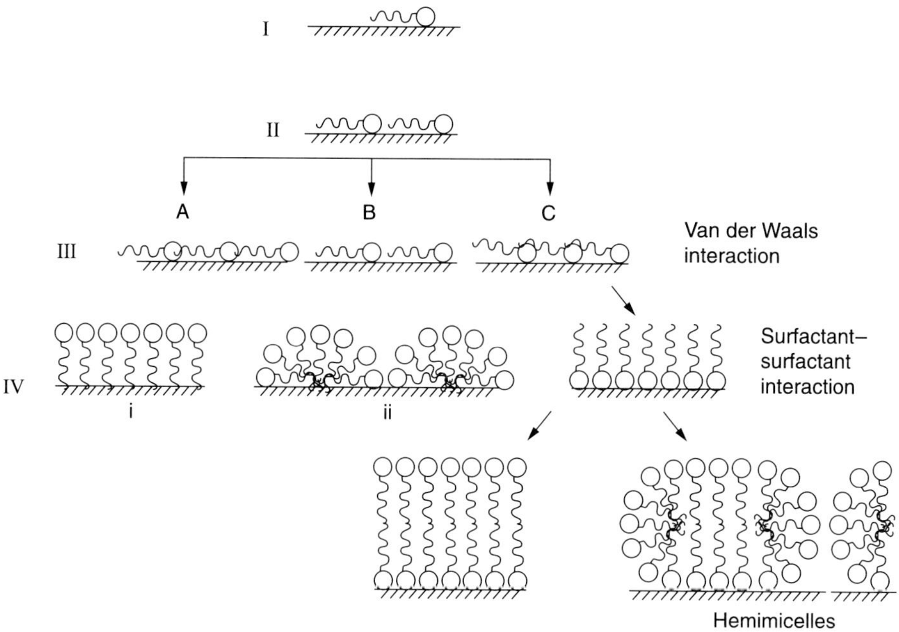

Fig. 5.4. Model for the adsorption of nonionic surfactants, showing orientation of surfactant molecules at the surface. I–V are the successive stages of adsorption, and sequence A–C corresponds, respectively, to situations where there are relatively weak, intermediate, and strong interactions between the adsorbent and the hydrophilic moiety of the surfactant.

by the hydrophobic portion of the surfactant molecule. This is mostly the case with agrochemicals that have hydrophobic surfaces. However, if the chemical is hydrophilic, the interaction will be dominated by the EO chain. The approach to monolayer saturation with the molecules lying flat is accompanied by a gradual decrease in the slope of the adsorption isotherm (region II in Figure 5.3). Increasing the size of the surfactant molecule, e.g. increasing the length of the alkyl or EO chain, will decrease adsorption (when expressed in moles per unit area). Conversely, increasing temperature will increase adsorption as a result of desolvation of the EO chains, thus reducing their size. Moreover, increasing temperature reduces the solubility of the nonionic surfactant and this enhances adsorption.

The subsequent stages of adsorption (regions III and IV) are determined by surfactant–surfactant interaction, although surfactant–surface interaction initially determines adsorption beyond stage II. This interaction depends on the nature of the surface and the hydrophilic–lipophilic balance of the surfactant molecule (HLB). For a hydrophobic surface, adsorption occurs via the alkyl group of the surfactant. For a given EO chain, the adsorption will increase with increase in the alkyl chain length. Conversely, for a given alkyl chain length, adsorption increases with decreasing the PEO chain length.

As the surfactant concentration approaches the c.m.c., the alkyl groups tend to aggregate. This will cause vertical orientation of the surfactant molecules (stage IV) and compress the head group and for an EO chain, resulting in a less coiled more extended conformation. The larger the surfactant alkyl chain the greater the cohesive forces, and hence the smaller the cross sectional area. This may explain why saturation adsorption increases with increasing alkyl chain length.

Interactions occurring in the adsorption layer during the fourth and subsequent stages of adsorption are similar to those that occur in bulk solution. Aggregate units may be formed (Figure 5.4, hemi-micelles or micelles). This picture was supported by Kleminko et al. [14] who found close agreement between saturation adsorption and adsorption calculations based on the assumption that the surface is covered with close-packed hemi-micelles. Kleminko [15] developed a theoretical model for the three stages of adsorption of nonionic surfactants. In the first stage (flat orientation) a modified Langmuir adsorption equation was used. In the second stage of horizontal orientation, the surface concentration increases by an amount that is determined by the displacement of the ethoxy chain by the alkyl group. Finally, in the region of the hemi-micelle formation, the adsorption can be described by a simple Langmuir equation of the form,

$$C_2 K_a^* = \frac{\Gamma_2}{(\Gamma_2^\infty - \Gamma_2)} \tag{5.12}$$

where Γ_2^∞ is the maximum surface excess, i.e. the surface excess when the surface is covered with close-packed hemi-micelles, K_a^* is a constant that is inversely proportional to the c.m.c. and C_2 is the equilibrium concentration.

5.3 Adsorption of Polymeric Surfactants at the Solid/Liquid Interface

The simplest type of a polymeric surfactant is a homopolymer, that is formed from the same repeating units [16, 17]: poly(ethylene oxide) (PEO); poly(vinylpyrrolidone) (PVP). Homopolymers have little surface activity at the oil/water (O/W) interface. However, homopolymers may adsorb significantly at the solid/liquid (S/L) interface. Even if the adsorption energy per monomer segment is small (a fraction of kT, where k is the Boltzmann constant and T is the absolute temperature), the total adsorption energy per molecule may be sufficient (several segments are adsorbed at the surface) to overcome the unfavourable entropy loss of the molecule at the S/L interface. Homopolymers may also adsorb at the solid surface by some specific interaction, e.g. hydrogen bonding (for example, adsorption of PEO or PVP on silica). In general, homopolymers are not the most suitable emulsifiers or dispersants.

A small variant is to use polymers that contain specific groups that have high affinity for the surface, e.g. partially hydrolysed poly(vinyl acetate) (PVAc), which is technically referred to as poly(vinyl alcohol) (PVA) – commercially available PVA molecules contain 4–12% acetate groups. The acetate groups give the molecule its amphipathic character – on a hydrophobic surface (such as polystyrene) the polymer adsorbs with preferential attachment of the acetate groups on the surface, leaving the more hydrophilic vinyl alcohol segments dangling in the aqueous medium. Partially hydrolysed PVA molecules exhibit surface activity at the O/W interface.

The most convenient polymeric surfactants are those of the block and graft copolymer type. A block copolymer is a linear arrangement of blocks of varying composition (**5.1**) [16].

Diblock-Poly A–*block*-Poly B ~~A~~~~~ ~~~~~B~~

Triblock-Poly A–*block*-Poly B - Poly A

~~A~~~~~ ~~~~B~~~~ ~~~~~A~~

5.1

A graft copolymer is a nonlinear array of one B block on which several A polymers are grafted (**5.2**).

5.2

Two types of investigations are essential to unravel the behaviour of block and graft copolymers: (1) their properties in a solvent in which both the A and B blocks are soluble, giving information on their conformation; (2) properties in a solvent which is a non-solvent for one of the blocks but a good solvent for the other.

Block copolymers exhibit surface activity since one block is soluble in one of the phases and the other is miscible in the other phase; e.g. A-B block, where A is hydrophilic and B is hydrophobic (**5.3**).

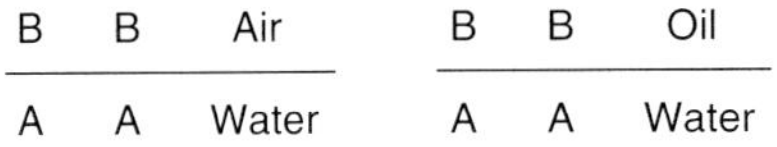

5.3

Since block copolymers are amphiphilic, they aggregate in solution to form micelles. A-B-A block copolymers may form micelles with smaller aggregation numbers (**5.4**), while A-B block copolymers can form simple micelles (**5.5**).

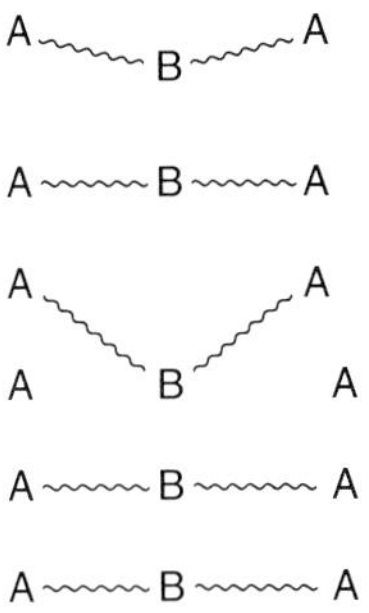

5.4

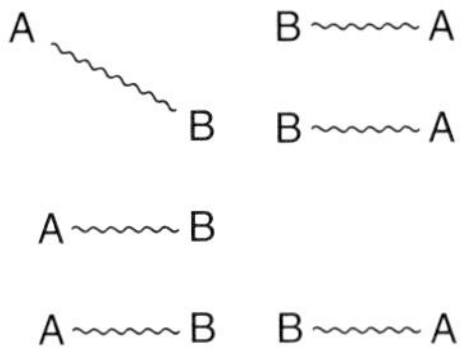

5.5

Graft copolymers also aggregate in solution to form micelles – again with small aggregation numbers. A dimer may be the form of aggregation.

Most block and graft copolymers have low critical micelle concentrations (c.m.c.s) and in many cases it is not easy to measure their c.m.c. The aggregation process is also affected by temperature and solvency of the medium for the A chains. One of the most useful methods to follow the aggregation of block and graft copolymers is to use time-averaged light scattering. By measuring the intensity as a function of concentration one can extrapolate the results to zero concentration and obtain the molecular weight of the micelle. This allows one to obtain the aggregation number from a knowledge of the molecular weight of the monomer.

Several examples of block and graft copolymers may be quoted. Triblock polymeric surfactants ["Pluronics" (BASF) or "Synperonic PE" (ICI)] – two poly-A blocks of PEO and one block poly-B of poly(propylene oxide) (PPO); several chain lengths of PEO and PPO are available. Triblocks of PPO-PEO-PEO (inverse "Pluronics") are also available. Polymeric triblock surfactants can be applied as emulsifiers and dispersants. The hydrophobic PPO chain resides at the hydrophobic surface, leaving the two PEO chains dangling in aqueous solution (providing steric stabilisation).

The above triblocks are not the most efficient emulsifiers or dispersants – the PPO chain is not sufficiently hydrophobic to provide a strong "anchor" to a hydrophobic surface or to an oil droplet. The reason for the surface activity of the PEO-PPO-PEO triblock at the O/W interface is probably due to "rejection" anchoring – the PPO chain is not soluble in water or most oils.

Several other di- and triblock copolymers have been synthesised: diblocks, polystyrene-*block*-poly(vinyl alcohol); triblocks, poly(methyl methacrylate)-*block*-poly(ethylene oxide)-*block*-poly(methyl methacrylate); diblocks, polystyrene-*block*-poly(ethylene oxide); triblocks, poly(ethylene oxide)-*block*-polystyrene-*block*-poly(ethylene oxide).

An alternative (and perhaps more efficient) polymeric surfactant is the amphipathic graft copolymer consisting of a polymeric backbone B (polystyrene or poly(methyl methacrylate)) and several A chains ("teeth") such as poly(ethylene oxide). The graft copolymer is referred to as a "comb" stabiliser – the polymer forms a "brush" at the solid/liquid interface. The copolymer is usually prepared by grafting a macromonomer such as methoxy poly(ethylene oxide) methacrylate with poly(methyl methacrylate). In most cases, some poly(methacrylic acid) is incorporated with the poly(methyl methacrylate) backbone – this leads to reduction of the glass transition of the backbone, making the chain more flexible for adsorption at the solid/liquid interface. Typical commercially available graft copolymers are Atlox 4913 and Hypermer CG-6 supplied by ICI.

The "grafting into" technique has also been used to synthesise polystyrene-poly(ethylene oxide) graft copolymers – these molecules are not commercially available.

Recently, a novel graft copolymer based on a naturally occurring polysaccharide, namely Inulin (polyfructose), has been synthesised [17]. Inulin is a polydisperse polysaccharide consisting mainly, if not exclusively, of $\beta(2 \rightarrow 1)$ fructosyl fructose units (F_m) with normally, but not necessarily, one glucopyranose unit at the reducing end (GF_n) [18, 19]. To produce the amphipathic graft copolymer, the chains were modified by introducing alkyl groups (C_4–C_{18}) on the polyfructose backbone through isocyanates. The structure of the molecule (Inulin carbamate) is illustrated below (**5.6**).

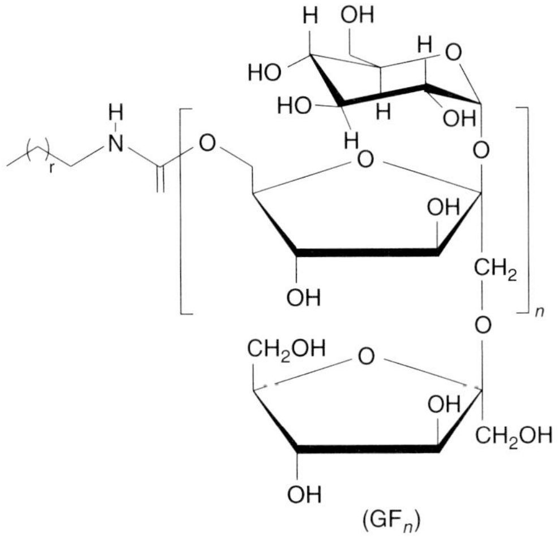

5.6

In the structure of GF_n, the alkyl groups represent the B chains (randomly distributed on the sugar backbone on primary hydroxyl functions as well as on the secondary ones) that become strongly adsorbed on a hydrophobic solid such as carbon black, polystyrene or an oil droplet. The sugar chain forms the stabilising chain as this is highly water soluble. These graft copolymers are surface active and they lower the surface tension of water and the interfacial tension at the oil/water interface. They will also adsorb on hydrophobic surfaces with the alkyl groups strongly attached (multipoint anchoring), leaving the polyfructose chains dangling in solution and probably forming large loops. These graft copolymers can produce highly stable suspensions and emulsions, in particular at high electrolyte concentrations [20].

5.4
Adsorption and Conformation of Polymeric Surfactants at Interfaces

Understanding the adsorption and conformation of polymeric surfactants at interfaces is key to knowing how these molecules act as stabilizers. Most basic ideas on adsorption and conformation of polymers have been developed for the solid/liquid interface [21]. The same concepts may be applied to the liquid/liquid interface, with some modification whereby some parts of the molecule may reside within the oil phase, rather than simply staying at the interface. Such modification does not alter the basic concepts, particularly when dealing with stabilization by these molecules.

Polymer adsorption involves several interactions that must be considered separately. Three main interactions must be taken into account, namely that between solvent molecules and the surface (or oil for o/w emulsions, which needs to be displaced for the polymer segments to adsorb), between the chains and the solvent, and between the polymer and the surface. In addition, one of the most fundamental considerations is the conformation of the polymer molecule at the interface. These molecules adopt various conformations, depending on their structure. The simplest case is that of a homopolymer that consists of identical segments [e.g. poly(ethylene oxide)], which shows a sequence of loops, trains and tails (Figure 5.5a). Notably, for such a polymer to adsorb, the reduction in entropy of the chain as it approaches the interface must be compensated by an energy of adsorption between the segments and the surface. In other words, the chain segments must have a minimum adsorption energy, χ^s, otherwise no adsorption occurs. With polymers that are highly water soluble, such as poly(ethylene oxide) (PEO), the interaction energy with the surface may be too small for adsorption to occur, and so the whole molecule may not be strongly adsorbed to the surface. For this reason, many commercially available polymers that are described as homopolymers, such as poly(vinyl alcohol) (PVA) contain some hydrophobic groups or short blocks (vinyl acetate in the case of PVA) that ensure their adsorption to hydrophobic surfaces (Figure 5.5b). Clearly, if all the segments have a high affinity to the surface, the whole molecule may lie flat on the surface (Figure 5.5c). This is rarely the case, since the molecule will have very low solubility in the continuous medium.

The most favourable structures for polymeric surfactants are those represented in Figure 5.5d, 5e and 5f, referred to as block and graft copolymers. Figure 5.5d shows an A-B block, consisting of a B chain that has a high affinity for the surface (or is soluble in the oil phase), referred to as the "anchoring" chain, and an A chain that has very low affinity for the surface and is strongly solvated by the medium. As will be discussed in the section on stabilization, this is the most convenient structure, since the forces that ensure strong adsorption are opposite to those that ensure stability. A variance on the structure shown in Figure 5.5d is the A-B-A block copolymer shown in Figure 5.5e. Here, the anchor chain B contains two stabilizing chains (tails). Figure 5.5f shows another variation, which is described as a graft copolymer ("comb" type structure), with one B chain and several A chains (tails or "teeth").

From this description of polymer configurations, a full characterization of the adsorption process clearly requires a knowledge of the amount of polymer adsorbed per unit area of the surface, Γ (mol m^{-2} or mg m^{-2}), the fraction of segments in close contact with the surface, p, and the distribution of polymer segments, $\rho(z)$, from the surface towards the bulk solution. We also must know how far the seg-

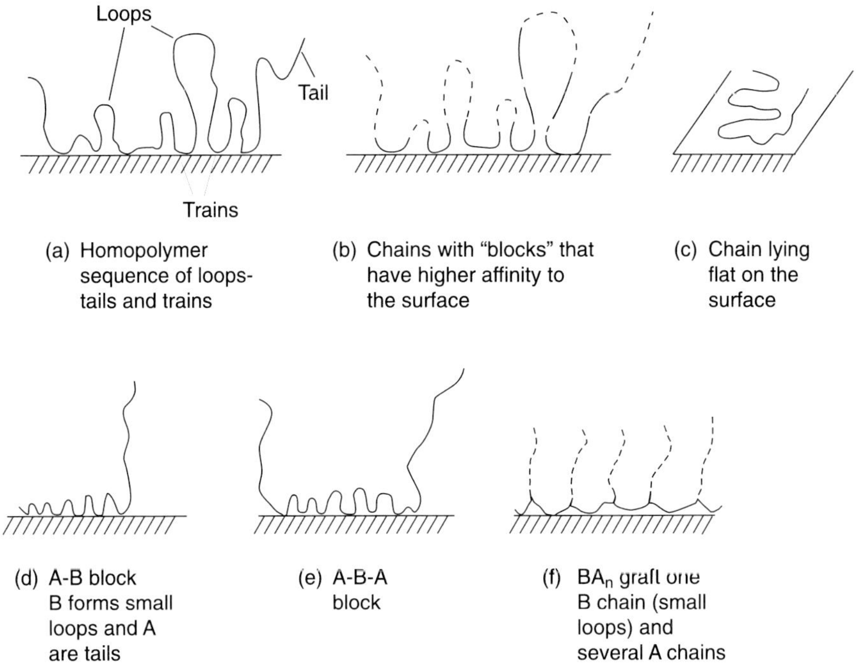

Fig. 5.5. Various conformations of polymeric surfactants adsorbed on a plane surface: (a) Random conformation of loops-trains-tails (homopolymer); (b) preferential adsorption of "short blocks"; (c) chain lying flat on the surface; (d) AB block copolymer with loop-train configuration of B and long tail of A; (e) ABA block as in (d); (f) BA_n graft with backbone B forming small loops and several tails of A ("teeth").

ments extend into solution, i.e. the adsorbed layer thickness δ, and how all these parameters change with (1) polymer coverage (concentration), the structure of the polymer and its molecular weight and (2) the environment such as solvency of the medium for the chains and temperature.

Several theories that describe polymer adsorption have been developed either using a statistical mechanical approach or quasi-lattice models. In the former, the polymer is considered to consist of three types of structures with different energy states, trains, loops and tails [22, 23]. The structures close to the surface (trains) are adsorbed with an internal partition function determined by short-range forces between the segment and surface (assigned an adsorption energy per segment χ^s). The segments in loops and tails are considered to have an internal partition function equivalent to that of segments in bulk solution and these are assigned a segment–solvent interaction parameter χ (Flory–Huggins interaction parameter). By equating the chemical potential of the macromolecule in the adsorbed state and in bulk solution, the adsorption isotherm can be determined. In earlier theories, the case of an isolated chain on the surface (low coverage) was consid-

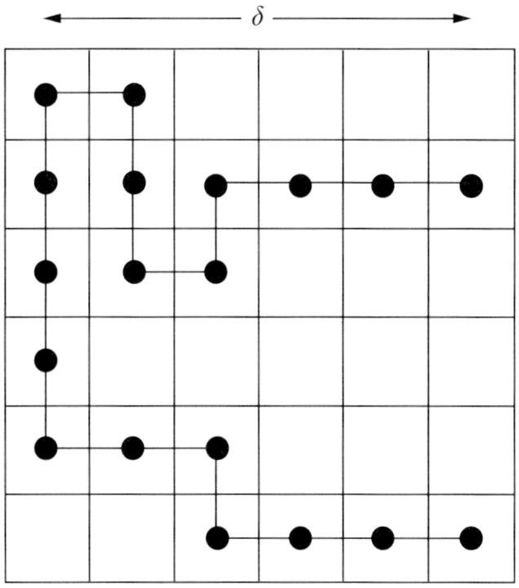

Fig. 5.6. Possible conformation of a polymer molecule at an interface.

ered, but later the theories were modified to take into account the lateral interaction between the chains, i.e. at high coverage.

The quasi-lattice model was developed by Roe [24] and by Scheutjens and Fleer [25, 26]. The basic procedure was to describe all chain conformations as step-weighted random walks on a quasi-crystalline lattice that extends in parallel layers away from the surface. This is illustrated in Figure 5.6, which shows a possible conformation of a polymer molecule at a surface.

The partition function is written in terms of a number of chain configurations that are treated as connected sequences of segments. In each layer, random mixing (Bragg–William or mean field approximation) between segments and solvent molecules is assumed. Each step in the random walk is assigned a weighting factor p_i that is considered to consist of three contributions, namely the adsorption energy χ^s, the configurational entropy of mixing and the segment–solvent interaction parameter χ.

The above theories gave several predictions for polymeric surfactant adsorption. Figure 5.7 shows typical adsorption isotherms plotted as surface coverage θ (in equivalent monolayers) versus polymer volume fraction ϕ_* in bulk solution (ϕ_* was taken to vary between 0 and 10^{-3}, which is the normal experimental range). Figure 5.7 shows the effect of increasing the chain length r and the effect of solvency (using two values for the Flory–Huggins interaction parameter, i.e. $\chi = 0$ (athermal solvent) and $\chi = 0.5$ (theta solvent). As the number of segments in the chain increases from a low (with few segments) to high (many segments) values, the adsorption isotherm changes from a Langmuirian type (characteristic for surfactant adsorption) to a high-affinity type.

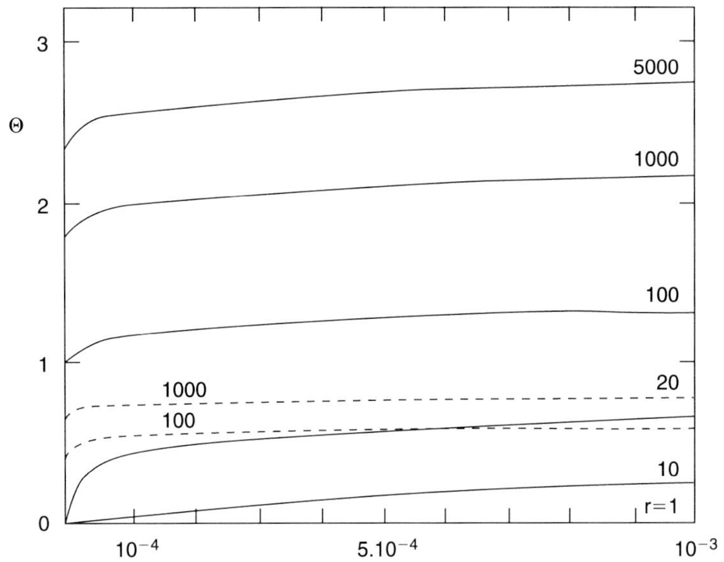

Fig. 5.7. Adsorption isotherms for oligomers and polymers in the dilute range; —— $\chi = 0.5$; - - - $\chi = 0$; $\chi_s = 1$; hexagonal lattice.

In the latter case, the first addition of polymer chains to the solution results in their virtual complete adsorption. The adsorption isotherms for chains with $r = 100$ and above are typical of those obtained experimentally for most polymers that are not too polydisperse, i.e. showing a steep rise followed by a nearly horizontal pseudo-plateau (which only increases a few percent per decade of ϕ_*). Adsorption in this case is described as being "irreversible", i.e. the equilibrium between adsorbed and free polymer is shifted towards the surface. This explains the strong anchoring of the polymer chains to the surface. As the solvency of the medium for the chains decreases, the amount of polymer adsorbed increases. This is clearly illustrated in Figure 5.7 when comparing the results obtained when $\chi = 0$ (very good solvent) with those obtained using a poor solvent with $\chi = 0.5$. In good solvents (dashed lines in Figure 5.7) θ is much smaller and levels off for long chains to attain an adsorption plateau that is essentially independent of molecular weight. This explains the relatively "weaker" adsorption of homopolymers that are highly solvated by the medium. It is now clear from these theories why block and graft copolymers are preferred for stabilization of dispersions. The poor solubility of the anchor chain B in the medium and its strong affinity to the surface ensures the strong adsorption of the molecule. In contrast, the high solubility of the stabilizing chain A ensures effective steric stabilization. Another prediction from the theories is that the higher the molecular weight of the polymer, the higher the amount of adsorption, when the latter is expressed in mg m^{-2}.

Some general features of the adsorption isotherms over a wide concentration range can be illustrated by using logarithmic scales for both θ and ϕ_*, which high-

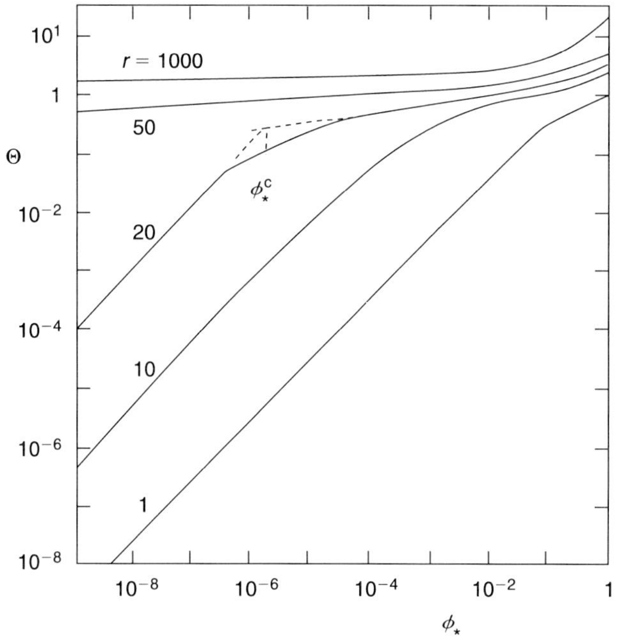

Fig. 5.8. Log–log presentation of adsorption isotherms of various r values, $\chi_s = 1$; $\chi = 0.5$; hexagonal lattice.

light the behaviour in extremely dilute solutions. Such a presentation [27] is shown in Figure 5.8. These results show a linear Henry region followed by a pseudo-plateau region. A transition concentration, ϕ_*^c, can be defined by extrapolation of the two linear parts; ϕ_*^c decreases exponentially with increasing chain length and when $r = 50$, ϕ_*^c is so small (10^{-12}) that it does not appear within the scale shown in Figure 5.8. With $r = 1000$, ϕ_*^c reaches the ridiculously low value of 10^{-235}. The region below ϕ_*^c is the Henry region, where the adsorbed polymer molecules behave essentially as isolated molecules. The representation in Figure 5.8 also answers the question of reversibility versus irreversibility for polymer adsorption. When $r > 50$, the pseudo-plateau region extends down to very low concentrations ($\phi_*^c = 10^{-12}$), which explains why one cannot easily detect any desorption upon dilution. Clearly, if such extremely low concentration can be reached, desorption of the polymer may take place. Thus, the lack of desorption (sometimes referred to as irreversible adsorption) is because the equilibrium between adsorbed and free polymer is shifted far in favour of the surface due to the high number of possible attachments per chain.

Another to emerge from Scheutjens and Fleer's theory [28] is the difference in shape between experimental and theoretical adsorption isotherms in the low concentration region. Experimental isotherms are usually rounded, whereas those predicted from theory are flat. This is accounted for in terms of the molecular weight distribution (polydispersity) that is encountered in many practical systems. This effect has been explained by Cohen-Stuart et al. [28]. With polydisperse polymer fractions, the larger molecules adsorb preferentially over the smaller ones. At low polymer concentrations, nearly all molecular weights are adsorbed, leaving only a small fraction of polymer with the lowest molecular weight in solution. As the polymer concentration is increased, the higher molecular weight fractions displace the lower ones on the surface, releasing them into solution, thus shifting the molecular weight distribution of the polymer in the bulk solution to lower values. This process continues with further increase in polymer concentration, leading to a fractionation process whereby the higher molecular weight fractions are adsorbed at the expense of the lower fractions that are released in the bulk. However, in very concentrated solutions, monomers adsorb preferentially with respect to polymers and short chains with respect to larger ones. This is because, in this region, the conformational entropy term dominates the free energy, disfavouring the adsorption of long chains.

The bound fraction, p, is high at low polymer concentrations ($\phi_* < \phi_*^c$), approaching unity, and it is relatively independent of molecular weight when $r > 20$. In addition, p also increases with increasing adsorption energy, χ^s, but it decreases with increasing surface coverage and increasing molecular weight of the polymer. The structure of the adsorbed layer is described in terms of the segment density distribution, $\rho(z)$. As an illustration, Figure 5.9 shows some calculations by Scheutjens and Fleer [28] for loops and tails with $r = 1000$, $\phi_* = 10^{-6}$ and $\chi = 0.5$. In this example, 38% of the segments are in trains, 55.5% in loops and 6.5% in tails. This theory demonstrates the importance of tails, which dominate the total distribution in the outer region of the adsorbed layer. As we will discuss in the next section, the segment density distribution is not easily determined, and it usually assigns a value for the adsorbed layer thickness (δ). This increases with increase of the molecular weight of the polymer and increase of solvency of the medium for the chains.

5.5 Experimental Methods for Measurement of Adsorption Parameters for Polymeric Surfactants

5.5.1 Amount of Polymer Adsorbed Γ – Adsorption Isotherms

The amount of polymer adsorbed, Γ, can be directly determined in a similar way as described for surfactants, except in this case one has to consider the relatively slow adsorption process, which may take several hours or even days to reach equilibrium. In addition, one needs very sensitive analytical methods to determine the polymer concentration in the early stages of adsorption (which can be in the ppm range). As mentioned before, the amount of adsorption Γ can be calculated from a

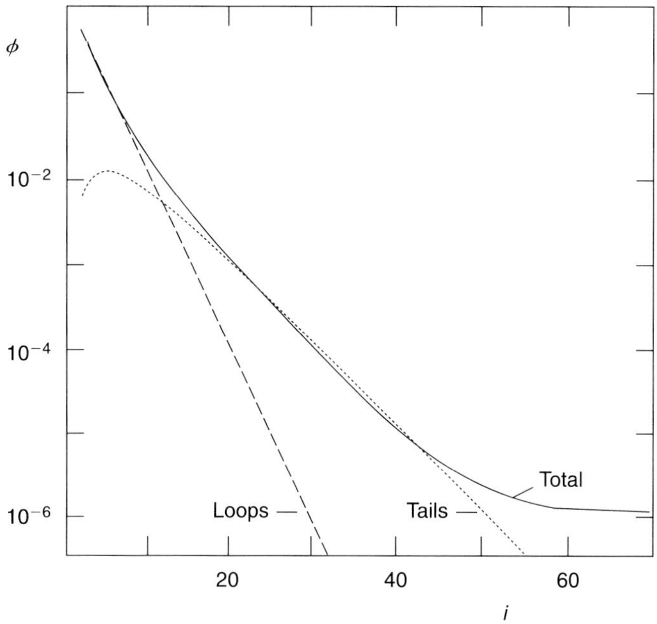

Fig. 5.9. Loop, tail and total segment concentration profiles according to Scheutjens and Fleer's theory [23], $\chi = 0.5$, $\chi_s = 1$, $r = 1000$, $\phi_* = 10^{-6}$.

knowledge of the initial polymer concentration C_1 and that after reaching equilibrium C_2, the mass of the solid m and the specific surface area A_s as given by Eq. (5.11).

Figure 5.10 illustrates this, showing the adsorption isotherms at 25 °C for poly(vinyl alcohol) (PVA) (containing 12% acetate groups) on polystyrene latex [29]. The polymer was fractionated using preparative gel permeation chromatography [29] or by a sequential precipitation technique using acetone [30]. The fractions were characterised for their molecular weight using ultracentrifugation and later by intrinsic viscosity measurements. The intrinsic viscosity $[\eta]$ could be related to the weight average molecular weight of the polymer (determined by ultracentrifugation) using the Mark–Houwink relationship,

$$[\eta] = KM^{\alpha} \tag{5.13}$$

The constants K and α were established from knowledge of $[\eta]$ and M. The latter values could also be used to calculate the molecular dimensions (radius of gyration), and the polymer–solvent interaction parameter χ was also determined. The polystyrene latex used for the adsorption measurements was a model system prepared using surfactant-free polymerisation and the particles were fairly monodis-

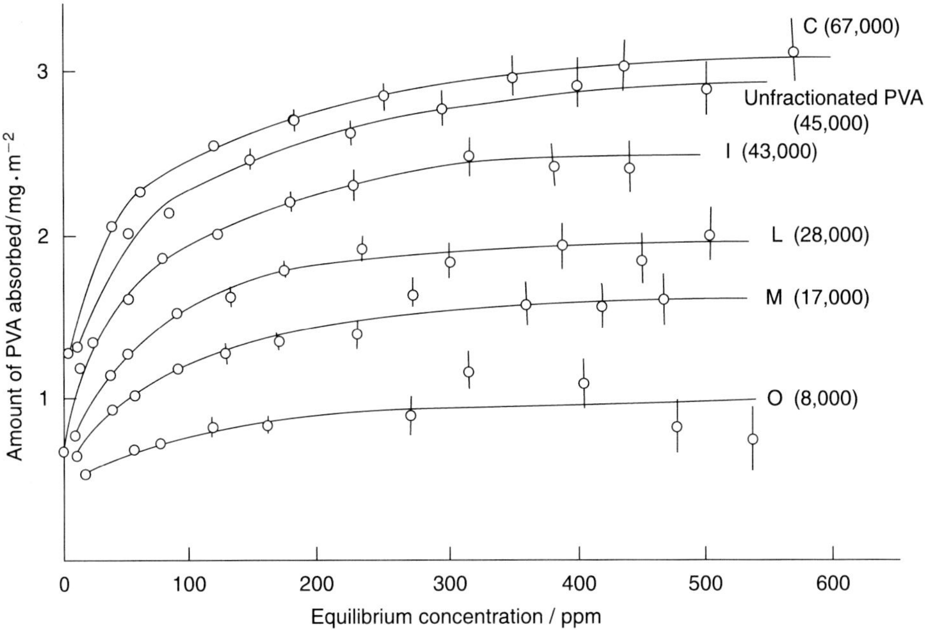

Fig. 5.10. Adsorption isotherms of poly(vinyl alcohol) (PVA) at 25 °C.

perse. Hence, the specific surface area of the particles could be estimated from simple geometry using electron microscopy.

Figure 5.10 shows the high affinity isotherms for the polymers and the increase in adsorption of the polymer with increasing molecular weight. Similar isotherms are expected for the adsorption of the polymer on oil droplets. However, in the latter case the full isotherm can not be obtained since to produce the emulsion one requires a minimum amount of polymer. In addition, the surface area of the emulsion has to be determined at each point from the droplet size distribution.

To study the effect of solvency on adsorption, measurements were carried out as a function of temperature [30] and addition of electrolyte (KCl or Na_2SO_4) [31]. Increasing temperature and/or addition of electrolyte reduces the solvency of the medium for the PVA chains (due to break down of the hydrogen bonds between the vinyl alcohol units and water). Figure 5.11 shows the adsorption isotherms for PVA with $M = 65\,100$ as a function of temperature. This shows a systematic increase of adsorption with rising temperature, i.e. with reduction of solvency (increase in χ), as expected from theory. The results obtained in the presence of electrolyte are shown in Figures 5.12 and 5.13. In both cases, addition of electrolyte increases adsorption of PVA, again due to the reduction of solvency of the medium for the chains.

The above polymer (PVA) is a "blocky" copolymer (containing short vinyl acetate blocks) and hence it does not represent the case for adsorption of homopoly-

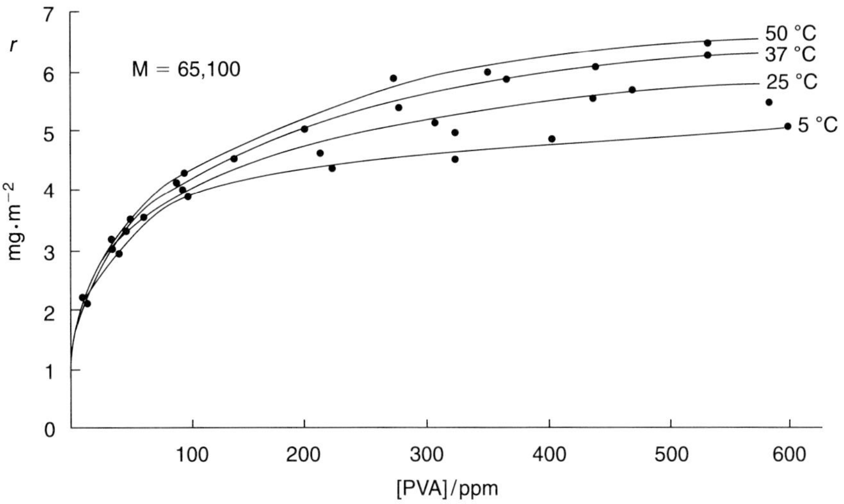

Fig. 5.11. Adsorption isotherms for PVA ($M_r = 65000$) on polystyrene latex 5 °C (lower curve) to 50 °C (upper curve).

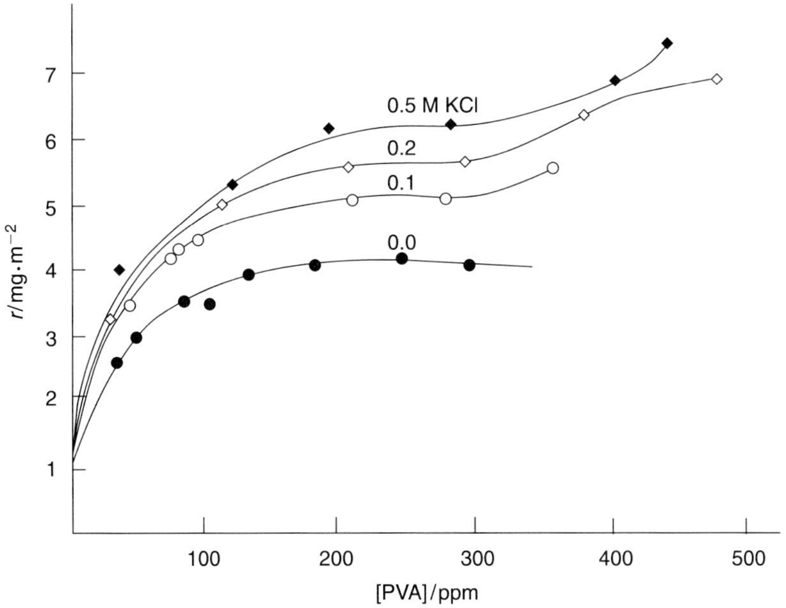

Fig. 5.12. Adsorption isotherms for PVA on polystyrene latex particles at various KCl concentrations.

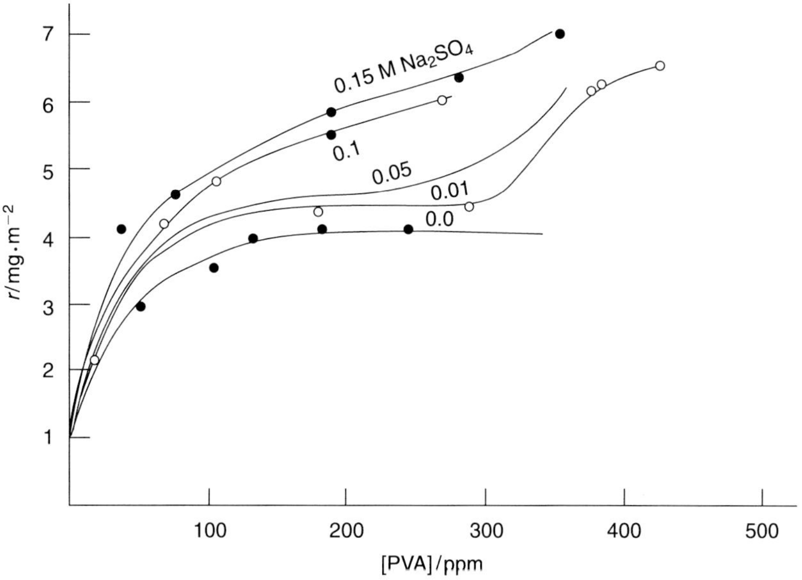

Fig. 5.13. Adsorption isotherms for PVA on polystyrene latex at various Na_2SO_4 concentrations.

mers. The latter case is exemplified by poly(ethylene oxide) (PEO) [32] as is illustrated in Figure 5.14 for adsorption on polystyrene latex using three different molecular weight PEO. As with PVA, the isotherms are of the high affinity type and the adsorbed amount increases with increasing molecular weight of the polymer. However, the amount of adsorption is much lower than that obtained using PVA, reflecting the difference between the two polymers.

5.5.2
Polymer Bound Fraction *p*

The bound fraction p represents the ratio of the number of segments in close contact with the surface (i.e. in trains) to the total number of segments in the polymer chain. The value of p can be directly determined using spectroscopic methods such as infrared (IR), electron spin resonance (ESR) and nuclear magnetic resonance (NMR). The IR method depends on measuring the shift in some absorption peak for a polymer and/or surface group [33, 34]. ESR and NMR methods depend on the reduction in mobility of the segments that are in close contact with the surface (larger rotational correlation time for trains when compared to loops). By using a pulsed NMR technique one can estimate p [35, 36]. An indirect method for estimating p is to use microcalorimetry. Basically one compares the enthalpy of adsorption per molecule with that per segment [37]. The latter may be obtained by using small molecules of similar structure to a polymer segment.

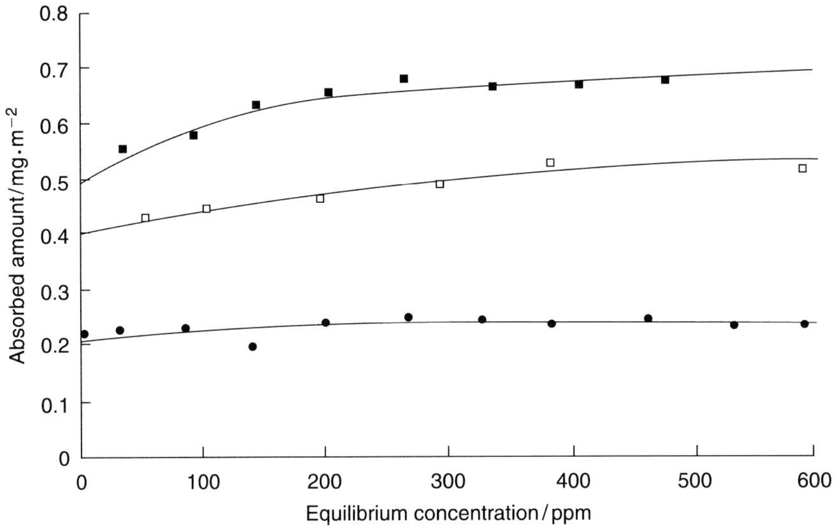

Fig. 5.14. Adsorption isotherms for PEO with various molecular weights on polystyrene latex. Molecular weights in order of decreasing adsorbed amounts are 930, 114 and 10.3 K.

5.5.3
Adsorbed Layer Thickness δ and Segment Density Distribution $\rho(z)$

Three direct methods can be applied for determination of adsorbed layer thickness: ellipsometry, attenuated total reflection (ATR) and neutron scattering. The first two [38] depend on the difference between refractive indices between the substrate, the adsorbed layer and bulk solution and require a flat reflecting surface. Ellipsometry [38] is based on the principle that light undergoes a change in polarizability when it is reflected at a flat surface (whether covered or uncovered with a polymer layer).

The above limitations when using ellipsometry or ATR are overcome by the application of neutron scattering, which can be applied to both flat surfaces and particulate dispersions. The basic principle is to measure the scattering due to the adsorbed layer, when the scattering length density of the particle is matched to that of the medium (the so-called "contrast-matching" method). Contrast matching of particles and medium can be achieved by changing the isotopic composition of the system (using deuterated particles and mixture of D_2O and H_2O). It has also used to measure the adsorbed layer thickness of polymers, e.g. PVA or poly(ethylene oxide) (PEO) on polystyrene latex [39]. Apart from obtaining δ, one can also determine the segment density distribution $\rho(z)$. Figure 5.15 illustrates this with the normalised density distribution for PVA ($M = 37\,000$) on a polystyrene (PS) latex.

The results show a monotonic decay of $\rho(z)$ with distance z from the surface and several regions may be distinguished. Close to the surface ($0 < z < 3$ nm), the decay in $\rho(z)$ is rapid, and, assuming a thickness of 1.3 nm for the bound layer, p was

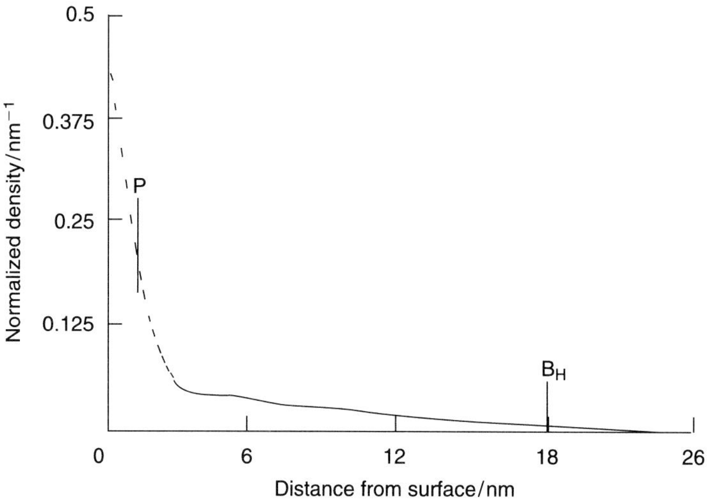

Fig. 5.15. Density $\rho(z)$ against distance z from the surface for PVA ($M_r = 37000$) adsorbed on deuterated PS latex in D_2O–H_2O.

calculated to be 0.1, which is in close agreement with the results obtained using NMR. In the middle region, $\rho(z)$ shows a shallow maximum followed by a slow decay, which extends to 18 nm, i.e. close to the hydrodynamic layer thickness δ_h of the polymer chain (see below); δ_h is determined by the longest tails and is about 2.5 times the radius of gyration in bulk solution (~7.2 nm). This slow decay of $\rho(z)$ with z at long distances is in qualitative agreement with Scheutjens and Fleers' theory [23], which predicts the presence of long tails. The shallow maximum at intermediate distances suggests that the observed segment density distribution is a summation of a fast monotonic decay due to loops and trains together with the segment density for tails which have a maximum density away from the surface. The latter maximum was clearly observed for a sample that had PEO grafted to a deuterated polystyrene latex [39] (where the configuration is represented by tails only).

The hydrodynamic thickness of block copolymers behaves differently from that of homopolymers (or random copolymers). Figures 5.16 and 5.17 illustrate this for an ABA block copolymer of poly(ethylene oxide)-poly(propylene oxide)-poly(ethylene oxide) (PEO-PPO-PEO) [30], showing the adsorbed amount (Figure 5.16) and the hydrodynamic thickness (Figure 5.17) versus fraction of anchor segment. The theoretical (Scheutjens and Fleer) prediction of adsorbed amount and layer thickness versus fraction of anchor segment are shown in the inserts of the figures. When there are two buoy blocks and a central anchor block, as in the above example, the A-B-A block shows similar behaviour to that of an A-B block. However, if there are two anchor blocks and a central buoy block, the polymer molecule precipitates at the particle surface, which is reflected in a continuous increase of adsorption with increasing polymer concentration, as has been shown for an A-B-A block of PPO-PEO-PPO [30].

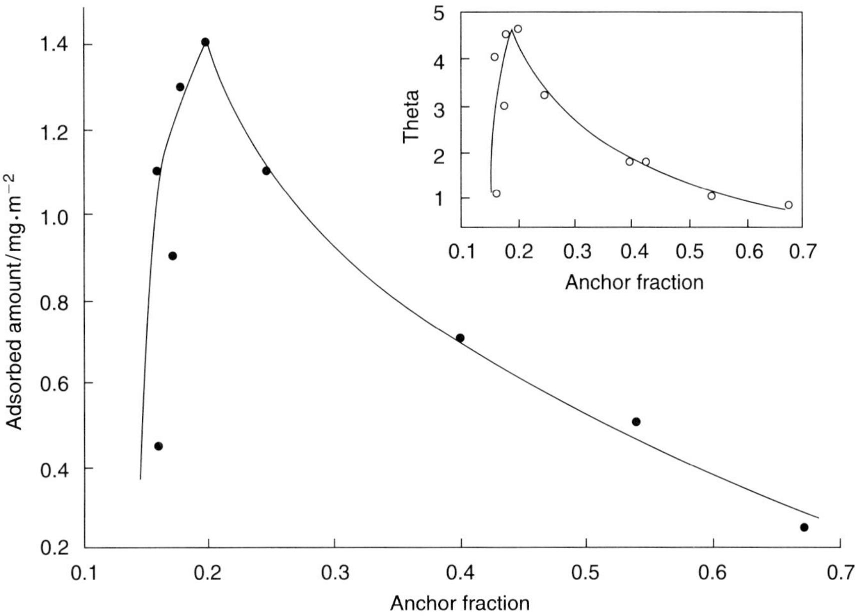

Fig. 5.16. Adsorbed amount versus fraction of anchor segment ν_A for PEO-PPO-PEO block copolymer. Insert: theoretical predictions.

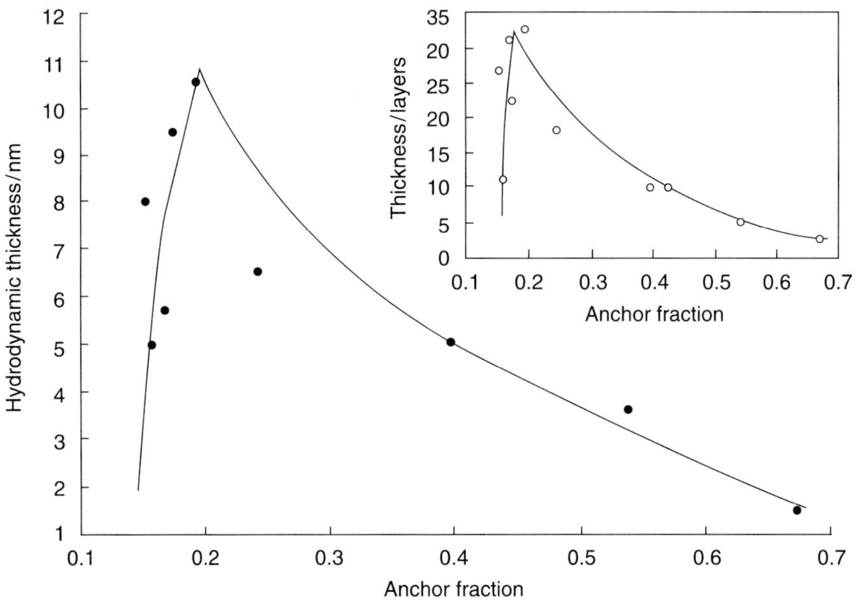

Fig. 5.17. Hydrodynamic thickness versus fraction of anchor segment ν_A for PEO-PPO-PEO block copolymer. Insert: theoretical predictions.

The above technique of neutron scattering clearly gives a quantitative picture of the adsorbed polymer layer. However, its practical application is limited since one needs to prepare deuterated particles or polymers for the contrast matching procedure. Practical methods for determination of the adsorbed layer thickness are mostly based on hydrodynamic methods described below.

5.5.4 Hydrodynamic Thickness Determination

Several methods may be applied to determine the hydrodynamic thickness of adsorbed polymer layers, of which viscosity, sedimentation coefficient (using an ultracentrifuge) and dynamic light scattering measurements are the most convenient. A less accurate method is from zeta potential measurements. These techniques are based on hydrodynamic techniques and these are discussed below.

The viscosity method [40] depends on measuring the increase in the volume fraction of the particles as a result of the presence of an adsorbed layer of thickness δ_h. The volume fraction of the particles (ϕ) plus the contribution of the adsorbed layers is usually referred to as the effective volume fraction ϕ_{eff}. Assuming the particles behave as hard-spheres, then the measured relative viscosity η_r is related to the effective volume fraction by the Einstein's equation, i.e.

$$\eta_r = 1 + 2.5\phi_{eff} \tag{5.14}$$

ϕ_{eff} and ϕ are related from simple geometry by

$$\phi_{eff} = \phi\left[1 + \left(\frac{\delta_h}{R}\right)\right]^3 \tag{5.15}$$

where R is the particle radius. Thus, from a knowledge of η_r and ϕ one can obtain δ_h using the above equations.

The sedimentation method depends on measuring the sedimentation coefficient (determined using an ultracentrifuge) of the particles S_0' (extrapolated to zero concentration) in the presence of the polymer layer [41]. Assuming the particles obey Stokes' law, S_0' is given by Eq. (5.16),

$$S_0' = \frac{\frac{4}{3}\pi R^3(\rho - \rho_s) + \frac{4}{3}\pi[(R + \delta_h)^3 - R^3](\rho_s^{ads} - \rho_s)}{6\pi\eta(R + \delta_h)} \tag{5.16}$$

where ρ and ρ_s are the mass density of the solid and solution phase, respectively, and ρ^{ads} is the average mass density of the adsorbed layer, which may be obtained from the average mass concentration of the polymer in the adsorbed layer.

To apply the above methods one should use a dispersion with monodisperse particles with a radius that is not much larger than δ_h. Small model particles of polystyrene may be used.

A relatively simple sedimentation method for determining δ_h is the slow speed centrifugation applied by Garvey et al. [41]. Basically, a stable monodisperse dispersion is slowly centrifuged at low g (< 50g) to form a close-packed (hexagonal or cubic) lattice in the sediment. From a knowledge of ϕ and the packing fraction (0.74 for hexagonal packing), the separation between the centre of two particles R_{-} may be obtained, i.e.,

$$R_{\delta} = R + \delta_{\mathrm{h}} = \left(\frac{0.74 V \rho_1 R^3}{W}\right) \tag{5.17}$$

where V is the sediment volume, ρ_1 is the density of the particles and W their weight.

The most rapid technique for measuring δ_{h} is photon correlation spectroscopy (PCS) (sometimes referred to as quasi-elastic light scattering), which allows one to obtain the diffusion coefficient of the particles with and without the adsorbed layer ($D_{\delta_{-}}$ and D respectively). This is obtained from measurement of the intensity fluctuation of scattered light as the particles undergo Brownian diffusion [42]. When a light beam (e.g. monochromatic laser beam) passes through a dispersion an oscillating dipole is induced in the particles, thus re-radiating the light. Owing to the random arrangement of the particles (which are separated by a distance comparable to the wavelength of the light beam, i.e. the light is coherent with the interparticle distance), the intensity of the scattered light will, at any instant, appear as a random diffraction or "speckle" pattern. As the particles undergo Brownian motion, the random configuration of the speckle pattern changes. The intensity at any one point in the pattern will, therefore, fluctuate such that the time taken for an intensity maximum to become a minimum (i.e. the coherence time) corresponds approximately to the time required for a particle to move one wavelength. Using a photomultiplier of active area about the size of a diffraction maximum, i.e. approximately one coherence area, this intensity fluctuation can be measured. A digital correlator is used to measure the photocount or intensity correlation function of the scattered light. The photocount correlation function can be used to obtain the diffusion coefficient D of the particles. For monodisperse non-interacting particles (i.e. at sufficient dilution), the normalised correlation function $[g^{(1)}(\tau)]$ of the scattered electric field is given by the equation,

$$[g^{(1)}(\tau)] = \exp[-(\Gamma\tau)] \tag{5.18}$$

where τ is the correlation delay time and Γ is the decay rate or inverse coherence time. Γ is related to D by Eq. (5.19),

$$\Gamma = DK^2 \tag{5.19}$$

where K is the magnitude of the scattering vector that is given by

$$K = \left(\frac{4n}{\lambda_0}\right) \sin\left(\frac{\theta}{2}\right) \tag{5.20}$$

where n is the refractive index of the solution, λ is the wavelength of light in vacuum and θ is the scattering angle.

From D, the particle radius R is calculated using the Stokes–Einstein equation,

$$D = \frac{kT}{6\pi\eta R} \tag{5.21}$$

where k is the Boltzmann constant and T is the absolute temperature. For a polymer-coated particle R is denoted R_δ, which is equal to $R + \delta_h$. Thus, by measuring D_δ and D, one can obtain δ_h. Notably, the accuracy of the PCS method depends on the ratio of δ_h/R, since δ_h is determined by difference. Since the accuracy of the measurement is $\pm 1\%$, δ_h should be at least 10% of the particle radius. This method can only be used with small particles and reasonably thick adsorbed layers.
Electrophoretic mobility, u, measurements can also be applied to measure δ_h [34]. From u, the zeta potential ζ, i.e. the potential at the slipping (shear) plane of the particles, can be calculated. Adsorption of a polymer causes a shift in the shear plane from its value in the absence of a polymer layer (which is close to the Stern plane) to a value that depends on the thickness of the adsorbed layer. Thus by measuring ζ in the presence (ζ_δ) and absence (ζ) of a polymer layer one can estimate δ_h. Assuming that the thickness of the Stern plane is Δ, then $\zeta_{\delta\_}$ may be related to the ζ (which may be assumed to be equal to the Stern potential Ψ_d) by Eq. (5.22),

$$\tanh\left(\frac{e\Psi_\delta}{4kT}\right) = \tanh\left(\frac{e\zeta}{4kT}\right) \exp[-\kappa(\delta_h - \Delta)] \tag{5.22}$$

where κ is the Debye parameter that is related to electrolyte concentration and valency.

Notably, δ_h calculated using the above simple equation shows a dependence on electrolyte concentration and hence the method cannot be used in a straightforward manner. Cohen-Stuart et al. [43] showed that the measured electrophoretic thickness δ_e approaches δ_h only at low electrolyte concentrations. Thus, to obtain δ_h from electrophoretic mobility measurements, results should be obtained at various electrolyte concentrations and δ_e should be plotted versus the Debye length $(1/\kappa)$ to obtain the limiting value at high $(1/\kappa)$ (i.e. low electrolyte concentration), which now corresponds to δ_h.

References

1 D. B. Hough, H. Randall: *Adsorption from Solution at the Solid/Liquid Interface*, G. D. Parfitt, C. H. Rochester (eds.), Academic Press, London, 1983, 247.

2 D. W. Fuerstenau, T. Healy: *Adsorptive Bubble Seperation Techniques*, R. Lemlich, ed., Academic Press, London, 1972, 91.

3 P. Somasundaran, E. D. Goddard, *Modern Aspects Electrochem.*, **1979**, *13*, 207.

4 T. W. Healy, *J. Macromol. Sci. Chem.*, **1974**, *118*, 603.

5 P. Somasundaran, H. Hannah: *Improved Oil Recovery by Surfactant and Polymer Flooding*, D. O. Shah, R. S. Schechter (eds.), Academic Press, London, 1979, 205.

6 F. G. Greenwood, G. D. Parfitt, N. H. Picton, D. G. Wharton, *Adv. Chem. Ser.*, **1968**, *79*, 135.

7 R. E. Day, F. G. Greenwood, G. D. Parfitt, *4th Int. Congr. Surf. Act. Subst.*, **1967**, *18*, 1005.

8 F. Z. Saleeb, J. A. Kitchener, *J. Chem. Soc.*, **1965**, 911.

9 P. Conner, R. H. Ottewill, *J. Colloid Interface Sci.*, **1971**, *37*, 642.

10 D. Fuerestenau: *The Chemistry of Biosurfaces*, M. L. Hair (eds.), Marcel Dekker, New York, 1971, 91.

11 T. Wakamatsu, D. W. Fuerstenau, *Adv. Chem. Ser.*, **1968**, *71*, 161.

12 A. M. Gaudin, D. W. Fuerstenau, *Trans. AIME*, **1955**, *202*, 958.

13 J. S. Clunie, B. Ingram: *Adsorption from Solution at the Solid/Liquid Interface*, G. D. Parfitt, C. H. Rochester (eds.), Academic Press, London, 1983, 105.

14 N. A. Kleminko, Tryasorukova, Permilouskayan, *Kolloid Zh.*, **1974**, *36*, 678.

15 N. A. Kleminko, *Kolloid Zh.*, **1978**, *40*, 1105; **1979**, *41*, 78.

16 I. Piirma: *Polymeric Surfactants*, Marcel Dekker, New York, 1992, Surfactant Science Series No. 42.

17 C. V. Stevens, A. Meriggi, M. Peristerpoulou, P. P. Christov, K. Booten, B. Levecke, A. Vandamme, N. Pittevils, T. F. Tadros, *Biomacromolecules*, **2001**, *2*, 1256.

18 E. L. Hirst, D. I. McGilvary, E. G. Percival, *J. Chem. Soc.*, **1950**, 1297.

19 M. Suzuki: *Science and Technology of Fructans*, M. Suzuki, N. J. Chatterton (eds.), CRC Press, Boca Raton, FL, 1993, 21.

20 T. F. Tadros, K. Booten, B. Levecke, Vandamme, to be published.

21 T. Tadros: *Polymer Colloids*, R. Buscall, T. Corner, and Stageman (eds.), Elsevier Applied Sciences, London, 1985, 105.

22 A. Silberberg, *J. Chem. Phys.*, **1968**, *48*, 2835.

23 C. A. Hoeve, *J. Polym. Sci.*, **1970**, *30*, 361; **1971**, *34*, 1.

24 R. J. Roe, *J. Chem. Phys.*, **1974**, *60*, 4192.

25 J. M. H. M. Scheutjens, G. J. Fleer, *J. Phys. Chem.*, **1979**, *83*, 1919.

26 J. M. H. M. Scheutjens, G. J. Fleer, *J. Phys. Chem.*, **1980**, *84*, 178.

27 J. M. H. M. Scheutjens, G. J. Fleer, *Adv. Colloid Interface Sci.*, **1982**, *16*, 341.

28 G. J. Fleer, M. A. Cohen-Stuart, J. M. H. M. Scheutjens, T. Cosgrove, B. Vincent: *Polymers of Interfaces* Chapman & Hall, London, 1993.

29 M. J. Garvey, T. F. Tadros, B. Vincent, *J. Colloid Interface Sci.*, **1974**, *49*, 57.

30 T. Boomgaard van den, T. A. King, T. F. Tadros, H. Tang, B. Vincent, *J. Colloid Interface Sci.*, **1978**, *61*, 68.

31 T. F. Tadros, B. Vincent, *J. Colloid Interface Sci.*, **1978**, *72*, 505.

32 T. M. Obey, P. Griffiths: *Principles of Polymer Science and Technology in Cosmetics and Personal Care*, E. D. Goddard, J. V. Gruber (eds.), Marcel Dekker, New York, 1999, Chapter 2.

33 E. Killmann, E. Eisenlauer, M. J. Korn, *Polym. Sci. Polym. Symp.*, **1977**, *61*, 413.

34 B. J. Fontana, J. R. Thomas, *J. Phys. Chem.*, **1961**, *65*, 480.

35 I. D. Robb, R. Smith, *Eur. Polym. J.*, **1974**, *10*, 1005.

36 K. G. Barnett, T. Cosgrove, B. Vincent, A. Burgess, T. L. Crowley, J. Kims, J. D. Turner, T. F. Tadros, *Polymer* **1981**, *22*, 283.

37 M. A. Cohen-Staurt, G. J. Fleer, J. Bijesterbosch, *Colloid Interface Sci.*, **1982**, *90*, 321.

38 F. Abeles: *Ellipsometry in the Measurement of Surfaces and Thin Films*, E. Passaglia, R. R. Stromberg, J. Kruger (eds.), Nat. Bur. Stand. Misc. Publ., 1964, 41, Volume 256.

39 T. Cosgrove, T. L. Crowley, T. Ryan, *Macromolecules*, **1987**, *20*, 2879.

40 A. Einstein: *Investigations on the Theory of the Brownian Movement*, Dover, New York, 1906.

41 M. J. Garvey, T. F. Tadros, B. Vincent, *J. Colloid Interface Sci.*, **1976**, *55*, 440.

42 P. N. Pusey: *Industrial Polymers: Characterisation by Molecular Weights*, J. H. S. Green, R. Dietz (eds.), Transcripta Books, London, 1973.

43 M. A. Cohen-Stuart, J. W. Mulder, *Colloids Surf.*, **1985**, *15*, 49.

6
Applications of Surfactants in Emulsion Formation and Stabilisation

6.1
Introduction

Emulsions are a class of disperse systems consisting of two immiscible liquids. The liquid droplets (the disperse phase) are dispersed in a liquid medium (the continuous phase) [1]. Several classes may be distinguished: oil-in-water (O/W), water-in-oil (W/O) and oil-in-oil (O/O). The latter class are exemplified by an emulsion consisting of a polar oil (e.g. propylene glycol) dispersed in a non-polar oil (paraffinic oil) and vice versa.

To disperse two immiscible liquids one needs a third component: the emulsifier. Emulsions may be classified according to the nature of the emulsifier or the structure of the system (Table 6.1) [1].

On storage, several breakdown processes may occur that depend on the particle size distribution and the density difference between droplets and the medium. It is the magnitude of the attractive versus repulsive forces that determines flocculation. The solubility of the disperse droplets and the particle size distribution determines Ostwald ripening. The stability of the liquid film between the droplets determines coalescence: phase inversion [1]. The various breakdown processes are illustrated in the Figure 6.1.

The physical phenomena involved in each breakdown process are not simply described, requiring analysis of the various surface forces involved. In addition, the

Tab. 6.1. Classification of emulsion types.

Nature of emulsifier	*Structure of the system*
Simple molecules and ions	Nature of internal and external phases
Nonionic surfactants	O/W, W/O
Ionic surfactants	Micellar emulsions
Surfactant mixtures	(microemulsions)
Nonionic polymers	Macroemulsions
Polyelectrolytes	Bilayer droplets
Mixed polymers and surfactants	Double and multiple emulsions
Liquid crystalline phases	Mixed emulsions
Solid particles	
(Pickering emulsions)	

Applied Surfactants: Principles and Applications. Tharwat F. Tadros

ISBN: 3-527-30629-3

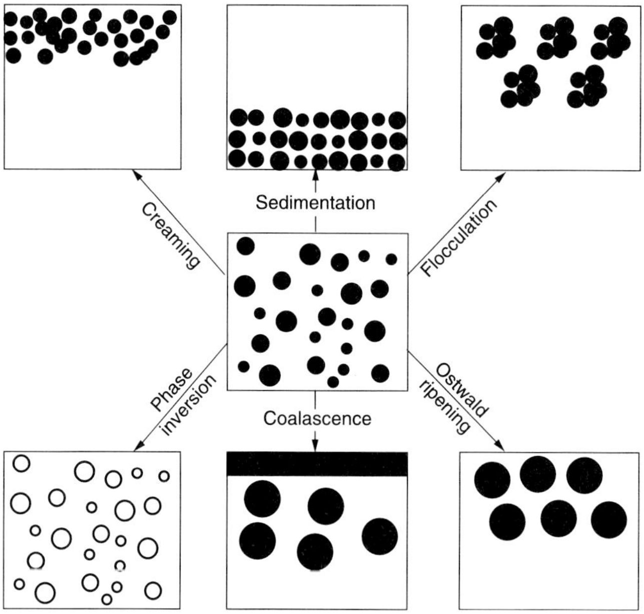

Fig. 6.1. Schematic of the emulsion breakdown processes.

above processes may take place simultaneously rather than consecutively, thereby complicating the analysis. Model emulsions with monodisperse droplets cannot be easily produced and hence any theoretical treatment must take into account the effect of droplet size distribution. Surfactant and polymer adsorption in an emulsion are not easily measured; one has to extract such information from measurement at a planar interface.

6.1.1
Industrial Applications of Emulsions

Among several applications of emulsions the most important are listed here: Food emulsion, e.g. mayonnaise, salad creams, deserts, beverages, etc. Personal care and cosmetics, e.g. hand creams, lotions, hair sprays, sunscreens, etc. Agrochemicals, e.g. self-emulsifiable oils which produce emulsions on dilution with water, emulsion concentrates (EWs) and crop oil sprays. Pharmaceuticals, e.g. anaethetics of O/W emulsions, lipid emulsions, double and multiple emulsions, etc. Paints, e.g. emulsions of alkyd resins, latex emulsions, etc. Dry cleaning formulations – these may contain water droplets emulsified in the dry cleaning oil that is necessary to remove soils and clays. Bitumen emulsions – emulsions prepared stable in the containers but when applied to the road chippings they must coalesce to form a uniform film of bitumen. Emulsions in the oil industry – many crude oils contain water droplets (e.g. North Sea oil) and these must be removed by coalescence followed by separation. Oil slick dispersions – oil spilled from tankers must be emulsified and then separated. Emulsification of unwanted oil – this is an important process for pollution control.

Such importance of emulsion in industry justifies a great deal of basic research to understand the origin of instability and methods to prevent their break down. Unfortunately, fundamental research on emulsions is difficult since model systems (e.g. with monodisperse droplets) are hard to produce.

6.2
Physical Chemistry of Emulsion Systems

6.2.1
Thermodynamics of Emulsion Formation and Breakdown

Consider a system in which an oil is represented by a large drop (2) of area A_1 immersed in a liquid 2, which is now subdivided into many smaller droplets (1) with total area A_2 ($A_2 \gg A_1$) (Figure 6.2). The interfacial tension γ_{12} is the same for the large and smaller droplets since the latter are generally in the region of 0.1 to few μm.

The change in free energy in going from state I to state II is made from two contributions: a surface energy term (which is positive) that is equal to $\Delta A\gamma_{12}$ (where $\Delta A = A_2 - A_1$). An entropy of dispersions term that is also positive (since producing a large number of droplets is accompanied by an increase in configurational entropy), which is equal to $T\Delta S^{\text{conf}}$.

From the second law of thermodynamics,

$$\Delta G^{\text{form}} = \Delta A\gamma_{12} - T\Delta S^{\text{conf}} \tag{6.1}$$

In most cases, $\Delta A\gamma_{12} \gg -T\Delta S^{\text{conf}}$, which means that ΔG^{form} is positive, i.e. emulsion formation is non-spontaneous and the system is thermodynamically unstable. In the absence of any stabilization mechanism, the emulsion will breakdown by flocculation, coalescence, Ostwald ripening or a combination of all these processes – This is illustrated in Figure 6.3, which shows several paths for emulsion breakdown [1].

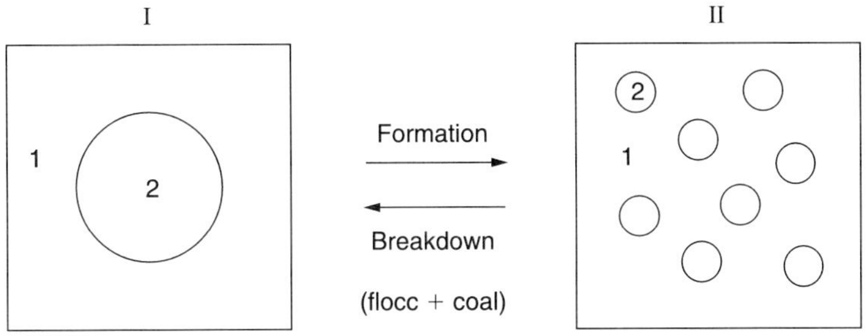

Fig. 6.2. Schematic of emulsion formation and breakdown.

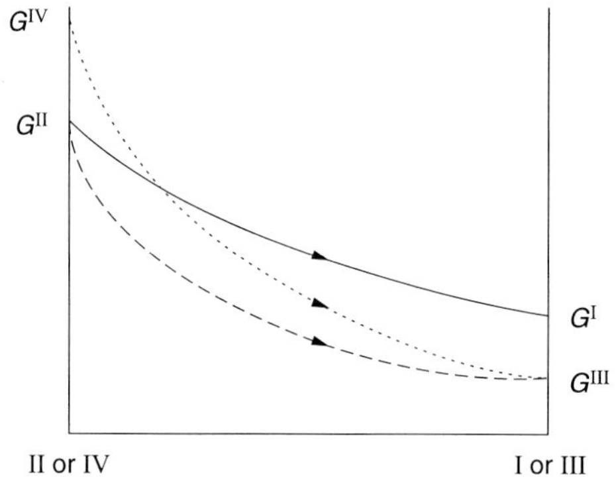

Fig. 6.3. Free energy path in emulsion breakdown (—) Flocc. + coal.; (– – –) Flocc. + coal. + Sed.; (- - - -) Flocc. + coal. + sed. + Ostwald ripening.

In the presence of a stabilizer (surfactant and/or polymer), an energy barrier is created between the droplets and, therefore, the reversal from state II to state I becomes non-continuous due to the presence of these energy barriers (Figure 6.4). In the presence of energy barriers, the system becomes kinetically stable [1]. As we will see later, the energy barrier can be created by electrostatic and/or steric repulsion that will overcome the van der Waals attraction.

6.2.2
Interaction Energies (Forces) Between Emulsion Droplets and their Combinations

Generally, there are three main interaction energies (forces) between emulsion droplets, which are discussed below.

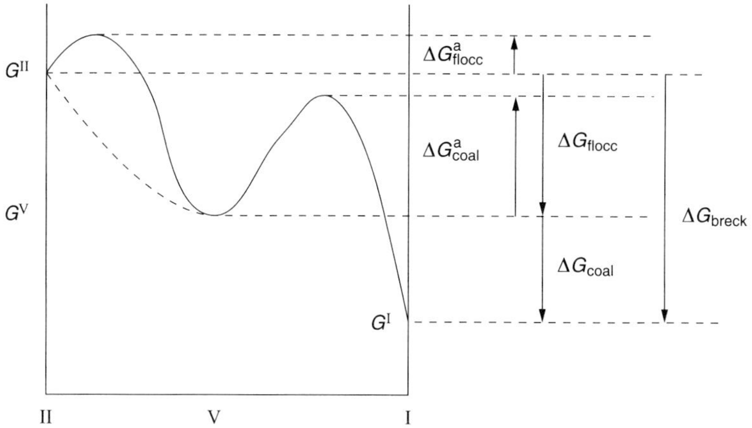

Fig. 6.4. Schematic free energy path for breakdown (flocculation and coalescence) for systems containing an energy barrier.

6.2.2.1 Van der Waals Attraction

There are three types of van der Waals attraction between atoms or molecules [2, 3]: dipole–dipole (Keesom), dipole-induced dipole (Debye) and dispersion (London) interactions. The most important are the London dispersion interactions, which arise from charge fluctuations.

Hamaker [2] suggested that the London dispersion interactions between atoms or molecules in macroscopic bodies (such as emulsion droplets) can be added, resulting in a strong van der Waals attraction, particularly at short separations between the droplets. For two droplets with equal radii R, at a separation distance h, the van der Waals attraction G_A is given by the following equation (due to Hamaker),

$$G_A = -\frac{AR}{12h} \tag{6.2}$$

where A is the effective Hamaker constant,

$$A = (A_{11}^{1/2} - A_{22}^{1/2})^2 \tag{6.3}$$

The Hamaker constant of any material depends on the number of atoms or molecules per unit volume q and the London dispersion constant β,

$$A = \pi q^2 \beta \tag{6.4}$$

G_A increases very rapidly with decreasing h. This is illustrated in Figure 6.5, which shows the van der Waals energy–distance curve for two emulsion droplets with separation distance h.

In the absence of repulsion, flocculation rapidly leads to large clusters. To counteract the van der Waals attraction, it is necessary to create a repulsive force. The two main types of repulsion can be distinguished, depending on the nature of the emulsifier used, are electrostatic (due to the creation of double layers) and steric (due to the presence of adsorbed surfactant or polymer layers).

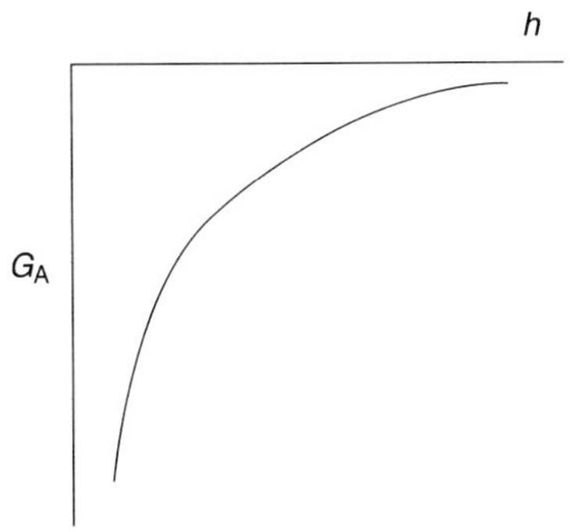

Fig. 6.5. Variation of van der Waals attraction with separation h between emulsion droplets.

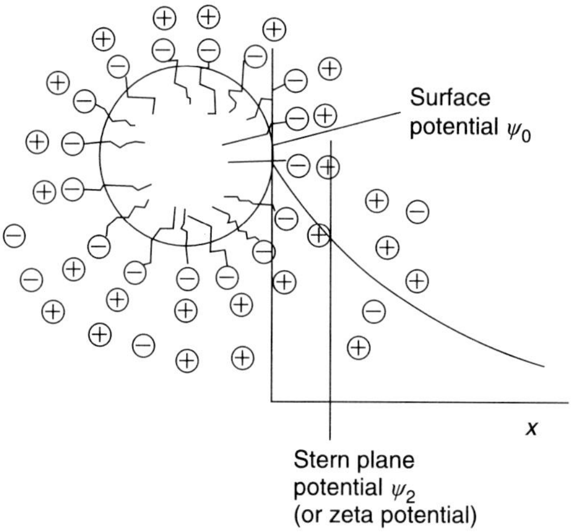

Fig. 6.6. Schematic of double layers produced by adsorption of an ionic surfactant.

6.2.2.2 **Electrostatic Repulsion**

This can be produced by adsorption of an ionic surfactant (Figure 6.6).

The surface potential Ψ_0 decreases linearly to Ψ_d (Stern or zeta potential) and then exponentially with increasing distance x [4].

When two droplets approach to a distance h that is smaller than the double layer extension, double layer overlap occurs and this leads to repulsion (the double layers cannot be fully developed) [5]. The double layer extension depends on electrolyte concentration and valency (the lower the electrolyte concentration and the lower the valency the more extended the double layer is).

The repulsive interaction G_{el} is given by the following expression,

$$G_{el} = 2\pi R \varepsilon_r \varepsilon_0 \Psi_0^2 \ln[1 + \exp - (\kappa h)] \tag{6.5}$$

ε_r is the relative permittivity and ε_0 is the permittivity of free space.

κ is the Debye–Hückel parameter, and $1/\kappa$ is the extension of the double layer (double layer thickness) that is given by the expression,

$$\left(\frac{1}{\kappa}\right) = \left(\frac{\varepsilon_r \varepsilon_0 kT}{2 n_0 Z_i^2 e^2}\right) \tag{6.6}$$

where n_0 is the number of ions per unit volume of each type present in bulk solution, Z_i is the valency of the ions and e is the electronic charge.

Values of $(1/\kappa)$ at various 1:1 electrolyte concentrations are tabulated below.

C (mol dm^{-3})	10^{-5}	10^{-4}	10^{-3}	10^{-2}	10^{-1}
$(1/\kappa)$ (nm)	100	33	10	3.3	1

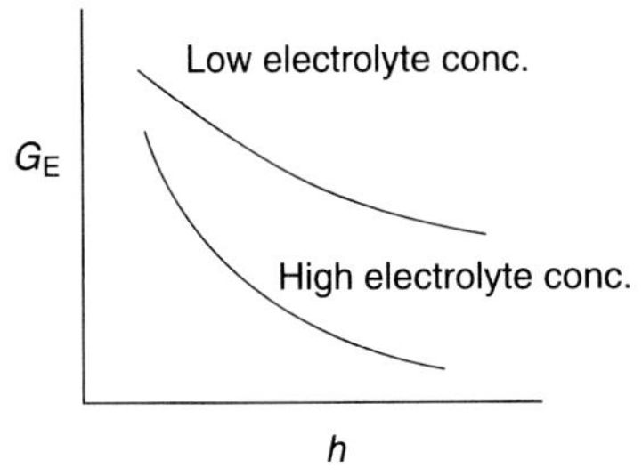

Fig. 6.7. Variation of G_{el} with h at low and high electrolyte concentrations.

The double layer extension decreases with increasing electrolyte concentration, meaning that the repulsion decreases with decreasing electrolyte concentration (Figure 6.7).

The combination of van der Waals attraction and double layer repulsion results in the well-known theory of colloid stability due to Deryaguin, Landau, Verwey and Overbeek (DLVO theory) [6, 7].

$$G_T = G_{el} + G_A \tag{6.7}$$

Figure 6.8 gives a schematic representation of the force (energy)–distance curve according to DLVO theory.

The above presentation is for a system at low electrolyte concentration. At large h, attraction prevails, resulting in a shallow minimum (G_{sec}) of the order of few kT units. At very short h, $V_A \gg G_{el}$, which gives a deep primary minimum (several

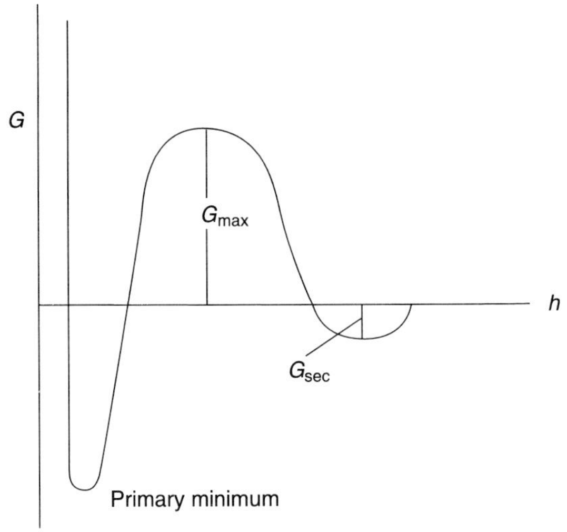

Fig. 6.8. Total energy–distance curve according to the DLVO theory.

hundred kT units). At intermediate h, $G_{el} > G_A$, affording a maximum (energy barrier) whose height depends on Ψ_0 (or ζ) and electrolyte concentration and valency – the energy maximum is usually kept at $>25kT$ units.

The energy maximum precludes close approach of the droplets and flocculation into the primary minimum is prevented. The higher Ψ_0 is and the lower the electrolyte concentration and valency, the higher the energy maximum. At intermediate electrolyte concentrations, weak flocculation into the secondary minimum may occur.

6.2.2.3 **Steric Repulsion**

This is produced by using nonionic surfactants or polymers, e.g. alcohol ethoxylates, or A-B-A block copolymers PEO-PPO-PEO (Figure 6.9).

The "thick" hydrophilic chains (PEO in water) produce repulsion due to two main effects [8]:

(a) Unfavourable mixing of the PEO chains, when these are in good solvent conditions (moderate electrolyte and low temperatures) – this is referred to as the osmotic or mixing free energy of interaction that is given by Eq. (6.8).

$$\frac{G_{mix}}{kT} = \left(\frac{4\pi}{V_1}\right)\phi_2^2 N_{av}\left(\frac{1}{2}-\chi\right)\left(3R+2\delta+\frac{h}{2}\right)\left(\delta-\frac{h}{2}\right)^2 \tag{6.8}$$

V_1 is the molar volume of the solvent, ϕ_2 is the volume fraction of the polymer chain with a thickness δ and χ is the Flory–Huggins interaction parameter.

When $\chi < 0.5$, G_{mix} is positive and the interaction is repulsive. When $\chi > 0.5$, G_{mix} is negative and the interaction is attractive. When $\chi = 0.5$, $G_{mix} = 0$ and this is referred to as the θ-condition.

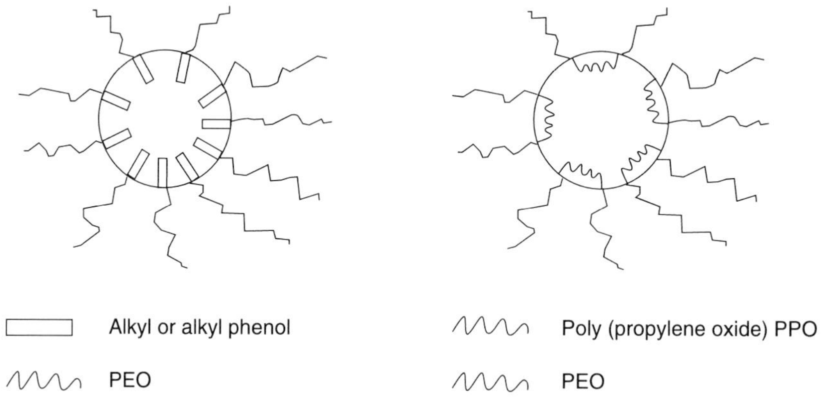

Fig. 6.9. Schematic of adsorbed layers.

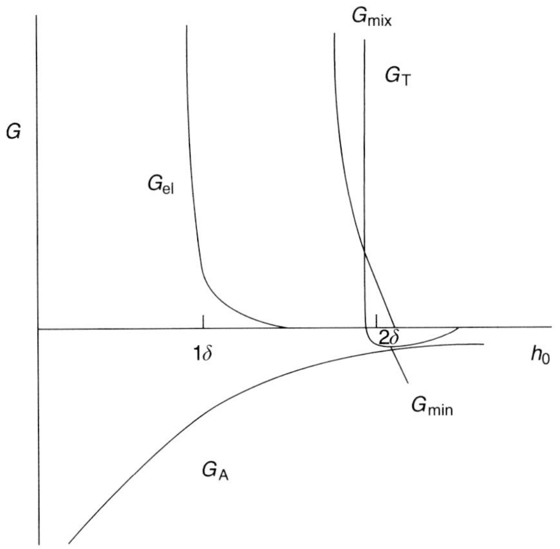

Fig. 6.10. Representation of the energy–distance curve for a sterically stabilised emulsion.

(b) Entropic, volume restriction or elastic interaction (G_{el}) results from the loss in configurational entropy of the chains on significant overlap. Entropy loss is unfavourable and, therefore, G_{el} is always positive.

Combination of G_{mix}, G_{el} with G_A gives the total energy of interaction G_T (theory of steric stabilisation) (Eq. 6.9).

$$G_T = G_{mix} + G_{el} + G_A \tag{6.9}$$

The schematic representation of the variation of G_{mix}, G_{el} and G_A with h given in Figure 6.10 shows that there is only one minimum (G_{min}), whose depth depends on R, δ and A. When $h_0 < 2\delta$, strong repulsion occurs and it increases very sharply with further decrease in h_0. At a given particle size and Hamaker constant, the larger the adsorbed layer thickness, the smaller the depth of the minimum. If G_{min} is made sufficiently small (large δ and small R), one may approach thermodynamic stability. This explains the case with nanoemulsions, which will be discussed in a separate chapter.

6.3 Mechanism of Emulsification

To prepare an emulsion, oil, water, surfactant and energy are needed [9, 10]. This can be considered from an examination of the energy required to expand the interface, $\Delta A\gamma$ (where ΔA is the increase in interfacial area when the bulk oil with area

A_1 produces numerous droplets with area A_2; $A_2 \gg A_1$, γ is the interfacial tension). Since γ is positive, the energy required to expand the interface is large and positive. This energy term cannot be compensated by the small entropy of dispersion $T\Delta S$ (which is also positive), and, as already discussed, the total free energy of formation of an emulsion, ΔG is positive.

Thus, emulsion formation is non-spontaneous and energy is required to produce the droplets. The formation of large droplets (few μm) as is the case for macroemulsions is fairly easy and hence high speed stirrers such as the Ultraturrax or Silverson Mixer are sufficient to produce the emulsion. In contrast, small drops (submicron, as is the case with nanoemulsions) are difficult to produce, requiring a large amount of surfactant and/or energy. The high energy required to form nanoemulsions can be understood from a consideration of the Laplace pressure p (the difference in pressure between inside and outside the droplet) [9, 10],

$$\Delta p = \gamma\left(\frac{1}{R_1} + \frac{1}{R_2}\right) \tag{6.10}$$

where R_1 and R_2 are the principal radii of curvature of the drop.

For a spherical drop, $R_1 = R_2 = R$ and

$$\Delta p = \frac{\gamma}{2R} \tag{6.11}$$

To break a drop into smaller ones, it must be strongly deformed and this deformation increases p [9, 10]. Figure 6.11 illustrates this, showing the situation when a spherical drop deforms into a prolate ellipsoid.

Near 1 there is only one radius of curvature R_a, whereas near 2 there are two radii of curvature $R_{b,1}$ and $R_{b,2}$. Consequently, the stress needed to deform the

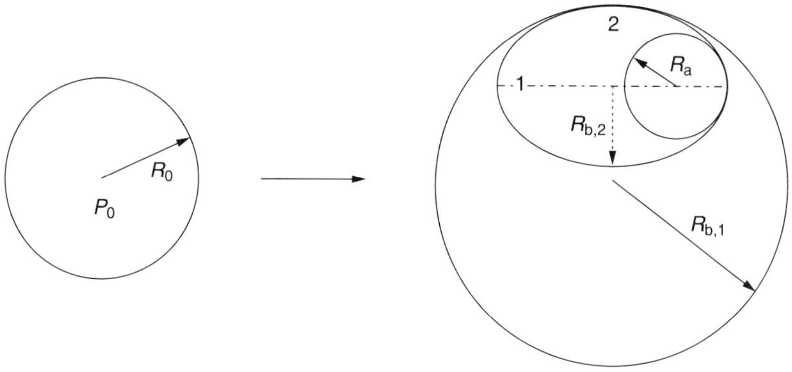

Fig. 6.11. Illustration of increase in Laplace pressure when a spherical drop is deformed to a prolate ellipsoid [9, 10].

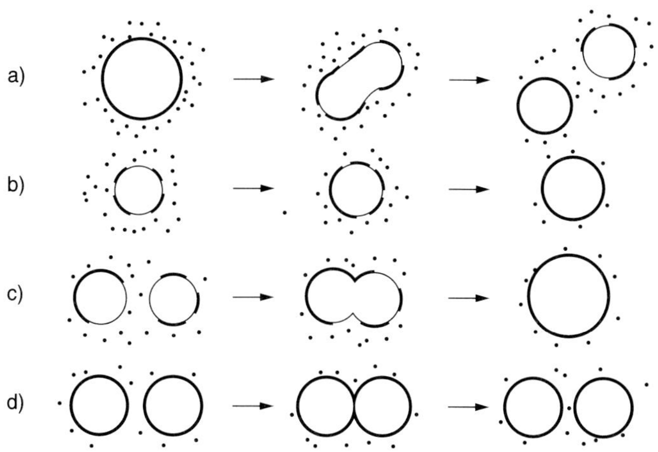

Fig. 6.12. Schematic of the various processes occurring during emulsion formation. Drops are depicted by thin lines and the surfactant by heavy lines and dots.

drop is higher for a smaller drop. Since the stress is generally transmitted by the surrounding liquid via agitation, higher stresses need more vigorous agitation, hence more energy is needed to produce smaller drops.

Surfactants play major roles in the formation of emulsions: by lowering the interfacial tension, p is reduced and hence the stress needed to break up a drop is reduced [9, 10]. Surfactants prevent coalescence of newly formed drops.

Figure 6.12 illustrates the various processes occurring during emulsification – break up of droplets, adsorption of surfactants and droplet collision (which may or may not lead to coalescence) [9, 10].

Each of the above processes occurs numerous times during emulsification and the time scale of each process is very short, typically a microsecond. This shows that emulsification is a dynamic process, and that events occurring in the microsecond domain could be very important [9, 10].

To describe emulsion formation one has to consider two main factors: hydrodynamics and interfacial science. To assess emulsion formation, one usually measures the droplet size distribution, using, for example, laser diffraction techniques – a useful average diameter d is

$$d_{\mathrm{nm}} = \left(\frac{S_{\mathrm{m}}}{S_{\mathrm{n}}}\right)^{1/(\mathrm{n}-\mathrm{m})} \tag{6.12}$$

In most cases d_{32} (the volume/surface average or Sauter mean) is used. The width of the size distribution can be given as the variation coefficient c_{m}, which is the standard deviation of the distribution weighted with d^{m} divided by the corresponding average d. Generally C_2 will be used that corresponds to d_{32}.

An alternative description of emulsion quality uses the specific surface area A (surface area of all emulsion droplets per unit volume of emulsion),

$$A = \pi s^2 = \frac{6\phi}{d_{32}} \tag{6.13}$$

6.4
Methods of Emulsification

Several procedures [9, 10] may be applied for emulsion preparation, ranging from simple pipe flow (low agitation energy L), static mixers and general stirrers (low to medium energy, L-M), high speed mixers such as the Ultraturrax (M), colloid mills and high pressure homogenizers (high energy, H), and ultrasound generators (M-H). Preparation methods can be continuous (C) or batch-wise (B): pipe flow and static mixers (C); stirrers and Ultraturrax (B, C); colloid mill and high pressure homogenizers (C); ultrasound (B, C).

In all methods, there is liquid flow [11, 12]: both unbounded and strongly confined flow. In the unbounded flow any droplets are surrounded by a large amount of flowing liquid (the confining walls of the apparatus are far away from most of the droplets). The forces can be frictional (mostly viscous) or inertial. Viscous forces cause shear stresses to act on the interface between the droplets and the continuous phase (primarily in the direction of the interface). The shear stresses can be generated by laminar flow (LV) [13] or turbulent flow (TV) [14] – this depends on the Reynolds number R_e,

$$R_e = \frac{vl\rho}{\eta} \tag{6.14}$$

where v is the linear liquid velocity, ρ is the liquid density and η is its viscosity; l is a characteristic length given by the diameter of flow through a cylindrical tube and by twice the slit width in a narrow slit.

For laminar flow $R_e < \sim 1000$, whereas for turbulent flow $R_e > \sim 2000$; thus whether the regime is linear or turbulent depends on the scale of the apparatus, the flow rate and the liquid viscosity. Turbulent eddies that are much larger than the droplets exert shear stresses on the droplets. If the turbulent eddies are much smaller than the droplets, inertial forces will cause disruption (TI). In bounded flow other relations hold – if the smallest dimension of the part of the apparatus in which the droplets are disrupted (say a slit) is comparable to droplet size, other relations hold (the flow is always laminar) [9, 10].

A different regime prevails if the droplets are directly injected through a narrow capillary into the continuous phase (injection regime), e.g. membrane emulsification.

Within each regime, an essential variable is the intensity of the forces acting:

$$\text{Viscous stress during laminar flow} = \eta G \tag{6.15}$$

where G is the velocity gradient.

The intensity in turbulent flow [9, 10] is expressed by the power density ε (the amount of energy dissipated per unit volume per unit time (for laminar flow $\varepsilon = \eta G^2$).

The most important regimes are laminar/viscous (LV), turbulent/viscous (TV) and turbulent/inertial (TI).

For water as the continuous phase, the regime is always TI. For higher viscosities of the continuous phase ($\eta_C = 0.1$ Pa s), the regime is TV. For still higher viscosities or a small apparatus (small l), the regime is LV. For very small apparatus (as with most laboratory homogenizers), the regime is nearly always LV.

For the above regimes, a semi-quantitative theory is available that can give the time scale and magnitude of the local stress σ_{ext}, the droplet diameter d, time scale of droplets deformation τ_{def}, time scale of surfactant adsorption, τ_{ads} and mutual collision of droplets [9, 10].

An important parameter that describes droplet deformation is the Weber number W_e (which gives the ratio of the external stress over the Laplace pressure) (Eq. 6.16).

$$W_e = \frac{G\eta_C R}{2\gamma} \tag{6.16}$$

The viscosity of the oil is important in the break-up of droplets – the higher the viscosity, the longer it takes to deform a drop. The deformation time τ_{def} is given by the ratio of oil viscosity to the external stress acting on the drop (Eq. 6.17).

$$\tau_{def} = \frac{\eta_D}{\sigma_{ext}} \tag{6.17}$$

The viscosity of the continuous phase η_C plays an important role in some regimes. For the turbulent inertial regime, η_C has no effect on droplet size. For the turbulent viscous regime, larger η_C leads to smaller droplets. For laminar viscous, the effect is even stronger.

6.5 Role of Surfactants in Emulsion Formation

Surfactants lower the interfacial tension γ and this causes a reduction in droplet size. The latter decrease with decrease in γ. For Turbulent Inertial regime, the

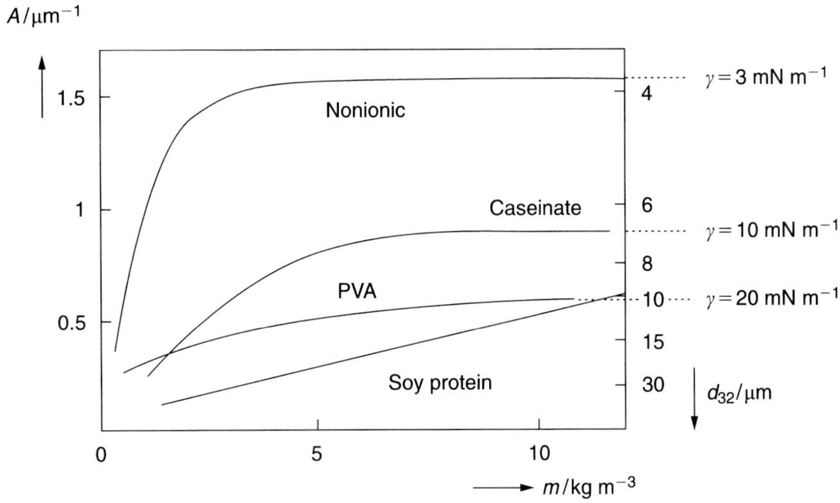

Fig. 6.13. Variation of A and d_{32} with m for various surfactant systems.

droplet diameter is proportional to $\gamma^{3/5}$. Figure 6.13 illustrates the effect of reducing γ on droplet size, showing a plot of droplet surface area A and mean drop size d_{32} as a function of surfactant concentration m for various systems [9, 10].

The amount of surfactant required to produce the smallest drop size will depend on its activity a (concentration) in the bulk, which determines the reduction in γ, as given by the Gibbs adsorption equation,

$$-\mathrm{d}\gamma = RT\Gamma \mathrm{d} \ln a \tag{6.18}$$

where R is the gas constant, T is the absolute temperature and Γ is the surface excess (number of moles adsorbed per unit area of the interface). Γ increases with increasing surfactant concentration and, eventually, reaches a plateau value (saturation adsorption). This is illustrated in Figure 6.14 for various emulsifiers.

The value of γ obtained depends on the nature of the oil and surfactant used – small molecules such as nonionic surfactants lower γ more than do polymeric surfactants such as PVA. Another important role of the surfactant is its effect on the interfacial dilational modulus ε (Eq. 6.19) [15].

$$\varepsilon = \frac{\mathrm{d}\gamma}{\mathrm{d} \ln A} \tag{6.19}$$

During emulsification, the interfacial area A increases, causing a reduction in Γ. The equilibrium is restored by adsorption of surfactant from the bulk, but this takes time (shorter times occur at higher surfactant activity). Thus, ε is small at both small and large a. Because of the lack or slowness of equilibrium with poly-

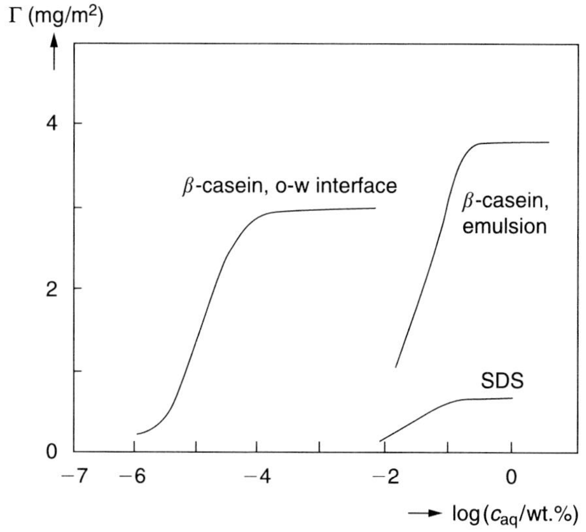

Fig. 6.14. Variation of Γ (mg m^{-2}) with log C_{eq} (wt%). Oils are β-casein (O-W interface) toluene, β-casein (emulsions) soybean, SDS benzene [9, 10].

meric surfactants, ε will not be the same for expansion and compression of the interface.

In practice, surfactant mixtures are used and these have pronounced effects on γ and ε. Some specific surfactant mixtures give lower γ than either of the two individual components [9, 10]. The presence of more than one surfactant molecule at the interface tends to increase ε at high surfactant concentrations. The various components vary in surface activity. Those with the lowest γ tend to predominate at the interface, but, if present at low concentrations, it may take a long time to reach the lowest value. Polymer–surfactant mixtures may show some synergetic surface activity.

6.5.1
Role of Surfactants in Droplet Deformation

Apart for their effect on reducing γ, surfactants play major roles in deformation and break-up of droplets [9, 10]. This is summarised as follows. Surfactants allow the existence of interfacial tension gradients, which is crucial for formation of stable droplets. In the absence of surfactants (clean interface), the interface cannot withstand a tangential stress; the liquid motion will be continuous (Figure 6.15a). If a liquid flows along the interface with surfactants, the latter will be swept downstream, causing an interfacial tension gradient (Figure 6.15b). A balance of forces will be established,

$$\eta\left[\frac{\mathrm{d}V_x}{\mathrm{d}y}\right]_{y=0} = -\frac{\mathrm{d}y}{\mathrm{d}x} \tag{6.20}$$

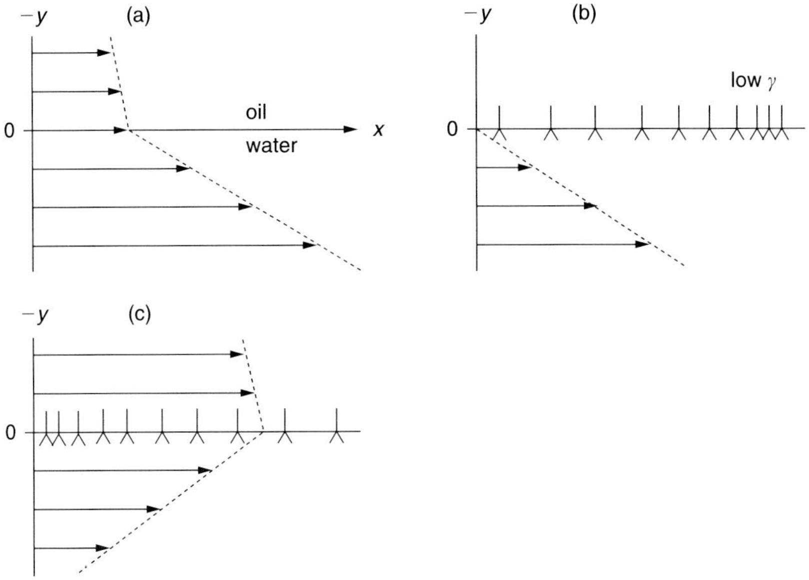

Fig. 6.15. Interfacial tension gradients and flow near an oil–water interface: (a) no surfactants; (b) velocity gradient causes an interfacial tension gradient; (c) interfacial tension gradient causes flow (Marangoni effect).

If the γ-gradient can become large enough, it will arrest the interface. If the surfactant is applied at one site of the interface, a γ-gradient is formed that will cause the interface to move roughly at a velocity given by Eq. (6.21).

$$v = 1.2[\eta\rho z]^{-1/3}|\Delta\gamma|^{2/3} \tag{6.21}$$

The interface will then drag some of the bordering liquid with it (Figure 6.15c).

Interfacial tension gradients [16–19] are very important in stabilising the thin liquid film between the droplets that is very important during the beginning of emulsification (films of the continuous phase may be drawn through the disperse phase and collision is very large). The magnitude of the γ-gradients and of the Marangoni effect depends on the surface dilational modulus ε, which for a plane interface with one surfactant-containing phase is given by the expression

$$\varepsilon = \frac{-\mathrm{d}\gamma/\mathrm{d}\ln\Gamma}{(1 + 2\zeta + 2\zeta^2)^{1/2}} \tag{6.22}$$

$$\zeta = \frac{\mathrm{d}m_C}{\mathrm{d}\Gamma}\left(\frac{D}{2\omega}\right)^{1/2} \tag{6.23}$$

$$\omega = \frac{\mathrm{d}\ln A}{\mathrm{d}t} \tag{6.24}$$

where D is the diffusion coefficient of the surfactant and ω represents a time scale (time needed for doubling the surface area) that is roughly equal to τ_{def}.

During emulsification, ε is dominated by the magnitude of the denominator in Eq. (6.22) because ζ remains small. The value of $dm_C/d\Gamma$ tends to go to very high values when Γ reaches its plateau value; ε goes to a maximum when m_C is increased.

For conditions that prevail during emulsification, ε increases with m_C and it is given by the relationship,

$$\varepsilon \approx \frac{d\pi}{d \ln \Gamma} \tag{6.25}$$

where π is the surface pressure ($\pi = \gamma_0 - \gamma$). Figure 6.16 shows the variation of π with ln Γ; ε is given by the slope of the line.

SDS shows a much higher ε than β-casein and lysozome, because Γ is higher for SDS. The two proteins show difference in their ε, which may be attributed to the conformational changes that occur upon adsorption [9, 10].

The presence of a surfactant means that, during emulsification, the interfacial tension need not be the same everywhere (Figure 6.15). This has two consequences: (1) the equilibrium shape of the drop is affected; (2) any γ-gradient formed will slow down the motion of the liquid inside the drop (this diminishes the amount of energy needed to deform and break-up the drop).

Another important role of the emulsifier is to prevent coalescence during emulsification. This is certainly not due to the strong repulsion between the droplets,

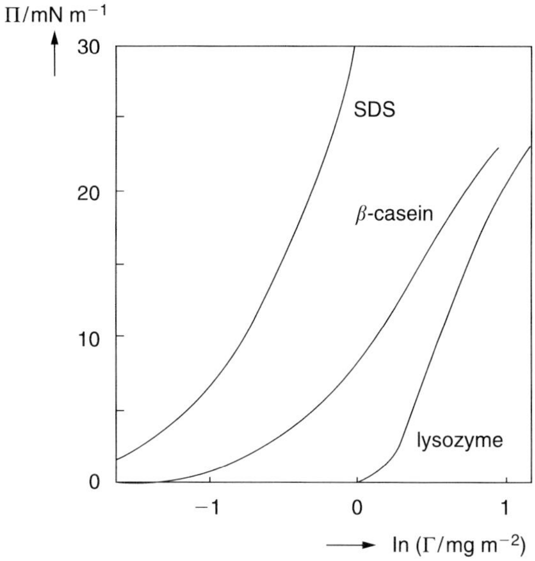

Fig. 6.16. Π versus ln Γ for three emulsifiers.

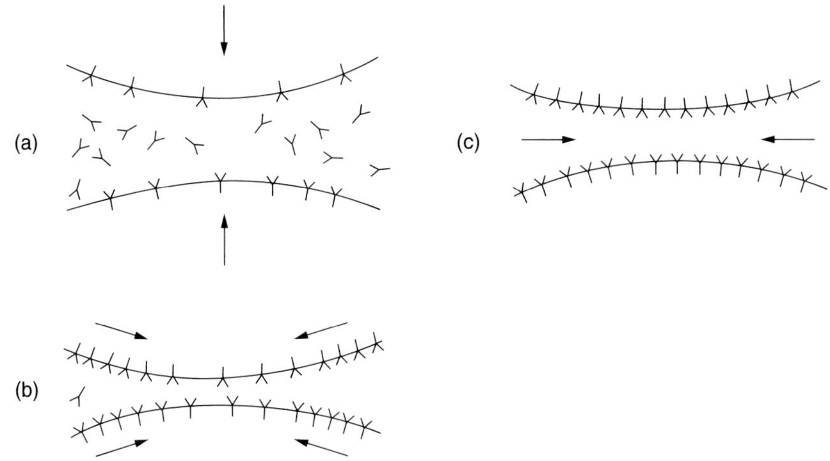

Fig. 6.17. Representation of the Gibbs–Marangoni effect for two approaching drops.

since the pressure at which two drops are pressed together is much greater than the repulsive stresses. The counteracting stress must be due to the formation of γ-gradients. When two drops are pushed together, liquid will flow out from the thin layer between them, and the flow will induce a γ-gradient (Figure 6.15c). This produces a counteracting stress (Eq. 6.21) given by,

$$\tau_{\Delta\gamma} \approx \frac{2|\Delta\gamma|}{(1/2)d} \tag{6.26}$$

The factor 2 follows from the fact that two interfaces are involved. Taking $\Delta\gamma = 10$ mN m^{-1}, the stress amounts to 40 kPa (which is of the same order of magnitude as the external stress).

Closely related to the above mechanism, is the Gibbs–Marangoni effect (Figure 6.17). Depletion of surfactant in the thin film between approaching drops results in γ-gradient without liquid flow being involved. This produces an inward flow of liquid that tends to drive the drops apart [9, 10].

The Gibbs–Marangoni effect also explains the Bancroft rule, which states that the phase in which the surfactant is most soluble forms the continuous phase. If the surfactant is in the droplets, a γ-gradient cannot develop and the drops would be prone to coalescence. Thus, surfactants with HLB > 7 tend to form O/W emulsions and those with HLB < 7 tend to form W/O emulsions. The Gibbs–Marangoni effect also explains the difference between surfactants and polymers for emulsification. Polymers give larger drops than surfactants, and they also give a smaller ε at small concentrations than surfactants do (Figure 6.16).

Various other factors should also be considered for emulsification: the disperse phase volume fraction ϕ. An increase in ϕ leads to an increase in droplet collision and, hence, to coalescence during emulsification. With increasing ϕ, the viscosity of the emulsion increases and could change the flow from turbulent to laminar (LV regime).

The presence of many particles results in a local increase in velocity gradients. This means that G increases. In turbulent flow, an increase in ϕ will induce turbulence depression. This will result in larger droplets – turbulence depression by added polymers tends to remove the small eddies, resulting in the formation of larger droplets.

If the mass ratio of surfactant to continuous phase is kept constant, increasing ϕ results in decreasing surfactant concentration and, hence, to an increase in γ_{eq}, resulting in larger droplets. If the mass ratio of surfactant to disperse phase is kept constant, the above changes are reversed.

General conclusions cannot be drawn since several of the above-mentioned mechanism may come into play. Experiments using a high pressure homogenizer at various ϕ at constant initial m_C (regime TI changing to TV at higher ϕ) showed that with increasing ϕ (> 0.1) the resulting droplet diameter increased and the dependence on energy consumption became weaker. Figure 6.18 shows a comparison of the average droplet diameter versus power consumption using different

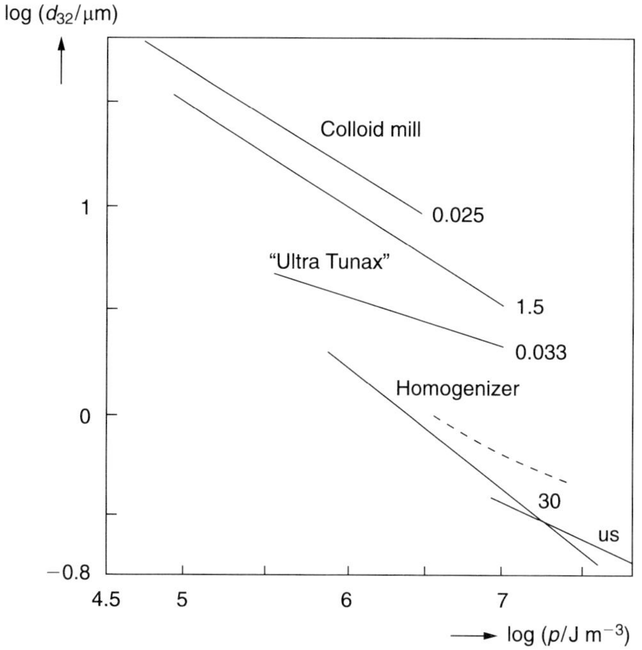

Fig. 6.18. Average droplet diameters obtained in various emulsifying machines as a function of energy consumption p. The number near the curves denotes the viscosity ratio λ; the results for the homogeniser are for $\phi = 0.04$ (solid line) and $\phi = 0.3$ (broken line) ('us' = ultrasonic generator).

emulsifying machines. The smallest droplet diameters are obtained with high-pressure homogenizers [9, 10].

6.6 Selection of Emulsifiers

6.6.1 Hydrophilic-Lipophilic Balance (HLB) Concept

The selection of different surfactants in the preparation of either O/W or W/O emulsions is often still made on an empirical basis. A semi-empirical scale for selecting surfactants is the hydrophilic–lipophilic balance (HLB number) developed by Griffin [20, 21]. This scale is based on the relative percentage of hydrophilic to lipophilic (hydrophobic) groups in the surfactant molecule(s). For an O/W emulsion droplet the hydrophobic chain resides in the oil phase whereas the hydrophilic head group resides in the aqueous phase. For a W/O emulsion droplet, the hydrophilic group(s) reside in the water droplet, whereas the lipophilic groups reside in the hydrocarbon phase. Table 6.2 summarizes HLB ranges and their application.

Table 6.2 gives a guide to the selection of surfactants for a particular application. The HLB number depends on the nature of the oil [21, 22]. As an illustration Table 6.3 gives the required HLB numbers to emulsify various oils.

The relative importance of the hydrophilic and lipophilic groups was first recognised when using mixtures of surfactants containing varying proportions of a low and high HLB numbers [20, 21]. The efficiency of any combination (as judged by

Tab. 6.2. Summary of HLB ranges and their applications.

HLB range	*Application*
3–6	W/O emulsifier
7–9	Wetting agent
8–18	O/W emulsifier
13–15	Detergent
15–18	Solubiliser

Tab. 6.3. Required HLB numbers to emulsify various oils.

Oil	*W/O emulsion*	*O/W emulsion*
Paraffin oil	4	10
Beeswax	5	9
Linolin, anhydrous	8	12
Cyclohexane	–	15
Toluene	–	15

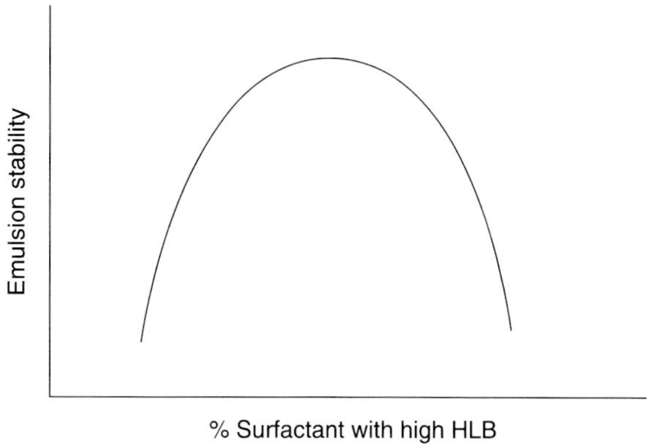

Fig. 6.19. Variation of emulsion stability with HLB.

phase separation) was found to pass a maximum when the blend contained a particular proportion of the surfactant with the higher HLB number (Figure 6.19).

The average HLB number may be calculated by additivity,

$$\mathrm{HLB} = x_1 \mathrm{HLB}_1 + x_2 \mathrm{HLB}_2 \tag{6.27}$$

x_1 and x_2 are the weight fractions of the two surfactants with HLB_1 and HLB_2.

Griffin [20, 21] developed simple equations to calculate the HLB number of relatively simple nonionic surfactants. For a polyhydroxy fatty acid ester

$$\mathrm{HLB} = 20\left(1 - \frac{S}{A}\right) \tag{6.28}$$

S is the saponification number of the ester and A is the acid number.

For a glyceryl monostearate, $S = 161$ and $A = 198$ – The HLB is 3.8 (suitable for W/O emulsion). For a simple alcohol ethoxylate, the HLB number can be calculated from the weight percent of ethylene oxide (E) and Polyhydric alcohol (P),

$$\mathrm{HLB} = \frac{E + P}{5} \tag{6.29}$$

If the surfactant contains PEO as the only hydrophilic group, the contribution from one OH group can be neglected,

$$\mathrm{HLB} = \frac{E}{5} \tag{6.30}$$

For the nonionic surfactant $C_{12}H_{25}$–O–$(CH_2$–CH_2–$O)_6$ HLB is 12 (suitable for O/W emulsion).

Tab. 6.4. HLB group numbers.

Group	***HLB number***
Hydrophilic	
$-SO_4Na^+$	38.7
$-COO^-$	21.2
–COONa	19.1
N(tertiary amine)	9.4
Ester (sorbitan ring)	6.8
–O–	1.3
CH– (sorbitan ring)	0.5
Lipophilic	
(–CH–), ($-CH_2-$), CH_3	0.475
Derived	
$-CH_2-CH_2-O$	0.33
$-CH_2-CH_2-CH_2-O-$	−0.15

The above simple equations cannot be used for surfactants containing propylene oxide or butylene oxide. In addition, they cannot be applied for ionic surfactants. Davies [23] devised a method for calculating the HLB number for surfactants from their chemical formulae, using empirically determined group numbers. A group number is assigned to various component groups. Table 6.4 summarises the group numbers for some surfactants.

The HLB is given by the following empirical equation,

$$\text{HLB} = 7 + \sum(\text{hydrophilic group nos}) - \sum(\text{lipohilic group nos}) \tag{6.31}$$

Davies [23] has shown that the agreement between HLB numbers calculated from the above equation and those determined experimentally is quite satisfactory.

Various other procedures have been developed to obtain a rough estimate of the HLB number. Griffin found a good correlation between the cloud point of a 5% solution of various ethoxylated surfactants and their HLB number (see Figure 6.20).

Davies [23] attempted to relate the HLB values to the selective coalescence rates of emulsions. Such correlations were not realised since emulsion stability and even its type were found to largely depend on the method of dispersing the oil into the water and vice versa. At best the HLB number can only be used as a guide for selecting optimum compositions of emulsifying agents.

One may take any pair of emulsifying agents, which fall at opposite ends of the HLB scale, e.g. Tween 80 (sorbitan monooleate with 20 moles EO, HLB = 15) and Span 80 (sorbitan monooleate, HLB = 5) and use them in various proportions to cover a wide range of HLB numbers. The emulsions should be prepared in the same way, with a few percent of the emulsifying blend. The stability of the emulsions is then assessed at each HLB number, either from the rate of coalescence or qualitatively by measuring the rate of oil separation. In this way one may be able to

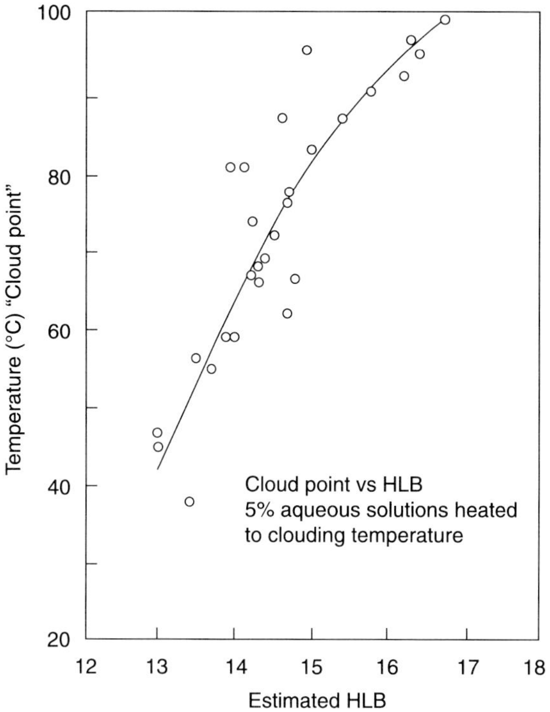

Fig. 6.20. Correlation of cloud point with HLB.

find the optimum HLB number for a given oil. Having found the most effective HLB, various other surfactant pairs are compared at this HLB to find the most effective pair.

6.6.2
Phase Inversion Temperature (PIT) Concept

This concept, developed by Shinoda [24, 25], is closely related to the HLB balance concept described above. Shinoda and co-workers found that many O/W emulsions stabilised with nonionic surfactants undergo a process of inversion at a critical temperature (PIT). The PIT can be determined by following the emulsion conductivity (a small amount of electrolyte is added to increase the sensitivity) as a function of temperature. The conductivity of the O/W emulsion increases with rising temperature till the PIT is reached, above which there will be a rapid reduction in conductivity (W/O emulsion is formed).

The PIT is influenced by the HLB number of the surfactant [24, 25]. Figure 6.21 shows this for cyclohexane/water emulsions stabilised by various nonionic surfactants. The size of the emulsion droplets was found to depend on the temperature and HLB number of the emulsifiers. The droplets are less stable towards coales-

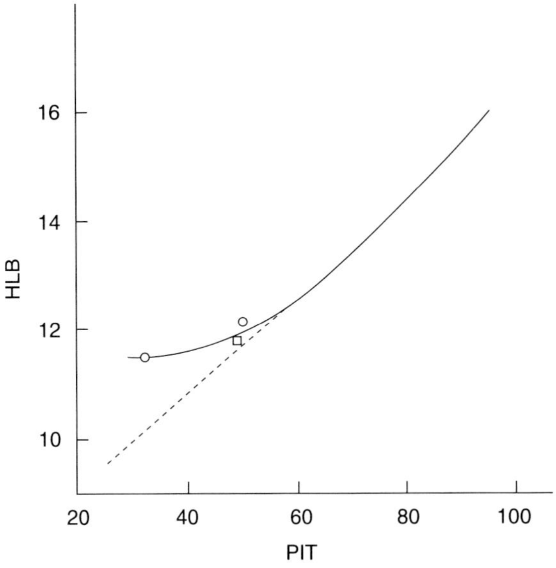

Fig. 6.21. Correlation of PIT with HLB of surfactants.

cence close to the PIT. However, by rapid cooling of the emulsion a stable system may be produced. Relatively stable O/W emulsions were obtained when the PIT of the system was 20–65 °C higher than the storage temperature. Emulsions prepared at a temperature just below the PIT followed by rapid cooling generally have smaller droplet sizes. This can be understood if one considers the change of interfacial tension with temperature (Figure 6.22). The interfacial tension decreases with increasing temperature, reaching a minimum close to the PIT, after which it increases. Thus, droplets prepared close to the PIT are smaller than those prepared at lower temperatures. These droplets are relatively unstable towards co-

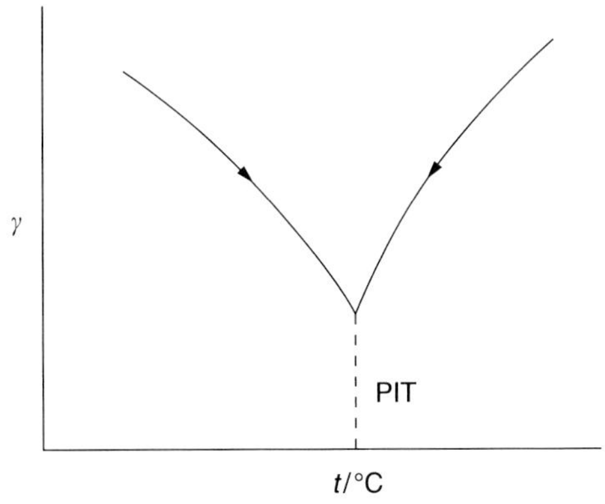

Fig. 6.22. Variation of interfacial tension with temperature.

alescence near the PIT, but by rapid cooling of the emulsion one can retain the smaller size. The above procedure may be applied to prepare mini (nano)emulsions.

The optimum stability of the emulsion is relatively insensitive to changes in the HLB or the PIT of the emulsifier, but instability is very sensitive to the PIT of the system. It is essential, therefore, to measure the PIT of the emulsion as a whole (with all other ingredients). At a given HLB, the stability of the emulsions against coalescence increases markedly as the molar mass of both the hydrophilic and lipophilic components increases.

The enhanced stability using high molecular weight surfactants (polymeric surfactants) can be understood by considering the steric repulsion, which produces more stable films. Films produced using macromolecular surfactants resist thinning and disruption, thus reducing the possibility of coalescence.

The emulsions showed maximum stability when the distribution of the PEO chains was broad. The cloud point is lower but the PIT is higher than in the corresponding case for narrow size distributions. The PIT and HLB number are directly related parameters.

Addition of electrolytes reduces the PIT and hence an emulsifier with a higher PIT is required when preparing emulsions in the presence of electrolytes. Electrolytes cause dehydration of the PEO chains and in effect this reduces the cloud point of the nonionic surfactant. One needs to compensate for this effect by using a surfactant with higher HLB. The optimum PIT of the emulsifier is fixed if the storage temperature is fixed. In view of the above correlation between PIT and HLB and the possible dependence of the kinetics of droplet coalescence on the HLB number, Sherman and co-workers suggested the use of PIT measurements as a rapid method for assessing emulsion stability. Figure 6.23 illustrates this with a plot of the rate of coalescence of paraffin oil/water emulsions prepared using blends of Tween and Span surfactants of various HLB numbers.

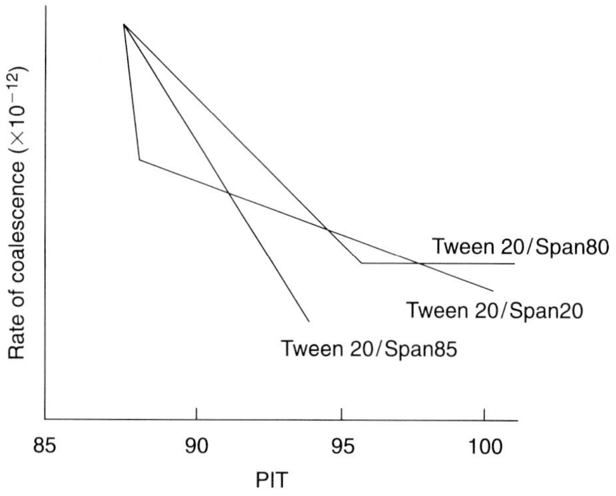

Fig. 6.23. Variation of rate of coalescence with PII.

However, one should be careful in using such methods for assessment of the long-term stability since the correlations were based on a very limited number of surfactants and oils. Measurement of the PIT can at best be used as a guide for preparation of stable emulsions. Assessment of the stability should be evaluated by following the droplet size distribution as a function of time using a Coulter counter or light diffraction techniques. Following the rheology of the emulsion as a function of time and temperature may also be used to assess the stability against coalescence. Care should be taken in analysing the rheological data.

The above results suggest a correlation between emulsion stability against coalescence and the PIT. Coalescence results in an increase in the droplet size, which is usually followed by a reduction in the viscosity of the emulsion. This trend is only observed if the coalescence is not accompanied by flocculation of the emulsion droplets (which results in an increase in the viscosity). Ostwald ripening can also complicate the analysis of the rheological data.

6.7 Cohesive Energy Ratio (CER) Concept for Emulsifier Selection

Beerbower and Hills [26] considered the dispersing tendency on oil and water interfaces of the surfactant or emulsifier in terms of the ratio of the cohesive energies of the mixtures of oil with the lipophilic portion of the surfactant and the water with the hydrophilic portion. They used the Winsor R_0 concept, which is the ratio of the intermolecular attraction of oil molecules (O) and the lipophilic portion of surfactant (L), C_{LO}, to that of water (W) and the hydrophilic portion (H), C_{HW},

$$R_0 = \frac{C_{\mathrm{LO}}}{C_{\mathrm{HW}}} \tag{6.32}$$

Several interaction parameters may be identified at the oil and water sides of the interface. One can identify at least nine interaction parameters,

$C_{\mathrm{LL}}, C_{\mathrm{OO}}, C_{\mathrm{LO}}$ (at oil side)
$C_{\mathrm{HH}}, C_{\mathrm{WW}}, C_{\mathrm{HW}}$ (at water side)
$C_{\mathrm{LW}}, C_{\mathrm{HO}}, C_{\mathrm{LH}}$ (at the interface)

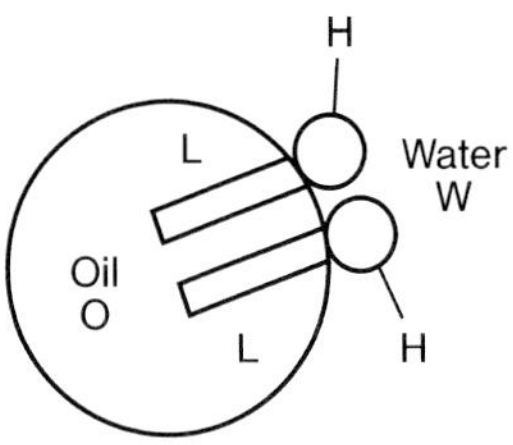

In the absence of emulsifier, there will be only three interaction parameters: $C_{\mathrm{OO}}, C_{\mathrm{WW}}, C_{\mathrm{OW}}$. If $C_{\mathrm{OW}} \ll C_{\mathrm{WW}}$, the emulsion breaks.

The above interaction parameters may be related to the Hildebrand Solubility parameter δ [27] (at the oil side of the interface) and the Hansen [28] non-polar, hydrogen bonding and polar contributions to δ at the water side of the interface. The solubility parameter of any component is related to its heat of vaporization ΔH by Eq. (6.33), where V_M is the molar volume.

$$\delta_2 = \frac{\Delta H - RT}{V_M} \quad (6.33)$$

Hansen [28] considered δ (at the water side of the interface) to consist of three main contributions, a dispersion contribution, δ_d, a polar contribution, δ_p and a hydrogen bonding contribution, δ_h. These contributions have different weighting factors,

$$\delta^2 = \delta_d^2 + 0.25\,\delta_p^2 + 0.25\,\delta_h^2 \quad (6.34)$$

Beerbower and Hills [26] used the following expression for the HLB number,

$$\text{HLB} = 20\left(\frac{M_H}{M_L + M_H}\right) = 20\left(\frac{V_H \rho_H}{V_L \rho_L + V_H \rho_H}\right) \quad (6.35)$$

where M_H and M_L are the molecular weights of the hydrophilic and lipophilic portions of the surfactants. V_L and V_H are their corresponding molar volumes whereas ρ_H and ρ_L are the respective densities.

The cohesive energy ratio was originally defined by Winsor, Eq. (6.32).

When $C_{LO} > C_{HW}$, $R > 1$ and a W/O emulsion forms. If $C_{LO} < C_{HW}$, $R < 1$ and an O/W emulsion forms. If $C_{LO} = C_{HW}$, $R = 1$ and a planar system results; this denotes the inversion point.

R_0 can be related to V_L, δ_L and V_H, δ_H by the expression,

$$R_0 = \frac{V_L \delta_L^2}{V_H \delta_H^2} \quad (6.36)$$

Using Eq. (6.34),

$$R_0 = \frac{V_L(\delta_d^2 + 0.25\delta_p^2 + 0.25\delta_h^2)_L}{V_h(\delta_d^2 + 0.25\delta_p^2 + 0.25\delta_h^2)_H} \quad (6.37)$$

Combining Eqs. (6.36) and (6.37) one obtains Eq. (6.38) as a general expression for the cohesive energy ratio.

$$R_0 = \left(\frac{20}{\text{HLB}} - 1\right)\frac{\rho_h(\delta_d^2 + 0.25\delta_p^2 + 0.25\delta_h^2)_L}{\rho_L(\delta_d^2 + 0.25\delta_p^2 + 0.25\delta_p^2)_L} \quad (6.38)$$

For an O/W system, HLB = 12–15 and $R_0 = 0.58$–0.29 ($R_0 < 1$). For a W/O system, HLB = 5–6 and $R_0 = 2.3$–1.9 ($R_0 > 1$). For a planar system, HLB = 8–10 and $R_0 = 1.25$–0.85 ($R_0 \sim 1$).

The R_0 equation combines both the HLB and cohesive energy densities – it gives a more quantitative estimate of emulsifier selection. R_0 considers HLB, molar volume and chemical match. The success of the above approach depends on the availability of data on the solubility parameters of the various surfactant portions. Some values are tabulated in the book by Barton [29].

6.8
Critical Packing Parameter (CPP) for Emulsifier Selection

The critical packing parameter (CPP) is a geometric expression relating the hydrocarbon chain volume (v) and length (l) and the interfacial area occupied by the head group (a) [30],

$$\text{CPP} = \frac{v}{l_c a_0} \qquad (6.39)$$

a_0 is the optimal surface area per head group, l_c is the critical chain length.

Regardless of the shape of any aggregated structure (spherical or cylindrical micelle or a bilayer), no point within the structure can be farther from the hydrocarbon–water surface than l_c. The critical chain length, l_c, is roughly equal to, but less than, the fully extended length of the alkyl chain.

The CPP for any micelle shape can be calculated from simple packing constraints.

Consider a spherical micelle (**6.1** and **6.2**).

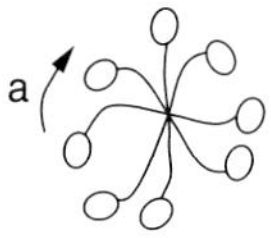

6.1

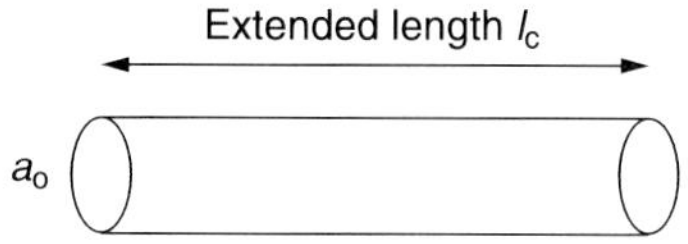

6.2

$$\text{Volume of the micelle} = \tfrac{4}{3}\pi r^3 = nv \qquad (6.40)$$

$$\text{Area of the micelle} = 4\pi r^2 = na \qquad (6.41)$$

v is the volume of the hydrocarbon chain and n is the aggregation number. The cross sectional area of the hydrocarbon chain is given by Eq. (6.42).

$$a_0 = \frac{v}{l_c} \qquad (6.42)$$

From Eqs. (6.40) and (6.41)

$$a = \frac{3v}{r} \tag{6.43}$$

Since r has to be less than l_c, packing constraints imply that $a > 3a_0$ or $a_0/a < \frac{1}{3}$. Thus, the CPP for a spherical micelle is $< \frac{1}{3}$.

Surfactants that form spherical micelles with the above packing constraints are more suitable for O/W emulsions.

For a cylindrical micelle,

$$\text{Volume} = \pi r^2 l = nv \tag{6.44}$$

$$\text{Area} = 2\pi rl = na \tag{6.45}$$

$$a = \frac{2v}{r} \tag{6.46}$$

Since r has to be less than the extended length of the hydrocarbon chain l_c, then packing constraints imply that $a > 2a_0$ or $a_0/a < \frac{1}{2}$. Thus the CPP for a cylindrical micelle is $< \frac{1}{2}$.

When the CPP exceeds $\frac{1}{2}$, but is less than 1, spherical bilayers (vesicles) can be produced. When the CPP is ~ 1, the bilayers may remain planar. With CPP > 1, inverted micelles are produced. Surfactants that produce these structures are suitable for forming W/O emulsions.

Table 6.5 gives predictions of the aggregates produced based on the above CPP concept.

6.9 Creaming or Sedimentation of Emulsions

This is the result of gravity, when the density of the droplets and the medium are not equal. Figure 6.24 gives a schematic picture for creaming or sedimentation for three cases [1].

Case (a) represents the situation for small droplets (< 0.1 μ, i.e. nanoemulsions) whereby the Brownian diffusion kT (where k is the Boltzmann constant and T is the absolute temperature) exceeds the force of gravity (mass $\times$ acceleration due to gravity, g),

$$kT \gg \tfrac{4}{3}\pi R^3 \Delta\rho g L \tag{6.47}$$

where R is the droplet radius, $\Delta\rho$ is the density difference between the droplets and the medium, and L is the height of the container.

Figure 6.24b represents emulsions consisting of "monodisperse" droplets with radius > 1 μm. Here, the emulsion separates into two distinct layers with the droplets forming a cream or sediment, leaving the clear supernatant liquid. This situation is seldom observed in practice.

Tab. 6.5.

Lipid	*Critical packing parameter* **vianlc**	*Critical packing shape*	*Structures formed*
Single-chained lipids (surfactants) with large head-group areas: • *SDS in low salt*	<1/3	Cone 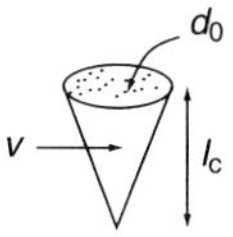	Spherical micelles 
Single-chained lipids with small head-group areas: • *SDS and CTAB in high salt* • *nonionic lipids*	1/3–1/2	Truncated cone 	Cylindrical micelles 
Double-chained lipids with large head-group areas, fluid chains: • *Phosphatidyl choline (lecithin)* • *phosphatidyl serine* • *phosphatidyl glycerol* • *phosphatidyl inositol* • *phosphatidic acid* • *sphingomyelin, DGDG*[a] • *dihexadecyl phosphate* • *dialkyl dimethyl ammonium* • *salts*	1/2–1	Truncated cone 	Flexible bilayers, vesicles 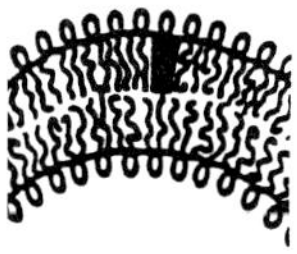
Double-chained lipids with small head-group areas, anionic lipids in high salt, saturated frozen chains: • *phosphatidyl ethanaiamine* • *phosphatidyl serine* + Ca^{2+}	~1	Cylinder 	Planar bilayers 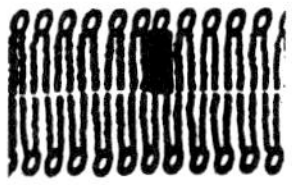
Double-chained lipids with small head-group areas, nonionic lipids, poly(cis) unsaturated chains, high *T*: • *unsat. phosphatidyl ethanolamine* • *cardiolipin* + Ca^{2+} • *phosphatidic acid* + Ca^{2+} • *cholesterol, MGDG*[b]	>1	Inverted truncated cone or wedge 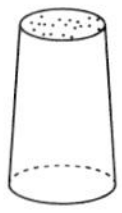	Inverted micelles 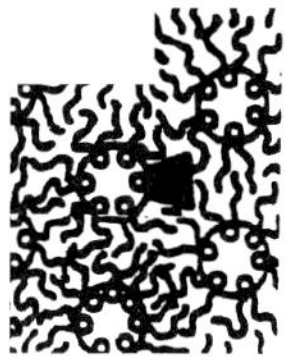

[a] DGDG digalactosyl diglyceride, diglucosyldiglyceride.
[b] MGDG monogalactosyl diglyceride, monoglucosyl diglyceride.

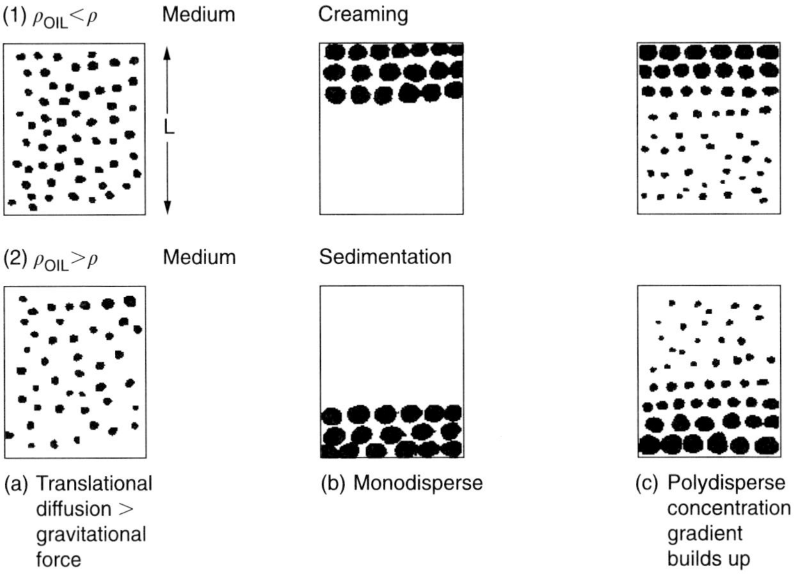

Fig. 6.24. Representation of creaming or sedimentation (see text for details).

Figure 6.24c represents polydisperse (practical) emulsions, in which case the droplets will cream or sediment at various rates. In the last case, a concentration gradient builds up, with the larger droplets staying at either the top of the cream layer or the bottom,

$$C(h) = C_0 \exp\left(-\frac{mgh}{kT}\right) \tag{6.48}$$

$$m = \tfrac{4}{3}\pi R^3 \Delta\rho \tag{6.49}$$

$C(h)$ is the concentration (or volume fraction ϕ) of droplets at height h, whereas C_0 is the concentration at zero time, which is the same at all heights.

6.9.1 Creaming or Sedimentation Rates

6.9.1.1 Very Dilute Emulsions ($\phi < 0.01$)

In this case the rate could be calculated using Stokes' law, which balances the hydrodynamic force with gravity force,

$$\text{Hydrodynamic force} = 6\pi\eta R v_0 \tag{6.50}$$

$$\text{Gravity force} = \tfrac{4}{3}\pi R^3 \Delta\rho g \tag{6.51}$$

$$v_0 = \frac{2}{9}\frac{\Delta\rho g R^2}{\eta_0} \tag{6.52}$$

v_0 is the Stokes' velocity and η_0 is the viscosity of the medium.

For an O/W emulsion with $\Delta\rho = 0.2$ in water ($\eta_0 \sim 10^{-3}$ Pa s), the rate of creaming or sedimentation is $\sim 4.4 \times 10^{-5}$ ms^{-1} for 10 μm droplets and $\sim 4.4 \times 10^{-7}$ m s^{-1} for 1 μm droplets. This means that in a 0.1 m container creaming or sedimentation of the 10 μm droplets is complete in ~0.6 hour and for the 1 μm droplets this takes ~60 hours.

6.9.1.2 Moderately Concentrated Emulsions ($0.2 < \phi < 0.1$)

Here one has to take into account the hydrodynamic interaction between the droplets, which reduces the Stokes velocity to a v given by the following expression,

$$v = v_0(1 - k\phi) \tag{6.53}$$

where k is a constant that accounts for hydrodynamic interaction; k is of the order of 6.5, which means that the rate of creaming or sedimentation is reduced by about 65%.

6.9.1.3 Concentrated Emulsions ($\phi > 0.2$)

The rate of creaming or sedimentation becomes a complex function of ϕ, as is illustrated in Figure 6.25, which also shows the change of relative viscosity η_r with ϕ.

As seen in figure, v decreases with increasing ϕ and ultimately approaches zero when ϕ exceeds a critical value (ϕ_p), which is the so-called "maximum pack-

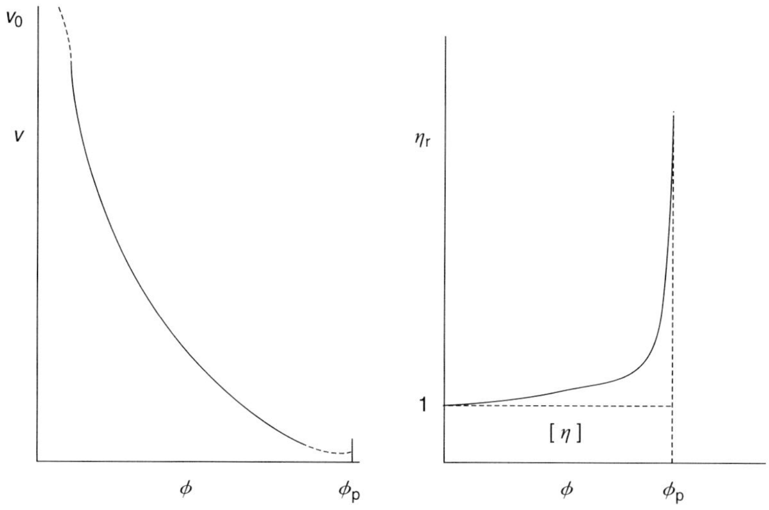

Fig. 6.25. Variation of v_0 and η_r with ϕ.

ing fraction". For monodisperse "hard-spheres", ϕ_p ranges from 0.64 (for random packing) to 0.74 for hexagonal packing – it exceeds 0.74 for polydisperse systems. Also, for emulsions that are deformable, ϕ_p can be much larger than 0.74.

Figure 6.25 also shows that when ϕ approaches ϕ_p, η_r approaches ∞. In practice most emulsions are prepared at ϕ well below ϕ_p, usually in the range 0.2–0.5, and under these conditions creaming or sedimentation is the rule rather than the exception.

Several procedures that may be applied to reduce or eliminate creaming or sedimentation are discussed below.

6.9.2 Prevention of Creaming or Sedimentation

6.9.2.1 Matching Density of Oil and Aqueous Phases

Clearly, if $\Delta\rho = 0$, $v = 0$; however, this method is seldom practical. Density matching, if possible, only occurs at one temperature.

6.9.2.2 Reduction of Droplet Size

Since the gravity force is proportional to R^3, if R is reduced by a factor of 10 the gravity force is reduced by 1000. Below a certain droplet size (which also depends on the density difference between oil and water), the Brownian diffusion may exceed gravity and creaming or sedimentation is prevented. This is the principle of formulation of nanoemulsions (with size range 50–200 nm), which may show very little or no creaming or sedimentation. The same applies for microemulsions (size range 5–50 nm).

6.9.2.3 Use of "Thickeners"

These are high molecular weight polymers, natural or synthetic such as xanthan gum, hydroxyethyl cellulose, alginates, carrageenans, etc. To understand the role of these "thickeners", let us consider the gravitational stresses exerted during creaming or sedimentation,

$$\text{Stress} = \text{mass of drop} \times \text{acceleration of gravity} = \tfrac{4}{3}\pi R^3 \Delta\rho g \tag{6.54}$$

To overcome such stress one needs a restoring force,

$$\text{Restoring force} = \text{Area of drop} \times \text{stress of drop} = 4\pi R^2 \sigma_p \tag{6.55}$$

Thus, the stress exerted by the droplet σ_p is given by,

$$\sigma_p = \frac{\Delta\rho R g}{3} \tag{6.56}$$

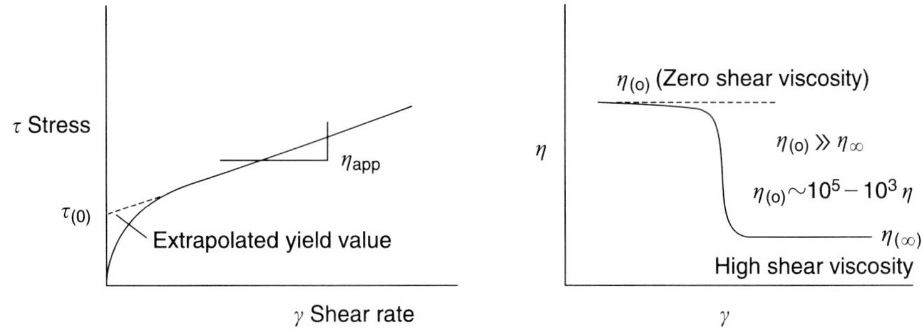

Fig. 6.26. Variation of (stress) τ and viscosity η with shear rate γ.

Simple calculation shows that σ_p is in the range 10^{-3}–10^{-1} Pa, implying that for prediction of creaming or sedimentation one needs to measure the viscosity at such low stresses. This can be obtained by using constant stress or creep measurements.

The above described "thickeners" satisfy the criteria for obtaining very high viscosities at low stresses or shear rates. This can be illustrated from plots of shear stress τ and viscosity η versus shear rate (or shear stress) (Figure 6.26).

These systems are described as "pseudo-plastic" or shear thinning. The low shear (residual or zero shear rate) viscosity $\eta(o)$ can reach several thousand Pa s and such high values prevent creaming or sedimentation.

The above behaviour is obtained above a critical polymer concentration (C^*) which can be located from plots of log η versus log C (illustrated in Figure 6.27). Below C^* the log η–log C curve has a slope in the region of 1, whereas above C^* the slope of the line exceeds 3.

In most cases, good correlation between the rate of creaming or sedimentation and $\eta(o)$ is obtained.

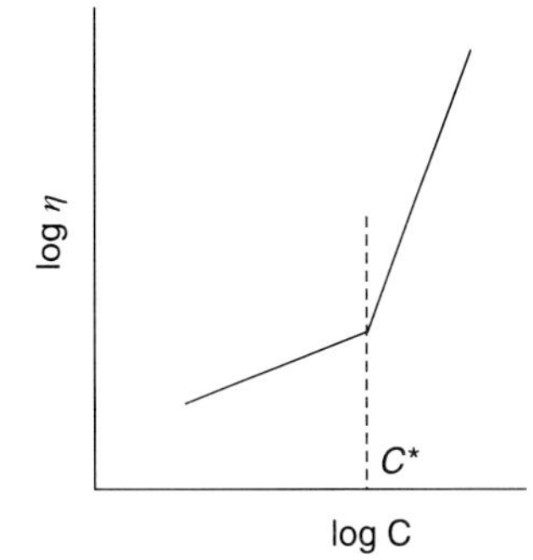

Fig. 6.27. Variation of log η with log C for polymer solutions.

6.9.2.4 **Controlled Flocculation**

As will be described in the section on flocculation, the total energy–distance of separation curve for electrostatically stabilised shows a shallow minimum (secondary minimum) at relatively long separation between the droplets. By addition of small amounts of electrolyte such minimum can be made sufficiently deep for weak flocculation to occur. The same applies for sterically stabilised emulsions, which show only one minimum, whose depth can be controlled by reducing the thickness of the adsorbed layer. This can be achieved by reducing the molecular weight of the stabiliser and/or addition of a non-solvent for the chains (e.g. electrolyte).

The above phenomenon of weak flocculation may be applied to reduce creaming or sedimentation, although in practice this is not easy since one has also to control the droplet size.

6.9.2.5 **Depletion Flocculation**

This is obtained by addition of "free" (non-adsorbing) polymer in the continuous phase [31]. At a critical concentration, or volume fraction of free polymer, ϕ_p^+, weak flocculation occurs, since the free polymer coils become "squeezed-out" from between the droplets. Figure 6.28 illustrates this, showing the situation when the polymer volume fraction exceeds the critical concentration.

The osmotic pressure outside the droplets is higher than that between the droplets and this results in an attraction whose magnitude depends on the concentration of the free polymer and its molecular weight, as well as the droplet size and ϕ. The value of ϕ_p^+ decreases with increasing molecular weight of the free polymer. It also decreases as the volume fraction of the emulsion increases.

The above weak flocculation can be applied to reduce creaming or sedimentation although it suffers from the following drawbacks: Temperature dependence – as the temperature increases, the hydrodynamic radius of the free polymer decreases

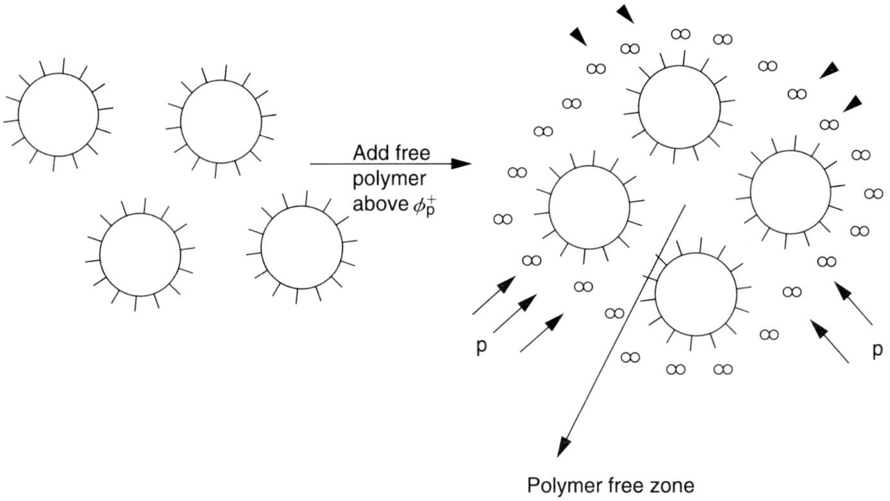

Fig. 6.28. Schematic representation of depletion flocculation.

(due to dehydration) and hence more polymer will be required to achieve the same effect at lower temperatures. If the free polymer concentration is increased above a certain limit, phase separation may occur and the flocculated emulsion droplets may cream or sediment faster than in the absence of the free polymer.

6.10
Flocculation of Emulsions

Flocculation is the result of the van der Waals attraction that is universal for all disperse systems. The van der Waals attraction G_A is described in detail in the section on physical chemistry of emulsion systems. This showed that G_A is inversely proportional to the droplet–droplet separation h and it depends on the effective Hamaker constant A of the emulsion system. One way to overcome the van der Waals attraction is by electrostatic stabilisation using ionic surfactants, which results in the formation of electrical double layers that introduce a repulsive energy that overcomes the attractive energy. Emulsions stabilised by electrostatic repulsion become flocculated at intermediate electrolyte concentrations (see below). The second and most effective method of overcoming flocculation is by "steric stabilisation", using nonionic surfactants or polymers. Stability may be maintained in electrolyte solutions (as high as 1 mol dm^{-3}, depending on the nature of the electrolyte) and up to high temperatures (in excess of 50 °C) provided the stabilising chains (e.g. PEO) are still in better than θ-conditions ($\chi < 0.5$).

6.10.1
Mechanism of Emulsion Flocculation

This can occur if the energy barrier is small or absent (for electrostatically stabilised emulsions) or when the stabilising chains reach poor solvency (for sterically stabilised emulsions, i.e. $\chi > 0.5$). For convenience, I will discuss the flocculation of electrostatically and sterically stabilised emulsions separately.

6.10.1.1 Flocculation of Electrostatically Stabilised Emulsions

As discussed in the section on physical chemistry of emulsion systems, the condition for kinetic stability is $G_{max} > 25kT$; when $G_{max} < 5kT$, flocculation occurs. Two types of flocculation kinetics may be distinguished: Fast flocculation with no energy barrier and slow flocculation when an energy barrier exists.

Fast flocculation kinetics have been treated by Smoluchowski [32], who considered the process to be represented by second-order kinetics and the process is simply diffusion controlled. The number of particles n at any time t may be related to the initial number (at $t = 0$) n_0 by the following expression,

$$n = \frac{n_0}{1 + kn_0 t} \tag{6.57}$$

where k is the rate constant for fast flocculation that is related to the diffusion coefficient of the particles D, i.e.

$$k = 8\pi DR \tag{6.58}$$

D is given by the Stokes–Einstein equation,

$$D = \frac{kT}{6\pi\eta R} \tag{6.59}$$

Combining Eqs. (6.58) and (6.59) gives

$$k = \frac{4}{3}\frac{kT}{\eta} = 5.5 \times 10^{-18}\ \mathrm{m^3\ s^{-1}}\ \text{for water at } 25\ ^\circ\mathrm{C} \tag{6.60}$$

The half-life $t_{1/2}$ ($n = \frac{1}{2}n_0$) can be calculated at various n_0 or volume fraction ϕ as give in Table 6.6.

Tab. 6.6. Half-life of emulsion flocculation.

***R* (μm)**		ϕ		
	10^{-5}	10^{-2}	10^{-1}	5×10^{-1}
0.1	765 s	76 ms	7.6 ms	1.5 ms
1.0	21 h	76 s	7.6 s	1.5 s
10.0	4 month	21 h	2 h	25 m

Slow flocculation kinetics have been treated by Fuchs [33] who related the rate constant k to the Smoluchowski rate by the stability constant W,

$$W = \frac{k_0}{k} \tag{6.61}$$

W is related to G_{max} by the following expression,

$$W = \tfrac{1}{2}k \exp\left(\frac{G_{max}}{kT}\right) \tag{6.62}$$

Since G_{max} is determined by the salt concentration C and valency, one can derive an expression relating W to C and Z [34],

$$\log W = -2.06 \times 10^9 \left(\frac{R\gamma^2}{Z^2}\right) \log C \tag{6.63}$$

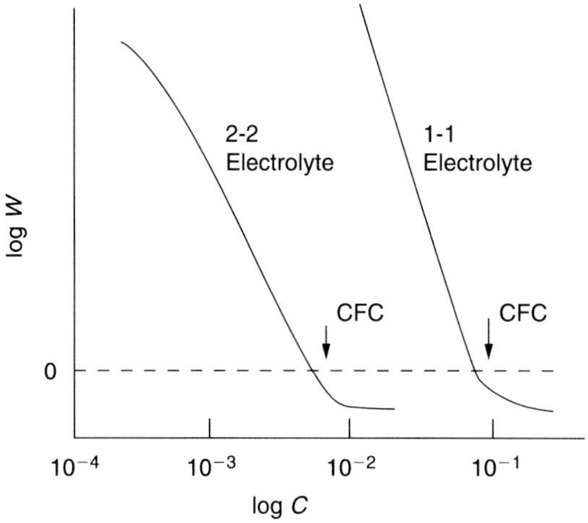

Fig. 6.29. Log W versus log C curves for electrostatically stabilised emulsions.

where γ is a function that is determined by the surface potential Ψ_0,

$$\gamma = \left[\frac{\exp(Ze\Psi_0/kT) - 1}{\exp(Ze\Psi_0/kT) + 1}\right] \tag{6.64}$$

Figure 6.29 shows plots of log W versus log C. The condition log $W = 0$ ($W = 1$) is the onset of fast flocculation – the electrolyte concentration at this point defines the critical flocculation concentration CFC. Above the CFC, $W < 1$, due to the contribution of van der Waals attraction, which accelerates the rate above the Smoluchowski value. Below the CFC, $W > 1$ and it increases with decreasing electrolyte concentration. The figure also shows that the CFC decreases with increasing valency in accordance to the Scultze–Hardy rule.

Another mechanism of flocculation is that involving the secondary minimum (G_{min}), which is few kT units. In this case, flocculation is weak and reversible and hence one must consider both the rate of flocculation (forward rate k_{f}) and deflocculation (backward rate k_{b}). The rate or decrease of particle number with time is given by

$$-\frac{\mathrm{d}n}{\mathrm{d}t} = -k_{\mathrm{f}}n^2 + k_{\mathrm{b}}n \tag{6.65}$$

The backward reaction (break-up of weak flocs) reduces the overall rate of flocculation.

6.10.1.2 Flocculation of Sterically Stabilised Emulsions

This occurs when the solvency of the medium for the chain becomes worse than a θ-solvent ($\chi > 0.5$). Under these conditions G_{mix} becomes negative, i.e. attractive,

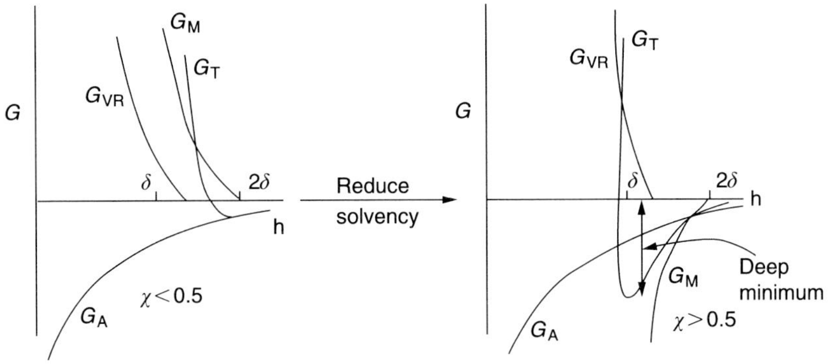

Fig. 6.30. Schematic representation of flocculation of sterically stabilised emulsions.

and a deep minimum is produced, resulting in catastrophic flocculation (referred to as incipient flocculation) [8]. This is schematically represented in Figure 6.30.

With many systems, good correlation between the flocculation point and the θ point is obtained [8]. For example, the emulsion will flocculate at a temperature (referred to as the critical flocculation temperature, CFT) that is equal to the θ-temperature of the stabilising chain. The emulsion may flocculate at a critical volume fraction of a non-solvent (CFV) that is equal to the volume of non-solvent that brings it to a θ-solvent.

6.10.2 General Rules for Reducing (Eliminating) Flocculation

This section summarises the criteria required to reduce (eliminate) flocculation.

6.10.2.1 Charge Stabilised Emulsions, e.g. Using Ionic Surfactants

The most important criterion is to make G_{max} as high as possible; this is achieved by three main conditions: High surface or zeta potential; low electrolyte concentration; and low valency of ions.

6.10.2.2 Sterically Stabilised Emulsions

Four main criteria are necessary here:

(1) Complete coverage of the droplets by the stabilising chains.
(2) Firm attachment (strong anchoring) of the chains to the droplets. This requires the chains to be insoluble in the medium and soluble in the oil; however, this is incompatible with stabilisation, which requires a chain that is soluble in the medium and strongly solvated by its molecules. These conflicting requirements are solved by the use of A-B, A-B-A block or BA_n graft copolymers (B is the "anchor" chain and A is the stabilising chain(s)) (**6.3**).

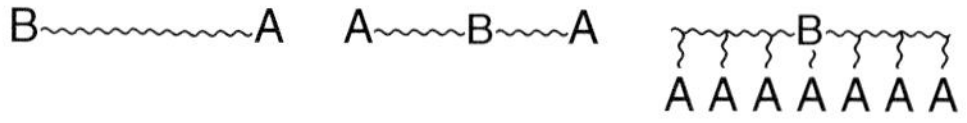

6.3

Examples of B chains for O/W emulsions are polystyrene, poly(methyl methacrylate), poly(propylene oxide) and alkyl poly(propylene oxide). For the A chain(s), poly(ethylene oxide) (PEO) or poly(vinyl alcohol) are good examples. For W/O emulsions, PEO can form the B chain, whereas the A chain(s) could be poly(hydroxy stearic acid) (PHS), which is strongly solvated by most oils.

(3) Thick adsorbed layers: the adsorbed layer thickness should be in the region of 5–10 nm – this means that the molecular weight of the stabilising chains could be in the region of 1000–5000.

(4) The stabilising chain should be maintained in good solvent conditions ($\chi < 0.5$) under all conditions of temperature changes on storage.

6.11 Ostwald Ripening

The driving force for Ostwald ripening is the difference in solubility between small and large droplets (the smaller droplets have higher Laplace pressure and higher solubility than the larger ones). This is illustrated below, where R_1 decreases and R_2 increases because of diffusion of molecules from the smaller to the larger droplets (Scheme 6.1).

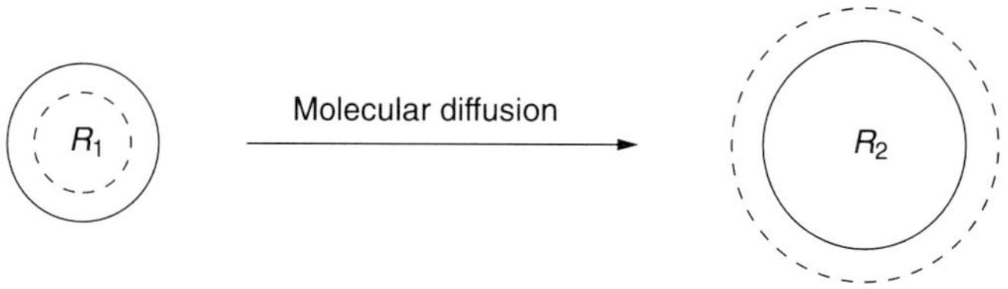

Scheme 6.1

The difference in chemical potential between different sized droplets was given by Lord Kelvin [35],

$$S(r) = S(\infty)\exp\left(\frac{2\gamma V_m}{rRT}\right) \tag{6.66}$$

where $S(r)$ is the solubility surrounding a particle of radius r, $S(\infty)$ is the bulk solubility, V_m is the molar volume of the dispersed phase, R is the gas constant and T is the absolute temperature. The quantity $2\gamma V_m/rRT$ is termed the characteristic length. It has an order of $\sim$1 nm or less, indicating that the difference in solubility of a 1 μm droplet is of the order of 0.1% or less. Theoretically, Ostwald ripening should lead to condensation of all droplets into a single drop [27]. This does not occur in practice since the rate of growth decreases with increasing droplet size.

For two droplets with radii r_1 and r_2 ($r_1 < r_2$),

$$\frac{RT}{V_m} \ln\left[\frac{S(r_1)}{S(r_2)}\right] = 2\gamma\left[\frac{1}{r_1} - \frac{1}{r_2}\right] \tag{6.67}$$

Eq. (6.67) shows that the larger the difference between r_1 and r_2, the higher the rate of Ostwald ripening.

Ostwald ripening can be quantitatively assessed from plots of the cube of the radius versus time t [36–38],

$$r^3 = \frac{8}{9}\left[\frac{S(\infty)\gamma V_m D}{\rho RT}\right] t \tag{6.68}$$

D is the diffusion coefficient of the disperse phase in the continuous phase.

Several methods may be applied to reduce Ostwald ripening:

(1) Addition of a second disperse phase component that is insoluble in the continuous medium (e.g. squalane) [39]. In this case partitioning between different droplet sizes occurs, with the component having low solubility expected to be concentrated in the smaller droplets. During Ostwald ripening in a two-component system, equilibrium is established when the difference in chemical potential between different size droplets (which results from curvature effects) is balanced by the difference in chemical potential resulting from partitioning of the two components – this effect reduces further growth of droplets.
(2) Modification of the interfacial film at the O/W Interface. According to Eq. (6.68), reduction in γ results in reduction of the Ostwald ripening rate. By using surfactants that are strongly adsorbed at the O/W interface (i.e. polymeric surfactants) and which do not desorb during ripening (by choosing a molecule that is insoluble in the continuous phase) the rate could be significantly reduced [40].

An increase in the surface dilational modulus ε ($= d\gamma/d \ln A$) and decrease in γ would be observed for the shrinking drop and this tends to reduce further growth.

A-B-A block copolymers such as PHS-PEO-PHS (which is soluble in the oil droplets but insoluble in water) can be used to achieve the above effect. This polymeric emulsifier enhances the Gibbs elasticity and reduces γ to very low values.

6.12 Emulsion Coalescence

When two emulsion droplets come in close contact in a floc or creamed layer or during Brownian diffusion, thinning and disruption of the liquid film may occur, resulting in eventual rupture. On close approach of the droplets, film thickness fluctuations may occur – alternatively, the liquid surfaces undergo some fluctuations forming surface waves (Figure 6.31).

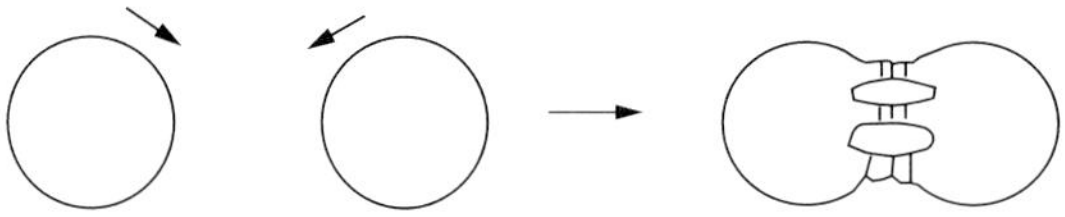

Fig. 6.31. Representation of surface fluctuations.

The surface waves may grow in amplitude and the apices may join as a result of the strong van der Waals attraction (at the apex, the film thickness is the smallest). The same applies if the film thins to a small value (critical thickness for coalescence).

Deryaguin [41] introduced a very useful concept in suggesting that a "Disjoining Pressure" $\pi(h)$ is produced in the film, which balances the excess normal pressure,

$$\pi(h) = P(h) - P_0 \tag{6.69}$$

where $P(h)$ is the pressure of a film with thickness h and P_0 is the pressure of a sufficiently thick film such that the net interaction free energy is zero.

$\pi(h)$ may be equated to the net force (or energy) per unit area acting across the film,

$$\pi(h) = -\frac{dG_T}{dh} \tag{6.70}$$

where G_T is the total interaction energy in the film.

$\pi(h)$ consists of three contributions, due to electrostatic repulsion (π_E), steric repulsion (π_s) and van der Waals attraction (π_A),

$$\pi(h) = \pi_E + \pi_S + \pi_A \tag{6.71}$$

To produce a stable film $\pi_E + \pi_s > \pi_A$ and this is the driving force for prevention of coalescence, which can be achieved by two mechanisms and their combination: (1) Increased repulsion both electrostatic and steric. (2) Dampening of the fluctuation by enhancing the Gibbs elasticity. In general, smaller droplets are less susceptible to surface fluctuations and hence coalescence is reduced. This explains the high stability of nanoemulsions.

Several methods may be applied to achieve the above effects:

(1) Use of mixed surfactant films. In many cases using mixed surfactants, say anionic and non-ionic or long-chain alcohols can reduce coalescence as a result of several effects: high Gibbs elasticity; high surface viscosity; hindered diffusion of surfactant molecules from the film.
(2) Formation of lamellar liquid crystalline phases at the O/W interface. This mechanism was proposed by Friberg and co-workers [42], who suggested that surfactant or mixed surfactant films can produce several bilayers that "wrap" the droplets. As a result of these multilayer structures, the potential drop is shifted to longer distances, thus reducing the van der Waals attraction.

Figure 6.32 gives a schematic representation of the role of liquid crystals, illustrating the difference between having a monomolecular layer and a multilayer (as with liquid crystals).

For coalescence to occur, these multilayers have to be removed "two-by-two" and this forms an energy barrier preventing coalescence.

6.12.1
Rate of Coalescence

Since film drainage and rupture is a kinetic process, coalescence is also a kinetic process. If one measures the number of particles n (flocculated or not) at time t,

$$n = n_t + n_v m \tag{6.72}$$

where n_t is the number of primary particles remaining, n is the number of aggregates consisting of m separate particles.

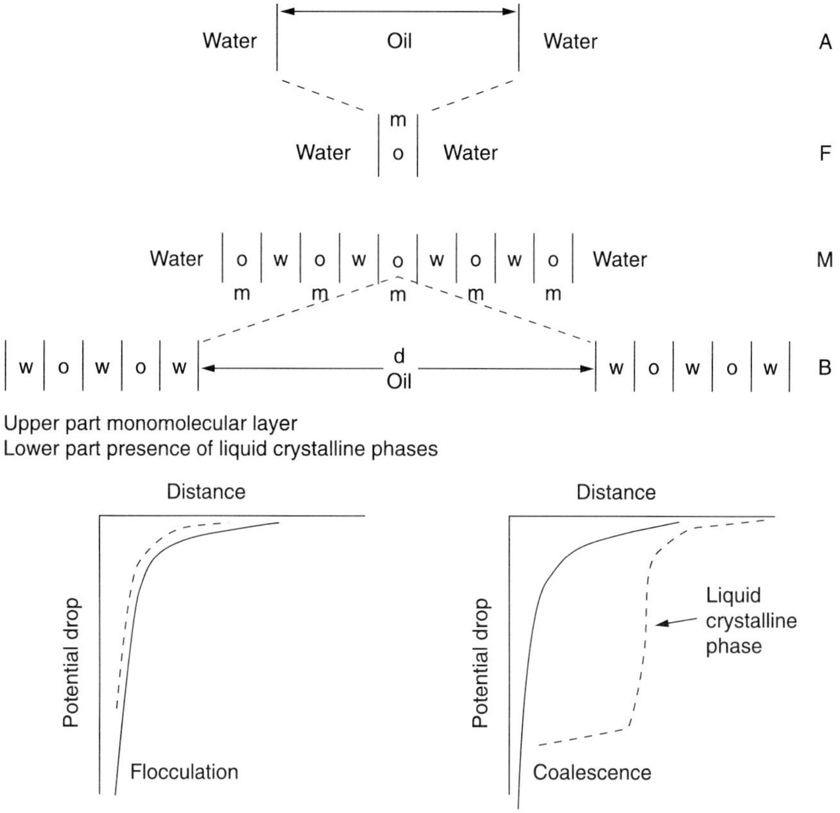

Fig. 6.32. Schematic representation of the role of liquid crystalline phases.

For studying emulsion coalescence, one should consider the rate constant of flocculation and coalescence. If coalescence is the dominant factor, then the rate k follows a first-order kinetics,

$$n = \frac{n_0}{kt}[1 + \exp -(kt)] \tag{6.73}$$

showing that a plot of log n versus t should give a straight line from which k can be calculated.

6.13
Phase Inversion

Phase inversion of emulsions can be one of two types: Transitional inversion induced by changing factors that affect the HLB of the system, e.g. temperature and/or electrolyte concentration, and catastrophic inversion, which is induced by increasing the volume fraction of the disperse phase [43].

Catastrophic inversion is illustrated in Figure 6.33, which shows the variation of viscosity and conductivity with the oil volume fraction ϕ. As can be seen, inversion occurs at a critical ϕ, which may be identified with the maximum packing fraction. At ϕ_{cr}, η suddenly decreases; the inverted W/O emulsion has a much lower volume fraction.

κ also decreases sharply at the inversion point since the continuous phase is now oil, which has very low conductivity.

Earlier theories of phase inversion were based on packing parameters – inversion occurs when ϕ exceeds the maximum packing (~0.64 for random packing and ~0.74 for hexagonal packing of monodisperse spheres; for polydisperse

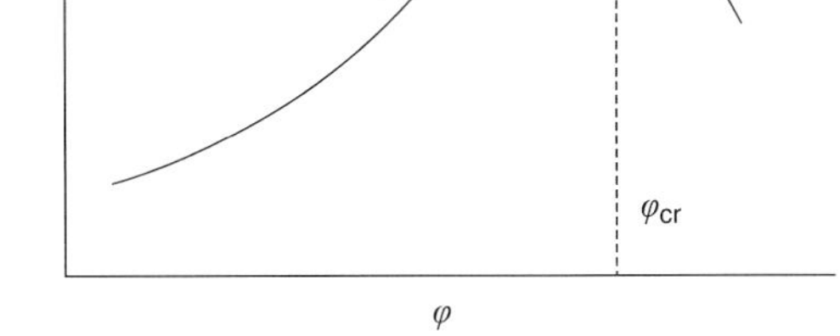

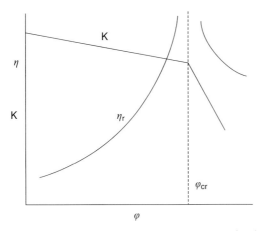

Fig. 6.33. Variation of conductivity and viscosity with volume fraction of oil.

systems, the maximum packing exceeds 0.74). However, these theories are not adequate, since many emulsions invert at ϕ well below the maximum packing due to the change in surfactant characteristics with variation of conditions. For example, when using a non-ionic surfactant based on PEO, the latter chain changes its solvation by increasing the temperature and/or addition of electrolyte. Many emulsions show phase inversion at a critical temperature (the phase inversion temperature) that depends on the HLB number of the surfactant as well as the presence of electrolytes. By increasing temperature and/or addition of electrolyte, the PEO chains become dehydrated and, finally, they become more soluble in the oil phase – under these conditions the O/W emulsion will invert to a W/O emulsion. This dehydration effect amounts to a decrease in the HLB number, and inversion will occur when the HLB reaches a value that is more suitable for a W/O emulsion. At present, there is no quantitative theory that accounts for the phase inversion of emulsions.

6.14 Rheology of Emulsions

The flow characteristics (rheology) of emulsions are of considerable importance, both from a fundamental and applied point of view [44]. At a fundamental level, the rheology of emulsions is a direct manifestation of the various interaction forces at work in the system [44]. The various processes that occur in emulsion systems, such as creaming and sedimentation, flocculation, coalescence, Ostwald ripening and phase inversion, may be investigated using various rheological techniques, such as measurements of shear stress as a function of shear rate (steady state), strain as a function of time at a constant applied stress (creep) and oscillatory techniques. The principle of each of these methods will be briefly described. In addition, the properties of the interfacial film (polymeric surfactant) can be studied from investigations of the interfacial rheology of the film, such as its viscosity and elasticity. Again, the principles of these measurements will be described. As we will see later, interfacial rheological investigations provide a fundamental understanding of the various breakdown processes in emulsions, such as thinning and disruption of liquid films and coalescence of droplets.

At an applied level, study of the rheology of emulsions is vital in many industrial applications of personal care products. It is perhaps useful to summarize the factors that affect emulsion rheology in a qualitative way. One of the most important factors is the volume fraction of the disperse phase, ϕ. In very dilute emulsions ($\phi < 0.01$), the relative viscosity, η_r, of the system may be related to ϕ using the simple Einstein equation (as for solid/liquid dispersions) [45], i.e.

$$\eta_r = 1 + 2.5\phi \tag{6.74}$$

As the volume fraction of the emulsion is gradually increased, the relative viscosity becomes a more complex function of ϕ and it is convenient to use a polynomial representing the variation of η_r with _, i.e.,

$$\eta_r = 1 + k_1\phi + k_2\phi^2 + k_3\phi^3 + \cdots \tag{6.75}$$

where k_1 is the Einstein's coefficient (2.5 for hard spheres), k_2 is a coefficient that accounts for hydrodynamic interaction between the droplets, which arises from the overlap of the associated flow pattern and their eventual overlap at appreciable _ values [46]; k_2 is equal to 6.2. The hydrodynamic interaction term is usually sufficient to describe the viscosity of dispersions up to _ = 0.2. Above this volume fraction, higher order interaction terms (k_3 _^{3}) are necessary. As we will see later, only semi-empirical equations are available to describe the variation of _$_r$ with _ over a wide range.

Another factor that may affect the rheology of emulsions is the viscosity of the disperse droplets. This is particularly the case when the viscosity of the droplets is comparable or lower than that of the dispersions medium. This problem was considered by Taylor [47], who extended Einstein's hydrodynamic treatment for suspensions to the case of droplets in a liquid medium. Taylor [47] assumed that the emulsifier film around the droplets would not prevent the transmission of tangential and normal stresses from the continuous phase to the disperse phase and that there was no slippage at the O/W interface. These stresses produce fluid circulation within the droplets, which reduces the flow patterns around them. Taylor derived the following expression for _$_r$,

$$\eta_r = 1 + 2.5\left(\frac{\eta_i + 0.4\eta_0}{\eta_i + \eta_0}\right)\phi \tag{6.76}$$

where η_i is the viscosity of the internal phase and η_0 is that of the external phase. Clearly, when $\eta_i \gg \eta_0$ (as with most O/W emulsions), the term between the brackets becomes equal to unity and Eq. (6.76) reduces to the Einstein's equation. Conversely, when $\eta_i \ll \eta_0$ (as with foams) the term between brackets becomes equal to 0.4 and the Einstein's coefficient becomes equal to 1. When η_i is comparable to η_0, Einstein's coefficient can assume values between 1 and 2.5 depending on the relative ratio of η_i/η_0. Notably, however, this analysis assumed a deformable droplet, which may not be the case when a surfactant or polymer film is present at the interface. In this case, interfacial tension gradients and/or surface viscosity will make these droplets appear as hard spheres and the emulsion behaves in a similar way to a solid/liquid dispersion.

The third factor that affects emulsion rheology is the droplet size distribution. This is particularly the case at high volume fractions. When $\phi > 0.6$, η_r is inversely proportional to the reciprocal of the mean droplet diameter [48]. The above equations do not show any dependence on droplet size and an account should be made for this effect by considering the average distance between the droplets in an emulsion. At high shear rate, the droplets are completely deflocculated (i.e. all structure is destroyed) and they are equidistance from each other. At a critical separation between the droplets, which depends on droplet size, the viscosity shows a rapid increase. The average distance of separation between the droplets, h_m, is related to the droplet diameter, d_m, by the simple expression,

$$h_m = d_m \left[\left(\frac{\phi_{max}}{\phi} \right)^{1/3} - 1 \right] \tag{6.77}$$

where ϕ_{max} is the maximum packing fraction, which is equal to 0.74 for hexagonally packed monodisperse spheres. With most emulsions, ϕ_{max} reaches a higher value than 0.74 as a result of polydispersity. Equation (6.77) indicates that, with small droplets, the critical value of h_m is reached at lower ϕ than with larger droplets.

Several other factors that affect the rheology of emulsions may be considered that are related to the properties of the continuous phase and the interfacial film. Three main properties of the continuous phase may be considered. The first and most important is the viscosity of the medium, which is affected by the additives present such as excess emulsifier and thickeners (e.g., polysaccharides) that are added in many personal care emulsions to prevent sedimentation or creaming as well as to produce the right consistency for application. The second property of the medium that affects emulsion rheology is the chemical composition such as polarity and pH, which affects the charge on the droplets and hence their repulsion. The viscosity of the emulsion is directly related to the magnitude of the repulsive forces. The latter are also affected by the nature and concentration of the electrolyte, which represents the third important property of the medium. The influence of charge and repulsion between droplets in an emulsion are sometimes referred to as electroviscous effects. Two such effects may be distinguished. The first arises from distortion of the double layers around the droplets as the latter are sheared. This effect is very small and it contributes a small increase in the relative viscosity. However, the second electroviscous effect arises from overlap of the double layers, which becomes significant in concentrated emulsions. The magnitude of the secondary electroviscous effect is proportional to ϕ^2. Thus, this effect can cause a large increase in viscosity. Clearly, by the addition of electrolytes, the double layers are compressed and this results in a large reduction of the electroviscous effects. This could find application in many practical systems, where a high viscosity is undesirable.

The rheology of emulsions may also be influenced by the interfacial rheology of the emulsifier film surrounding the droplets. When shear is applied to an interfacial film, its constituent molecules as well as the molecules of the oil and water phases in its immediate vicinity are displaced from their equilibrium positions [49]. The stress that develops depends on the associated molecular rearrangement. This will have an effect on the interfacial viscosity of the film, η_s. This will affect the bulk rheology of the emulsion, if the latter is formed from large deformable droplets. The viscosity of an emulsion in which the drops deform under shear increases more rapidly with ϕ than for emulsions of identical size but surrounded by an elastic film that prevents deformation. Clearly, when the droplets are very small deformation is less likely and interfacial rheology becomes less significant. In some cases the chemical nature of the emulsifier has an effect on the relative viscosity, particularly at high ϕ. Sherman [50, 51] has demonstrated this for food emulsions prepared using emulsifiers of different chemical nature. The nature and concentration of emulsifier can also have a dramatic effect on emulsion phase inversion [51, 52]. An excess emulsifier will have a pronounced effect on the viscosity of the continuous phase, resulting in an increase in the overall viscosity of the system. Under these conditions phase inversion may occur at a relatively lower disperse volume fraction compared with that at lower emulsifier concentration. The nature of the emulsifier, in particular its solubility and distribution in both phases, also has a large effect on the rheology of the system. Unfortunately, there are no systematic studies of these effects on the rheology of emulsions. Many personal care emulsions and creams are complex systems that are formulated to behave like "semisolids", being solid-like at ambient conditions and transformed into a liquid-like consistency when stressed during the application on the skin. The dominant colloidal structural elements of these semisolid preparations are three-dimensional colloidal solid networks in which a liquid is incorporated. Such a bi-coherent (sponge-like) structure may be referred to as a gel. These gel structures may be either in a crystalline or liquid crystalline state, with properties determined by the bulk rheology of the system. Bulk rheology is, in turn, determined by the structure of the liquid crystalline phases produced, which may be established using low-angle X-ray and freeze–fracture techniques [42]. The structure of the system and its rheological properties also determines the stability, interaction with the skin and release.

6.15 Interfacial Rheology

Interfacial rheology deals with the response of mobile interfaces to deformation [52]. Emulsions contain a molecular or macromolecular surfactant film at the fluid interface, which, apart from being necessary for the stabilization of the dispersion, also initiates additional interfacial stresses beyond that already contributed by a homogeneous interfacial tension, γ. If a non-uniformity of surfactant concentration develops within the fluid interface, an interfacial tension gradient $d\gamma/dA$, where A is the interfacial area, is produced. This gradient is sometimes defined by the Gibbs elasticity, ε, which is simply equal to $d\gamma/d \ln A$, which induces both area and volumetric liquid motion. The gradient-driven flow is the basis of the so-called "Marangoni effect" [53]. In addition to the possible existence of these interfacial tension gradients, other interfacial rheological stresses of a viscous nature may arise, such as those relating to interfacial shear and dilational viscosities [52]. Many surfactant and polymer films also exhibit non-Newtonian interfacial rheological behaviour that may be characterized by Bingham plastic models and interfacial viscoelasticity. The basic equations needed to describe these interfacial rheological parameters are summarised below, followed by a brief description of some of the essential techniques required to measure interfacial rheology. Finally, some results will be given to correlate interfacial rheology with emulsion stability.

6.15.1
Basic Equations for Interfacial Rheology

The interfacial shear viscosity, η_s, is the ratio between the shear stress, σ, and shear rate, γ, in the plane of the interface, i.e. it is a two-dimensional viscosity. The unit for surface viscosity is, therefore, N m^{-1} s (surface Pa s). A liquid/liquid (or liquid/vapour) interface with no adsorbed surfactant or polymer shows only a negligible interfacial shear viscosity. However, in the presence of an adsorbed surfactant or polymer layer, an appreciable interfacial shear viscosity is obtained (which can be orders of magnitude higher than the bulk viscosity of the film). This appreciable shear viscosity can be accounted for in terms of the orientation of the surfactant or polymer molecules at the interface. For example, surfactant molecules at the O/W interface usually form a monolayer of vertically oriented molecules with the hydrophobic portion pointing to (or dissolved in) the oil, leaving the polar head groups pointing in the aqueous phase. A two-dimensional surface pressure, π, may be defined, i.e.

$$\pi = \gamma_0 - \gamma \tag{6.78}$$

where γ_0 is the interfacial tension of the clean interface (i.e., before adsorption of surfactant or polymer) and γ is the corresponding value with the adsorbed film. Since γ_0 is of the order of 30–50 mN m^{-1}, whereas γ be as low as a fraction of mN m^{-1}, clearly π can be high, reaching values of the order of 30–50 mN m^{-1}. Thus, any shear field applied across the interface containing these adsorbed surfactants or polymers (with high surface pressure) results in a large viscous interaction between adjacent molecules. With macromolecules that form loops and tails at the interface, the film resists compression due to the lateral repulsion between the loops and tails.

The interfacial dilational elasticity, ε, arises from interfacial tension gradients, which are due to inhomogeneous surfactant or polymer films. The regions that are depleted from the film have higher interfacial tension than those containing the adsorbed film. Consequently, an interfacial tension gradient $d\gamma/dA$ is set up and the Gibbs dilational elasticity may be defined as

$$\varepsilon = \frac{d\gamma}{d\ln A} \tag{6.79}$$

The above situation may arise during emulsification or on approach of two emulsion droplets. As the interface is stretched, the film will no longer cover the whole interface, and regions depleted of surfactant or polymer are created. This results in interfacial tension gradients, with surfactant or polymer molecules tending to diffuse from the bulk to the interface to fill these depleted regions. During this process, liquid may be transported to the interface, a phenomenon usually referred to as the "Marangoni" effect [19]. This Gibbs–Marangoni effect is sometimes believed to be the driving force for stabilization of thin liquid films between droplets, thus preventing their coalescence.

The interfacial dilational viscosity, η_s^d can be simply defined if one considers a uniform expansion of the interface at a constant rate d ln A/dt, i.e.

$$\eta_s^d = \frac{d\gamma}{d \ln A \, dt} \tag{6.80}$$

As mentioned above, interfacial films exhibit non-Newtonian flow, which can be treated in the same manner as for dispersions and polymer solutions. The steady-state flow can be described using Bingham plastic models. Viscoelastic behaviour can be treated using stress relaxation or strain relaxation (creep) models as well as dynamic (oscillatory) models. The Bingham-fluid model of interfacial rheological behaviour [54] assumes the presence of a surface yield stress, σ_s, i.e.

$$\sigma = \sigma_s + \eta_s \dot{\gamma} \tag{6.81}$$

In stress relaxation experiments, a sudden strain is applied on the film, within a short period of time, and the stress σ followed as a function of time. If $\sigma(t)$ is the stress at time t and σ_0 is the instantaneous value at the moment when the constant strain γ is applied, then,

$$\ln \frac{\sigma(t)}{\sigma_0} = \frac{t}{t_r} \tag{6.82}$$

where t_r is the relaxation time that is given by the ratio η/G, where G is the relaxation modulus.

In strain relaxation (creep) experiments, a small constant stress is applied on the film and the strain or compliance J (where $J = \gamma/\sigma$) is followed as a function of time. The compliance at any time $t, J(t)$, is given by the expression

$$J(t) = \frac{\left[1 - \exp\left(-\frac{t}{t_r}\right)\right]}{G} \tag{6.83}$$

In dynamic (oscillatory) experiments, the stress or strain is varied periodically with a sinusoidal alteration at a frequency ω (rad s^{-1}) and the resulting strain or stress is compared with the applied values. For a viscoelastic material, the stress and strain show a time shift Δt between the sine waves of the stress and strain. The product of this time shift and the frequency ω gives the phase angle shift δ (note that for a viscoelastic material $0 < \delta < 90°$). The amplitude ratio of stress and strain gives the complex modulus G^*, which is split into two components through the phase angle shift δ: the in-phase component G' (the real part of the complex modulus, referred to as the storage or elastic modulus) and the out-of-phase component G'' (the imaginary part of the complex modulus, referred to as the loss or viscous modulus), i.e.

$$G' = |G^*| \cos \delta \tag{6.84}$$

$$G'' = |G^*| \sin \delta \tag{6.85}$$

and

$$|G^*| = G' + iG'' \tag{6.86}$$

The dilational modulus ε^* for a sinusoidally oscillating surface dilation can also be split into in-phase (referred to as the dilational elasticity) and out-of-phase components, i.e.

$$|\varepsilon^*| = \varepsilon' + i\varepsilon'' \tag{6.87}$$

6.15.2 Basic Principles of Measurement of Interfacial Rheology

The simplest procedure to measure the interfacial shear viscosity is to use a torsion pendulum surface viscometer [55]. This technique observes the damping of a torsion pendulum due to the viscous drag of a surface film. The shearing element can be in the form of a ring, a disc or knife-edged disc, which is suspended by a torsion wire and positioned at the place of the interface (Figure 6.34). Measurements are

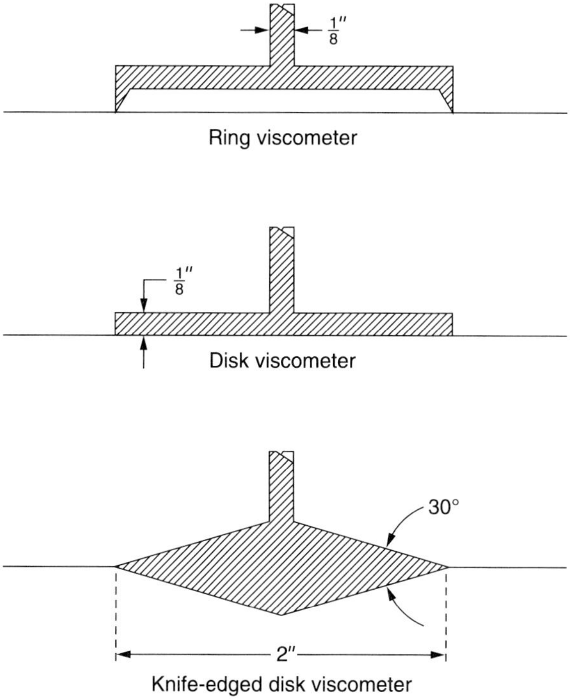

Fig. 6.34. Surface viscometer designs.

made of the period of the pendulum and the damping as the pendulum oscillates. The apparent surface viscosity, η_s, is given by

$$\eta_s = \eta_0 \left(\frac{\Delta/\Delta_0}{t/t_0} - 1 \right) \tag{6.88}$$

where η_0 is the sum of the bulk viscosities of the two phases forming the interface. Δ is the difference in logarithm of the amplitude of successive swings for the interface with adsorbed surfactant or polymer and Δ_0 is the corresponding value without film; t is the period of the pendulum for the film-covered interface and t_0 the corresponding value without film.

The surface viscosity may be related to the torsion modulus of the wire, C_w, the polar moment of inertia of the oscillating pendulum, I, and the dimensions of the viscometer by the expression [55],

$$\eta_s = \frac{C_w I}{2\pi} \frac{R_2^2 - R_1^2}{R_1^2 R_2^2} \left[\frac{\Delta}{7.4 + \Delta^2} - \frac{\Delta_0}{7.4 + \Delta_0} \right] \tag{6.89}$$

where R_1 is the radius of the surface viscometer and R_2 is the radius of the container.

One of the main drawbacks of the torsion pendulum viscometer is that it uses a range of shear rates and hence it is not suitable for measurement of non-Newtonian films. In the latter case, it is preferable to use rotational torsional viscometer, where the surface film is sheared between rotating concentric rings on a surface. The shear rate can be held constant by rotating one ring, while measuring the torque T on the other ring (Eq. 6.90), [55] where Ω is the angular velocity.

$$\eta_s = \frac{T}{4\pi\Omega} \frac{R_2^2 - R_1^2}{R_1^2 R_2^2} + \frac{\sigma_s}{\Omega} \ln \frac{R_2}{R_1} \tag{6.90}$$

Another convenient method for measuring the interfacial shear viscosity is to use the deep-channel surface viscometer (Figure 6.35) [52]. Basically, this consists of two concentric brass cylinders (separated by a distance y_0) lowered into a pool of liquid contained within a brass dish, to a depth at which the brass cylinders nearly touch the bottom of the dish. The dish is rotated with a known angular velocity, ω_0, and the midchannel or "centerline" surface motion of the interface within the channel (formed by the concentric cylinders) is measured using talc or Teflon particles placed within the fluid interface, i.e.

$$\frac{\eta_s \pi}{\eta y_0} = \frac{v_c^*}{v_c} - 1 \tag{6.91}$$

where η is the bulk shear viscosity, v_c^* is the center-line surface velocity in the presence of the film and v_c is the corresponding value in the absence of the film.

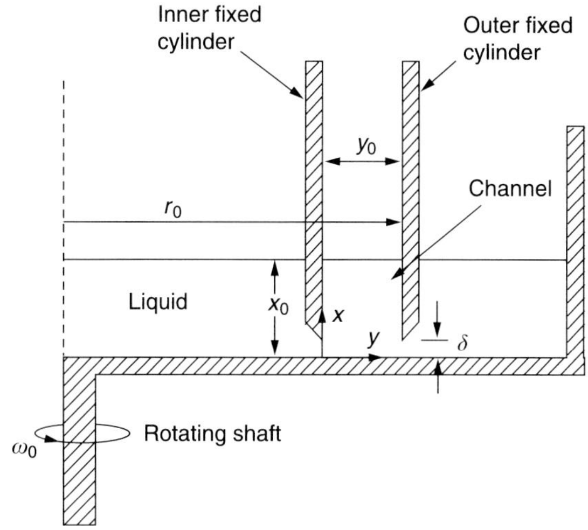

Fig. 6.35. Deep channel surface viscometer.

Three techniques may be applied to measure the dilational surface elasticity and viscosity [52]. The first method applies surface waves to the interface (with frequency ω). The dilational elasticity, ε', is given by

$$\varepsilon' = \frac{\varepsilon_0[1 + (\tau/\omega)^{1/2}]}{[1 + 2(\tau/\omega)^{1/2} + 2(\tau/\omega)]} \tag{6.92}$$

where ε_0 is the Gibbs elasticity and τ is a "diffusion parameter" that is related to the diffusion coefficient D of the surfactant molecule.

The relaxation elasticity, ε'', is given by

$$\varepsilon'' = -\frac{\varepsilon_0(\tau/\omega)^{1/2}}{[1 + 2(\tau/\omega)^{1/2} + 2(\tau/\omega)]} \tag{6.93}$$

The tangential bulk-phase stress component evaluated at the interface combines an elastic (interfacial tension gradient) effect, ε', and an apparent viscous effect, $(\eta_s^d + \eta_s) + \varepsilon''/\omega$. One of the most convenient methods of measuring capillary waves is to use light scattering [56], which can yield information on both the tension and dilational modulus of the interface.

The second method for measuring dilational elasticity and viscosity is based on rotation, translation or deformation of bubbles and droplets [52]. Agrawal and Wasan [57] have suggested that the translational velocity of bubbles or droplets in a quiescent liquid might be used to determine the apparent dilational viscosity. Unfortunately, this simple method is not suitable since the settling velocity is not sensitive enough to the magnitude of the apparent surface viscosity. Wei et al. [58]

suggested that measurements of the circulation velocity of a tracer particle in the equatorial plane of a spherical droplet rotating within a shear field might provide a means for deducing a combination of shear and dilational viscosities. The study of droplet shape deformation in a shear field provides another method for measuring the apparent interfacial dilational viscosity [59]. Such a method was applied by Phillips et al. [59], using the spinning drop technique.

The third method for measuring dilational elasticity and viscosity is the maximum bubble pressure method [60]. Although this method overcomes some of the problems encountered in the surface wave and droplet deformational methods, it can only be applied for measurements at the air/liquid interface.

Several methods have been suggested for measuring the non-Newtonian rheological behaviour of surfactant and polymer films. For example, Haydon et al. [61] constructed a special apparatus to measure the two-dimensional creep and stress relaxation of adsorbed protein film at the O/W interface. In creep experiments, a constant torque (in mN m^{-1}) was applied and the resulting deformation (in radians) was recorded as a function of time. In the stress relaxation experiments, a certain deformation γ was produced in the film by applying an initial stress, and the deformation was kept constant by gradually decreasing the stress.

The deep-channel viscometer could also be adapted for measurement of the non-linear interfacial rheological behaviour of the film [52]. In this case several small tracer particles are placed on the fluid interface at different radial positions and the angular velocities are determined from measurements of the period of revolution. When used to measure viscoelastic properties, the deep-channel viscometer is operated in an oscillatory mode, in which case the floor of the viscometer is oscillated sinusoidally. Simultaneous measurements of the phase angle between the surface motion and the oscillating motion of the bottom dish, and the "surface-to-floor" amplitude ratio, may permit determination of the viscoelastic properties of the fluid interface, presuming knowledge of an appropriate rheological model [52].

6.15.3
Correlation of Interfacial Rheology with Emulsion Stability

Several examples illustrate the correlation between interfacial rheology and emulsion stability. Cockbain and co-workers [62] made one of the first observations, finding that addition of an alcohol such as lauryl alcohol to an emulsion stabilized by an anionic surfactant increased the emulsion stability. This was attributed to an increase in the interfacial shear viscosity. Later, Prince et al. [63] found that the dilational elasticity of the film increased markedly in the presence of the alcohol and they, therefore, attributed the enhanced stability to such high surface elasticity. Other authors attributed the enhanced stability to a high interfacial viscosity [57], although Prince et al. [63] argued against this since they found that the film stability was not very sensitive to either temperature changes or the concentration of alcohol, which had a pronounced effect on η_s. Wasan et al. [64] claimed a correlation between interfacial shear viscosity and emulsion stability. One of the most convincing examples of the correlation of interfacial rheology with emulsion stability is the results of Haydon et al. [61]. These authors systematically investigated the rheolog-

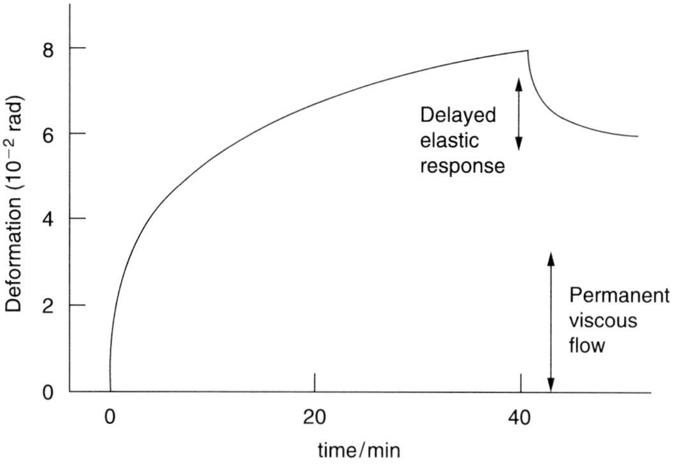

Fig. 6.36. Creep curve of an adsorbed bovine serum albumin film (pH 5.2) at a light petroleum–water interface, at a constant stress of 0.0116 N m^{-1}.

ical characteristics of various proteins, namely albumin, poly (ϕ-L-lysine) and arabinic acid, at the O/W interface and correlated these measurements with stability of oil droplets at a planar O/W interface. As mentioned above, the viscoelastic properties of the adsorbed films were studied using creep and stress relaxation measurements. Figure 6.36 shows a typical creep curve for bovine serum albumin at the petroleum ether/water interface. The curve shows an initial, instantaneous deformation characteristic of an elastic body, followed by a nonlinear flow that gradually declines and approaches the steady state of a viscous body. After 30 minutes, when the external force was withdrawn, the film tended to revert to its initial state, with an initial instantaneous recovery followed by a slow one. The original state was not obtained even after 20 hours and the film seemed to have undergone some flow. This behaviour illustrates the viscoelastic properties of the bovine serum albumin.

Biswas and Haydon [61] also found a striking effect of pH on the rigidity of the protein film. This is illustrated in Figure 6.37, where the shear modulus G and surface viscosity η_s are plotted as a function of pH. The elasticity of the film is seen to be at a maximum at the isoelectric point of the protein. Biswas and Haydon then determined the rate of coalescence of petroleum ether drops at a planar O/W interface by measuring the lifetime of a droplet resting beneath the interface. The half-life of the droplets was plotted as a function of pH as shown in Figure 6.37, which clearly illustrates the correlation with G and η_s.

Biswas and Haydon [61] derived an equation relating the time of coalescence (τ) with the viscoelasticity of the film, the thickness of the adsorbed film h and the critical distortion of the plane interface under the weight of the drop, i.e.

$$\tau = \eta_s \left[3C' \frac{h^2}{A} - \frac{1}{G} - \phi(t) \right] \tag{6.94}$$

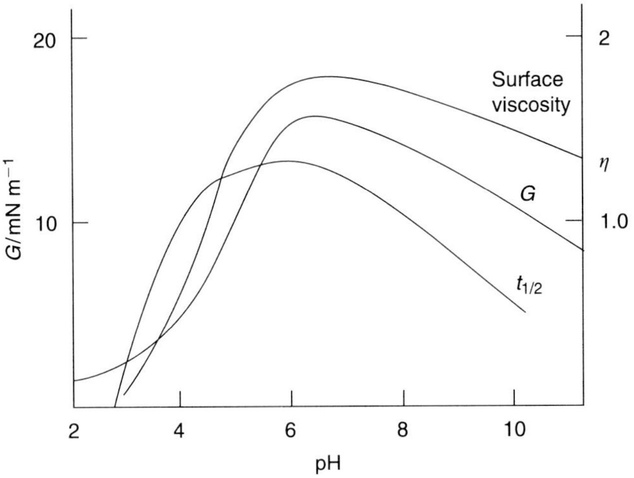

Fig. 6.37. Shear modulus, surface viscosity and half-life of petroleum ether drops beneath a plane light petroleum–0.1 mol dm^{-3} aqueous KCl interface, as a function of pH.

where G is the (instantaneous) elasticity, η_s the long time viscosity (i.e., for an infinite time of retardation), $\phi(t)$ is the elastic deformation per unit stress and $3C'$ is a critical deformation factor. Equation (6.94) predicts that: (1) the life time of the drop, τ, increases with increasing viscosity of the protective film; (2) the rate process of coalescence is not influenced by the instantaneous elasticity, but this quantity is likely to set a limit on the process through the critical deformation factor $3C'$; (3) the life time should depend on the film thickness and vary linearly with h^2 if the retarded elasticity $\phi(t)$ is neglected; (4) τ should be a fixed (not a fluctuating quantity).

The results of Biswas and Haydon [61] indicate clearly that no significant stabilization occurred with non-viscoelastic films. However, the presence of viscoelasticity is not sufficient to confer stability when drainage is rapid. For example, the highly viscoelastic films of bovine serum albumin or pepsin could not stabilize W/O emulsions; the same was found with pectin and gum arabic. In these cases the drainage was clearly rapid, even from rigid films, e.g. W/O droplets in the case of bovine serum albumin. In fact, as expected, it was only after solvent drainage, and the disperse phases were still separated by a film of high viscosity, that enhanced stabilization occurred. It was concluded from these investigations that, in agreement with the prediction of theory, a film with appreciable thickness is required for stability to coalescence. In addition, the main part of the film should be located on the continuous side of the interface.

Several other examples may be found in the literature [65], in which a correlation between the interfacial viscosity of macromolecular stabilized films with droplet stability was found. However, there are also several cases where stable emulsions could be prepared without any significant interfacial viscosity or elasticity. Therefore, one should be careful in using interfacial rheology as a predictive test for emulsion stability. Other factors, such as film drainage and thickness may be of more importance. Despite these limitations, interfacial rheology offers a powerful tool for understanding the properties of surfactant and macromolecular films at the liquid/liquid interface. In cases where a correlation between the interfacial viscosity and/or elasticity and emulsion stability is found, one could use these measurements to screen various other components that have marked effect on these parameters.

6.16 Investigations of Bulk Rheology of Emulsion Systems

As discussed above, the bulk rheology of emulsion systems can be investigated using steady state (shear stress as a function of shear rate), constant stress and oscillatory techniques. These methods are the same as those described for interfacial rheology. In this section, I will describe some results on various emulsion systems to illustrate the use of rheological measurements in investigating the interaction between emulsion droplets. Firstly, the viscosity–volume fraction relationship for O/W and W/O emulsions will be discussed to show the analogy with suspensions. This is followed by a section on the viscoelastic properties of concentrated emulsions. A third section deals with the case of flocculated emulsions, which may be produced, for example, by the addition of a "free" (non-adsorbing) polymer or by van der Waals attraction.

6.16.1 Viscosity-Volume Fraction Relationship for Oil/Water and Water/Oil Emulsions

Figure 6.38 gives relative viscosity (η_r)–volume fraction (ϕ) curves for paraffin oil/water emulsions.

Four emulsion systems were prepared using nonionic surfactants, namely Synperonic NPE 1800 and its analogues [66]. These surfactants have the structural formula: C_9H_{19}-C_6H_5-(CH_2-CH(CH_3)-O)$_m$-(CH_2-CH_2-O)$_n$-OH. They all contain the same hydrophobic chain (nonyl phenyl and 13 moles propylene oxide), but have different numbers of moles of ethylene oxide: 27 for Synperonic NPE 1800, 48 for NPE A, 80 for NPE B and 174 for NPE C. The average droplet diameter for each emulsion was determined using the Coulter counter and the volume mean diameter (VMD) is given in the legends of Figure 6.38. The molecular weight (M_w) and hydrodynamic thickness (δ_h) of these surfactant molecules were determined, and are tabulated below:

Synperonic	*NPE 1800*	*NPE A*	*NPE B*	*NPE C*
M_w	2180	3080	4460	8650
δ_h (nm)	5.8	6.4	8.5	11.6

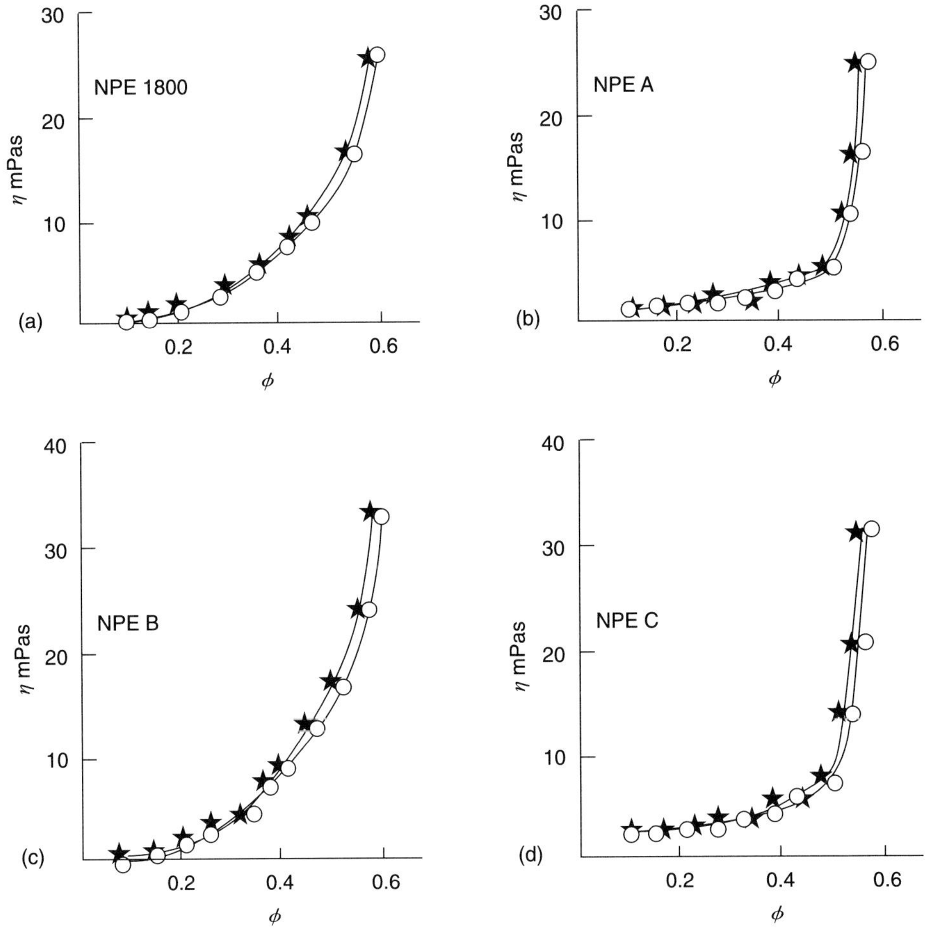

Fig. 6.38. η_r versus ϕ curves for paraffin oil–water emulsions: (a) Synperonic NPE 1800, VMD = 3.5 μm; (b) NPE A, VMD = 4 μm; (c) NPE B, VMD = 4.5 μm; (d) NPE C, VMD = 5 μm. (×) Experimental results; (○) theoretical values according to the Dougherty–Krieger equation.

The above emulsion system is fairly simple, since it is likely that the nonyl phenyl and propylene oxide chain is on the oil side of the interface, whereas the poly(ethylene oxide) chain is on the aqueous side of the interface. The hydrodynamic thickness of the surfactants is much less than the droplet radius and hence these sterically stabilized emulsion may approximate hard-sphere dispersions very closely (with an effective radius $R_{\text{eff}} = R + \delta_{\text{h}}$). This can be tested by fitting the data to the hard sphere model suggested by Dougherty and Krieger [67, 68].

By application of the theory of corresponding states, these authors derived the following equation for the relative viscosity,

$$\eta_r = \left[1 - \left(\frac{\phi}{\phi_p}\right)\right]^{-[\eta]\phi_p} \qquad (6.95)$$

where $[\eta]$ is the intrinsic viscosity, which has a theoretical value of 2.5 for rigid spheres and ϕ_p is the maximum packing fraction, which is equal to 0.64 for random packing and 0.74 for hexagonal close packing of monodisperse spheres. However, Krieger [68] showed that, with hard sphere dispersions, ϕ_p is close to 0.6. Since the emulsions are polydisperse, a higher ϕ_p is to be expected. The value of ϕ_p for each emulsion was estimated from plots of $\eta^{-1/2}$ versus ϕ, which gave a straight line. Extrapolation to $\eta^{-1/2} = 0$ (i.e., $\eta = \infty$) gave ϕ_p. The values obtained were 0.73, 0.73, 0.79 and 0.69 for Synperonic NPE 1800, NPE A, NPE B and NPE C respectively. Using the calculated ϕ_p, plots of η_r were constructed (Figure 6.38). These results show that the emulsions stabilized with the Synperonic NPE surfactants approximate very closely hard sphere dispersions.

Figure 6.39 gives the relative viscosity–volume fraction curve for water-in-oil emulsions [69]. Here, isoparaffinic oil (Isopar M) was used and the emulsions were prepared using an A-B-A block copolymer of PHS-PEO-PHS [Arlacel P135, supplied by ICI, where PHS refers to poly-12-hydroxystearic acid and PEO refers to poly(ethylene oxide)]. The weight average molecular weight of the polymer is 6809, while its number average is 3499. The emulsion had a narrow size distribution, with a z-average radius R of 183 nm, as determined by photon correlation spectroscopy.

As shown before [70], the viscosity–volume faction curve may be used to obtain the adsorbed layer thickness as a function of ϕ. Assuming that the W/O emulsion

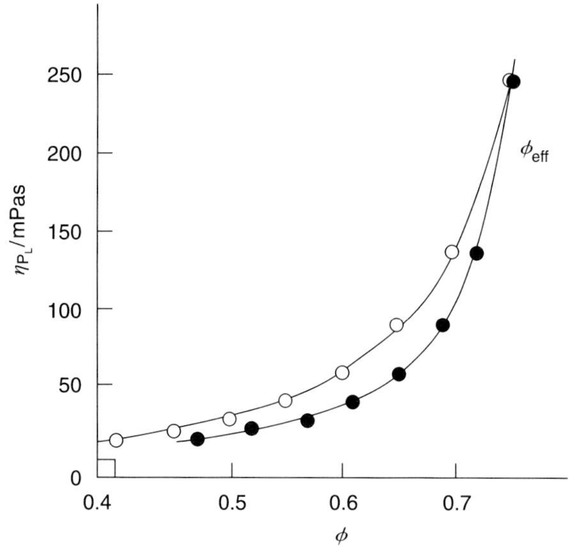

Fig. 6.39. Viscosity–volume fraction curve for W/O emulsions.

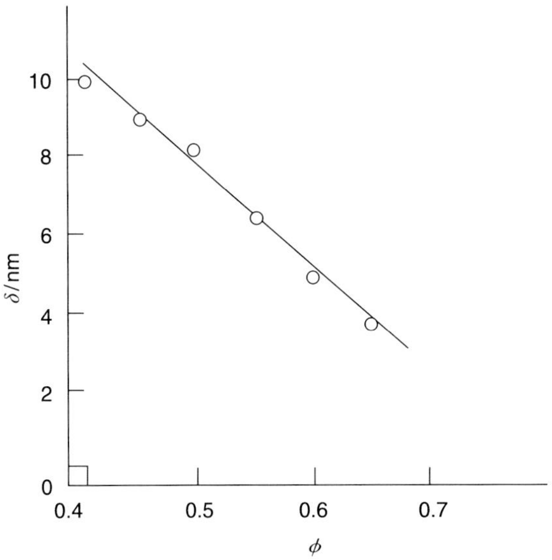

Fig. 6.40. Variation of δ with ϕ.

behaves as near hard sphere dispersions, the Dougherty–Krieger equation [67, 68] can be applied to obtain the effective volume fraction. Using Eq. (6.95), ϕ_{eff} can be calculated from η_r provided a reasonable estimate can be made of $[\eta]$ and ϕ_p. $[\eta]$ was taken to be equal to 2.5, whereas ϕ_p was estimated from a plot of $\eta^{-1/2}$ versus ϕ, as described above; ϕ_p was found to be 0.84, which is reasonable considering the polydispersity of the emulsions. Figure 6.39 shows the ϕ_{eff}s. From ϕ_{eff} and ϕ, the adsorbed layer thickness, δ, was calculated using the expression

$$\phi_{eff} = \phi\left[1 + \left(\frac{\delta}{R}\right)\right]^3 \tag{6.96}$$

The plot of δ versus ϕ in Figure 6.40 shows that δ decreases linearly with increase in ϕ. The value at $\phi = 0.4$ is 10 nm, which is a measure of the fully extended PHS chain. At such relatively low ϕ, there will be no interpenetration of the PHS chains since the distance between the droplets is relatively large. This δ obtained from rheology is in close agreement with the results recently obtained from thin liquid film measurements between two water droplets [71]. It also agrees closely with the results obtained by Ottewill and co-workers [72, 73] using compression cells and small-angle neutron scattering. The decrease in δ with increase in ϕ is similar to the results obtained using latex dispersions stabilized with grafted PEO chains [70]. This reduction in δ with increase in ϕ may be attributed to the interpenetration and/or compression of the chains on increasing ϕ. If complete interpenetration is possible, δ can be halved in dilute dispersions. Indeed, Figure 6.40 shows that δ is reduced to 4 nm at $\phi = 0.65$. This reduction in δ can also be attributed to compression of the chains on close approach, without the need of invoking any interpenetration. Probably, a combination of both mechanisms may occur on approach of the droplets in a concentrated dispersion.

6.16.2
Viscoelastic Properties of Concentrated O/W and W/O Emulsions

The viscoelastic properties of concentrated O/W and W/O emulsions have been investigated using dynamic (oscillatory) measurements. For that purpose a Bohlin VOR (Bohlin Reologie, Lund, Sweden) was used. Concentric cylinder platens were used in measurements carried out at 25 ± 0.1 °C. In oscillatory measurements, the response in stress of a viscoelastic material subjected to a sinusoidally varying strain is monitored as a function of strain amplitude and frequency. The stress amplitude is also a sinusoidally varying function in time, but for a viscoelastic material it is shifted out of phase with the strain. The phase angle shift between stress and strain, δ, is given by

$$\delta = \Delta t \omega \tag{6.97}$$

where ω is the frequency in radians s^{-1} ($\omega = 2\pi\nu$, where ν is the frequency in hertz). From measurement of the angular deflection (using a transducer) and the resulting torque on the detector shaft (the inner cylinder is connected to interchangeable torque bars) used to monitor the stress, the phase angle shift and stress and strain amplitudes (τ_0 and γ_0 respectively) are determined and one can obtain the rheological parameters G^* (complex modulus), G' (storage modulus), G'' (loss modulus) and η' (dynamic viscosity). G' is a measure of the energy stored elastically in the system (the elastic component of the complex modulus), whereas G'' is a measure of the energy dissipated as viscous flow.

In viscoelastic measurements, one measures the viscoelastic parameters as a function of strain amplitude (at a fixed frequency) to obtain the linear viscoelastic regions. The strain amplitude is gradually increased from the smallest possible value at which a measurement can be made, and the rheological parameters are monitored as a function of γ_0. Initially, the rheological parameters remain virtually constant and independent of the strain amplitude. However, above a critical value of strain amplitude (referred to as γ_{cr}), the rheological parameters show a change with further increase in γ_0. G^* and G' show a decrease with increase in γ_0 above γ_{cr}, whereas G'' usually shows an increase. This behaviour is attributed to the effect of strain amplitude on the structure of the concentrated dispersion. At $\gamma_0 < \gamma_{cr}$, the structure of the concentrated dispersion is not perturbed, whereas at $\gamma_0 > \gamma_{cr}$, some breakdown of the structure may occur. For example, in flocculated systems some flocs are broken down above γ_{cr}. Similarly, for a stable dispersion, the regular arrangement of the particles in the system is disturbed above γ_{cr}. Therefore the viscoelastic measurements have to be carried out in the linear region (i.e. at $\gamma_0 < \gamma_{cr}$) to obtain information on the structure of the system without any appreciable per-

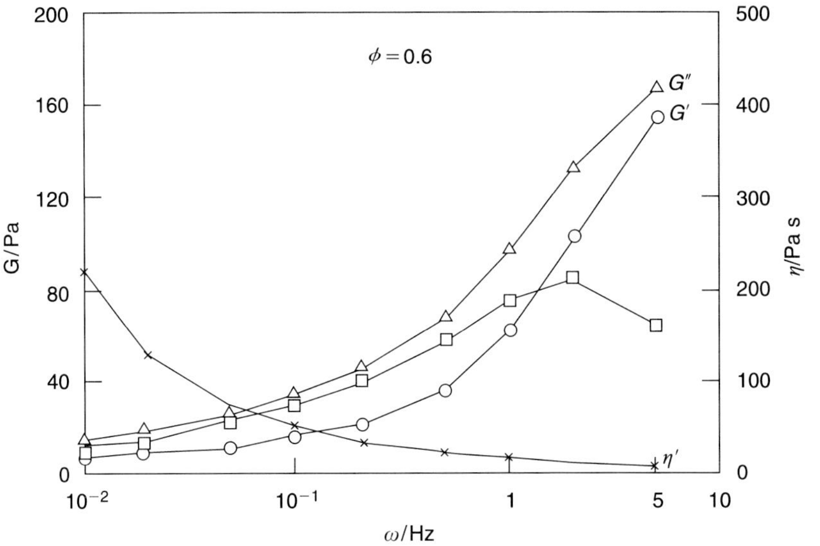

Fig. 6.41. Variation of G^*, G', G'' and η' with frequency (Hz) for an O/W emulsion, $\phi = 0.6$.

turbation of that structure. Once the linear region is established, measurements are made as a function of frequency. By fixing the frequency range, while changing the parameters of the system, such as its volume fraction, solvency of the medium for the stabilising chains, etc., one can obtain information of the interparticle interaction in the concentrated dispersion [74]. This is illustrated below.

Figure 6.41 shows typical plots of G^*, G', G'' and η' as a function of frequency (Hz) for an isoparaffinic O/W emulsion [75] with $\phi = 0.6$, which is stabilized using an ABA block copolymer of PEO-PPO-PEO (Synperonic PE), with an average of 47.3 poly(propylene oxide) (PPO) units and 41.6 poly(ethylene oxide) (PEO) units. The volume mean diameter of the droplets was 0.98 μm (as determined by the Coulter counter). Below a certain frequency $G'' > G'$, whereas above that frequency $G' > G''$ (Figure 6.41). This behaviour is typical of a viscoelastic system. In the low frequency regime, the system shows a more viscous than elastic response, since in this region (relatively long time scale) the energy dissipation is relatively more than the elastic energy stored in the system. In the high frequency regime (relatively short time scale), this energy dissipation is not significant and the system stores most of the energy, showing a predominantly elastic response. The characteristic frequency, ω^* (rad s^{-1}) at which $G' = G''$ (the cross over point) is related to the relaxation time of the system, i.e. $t_r = 1/\omega$. Thus, by carrying out oscillatory measurements as a function of frequency at various volume fractions, one can obtain the variation of relaxation time with ϕ.

As expected, ω^* shifts to lower frequencies as ϕ increases, i.e. t_r increases with increasing ϕ. This is illustrated in Figure 6.42, which clearly shows the rapid increase in t_r with ϕ when the latter exceeds 0.54. This increase in relaxation time is

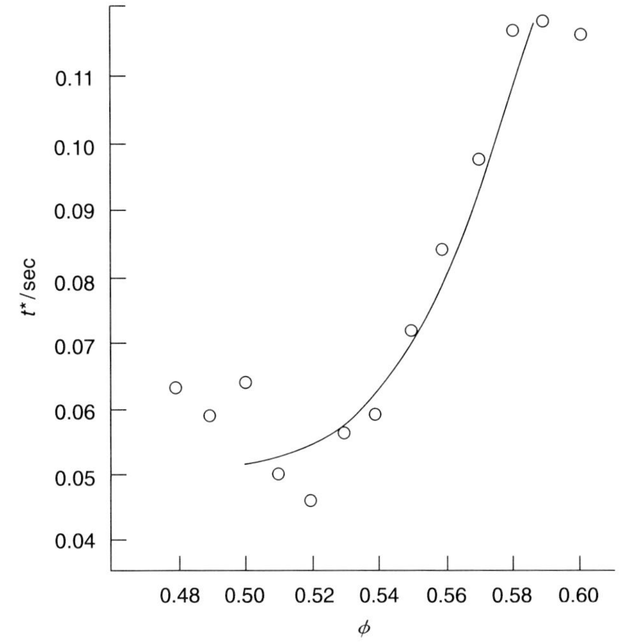

Fig. 6.42. Variation of t_r with ϕ for an O/W emulsion.

the result of the strong elastic interaction between the droplets, which increases in magnitude with increasing ϕ. In addition, by increasing ϕ, the diffusion coefficient of the droplets decreases. Both effects result in an increase in t_r.

The above trend is also observed if G^*, G' and G'' are plotted versus ϕ. Figure 6.43 illustrates this for the above emulsions at a frequency of 2 Hz. At $\phi < 0.56$, $G'' > G'$, whereas at $\phi > 0.56$, $G' > G''$. This reflects the rise in steric interaction with increase in ϕ. At $\phi < 0.56$, the droplet–droplet separation is probably larger than twice the adsorbed layer thickness and hence the adsorbed layers are not forced to overlap or compress. In this case, the repulsive interaction between the adsorbed layers is relatively weak and the emulsion shows predominantly viscous response. However, when $\phi > 0.56$, the droplet–droplet separation may become smaller than twice the adsorbed layer thickness and the chains are forced to interpenetrate and/or compress. This leads to strong steric repulsion and the emulsion shows a predominantly elastic response. The higher ϕ, the smaller the distance between the droplets and the stronger the steric interaction. This explains the rapid increase in G' as ϕ increases above 0.56 and the progressively larger value of G' relative to G''.

Similar results have been obtained for the W/O emulsions stabilized by the A-B-A block of PHS-PEO-PHS [70]. Figure 6.44 shows the variation of relaxation time t^* with ϕ, whereas Figure 6.45 shows the variation of G' and G'' (at $\omega = 1$ Hz) with ϕ. Figure 6.44 shows that t^* increases rapidly with increase ϕ, when the latter is

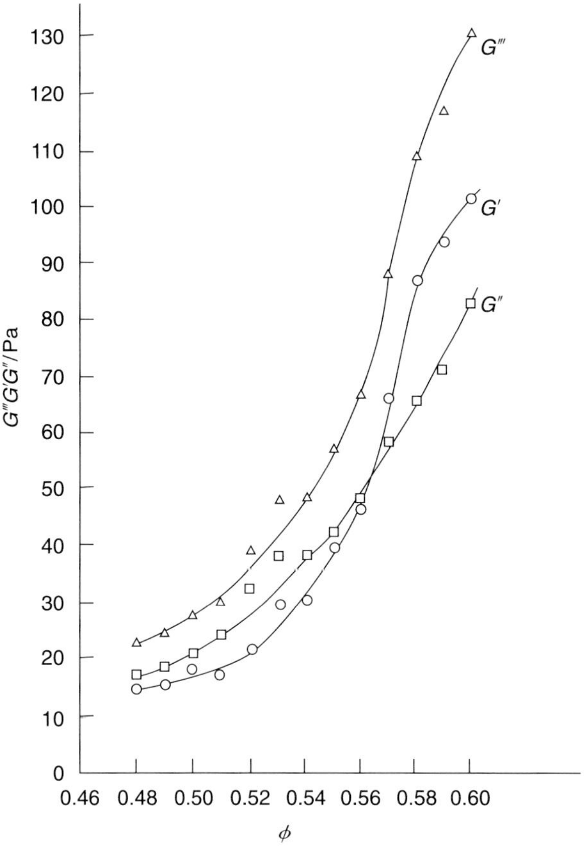

Fig. 6.43. Variation of G^*, G' and G'' (at $\omega = 2$ Hz) for an O/W emulsion.

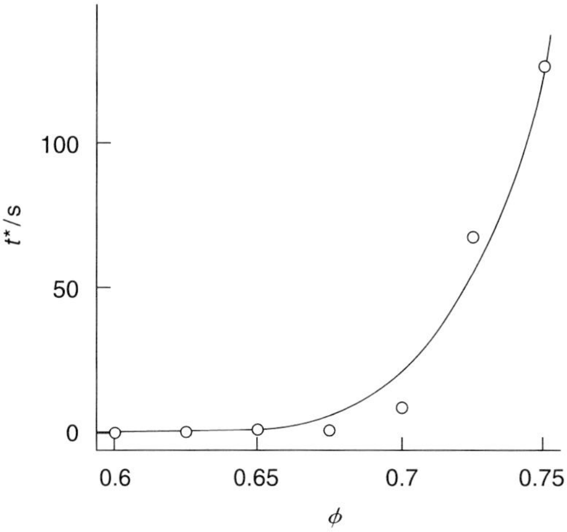

Fig. 6.44. Variation of t^* with ϕ for W/O emulsions.

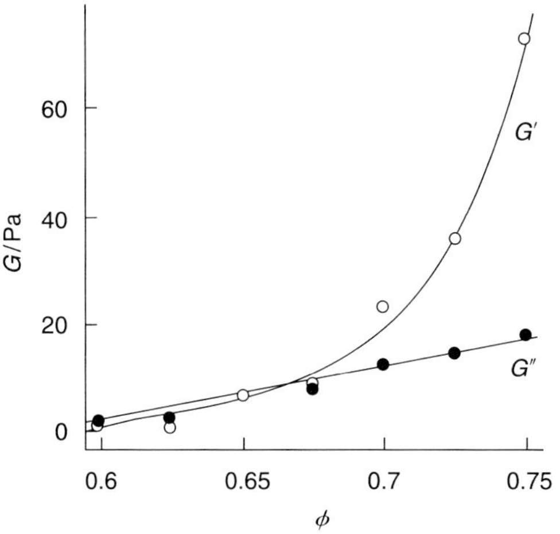

Fig. 6.45. Variation of G' (○) and G'' (●) with ϕ for W/O emulsions.

greater than 0.67. This, as mentioned above, is the result of a reduction in diffusion coefficient of the droplets and an increase in steric repulsion as the surface-to-surface separation between the droplets becomes less than twice the adsorbed layer thickness. Figure 6.44 also shows a transition from predominantly viscous to predominantly elastic response as ϕ exceeds 0.67. This is a direct manifestation of the strong elastic interaction that occurs at and above this critical ϕ.

Comparison of the results obtained using O/W versus W/O emulsions shows that the critical volume fraction, ϕ_{cr}, at which the emulsion shows change from predominantly viscous to predominantly elastic response is higher for the W/O emulsions when compared with the O/W water emulsions. Both systems are polydisperse, and the difference in droplet size distribution is unlikely to be the cause of this difference. The most likely reason is the difference in interaction between the two systems. With the W/O system, such interaction is short range and is governed by the adsorbed layer thickness, which, as mentioned above, is of the order of 5–10 nm, depending on the volume fraction. In this case, the PHS chains are in a medium of low permittivity (~2 for isoparaffinic oil) and there will be no contribution from double layers. The O/W emulsions, however, are stabilized by PEO and these chains are in an aqueous medium with a high permittivity (~78 for water). The PEO chains are relatively short (41.6 units for two A chains, i.e. ~21 EO units per chain) and, therefore, the adsorbed layer thickness is not large (probably of the order of 5 nm). However, in this case there will be some contribution from double layer repulsion (since no electrolyte was added to the system) and the interaction is longer in nature than that of the W/O system.

The effect of droplet size and its distribution and the adsorbed layer thickness may be inferred from a comparison of the results obtained with the O/W emulsions with those obtained using polystyrene latex dispersions containing grafted

PEO (with molecular weight 2000) [76]. As discussed above, the viscoelastic behaviour of the system (which reflects the steric interaction) is determined by the ratio of the adsorbed layer thickness to the particle radius (δ/R). The larger this ratio, the lower the volume fraction at which the system changes from predominantly viscous to predominantly elastic response. With relatively polydisperse systems, ϕ_{cr} shifts to higher values than for monodisperse systems with the same mean size.

6.16.3
Viscoelastic Properties of Weakly Flocculated Emulsions

A weakly flocculated emulsion may be produced by the addition of a "free" (non-adsorbing) polymer to a sterically stabilized emulsion. This has been illustrated by addition of PEO to an emulsion stabilized by Synperonic NPE 1800 [77]. The stabilizing chain in this case is also PEO and the added "free" polymer does not adsorb on the droplets. This system is, therefore, similar to polystyrene latex with grafted PEO to which "free" PEO is added [78, 79]. Above a critical volume fraction of the free polymer, ϕ_p^+ (which depends on the molecular weight), flocculation of the dispersion occurs. The origin of flocculation may be visualised from the schematic picture given in Fig. 6.28. As the free polymer concentration is gradually increased, a critical concentration is reached, whereby the polymer coils can no longer fit in between the particles or droplets. In other words, the polymer coils are "squeezed out" from between the particles or droplets. This results in a polymer-free zone between the particles and the osmotic pressure of the polymer solution outside the particles becomes higher than that in the regions between the particles (which are depleted from free polymer). This causes attraction of the particles or droplets in the dispersion, a phenomenon referred to as depletion flocculation. The free energy of attraction due to depletion is given by the following expression [80, 81],

$$G_{dep} = 2\pi R\left(\frac{\mu_i - \mu_1^\circ}{v_1}\right)\Delta^2\left(1 + \frac{2\Delta}{3R}\right) \tag{6.98}$$

where v_1 is the molecular volume of the solvent, μ_1 is the chemical potential of the solvent in the presence of the free polymer and μ_1° that before the addition of the free polymer. $(\mu_1 - \mu_1^\circ)/v_1$ is a measure of the osmotic pressure of the polymer solution. Since $\mu_1 < \mu_1^\circ$, then G_{dep} is negative (i.e., attractive). Δ is the thickness of the depletion layer, which is determined by the radius of gyration of the free polymer coil. Thus, G_{dep} has a magnitude that is determined by the osmotic pressure of the polymer solution and a range that is determined by the radius of gyration of the free polymer coil. Since G_{dep} is proportional to the osmotic pressure, it is clear that, for a given molecular weight, as the free polymer concentration is increased above ϕ_p^+, G_{dep} also increases. Since, G_{dep} is proportional to Δ^2, then the higher the molecular weight the lower the ϕ_p^+. As discussed before [72], ϕ_p^+ also decreases with increasing volume fraction of the dispersion.

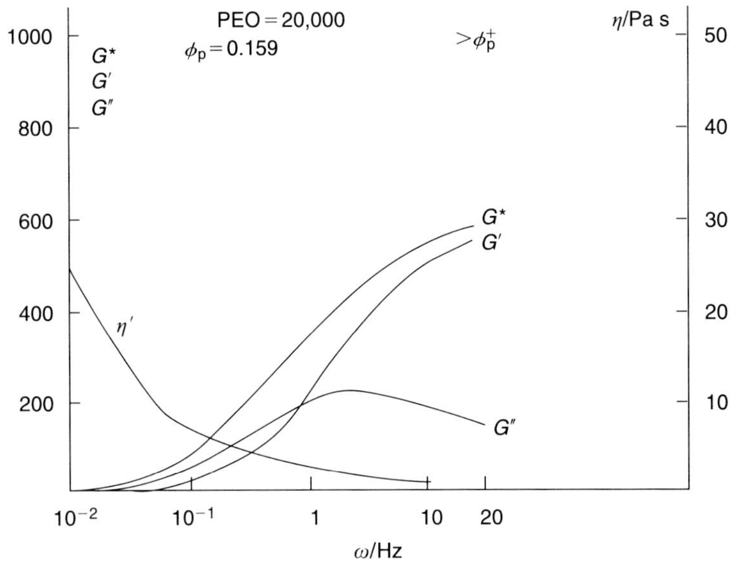

Fig. 6.46. Variation of G^*, G', G'' and η' with frequency (Hz) for an O/W emulsion ($\phi = 0.6$) with added PEO ($M_r = 20000$) at $\phi_p = 0.159$.

Weak flocculation of an emulsion, produced by the addition of a free polymer, is expected to affect the viscoelastic properties of the emulsion. This is illustrated in Figure 6.46, which shows the variation of G^*, G', G'' and η' with frequency for an emulsion with $\phi = 0.6$ and at a volume fraction of free polymer (PEO with $M = 20\,000$) that is above ϕ_p^+ (0.159).

The emulsion clearly behaves as a typical viscoelastic system. At a frequency < 1 Hz, $G'' > G'$, whereas above 1 Hz, $G' > G''$, and G' approaches G^* at high frequency. G'', however, becomes equal to G' at 1 Hz, reaches a maximum at 1–2 Hz, and then falls gradually with increasing frequency. The characteristic frequency (at which $G' = G''$) is $\sim$6 rad s^{-1}, giving a relaxation time of the order of 0.15 s. This reflects the flocculation of the emulsion, since, before addition of the free polymer, the emulsion was predominantly viscous within the whole frequency range. Similar results were obtained with the other flocculated systems. The plots of G^* and G' versus ϕ_p given in Figure 6.47 show that both G^* and G' increase rapidly at $\phi_p > \sim 0.03$. Similar trends were obtained with emulsions flocculated by PEO with $M = 35\,000$ and $90\,000$, but the free polymer concentration above which the moduli showed a rapid increase were lower. The ϕ_p^+ obtained were 0.03, 0.022 and 0.012 for PEO with $M = 20\,000, 35\,000$ and $90\,000$ respectively. These values are comparable to those obtained using polystyrene latex dispersions [78, 79]. Thus, viscoelastic measurements can be applied to study the flocculation of emulsions in the same manner as for suspensions.

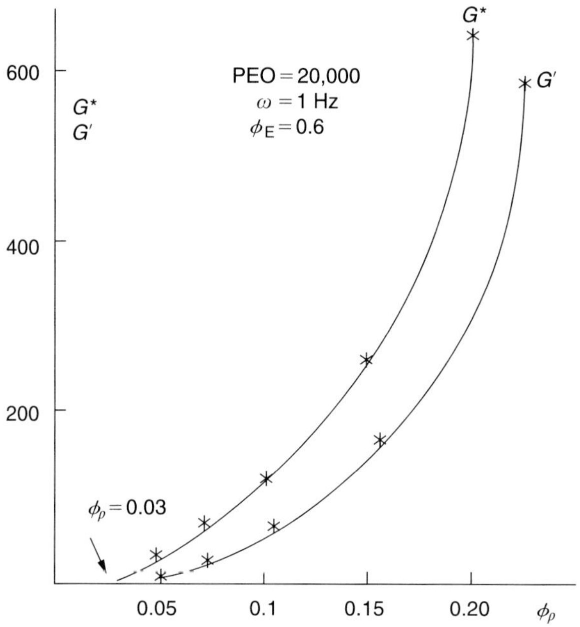

Fig. 6.47. Variation of G^* and G' with ϕ_p (PEO, $M_r = 20000$) for an O/W emulsion ($\phi = 0.6$).

6.17
Experimental Methods for Assessing Emulsion Stability

As mentioned in the introduction, the emulsion breakdown processes is far from understood at a molecular level. It is thus, necessary to develop methods of assessment of each process and to attempt to predict the long-term physical stability of emulsions.

6.17.1
Assessment of Creaming or Sedimentation

Several methods may be applied to assess the creaming or sedimentation of emulsion:

(1) Measurement of the rate by direct observation of emulsion separation using graduated cylinders that are placed at constant temperature. This method allows one to obtain the rate as well as the equilibrium cream or sediment volume.
(2) Turbidity measurements as a function of height at various times, using, for example, the Turboscan (which measures turbidity from the back scattering of near IR light).
(3) Ultrasonic velocity and absorption at various heights in the cream or sedimentation tubes.

Centrifugation may be applied to accelerate the rate of creaming or sedimentation, but one should be careful in the amount of g force that may be applied (g should not exceed the critical g force that causes deformation of the emulsion droplets and oil separation).

6.17.2
Assessment of Emulsion Flocculation

For dilute emulsions (which may be obtained by carefully diluting the concentrate in the supernatant liquid), the rate of flocculation can be determined by measuring turbidity, τ, as a function of time

$$\tau = A n_0 V_1^2 (1 + n_0 k t) \tag{6.99}$$

where A is an optical constant, n_0 is the number of droplets at time t = zero, V_1 is the volume of the droplets and k is the rate constant of flocculation.

Thus, a plot of τ versus t gives a straight line, in the initial time of flocculation, and k can be calculated from the slope of the line. Flocculation of emulsions can also be assessed by direct droplet counting using optical microscopy (with image analysis), using the Coulter counter and Photosedimentation methods.

6.17.3
Assessment of Ostwald Ripening

As mentioned above, the best procedure to follow Ostwald ripening is to plot R^3 versus time, following Eq. (6.68). This gives a straight line, from which the rate of Ostwald ripening can be calculated. In this way one can assess the effect of the various additives that may reduce Ostwald ripening, e.g. addition of highly insoluble oil and/or an oil-soluble polymeric surfactant.

6.17.4
Assessment of Coalescence

The rate of coalescence is measured by following the droplet number n or average droplet size d (diameter) as a function of time. Plots of log(droplet number) or average diameter versus time give straight lines (at least in the initial stages of coalescence), from which the rate of coalescence k can be estimated using Eq. (6.73). In this way, one can compare the different stabilisers, e.g. mixed surfactant films, liquid crystalline phase and macromolecular surfactants.

6.17.5
Assessment of Phase Inversion

The most common procedure to assess phase inversion is to measure the conductivity or the viscosity of the emulsion as a function of ϕ, increase of temperature and/or addition of electrolyte. For example, for an O/W emulsion phase, inversion to W/O is accompanied by a rapid decrease in conductivity and viscosity.

References

1 T. F. TADROS, B. VINCENT: *Encyclopedia of Emulsion Technology*, P. BECHER (ed.): Marcel Dekker, New York, 1983.
2 H. C. HAMAKER, *Physica (Utrecht)*, **1937**, *4*, 1058.
3 H. R. KRUYT: *Colloid Science*, Elsevier Science, Amsterdam, 1952, Volume I.
4 J. LYKLEMA: *Solid/Liquid Dispersions*, Th. F. TADROS (ed.): Academic Press, London, 1987.
5 B. BIJESTERBOSCH: *Solid/Liquid Dispersions*, Th. F. TADROS (ed.): Academic Press, London, 1987.
6 B. V. DERYAGUIN, L. LANDAU, *Acta Physicochem. USSR*, **1941**, *14*, 633.
7 E. J. W. VERWEY, J. T. G. OVERBEEK: *Theory of Stability of Lyophobic Colloids*, Elsevier Science, Amsterdam, 1948.
8 D. H. NAPPER, *Polymeric Stabilisation of Dispersions*, Academic Press, London, 1983.
9 P. WALSTRA: *Encyclopedia of Emulsion Technology*. P. BEECHER (ed.): Marcel Dekker, New York, 1983, Volume 1.
10 P. WALSTRA, P. E. A. SMOLDERS: *Modern Aspects of Emulsions*, B. P. BINKS (ed.): The Royal Society of Chemistry, Cambridge, 1998.
11 H. A. STONE, *Ann. Rev. Fluid. Mech.*, **1994**, *226*, 95.
12 J. A. WIERENGA, J. J. M. JANSSEN, W. G. M. AGTEROF, *Trans. Inst. Chem. Eng.*, **1996**, *74A*, 554.
13 V. G. LEVICH: *Physicochemical Hydrodynamics*, Prentice-Hall, Englewood Cliffs, 1962.
14 J. T. DAVIS: *Turbulent Phenomena*, Academic Press, London, 1972.
15 E. H. LUCASSES-REYNDERS: *Encyclopedia of Emulsion Technology*. P. BECHER (ed.): Marcel Dekker, New York, 1996.
16 D. E. GRAHAM, M. C. PHILLIPS, *J. Colloid Interface Sci.*, **1979**, *70*, 415.
17 E. H. LUCASSES-REYNDERS, *Collids Surf.*, **1994**, *A91*, 79.
18 J. LUCASSEN: *Anionic Surfactants*, E. H. LUCASSESN-REYNDERS (ed.): Marcel Dekker, New York, 1981.
19 M. VAN DEN TEMPEL, *Proc. Int. Congr. Surf. Activity*, **1960**, *2*, 573.
20 W. C. GRIFFIN, *J. Cosmet. Chem.*, **1954**, *1*, 311.
21 P. BECHER: *Nonionic Surfactants*, M. J. SCHICK (ed.): Marcel Dekker, New York, 1987, Surfactant Science Series, Volume 1.
22 L. MARZALL: *Nonionic Surfactants Editor*, M. J. SCHICK (ed.): Marcel Dekker, New York, 1987, Surfactant Science Series, Volume 23.
23 J. T. DAVIES: *Proc. Int. Congr. Surface Activity*, Academic Press, London, 1959, 426, Volume 1; J. T. DAVIES, E. K. RIDEAL, *Interfacial Phenomena*, Academic Press, New York, 1961.
24 K. SHINODA, *J. Colloid Interface Sci.*, **1967**, *25*, 396.
25 K. SHINODA, H. SAITO, *J. Colloid Interface Sci.*, **1969**, *30*, 258.
26 A. BEERBOWER, M. W. HILLS, *Am. Cosmet. Perfum.*, **1972**, *87*, 85.
27 J. H. HILDEBRAND: *Solubility of Non-Electrolytes*, 2nd edition, Reinhold, New York, 1936.
28 C. M. HANSEN, *J. Paint Technol.*, **1967**, *39*, 505.
29 A. F. M. BARTON: *Handbook of Solubility Prameters and Other Cohesive Parameters*, CRC Press, New York, 1983.
30 J. N. ISRAELACHVILI, J. N. MITCHELL, B. W. NINHAM, *J. Chem. Soc., Faraday Trans. II*, **1976**, *72*, 1525.
31 *Effect of Polymers on Dispersion Properties*, T. F. TADROS (ed.): Academic Press, London, 1982.
32 M. V. SMOLUCHOWSKI, *Z. Phys. Chem.*, **1927**, *92*, 129.
33 N. FUCHS, *Z. Physik.*, **1936**, *89*, 736.
34 H. REERINK, J. T. G. OVERBEEK, *Discuss. Faraday Soc.*, **1954**, *18*, 74.
35 W. THOMPSON (Lord Kelvin), *Phil. Mag.*, **1871**, *42*, 448.
36 A. S. KABALANOV, E. D. SHCHUKIN, *Adv. Colloid Interface Sci.*, **1992**, *38*, 69; A. S. KABALANOV, *Langmuir*, **1994**, *10*, 680.

37 I. M. Lifshitz, V. V. Slesov, *Sov. Phys. JETP*, **1959**, *35*, 331.
38 C. Wagner, *Z. Electrochem.*, **1961**, *35*, 581.
39 W. I. Higuchi, J. Misra, *J. Pharm. Sci.*, **1962**, *51*, 459.
40 P. Walstra: *Encyclopedia of Emulsion Technology*, P. Becher (ed.): Marcel Dekker, New York, 1996, Volume 4.
41 B. V. Deryaguin, R. L. Scherbaker, *Kolloid Zh.*, **1961**, *23*, 33.
42 S. Friberg, P. O. Jansson, E. Cederberg, *J. Colloid Interface Sci.*, **1976**, *55*, 614.
43 B. W. Brooks, H. N. Richmond, M. Zefra: *Modern Aspects of Emulsion Science*, P. B. Binks (ed.): The Royal Society of Chemistry, Cambridge, 2002.
44 T. F. Tadros, *Colloids Surf.*, **1994**, *91*, 39.
45 A. Einstein: *Investigations on the Theory of the Brownian Movement*, Dover, New York, 1906.
46 G. K. Batchelor, *J. Fluid Mech.*, **1972**, *52*, 245.
47 G. I. Taylor, *Proc. R. Soc.*, **1932**, *A138*, 41.
48 E. G. Richardson, *Kolloid Z.*, **1933**, *65*, 32; *J. Colloid Sci.*, **1950**, *5*, 404; **1953**, *8*, 367.
49 P. Shermam: *Emulsion Science*, P. Sherman (ed.): Academic Press, London, 1968, Chapter 4.
50 P. Sherman, *J. Colloid Sci.*, **1955**, *10*, 63.
51 P. Sherman, *J. Soc. Chem. Ind. (London)*, **1950**, *69*.
52 D. A. Edwards, H. Brenner, D. T. Wasan: *Interfacial Transport Processes and Rheology*, Butterworth-Heinemann, Oxford, 1991.
53 C. G. M. Marangoni, *Ann. Phys. (Poggendorff)*, **1871**, *3*, 337.
54 R. J. Mannheimer, R. S. Schechter, *J. Colloid Interface Sci.*, **1970**, *32*, 225.
55 D. W. Criddle: *Rheology, Theory, and Applications*. F. R. Eirich, eds., Academic Press, London, 1960, 429, Volume 3, Chapter 11.
56 D. Langevin, *J. Colloid Interface Sci.*, **1981**, *80*, 412.
57 S. K. Agrawal, D. T. Wasan, *Chem. Eng. J.*, **1979**, *18*, 215.
58 L. Wei, W. Schmidt, J. C. Slattery, *J. Colloid Interface Sci.*, **1974**, *48*, 1.
59 W. J. Phillips, R. W. Graves, R. W. Flumerfelt, *J. Colloid Interface Sci.*, **1980**, *76*, 350.
60 R. L. Bendure, *J. Colloid Interface Sci.*, **1971**, *35*, 238.
61 B. Biswas, D. A. Haydon, *Proc. R. Soc.*, **1963**, *A271*, 296.
62 E. G. Cockbain, T. S. McRoberts, *J. Colloid Sci.*, **1953**, *8*, 440.
63 A. Prince, C. Arcuri, M. van den Tempel, *J. Colloid Interface Sci.*, **1967**, *24*, 811.
64 D. T. Wasan, J. J. McNamara, S. M. Shah, K. Sampath, N. Aderangi, *J. Rheol.*, **1979**, *23*, 181.
65 J. R. Campanelli, D. G. Cooper, *J. Chem. Eng.*, **1989**, *67*, 851.
66 Th. F. Tadros, P. D. Winn, to be published.
67 I. M. Krieger, M. Dougherty, *Trans. Soc. Rheol.*, **1959**, *3*, 137.
68 I. M. Krieger, *Adv. Colloid Interface Sci.*, **1972**, *3*, 111.
69 T. F. Tadros, P. K. Thomas, to be published.
70 C. Prestidge, T. F. Tadros, *J. Colloid Interface Sci.*, **1988**, *124*, 660.
71 M. S. Aston, T. M. Herrington, T. F. Tadros, *Colloids Surf.*, **1989**, *40*, 49.
72 R. J. Cairns, R. H. Ottewill, *J. Colloid Interface Sci.*, **1976**, *54*, 45.
73 R. H. Ottewill, personal comm.
74 T. F. Tadros, *Langmuir*, **1990**, *6*, 28.
75 J. L. Cutler, Th. F. Tadros, to be published.
76 W. Liang, T. F. Tadros, P. F. Luckham, *J. Colloid Interface Sci.*, **1992**, *153*, 131.
77 T. F. Tadros, P. D. Winn, to be published.
78 C. Prestidge, T. F. Tadros, *Colloids Surf.*, **1988**, *31*, 325.
79 W. Liang, T. F. Tadros, P. F. Luckham, *J. Colloid Interface Sci.*, **1993**, *155*, 156.
80 S. Asakura, F. Oosawa, *J. Polym. Sci.*, **1958**, *33*, 245.
81 G. J. Fleer, J. H. M. H. Scheutjens, B. Vincent, *ACS Symp. Ser.*, **1984**, *240*, 245.

7 Surfactants as Dispersants and Stabilisation of Suspensions

7.1 Introduction

Surfactants are used as dispersants in solid/liquid dispersions (suspensions). For that reason, surfactants find application in almost every industrial preparation, e.g. paints, dyestuffs, paper coatings, printing inks, agrochemicals, pharmaceuticals, cosmetics, food products, detergents, ceramics, etc. The powder can be either hydrophobic, e.g. organic pigments, agrochemicals, ceramics, or hydrophilic, e.g. silica, titania, clays. The liquid can be aqueous or non-aqueous. The role of surfactants in dispersing solids in liquids can be understood from their accumulation at the solid/liquid interface (Chapter 5). It is essential to understand the process of dispersion at a fundamental level: "Dispersion is a process whereby aggregates and agglomerates of powders are dispersed into "individual" units, usually followed by a wet milling process (to subdivide the particles into smaller units) and stabilisation of the resulting dispersion against aggregation and sedimentation" [1, 2].

This next section will describe the role of surfactants in preparing solid/liquid dispersions (suspensions). Subsequent sections cover the origin of charge in suspension particles, the electrical double layer and the concept of zeta potential. The stabilisation of suspensions by surfactants both electrostatically and sterically will be briefly described; this has been described in some detail in Chapter 5. The different states of suspensions on standing and how three-dimensional structures formed can be accounted for in terms of the various interaction forces that occur between the particles are then discussed. A short section covers the rheology of suspensions and how this is affected by the presence of the surfactant. This is followed by a look at the settling of suspensions and prevention of formation of compact sediments (clays or cakes). Finally, the various procedures that may be used for characterisation of suspensions are summarised. Particular attention is given to the application of rheological techniques for the assessment and prediction of the long-term stability of suspensions. As far as possible the fundamental principles involved in each of the above processes will be described.

Applied Surfactants: Principles and Applications. Tharwat F. Tadros

ISBN: 3-527-30629-3

7.2
Role of Surfactants in Preparation of Solid/Liquid Dispersions

There are two main processes for the preparation of solid/liquid dispersions. The first depends on the "build-up" of particles from molecular units, i.e. the so-called condensation method, which involves two main processes, nucleation and growth. Here, it is necessary first to prepare a molecular (ionic, atomic or molecular) distribution of the insoluble substances; then by changing the conditions precipitation is caused, leading to the formation of nuclei that grow to the particles in question. In the second procedure, usually referred to as a dispersion process, larger "lumps" of the insoluble substances are subdivided by mechanical or other means into smaller units. The role of surfactants in the preparation of suspensions by these two methods will be described separately.

7.2.1
Role of Surfactants in Condensation Methods

7.2.1.1 Nucleation and Growth

To understand the role of surfactants in the condensation methods, it is essential to consider the major processes involved, namely nucleation and growth. Nucleation is the spontaneous appearance of a new phase from a metastable (supersaturated) solution of the material in question [3]. The initial stages of nucleation result in the formation of small nuclei where the surface-to-volume ratio is very large and hence the role of specific surface energy is very important. With the progressive increasing size of the nuclei, the ratio becomes smaller and, eventually, large crystals appear, with a corresponding reduction in the role played by the specific surface energy. As shown below, addition of surfactants can be used to control the process of nucleation and the size of the resulting nucleus.

According to Gibbs [4] and Volmer [5], the free energy of formation of a spherical nucleus, ΔG, is given by the sum of two contributions: a positive surface energy term ΔG_S which increases with increase in the radius of the nucleus (r) and a negative contribution ΔG_V due to the appearance of a new phase, which also increases with increasing r,

$$\Delta G = \Delta G_S + \Delta G_V \tag{7.1}$$

ΔG_S is given by the product of area of the nucleus and the specific surface energy (solid/liquid interfacial tension) γ; ΔG_V is related to the relative supersaturation (S/S_0),

$$\Delta G = 4\pi r^2 \gamma - \left(\frac{4\pi r^3 \rho}{3M}\right) RT \ln\left(\frac{S}{S_0}\right) \tag{7.2}$$

where ρ is the density, R is the gas constant and T is the absolute temperature.

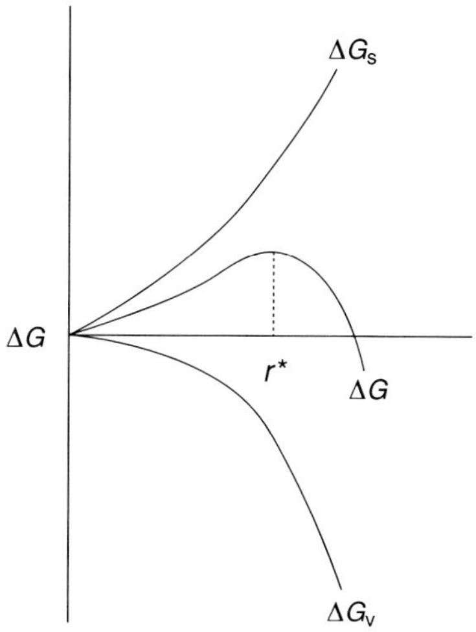

Fig. 7.1. Variation of free energy of formation of a nucleus with radius.

In the initial stages of nucleation, ΔG_S increases faster with increasing r than does ΔG_V, and ΔG remains positive, reaching a maximum at a critical radius r^*, after which it decreases and eventually becomes negative. This occurs since the second term in Eq. (7.2) rises faster with r than the first term (r^3 versus r^2). When ΔG becomes negative, growth becomes spontaneous and the clusters grow rapidly. This is illustrated in Figure 7.1, which shows the critical size of the nucleus r^* above which growth becomes spontaneous. The free energy maximum ΔG^* at the critical radius represents the barrier that has to be overcome before growth becomes spontaneous. Both r^* and ΔG^* can be obtained by differentiating Eq. (7.2) with respect to r and equating the result to zero. This gives the following expressions,

$$r^* = \frac{2\gamma M}{\rho RT \ln(S/S_0)} \tag{7.3}$$

$$\Delta G^* = \frac{16}{3} \frac{\pi \gamma^3 M^2}{(\rho RT)^2[(\ln(S/S_0)]^2} \tag{7.4}$$

Equations (7.1) to (7.4) clearly show that the free energy of formation of a nucleus and the critical radius r^* above which the cluster formation grows spontaneously depend on two main parameters, γ and (S/S_0) both of which are influenced by surfactants; γ is influenced directly, by adsorption of surfactant on the surface of the nucleus, which lowers γ and this reduces r^* and ΔG^*. In other words, spontaneous formation of clusters occurs at smaller critical radii. In addition, surfactant adsorption stabilises the nuclei against any flocculation. The presence of micelles in solution also affects the process of nucleation and growth, both directly and indirectly. The micelles can act as "nuclei" on which growth may occur. In addition, they may solublise the molecules of the material, thus affecting the relative supersaturation, which can effect both nucleation and growth.

7.2.1.2 Emulsion Polymerisation

In emulsion polymerisation, the monomer, e.g. styrene or methyl methacrylate that is insoluble in the continuous phase, is emulsified using a surfactant that adsorbs at the monomer/water interface [6]. The surfactant micelles in bulk solution solublise some of the monomer. A water-soluble initiator such as potassium persulphate ($K_2S_2O_8$) is added and this decomposes in the aqueous phase, forming free radicals that interact with the monomers to give oligomeric chains. It was long assumed that nucleation occurs in the "monomer swollen micelles", owing to the sharp increase in the rate of reaction above the critical micelle concentration and that the number of particles formed and their size depend largely extent on the nature of the surfactant and its concentration (which determines the number of micelles formed). However, this mechanism has been disputed by the suggestion that the presence of micelles means that excess surfactant is available and that molecules will readily diffuse to any interface.

The most accepted theory of emulsion polymerisation is referred to as the coagulative nucleation theory [7, 8]. A two-step coagulative nucleation model has been proposed by Napper and co-workers [7, 8]. In this process the oligomers grow by propagation, followed by a termination process in the continuous phase. A random coil is produced that is insoluble in the medium and this produces a precursor oligomer at the θ-point. The precursor particles subsequently grow primarily by coagulation to form true latex particles. Some growth may also occur by further polymerisation. The colloidal instability of the precursor particles may arise from their small size, and the slow rate of polymerisation can be due to reduced swelling of the particles by the hydrophilic monomer [7, 8]. Surfactants play a crucial role in these processes since they determine the stabilising efficiency, and the effectiveness of the surface active agent ultimately determines the number of particles formed. This was confirmed by using different surface active agents. The effectiveness of any surface active agent in stabilising the particles was the dominant factor, and the number of micelles formed was relatively unimportant.

According to the theory of Smith and Ewart [9] for the kinetics of emulsion polymerisation, the rate of propagation R_p is related to the number of particles N formed in a reaction by the equation,

$$-\frac{d[M]}{dt} = R_p k_p N n_{av}[M] \tag{7.5}$$

where [M] is the monomer concentration in the particles, k_p is the propagation rate constant and n_{av} is the average number of radicals per particle.

According to Eq. (7.5), the rate of polymerisation and the number of particles are directly related, i.e. an increase in the number of particles will increase the rate. This has been found for many polymerisations, although there are exceptions. The number of particles is related to the surfactant concentration [S] by Eq. (7.6).

$$N \approx [S]^{3/5} \tag{7.6}$$

Using the coagulative nucleation model, Napper et al. [7, 8] found that the final particle number increases with increasing surfactant concentration with a monotonically diminishing exponent. The slope of d(log N_c)/d(log t) varies from 0.4 to 1.2. At high surfactant concentration, the nucleation time will be long since the new precursor particles will be readily stabilised. As a result, more latex particles are formed and, eventually, will outnumber the very small precursor particles at long times. Precursor/particle collisions will become more frequent and fewer latex particles are produced. dN_c/dt will approach zero and at long times the number of latex particles remains constant. This shows the inadequacy of the Smith–Ewart theory, which predicts a constant exponent (3/5) at all surfactant concentrations. Thus, the coagulative nucleation mechanism has been accepted as the most probable theory for emulsion polymerisation. In all cases, the nature and concentration of surfactant used is crucial, and this is very important in the industrial preparation of latex systems.

Most reports on emulsion polymerisation have been limited to commercially available surfactants, which in many cases are relatively simple molecules, such as sodium dodecyl sulphate and simple nonionic surfactants. However, studies on the effect of surfactant structure on latex formation have revealed the importance of the structure of the molecule. As discussed in Chapter 6, block and graft copolymers (polymeric surfactants) are expected to be better stabilisers than simple surfactants. Studies on styrene polymerisation using an A-B block of polystyrene with poly(ethylene oxide) (PS-PEO), with various ratios of the molecular weight of the two blocks, showed that an optimum composition is required [10]. For efficient anchoring to the latex particles, the block length need not be more than 10 units and a PEO block with a molecular weight of 3000 is sufficient to stabilise the particles. The results also showed that using a higher molecular weight stabiliser could be counter-productive.

7.2.1.3 **Dispersion Polymerisation**

Here, the reaction mixture, consisting of monomer, initiator and solvent (aqueous or non-aqueous), is usually homogeneous; as polymerisation proceeds, polymer separates out and the reaction continues in a homogeneous manner [11]. A dispersant, sometimes referred to as "protective agent" is added to stabilise the particles once formed.

The above mechanism for the preparation of polymer particles is usually applied for preparation of non-aqueous dispersions (latex particles dispersed in a non-aqueous medium). As mentioned above, the two main criteria for this type of polymerisation are the insolubility of the formed polymer in the continuous phase and the solubility of the monomer and initiator in the dispersion medium. Initially, polymerisation starts as a homogeneous system, but after polymerisation proceeds to some extent, the insolubility of the formed polymer chains causes their precipitation. The process can be visualised as starting with the formation of polymer chains by free radical initiation, followed by formation of nuclei that then grow into polymer particles.

In the early production of non-aqueous latex dispersions, a hydrocarbon solvent was chosen as the continuous medium. However, later, mixed solvents with polar components were used. Indeed, the process of dispersion polymerisation has been applied in many cases using completely polar solvents such as alcohol, water or alcohol/water mixtures [11].

The mechanism of dispersion polymerisation has been discussed in detail in the book edited by Barrett [11]. A distinct difference between emulsion and dispersion polymerisation may be considered in terms of the rate of reaction. As mentioned above, with emulsion polymerisation the rate of reaction depends on the number of particles formed. However, with dispersion polymerisation, the rate is independent of the number of particles formed. This is to be expected, since in the latter case polymerisation initially occurs in the continuous phase, whereby both monomer and initiator are soluble, and the continuation of polymerisation after precipitation is questionable. Although in emulsion polymerisation the initial monomer initiation reaction also occurs in the continuous medium, the particles formed become swollen with the monomer and polymerisation may continue in these particles. A comparison of the rate of reaction for dispersion and solution polymerisation showed a much faster rate for the former process [11].

As mentioned above, to prevent aggregation of the formed polymer particles one needs a dispersant (polymer surfactant) that must satisfy several criteria. The most effective dispersants are those of the block (A-B or A-B-A) or graft (BA_n) type. The B chain is chosen to be insoluble in the medium and has high affinity to the surface of the polymer particles (or becomes incorporated within its matrix) (Chapter 5). This is usually referred to as the "anchor" chain. A chain(s) are chosen to be highly soluble in the medium and strongly solvated with its molecules. It should give a Flory–Huggins interaction parameter (χ) < 0.5 to ensure effective steric stabilisation.

The nature and concentration of the stabiliser determines the number of particles formed in dispersion polymerisation. In general, increasing dispersant concentration increases the number of particles formed (at any given monomer content), i.e. smaller latex particles are produced. This is not surprising, since smaller particles have larger surface areas, requiring a higher dispersant concentration.

The particles in dispersion polymerisation were considered to be formed by two main steps [10]: (1) initiation of monomer in the continuous phase and subsequent growth of the oligomeric chains until insolubility occurs; (2) grown oligomeric chains associate to form aggregates, which below a certain critical size are unstable but gain stability through dispersant adsorption. However, several other processes may take place, e.g. homocoagulation (collision with other precursor particles), growth by propagation, adsorption of stabiliser and swelling by monomer. Notably, however, that the number of particles in the final latex cannot be dependent on particle nucleation only, since there is another step involved that determines how many of the precursor particles created are involved in the formation of one colloidally stable particle. This step depends on the nature of the stabiliser and how many particles have to heterocoagulate to decrease the total surface area to a size that the stabiliser in the system can stabilise.

7.2.2
Role of Surfactants in Dispersion Methods

Dispersion methods are used to prepare suspensions of preformed particles. The term dispersion is used to refer to the complete process of incorporating the solid into a liquid such that the final product consists of fine particles distributed throughout the dispersion medium. The role of surfactants (or polymers) in the dispersion can be by considering the stages involved [1]. Three stages have been considered [3]: wetting of the powder by the liquid, breaking of the aggregates and agglomerates, and comminution (milling) of the resulting particles into smaller units. These three stages are considered below.

7.2.2.1 Powder Wetting

Wetting is a fundamental process in which one fluid phase is displaced completely or partially by another fluid phase from the surface of a solid. A useful parameter to describe wetting is the contact angle θ of a liquid drop on a solid substrate. If the liquid makes no contact with the solid, i.e. $\theta = 180^\circ$, the solid is referred to as non-wettable by the liquid in question. This may be the case for a perfectly hydrophobic surface with a polar liquid such as water. However, when $180^\circ > \theta > 90^\circ$, one may refer to a case of poor wetting. When $0^\circ < \theta < 90^\circ$, partial (incomplete) wetting is the case, whereas when $\theta = 0^\circ$ complete wetting occurs and the liquid spreads on the solid substrate, forming a uniform liquid film. The cases of partial and complete wetting are schematically shown in Figure 7.2, for a liquid on a perfectly smooth solid substrate.

The utility of contact angle measurements depends on equilibrium thermodynamic arguments (static measurements) using the well-known Young's equation (Eq. 7.12). The value depends on: (1) The history of the system; (2) whether the liquid is tending to advance across or recede from the solid surface (advancing angle θ_A, receding angle θ_R; usually $\theta_A > \theta_R$).

Under equilibrium, the liquid drop takes the shape that minimises the free energy of the system. Three interfacial tensions can be identified: γ_{SV}, solid/vapour area A_{SV}; γ_{SL}, solid/liquid area A_{SL}; γ_{LV}, liquid/vapour area A_{LV}. Figure 7.3 gives

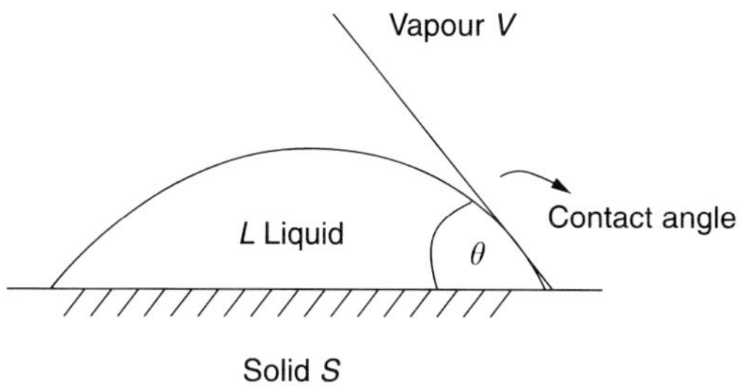

Fig. 7.2. Schematic of complete and incomplete wetting.

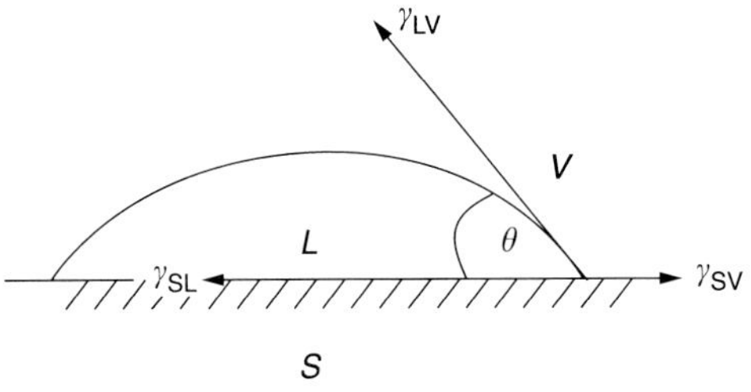

Fig. 7.3. Representation of the contact angle and wetting line.

a schematic representation of the balance of tensions at the solid/liquid/vapour interface. The contact angle is that formed between the planes tangent to the surfaces of the solid and liquid at the wetting perimeter. Here, solid and liquid are simultaneously in contact with each other and the surrounding phase (air or vapour of the liquid). The wetting perimeter is referred to as the three-phase line or wetting line. In this region, vapour, liquid and solid are in equilibrium.

$\gamma_{SV}A_{SV} + \gamma_{SL}A_{SL} + \gamma_{LV}A_{LV}$ should be a minimum at equilibrium, leading to the well-known Young's equation,

$$\gamma_{SV} = \gamma_{SL} + \gamma_{LV} \cos \theta \tag{7.7}$$

$$\cos \theta = \frac{\gamma_{SV} - \gamma_{SL}}{\gamma_{LV}} \tag{7.8}$$

The contact angle θ depends on the balance between the solid/vapour (γ_{SV}) and solid/liquid (γ_{SL}) interfacial tensions. The angle that a drop assumes on a solid surface is the result of the balance between the adhesion force between solid and liquid and the cohesive force in the liquid,

$$\gamma_{LV} \cos \theta = \gamma_{SV} - \gamma_{SL} \tag{7.9}$$

If there is no interaction between solid and liquid,

$$\gamma_{SL} = \gamma_{SV} + \gamma_{LV} \tag{7.10}$$

i.e., $\cos \theta = -1$ or $\theta = 180°$.

If there is strong interaction between solid and liquid (maximum wetting), the latter spreads until Young's equation is satisfied,

$$\gamma_{LV} = \gamma_{SV} - \gamma_{SL} \tag{7.11}$$

i.e., $\cos \theta = 1$ or $\theta = 0°$; the liquid is said to spread spontaneously on the solid surface.

When the surface of the solid is in equilibrium with the liquid vapour, we can consider the spreading pressure π (Figure 7.4).

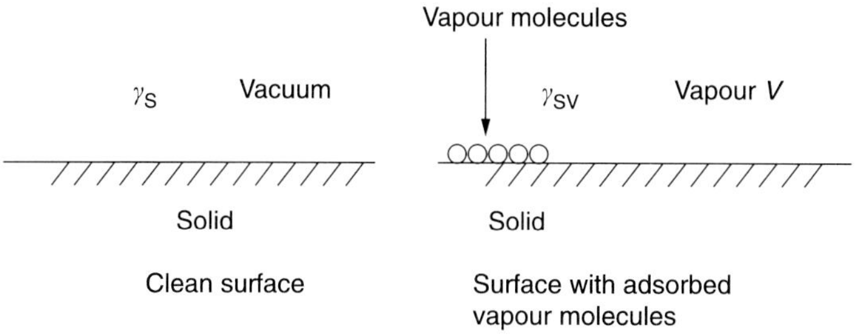

Fig. 7.4. Scheme of the spreading pressure.

The solid surface tension is lowered as a result of adsorption of vapour molecules,

$$\pi = \gamma_s - \gamma_{SV} \tag{7.12}$$

Young's equation can be written as:

$$\gamma_{LV} \cos\theta = \gamma_S - \gamma_{SL} - \pi \tag{7.13}$$

Adhesion Tension There is no direct way of measuring γ_{SV} or γ_{SL}. The difference between γ_{SV} and γ_{SL} can be obtained from contact angle measurements ($= \gamma_{LV} \cos\theta$). This difference is referred to as the wetting tension or adhesion tension

$$\text{Adhesion tension} = \gamma_{SV} - \gamma_{SL} = \gamma_{LV} \cos\theta \tag{7.14}$$

Gibbs defined the adhesion tension τ as the difference between the surface pressure of the solid/liquid and that between the solid/vapour interface,

$$\tau = \pi_{SL} - \pi_{SV} \tag{7.15}$$

$$\pi_{SV} = \gamma_s - \gamma_{SV} \tag{7.16}$$

$$\pi_{SL} = \gamma_s - \gamma_{SL} \tag{7.17}$$

$$\tau = \gamma_{SV} - \gamma_{SL} = \gamma_{LV} \cos\theta \tag{7.18}$$

Work of Adhesion, W_a The work of adhesion is a direct measure of the free energy of interaction between solid and liquid (Figure 7.5).

$$W_a = (\gamma_{LV} + \gamma_{SV}) - \gamma_{SL} \tag{7.19}$$

Using Young's equation,

$$W_a = \gamma_{LV} + \gamma_{SV} - \gamma_{LV} \cos\theta = \gamma_{LV}(\cos\theta + 1) \tag{7.20}$$

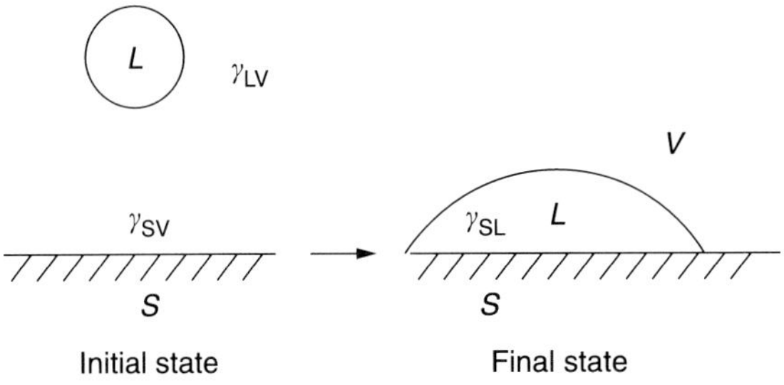

Fig. 7.5. Schematic of the work of adhesion.

The work of adhesion depends on γ_{LV}, the liquid/vapour surface tension and θ, the contact angle between liquid and solid.

The work of cohesion W_c is the work of adhesion when the two surfaces are the same. Consider a liquid cylinder with unit cross sectional area (Figure 7.6).

$$W_c = 2\gamma_{LV} \tag{7.21}$$

For adhesion of a liquid on a solid, $W_a \sim W_c$ or $\theta = 0°$ ($\cos\theta = 1$).

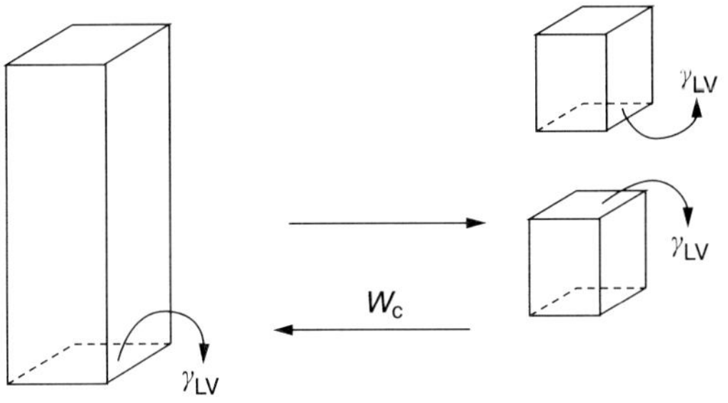

Fig. 7.6. Schematic of the work of cohesion.

Spreading Coefficient *S* Harkins [12] defined the spreading coefficient as the work required to destroy a unit area of SL and LV and leave a unit area of bare solid SV (Figure 7.7).

The spreading coefficient S = surface energy of final state – surface energy of the initial state.

$$S = \gamma_{SV} - (\gamma_{SL} + \gamma_{LV}) \tag{7.22}$$

Using Young's equation,

$$\gamma_{SV} = \gamma_{SL} + \gamma_{LV}\cos\theta \tag{7.23}$$

$$S = \gamma_{LV}(\cos\theta - 1) \tag{7.24}$$

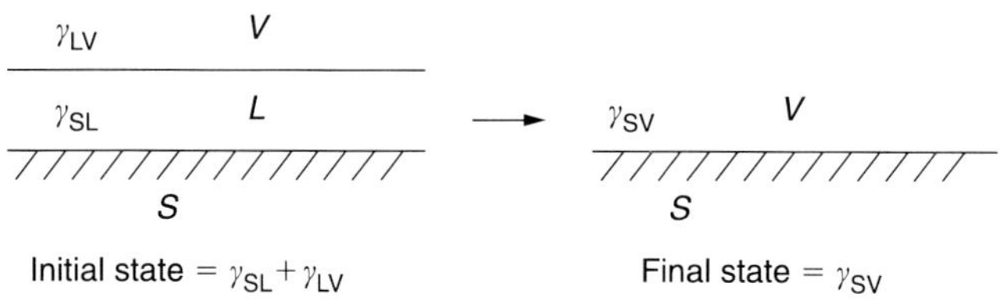

Fig. 7.7. Representation of the spreading coefficient.

If S is zero (or positive), i.e. $\theta = 0°$, the liquid will spread until it completely wets the solid. If S is negative, i.e. $\theta > 0°$, only partial wetting occurs. Alternatively, one can use the equilibrium (final) spreading coefficient.

For dispersion of powders into liquids, one usually requires complete spreading, i.e. θ should be zero.

Contact Angle Hysteresis For a liquid spreading on a uniform, non-deformable solid (idealised case), there is only one contact angle – the equilibrium value. With real systems (practical solids) several stable contact angles can be measured. Two relatively reproducible angles can be measured: largest – advancing angle θ_A; smallest – receding angle θ_R (Figure 7.8). θ_A is measured by advancing the periphery of a drop over a surface (e.g. by adding more liquid to the drop). θ_R is measured by pulling the liquid back. The difference $\theta_A - \theta_R$ is referred to as contact angle hysteresis.

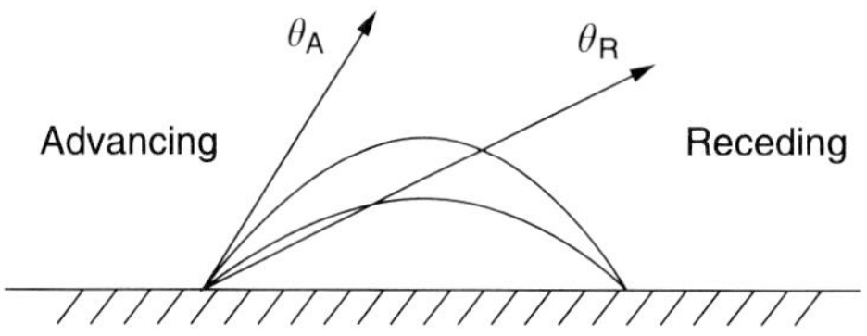

Fig. 7.8. Schematic of advancing and receding contact angles.

Reasons for Hysteresis

(1) Penetration of wetting liquid into pores during advancing contact angle measurements.
(2) Surface roughness: The first and rear edges both meet the liquid with some intrinsic angle θ_0 (microscopic contact angle). The macroscopic angles θ_A and θ_R vary significantly. This is best illustrated for a surface inclined at an angle α from the horizontal (Figure 7.9).

θ_0s are determined by contact of the liquid with the "rough" valleys (microscopic contact angle). Both θ_A and θ_R are determined by contact of liquid with arbitrary

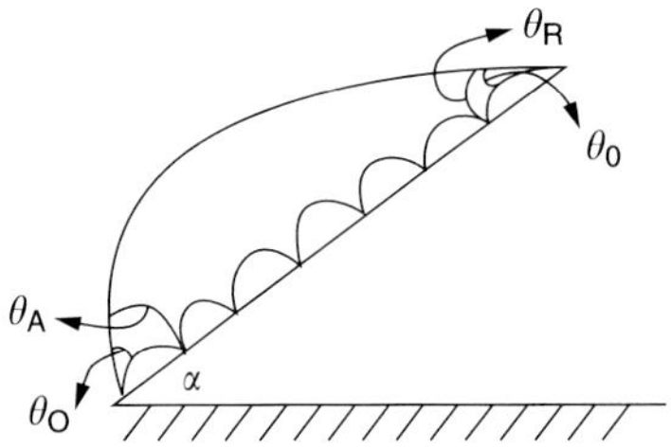

Fig. 7.9. Representation of contact angle hysteresis.

parts on the surface (peak or valley). Surface roughness can be accounted for by comparing the "real" area of the surface (A) with the apparent one,

$$r = \frac{A}{A'} \tag{7.25}$$

A = area of surface taking into account all peaks and valleys. A' = apparent area (same macroscopic dimension); $r > 1$.

$$\cos\theta = r\cos\theta_0 \tag{7.26}$$

Wenzel's Equation

θ = macroscopic contact angle
θ_0 = microscopic contact angle

$$\cos\theta = r\left[\frac{(\gamma_{SV} - \gamma_{SL})}{\gamma_{LV}}\right] \tag{7.27}$$

If $\cos\theta$ is negative on a smooth surface ($\theta > 90°$) it becomes more negative on a rough surface (θ is larger) and surface roughness reduces wetting. If $\cos\theta$ is positive on a smooth surface ($\theta < 90°$) it becomes more positive on a rough surface (θ is smaller) and roughness enhances wetting.

Surface Heterogeneity Most practical surfaces are heterogeneous, consisting of "islands" or "patches" with different surface energies. As the drop advances on such a surface, its edge tends to stop at the boundary of the "island". The advancing angle is associated with the intrinsic angle of the high contact angle region. The receding angle will be associated with the low contact angle region. If the heterogeneities are very small compared with the dimensions of the liquid drop, one can define a composite contact angle using Cassie's equation,

$$\cos\theta = Q_1\cos\theta_1 + Q_2\cos\theta_2 \tag{7.28}$$

Q_1 = fraction of surface having contact angle θ_1; Q_2 = fraction of surface having contact angle θ_2. Both θ_1 and θ_2 are the maximum and minimum possible angles.

Critical Surface Tension of Wetting Fox and Zisman introduced a systematic way of characterising the "wettability" of a surface [13]. For a given substrate and for a series of related liquids (e.g. n-alkanes, siloxanes and dialkyl ethers) a plot of $\cos\theta$ versus γ_{LV} gives a straight line (Figure 7.10).

Extrapolation of the straight line to $\cos\theta = 1$ ($\theta = 0$) gives the critical surface tension of wetting γ_c. Any liquid with $\gamma_{LV} < \gamma_c$ will give $\theta = 0$, i.e. it wets the surface completely; γ_c is the surface tension of a liquid that just spreads on the substrate to give complete wetting.

The above linear relationship can be represented by the following empirical equation,

$$\cos\theta = 1 + b(\gamma_{LV} - \gamma_c) \tag{7.29}$$

High-energy solids, e.g. glass, give high γ_c (> 40 mN m^{-1}). Low energy solids, e.g. hydrophobic surfaces, give lower γ_c ($\sim$30 mN m^{-1}). Very low energy solids such as Teflon [poly(tetrafluoroethylene), PTFE] give still lower γ_c (< 25 mN m^{-1}).

The above concept explains the degree of wettability of the above surfaces with water; glass being the easiest to wet and Teflon being the most difficult.

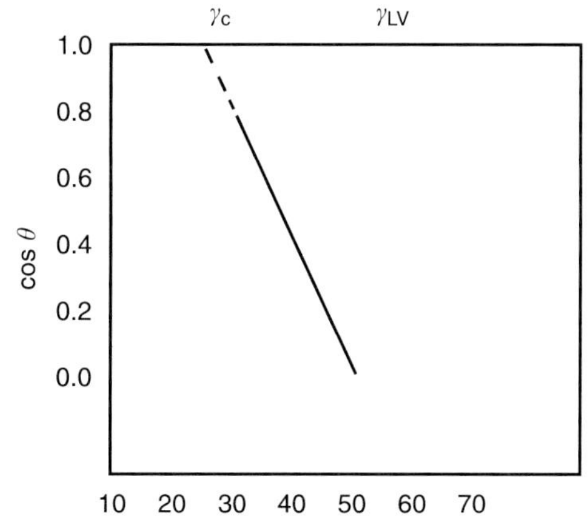

Fig. 7.10. Schematic of the critical surface tension of wetting.

7.3
Effect of Surfactant Adsorption

Surfactants lower the surface tension of water, γ, and they adsorb at the solid/liquid interface. A plot of γ_{LV} versus $\log C$ (where C is the surfactant concentration)

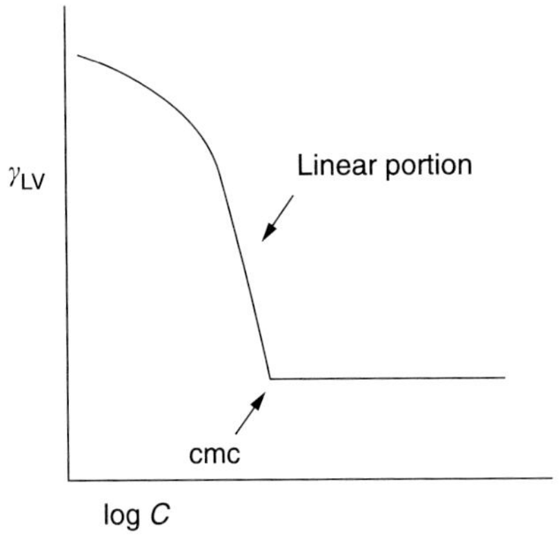

Fig. 7.11. Representative γ versus log C curve for surfactants.

results in a gradual reduction in γ_{LV}, followed by a linear decrease of γ_{LV} with log C (just below the critical micelle concentration, c.m.c.), and when the c.m.c. is reached γ_{LV} remains virtually constant (Figure 7.11).

From the slope of the linear portion of the $\gamma - \log C$ curve (just below the c.m.c.), one can obtain the surface excess (number of moles of surfactant per unit area at the L/A interface). Using the Gibbs adsorption isotherm,

$$\frac{d\gamma}{d \log C} = -2.303 RT\Gamma \tag{7.30}$$

Γ = surface excess (mol m^{-2}), R = gas constant, T = absolute temperature.

From Γ one can obtain the area per molecule,

$$\text{Area per molecule} = \frac{1}{\Gamma N_{av}} (m^2) = \frac{10^{18}}{\Gamma N_{av}} (nm^2) \tag{7.31}$$

Most surfactants produce a vertically oriented monolayer just below the c.m.c.

The area per molecule is usually determined by the cross sectional area of the head group. For ionic surfactants containing say $-OSO_3^-$ or $-SO_3^-$ head groups, the area per molecule is in the region of 0.4 nm^2. For nonionic surfactants containing several moles of ethylene oxide (8–10), the area per molecule can be much larger (1–2 nm^2). Surfactants will also adsorb at the solid/liquid interface. For hydrophobic surfaces, the main driving force for adsorption is by hydrophobic bonding. This results in lowering of the contact angle of water on the solid surface. For hydrophilic surfaces, adsorption occurs via the hydrophilic group, e.g. cationic surfactants on silica. Initially, the surface becomes more hydrophobic and the contact angle θ increases with increasing surfactant concentration. However, at higher cationic surfactant concentrations, a bilayer is formed by hydrophobic interaction between the alkyl groups, and the surface becomes more and more hydrophilic until, eventually, the contact angle reaches zero at high surfactant concentrations.

Smolders [14] suggested the following relationship for change of θ with C,

$$\frac{d\gamma_{LV}\cos\theta}{d\ln C} = \frac{d\gamma_{SV}}{d\ln C} - \frac{d\gamma_{SL}}{d\ln C} \tag{7.32}$$

Using the Gibbs equation,

$$\sin\theta\left(\frac{d\gamma}{d\ln C}\right) = RT(\Gamma_{SV} - \Gamma_{SL} - \gamma_{LV}\cos\theta) \tag{7.33}$$

since $\gamma_{LV}\sin\theta$ is always positive, then $(d\theta/d\ln C)$ will always have the same sign as the right-hand side of Eq. (7.33). Three cases may be distinguished: $(d\theta/d\ln C) < 0$, $\Gamma_{SV} < \Gamma_{SL} + \Gamma_{LV}\cos\theta$; addition of surfactant improves wetting; $(d\theta/d\ln C) = 0$, $\Gamma_{SV} = \Gamma_{SL} + \Gamma_{LV}\cos\theta$; surfactant has no effect on wetting; $(d\theta/d\ln C) > 0$, $\Gamma_{SV} > \Gamma_{SL} + \Gamma_{LV}\cos\theta$; surfactant causes dewetting.

7.4 Wetting of Powders by Liquids

Wetting of powders by liquids is very important in their dispersion, e.g. in the preparation of concentrated suspensions. The particles in a dry powder form either aggregate or agglomerate (Figure 7.12).

It is essential in the dispersion process to wet both external and internal surfaces and displace the air entrapped between the particles. Wetting is achieved by the use of surface active agents (wetting agents) of the ionic or nonionic type that can diffuse quickly (i.e. lower the dynamic surface tension) to the solid/liquid interface and displace the air entrapped by rapid penetration through the channels between the particles and inside any "capillaries". For wetting of hydrophobic powders into

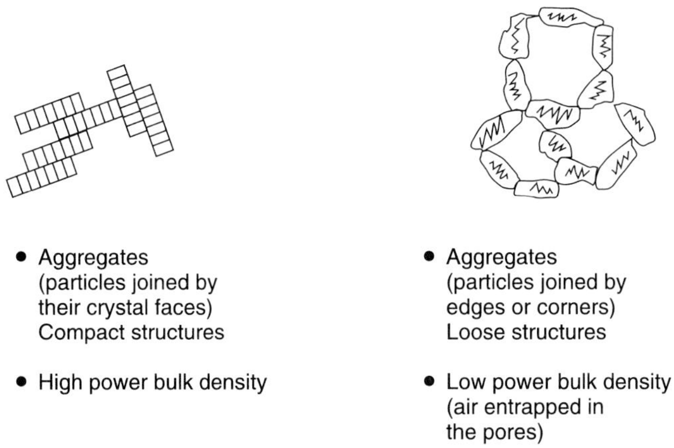

Fig. 7.12. Schematic of aggregates and agglomerates.

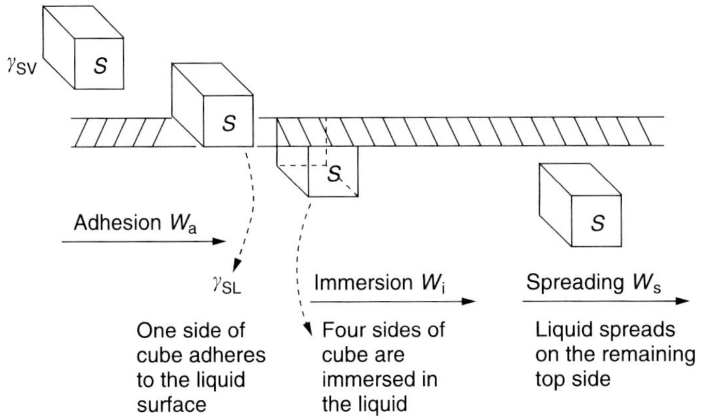

Fig. 7.13. Scheme showing the wetting of a cube of solid.

water, anionic surfactants, e.g. alkyl sulphates or sulphonates or nonionic surfactants of the alcohol or alkyl phenol ethoxylates are usually used.

A useful concept for choosing wetting agents of the ethoxylated surfactants is the hydrophilic–lipophilic balance (HLB) concept,

$$\text{HLB} = \frac{\%\text{ of hydrophilic groups}}{5} \tag{7.34}$$

Most wetting agents of this class have an HLB number in the range 7–9.

The process of wetting of a solid by a liquid involves three types of wetting: Adhesion wetting, W_a; immersion wetting W_i; spreading wetting W_s. This can be illustrated by considering a cube of solid, each side with a unit area of (Figure 7.13).

In every step one can apply the Young's equation,

$$\gamma_{SV} = \gamma_{SL} + \gamma_{LV} \cos\theta \tag{7.35}$$

$$W_a = \gamma_{SL} - (\gamma_{SV} + \gamma_{LV}) = -\gamma_{LV}(\cos\theta + 1) \tag{7.36}$$

$$W_i = 4\gamma_{SL} - 4\gamma_{SV} = -4\gamma_{LV} \cos\theta \tag{7.37}$$

$$W_s = (\gamma_{SL} + \gamma_{LV}) - \gamma_{SV} = -\gamma_{LV}(\cos\theta - 1) \tag{7.38}$$

The work of dispersion W_d is the sum of W_a, W_i and W_s,

$$W_d = W_a + W_i + W_s = 6\gamma_{SV} - \gamma_{SL} = -6\gamma_{LV} \cos\theta \tag{7.39}$$

Wetting and dispersion depends on γ_{LV} (liquid surface tension) and the contact angle θ between liquid and solid. W_a, W_i and W_s are spontaneous when $\theta < 90°$. W_d is spontaneous when $\theta = 0$. Since surfactants are added in sufficient amounts ($\gamma_{dynamic}$ is lowered sufficiently) spontaneous dispersion is the rule rather than the exception.

Wetting of the internal surface requires penetration of the liquid into channels between and inside the agglomerates. The process is similar to forcing a liquid through fine capillaries. To force a liquid through a capillary with radius r, a pressure p is required that is given by

$$p = -\frac{2\gamma_{LV}\cos\theta}{r} = \left[\frac{-2(\gamma_{SV} - \gamma_{SL})}{r\gamma_{LV}}\right] \tag{7.40}$$

γ_{SL} has to be made as small as possible; rapid surfactant adsorption to the solid surface, low θ. When $\theta = 0$, $p \propto \gamma_{LV}$. Thus, for penetration into pores one requires a high γ_{LV}. Wetting of the external surface requires a low contact angle (θ) and low surface tension γ_{LV}. Wetting of the internal surface (i.e. penetration through pores) requires low θ but high γ_{LV}. These two conditions are incompatible and a compromise has to be made: $\gamma_{SV} - \gamma_{SL}$ must be kept at a maximum; γ_{LV} should be kept as low as possible but not too low.

The above conclusions illustrate the problem of choosing the best dispersing agent for a particular powder. This requires measurement of the above parameters as well as testing the efficiency of the dispersion process.

7.5 Rate of Penetration of Liquids

7.5.1 Rideal–Washburn Equation

For horizontal capillaries (gravity neglected), the depth of penetration l in time t is given by the Rideal–Washburn equation [15, 16],

$$l = \left[\frac{rt\gamma_{LV}\cos\theta}{2\eta}\right]^{1/2} \tag{7.41}$$

To enhance the rate of penetration, γ_{LV} has to be made as high as possible, and both θ and η as low as possible.

For dispersion of powders into liquids one should use surfactants that lower θ while not reducing γ_{LV} too much. The viscosity of the liquid should also be kept at a minimum. Thickening agents (such as polymers) should not be added during the dispersion process. It is also necessary to avoid foam formation during the dispersion.

For a packed bed of particles, r may be replaced by K, which contains the effective radius of the bed and a turtuosity factor, which takes into account the complex path formed by the channels between the particles, i.e.,

$$l^2 = \frac{kt\gamma_{LV}\cos\theta}{2\eta} \tag{7.42}$$

Thus a plot of l^2 versus t gives a straight line, from the slope of which one can obtain θ.

The Rideal–Washburn equation can be applied to obtain the contact angle of liquids (and surfactant solutions) in powder beds. K should first be obtained using a liquid that produces zero contact angle. This is discussed below.

7.5.2 Measurement of Contact Angles of Liquids and Surfactant Solutions on Powders

A packed bed of powder is prepared, say, in a tube fitted with a sintered glass at the end (to retain the powder particles). The powder must be packed uniformly in the tube (a plunger may be used). The tube containing the bed is immersed in a liquid that gives spontaneous wetting (e.g. a lower alkane), i.e. the liquid gives a zero contact angle and $\cos\theta = 1$. By measuring the rate of penetration of the liquid (this can be carried out gravimetrically using, for example, a microbalance or a Kruss instrument) one can obtain K. The tube is then removed from the lower alkane liquid and left to stand for evaporation of the liquid. It is then immersed in the liquid in question and the rate of penetration is measured again as a function of time. Using Eq. (7.27), one can calculate $\cos\theta$ and hence θ.

7.6 Structure of the Solid/Liquid Interface

7.6.1 Origin of Charge on Surfaces

A great variety of processes occur to produce a surface charge.

7.6.1.1 Surface Ions

These are ions that have such a high affinity for the surface of the particles that they may be taken as part of the surface, e.g. Ag^+ and I^- for AgI. For AgI in a solution of KNO_3, the surface charge σ_0 is given by the expression

$$\sigma_0 = F(\Gamma_{Ag^+} - \Gamma_{I^-}) = F\Gamma_{AgNO_3} - \Gamma_{KI} \qquad (7.43)$$

where F is the Faraday constant (96 500 C mol^{-1}) and Γ is the surface excess of ions (mol m^{-2}).

Similarly, for an oxide such as silica or alumina in KNO_3, H^+ and OH^- may be taken as part of the surface,

$$\sigma_0 = F(\Gamma_{H^+} - \Gamma_{OH^-}) = F(\Gamma_{HCl} - \Gamma_{KOH}) \qquad (7.44)$$

The ions that determine the charge on the surface are termed potential-determining ions.

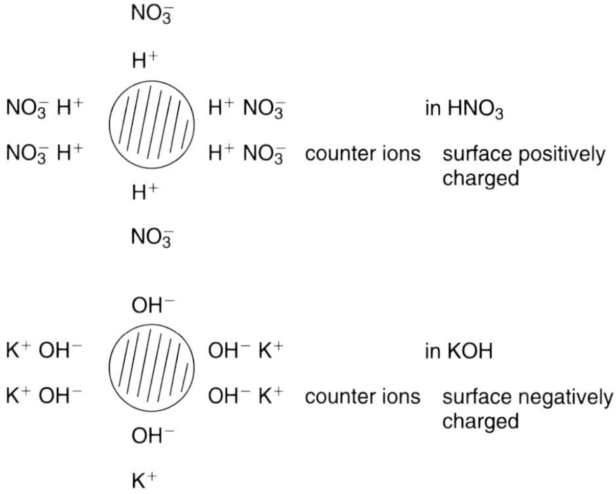

Fig. 7.14. Schematic of an oxide surface.

Consider an oxide surface (Figure 7.14).

The charge depends on the pH of the solution: Below a certain pH the surface is positive and above a certain pH it is negative. At a specific pH ($\Gamma_H = \Gamma_{OH}$) surface is uncharged; this is referred to as the point of zero charge (p.z.c.).

The p.z.c. depends on the type of the oxide: For an acidic oxide such as silica it is ca. pH 2–3. For a basic oxide such as alumina p.z.c. is ~pH 9. For an amphoteric oxide such as titania the p.z.c. is ~pH 6.

In some cases, specifically adsorbed ions (that have non-electrostatic affinity to the surface) "enrich" the surface but may not be considered as part of the surface, e.g. bivalent cations on oxides, cationic and anionic surfactants on most surfaces [17].

7.6.1.2 **Isomorphic Substitution**

This can be achieved with, for example, sodium montmorillonite, i.e. by replacement of cations inside the crystal structure by cations of lower valency, e.g. Si^{4+} with Al^{3+}. The deficit of one positive charge gives one negative charge. The surface of Na montmorillonite is negatively charged with Na^+ as counter ions (Figure 7.15). The surface charge + counter ions form the electrical double layer.

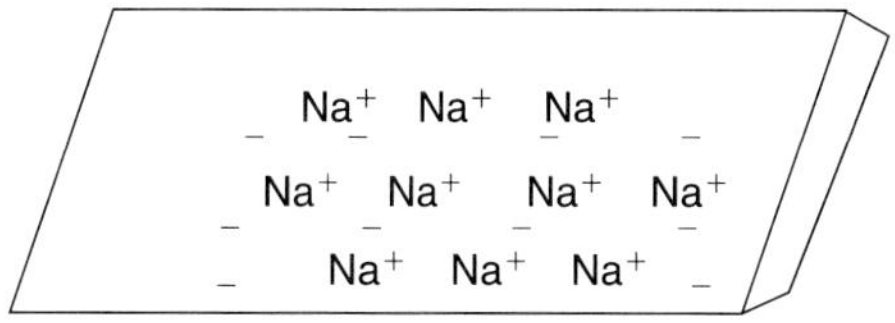

Fig. 7.15. Schematic of a clay particle.

7.7
Structure of the Electrical Double Layer

7.7.1
Diffuse Double Layer (Gouy and Chapman)

The surface charge σ_0 is compensated by an unequal distribution of counter ions (opposite in charge to the surface) and co-ions (same sign as the surface) that extend to some distance from the surface [17]. Figure 7.16 shows this schematically. The potential decays exponentially with distance x. At low potentials,

$$\Psi = \Psi_0 \exp -(\kappa x) \tag{7.45}$$

Note that when $x = 1/\kappa$, $\Psi_x = \Psi_0/e$; $1/\kappa$ is referred to as the "thickness" of the "double layer".

The double layer extension depends on electrolyte concentration and valency of the counter ions,

$$\left(\frac{1}{\kappa}\right) = \left(\frac{\varepsilon_r \varepsilon_0 kT}{2n_0 Z_i^2 e^2}\right)^{1/2} \tag{7.46}$$

ε_r is the permittivity (dielectric constant), which is 78.6 for water at 25 °C; ε_0 is the permittivity of free space. k is the Boltzmann constant and T is the absolute temperature; n_0 is the number of ions per unit volume of each type present in bulk solution and Z_i is the valency of the ions; e is the electronic charge.

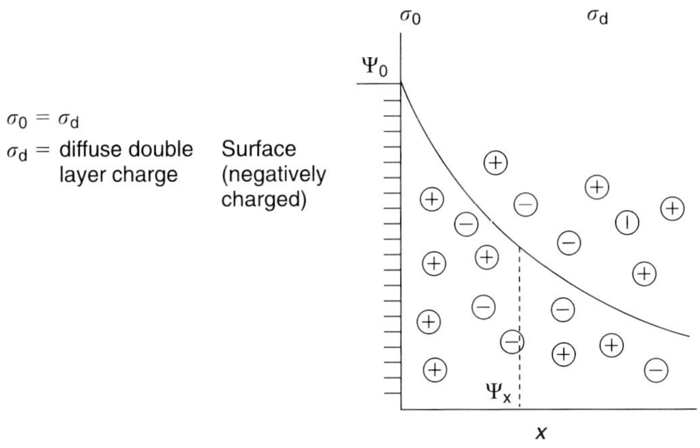

Fig. 7.16. Diffuse double layer according to Gouy and Chapman.

The data tabulated below for 1:1 electrolyte (e.g. KCl) show that the double layer extension increases with decreasing electrolyte concentration.

C (mol dm^{-3})	10^{-5}	10^{-4}	10^{-3}	10^{-2}	10^{-1}
$(1/\kappa)$ (nm)	100	33	10	3.3	1

7.7.2
Stern–Grahame Model of the Double Layer

Stern [17] introduced the concept of the non-diffuse part of the double layer for specifically adsorbed ions, the rest being diffuse in nature (Figure 7.17).

The potential drops linearly in the Stern region, and then exponentially. Grahame distinguished two types of ions in the Stern plane, physically adsorbed counter ions (outer Helmholtz plane) and chemically adsorbed ions (that lose part of their hydration shell) (inner Helmholtz plane).

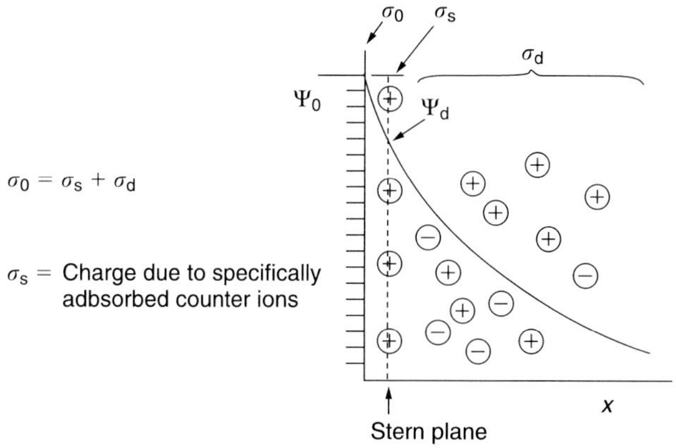

Fig. 7.17. Double layer according to Stern and Grahame.

7.8
Electrical Double Layer Repulsion

When charged colloidal particles in a dispersion approach each other such that the double layer begins to overlap (particle separation becomes less than twice the double layer extension), repulsion occurs. The individual double layers can no longer develop unrestrictedly, since the limited space does not allow complete potential decay [18].

Figure 7.18 illustrates this for two flat plates.

The potential $\Psi_{H/2}$ half-way between the plates is no longer zero (as would be the case for isolated particles at $x \to \infty$).

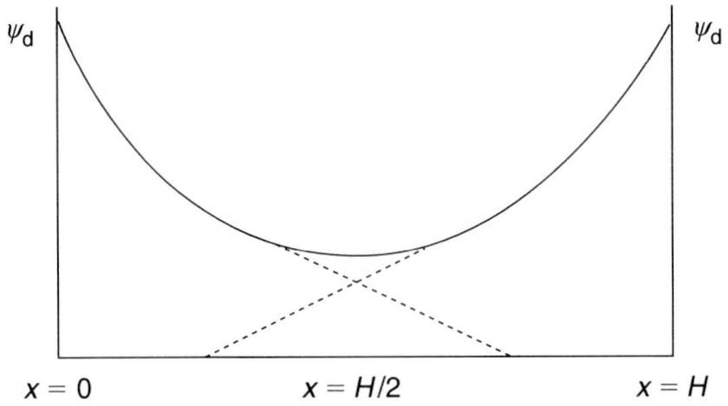

Fig. 7.18. Double layer interaction for two flat plates.

For two spherical particles of radius R and surface potential Ψ_0 and condition $\kappa R < 3$, the expression for the electrical double layer repulsive interaction is given by

$$G_{el} = \frac{4\pi\varepsilon_r\varepsilon_0 R^2\Psi_0^2 \exp -(\kappa h)}{2R + h} \quad (7.47)$$

where h is the closest separation between the surfaces.

The above expression shows the exponential decay of G_{el} with h. The higher κ is (i.e. the higher the electrolyte concentration), the steeper the decay (Figure 7.19).

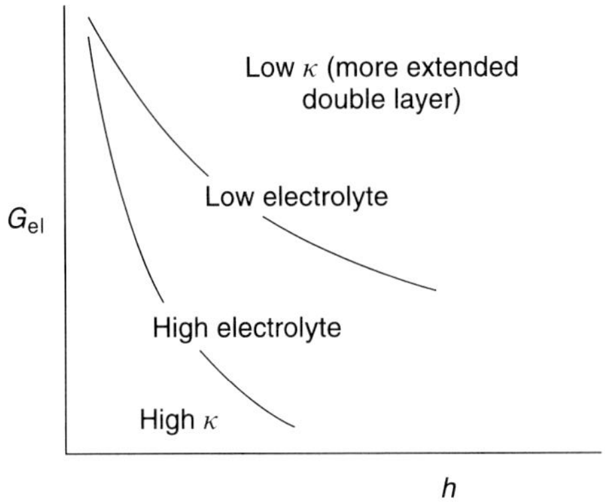

Fig. 7.19. Variation of G_{el} with h at different electrolyte concentrations. This means that at any given distance h, the double layer repulsion decreases with increasing electrolyte concentration.

7.9 Van der Waals Attraction

As is well known, atoms or molecules always attract each other at short separations. The attractive forces are of three different types: Dipole–dipole interaction

(Keesom), dipole-induced–dipole interaction (Debye) and the London dispersion force. The London dispersion force is the most important, since it occurs for polar and non-polar molecules. It arises from fluctuations in the electron density distribution.

At small distances of separation r in vacuum, the attractive energy between two atoms or molecules is given by

$$G_{aa} = -\frac{\beta_{11}}{r^6} \tag{7.48}$$

β_{11} is the London dispersion constant.

For colloidal particles made of atom or molecular assemblies, the attractive energies may be added, resulting in the following expression for two spheres (at small h),

$$G_A = -\frac{AR}{12h} \tag{7.49}$$

where A is the effective Hamaker constant,

$$A = (A_{11}^{1/2} - A_{22}^{1/2})^2 \tag{7.50}$$

A_{11} is the Hamaker constant between particles in a vacuum and A_{22} Hamaker constant for equivalent volumes of the medium.

$$A = \pi q^2 \beta_{ii} \tag{7.51}$$

q is number of atoms or molecules per unit volume.

G_A decreases with increasing h (Figure 7.20); at very short distances, the Born repulsion appears.

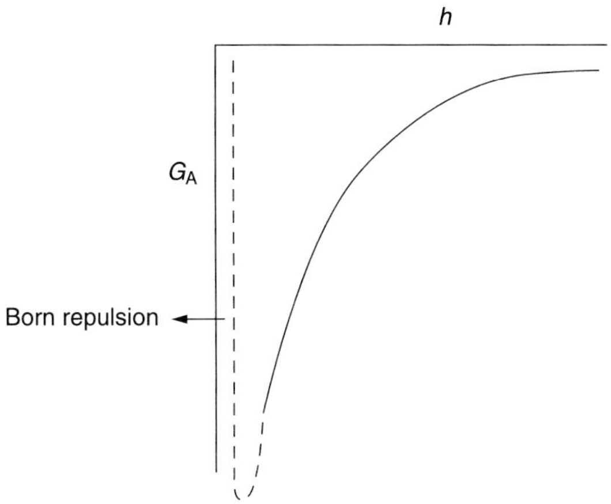

Fig. 7.20. Variation of G_A with h.

7.10
Total Energy of Interaction: Deryaguin–Landau–Verwey–Overbeek (DLVO) Theory [19, 20]

Combination of G_{el} and G_A results in the well-known theory of stability of colloids (DLVO theory) [19, 20],

$$G_T = G_{el} + G_A \tag{7.52}$$

Figure 7.21 shows a plot of G_T versus h, which represents the case at low electrolyte concentrations, i.e. strong electrostatic repulsion between the particles.

G_{el} decays exponentially with h, i.e. $G_{el} \rightarrow 0$ as h becomes large.

G_A is $\propto 1/h$, i.e. G_A does not decay to 0 at large h.

At long separations, $G_A > G_{el}$, resulting in a shallow minimum (secondary minimum). At very short distances, $G_A \gg G_{el}$, resulting in a deep primary minimum.

At intermediate distances, $G_{el} > G_A$, resulting in energy maximum, G_{max}, whose height depends on Ψ_0 (or Ψ_d) and the electrolyte concentration and valency.

At low electrolyte concentrations ($< 10^{-2}$ mol dm^{-3} for a 1:1 electrolyte), G_{max} is high ($> 25kT$) and this prevents particle aggregation into the primary minimum. The higher the electrolyte concentration (and the higher the valency of the ions), the lower the energy maximum.

Under some conditions (depending on electrolyte concentration and particle size), flocculation into the secondary minimum may occur. This flocculation is weak and reversible. By increasing the electrolyte concentration, G_{max} decreases

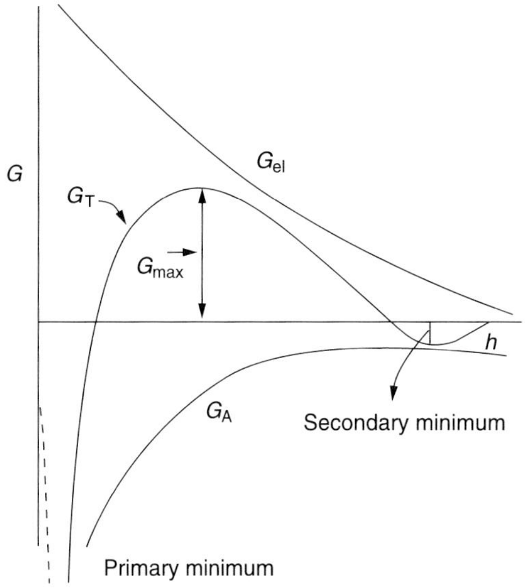

Fig. 7.21. Schematic of the variation of G_T with h according to DLVO theory.

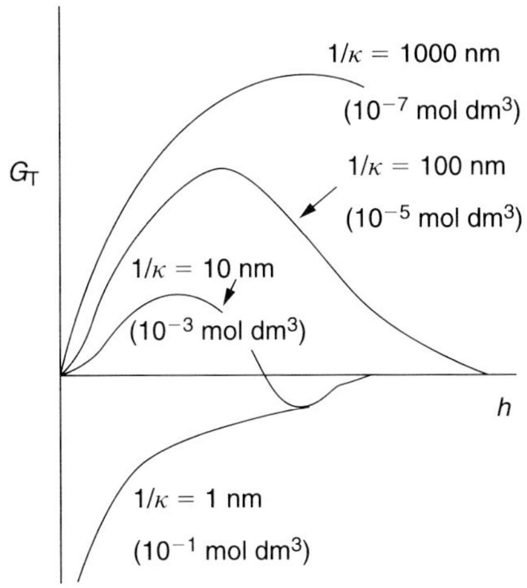

Fig. 7.22. Variation of G_T at various electrolyte concentrations.

until, at a given concentration, it vanishes and particle coagulation occurs. Figure 7.22 illustrates this, showing the variation of G_T with h at various electrolyte concentrations.

Coagulation occurs at a critical electrolyte concentration, the critical coagulation concentration (c.c.c.), which depends on the electrolyte valency. At low surface potentials, c.c.c. $\propto 1/Z^2$. This referred to as the Schultze–Hardy rule.

One can define a rate constant for flocculation: k_0 = rapid rate of flocculation (in the absence of an energy barrier) and k = slow rate of flocculation (in the presence of an energy barrier):

$$\frac{k_0}{k} = W \text{ (the Stability ratio)} \tag{7.53}$$

Note that W increases as G_{max} increases.

The stability of colloidal dispersions can be quantitatively assessed from plots of log W versus log C (Figure 7.23).

7.11
Criteria for Stabilisation of Dispersions with Double Layer Interaction

The two main criteria for stabilisation are: (1) High surface or Stern potential (zeta potential), high surface charge. (2) Low electrolyte concentration and low valency of counter and co-ions. One should ensure that an energy maximum in excess of $25kT$ exists in the energy–distance curve. When $G_{max} \gg kT$, the particles in the dispersion cannot overcome the energy barrier, thus preventing coagulation.

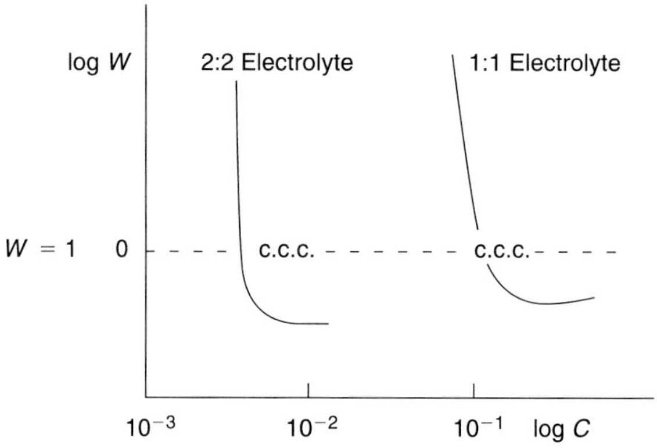

Fig. 7.23. Log *W* versus log *C* curves.

In some cases, particularly with large and asymmetric particles, flocculation into the secondary minimum may occur. This flocculation is usually weak and reversible and may be advantageous for preventing the formation of hard sediments.

7.12 Electrokinetic Phenomena and the Zeta Potential

As mentioned above, one of the main criteria for electrostatic stability is the high surface or zeta potential, which can be experimentally measured (vide infra). Before describing the experimental techniques for measuring the zeta potential it is essential to consider the electrokinetic effects in some detail, describing the theories that can be used to calculate the zeta potential from the particle electrophoretic mobility [21].

Electrokinetic effects are the direct result of charge separation at the interface between two phases (Figure 7.24).

Consider a negatively charged surface; positive ions (counter ions) are attracted to the surface, whereas negative ions (co-ions) are repelled (Figure 7.25).

The accumulation of excess positive ions causes a gradual reduction in the potential from its value Ψ_0 at the surface to 0 in bulk solution. At a point p from the surface, one can define a potential Ψ_x.

Electrokinetic effects arise when one of the two phases is caused to move tangentially past the second phase. Tangential motion can be caused by: Electric field;

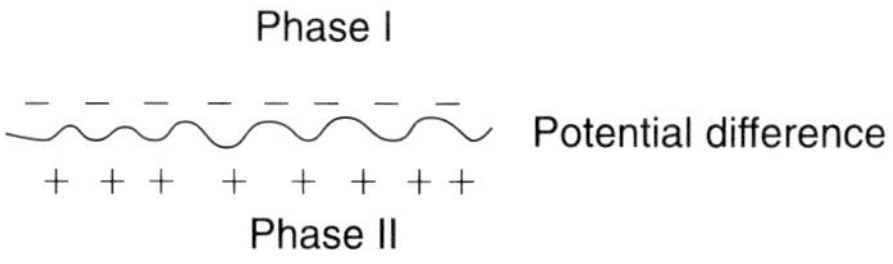

Fig. 7.24. Representation of charge separation.

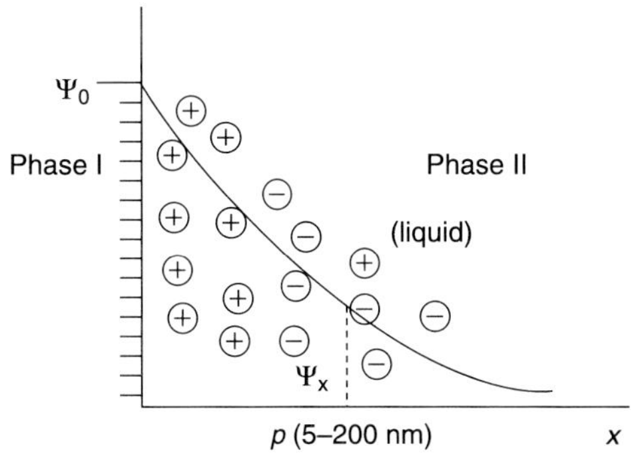

Fig. 7.25. Representation of charge accumulation at an interface.

forcing a liquid in a capillary; the gravitational field on the particles. This leads to the four types of electrokinetic phenomena described below.

(1) Electrophoresis: Movement of one phase is induced by application of an external electric field. One measures the particle velocity v from which the electrophoretic mobility u can be calculated,

$$u = \frac{v}{(E/l)} \ \text{m}^2\ \text{V}^{-1}\ \text{s}^{-1} \tag{7.54}$$

where E is the applied potential and l is the distance between the two electrodes; E/l is the field strength.

(2) Electro-osmosis: Here the solid is kept stationary (e.g. in the form of a glass tube) and the liquid is allowed to move under the influence of an electric field. The applied field acts on the charges (ions in the liquid) and when these move they drag liquid with them.

(3) Streaming potential: The liquid is forced through a capillary or a porous plug (containing the particles) under the influence of a pressure gradient. The excess charges near the wall (or the surface of particles in the plug) are carried along by the liquid flow, thus producing an electric field that can be measured by using electrodes and an electrometer.

(4) Sedimentation Potential (Dorn effect): In this case the particles are allowed to settle or rise through a fluid under the influence of gravity (or using a centrifugal force) – When the particles move they leave behind their ionic atmosphere and this creates a potential difference in the direction of motion which can be measured using electrodes and an electrometer.

Only electrophoresis will be discussed here since this is the most commonly used method for dispersions, allowing one to measure the particle mobility, which can be converted into the zeta potential using theoretical treatments.

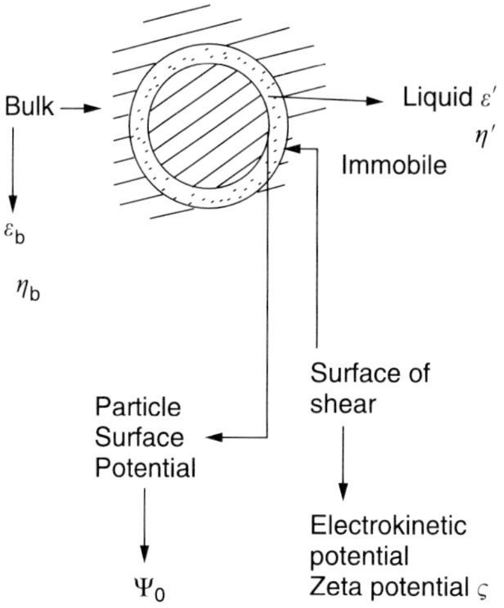

Fig. 7.26. Schematic of the shear plane.

In all electrokinetic phenomena [21], a fluid moves with respect to a solid surface. One needs to derive a relationship between fluid velocity (which varies with distance from the solid) and the electric field in the interfacial region.

The most important concept is the surface of shear, an imaginary surface close to the surface, within which the fluid is stationary. Figure 7.26 illustrates this, showing the position of the surface potential Ψ_0, the shear plane and zeta potential (that is close to the Stern potential Ψ_d).

Measurement of zeta potential (ζ) is valuable in determining the properties of dispersions. In addition, it has many other applications in various fields: Electrode kinetics, electro-dialysis, corrosion, adsorption of surfactants and polymers, crystal growth, mineral flotation and particle sedimentation.

Although measurement of particle mobility is fairly simple (particularly with the development of automated instruments), interpretation of the results is not simple. The calculation of zeta potential from particle mobility is not straightforward since this depends on the particle size and shape as well as the electrolyte concentration. For simplicity we will assume that the particles are spherical.

7.13 Calculation of Zeta Potential

7.13.1 Von Smoluchowski (Classical) Treatment [22]

This applies to the case where the particle radius R is much larger than the double layer thickness ($1/\kappa$), i.e. $\kappa R \gg 1$. This generally applies to particles that are greater than 0.5 μm (when the 1:1 electrolyte concentration is lower than 10^{-3} mol dm^{-3}, i.e. $\kappa R > 10$),

$$u = \frac{\varepsilon_r \varepsilon_0 \zeta}{\eta} \tag{7.55}$$

where ε_r is the relative permittivity of the medium (78.6 for water at 25 °C), ε_0 is the permittivity of free space (8.85×10^{-12} F m^{-1}) and η is the viscosity of the medium (8.9×10^{-4} Pa s for water at 25 °C) ζ is the zeta potential in volts.

For water at 25 °C,

$$\zeta = 1.282 \times 10^6 u \tag{7.56}$$

u is expressed in $m^2\ V^{-1}\ s^{-1}$.

7.13.2
Hückel Equation [23]

This applies for the case $\kappa R < 1$,

$$u = \frac{2}{3} \frac{\varepsilon_r \varepsilon_0 \zeta}{\eta} \tag{7.57}$$

Eq. (7.57) applies for small particles (< 100 nm) and thick double layers (low electrolyte concentration).

7.13.3
Henry's Treatment [24]

Henry accounted for the discrepancy between Smoluchowski and Hückel's treatment. Hückel disregarded the deformation of the electric field by the particle, whereas Smoluchowski assumed the field to be uniform and everywhere parallel to the particle surface. These two assumptions are justified in the extreme cases of $\kappa R \ll 1$ and $\kappa R \gg 1$ respectively.

For intermediate cases where κR is not too small or too large, Henry derived the following expression (which can be applied at all κR values),

$$u = \frac{2}{3} \frac{\varepsilon_r \varepsilon_0 \zeta}{\eta} f(\kappa R) \tag{7.58}$$

The function $f(\kappa R)$, Henry's correction factor, depends also on the particle shape. Values of $f(\kappa R)$ at various values of κR are tabulated below.

κR	0	1	2	3	4	5	10	25	100	∞
$f(\kappa R)$	1.0	1.027	1.066	1.101	1.133	1.160	1.239	1.370	1.460	1.500

Henry's calculations are based on the assumption that the external field can be superimposed on the field due to the particle, and hence it can only be applied for low potentials ($\zeta < 25$ mV). It also does not take into account the distortion of the field induced by the movement of the particle (relaxation effect).

Wiersema, Loeb and Overbeek [25] introduced two corrections to Henry's treatment, namely the relaxation and retardation (movement of the liquid with the double layer ions) effects and Ottewill and Shaw have compiled a numerical tabulation of the relation between mobility and zeta potential [26]. Such tables are useful for converting u into ζ at all practical values of κR.

7.14 Measurement of Electrophoretic Mobility

7.14.1 Ultramicroscopic Technique (Microelectrophoresis)

This is the most commonly used method since it allows direct observation of the particles using an ultramicroscope (suitable for particles larger than 100 nm). Basically, a dilute suspension is placed in a cell consisting of a thin-walled (~100 μm) glass tube that is attached to two larger bore tubes with sockets for placing the electrodes. The cell is immersed in a thermostated bath (accurate to ±0.1 °C) that contains an attachment for illumination and a microscope objective for observing the particles. It is also possible to use a video camera to observe directly the particles.

Since the glass walls are charged (usually negative at practical pH measurements), the solution in the cell will, in general, experience electro-osmotic flow. Only where the electro-osmotic flow is zero, i.e. at the stationary level, can the electrophoretic mobility of the particles be measured. The stationary level is located at a 0.707 of the radius from the centre of the tube or 0.146 of the internal diameter from the wall.

By focusing the microscope objective at the top and bottom of the walls of the tube, one can easily locate the position of the stationary levels.

The average particle velocity is measured at the top and bottom stationary levels by averaging at least 20 measurements in each direction (the eye piece of the microscope is fitted with a graticule).

Several commercial instruments are available (e.g. Rank Brothers, Bottisham Cambridge England and Pen Kem in USA).

For large particles (> 1 μm and high density) sedimentation may occur during the measurement. In this case one can use a rectangular cell and observe the particles horizontally from the side of the glass cell.

Microelectrophoresis has many advantages since the particles can be measured in their normal environment. It is preferable to dilute the suspension with the supernatant liquid, which can be produced by centrifugation.

7.14.2
Laser Velocimetry Technique

This method is suitable for small particles that undergo Brownian motion. The light scattered by small particles will show intensity fluctuations as a result of the Brownian diffusion (Doppler shift). By application of an electric field as the particles undergo Brownian motion and measuring the fluctuation in intensity of the scattered light (using a correlator) one can measure the particle mobility.

Two laser beams of equal intensity are allowed to cross at a particular point within the cell containing the suspension of particles. At the intersection of the beam, which is focused at the stationary level, interferences of known spacing are formed. The particles moving through the fringes under the influence of the electric field scatter light whose intensity fluctuates with a frequency that is related to the mobility of the particles.

The photons are detected by a photomultiplier and the signal is fed to the correlator. The resulting correlation function is analysed to determine the frequency (Doppler) spectrum and this is converted into the particle velocity V,

$$V = \Delta \nu s \tag{7.59}$$

where $\Delta\nu$ is the Doppler shift frequency and s is the spacing between the interference fringes in the region where the beams cross; s is given by the relationship

$$s = \frac{\lambda}{2\sin(\alpha/2)} \tag{7.60}$$

where λ is the laser wavelength and α is the angle between the crossing laser beams.

The velocity spectrum is then converted into a mobility spectrum (allowing one to obtain the mobility distribution) and the mobility is converted into zeta potential using Hückel's equation.

Several commercial instruments are available: Malvern Zeta Sizer, Coulter Delsa Sizer.

7.15
General Classification of Dispersing Agents

As mentioned above, for dispersing powders into liquids, one usually requires the addition of a dispersing agent that satisfies the following requirements: Lowers the surface tension of the liquid to aid wetting of the powder; adsorbs at the solid/liquid interface to lower the solid/liquid interfacial tension; lowers the contact angle of the liquid on the solid surface (zero contact angle is very common); helps to break-up the aggregates and agglomerates as well as in subdivision of the particles into smaller units; and stabilises the particles formed against any aggregation (or rejoining).

All dispersing agents are surface active and they can be simple surfactants (anionic, cationic, zwitterionic or nonionic), polymers or polyelectrolytes. The dispersing agent should be soluble (or at least dispersible) in the liquid medium and it should adsorb at the solid/liquid interface.

In this section we describe the general classification of dispersing agents. The adsorption of surfactants and polymers at the solid/liquid interface was treated in Chapter 5. The various classes can be summarised as follows.

7.15.1 Surfactants

Ionic, anionic, e.g. sodium dodecyl sulphate ($C_{12}H_{25}OSO_3^-Na^+$); cationic, e.g. cetyltrimethylammonium chloride $C_{16}H_{33}-N^+(CH_3)_3Cl^-$); zwitterionic, e.g. 3-dimethyldodecylamine propane sulphonate (Betaine $C_{12}H_{25}-N^+(CH_3)_2-CH_2-CH_2-CH_2-SO_3$); nonionic, alcohol ethoxylates [$C_nH_{2n+1}-O-(CH_2-CH_2-O)_n-H$], alkyl phenol ethoxylates [$C_nH_{2n+1}-C_6H_4-O-(CH_2-CH_2-O)_n-H$], amine oxides, e.g. decyl dimethyl amine oxide [$C_{10}H_{21}-N(CH_3)_2 \rightarrow O$], amine ethoxylates.

7.15.2 Nonionic Polymers

Poly(vinyl alcohol) [with poly(vinyl acetate) blocks – usually 4–12%], $-(CH_2-CH-OH)_x-(CH_2-CH(OCOCH_3)-OH)_y-(CH_2-CH-OH)_x-$. Block copolymers of ethylene oxide–propylene oxide (ABA block of PEO-PPO-PEO), e.g. Pluronics (BASF), Synperonic PE (ICI), $H-(O-CH_2-CH_2)_n-(CH_2-CH(CH_3)-O)_m-(CH_2-CH_2-O)_n-H$. Graft copolymers, e.g. poly(methyl methacrylate) (PMMA) backbone [with some poly(methacrylic acid)] with grafted PEO chains, e.g. Atlox 4913 Hypermer CG6 (ICI).

7.15.3 Polyelectrolytes

Poly(acrylic acid), $-(CH_2-CH-COOH)_n-$, at pH > 5, the carboxylic acid groups ionise to form an anionic polyelectrolyte; poly(acrylic acid)/poly(methacrylic acid), $(CH_2-CH-COO^-)_n-(CH_2-C(CH_3)-COO^-)_m$; naphthalene formaldehyde sulphonated condensates, lignosulphonates.

7.16 Steric Stabilisation of Suspensions

The use of natural and synthetic polymers (referred to as polymeric surfactants) to stabilise solid/liquid dispersions plays an important role in industrial applications [27] such as in paints, cosmetics, agrochemicals and ceramics.

Polymers are particularly important for the preparation of concentrated dispersions, i.e. at high volume fraction ϕ of the disperse phase; ϕ = (volume of all particles)/(total volume of dispersion).

Polymers are also essential for the stabilisation of non-aqueous dispersions, since in this case electrostatic stabilisation is not possible (due to the low dielectric constant of the medium).

To understand the role of polymers in dispersion stability, it is essential to consider the adsorption and conformation of the macromolecule at the solid/liquid interface [1].

7.17 Interaction Between Particles Containing Adsorbed Polymer Layers

When two particles, each with a radius R and containing an adsorbed polymer layer with a hydrodynamic thickness δ_h, approach each other to a surface–surface separation distance h that is smaller than $2\delta_h$ the polymer layers interact, resulting in two main situations [27, 28]: either the polymer chains may overlap or the polymer layer may undergo some compression. In both cases, there will be an increase in the local segment density of the polymer chains in the interaction region (Figure 7.27).

The real situation perhaps lies between the above two cases, i.e. the polymer chains may undergo some interpenetration and some compression. Provided the dangling chains (the A chains in A-B, A-B-A block or BA_n graft copolymers) are in a good solvent, this local increase in segment density in the interaction zone will result in strong repulsion as a result of two main effects:

(1) Increase in the osmotic pressure in the overlap region as a result of the unfavourable mixing of the polymer chains, when these are in good solvent conditions [1, 2]. This is referred to as osmotic repulsion or mixing interaction and it is described by a free energy of interaction G_{mix}.
(2) Reduction of the configurational entropy of the chains in the interaction zone – this entropy reduction results from the decrease in the volume available for the chains when these are either overlapped or compressed. This is referred to as volume restriction interaction, entropic or elastic interaction and it is described by a free energy of interaction G_{el}.

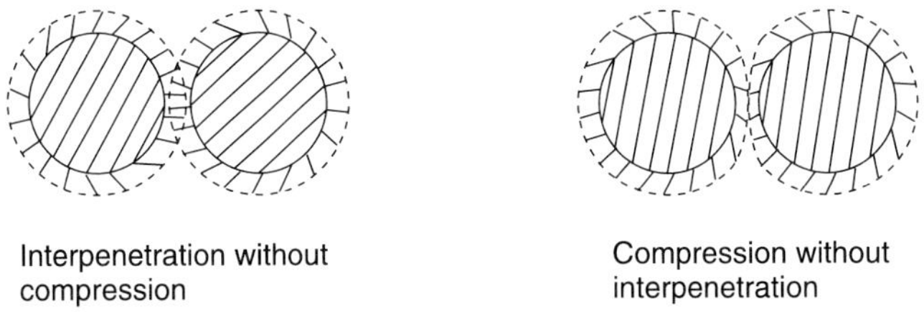

Fig. 7.27. Scheme of the interaction of two polymer layers.

Combination of G_{mix} and G_{el} is usually referred to as the steric interaction free energy, G_s, i.e.

$$G_s = G_{mix} + G_{el} \tag{7.61}$$

The sign of G_{mix} depends on the solvency of the medium for the chains. If in a good solvent, i.e. the Flory–Huggins interaction parameter χ is less than 0.5, then G_{mix} is positive and the mixing interaction leads to repulsion (see below). In contrast, if $\chi > 0.5$ (i.e. the chains are in a poor solvent condition) then G_{mix} is negative and the mixing interaction becomes attractive.

G_{el} is always positive and, hence, in some cases one can produce stable dispersions in a relatively poor solvent (enhanced steric stabilisation) [2].

7.17.1 Mixing Interaction G_{mix}

This results from the unfavourable mixing of the polymer chains, when under good solvent conditions (Figure 7.28). Consider two spherical particles with the same radius and each containing an adsorbed polymer layer with thickness δ. Before overlap, one can define in each polymer layer a chemical potential for the solvent μ_i^α and a volume fraction for the polymer in the layer ϕ_2. In the overlap region (volume element dV), the chemical potential of the solvent is reduced to μ_i^β; this results from the increase in polymer segment concentration in this overlap region.

In the overlap region, the chemical potential of the polymer chains is now higher than in the rest of the layer (with no overlap). This amounts to an increase in the osmotic pressure in the overlap region. As a result solvent will diffuse from the bulk to the overlap region, thus separating the particles and, hence, a strong repulsive energy arises from this effect.

The above repulsive energy can be calculated by considering the free energy of mixing of two polymer solutions, as for example treated by Flory and Krigbaum [29]. The free energy of mixing is given by two terms: an entropy term that de-

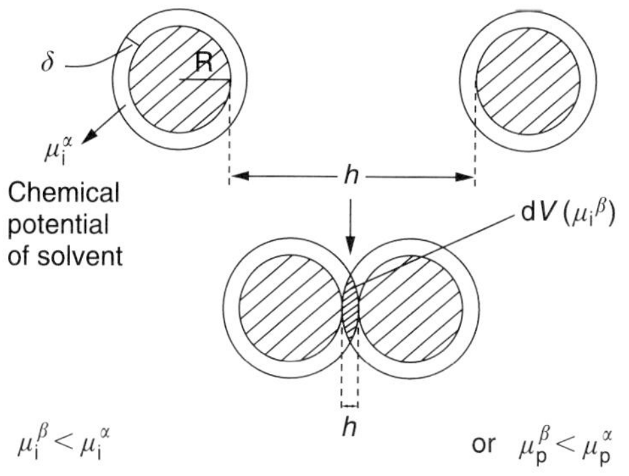

Fig. 7.28. Representation of polymer layer overlap.

pends on the volume fraction of polymer and solvent and an energy term that is determined by the Flory–Huggins interaction parameter χ,

$$\delta(G_{mix}) = kT(n_1 \ln \phi_1 + n_2 \ln \phi_2 + \chi n_1 \phi_2) \quad (7.62)$$

where n_1 and n_2 are the respective number of moles of solvent and polymer with volume fractions ϕ_1 and ϕ_2, k is the Boltzmann constant and T is the absolute temperature.

The total change in free energy of mixing for the whole interaction zone, V, is obtained by summing over all the elements in V,

$$G_{mix} = \frac{2kTV_2^2}{V_1} \nu_2 \left(\frac{1}{2} - \chi\right) R_{mix}(h) \quad (7.63)$$

where V_1 and V_2 are the molar volumes of solvent and polymer respectively, ν_2 is the number of chains per unit area, and $R_{mix}(h)$ is a geometric function that depends on the form of the segment density distribution of the chain normal to the surface, $\rho(z)$; k is the Boltzmann constant and T is the absolute temperature. Using the above theory one can derive an expression for the free energy of mixing of two polymer layers (assuming a uniform segment density distribution in each layer) surrounding two spherical particles as a function of the separation distance h between the particles [30].

The expression for G_{mix} is,

$$G_{mix} = \left(\frac{2V_2^2}{V_1}\right) \nu_2 \left(\frac{1}{2} - \chi\right) \left(3R + 2\delta + \frac{h}{2}\right) \left(\delta - \frac{h}{2}\right)^2 \quad (7.64)$$

The sign of G_{mix} depends on the value of the Flory–Huggins interaction parameter χ: if $\chi < 0.5$, G_{mix} is positive and the interaction is repulsive; if $\chi > 0.5$ then G_{mix} is negative and the interaction is attractive. The condition $\chi = 0.5$ and $G_{mix} = 0$ is termed the θ-condition. The latter corresponds to the case where the polymer mixing behaves as ideal, i.e. mixing of the chains does not lead to either an increase or decrease of the free energy of the system.

7.17.2 Elastic Interaction, G_{el}

This arises from the loss in configurational entropy of the chains on the approach of a second particle (Figure 7.29). As a result of this approach, the volume available for the chains becomes restricted, resulting in a loss of the number of configurations. This can be illustrated by considering a simple molecule, represented by a rod that rotates freely in a hemisphere across a surface [31].

When the two surfaces are separated by an infinite distance (∞) the number of configurations of the rod is $\Omega(\infty)$, which is proportional to the volume of the hemisphere. When a second particle approaches to a distance h such that it cuts the

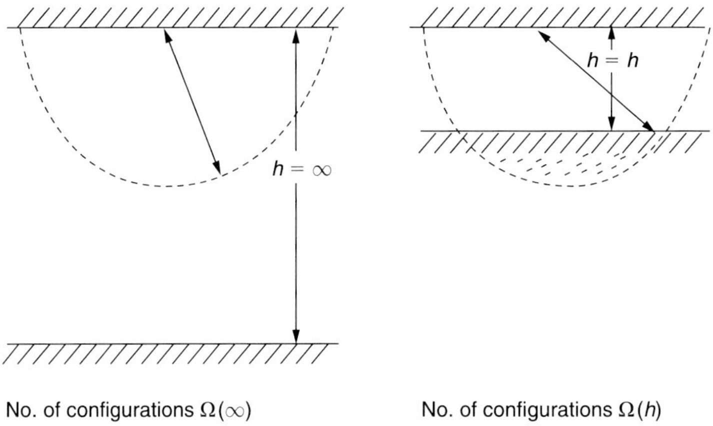

Fig. 7.29. Scheme of configurational entropy loss on approach of a second particle.

hemisphere (losing some volume), the volume available to the chains is reduced and the number of configurations become $\Omega(h)$ which is less than $\Omega(\infty)$.

For two flat plates, G_{el} is given by the expression

$$\frac{G_{el}}{kT} = 2\nu_2 \ln\left[\frac{\Omega(h)}{\Omega(\infty)}\right] = 2\nu_2 R_{el}(h) \tag{7.65}$$

where $R_{el}(h)$ is a geometric function whose form depends on the segment density distribution. G_{el} is always positive and could play a major role in steric stabilisation. It becomes very strong when the separation between the particles becomes comparable to the adsorbed layer thickness δ.

Combination of G_{mix} and G_{el} with G_A gives the total energy of interaction G_T (assuming there is no contribution from any residual electrostatic interaction) [6], i.e.

$$G_T = G_{mix} + G_{el} + G_A \tag{7.66}$$

Figure 7.30 gives a schematic representation of the variation of G_{mix}, G_{el}, G_A and G_T with surface–surface separation h.

G_{mix} increases very sharply with decreasing h, when $h < 2\delta$. G_{el} increases very sharply with decreasing h, when $h < \delta$. G_T versus h shows a minimum, G_{min}, at separations comparable to 2δ; when $h < 2\delta$, G_T shows a rapid increase with further decrease in h [32].

Unlike the G_T–h curve predicted by the DLVO theory (which shows two minima and one energy maximum), the G_T–h for systems that are sterically stabilised show only one minimum, G_{min}, followed by a sharp increase in G_T with decreasing h (when $h, 2\delta$). The depth of the minimum depends on the Hamaker constant A, the particle radius R and adsorbed layer thickness δ; G_{min} increases with increasing

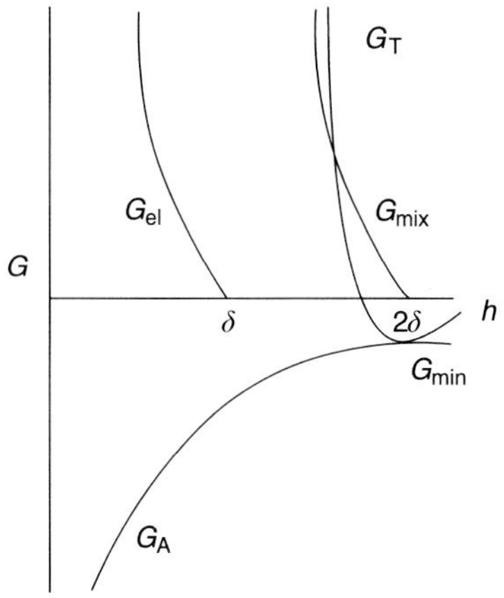

Fig. 7.30. Variation of G_{mix}, G_{el}, G_A and G_T with surface–surface distance (h) between the particles.

A and R. At a given A and R, G_{min} increases with decrease in δ (i.e. with decrease of the molecular weight, M_w, of the stabiliser). Figure 7.31 illustrates this with the energy–distance curves for poly(vinyl alcohol) with various molecular weights (1). M_w varies with δ as tabulated below.

M_w	67000	43000	28000	17000	8000
δ (nm)	25.5	19.7	14.0	9.8	3.3

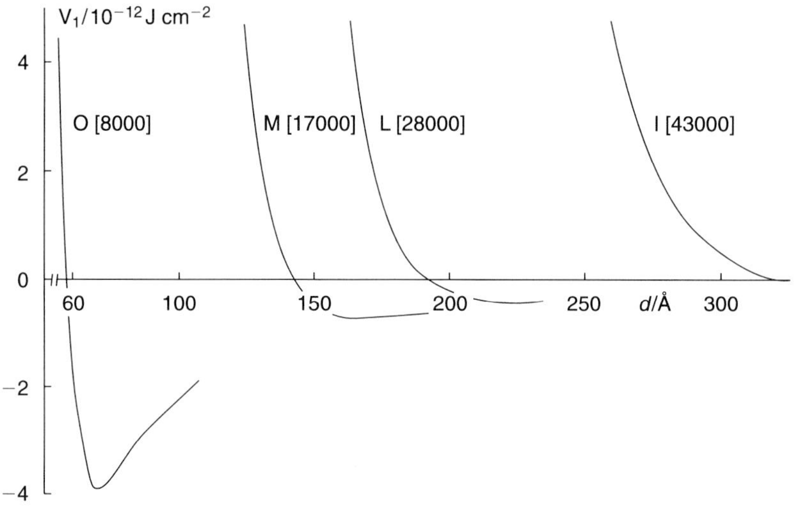

Fig. 7.31. Total interaction energy versus separation for particles with adsorbed layers of poly(vinyl alcohol) (PVA) of various thicknesses.

Figure 7.31 shows that G_{min} increases with decreasing molecular weight of PVA. When the molecular weight of the polymer is greater than 43 000, G_{min} is so small that it does not appear in the energy–distance curve. In this case the dispersion will approach thermodynamic stability (particularly at low volume fractions). When the molecular weight of the polymer becomes very small, as is the case with 8000, G_{min} become sufficiently large to cause weak flocculation.

7.18 Criteria for Effective Steric Stabilisation

(1) The particles should be completely covered by the polymer (the amount of polymer should correspond to the plateau value). Any bare patches may cause flocculation either by van der Waals attraction (between the bare patches) or by bridging flocculation (whereby a polymer molecule will become simultaneously adsorbed on two or more particles).
(2) The polymer should be strongly "anchored" to the particle surfaces, to prevent any displacement during particle approach. This is particularly important for concentrated suspensions. For this purpose, A-B, A-B-A block and BA_n graft copolymers are the most suitable, where the chain B is chosen to be highly insoluble in the medium and has a strong affinity to the surface. Examples of B groups for hydrophobic particles in aqueous media are polystyrene and poly(methyl methacrylate).
(3) The stabilising chain A should be highly soluble in the medium and strongly solvated by its molecules. Examples of A chains in aqueous media are poly(ethylene oxide) and poly(vinyl alcohol).
(4) δ should be sufficiently large (> 10 nm) to prevent weak flocculation.

7.19 Flocculation of Sterically Stabilised Dispersions

Two main types of flocculation may be distinguished:

(1) Weak flocculation: This occurs when the thickness of the adsorbed layer is small (usually <5 nm), particularly when the particle radius and Hamaker constant are large.
(2) Incipient flocculation: This occurs when the solvency of the medium is reduced to become worse than a θ-solvent (i.e. $\chi > 0.5$). Figure 7.32 illustrates this, where χ was changed from <0.5 (good solvent) to >0.5 (poor solvent).

When $\chi > 0.5$, G_{mix} becomes negative (attractive), which, when combined with the van der Waals attraction at this separation distance, gives a deep minimum, causing flocculation. In most cases, there is a correlation between the critical floc-

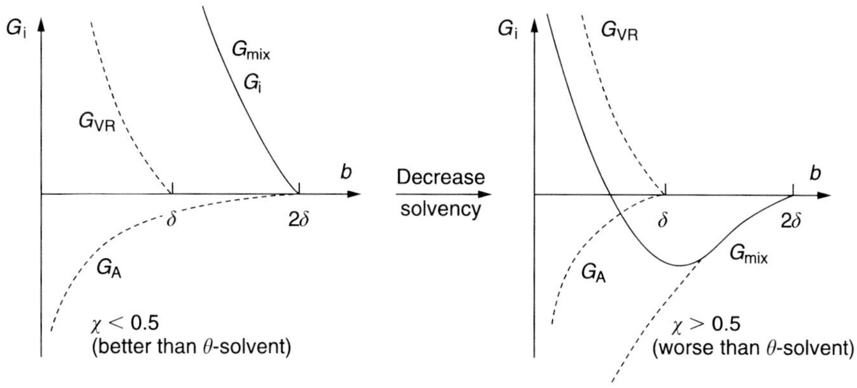

Fig. 7.32. Influence of reduction in solvency on the energy–distance curves for sterically stabilised dispersions.

culation point and the θ condition of the medium. Good correlation is found in many cases between the critical flocculation temperature (CFT) and θ-temperature of the polymer in solution (with block and graft copolymers one should consider the θ-temperature of the stabilising chains A) [28]. Good correlation is also found between the critical volume fraction (CFV) of a non-solvent for the polymer chains and their θ-point under these conditions. However, in some cases such correlation may break down, particularly the case for polymers that adsorb by multi-point attachment. This situation has been described by Napper [28], who referred to it as "enhanced" steric stabilisation.

Thus by measuring the θ-point (CFT or CFV) for the polymer chains (A) in the medium under investigation (which could be obtained from viscosity measurements) one can establish the stability conditions for a dispersion, before its preparation. This procedure also helps in designing effective steric stabilisers such as block and graft copolymers.

7.20 Properties of Concentrated Suspensions

One of the main features of concentrated suspensions is the formation of three-dimensional structure units, which determine their properties and in particular their rheology. The formation of these units is determined by the interparticle interactions, which need to be clearly defined and quantified. It is useful to define the concentration range above which a suspension may be considered concentrated. The particle number concentration and volume fraction, ϕ, above which a suspension may be considered concentrated is best defined in terms of the balance between the particle translational motion and interparticle interaction. At one extreme, a suspension may be considered dilute if the thermal motion (Brownian diffusion) of the particles predominates over the imposed interparticle interaction [33–35]. In this case, the particle translational motion is large and only occasional contacts occur between the particles, i.e. the particles do not "see" each other until collision occurs, giving a random arrangement of particles. In this case, the particle interactions can be represented by two-body collisions. In such "dilute" systems, gravity effects may be neglected and, if the particle size range is within the colloid range (1 nm–1 μm), no settling occurs. The properties of the suspension are time-independent and, therefore, any time-average quantity such as viscosity or scattering may be extrapolated to infinite dilution.

As the particle number concentration is increased in a suspension, the volume of space occupied by the particles increases relative to the total volume. Thus, a proportion of the space is excluded in terms of its occupancy by a single particle. Moreover, the particle–particle interaction increases and the forces of interaction between the particles play a dominant role is determining the properties of the system. With further increase in particle number concentration, the interactive contact between the particles increases until a situation is reached where the interaction produces a specific order between the particles, and a highly developed structure is reached. With solid in liquid dispersions, such a highly ordered structure, which is close to the maximum packing fraction ($\phi = 0.74$ for hexagonally closed packed array of monodisperse particles) is referred to as "solid" suspension. In such a system, any particle interacts with many neighbours and the vibrational amplitude is small relative to particle size; the properties of the system are essentially time independent [33–35].

Between the random arrangement of particles in "dilute" suspensions and the highly ordered structure of "solid" suspensions one may easily define "concentrated" suspensions. In this case, the particle interactions occur by many-body collisions and the translational motion of the particles is restricted. However, this reduced translational motion is not as great as with "solid" suspensions, i.e. the vibrational motion of the particles is large compared with particle size. A time-dependent system arises in which there will be spatial and temporal correlations.

On standing, concentrated suspensions reach various states (structures) that are determined by: (1) Magnitude and balance of the various interaction forces, electrostatic repulsion, steric repulsion and van der Waals attraction. (2) Particle size and shape distribution. (3) Density difference between disperse phase and medium, which determines the sedimentation characteristics. (4) Conditions and prehistory of the suspension, e.g. agitation, which determines the structure of the flocs formed (chain aggregates, compact clusters, etc.). (5) Presence of additives, e.g. high molecular weight polymers that may cause bridging or depletion flocculation.

Some of the various states that may be produced are shown in Figure 7.33. These states may be described in terms of three different energy–distance curves may: (a) Electrostatic, produced, for example, by the presence of ionogenic groups on the surface of the particles, or adsorption of ionic surfactants. (b) Steric, produced, for example, by adsorption of nonionic surfactants or polymers. (c) Electrostatic + steric (electrosteric) as, for example, produced by polyelectrolytes. These are illus-

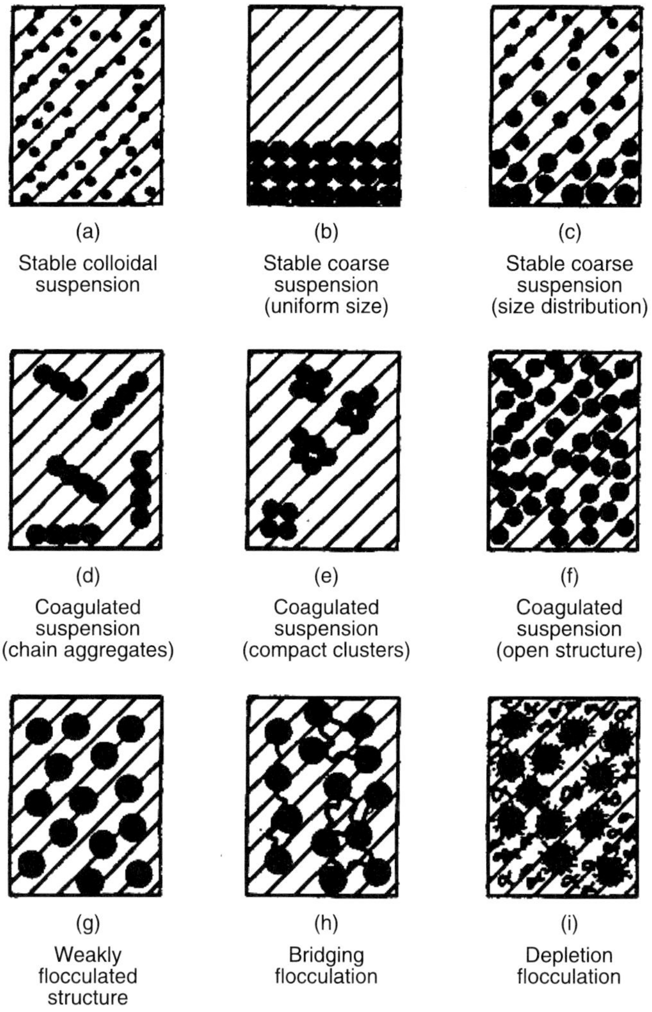

Fig. 7.33. Scheme of the various structures produced in concentrated suspensions.

trated in Figure 7.34. A brief description of the various states shown in Figure 7.33 is given below [36, 37].

States (a) to (c) in Figure 7.33 correspond to a suspension that is stable in the colloid sense. The stability is obtained as a result of net repulsion due to the presence of extended double layers (i.e. at low electrolyte concentration), the result of steric repulsion produced adsorption of nonionic surfactants or polymers, or the result of a combination of double layer and steric repulsion (electrosteric). State (a) represents a suspension with small particle size (submicron) whereby the Brownian diffusion overcomes the gravity force, producing a uniform distribution of the particles in the suspension, i.e.

$$kT \gg \tfrac{4}{3}\pi R^3 \Delta\rho g h \tag{7.67}$$

where k is the Boltzmann constant, T is the absolute temperature, R is the particle radius, $\Delta\rho$ is the buoyancy (difference in density between the particles and the medium), g is the acceleration due to gravity and h is the height of the container.

A good example of the above case is a latex suspension with a particle size well below 1 μm that is stabilised by ionogenic groups, by an ionic surfactant or nonionic surfactant or polymer. This suspension will show no separation on storage for long periods.

States (b) and (c) represent suspensions whereby the particle size range is outside the colloid range (> 1 μm). In this case, gravity exceeds the Brownian diffusion,

$$\tfrac{4}{3}\pi R^3 \Delta\rho g \gg kT \tag{7.68}$$

With state (b), the particles are uniform and they will settle under gravity, forming a hard sediment (technically referred to as "clay" or "cake"). The repulsive forces between the particles allow them to move past each other until they reach small distances of separation (that are determined by the location of the repulsive barrier). Owing to the small distances between the particles in the sediment it is very difficult to redisperse the suspension by simple shaking.

With case (c), consisting of a wide distribution of particle sizes, the sediment may contain larger proportions of the larger size particles, but a hard "clay" is still produced. These "clays" are dilatant (i.e. shear thickening) and they can be detected easily by inserting a glass rod in the suspension. Penetration of the glass rod into these hard sediments is very difficult.

States (d) to (f) in Figure 7.33 represent coagulated suspensions that have either a small or no repulsive energy barrier. State (d) represents coagulation under no stirring conditions, in which case chain aggregates are produced that will settle under gravity, forming a relatively open structure. State (e) represents coagulation under stirring conditions whereby compact aggregates are produced that will settle faster than the chain aggregates, and the sediment produced is more compact. State (f) represents coagulation at high volume fraction of the particles, ϕ. Here, the whole particles form a "one-floc" structure from chains and cross chains that extend from one wall to the other in the container. Such a coagulated structure may undergo some compression (consolidation) under gravity, leaving a clear supernatant liquid layer at the top of the container. This phenomenon is referred to as syneresis.

State (g) in Figure 7.33 represents the case of weak and reversible flocculation. This occurs when the secondary minimum in the energy–distance curve (Figure 7.34a) is deep enough to cause flocculation. This can occur at moderate electrolyte concentrations, in particular with larger particles. The same occurs with sterically and electrosterically stabilised suspensions (Figure 7.34b and c). This takes place when the adsorbed layer thickness is not very large, particularly with large par-

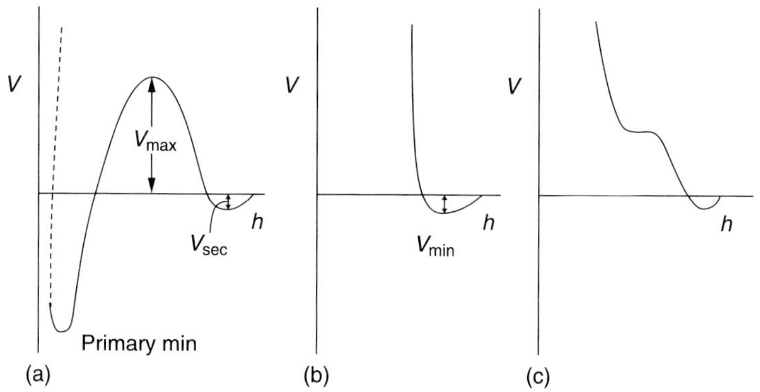

Fig. 7.34. Total energy of interaction for three different stabilisation mechanisms: (a) electrostatic; (b) steric; (c) electrosteric.

ticles. The minimum depth required to cause weak flocculation depends on the volume fraction of the suspension. The higher the volume fraction, the lower the minimum depth required for weak flocculation. This can be understood if one considers the free energy of flocculation, which consists of two terms, an energy term determined by the depth of the minimum (G_{min}) and an entropy term determined by reduction in configurational entropy on aggregation of particles,

$$\Delta G_{flocc} = \Delta H_{flocc} - T\Delta S_{flocc} \tag{7.69}$$

With dilute suspension, the entropy loss on flocculation is larger than with concentrated suspensions. Hence, for flocculation of a dilute suspension, a higher energy minimum is required than for concentrated suspensions.

The above flocculation is weak and reversible, i.e. on shaking the container redispersion of the suspension occurs. On standing, the dispersed particles aggregate to form a weak "gel". This process (referred to as sol–gel transformation) leads to reversible time dependence of viscosity (thixotropy). On shearing the suspension, the viscosity decreases and when the shear is removed, the viscosity is recovered. This phenomenon is applied in paints. On application of the paint (by a brush or roller), the gel is fluidised, allowing a uniform coating of the paint. When shearing is stopped, the paint film recovers its viscosity, avoiding any dripping.

State (h) represents the case whereby the particles are not completely covered by the polymer chains. Here, simultaneous adsorption of one polymer chain on more than one particle occurs, leading to bridging flocculation. If the polymer adsorption is weak (low adsorption energy per polymer segment), the flocculation could be weak and reversible. In contrast, if the adsorption of the polymer is strong, tough flocs are produced and the flocculation is irreversible. The last phenomenon is used for solid/liquid separation, e.g. in water and effluent treatment.

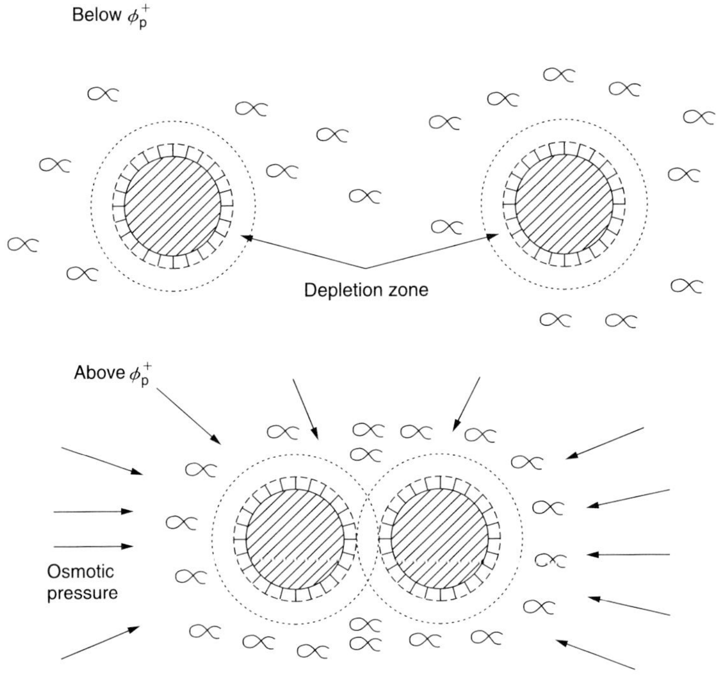

Fig. 7.35. Representation of depletion flocculation.

Case (1) represents a phenomenon, referred to as depletion flocculation, produced by addition of "free" non-adsorbing polymer [38]. In this case, the polymer coils cannot approach the particles to a distance Δ (which is determined by the radius of gyration of free polymer, R_G), since the reduction of entropy on close approach of the polymer coils is not compensated by an adsorption energy. The suspension particles will be surrounded by a depletion zone of thickness Δ. Above a critical volume fraction of the free polymer, ϕ_p^+, the polymer coils are "squeezed out" from between the particles and the depletion zones begin to interact. The interstices between the particles are now free from polymer coils and hence an osmotic pressure is exerted outside the particle surface (the osmotic pressure outside is higher than between the particles), resulting in weak flocculation [4]. A schematic representation of depletion flocculation is shown in Figure 7.35.

The magnitude of the depletion attraction free energy, G_{dep}, is proportional to the osmotic pressure of the polymer solution, which in turn is determined by ϕ_p and the molecular weight M. The range of depletion attraction is proportional to the thickness of the depletion zone, Δ, which is roughly equal to the radius of gyration, R_G, of the free polymer. A simple expression for G_{dep} is [38]

$$G_{dep} = \frac{2\pi R\Delta^2}{V_1}(\mu_1 - \mu_1^\circ)\left(1 + \frac{2\Delta}{R}\right) \qquad (7.70)$$

where V_1 is the molar volume of the solvent, μ_1 is the chemical potential of the solvent in the presence of free polymer with volume fraction ϕ_p and μ_1° is the chemical potential of the solvent in the absence of free polymer; $(\mu_1 - \mu_1^{\circ})$ is proportional to the osmotic pressure of the polymer solution.

7.21 Characterisation of Suspensions and Assessment of their Stability

A full assessment of the properties of suspensions requires three main types of investigations:

(1) Fundamental investigation of the system at a molecular level.
(2) Investigations into the state of the suspension on standing.
(3) Bulk properties of the suspension.

All three investigations require several sophisticated techniques, such as zeta potential measurements, surfactant and polymer adsorption and their conformation at the solid/liquid interface, measurement of the rate of flocculation and crystal growth, and several rheological measurements.

Apart from the above practical methods, which are present in most industrial laboratories, more fundamental information can be obtained using modern sophisticated techniques such as small angle X-ray and neutron scattering measurements, ultrasonic absorption techniques, etc.

Several other modern techniques are also available to investigate the state of the suspension: Freeze–fracture and electron microscopy, atomic force microscopy, scanning tunneling microscopy and confocal laser microscopy.

In all the above methods, care should be taken in sampling the suspension, which should cause as little disturbance as possible for the "structure" to be investigated. For example, when one investigates the flocculation of a concentrated suspension, dilution of the system for microscopic investigation may lead to break down of the flocs and a false assessment is obtained. The same applies to examinations of the rheology of a concentrated suspension, since transfer of the system from its container to the rheometer may lead to a break down of the structure.

For the above reasons one must establish well-defined procedures for every technique and this requires a great deal of skill and experience. It is advisable in all cases to develop standard operation procedures for the above investigations.

7.21.1 Assessment of the Structure of the Solid/Liquid Interface

7.21.1.1 Double Layer Investigations

Two procedures may be applied to investigate the charge and potential distribution at the solid/liquid interface, which is important in assessing electrostatic stabilisation: (1) Electrokinetic studies – this allows one to obtain the particle mobility as function of the system variables such as pH, electrolyte concentration, etc. From the electrophoretic mobility one can calculate the zeta potential, provided information is available on particle size and electrolyte concentration. When the above information is not available (as with many practical systems) one should use the electrophoretic mobility for relative comparison between various systems; the assumption can be made that the higher the mobility, the higher the surface charge and the more likely the system is stable against flocculation, if the charge is the main stabilising factor.

Clearly for systems stabilised by nonionic surfactants and polymers, electrophoretic mobility measurements are less informative. However, zeta potential measurements can be qualitatively used to obtain information on the adsorbed layer thickness for nonionic surfactants and polymers, as discussed before. When a nonionic surfactant or polymer adsorbs at the solid/liquid interface, a shift in the shear plane occurs and this results in reduction in the zeta potential. If the zeta potential of the particles is measured in the presence and absence of nonionic surfactant or polymer, then the adsorbed layer thickness can be roughly estimated from the reduction in zeta potential.

The above procedure requires measurement at various electrolyte concentrations, and extrapolation of the results to infinitely dilute electrolyte concentration.

Electrophoretic mobility measurements can also be used to investigate specifically adsorbed ions, which lead to a significant change in zeta potential and in some cases, with chemisorbed counter ions, charge reversal may occur.

7.21.1.2 Analytical Determination of Surface Charge

This can be applied for particles that contain ionogenic groups such as oxides. The oxide (such as silica or alumina) is covered with –OH groups that can be titrated with acid or alkali using a cell of the type:

$$E_1|\text{Oxide suspension in inert electrolyte at a given concentration}|E_2$$

where E_1 is an electrode reversible to H^+ and OH^- ion, such as a glass electrode, and E_2 is a reference electrode.

From a knowledge of the amount of H^+ and OH^- added and the amount remaining in solution, Γ_H and Γ_{OH} (the number of moles of H^+ and OH^- adsorbed) can be calculated and, hence, one can establish surface charge–pH isotherms.

Measurements are usually made at various electrolyte concentrations. In addition, the nature of the electrolyte is changed to investigate any possible specific adsorption.

7.21.1.3 Surfactant and Polymer Adsorption

A representative sample of the solid with known mass m and surface area A per gram is equilibrated with a surfactant or polymer concentration (C_1). After equilibrium is reached (at a given constant temperature), the solid is removed

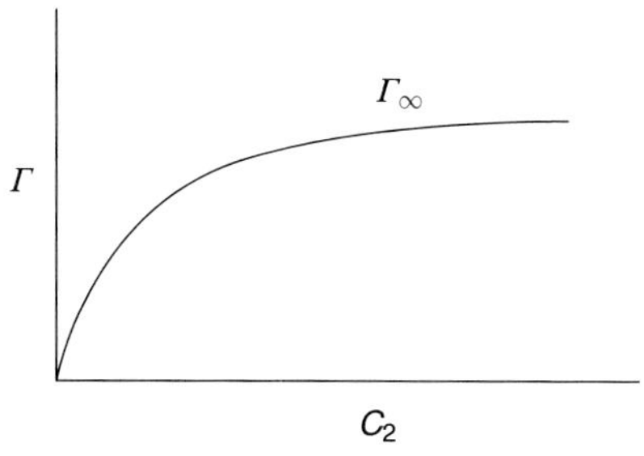

Fig. 7.36. Typical Langmuir isotherm for surfactant adsorption.

by centrifugation and the equilibrium concentration C_2 is determined analytically. The amount of adsorption Γ (mol m^{-2}) is given by,

$$\Gamma = \frac{(C_1 - C_2)}{mA} = \frac{\Delta C}{mA} \tag{7.71}$$

In most cases (particularly with surfactants) a plot of Γ versus C_2 gives a Langmuir-type isotherm (Figure 7.36).

The data can be fitted using the Langmuir equation,

$$\Gamma = \frac{\Gamma_\infty b C_2}{(1 + bC_2)} \tag{7.72}$$

where b is a constant that is related to the free energy of adsorption,

$$b \propto \left(-\frac{\Delta G_{\text{ads}}}{RT}\right) \tag{7.73}$$

Most polymers (particularly those with high molecular weight) give a high-affinity isotherm (Figure 7.37).

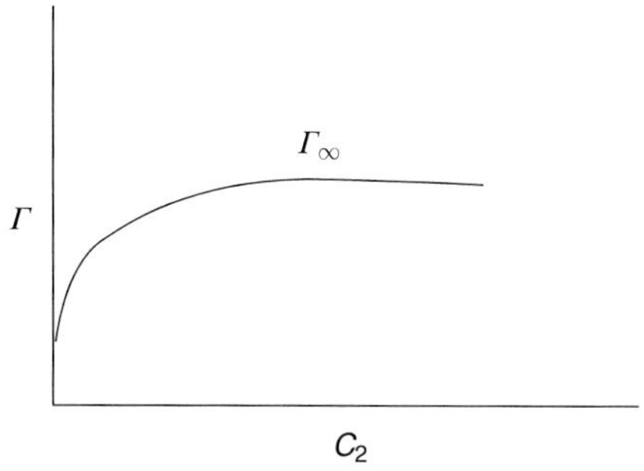

Fig. 7.37. Typical high-affinity isotherm for polymer adsorption.

7.21.2
Assessment of the State of the Dispersion

7.21.2.1 Measurement of Rate of Flocculation

Two general techniques may be applied for measuring the rate of flocculation of suspensions, both of which can only be applied for dilute systems. The first method is based on measuring the scattering of light by the particles. For monodisperse particles with a radius that is less than $\lambda/20$ (where λ is the wavelength of light) one can apply the Rayleigh equation, whereby the turbidity τ_0 is given by

$$\tau_0 = A' n_0 V_1^2 \tag{7.74}$$

in which A' is an optical constant (which is related to the refractive index of the particle and medium and the wavelength of light) and n_0 is the number of particles, each with a volume V_1.

By combining the Rayleigh theory with the Smoluchowski–Fuchs theory of flocculation kinetics [18] one can obtain Eq. (7.75) for the variation of turbidity with time, where k is the rate constant of flocculation.

$$\tau = A' n_0 V_1^2 (1 + 2n_0 kt) \tag{7.75}$$

The second method for obtaining the rate constant of flocculation is by direct particle counting as a function of time. For this purpose optical microscopy or image analysis may be used, provided the particle size is within the resolution limit of the microscope. Alternatively, the particle number may be determined using electronic devices such as a Coulter counter or a flow ultramicroscope.

The rate constant of flocculation is determined by plotting $1/n$ versus t, where n is the number of particles after time t, i.e.

$$\left(\frac{1}{n}\right) = \left(\frac{1}{n_0}\right) + kt \tag{7.76}$$

The rate constant k of slow flocculation is usually related to the rapid rate constant k_0 (the Smoluchowski rate) by the stability ratio W,

$$W = \left(\frac{k}{k_0}\right) \tag{7.77}$$

One usually plots log W versus log C (where C is the electrolyte concentration) to obtain the critical coagulation concentration (c.c.c.), which is the point at which log $W = 0$.

7.21.2.2 Measurement of Incipient Flocculation

This can be done for sterically stabilised suspensions, when the medium for the chains becomes a θ-solvent. This occurs, for example, on heating an aqueous suspension stabilised with poly(ethylene oxide) (PEO) or poly(vinyl alcohol) chains. Above a certain temperature (the θ-temperature), which depends on electrolyte concentration, the suspension flocculates. The temperature at which this occurs is defined as the critical flocculation temperature (CFT).

This process of incipient flocculation can be followed by measuring the turbidity of the suspension as a function of temperature. Above the CFT, the turbidity of the suspension rises very sharply.

For the above purpose, the cell in the spectrophotometer that is used to measure the turbidity is placed in a metal block that is connected to a temperature-programming unit (which allows one to increase the temperature rise at a controlled rate).

7.21.2.3 Measurement of Crystal Growth (Ostwald Ripening)

As discussed in Chapter 5, Ostwald ripening is the result of the difference in solubility S between small and large particles. The smaller particles are more soluble than the larger ones,

$$S \propto \frac{2\sigma}{r} \tag{7.78}$$

where σ is the solid/liquid interfacial tension and r is the particle radius.

For two particles with radii r_1 and r_2,

$$\frac{RT}{M} \ln\left(\frac{S_1}{S_2}\right) = \left(\frac{2\sigma}{\rho}\right)\left(\frac{1}{r_1} - \frac{1}{r_2}\right) \tag{7.79}$$

where R is the gas constant, T is the absolute temperature, M is the molecular weight and ρ is the density of the particles.

To obtain a measure of the rate of crystal growth, the particle size distribution of the suspension is followed as a function of time, using either a Coulter counter, a Master sizer or an optical disc centrifuge. One usually plots the cube of the average radius versus time, which gives a straight line, from which the rate of crystal growth can be determined (the slope of the linear curve).

7.22 Bulk Properties of Suspensions

7.22.1 Equilibrium Sediment Volume (or Height) and Redispersion

For a "structured" suspension, obtained by "controlled" flocculation or addition of "thickeners" (such as polysaccharides, clays or oxides), the "flocs" sediment at a rate dependent on the size and porosity of the aggregated mass. After this initial sedimentation, compaction and rearrangement of the floc structure occurs, a phenomenon referred to as consolidation.

Normally, in sediment volume measurements, one compares the initial volume V_0 (or height H_0) with the ultimately reached value V (or H). A colloidally stable suspension gives a "close-packed" structure with relatively small sediment volume (dilatant sediment referred to as clay). A weakly "flocculated" or "structured" suspension gives a more open sediment and hence a higher sediment volume. Thus by comparing the relative sediment volume V/V_0 or height H/H_0, one can distinguish between a clayed and flocculated suspension.

7.22.2 Rheological Measurements

Three different rheological measurements may be applied

(1) Steady-state shear stress–shear rate measurements (using a controlled shear rate instrument).
(2) Constant stress (creep) measurements (carried out using a constant stress instrument).
(3) Dynamic (oscillatory) measurements (preferably carried out using a constant strain instrument).

The above rheological techniques can be used to assess sedimentation and flocculation of suspensions. This will be discussed in detail below.

7.22.3 Assessment of Sedimentation

As will be shown in the next section, the rate of sedimentation decreases with increasing volume fraction of the disperse phase, ϕ, and ultimately approaches zero at a critical volume fraction ϕ_p (the maximum packing fraction). However, at $\phi \sim \phi_p$, the viscosity of the system approaches ∞. Thus, for most practical emulsions, the system is prepared at ϕ below ϕ_p and "thickeners" are added to reduce sedimentation. These "thickeners" are usually high molecular weight polymers (such as xanthan gum, hydroxyethyl cellulose or associative thickeners), finely divided inert solids (such as silica or swelling clays) or a combination of the two.

In all cases, a "gel" network is produced in the continuous phase that is shear thinning (i.e. its viscosity decreases with increasing shear rate) and viscoelastic (i.e. it has viscous and elastic components of the modulus). If the viscosity of the elastic network, at shear stresses (or shear rates) comparable to those exerted by the particles, exceeds a certain value, then sedimentation is completely eliminated.

The shear stress, σ_p, exerted by a particle (force/area) can be simply calculated,

$$\sigma_p = \frac{(4/3)\pi R^3 \Delta\rho g}{4\pi R^2} = \frac{\Delta\rho R g}{3} \tag{7.80}$$

For a 10 μm radius particle with density difference $\Delta\rho = 0.2$, σ_p is equal to

$$\sigma_p = \frac{0.2 \times 10^3 \times 10 \times 10^{-6} \times 9.8}{3} \approx 6 \times 10^{-3} \quad (7.81)$$

For smaller droplets smaller stresses are exerted.

Thus, to predict sedimentation, one has to measure the viscosity at very low stresses (or shear rates). These measurements can be carried out using a constant stress rheometer (Carrimed, Bohlin, Rheometrics or Physica). A constant stress σ (using for example a drag cup motor that can apply very small torques and using an air bearing system to reduce the frictional torque) is applied on the system (which may be placed in the gap between two concentric cylinders or a cone–plate geometry) and the deformation [strain γ or compliance $J = (\gamma/\sigma)$ Pa^{-1}] is followed as a function of time [39–41].

For a viscoelastic system, the compliance shows a rapid elastic response J_0 at $t \to 0$ [instantaneous compliance $J_0 = 1/G_0$, where G_0 is the instantaneous modulus, which is a measure of the elastic (i.e. "solid-like") component]. At $t \to 0$, J increases slowly with time and this corresponds to the retarded response ("bonds" are broken and reformed but not at the same rate). Above a certain time period (which depends on the system), the compliance shows a linear increase with time (i.e. the system reaches a steady state with constant shear rate). If, after the steady state is reached, the stress is removed elastic recovery occurs and the strain changes sign.

The above behaviour (usually referred to as "creep") is schematically represented in Figure 7.38.

The slope of the linear part of the creep curve gives the value of the viscosity at the applied stress, η_σ,

$$\frac{J}{t} = \frac{Pa^{-1}}{s} = \frac{1}{Pa\ s} = \eta_\sigma \quad (7.82)$$

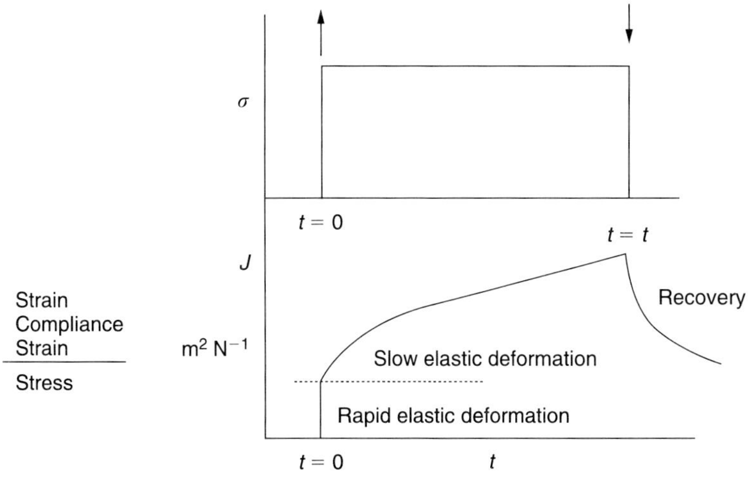

Fig. 7.38. Typical creep curve for a viscoelastic system.

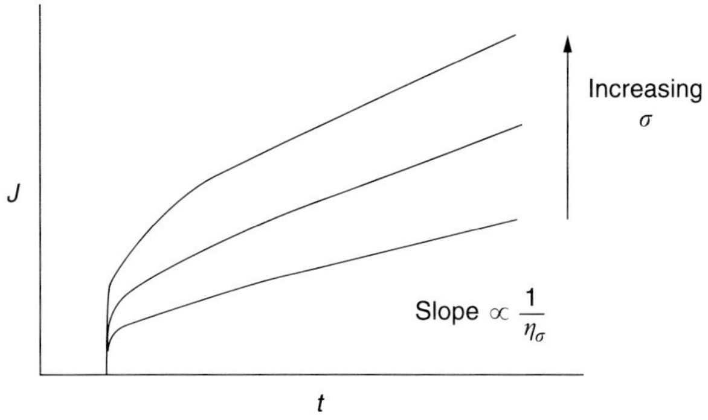

Fig. 7.39. Creep curves as a function of applied stress.

The recovery curve will only give the elastic component, which if superimposed on the ascending part of the curve will give the viscous component.

Thus, one measures creep curves as a function of the applied stress (starting from a very small stress of the order of 0.01 Pa). This is illustrated in Figure 7.39.

The viscosity η_σ (which is equal to the reciprocal of the slope of the straight portion of the creep curve) is plotted as a function of the applied stress (Figure 7.40).

Below a critical stress, σ_{cr}, the viscosity reaches a limiting value, $\eta_{(0)}$ namely the residual (or zero shear) viscosity; σ_{cr} may be denoted as the "true yield stress" of the emulsion, i.e. the stress above which the "structure" of the system is broken down. Above σ_{cr}, η_σ decreases rapidly with further increase of the shear stress (the shear thinning regime). It reaches another Newtonian value η_{∞}, which is the high shear limiting viscosity.

$\eta_{(0)}$ could be several orders of magnitudes (10^4–10^8) higher than η_∞. Usually, one obtains a good correlation between the rate of sedimentation v and the residual viscosity $\eta_{(0)}$ (Figure 7.41) [42].

Above a certain $\eta_{(0)}$, v becomes 0. Clearly, to minimise sedimentation one has to increase $\eta_{(0)}$; an acceptable level for the high shear viscosity η_∞ must be achieved, depending on the application. In some cases, a high $\eta_{(0)}$ may be accompanied by a

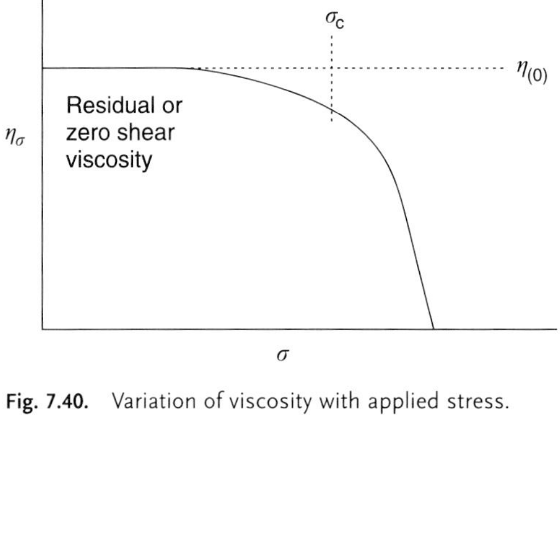

Fig. 7.40. Variation of viscosity with applied stress.

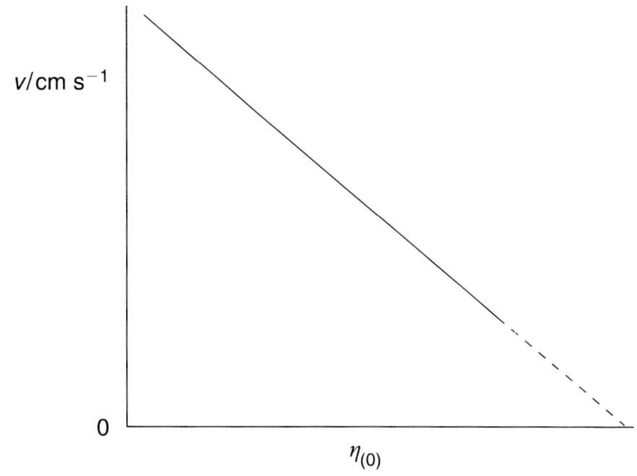

Fig. 7.41. Variation of sedimentation rate with residual viscosity.

high η_∞ (which may not be acceptable for application, for example if spontaneous dispersion on dilution is required). If this is the case, the formulation chemist should look for an alternative thickener.

Another problem encountered with many suspensions is that of "syneresis", i.e. the appearance of a clear liquid film at the top of the suspension. "Syneresis" occurs with most "flocculated" and/or "structured" (i.e. those containing a thickener in the continuous phase) suspensions. "Syneresis" may be predicted from measurement of the yield value (using steady-state measurements of shear stress as a function of shear rate) as a function of time or using oscillatory techniques (whereby the storage and loss modulus are measured as a function of strain amplitude and frequency of oscillation). These techniques will be discussed in detail below.

It is sufficient to state here that when a network of the suspension particles (either alone or combined with the thickener) is produced, the gravity force will cause some contraction of the network (which behaves as a porous plug), thus causing some separation of the continuous phase that is entrapped between the droplets in the network.

7.22.4 Assessment of Flocculation

As mentioned before, flocculation of suspensions is the result of the long-range van der Waals attraction. Flocculation can be weak (and reversible) or strong, depending on the magnitude of the net attractive forces. Weak flocculation may result in reversible time dependence of the viscosity, i.e. on shearing the emulsion at a given shear rate the viscosity decreases, and on standing the viscosity recovers to its original value. This phenomenon is referred to as thixotropy (sol–gel transformation).

Rheological techniques are most convenient to assess suspension flocculation without the need of any dilution (which in most cases result in breakdown of the floc structure). In steady-state measurements the suspension is carefully placed in the gap between concentric cylinder or cone-and-plate platens. For the concentric cylinder geometry, the gap width should be at least 10× larger than the largest droplet (a gap width that is greater than 1 mm is usually used). For the cone-and-plate geometry a cone angle of 4° or smaller is usually employed.

A controlled rate instrument is usually used for the above measurements; the inner (or outer) cylinder, the cone (or the plate) is rotated at various angular velocities (allowing one to obtain the shear rate γ) and the torque is measured on the other element (allowing one to obtain the stress σ).

For a Newtonian system (such as the case of a dilute suspension, with a volume fraction ϕ less than 0.1) σ is related to $\dot{\gamma}$ by the equation,

$$\sigma = \eta\dot{\gamma} \tag{7.83}$$

where η is the Newtonian viscosity (which is independent of the applied shear rate).

For most practical suspensions (with $\phi > 0.1$ and containing thickeners to reduce sedimentation) a plot of σ versus $\dot{\gamma}$ is not linear (i.e. the viscosity depends on the applied shear rate). The most common flow curve is shown in Figure 7.42 (usually described as a pseudo-plastic or shear thinning system). In this case, the viscosity decreases with increasing shear rate, reaching a Newtonian value above a critical shear rate.

Several models may be applied to analyse the results of Figure 7.42.

(a) Power Law Model

$$\sigma = k\dot{\gamma}^n \tag{7.84}$$

where k is the consistency index of the emulsion and n is the power (shear thinning) index ($n < 1$); the lower n the more shear thinning the suspension is.

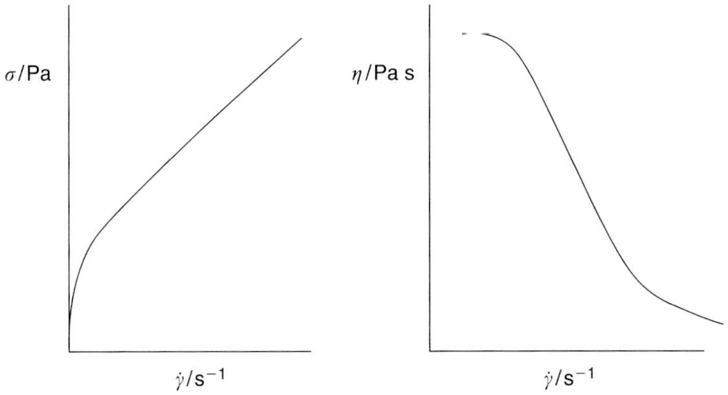

Fig. 7.42. Shear stress and viscosity versus shear rate for a pseudo-plastic system.

This is usually the case with weakly flocculated suspensions or those to which a "thickener" is added.

By fitting the results of Figure 7.42 to Eq. (7.84) (this is usually in the software of the computer connected to the rheometer) one can obtain the viscosity of the emulsion at a given shear rate,

$$\eta \text{ (at a given shear rate)} = \frac{\sigma}{\dot{\gamma}} = k\dot{\gamma}^{n-1} \tag{7.85}$$

(b) Bingham Model

$$\sigma = \sigma_\beta + \eta_{\text{pl}}\dot{\gamma} \tag{7.86}$$

where σ_β is the extrapolated yield value (obtained by extrapolation of the shear stress–shear rate curve to $\dot{\gamma} = 0$). Again this is provided in the software of the rheometer. η_{pl} is the slope of the linear portion of the σ–$\dot{\gamma}$ curve (usually referred to as the plastic viscosity).

Both σ_β and η_{pl} may be related to the flocculation of the suspension. At any given volume fraction of the emulsion and at a given particle size distribution, the higher the value of σ_β and η_{pl} the more the flocculated the suspension is. Thus, if one stores a suspension at any given temperature and makes sure that the particle size distribution remains constant (i.e. no Ostwald ripening occurs), an increase in the above parameters indicates flocculation of the suspension on storage. Clearly, if Ostwald ripening occurs simultaneously, σ_β and η_{pl} may change in a complex manner with storage time. Ostwald ripening results in a shift of the particle size distribution to higher diameters; this has the effect of reducing both σ_β and η_{pl}. If flocculation occurs simultaneously (having the effect of increasing these rheological parameters), the net effect may be an increase or decrease of the rheological parameters.

The above trend depends on the extent of flocculation relative to Ostwald ripening. Therefore, following σ_β and η_{pl} with storage time requires knowledge of Ostwald ripening and/or coalescence. Only in the absence of this latter breakdown process can one use rheological measurements as a guide of assessment of flocculation.

(c) Herschel–Buckley Model [43]

In many cases, the shear stress–shear rate curve may not show a linear portion at high shear rates. In this case, the data may be fitted with a Hershel–Buckley model,

$$\sigma = \sigma_\beta + k\dot{\gamma}^n \tag{7.87}$$

(d) Casson's Model [44]

This is another semi-empirical model that may be used to fit the data of Figure 7.42,

$$\sigma^{1/2} = \sigma_C^{1/2} + \eta_C^{1/2}\dot{\gamma}^{1/2} \qquad (7.88)$$

Note that σ_β is not equal to σ_C.

Eq. (7.88) shows that a plot of $\sigma^{1/2}$ versus $\dot{\gamma}^{1/2}$ gives a straight line, from which σ_C and η_C can be evaluated.

In all the above analyses, the assumption was made that a steady state was reached. In other words, no time effects occurred during the flow experiment.

7.22.5 Time Effects during Flow – Thixotropy

Many suspensions (particularly those that are weakly flocculated or "structured" to reduce sedimentation) show time effects during flow. At any given shear rate, the viscosity of the suspension continues to decrease with increasing the time of shear; on stopping the shear, the viscosity recovers to its initial value. This reversible decrease of viscosity is referred to as thixotropy.

The most common procedure of studying thixotropy is to apply a sequence of shear stress – shear rate regimes within controlled periods. If the flow curve is carried out within a very short time (say increasing the rate from 0 to say 500 s^{-1} in 30 s and then reducing it again from 500 to 0 s^{-1} within the same period), one finds that the descending curve is below the ascending one.

The above behaviour can be explained from consideration of the structure of the system. If, for example, the suspension is weakly flocculated, then on applying a shear force on the system this flocculated structure is broken down (and this is the cause of the shear thinning behaviour). On reducing the shear rate back to zero the structure builds up only in part within the duration of the experiment (30 s).

The ascending and descending flow curves show hysteresis that is usually referred to as "thixotropic loop". If the same experiment is now repeated over a longer time (say 120 s for the ascending and 120 s for the descending curves), the hysteresis decreases, i.e. the "thixotropic loop" becomes smaller.

By repeating the above experiments within various time periods one obtains a series of thixotropic loops (Figure 7.43).

The above study may be used to investigate the state of flocculation of a suspension. Weakly flocculated suspensions usually show thixotropy and the change of thixotropy with applied time may be used as an indication of the strength of this weak flocculation.

The above analysis is only qualitative and one cannot use the results in quantitatively. This is due to the possible breakdown of the structure on transferring the suspension to the rheometer and also during the uncontrolled shear experiment.

A very important point to consider during rheological measurements is the possibility of "slip" during the measurements. This is particularly the case with highly concentrated suspensions, whereby the flocculated system may form a "plug" in the gap of the platens, leaving a thin liquid film at the walls of the concentric cylinder or cone-and-plate geometry. To reduce "slip" one should use roughened walls for the platens.

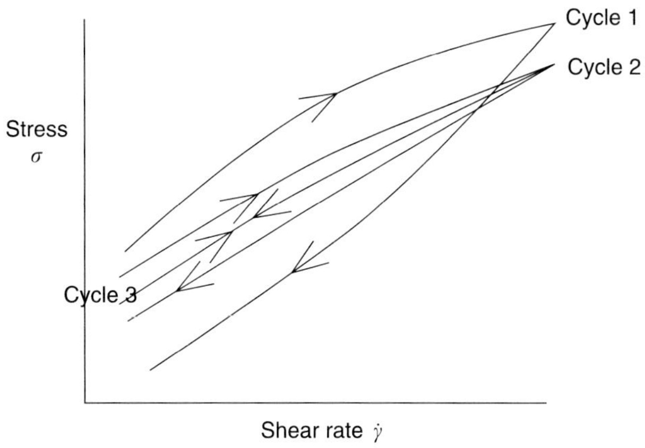

Fig. 7.43. Flow curves for a thixotropic system.

Strongly flocculated suspensions usually show much less thixotropy than weakly flocculated systems. Again, one must be careful in drawing definite conclusions without other independent techniques (e.g. microscopy).

7.22.6 Constant Stress (Creep) Experiments

This method has been described in detail in the section on sedimentation. Basically a constant stress σ is applied on the system and the compliance J (Pa^{-1}) is plotted as a function of time (creep curve, Figure 7.38).

The above experiment is repeated several times, increasing the stress from the smallest possible value (that can be applied by the instrument) in small increments. A set of creep curves are produced at various applied stresses (Figure 7.39).

From the slope of the linear portion of the creep curve (after the system reaches a steady state), the viscosity at each applied stress, η_σ, is calculated. A plot of η_σ versus σ (Figure 7.40) allows one to obtain the limiting (or zero shear) viscosity $\eta_{(0)}$ and the critical stress σ_{cr} (which may be identified with the "true" yield stress of the system).

The values of $\eta_{(0)}$ and σ_{cr} may be used to assess the flocculation of the suspension on storage. If flocculation occurs on storage (without any Ostwald ripening or coalescence), $\eta_{(0)}$ and σ_{cr} may show a gradual increase with increasing storage time.

As discussed in the previous section (on steady-state measurements), the trend becomes complicated if Ostwald ripening occurs simultaneously (both have the effect of reducing $\eta_{(0)}$ and σ_{cr}).

The above measurements should be supplemented by particle size distribution measurements of the diluted suspension (making sure that no flocs are present after dilution) to assess the extent of Ostwald ripening.

Another complication may arise from the nature of the flocculation. If the latter occurs in an irregular way (producing strong and tight flocs), $\eta_{(0)}$ may increase, while σ_{cr} may show some decrease, thus complicating the analysis of the results.

Despite the above complications, constant stress measurements may provide valuable information on the state of the suspension on storage. Carrying out creep experiments and ensuring that a steady state is reached can be time consuming. One usually carries out a stress sweep experiment, whereby the stress is gradually increased (within a predetermined time period to ensure that one is not too far from reaching the steady state) and plots of η_σ versus σ are established.

The above experiments are carried out at various storage times (say every two weeks) and temperatures. From the change of $\eta_{(0)}$ and σ_{cr} with storage time and temperature, one may obtain information on the degree and the rate of flocculation of the system.

Clearly, interpretation of the rheological results requires expert knowledge of rheology and measurement of the particle size distribution as a function of time.

One main problem in carrying the above experiments is sample preparation. When a flocculated emulsion is removed from the container, care should be taken not to cause much disturbance to that structure (minimum shear should be applied on transferring the emulsion to the rheometer). It is also advisable to use separate containers for assessment of the flocculation. A relatively large sample is prepared and this is then transferred to a number of separate containers.

7.22.7
Dynamic (Oscillatory) Measurements

These are by far the most commonly used method to obtain information on the flocculation of a suspension. A strain is applied in a sinusoidal manner, with an amplitude γ_0 and a frequency ν (cycles s^{-1} or Hz) or ω (rad s^{-1}). This is usually carried out by moving one of the platens say the cup (in a concentric cylinder geometry) or the plate (in a cone-and-plate geometry) back and forth in a sinusoidal manner. The stress on the other platen, the bob or the cone is simultaneously measured. These platens are usually connected to interchangeable torque bars, whereby the stress can be directly measured. The stress amplitude σ_0 is simultaneously measured.

In a viscoelastic system (such as the case with a flocculated suspension), the stress oscillates with the same frequency, but out-of-phase from the strain. From measurement of the time shift between strain and stress amplitudes (Δt) one can obtain the phase angle shift δ,

$$\delta = \Delta t\omega \tag{7.89}$$

Figure 7.44 gives a schematic representation of the variation of strain and stress with Δt.

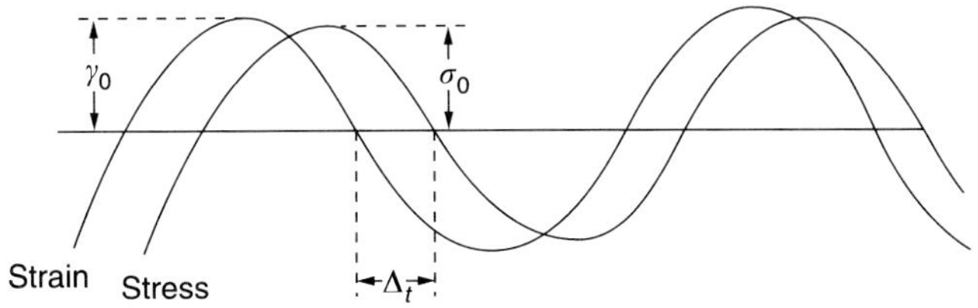

Fig. 7.44. Stress–strain relationship for a viscoelastic system.

From the amplitudes of stress and strain and the phase-angle shift one can obtain the various viscoelastic parameters: The complex modulus G^*, the storage modulus (the elastic component of the complex modulus) G', the loss modulus (the viscous component of the complex modulus) G'', $\tan\delta$ and the dynamic viscosity η'.

Complex modulus $$|G^*| = \frac{\sigma_0}{\gamma_0} \tag{7.90}$$

Storage modulus $$G' = |G^*| \cos\delta \tag{7.91}$$

Loss modulus $$G'' = |G^*| \sin\delta \tag{7.92}$$

$$\tan\delta = \frac{G''}{G'} \tag{7.93}$$

Dynamic viscosity $$\eta' = \frac{G''}{\omega} \tag{7.94}$$

G' is a measure of the energy stored in a cycle of oscillation; G'' is a measure of the energy dissipated as viscous flow in a cycle of oscillation; $\tan\delta$ is a measure of the relative magnitudes of the viscous and elastic components. Clearly, the smaller $\tan\delta$ is the more elastic the system is and vice versa.

η', the dynamic viscosity, shows a decrease with increase of frequency ω, reaching a limiting value as $\omega \to 0$; the value of η' in this limit is identical to the residual (or zero shear) viscosity $\eta_{(0)}$. This is referred to as the Cox–Mertz rule.

In oscillatory measurements one carries out two sets of experiments, strain sweep and oscillatory sweep, which are detailed below.

7.22.7.1 Strain Sweep Measurements

Here, the oscillation is fixed (say at 0.1 or 1 Hz) and the viscoelastic parameters are measured as a function of strain amplitude. This allows one to obtain the linear viscoelastic region. In this region all moduli are independent of the applied strain amplitude and become only a function of time or frequency. This is illustrated in Figure 7.45, which shows a schematic representation of the variation of G^*, G' and G'' with strain amplitude (at a fixed frequency).

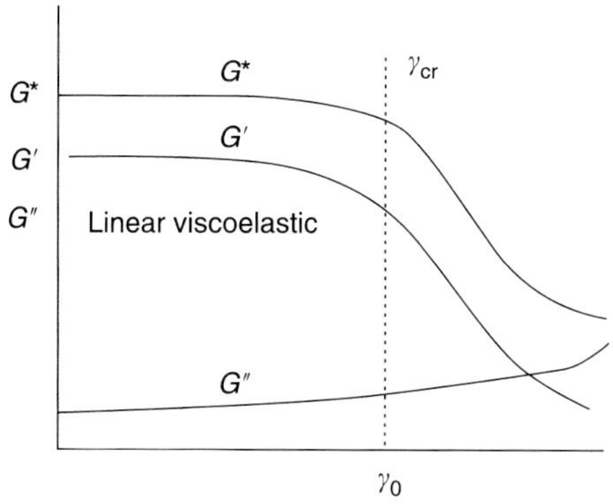

Fig. 7.45. Strain sweep results.

Figure 7.45 shows that G^*, G' and G'' remain virtually constant up to a critical strain value, γ_{cr}. This region is the linear viscoelastic region. Above γ_{cr}, G^* and G' starts to fall, whereas G'' starts to increase. This is the nonlinear region. The value of γ_{cr} may be identified with the minimum strain above which the "structure" of the emulsion starts to break down (for example breakdown of flocs into smaller units and/or breakdown of a "structuring" agent).

From γ_{cr} and G', one can obtain the cohesive energy E_c (J m^{-3}) of the flocculated structure,

$$E_c = \int_0^{\gamma_{cr}} \sigma \, d\gamma = \int_0^{\gamma_{cr}} G'\gamma \, d\gamma = \tfrac{1}{2} G' \gamma_{cr}^2 \tag{7.95}$$

E_c may be used in a quantitative manner as a measure of the extent and strength of the flocculated structure in a suspension. The higher E_c is, the more flocculated the structure. Clearly E_c depends on the volume fraction of the suspension as well as the particle size distribution (which determines the number of contact points in a floc). Therefore, for quantitative comparison between various systems, one has to make sure that the volume fraction of the disperse particles is the same and the suspensions have very similar particle size distributions.

E_c also depends on the strength of the flocculated structure, i.e. the energy of attraction between the particles [45, 46]. This depends on whether the flocculation is in the primary or secondary minimum. Flocculation in the primary minimum is associated with a large attractive energy and this leads to higher E_cs than are obtained for secondary minimum flocculation. For a weakly flocculated suspension, such as the case with secondary minimum flocculation of an electrostatically stabilised suspension, the deeper the secondary minimum, the higher the value of E_c (at any given volume fraction and particle size distribution of the suspension).

With a sterically stabilised suspension, weak flocculation can also occur when the thickness of the adsorbed layer decreases. Again E_c can be used as a measure of the flocculation; the higher the value of E_c, the stronger the flocculation.

If incipient flocculation occurs (on reducing the solvency of the medium for the change to worse than θ-condition) a much deeper minimum is observed and this is accompanied by a much larger increase in E_c.

To apply the above analysis, one must have an independent method for assessing the nature of the flocculation. Rheology is a bulk property that can give information on the interparticle interaction (whether repulsive or attractive) and to apply it in a quantitative manner one must know the nature of these interaction forces. However, rheology can be used in a qualitative manner to follow the change of the suspension on storage. Providing the system does not undergo Ostwald ripening, the change of the moduli with time and, in particular, the change of the linear viscoelastic region may be used as an indication of flocculation. Strong flocculation is usually accompanied by a rapid increase in G' and this may be accompanied by a decrease in the critical strain above which the "structure" breaks down. This may be used as an indication of formation of "irregular" flocs that become sensitive to the applied strain. The floc structure will entrap a large amount of the continuous phase, leading to an apparent increase in the volume fraction of the suspension and hence an increase in G'.

7.22.1.2 **Oscillatory Sweep**

In this case, the strain amplitude is kept constant in the linear viscoelastic region (one usually takes a point far from γ_{cr} but not too low (i.e. in the mid-point of the linear viscoelastic region) and measurements are carried out as a function of frequency. This is schematically represented in Figure 7.46 for a viscoelastic liquid system.

Both G^* and G' increase with increasing frequency and, ultimately, above a certain frequency they reach a limiting value and show little dependence on frequency.

G'' is higher than G' in the low frequency regime; it also increases with increasing frequency and at a certain characteristic frequency ω^* (which depends on the

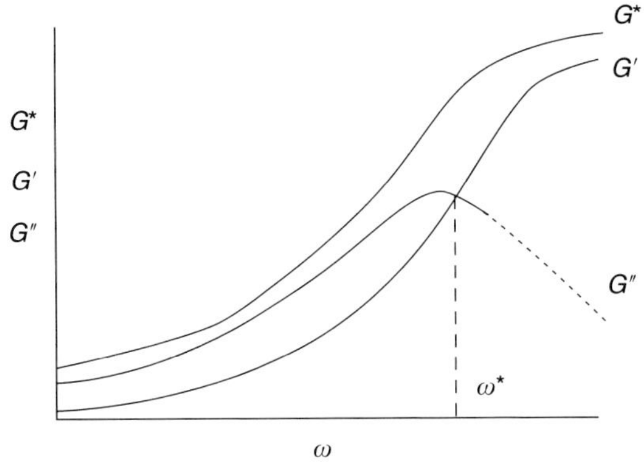

Fig. 7.46. Oscillatory sweep results.

system) it becomes equal to G' (usually referred to as the cross-over point), after which it reaches a maximum and then shows a reduction with further increase in frequency.

In the low frequency regime, i.e. below ω^*, $G'' > G'$; this regime corresponds to longer times (remember that the time is reciprocal of frequency) and under these conditions the response is more viscous than elastic. In the high frequency regime, i.e. above ω^*, $G' > G''$; this regime corresponds to short times and under these conditions the response is more elastic than viscous.

At sufficiently high frequency, G'' approaches zero and G' becomes nearly equal to G^*; this corresponds to very short time scales, whereby the system behaves as a near elastic solid. Very little energy dissipation occurs at such high frequency.

The characteristic frequency ω^* can be used to calculate the relaxation time of the system (t^*),

$$t^* = \frac{1}{\omega^*} \tag{7.96}$$

The relaxation time may be used as a guide for the state of the suspension. For a colloidally stable suspension (at a given particle size distribution), t^* increases with increase of the volume fraction of the solid phase, ϕ. In other words, the cross-over point shifts to lower frequency with increase in ϕ. For a given suspension, t^* increases with increasing flocculation, providing the particle size distribution remains the same (i.e. no Ostwald ripening).

G' also increases with increasing flocculation, since aggregation of particles usually results in liquid entrapment and the effective volume fraction of the suspension shows an apparent increase. With flocculation, the net attraction between the droplets also increases and this results in an increase in G'. G' is determined by the number of contacts between the particles and the strength of each contact (which is determined by the attractive energy).

Notably, in practice one may not obtain the full curve, due to the frequency limit of the instrument and, also, measurement at low frequency is time consuming. Usually one obtains part of the frequency dependence of G' and G''. In most cases, one has a more elastic than viscous system. Most suspension systems used in practice are weakly flocculated and they also contain "thickeners" or "structuring" agents to reduce sedimentation and to acquire the right rheological characteristics for application.

The exact values of G' and G'' required depend on the system and its application. In most cases a compromise has to be made between acquiring the right rheological characteristics for application and the optimum rheological parameters for long-term physical stability.

Application of rheological measurements to achieve the above conditions requires a great deal of skill and understanding of the factors that affect rheology.

7.23
Sedimentation of Suspensions and Prevention of Formation of Dilatant Sediments (Clays)

As discussed before, most suspensions undergo separation on standing as a result of the density difference between the particles and the medium, unless the particles are small enough for Brownian motion to overcome gravity. Figure 7.47 illustrates this for three cases of suspensions.

The most practical situation is that represented by Figure 7.47(C), whereby a concentration gradient of the particles occurs across the container. The concentration of particles C can be related to that before any settling C_0 by the equation

$$C = C_0 \exp\left(-\frac{mgh}{kT}\right) \tag{7.97}$$

where m is the mass of the particles that is given by $(4/3)\pi R^3 \Delta\rho$ (R is the particle radius and $\Delta\rho$ is the density difference between particle and medium), g is the acceleration due to gravity and h is the height of the container.

For a very dilute suspension of rigid non-interacting particles, the rate of sedimentation v_0 can be calculated by application of Stokes' law, whereby the hydrodynamic force is balanced by the gravitational force,

$$\text{Hydrodynamic force} = 6\pi\eta R v_0 \tag{7.98}$$

$$\text{Gravity force} = \tfrac{4}{3}\pi R^3 \Delta\rho g \tag{7.99}$$

$$v_0 = \frac{2}{9}\frac{R^2 \Delta\rho g}{\eta} \tag{7.100}$$

where η is the viscosity of the medium (water).

v_0 was calculated for three particle sizes (0.1, 1 and 10 μm) for a suspension with density difference $\Delta\rho = 0.2$. The values of v_0 are 4.4×10^{-9}, 4.4×10^{-7} and 4.4×10^{-5} m s^{-1} respectively. The time needed for complete sedimentation in a 0.1 m container is 250 days, 60 hours and 40 minutes respectively.

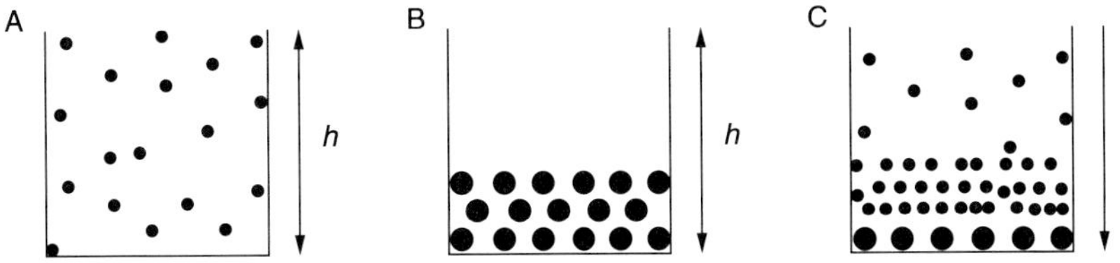

Fig. 7.47. Schematic representation of particle sedimentation: (A) Submicron particles, Brownian diffusion > gravity. (B) Coarse particles (> 1 μm) with uniform size. (C) Coarse particles with size distribution.

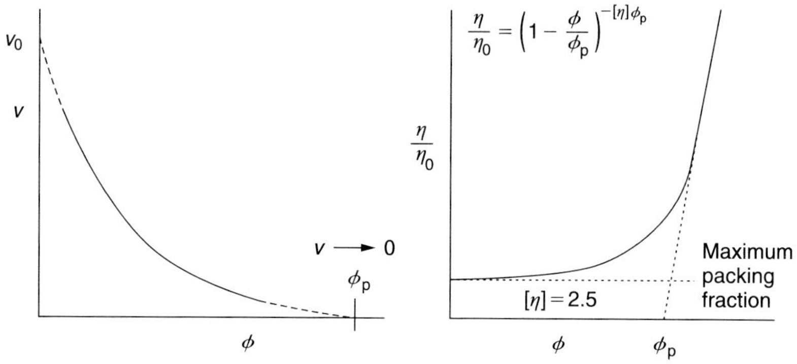

Fig. 7.48. Variation of sedimentation velocity and relative viscosity with ϕ.

For moderately concentrated suspensions, $0.2 > \phi > 0.01$, sedimentation is reduced as a result of hydrodynamic interaction between the particles, which no longer sediment independently. The sedimentation velocity, v, can be related to the Stokes' velocity v_0 by Eq. (7.101).

$$v = v_0(1 - 6.55\phi) \tag{7.101}$$

This means that for a suspension with $\phi = 0.1$, $v = 0.345v_0$, i.e. the rate is reduced by a factor of $\sim$3.

For more concentrated suspensions ($\phi > 0.2$), the sedimentation velocity becomes a complex function of ϕ. At $\phi > 0.4$, one usually enters the hindered settling regime, whereby all the particles sediment at the same rate (independent of size).

A schematic representation for the variation of v with ϕ is shown in Figure 7.48, which also shows the variation of relative viscosity with ϕ. Clearly, v decreases exponentially with increase in ϕ and ultimately approaches zero when ϕ approaches a critical value ϕ_p (the maximum packing fraction). The relative viscosity shows a gradual increase with increasing ϕ and, when $\phi = \phi_p$ the relative viscosity approaches infinity.

The maximum packing fraction ϕ_p can be calculated easily for monodisperse rigid spheres. For hexagonal packing $\phi_p = 0.74$, whereas for random packing $\phi_p = 0.64$. The maximum packing fraction increases with polydisperse suspensions. For example, for a bimodal particle size distribution (with a ratio of $\sim$10:1) $\phi_p > 0.8$.

The relative sedimentation rate (v/v_0) can be related to the relative viscosity η/η_0 as

$$\left(\frac{v}{v_0}\right) = \alpha\left(\frac{\eta_0}{\eta}\right) \tag{7.102}$$

The relative viscosity is related to the volume fraction ϕ by the Dougherty–Krieger equation [21] for hard spheres,

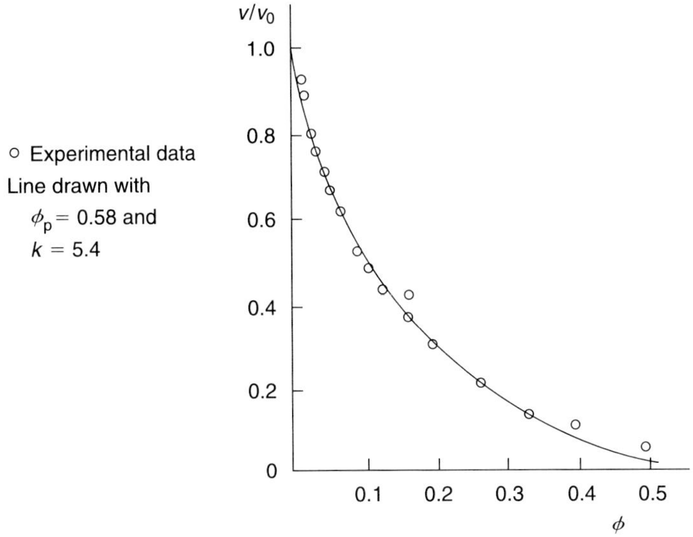

Fig. 7.49. Variation of v/v_0 with ϕ for polystyrene latex suspensions ($R = 1.55$ μm).

$$\frac{\eta}{\eta_0} = \left(1 - \frac{\phi}{\phi_p}\right)^{-[\eta]\phi_p} \tag{7.103}$$

where $[\eta]$ is the intrinsic viscosity (2.5 for hard spheres).

Combining Eqs. (7.102) and (7.103) gives

$$\frac{v}{v_0} = \left(1 - \frac{\phi}{\phi_p}\right)^{\alpha[\eta]\phi_p} = \left(1 - \frac{\phi}{\phi_p}\right)^{k\phi_p} \tag{7.104}$$

The above empirical relationship was tested for sedimentation of polystyrene latex suspensions with $R = 1.55$ μm in 10^{-3} mol dm^{-3} NaCl. The results are shown in Figure 7.49; the open circles are the experimental points, whereas the solid line is calculated using Eq. (7.103) with ϕ_p and $k = 5.4$.

The sedimentation of particles in non-Newtonian fluids, such as aqueous solutions containing high molecular weight compounds (e.g. hydroxyethyl cellulose or xanthan gum), is not simple since these non-Newtonian solutions are shear thinning with the viscosity decreasing with increase in shear rate. As discussed above, these solutions show a Newtonian region at low shear rates or shear stresses, usually referred to as the residual or zero shear viscosity $\eta(0)$.

As discussed above, the stress exerted by the particles is very small, in the region of 10^{-3}–10^{-1} Pa, depending on the particle size and the density of the particles.

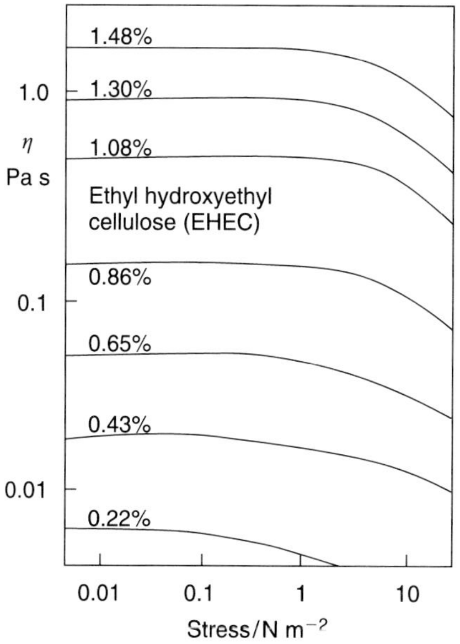

Fig. 7.50. Variation of viscosity with applied stress for solutions of EHEC at various concentrations.

Clearly, to predict sedimentation one needs to measure the viscosity at such low stresses. This is illustrated for solutions of ethyl hydroxy ethyl cellulose (EHEC) in Figure 7.50.

Good correlation is found between the rate of sedimentation for polystyrene latex and $\eta(0)$ [42] (Figure 7.51). For this suspension, no sedimentation occurred when $\eta(0)$ was greater than 10 Pa s.

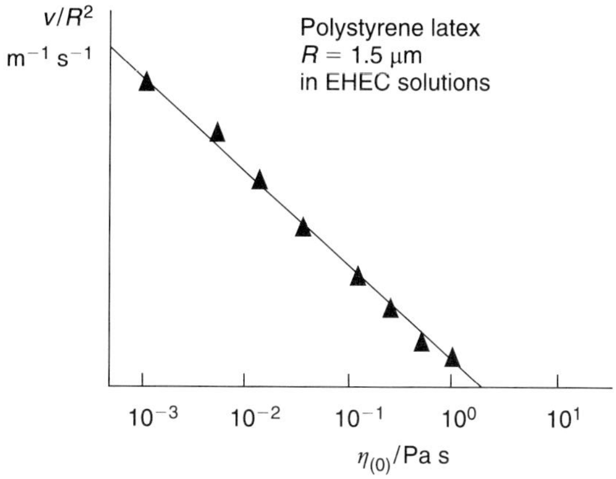

Fig. 7.51. Variation of v/R^2 with $\eta(0)$ for polystyrene latex suspensions ($R = 1.55$).

The situation with more practical dispersions is more complex due to the interaction between the thickener and the particles. Most practical suspensions show some weak flocculation and the "gel" produced between the particles and thickener may undergo some contraction as a result of the gravity force exerted on the whole network. A useful method to describe separation in these concentrated suspensions is to follow the relative sediment volume V_t/V_0 or relative sediment height h_t/h_0 (where the subscripts t and o refer to time t and zero time respectively) with storage time. For good physical stability the V_t/V_0 or h_t/h_0 should be as close as possible to unity (i.e. minimum separation). This can be achieved by balancing the gravitational force exerted by the gel network with the bulk "elastic" modulus of the suspension. The latter is related to the high frequency modulus G'.

7.24 Prevention of Sedimentation and Formation of Dilatant Sediments

Several methods may be applied to prevent sedimentation and formation of clays or cakes in a suspension and these are summarised below.

7.24.1 Balance of the Density of the Disperse Phase and Medium

It is clear from Stokes' law that if $\Delta\rho = 0$, $v_0 = 0$. This method can be applied only when the density of the particles is not much larger than that of the medium (e.g. $\Delta\rho \sim 0.1$). By dissolving an inert substance in the continuous phase one may achieve density matching. However, apart from its limitation to particles with density not much larger than the medium, the method is not very practical since density matching can only occur at one temperature.

7.24.2 Reduction of Particle Size

As mentioned above, if R is significantly reduced (to values below 0.1 μm) the Brownian diffusion can overcome the gravity force and no sedimentation occurs. This is the principle of formation of nano-suspensions.

7.24.3 Use of High Molecular Weight Thickeners

As discussed above, high molecular weight materials such as hydroxyethyl cellulose or xanthan gum will, when added above a critical concentration (at which polymer coil overlap occurs), produce very high viscosity at low stresses or shear rates (usually in excess of several hundred Pa s) and this will prevent sedimentation of the particles.

7.24.4
Use of "Inert" Fine Particles

Several fine particulate inorganic materials produce "gels" when dispersed in aqueous media, e.g. sodium montmorillonite or silica. These particulate materials produce three-dimensional structures in the continuous phase as a result of interparticle interaction. For example, sodium montmorillonite (referred to as a swellable clay) forms gels at low electrolyte concentrations by simple double layer interaction. At intermediate electrolyte concentrations, the clay particles produce gels by "face-to-edge" association since the faces of the platelets are negatively charged whereas the edges are positively charged. At sufficient particle concentration, the T-junctions produce a continuous gel network in the continuous phase, preventing sedimentation of the coarse suspension particles. Finely divided silica, such as Aerosil 200 (produced by Degussa), produces gel structures by simple association (by van der Waals attraction) of the particles into chains and cross chains. When incorporated in the continuous phase of a suspension, these gels prevent sedimentation.

7.24.5
Use of Mixtures of Polymers and Finely Divided Particulate Solids

By combining the thickeners such as hydroxyethyl cellulose or xanthan gum with particulate solids such as sodium montmorillonite, a more robust gel structure could be produced. This gel structure may be less temperature dependent and could be optimised by controlling the ratio of the polymer and the particles.

7.24.6
Depletion Flocculation

As discussed before, addition of free non-adsorbing polymer can produce weak flocculation above a critical volume fraction of the free polymer, ϕ_p. This weak flocculation produces a "gel" structure that reduces sedimentation. As an illustration, results were obtained for a sterically stabilised suspension [using a graft copolymer of poly(methyl methacrylate) with poly(ethylene oxide) side chains] to which hydroxyethyl cellulose with various molecular weights was added to the suspension. The weak flocculation was studied using oscillatory measurements. Figure 7.52 shows the variation of the complex modulus G^* with ϕ_p.

Above a critical ϕ_p (which depends on the molecular weight of HEC), G^* increases very rapidly with further increase in ϕ_p. When ϕ_p reaches an optimum concentration, sedimentation is prevented. Figure 7.53 illustrates this with the sediment volume in 10 cm cylinders as a function ϕ_p for various volume fractions of the suspension ϕ_s.

At sufficiently high volume fraction of the suspensions ϕ_s and high volume fraction of free polymer ϕ^p a 100% sediment volume is reached and this is effective in eliminating sedimentation and formation of dilatant sediments.

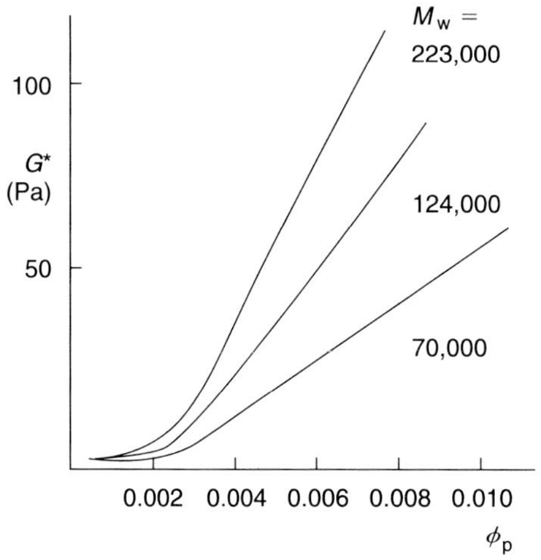

Fig. 7.52. Variation of G^* with ϕ_p for HEC with various molecular weights.

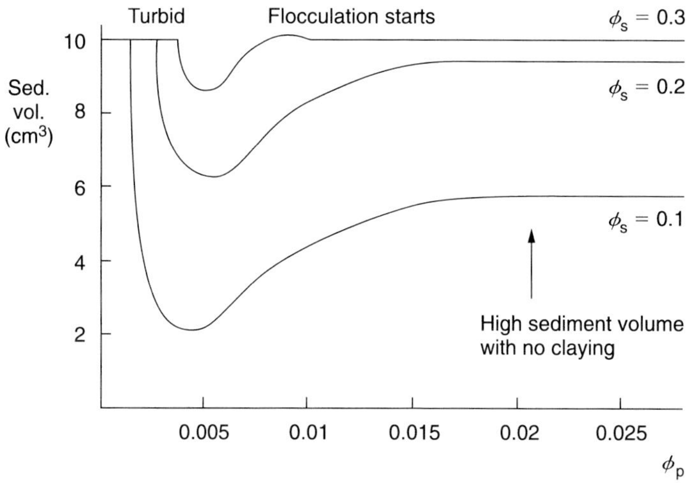

Fig. 7.53. Variation of sediment volume with ϕ_p.

7.24.7
Use of Liquid Crystalline Phases

As discussed in the chapter on phase behaviour of surfactants, the latter produce liquid crystalline phases at high concentrations. Three main types of liquid crystals can be identified: Hexagonal phase (sometimes referred to as middle phase), cubic phase and lamellar (neat phase). All these structures are highly viscous and they also show an elastic response. If produced in the continuous phase of suspensions, they can eliminate sedimentation of the particles. These liquid crystalline phases are particularly useful for application in liquid detergents that contain high surfactant concentrations. Their presence reduces sedimentation of the coarse builder particles (phosphates and silicates).

References

1 *Dispersion of Powders in Liquids*, G. D. Parfitt (ed.): Applied Science Publishers Ltd., London, 1977.
2 *Solid/Liquid Dispersions*, Th. F. Tadros (ed.): Academic Press, London, 1987.
3 G. D. Parfitt: *Dispersion of Powders in Liquids*, G. D. Parfitt (ed.), Applied Science Publishers, London, 1973.
4 J. W. Gibbs: *Scientific Papers*, Longman Green, London, 1906, Volume 1.
5 M. Volmer: *Kinetik der Phase Buildung*, Stemkopf, Dresden, 1939.
6 D. Blakely: *Emulsion Polymerisation*, Applied Science Publication, London, 1975.
7 G. Litchi, R. G. Gilbert, D. H. Napper, *J. Polym. Sci.*, **1983**, *21*, 269.
8 P. J. Feeney, D. H. Napper, R. G. Gilbert, *Macromolecules*, **1984**, *17*, 2520.
9 W. V. Smith, R. H. Ewart, *J. Chem. Phys.*, **1948**, *16*, 592.
10 I. Piirma: *Polymeric Surfactants*, Marcel Dekker, New York, 1992, Surfactant Science Series, Volume 42.
11 *Dispersion Polymerisation in Non-Aqueous Media*, K. E. J. Barrett (ed.): John Wiley & Sons, Chichester, 1975.
12 T. Blake: *Surfactants*, Th. F. Tadros (ed.): Academic Press, London, 1984.
13 W. A. Zisman, *Contact Angles, Wettability and Adhesion*, American Chemical Society, Washington, 1964, 1, Advances in Chemistry Series, No. 43.
14 C. A. Smolders, *Rec. Trav. Chim.*, **1961**, *80*, 650.
15 E. K. Rideal, *Phil. Mag*, **1922**, *44*, 1152.
16 E. D. Washburn, *Phys. Rev.*, **1921**, *17*, 273.
17 J. Lyklema: *Solid/Liquid Dispersions*, Th. F. Tadros (ed.): Academic Press, London, 1987.
18 B. H. Bijesterbosch: *Solid/Liquid Dispersions*, Th. F. Tadros (ed.): Academic Press, London, 1987.
19 *Colloid Science*, H. R. Kruyt (ed.): Elsevier Science, Amsterdam, 1952, Volume I.
20 E. J. W. Verwey, J. T. G. Overbeek: *Theory of Stability of Lyophobic Colloids*, Elsevier Science, Amsterdam, 1948.
21 R. J. Hunter: *Zeta Potential in Colloid Science; Principles and Applications*, Academic Press, London, 1981.
22 M. V. von Smoluchowski: *Handbuch der Electrizität und des Magnetismus*, Barth, Leipzig, 1914, Volume II.
23 E. Hückel, *Z. Phys.*, **1924**, *25*, 204.
24 D. C. Henry, *Proc. Royal Soc. London*, **1948**, *A133*, 106.
25 P. H. Wiersema, A. L. Loeb, J. T. G. Overbeek, *J. Colloid Interface Sci.*, **1967**, *22*, 78.
26 R. H. Ottewill, J. N. Shaw, *J. Electroanal. Interfacial Electrochem.*, **1972**, *37*, 133.
27 T. F. Tadros: *The Effect of Polymers on Dispersion Properties*, T. F. Tadros (ed.): Academic Press, London, 1981.
28 D. H. Napper: *Polymeric Stabilisation of Colloidal Dispersions*, Academic Press, London, 1981.
29 P. J. Flory, W. R. Krigbaum, *J. Chem. Phys.*, **1950**, *18*, 1086.
30 E. W. Fischer, *Z. Kolloid*, **1958**, *160*, 120.
31 E. L. Mackor, J. H. Waals van der, *J. Colloid Sci.*, **1951**, *7*, 535.
32 F. T. Hesselink, A. Vrij, J. T. G. Overbeek, *J. Phys. Chem.*, **1971**, *75*, 2094.
33 R. H. Ottewill: *Concentrated Dispersions*, J. W. Goodwin (ed.): Royal Scoiety of Chemistry, London, 1982, 43, Special Publication No. 43, Chapter 9.
34 R. H. Ottewill: *Science and Technology of Polymer Colloids*, G. W. Poehlein,

R. H. Ottewill, J. W. Goodwin, eds., Martinus Nishof Publishing, Boston, The Hague, 1983, 503, Volume II.

35 R. H. Ottewill: *Solid/Liquid Dispersions*, Th. F. Tadros (ed.): Academic Press, London, 1987.

36 T. F. Tadros, *Adv. Colloid Interface Sci.*, **1980**, *12*, 141.

37 Th. F. Tadros: *Science and Technology of Polymer Colloids*, G. W. Poehlein, R. H. Ottewill, eds., Marinus Nishof Publishing, Boston, The Hauge, 1983, Volume II.

38 A. Asakura, F. Oosawa, *J. Chem. Phys.*, **1954**, *22*, 1235; *J. Polym. Sci.* **1958**, *93*, 183.

39 J. D. Ferry: *Viscoelastic Properties of Polymers*, John Wiley & Sons, Chichester, 1980.

40 R. W. Whorlow: *Rheological Techniques*, Ellis Horwood, Chester, 1980.

41 J. W. Goodwin, R. Hughes: *Rheology for Chemists*, Royal Society of Chemistry, Cambridge, 2000.

42 R. Buscall, J. W. Goodwin, R. H. Ottewill, T. F. Tadros, *J. Colloid Interface Sci.*, **1982**, *85*, 78.

43 W. H. Herschel, R. Buckley, *Proc. Am. Soc. Test Mater.*, **1926**, *26*, 621; *Kolloid Z.*, **1926**, *39*, 291.

44 N. Casson, *Rheology of Disperse Systems*, C. C. Mill (ed.): Pergamon Press, Oxford, 1959, 84–104.

45 T. F. Tadros, *Adv. Colloid Interface Sci.*, **1993**, *46*, 1.

46 T. F. Tadros, *Adv. Colloid Interface Sci.*, **1996**, *68*, 91.

8 Surfactants in Foams

8.1 Introduction

Foam is a disperse system, consisting of gas bubbles separated by liquid layers. Because of the significant density difference between the gas bubbles and the medium, the system quickly separates into two layers with the gas bubbles rising to the top, which may undergo deformation to form polyhedral structures. This is discussed in detail below.

Pure liquids cannot foam unless a surface active material is present. When a gas bubble is introduced below the surface of a liquid, it bursts almost immediately as a soon as the liquid has drained away. With dilute surfactant solutions, as the liquid/air interface expands and the equilibrium at the surface is disturbed, a resorting force is set up that tries to establish the equilibrium.

The restoring force arises from the Gibb–Marangoni effect, which was discussed in detail in Chapter 5. Owing to the presence of surface tension gradients $d\gamma$ (due to incomplete coverage of the film by surfactant), a dilational elasticity ε is produced (Gibbs elasticity). This surface tension gradient induces a flow of surfactant molecules from the bulk to the interface and these molecules carry liquid with them (the Marangoni effect). The Gibbs–Marangoni effect prevents thinning and disruption of the liquid film between the air bubbles and this stabilises the foam. This process is also discussed in detail below.

Several surface active foaming materials may be distinguished – surfactants: ionic, nonionic and zwitterionic, polymers (polymeric surfactants), particles that accumulate at the air/solution interface, and specifically adsorbed cations or anions from inorganic salts.

Many of the above substances can cause foaming at extremely low concentrations (as low as 10^{-9} mol dm^{-3}).

In kinetic terms, foams may be classified into (1) unstable, transient foams (lifetime of seconds) and (2) metastable, permanent foams (lifetimes of hours or days).

Applied Surfactants: Principles and Applications. Tharwat F. Tadros

ISBN: 3-527-30629-3

8.2
Foam Preparation

Like most disperse systems, foams can be obtained by condensation and dispersion methods. The condensation methods for generating foam involve the creation of gas bubbles in the solution by decreasing the external pressure, by increasing temperature or as a result of chemical reaction. Thus, bubble formation may occur through homogeneous nucleation at high supersaturation or heterogeneous nucleation (e.g. from catalytic sites) at low supersaturation.

The most applied technique for generating foam is by a simple dispersion technique (mechanical shaking or whipping). This method is unsatisfactory since it is difficult to accurately control the amount of air incorporated. The most convenient method is to pass a flow of gas (sparging) through an orifice with a well-defined radius r_0.

The size of the bubbles (produced at an orifice) r may be roughly estimated from the balance of the buoyancy force F_b with the surface tension force F_s [1],

$$F_b = \tfrac{4}{3}\pi r^3 \rho g \tag{8.1}$$

$$F_s = 2\pi r_0 \gamma \tag{8.2}$$

$$r = \left(\frac{3\gamma r_0}{2\rho g}\right)^{1/3} \tag{8.3}$$

where r and r_0 are the radii of the bubble and orifice and ρ is the specific gravity of liquid.

Since the dynamic surface tension of the growing bubble is higher than the equilibrium tension, the contact base may spread, depending on the wetting conditions. Thus, the main problem is the value of γ to be used in Eq. (8.3). Another important factor that controls bubble size is the adhesion tension $\gamma \cos\theta$, where θ is the dynamic contact angle of the liquid on the solid of the orifice. With a hydrophobic surface, a bubble develops with a greater size than thc hole. One should always distinguish between the equilibrium contact angle θ and the dynamic contact angle, θ_{dyn} during bubble growth.

As the bubble detaches from the orifice, the dimensions of the bubble will determine the velocity of the rise. The rise of the bubble through the liquid causes a redistribution of surfactant on the bubble surface, with the top having a reduced concentration and the polar base having a higher concentration than the equilibrium value. This unequal distribution of surfactant on the bubble surface has an important role in foam stabilisation (due to the surface tension gradients). When the bubble reaches the interface, a thin liquid film is produced on its top. The life time of this thin film depends on many factors, e.g. surfactant concentration, rate of drainage, surface tension gradient, surface diffusion and external disturbances.

8.3 Foam Structure

Two main types of foams may be distinguished: (1) spherical foam ("Kugel Schaum"), consisting of gas bubbles separated by thick films of viscous liquid produced in freshly prepared systems. This may be considered as a temporary dilute dispersion of bubbles in the liquid. (2) Polyhedral gas cells produced on aging; thin flat "walls" are produced with junction points of the interconnecting channels (plateau borders). Due to the interfacial curvature, the pressure is lower and the film is thicker in the plateau border. A capillary suction effect of the liquid occurs from the centre of the film to its periphery.

The pressure difference between neighbouring cells, Δp, is related to the radius of curvature (r) of the plateau border by,

$$\Delta p = \frac{2\gamma}{r} \qquad (8.4)$$

In a foam column, several transitional structures may be distinguished (Figure 8.1). Near the surface, a high gas content (polyhedral foam) is formed, with a much lower gas content structure near the base of the column (bubble zone). A transition state may be distinguished between the upper and bottom layers.

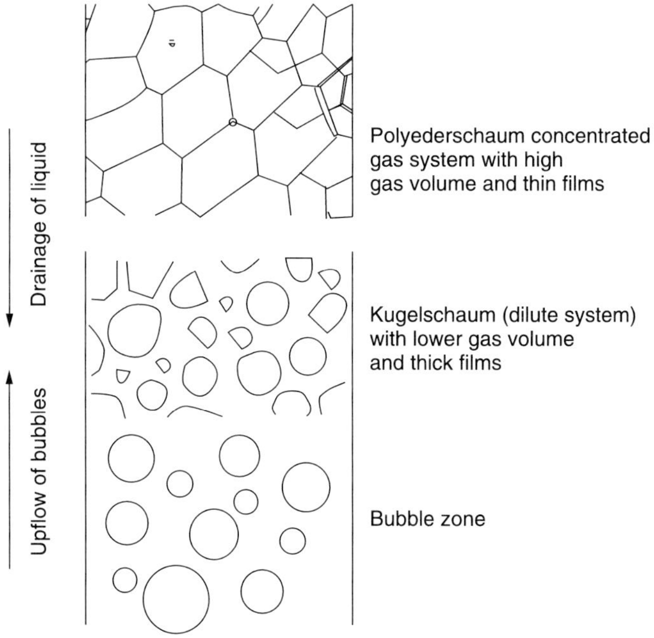

Fig. 8.1. Schematic of foam structure in a column.

The drainage of excess liquid from the foam column to the underlying solution is, initially, driven by hydrostatic, causing the bubble to become distorted. Foam collapse usually occurs from top to bottom of the column. Films in the polyhedral foam are more susceptible to rupture by shock, temperature gradient or vibration.

Another mechanism of foam instability is due to Ostwald ripening (disproportionation). The driving force for this process is the difference in Laplace pressure between the small and larger foam bubbles. The smaller bubbles have a higher Laplace pressure than the larger ones. The gas solubility increases with pressure and, hence, gas molecules will diffuse from the smaller to the larger bubbles. This process only occurs with spherical foam bubbles. It may be opposed by the Gibbs elasticity effect. Alternatively, rigid films produced using polymers may resist Ostwald ripening as a result of the high surface viscosity.

With polyhedral foam with planar liquid lamella, the pressure difference between the bubbles is not large, and hence Ostwald ripening is not responsible for the foam instability. With polyhedral foams, the main driving force for foam collapse is the surface forces that act across the liquid lamella.

To keep the foam stable (i.e. to prevent complete rupture of the film), this capillary suction effect must be prevented by an opposing "disjoining pressure" that acts between the parallel layers of the central flat film (see below).

The generalised model for drainage involves the plateau borders forming a "network" through which the liquid flows due to gravity.

8.4
Classification of Foam Stability

All foams are thermodynamically unstable (due to the high interfacial free energy). For convenience foams are classified according to the kinetics of their breakdown: (1) Unstable (transient) foams, lifetime of seconds. These are generally produced using "mild" surfactants, e.g. short-chain alcohols, aniline, phenol, pine oil, short-chain undissociated fatty acid. Most of these compounds are sparingly soluble and may produce a low degree of elasticity. (2) Metastable ("permanent") foams, lifetime hours or days. These metastable foams can withstand ordinary disturbances (thermal or Brownian fluctuations). They can collapse from abnormal disturbances (evaporation, temperature gradients, etc.).

The above metastable foams are produced from surfactant solutions near or above the critical micelle concentration (c.m.c.). The stability is governed by the balance of surface forces (see below). Film thickness is comparable to the range of intermolecular forces. In the absence of external disturbances, these foams may stay stable indefinitely. They are produced using proteins, long-chain fatty acids or solid particles.

Gravity is the main driving force for foam collapse, directly or indirectly through the plateau border. Thinning and disruption may be opposed by surface tension gradients at the air/water interface. Alternatively, the drainage rate may be decreased by increasing the bulk viscosity of the liquid (e.g. addition of glycerol or polymers). Stability may be increased in some cases by the addition of electrolytes that produce a "gel network" in the surfactant film. Foam stability may also be enhanced by increasing the surface viscosity and/or surface elasticity. High packing of surfactant films (high cohesive forces) may also be produced using mixed surfactant films or surfactant/polymer mixtures.

To investigate foam stability one must consider the role of the plateau border under dynamic and static conditions. One should also consider foam films with intermediate lifetimes, i.e. between unstable and metastable foams.

8.5 Drainage and Thinning of Foam Films

As mentioned above, gravity is the main driving force for film drainage. Gravity can act directly on the film or through capillary suction in the plateau borders. As a general rule, the rate of drainage of foam films may be decreased by increasing the bulk viscosity of the liquid from which the foam is prepared. This can be achieved by adding glycerol or high molecular weight poly(ethylene oxide). Alternatively, the viscosity of the aqueous surfactant phase can be increased by addition of electrolytes that form a "gel" network (liquid crystalline phases may be produced).

Film drainage can also be decreased by increasing the surface viscosity and surface elasticity. This can be achieved, for example, by addition of proteins, polysaccharides and even particles. These systems are applied in many food foams.

For convenience, the drainage of horizontal and vertical films will be treated separately.

8.5.1 Drainage of Horizontal Films

Most quantitative studies on film drainage have been carried out using small, horizontal films, as described in detail by Scheludko and co-workers [2–4]. Figure 8.2 illustrates the measuring cell for studying microscopic foam films.

The foam film c is formed in the middle of a biconcave drop b, situated in a glass tube of radius R, by withdrawing liquid from it (A and B) and in the hole of a porous plate g (C) (Figure 8.2). A suitable tube diameter in A and B is 0.2–0.6 mm and the film radius ranges from 100 to 500 nm. In C, the hole radius can considerably smaller, in the range of 120 μm and the film radius is 10 μm. When the film thins to form the so-called "black" film, black spots can be observed under the microscope.

Film thickness is determined by interferometry, which is based on comparison between the intensities of the light falling on the film and that reflected from it [4].

The drainage time T is determined and compared with the theoretical value for a flat film calculated from the Reynolds Eq. (8.5),

$$T = \int_{h_t}^{h_0} \frac{dh}{V} \tag{8.5}$$

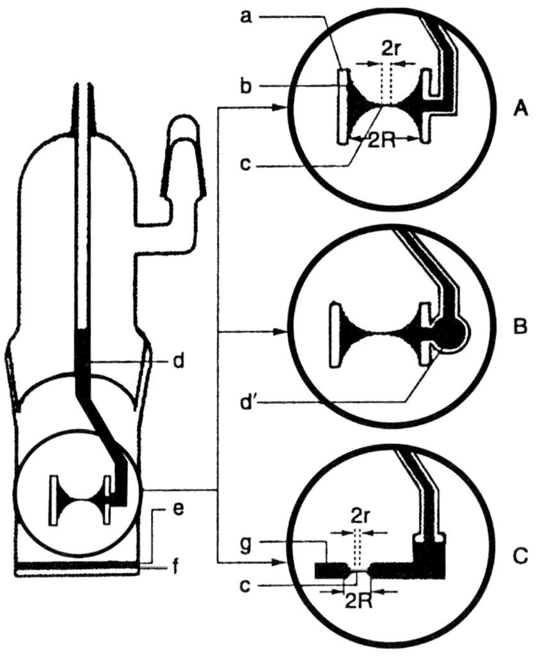

Fig. 8.2. Cell used for studying microscopic foam films: (A) in a glass tube; (B) with a reservoir of surfactant solution d′; (C) in a porous plate. a, glass tube film holder; b, biconcave drop; c, microscopic foam film; d, glass capillary; e, surfactant solution; f, optically flat glass; g, porous plate.

where h_0 is the initial film thickness and h_t is the value after time t; V is the velocity of thinning $V = -dh/dt$.

For horizontal, fairly thick films (> 100 nm), Scheludko [3] derived an expression for the thinning between two disc surfaces under the influence of a uniform external pressure. The change in film thickness with drainage time, V_{re}, is given by

$$V_{re} = -\frac{dh}{dt} = \frac{2h^3 \Delta P}{3\eta R^2} \tag{8.6}$$

where R is the radius of the disc, η is the viscosity of the liquid and ΔP is the difference in pressure between the film and bulk solution. ΔP was taken to be equal to the capillary pressure in the plateau border. For very thin films, the pressure gradient also includes the disjoining pressure (see below).

Eq. (8.6) applies for the following conditions: (1) The liquid flows between parallel plane surfaces; (2) film surfaces are tangentially immobile; and (3) the rate of thinning due to evaporation is negligible compared with the thinning due to drainage.

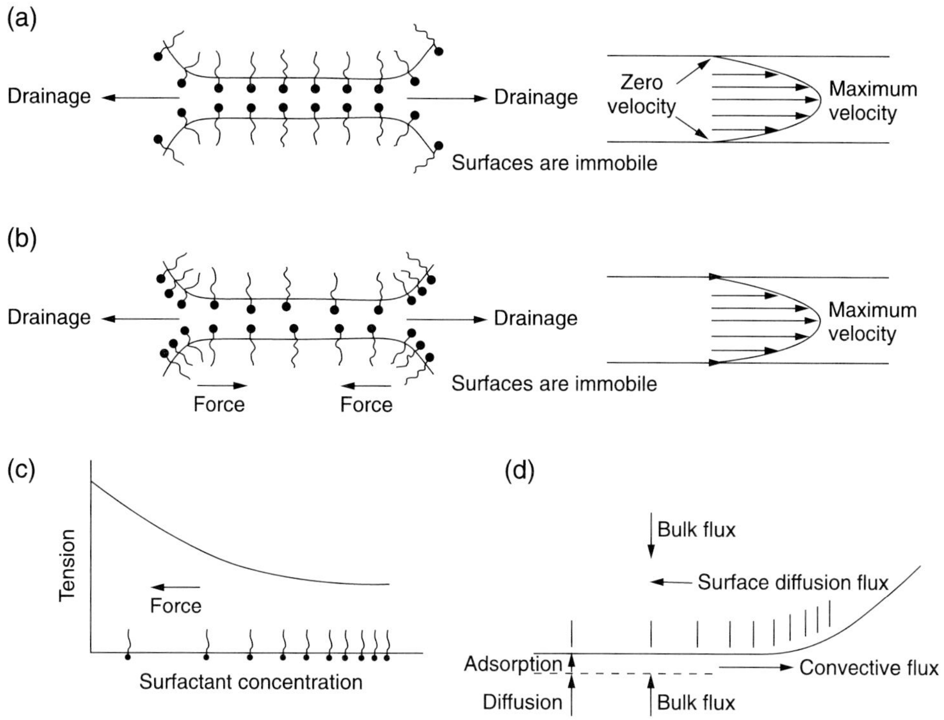

Fig. 8.3. Schematic of film drainage: (a) thick rigid films; (b) mobile surfactant films; (c) interfacial tension gradients; and (d) diffusion along the surface and from bulk solution.

Experimental results obtained by Scheludko and co-workers [2–4] with comparatively thick rigid films, produced from dilute solutions of sodium oleate and isoamyl alcohol, gave reasonably good agreement with the drainage equation. However, deviation from the Reynolds equation was observed in many cases due to tangential surface mobility. Surface viscosity can also slow down the drainage.

Figure 8.3 gives a schematic of film drainage [6], representing the cases of (a) a thick film (> 100 nm), where the drainage velocity can be determined from Reynold's equation. (b) The case for most surfactant films, where the surfaces are not rigid and the tangential velocity at the surface is not zero. (c) Surface mobility during drainage causes interfacial tension gradients. (d) Surface diffusion along the surface and from the bulk solution occurs with both adsorption and convective flow.

For thinner films, large electrostatic repulsive interactions can reduce the driving force for drainage and may lead to stable films. Also, for thick films that contain high surfactant concentrations (> c.m.c.), the micelles present in the film can cause a repulsive structural mechanism. The effect of deformation of the film surface during thinning is also extremely complicated.

8.5.2
Drainage of Vertical Films

Foam films can be produced by pulling a frame out of a reservoir containing a surfactant solution. Three stages can be identified: (1) initial formation of the film that is determined by the withdraw velocity; (2) drainage of the film within the lamella, which causes thinning with time; and (3) aging of the film, which may result in the formation of a metastable film.

Assuming that the monolayer of the surfactant film at the boundaries of the film is rigid, film drainage may be described by the viscous flow of the liquid under gravity between two parallel plates, as given by the Poiseille's equation,

$$V_{av} = \frac{\rho g h^2}{8\eta} \tag{8.7}$$

where h is the film thickness, ρ is the liquid density in the film, η is the viscosity of the liquid and g is the acceleration due to gravity.

As the process proceeds, thinning can also occur by a horizontal mechanism known as marginal regeneration [7, 8], in which the liquid is drained from the film near the border region and exchanged from within the low pressure plateau border. In this exchange, the total area of the film does not change significantly.

Marginal regeneration is shown schematically in Figure 8.4. The thicker film is drawn into the border by the negative excess pressure ΔP and the thinner film is pulled out of the border (Figure 8.4a). Figure 8.4b shows a schematic view of the

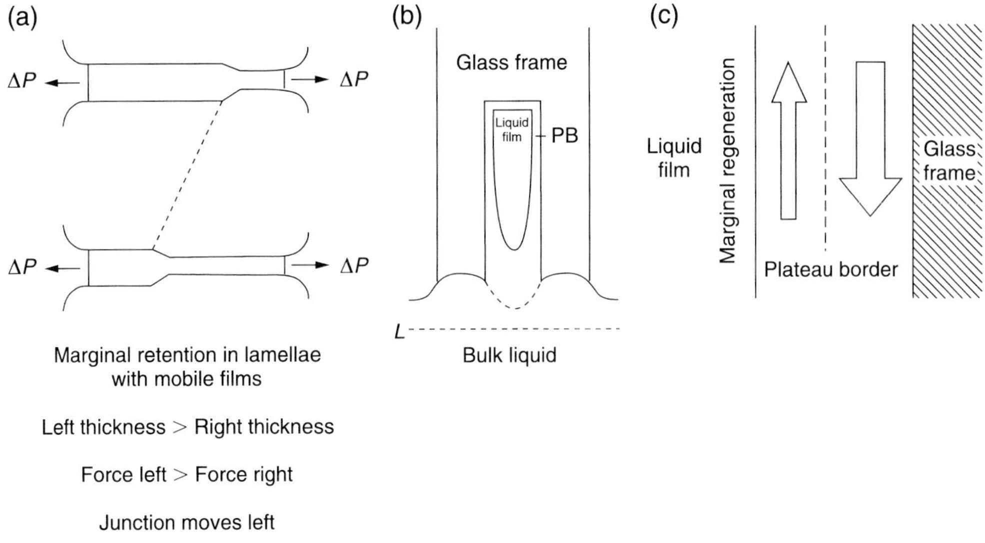

Fig. 8.4. Representation of marginal regeneration: (a) film drawn from thicker to thinner film; (b) schematic view of the film; (c) close-up of flows in the plateau border of the film.

film, whereas Figure 8.4c shows a close-up of flows in the plateau border of the vertical film. The liquid flow can be visualised by addition of small glass particles. In the film-near region, the liquid flows upwards, while near the frame the liquid flows downward. In between, particles show no or only restricted, more or less circular, motion [9].

The above regeneration mechanism results in the formation of patches of thin film at the border, with the excess fluid flowing into the border channel. The edge effects determine the drainage, with the rate of thinning varying inversely with film width [7, 8]. This results in thickness fluctuations caused by capillary waves.

Marginal regeneration is probably the most important cause of drainage in vertical films with mobile surfaces, i.e. with surfactant solutions at concentrations above the c.m.c.

8.6 Theories of Foam Stability

There is no single theory that can explain foam stability in a satisfactory manner. Several approaches have been considered and these are summarised below.

8.6.1 Surface Viscosity and Elasticity Theory

The adsorbed surfactant film is assumed to control the mechanical-dynamical properties of the surface layers by virtue of its surface viscosity and elasticity. This concept may be true for thick films (>100 nm) whereby intermolecular forces are less dominant (i.e. foam stability under dynamic conditions). Surface viscosity reflects the speed of the relaxation process that restores the equilibrium in the system after imposing a stress on it. Surface elasticity is a measure of the energy stored in the surface layer as a result of an external stress.

The viscoelastic properties of the surface layer are an important parameter. Surface scattering methods are the most useful techniques for studying the viscoelastic properties of surfactant monolayers. When transversal ripples occur, periodic dilation and compression of the monolayer arises and this can be accurately measured. This enables one to obtain the viscoelastic behaviour of monolayers under equilibrium and non-equilibrium conditions, without disturbing the original state of the adsorbed layer.

Some correlations have been found between surface viscosity and elasticity and foam stability, e.g. when adding lauryl alcohol to sodium lauryl sulphate, which tends to increase the surface viscosity and elasticity [10].

8.6.2 Gibbs–Marangoni Effect Theory

The Gibbs coefficient of elasticity, ε, was introduced as a variable resistance to surface deformation during thinning:

$$\varepsilon = 2\left(\frac{d\gamma}{d \ln A}\right) = -2\left(\frac{d\gamma}{d \ln h}\right) \tag{8.8}$$

$d \ln h$ = relative change in lamella thickness. ε is the "Film Elasticity of Compression Modulus" or "Surface Dilational Modulus", and is a measure of the ability of the film to adjust its surface tension in an instant stress. In general, the higher ε the more stable the film is; ε depends on surface concentration and film thickness. For a freshly produced film to survive, a minimum ε is required.

The main deficiency of the early studies on Gibbs elasticity was that it was applied to thin films and diffusion from the bulk solution was neglected. In other words, Gibbs theory applies to the case where there are insufficient surfactant molecules in the film to diffuse to the surface and lower the surface tension. This is clearly not the case with most surfactant films. For thick lamella under dynamic conditions, one should consider diffusion from the bulk solution, i.e. the Marangoni effect. The Marangoni effect tends to oppose any rapid displacement of the surface (Gibbs effect) and may provide a temporary restoring force to "dangerous" thin films. In fact, the Marangoni effect is superimposed on the Gibbs elasticity, so that the effective restoring force is a function of the rate of extension, as well as the thickness. When the surface layers behave as insoluble monolayers, then the surface elasticity has its greatest value and is referred to as the Marangoni dilational modulus, ε_m.

The Gibbs–Marangoni effect explains the maximum foaming behaviour at intermediate surfactant concentration [5]. This is illustrated in Figure 8.5. At low surfactant concentrations (well below the c.m.c.), the greatest possible differential surface tension will only be relatively small (Figure 8.5a) and little foaming will occur. At very high surfactant concentrations (well above the c.m.c.), the differential tension relaxes too rapidly because of the supply of surfactant that diffuses to the surface (Figure 8.5c). This causes the restoring force to have time to counteract the disturbing forces that produce a dangerously thinner film, and foaming is poor. It is the intermediate surfactant concentration range that produces maximum foaming (Figure 8.5b).

8.6.3
Surface Forces Theory (Disjoining Pressure)

This theory operates under static (equilibrium) conditions in relatively dilute surfactant solutions ($h < 100$ nm). In the early stages of formation, foam films drain under the action of gravitation or capillary forces. Provided the films remain stable during this drainage stage, they may approach a thickness in the range of 100 nm. At this stage, surface forces come into play, i.e. the range of the surface forces becomes now comparable to the film thickness. Deryaguin and co-workers [11, 12] introduced the concept of disjoining pressure, which should remain positive to slow down further drainage and film collapse. This is the principle of formation of thin metastable (equilibrium) films.

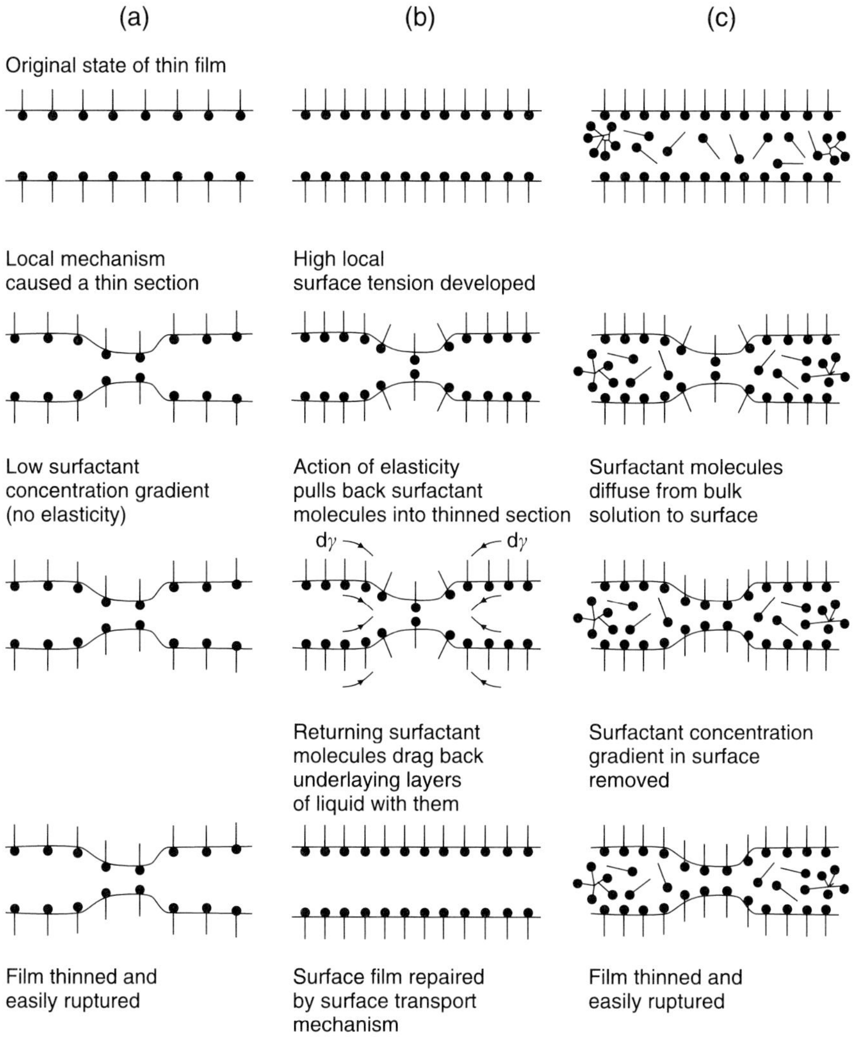

Fig. 8.5. Gibbs–Marangoni effect: (a) low surfactant concentration (< c.m.c.); (b) intermediate surfactant concentration; (c) high surfactant concentration (> c.m.c.).

In addition to the Laplace capillary pressure, three additional forces can operate at surfactant concentrations below the c.m.c.: Electrostatic double layer repulsion (π_e)l, van der Waals attraction (π_{vdW}), and steric (short-range) forces (π_{st}),

$$\pi = \pi_{el} + \pi_{vdW} + \pi_{st} \tag{8.9}$$

In the original definition of disjoining pressure, Deryaguin [11, 12] only considered the first two terms on the right-hand side of Eq. (8.9). At low electrolyte concentrations, double layer repulsion predominates and π_{el} can compensate the capillary pressure, i.e. $\pi_{el} = P_c$. This results in the formation of an equilibrium free film that is usually referred to as the thick common film CF (~50 nm thick). This equilibrium metastable film persists until thermal or mechanical fluctuations cause rupture. The stability of the CF can be described in terms of the theory of colloid stability due to Deryaguin, Landau [13] and Verwey and Overbeek [14] (DLVO theory).

The critical thickness at which the CF ruptures (due to thickness perturbations) fluctuates and an average value h_{cr} may be defined. However, an alternative situation may occur as h_{cr} is reached and, instead of rupturing, a metastable film (high stability) may be formed with a thickness $h < h_{cr}$. The formation of this metastable film can be experimentally observed through the formation of "islands of spots", which appear black in light reflected from the surface. This film is often referred to a "first black" or "common black" film. The surfactant concentration at which this "first black" film is produced can be 1–2 orders of magnitude lower than the c.m.c.

Further thinning can cause an additional transformation into a thinner stable region (a stepwise transformation). This usually occurs at high electrolyte concentrations, which leads to a second, very stable, thin black film, usually referred to as Newton secondary black film, with a thickness in the region of 4 nm. Under these conditions, short-range steric or hydration forces control the stability, providing the third contribution to the disjoining press, π_{st} in Eq. (8.9).

Figure 8.6 shows a schematic representation of the variation of disjoining pressure π with film thickness h, illustrating the transition from the common film to

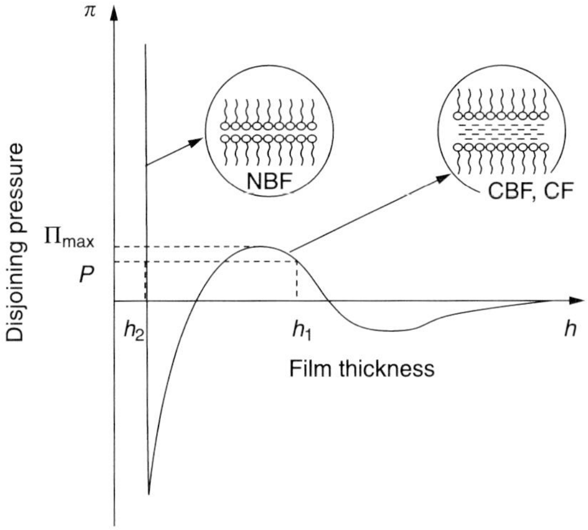

Fig. 8.6. Disjoining pressure versus film thickness showing the transition from common film (CF) to common black film (CBF) to Newton black film (NBF).

the common black film and to the Newton black film. The common black film is around 30 nm thick, whereas the Newton black film is ca. 4–5 nm thick, depending on electrolyte concentration.

Several investigations were carried out to study the above transitions from common film to common black film and finally to Newton black film. For sodium dodecyl sulphate, the common black films have thicknesses ranging from 200 nm in very dilute systems to about 5.4 nm. The thickness depends strongly on electrolyte concentration and the stability may be considered to be caused by the secondary minimum in the energy distance curve (see Chapter 7). In cases where the film thins further and overcomes the primary energy maximum, it will fall into the primary minimum potential energy sink, where very thin Newton black films are produced. The transition from common black films to Newton black films occurs at a critical electrolyte concentration that depends on the type of surfactant.

The rupture mechanisms of thin liquid films were considered by de Vries [15] and by Vrij and Overbeek [16]. It was assumed that thermal and mechanical disturbances (having a wave like nature) cause film thickness fluctuations (in thin films), leading to rupture or coalescence of bubbles at a critical thickness. Vrij and Overbeek [16] carried out a theoretical analysis of the hydrodynamic interfacial force balance, and expressed the critical thickness of rupture in terms of the attractive van der Waals interaction (characterised by the Hamaker constant A), the surface or interfacial tension γ, and disjoining pressure. The critical wavelength, λ_{crit}, for the perturbation to grow (assuming the disjoining pressure just exceeds the capillary pressure) was determined. Film collapse occurs when the amplitude of the fast growing perturbation was equal to the thickness of the film. The critical thickness of rupture, h_{cr}, was defined by Eq. (8.10), where a_f is the area of the film.

$$h_{crit} = 0.267\left(\frac{a_f A^2}{6\pi\gamma\Delta p}\right)^{1/7} \tag{8.10}$$

Many poorly foaming liquids with thick film lamella are easily ruptured, e.g. pure water and ethanol films (between 110 and 453 nm thick). Under these conditions, rupture occurs by growth of disturbances that may lead to thinner sections [17]. Rupture can also be caused by spontaneous nucleation of vapour bubbles (forming gas cavities) in the structured liquid lamella [18]. An alternative explanation for rupture of relatively thick aqueous films containing low level of surfactants is the hydrophobic attractive interaction between the surfaces, which may be caused by bubble cavities [19, 20].

8.6.4 Stabilisation by Micelles (High Surfactant Concentrations > c.m.c.)

At high surfactant concentrations (above the c.m.c.), micelles of ionic or nonionic surfactants can produce organised molecular structures within the liquid film [21, 22]. This will provide an additional contribution to the disjoining pressure. Thinning of the film occurs through a stepwise drainage mechanism, referred to as

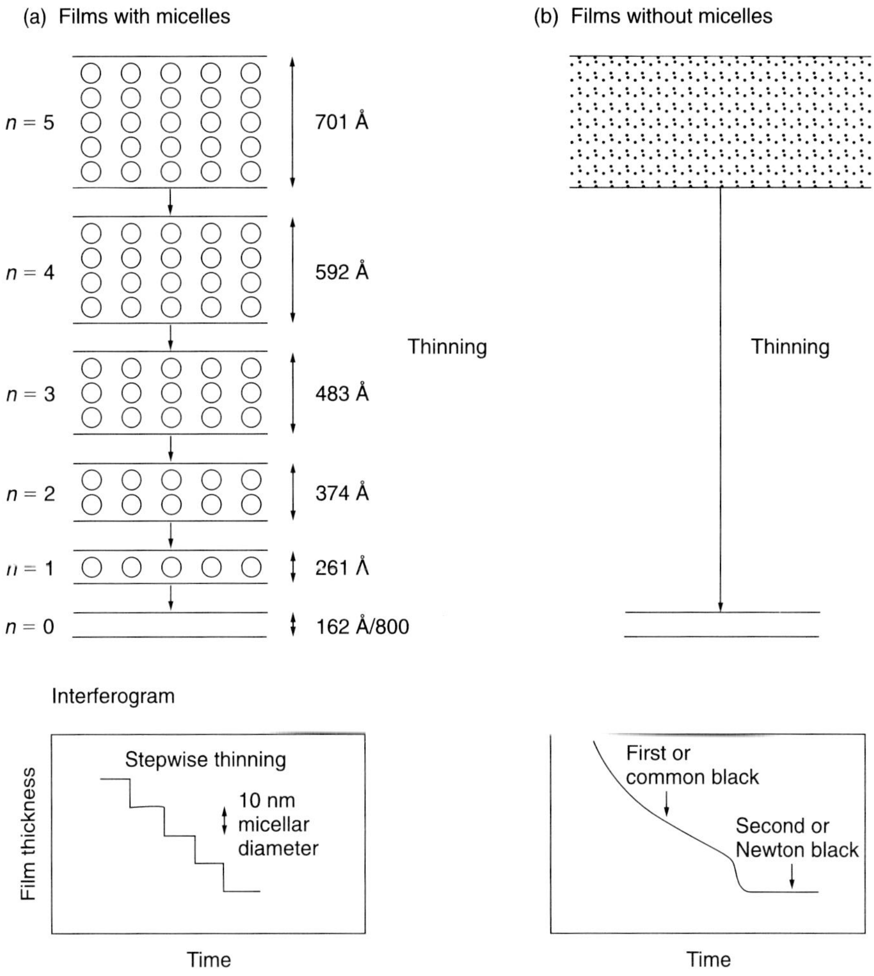

Fig. 8.7. Films with micelles (a), showing the stratification mechanism, and without (b).

stratification [23]. This is illustrated in Figure 8.7, which also shows the thinning of films without micelles for comparison.

The ordering of surfactant micelles (or colloidal particles) in the liquid film due to the repulsive interaction provides an additional contribution to the disjoining pressure and this prevents the thinning of the liquid film. Figure 8.8 summarises the different mechanisms in the main stages in the evolution of a thin liquid film containing a surfactant below and above the c.m.c. The general forms of the force curves for non-structured (below the c.m.c.) and structured (above the c.m.c.) are schematically illustrated in Figure 8.9. The force curve for the non-structured films follows the same trend as that described by DLVO theory.

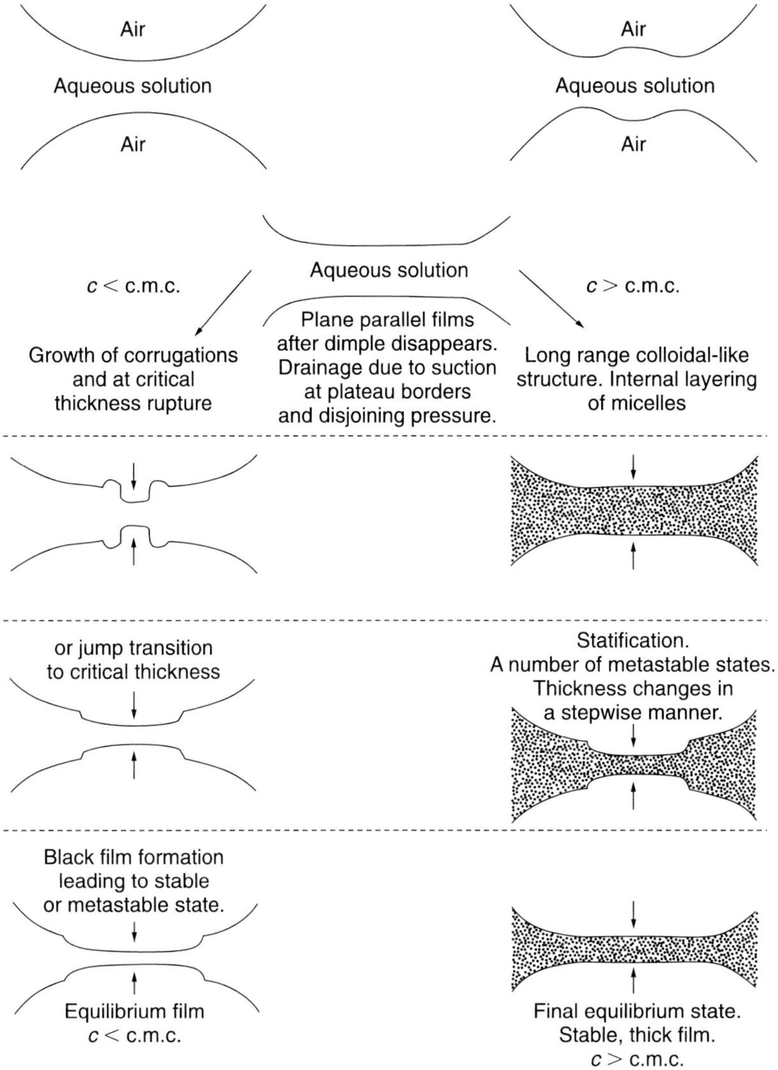

Fig. 8.8. Schematic of the evolution of a thin aqueous film at $c <$ c.m.c. and at $c >$ c.m.c.

8.6.5
Stabilization by Lamellar Liquid Crystalline Phases

This is particularly the case with nonionic surfactants that produce a lamellar liquid crystalline structure in the film between the bubbles [24, 25]. These liquid crystals reduce film drainage as a result of the increase in viscosity of the film. In addition, the liquid crystals act as a reservoir of surfactant of optimal composition to stabilise the foam.

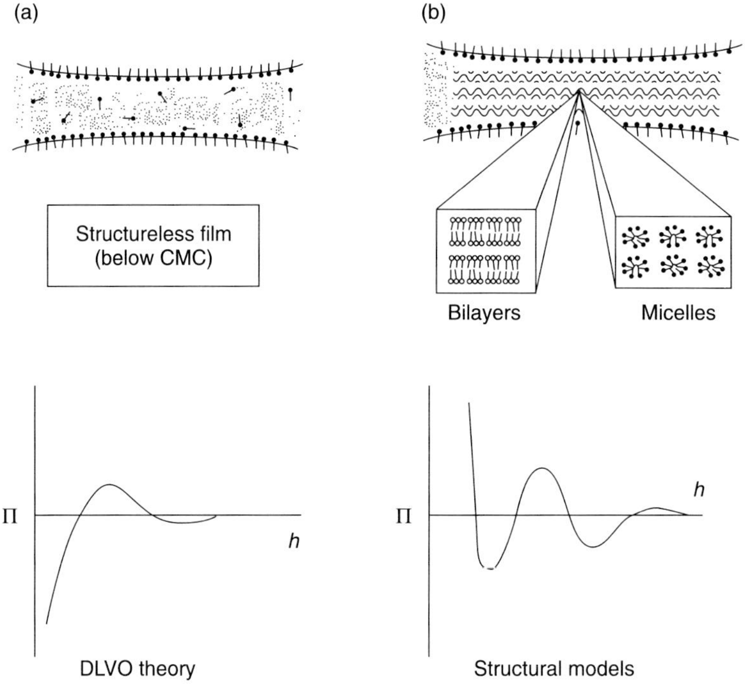

Fig. 8.9. General forms for the force–distance curves for a thin film at low surfactant concentration (structureless film, below the c.m.c.) (a) and a thin film with micelles or bilayer (supramolecular structuring) (b).

8.6.6
Stabilisation of Foam Films by Mixed Surfactants

A combination of surfactants gives slower drainage and improved foam stability. For example, mixtures of anionic and nonionic surfactants or anionic surfactant and long-chain alcohol produce much more stable films than the single components. This could be attributed to several factors. For example, the addition of a nonionic to an anionic surfactant reduces the c.m.c. of the anionic surfactant. The mixture can also produce a lower surface tension than the individual components. The combined surfactant system also has a high surface elasticity and viscosity when compared with the single components.

8.7
Foam Inhibitors

Two main types of inhibition may be distinguished: Antifoamers that are added to prevent foam formation and deformers that are added to eliminate an existing foam. For example, alcohols such as octanol are effective as defoamers but ineffective as antifoamers. Since the drainage and stability of liquid films is far from fully understood, it is very difficult to explain the antifoaming and foam-breaking action obtained by addition of substances. This is also complicated by the fact that in many industrial processes foams are produced by unknown impurities. For these reasons, the mechanism of action of antifoamers and defoamers is far from being understood [26]. The various methods that can be applied for foam inhibition and foam breaking are summarised below.

8.7.1
Chemical Inhibitors that Both Lower Viscosity and Increase Drainage

Chemicals that reduce the bulk viscosity and increase drainage can cause a decrease in foam stability. The same applies to materials that reduce surface viscosity and elasticity (swamping the surface layer with excess compound of lower viscosity).

It has been suggested that a spreading film of antifoam may simply displace the stabilising surfactant monolayer. As the oil lens spreads and expands on the surface, the tension will be gradually reduced to a lower uniform value. This will eliminate the stabilising effect of the interfacial tension gradients, i.e. elimination of surface elasticity.

Reduction of surface viscosity and elasticity may be achieved by low molecular weight surfactants. This will reduce the coherence of the layer, e.g. by addition of small amounts of nonionic surfactants. These effects depend on the molecular structure of the added surfactant. Other materials, which are not surface active, can also destabilise the film by acting as cosolvents that reduce the surfactant concentration in the liquid layer. Unfortunately, these non-surface-active materials, such as methanol or ethanol, need to be added in large quantities ($> 10\%$).

8.7.2
Solubilised Chemicals that Cause Antifoaming

Solubilised antifoamers such as tributyl phosphate and methyl isobutyl carbinol, when added to surfactant solutions such as sodium dodecyl sulphate and sodium oleate, reduce foam formation [27]. In cases where the oils exceed the solubility limit, the emulsifier droplets of oil can greatly influence the antifoam action. It has been claimed [27] that the oil solubilised in the micelle causes a weak defoaming action. Mixed micelle formation with extremely low concentrations of surfactant may explain the actions of insoluble fatty acid esters, alkyl phosphate esters and alkyl amines.

8.7.3
Droplets and Oil Lenses that Cause Antifoaming and Defoaming

Undissolved oil droplets form in the surface of the film and this can lead to film rupture. Several examples of oils may be used: Alkyl phosphates, diols, fatty acid esters and silicone oils [poly(dimethyl siloxane)].

A widely accepted mechanism for the antifoaming action of oils considers two steps: The oil drops enter the air/water interface, and the oil then spreads over the film, causing rupture.

The antifoaming action can be rationalised [28] in terms of the balance between the entering coefficient E and the Harkins [29] spreading coefficient S, which are given by the following equations,

$$E = \gamma_{W/A} + \gamma_{W/O} - \gamma_{O/A} \tag{8.11}$$

$$S = \gamma_{W/A} - \gamma_{W/O} - \gamma_{O/A} \tag{8.12}$$

where $\gamma_{W/A}$, $\gamma_{O/A}$ and $\gamma_{W/O}$ are the macroscopic interfacial tensions of the aqueous phase, oil phase and interfacial tension of the oil/water interface, respectively.

Ross and McBain [30] have suggested that, for efficient defoaming, the oil drop must enter the air/water interface and spread to form a duplex film at both sides of the original film. This leads to displacement of the original film, leaving an unstable oil film that can easily break. Ross used the spreading coefficient (Eq. 8.12) as a defoaming criterion [28].

For antifoaming, both E and S should be >0 for entry and spreading. Figure 8.10 give a schematic representation of oil entry and the balance of the relevant tensions [5]. A typical example of such spreading/breaking is illustrated for a hydrocarbon surfactant stabilised film. For most surfactant systems, $\gamma_{AW} = 35–45$ mN m^{-1} and $\gamma_{OW} = 5–10$ mN m^{-1} and, hence, for an oil to act as an antifoaming agent γ_{OA} should be less than 25 mN m^{-1}. This shows why low surface tension silicone oils, which have surface tensions as low as 10 mN m^{-1}, are effective.

8.7.4
Surface Tension Gradients (Induced by Antifoamers)

Some antifoamers are thought to act by eliminating the structure tension gradient effect in foam films by reducing the Marangoni effect. Since spreading is driven by a surface tension gradient between the spreading front and the leading edge of the spreading front, then the thinning and foam rupture can occur by this surface tension gradient acting as a shear force (dragging the underlying liquid away from the source). This could be achieved by solids or liquids containing surfactant other than that stabilising the foam. Alternatively, liquids that contain foam stabilisers at higher concentrations than that present in the foam may also act by this mechanism. A third possibility is the use of adsorbed vapours of surface active liquids.

8.7.5
Hydrophobic Particles as Antifoamers

Many solid particles with some degree of hydrophobicity cause destabilisation of foams, e.g. hydrophobic silica, PTFE particles. These particles exhibit a finite contact angle when adhering to the aqueous interface. It has been suggested that

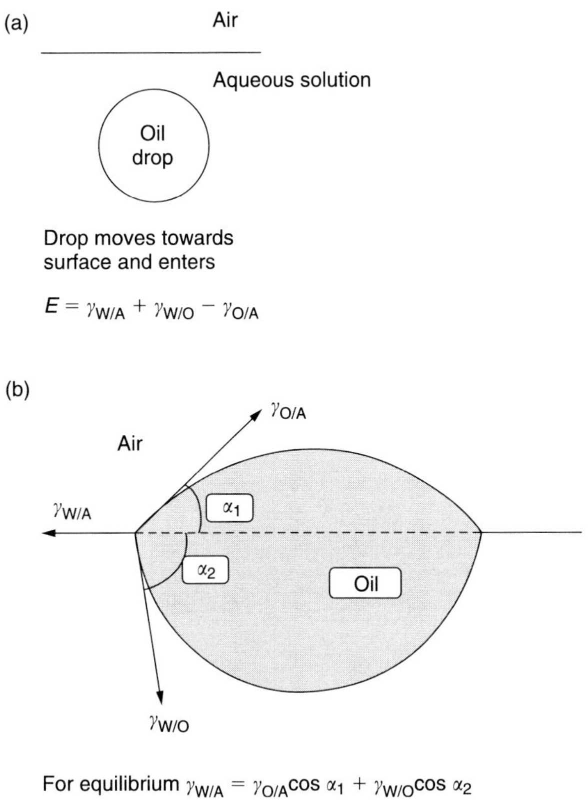

Fig. 8.10. Representation of entry of an oil droplet into the air–water interface (a) and its further spreading (b).

many of these hydrophobic particles can deplete the stabilising surfactant film by rapid adsorption and can cause weak spots in the film.

A further proposed mechanism, based on the degree of wetting of the hydrophobic particles [31], led to the idea of particle bridging. For large smooth particles (large enough to touch both surface and with a contact angle $\theta > 90°$) dewetting can occur. Initially, the Laplace pressure in the film adjacent to the particle becomes positive and causes liquid to flow away from the particle, leading to enhanced drainage and formation of a "hole". When $\theta < 90°$, then, initially, the situation is the same as for $\theta > 90°$, but as the film drains it attains a critical thickness where the film is planar and the capillary pressure becomes zero. At this point, further drainage reverses the sign of the radii of curvature, causing unbalanced capillary forces, which prevent drainage. This can cause a stabilising effect for certain types of particles. This means that a critical receding contact angle is required for efficient foam breaking.

With particles containing rough edges, the situation is more complex, as demonstrated by Johansson and Pugh [32], using finely ground quartz particles of different size fractions. The particle surfaces were hydrophobised by methylation. These and other reported studies confirmed the importance of size, shape and hydrophobicity of the particles on foam stability.

8.7.6 Mixtures of Hydrophobic Particles and Oils as Antifoamers

The synergetic antifoaming effect of mixtures of insoluble hydrophobic particles and hydrophobic oils when dispersed in aqueous medium has been well established in the patent literature. These mixed antifoamers are very effective at very low concentrations (10–100 ppm). The hydrophobic particles could be hydrophobised silica and the oil is poly(dimethyl siloxane) (PDMS).

One possible explanation of the synergetic effect is that the spreading coefficient of PDMS oil is modified by the addition of hydrophobic particles. The oil-particle mixtures are suggested to form composite entities where the particles can adhere to the oil–water interface. The presence of particles adhering to the oil–water interface may facilitate the emergence of oil droplets into the air–water interface to form lenses, leading to rupture of the oil–water–air film.

8.8 Physical Properties of Foams

8.8.1 Mechanical Properties

The compressibility of a foam is determined by the ability of the gas to compress, its wetting power is determined by the properties of the foaming solution [4]. As in any disperse system, a foam may acquire the properties of a solid body, i.e. it can maintain its shape and it possesses a shear modulus (see below).

One of the basic mechanical properties of foams is its compressibility [4] (elasticity) and a bulk modulus E_v may be defined by

$$E_v = -\frac{dp_0}{d \ln V} \tag{8.13}$$

where p_0 is the external pressure, causing deformation, and V is the volume of the deforming system.

By taking into account the liquid volume V_L, the modulus of bulk elasticity of the "wet" foam E'_v is given by

$$E_v = \frac{dp_0}{d \ln V_F} = \frac{V_F dp_0}{d(V_L + V_G)} = E_v\left(1 + \frac{V_L}{V_G}\right) \tag{8.14}$$

Thus, the real modulus of bulk elasticity ("wet" foam) is higher than E_v ("dry" foam).

8.8.2
Rheological Properties

Like any disperse system, foams produce non-Newtonian systems and to characterise its rheological properties one need to obtain information on the elasticity modulus (modulus of compressibility and expansion), the shear modulus, yield stress and effective viscosity, elastic recovery, etc.

It is difficult to study the rheological properties of a foam since on deformation its properties change. The most convenient geometry to measure foam rheology is a parallel plate. The rheological properties could be characterised by a variable viscosity (4),

$$\eta = \eta^*(\dot{\gamma}) + \frac{\tau_\beta}{\dot{\gamma}} \tag{8.15}$$

where $\dot{\gamma}$ is the shear rate.

The shear modulus of a foam is given by

$$G = \frac{\tau_\beta \Delta l}{H} \tag{8.16}$$

where Δl is the shear deformation and H is the distance between parallel plates in the rheometer.

Deryaguin [33] obtained the following expression for the shear modulus,

$$G = \frac{2}{5} p_\gamma = \frac{2}{5}\left(\frac{2}{3}\gamma\varepsilon\right) \approx \frac{4\gamma}{3R_v} \tag{8.17}$$

where R_v is the average volume of the bubble and ε is the specific surface area.

Bikerman [34] obtained Eq. (8.18) for the yield stress of a foam,

$$\tau_\beta = 0.5\frac{N_f}{N_F - 1} p_\gamma \cos\theta \approx \frac{\gamma}{R}\cos\theta \tag{8.18}$$

where N_f is the number of films contacting the plate per unit area and θ is the average angle between the plate and the film.

Princen [35] used a two-dimensional hexagonal package model to derive an expression for the shear modulus and yield stress of a foam, taking into account the foam expansion ratio and the contact angles,

$$G = 0.525\frac{\gamma\cos\theta}{R}\phi^{1/2} \tag{8.19}$$

$$\tau_\beta = 1.05\frac{\gamma \cos\theta}{R}\phi^{1/2}F_{max} \tag{8.20}$$

where F_{max} is a coefficient, equal to 0.1–0.5, depending on the gas volume fraction ϕ. For a "dry" foam ($\phi \to 1$), the yield stress can be calculated from

$$\tau_\beta = 0.525\gamma \cos\frac{\theta}{R} \tag{8.21}$$

For real foams, τ_β can be expressed by the general expression

$$\tau_\beta = C\frac{\gamma \cos\theta}{R}\phi^{1/3}F_{max} \tag{8.22}$$

where C is a coefficient that is approximately equal to 1.

8.8.3
Electrical Properties

Only the liquid phase in a foam possesses electrical conductivity. The specific conductivity of a foam, κ_F, depends on the liquid content and its specific conductivity, κ_L,

$$\kappa_F = \frac{\kappa_L}{nB} \tag{8.23}$$

where n is the foam expansion ratio and B is a structural coefficient that depends on the foam expansion ratio and the liquid phase distribution between the plateau borders; B changes monotonically from 1.5 to 3 with increasing foam expansion factor.

8.8.4
Electrokinetic Properties

In foams with a charged gas/liquid interface, one can obtain various electrokinetic parameters, such as the streaming and zeta potentials. For example, the relation between the volumetric flow of a liquid flowing through a capillary or membrane and zeta potential can be given by the Smoluchowski equation (see Chapter 7),

$$Q = \frac{\varepsilon\varepsilon_0 l}{\eta\kappa} = \frac{\varepsilon\varepsilon_0 r^2}{\eta}\frac{\Delta V}{L} \tag{8.24}$$

where ε is the permittivity of the liquid and ε_0 is the permittivity of free space, I is the electric current, η is the viscosity of the liquid; r is the capillary radius, L is its length and ΔV is the potential distance between the electrodes placed at the capillary ends.

The interpretation of electrokinetic results is complicated because of surface mobility and border and film elasticity, which cause large non-homogeneities in density and border radii at hydrostatic equilibrium and liquid motion.

8.8.5
Optical Properties

The extinction of the luminous flux passing through a foam layer occurs by light scattering (reflection, refraction, interference and diffraction from the foam elements) and light absorption by the solution [4]. In polyhedral foams, there are three structural elements that have clearly distinct optical properties: films, plateau border and vertexes.

The optical properties of single foam films have been extensively studied, but those of the foam as disperse system are poorly considered. The extinction of luminous flux (I/I_0, where I is the intensity of the light passing through the foam and I_0 is the intensity of the incident light) is concluded to be a linear function of the specific foam area. This could be used to determine the specific surface area of a foam.

8.9
Experimental Techniques for Studying Foams

8.9.1
Techniques for Studying Foam Films

Most quantitative studies on foams have been carried out using foam films. As discussed above, microscopic horizontal films were studied by Scheludko and coworkers [2–4]. A schematic representation of the set-up used to study horizontal foam films is given in Figure 8.2. The foam thickness was determined by interferometry. Studies on vertical films were carried out by Mysels and collaborators [5, 7].

One of the most important characteristics of foam films is the contact angle θ appearing at the contact of the film with the bulk phase (solution) from which it is formed. This could be obtained by a topographic technique (which is suitable for small contact angles) that is based on determination of the radii of the interference Newton rings when the film is observed in a reflected monochromatic light.

Another technique for studying foam films is to use α-particle irradiation, which can destroy the film. Depending on the intensity of the α-source, the film either ruptures instantaneously or lives for a much shorter time than required for its spontaneous rupture. The life time τ_a of a black film subjected to irradiation is considered as a parameter characterising the destructive effect of α-particles.

A third technique for studying foam films is fluorescence recovery after photobleaching (FRAP). This technique has been applied by Clarke et al. [36] for lateral diffusion in foam films. It involves irreversible photobleaching by intense laser light of fluorophore molecules in the sample. The time of redistribution of probe molecules (assumed to be randomly distributed within the constitutive membrane lipids in the film) is monitored. The lateral diffusion coefficient, D, is calculated from the rate of recovery of fluorescence in the bleaching region due to the entry of unbleaching fluoroprobes of adjacent parts of the membranes.

Deryaguin and Titijevskaya [37] measured the isotherms of disjoining pressure of microscopic foam films (common thin films) in a narrow range of pressures. At equilibrium, the capillary pressure p_σ in the flat horizontal foam film is equal to the disjoining pressure (π) in it,

$$p_\sigma = \pi = p_{\mathrm{g}} - p_{\mathrm{L}} \tag{8.25}$$

where p_{g} is the pressure in the gas phase and p_{L} is the pressure in the liquid phase.

Several other techniques have been applied to measure foam films, e.g. ellipsometry, FT-IR spectroscopy, X-ray reflection and measurement of gas permeability through the film. These techniques are described in detail in the text by Exerowa and Kruglyakov [4] to which the reader is referred.

8.9.2
Techniques for Studying Structural Parameters of Foams

The polyhedral foam consists of polyhedral gas bubbles, the faces of which are flat or lightly bent liquid films, while the edges are the plateau borders and the edge cross points are the vortexes. Several techniques can be applied to obtain the analytical dependence of these characteristics and the structural parameters of the foam [4].

The foam expansion ratio can be characterised by the liquid volume fraction in the foam, which is the sum of the volume fractions of the films, plateau borders and vertexes. Alternatively, one can use the foam density as a measure of the foam expansion ratio. The reduced pressure in the foam plateau border can be measured using a capillary manometer [4]. The bubble size and shape distribution in a foam can be determined by microphotography of the foam. Information about the liquid distribution between films and plateau borders is obtained from the data on the border radius of curvature, the film thickness and the film to plateau border number ratio obtained in an elementary foam cell.

8.9.3
Measurement of Foam Drainage

After foam formation the liquid starts to drain out of the foam. The "excess" liquid in the foam film drains into the plateau borders, then through them, flowing down from the upper to the lower foam layers, following the gravity direction until the gradient of the capillary pressure equalises the gravity force,

$$\frac{dp_\sigma}{dl} = \rho g \tag{8.26}$$

where l is a co-ordinate in opposite direction to gravity.

Simultaneously with drainage from films into borders the liquid begins to flow out from the foam, when the pressure in the lower foam films outweighs the external pressure. This process is similar to gel syneresis and it is sometimes referred to as "foam syneresis" and "foam drainage".

The rate of foam drainage is determined by the hydrodynamic characteristics of the foam as well as the rate of internal foam collapse and breakdown of the foam column. The foam drainage is determined by measuring the quantity of liquid that drains from the foam per unit time. Various types of vessels and graduated tubes can be used to measure the liquid quantity draining from a foam. Alternatively, one can measure the change in electrical conductivity of the layer at the vessel mouth compared with the electrical conductivity of the foaming solution [4].

8.9.4 Measurement of Foam Collapse

This can be followed by measuring the bubble size distribution as function of time, by, for example, microphotography or by the counting number of bubbles. Alternatively, one can measure the specific surface area or average bubble size as a function of time. Other techniques such as light scattering or ultrasound can also be applied.

References

1 E. Dickinson: *Introduction to Food Colloids*, Oxford University Press, Oxford, 1992.
2 A. Scheludko: *Colloid Science*, Elsevier Science, Amsterdam, 1966.
3 A. Scheludko, *Adv. Colloid Interface Sci.*, **1971**, *1*, 391.
4 D. Exerowa, P. M. Kruglyakov: *Foam and Foam Films*, Elsevier, Amsterdam, 1997.
5 R. J. Pugh, *Adv. Colloid Interface Sci.*, **1996**, *64*, 67.
6 O. Reynolds, *Phil. Trans. Royal Soc. London Ser. A*, **1886**, *177*, 157.
7 K. J. Mysels, *J. Phys. Chem.*, **1964**, *68*, 3441.
8 J. Lucassen: *Anionic Surfactants.* E. H. Lucassen-Reynders (ed.): Marcel Dekker, New York, 1981, 217.
9 H. N. Stein, *Adv. Colloid Interface Sci.*, **1991**, *34*, 175.
10 J. T. Davies: *Proceedings of the Second International Congress of Surface Activity.* J. H. Schulman (ed.): Butterworths, London, 1957, Volume 1.
11 B. V. Deryaguin, N. V. Churaev, *Kolloid Zh.*, **1976**, *38*, 438.
12 B. V. Deryaguin: *Theory of Stability of Colloids and Thin Films*, Consultant Bureau, New York, 1989.
13 B. V. Deryaguin, L. D. Landau, *Acta Physicochim. USSR*, **1941**, *14*, 633.

14 E. J. Verwey, J. T. G. Overbeek: *Theory of Stability of Lyophobic Colloids*, Elsevier Science, Amsterdam, 1948.

15 A. J. Vries de, *Discuss. Faraday Soc.*, **1966**, *42*, 23.

16 A. Vrij, J. T. G. Overbeek, *J. Am. Chem. Soc.*, **1968**, *90*, 3074.

17 B. Radoev, A. Scheludko, E. Manev, *J. Colloid Interface Sci.*, **1983**, *95*, 254.

18 V. G. Gleim, I. V. Shelomov, B. R. Shidlovskii, *J. Appl. Chem. USSR*, **1959**, *32*, 1069.

19 R. J. Pugh, R. H. Yoon, *J. Colloid Interface Sci.*, **1994**, *163*, 169.

20 P. M. Claesson, H. K. Christensen, *J. Phys. Chem.*, **1988**, *92*, 1650.

21 E. S. Johnott, *Philos. Mag.*, **1906**, *11*, 746.

22 J. Perrin, *Ann. Phys.*, **1918**, *10*, 160.

23 L. Loeb, D. T. Wasan, *Langmuir*, **1993**, *9*, 1668.

24 S. Frieberg, *Mol. Cryst. Liq. Cryst.*, **1977**, *40*, 49.

25 J. E. Perez, J. E. Proust, Ter-Minassian Saraga: *Thin Liquid Films*, I. B. Ivanov (ed.): Marcel Dekker, New York, 1988, 70.

26 *Defoaming*, P. R. Garrett (ed.): Marcel Dekker, New York, 1993, Surfactant Science Series, Volume 45.

27 S. Ross, R. M. Haak, *J. Phys. Chem.*, **1958**, *62*, 1260.

28 J. V. Robinson, W. W. Woods, *J. Soc. Chem. Ind.*, **1948**, *67*, 361.

29 W. D. Harkins, *J. Phys. Chem.*, **1941**, *9*, 552.

30 S. Ross, J. W. McBain, *Ind. Chem. Eng.*, **1944**, *36*, 570.

31 P. R. Garett, *J. Colloid Interface Sci.*, **1979**, *69*, 107.

32 G. Johansson, R. J. Pugh, *Int. J. Miner. Process.*, **1992**, *34*, 1.

33 B. V. Deryaguin, *Kollod. Z.*, **1933**, *64*, 1.

34 J. J. Bickerman, *Foams*, Springer-Verlag, New York, 1973.

35 H. Princen, *J. Colloid Interface Sci.*, **1983**, *91*, 160.

36 D. Clark, R. Dann, A. Mackie, J. Mingins, A. Pinder, P. Purdy, E. Russel, L. Smith, D. Wilson, *J. Colloid Interface Sci.*, **1990**, *138*, 195.

37 B. V. Deryaguin, A. S. Titijevskaya, *Kolloid Zh.*, **1953**, *15*, 416.

9
Surfactants in Nano-Emulsions

9.1
Introduction

Nano-emulsions are transparent or translucent systems mostly covering the size range 50–200 nm [1, 2]. They were also referred to as mini-emulsions [3, 4]. Unlike microemulsions (which are also transparent or translucent and thermodynamically stable, see Chapter 10), nano-emulsions are only kinetically stable. However, their long-term physical stability (with no apparent flocculation or coalescence) makes them unique and they are sometimes referred to as "Approaching Thermodynamic Stability".

The inherently high colloid stability of nano-emulsions can be well understood from a consideration of their steric stabilisation (when using nonionic surfactants and/or polymers) and how this is affected by the ratio of the adsorbed layer thickness to droplet radius, as will be discussed below.

Unless adequately prepared (to control the droplet size distribution) and stabilised against Ostwald ripening (which occurs when the oil has some finite solubility in the continuous medium), nano-emulsions may lose their transparency with time as a result of increasing droplet size.

Nano-emulsions are attractive for application in personal care and cosmetics as well as in health care due to the following advantages:

- The very small droplet size causes a large reduction in the gravity force and Brownian motion may be sufficient to overcome gravity. This means that no creaming or sedimentation occurs on storage.
- The small droplet size also prevents any flocculation of the droplets. Weak flocculation is prevented and this enables the system to remain dispersed with no separation.
- The small droplets size also prevents their coalescence, since these droplets are non-deformable and hence surface fluctuations are prevented. In addition, the significant surfactant film thickness (relative to droplet radius) prevents any thinning or disruption of the liquid film between the droplets.
- Nano-emulsions are suitable for efficient delivery of active ingredients through the skin. The large surface area of the emulsion system allows rapid penetration of actives.

Applied Surfactants: Principles and Applications. Tharwat F. Tadros

ISBN: 3-527-30629-3

- Due to their small size, nano-emulsions can penetrate through the "rough" skin surface and this enhances penetration of actives.
- The transparent nature of the system, their fluidity (at reasonable oil concentrations) as well as the absence of any thickeners may give them a pleasant aesthetic character and skin feel.
- Unlike microemulsions (which require a high surfactant concentration, usually in the region of 20% and higher), nano-emulsions can be prepared using reasonable surfactant concentrations. For a 20% O/W nano-emulsion, a surfactant concentration in the region of 5–10% may be sufficient.
- The small size of the droplets allow them to deposit uniformly on substrates – wetting, spreading and penetration may be also enhanced because of the low surface tension of the whole system and the low interfacial tension of the O/W droplets.
- Nano-emulsions can be applied for delivery of fragrants, which may be incorporated in many personal care products. This could also be applied in perfumes, which are desirable to be formulated alcohol free.
- Nano-emulsions may be applied as a substitute for liposomes and vesicles (which are much less stable) and it is possible in some cases to build lamellar liquid crystalline phases around the nano-emulsion droplets.

Despite the above advantages, nano-emulsions have only attracted interest in recent years because:

- Their preparation requires, in many cases, special application techniques such as the use of high-pressure homogenisers as well as ultrasonics. Such equipment (such as the Microfluidiser) has became available only in recent years.
- There is a perception in the Personal Care and Cosmetic Industry that nano-emulsions are expensive to produce. Expensive equipment is required as well as the use of high concentrations of emulsifiers.
- Lack of understanding of the mechanism of production of submicron droplets and the role of surfactants and cosurfactants.
- Lack of demonstration of the benefits that can be obtained from using nano-emulsions when compared with classical macroemulsion systems.
- Lack of understanding of the interfacial chemistry involved in production of nano-emulsions. For example, few formulation chemists are aware of the use of the phase inversion temperature (PIT) concept and how this can be usefully applied for the production of small emulsion droplets.
- Lack of knowledge on the mechanism of Ostwald ripening, which is perhaps the most serious instability problem with nano-emulsions.
- Lack of knowledge of the ingredients that may be incorporated to overcome Ostwald ripening. For example, addition of a second oil phase with very low solubility and/or incorporation of polymeric surfactants that strongly adsorb at the O/W interface (which are also insoluble in the aqueous medium).
- Fear of introduction of new systems without full evaluation of the cost and benefits.

However, despite these difficulties, several companies have introduced nano-emulsions in the market and, within the next few years, the benefits will be evaluated. Nano-emulsions have been used in the pharmaceutical field as drug delivery systems [5].

The acceptance of nano-emulsions as a new type of formulation depends on customer perception and acceptability. With the advent of new high-pressure homogenizers and the competition between various manufacturers, the cost of production of nano-emulsions will decrease and may approach that of classical macroemulsions.

Fundamental research into the role of surfactants in the process [6, 7] will lead to optimized emulsifier systems and more economic use of surfactants will emerge.

This chapter discusses the following topics:

(1) Fundamental principles of emulsification and the role of surfactants.
(2) Production of nano-emulsions using: (a) High-pressure homogenizers and (b) the phase inversion temperature (PIT) principle.
(3) Theory of steric stabilization of emulsions; the role of the relative ratio of adsorbed layer thickness to the droplet radius.
(4) Theory of Ostwald ripening and methods of reduction of the process, i.e. (a) incorporation of a second oil phase with very low solubility and (b) use of strongly adsorbed polymeric surfactants.
(5) Examples of recently prepared nano-emulsions and investigation of the above effects.

9.2 Mechanism of Emulsification

As mentioned in Chapter 6, oil, water, surfactant and energy are needed to prepare emulsions. This can be considered from an examination of the energy required to expand the interface, $\Delta A\gamma$ (where ΔA is the increase in interfacial area when the bulk oil with area A_1 produces a large number of droplets with area A_2; $A_2 \gg A_1$, γ is the interfacial tension). Since γ is positive, the energy to expand the interface is large and positive. This energy term cannot be compensated by the small entropy of dispersion $T\Delta S$ (which is also positive) and so the total free energy of formation of an emulsion, ΔG is positive,

$$\Delta G = \Delta A\gamma - T\Delta S \qquad (9.1)$$

Thus, emulsion formation is non-spontaneous and energy is required to produce the droplets. The formation of large droplets (few μm) as is the case for macro-emulsions is fairly easy and hence high speed stirrers such as the Ultraturrax or Silverson Mixer are sufficient to produce the emulsion. In contrast the formation of small drops (submicron, as is the case with nano-emulsions) is difficult, requiring a large amount of surfactant and/or energy.

The high energy required to form nano-emulsions can be understood in terms of the Laplace pressure p (the difference in pressure between inside and outside the droplet),

$$p = \gamma\left(\frac{1}{R_1} + \frac{1}{R_2}\right) \tag{9.2}$$

where R_1 and R_2 are the principal radii of curvature of the drop. For a spherical drop, $R_1 = R_2 = R$ and

$$p = \frac{2\gamma}{R} \tag{9.3}$$

To break up a drop into smaller ones, it must be strongly deformed and this deformation increases p. This is illustrated before in Figure 6.11 (see Chapter 6), showing the situation when a spherical drop deforms into a prolate ellipsoid. Near 1 there is only one radius of curvature R_a, whereas near 2 there are two radii of curvature $R_{b,1}$ and $R_{b,2}$. Consequently, the stress needed to deform the drop is higher for a smaller drop. Since the stress is generally transmitted by the surrounding liquid via agitation, higher stresses need more vigorous agitation, hence more energy is needed to produce smaller drops [8].

Surfactants play major roles in the formation of nano-emulsions: By lowering the interfacial tension, p is reduced and hence the stress needed to break up a drop is reduced. Surfactants prevent coalescence of newly formed drops.

Figure 6.12 illustrates the various processes that occur during emulsification: Break up of droplets, adsorption of surfactants and droplet collision (which may or may not lead to coalescence) [8]. Each of these processes occurs numerous times during emulsification and the time scale of each process is very short, typically a microsecond. This shows that the emulsification is a dynamic process and events that occur in a microsecond range could be very important.

As mentioned in Chapter 6, to describe emulsion formation one has to consider two main factors: Hydrodynamics and interfacial science

To assess nano-emulsion formation, one usually measures the droplet size distribution using dynamic light scattering techniques (photon correlation spectroscopy, PCS). In this technique, one measures the intensity fluctuation of scattered light by the droplets as they undergo Brownian motion [9]. When a light beam passes through a nano-emulsion, an oscillating dipole moment is induced in the droplets, thereby reradiating the light. Due to the random position of the droplets, the intensity of scattered light at any instant appears as a random diffraction or "speckle" pattern. As the droplets undergo Brownian motion, the random configuration of the pattern will, therefore, fluctuate such that the time taken for an intensity maximum to become a minimum, i.e. the coherence time, corresponds exactly to the time required for the droplet to move one wavelength. Using a photomultiplier of active area about the diffraction maximum, i.e. one coherence area, this intensity fluctuation can be measured. The analogue output is digitised using a digital correlator that measures the photocount (or intensity) correlation function of the scattered light. The photocount correlation function $G^{(2)}(\tau)$ is given by

$$G^{(2)}(\tau) = B(1 + \gamma^2[g^{(1)}(\tau)]^2) \tag{9.4}$$

where τ is the correlation delay time. The correlator compares $G^{(2)}(\tau)$ for many values of τ. B is the background value to which $G^{(2)}(\tau)$ decays at long delay times; $g^{(1)}(\tau)$ is the normalised correlation function of the scattered electric field and γ is a constant (~ 1).

For monodisperse non-interacting droplets,

$$g^{(1)} = \exp(-\Gamma\tau) \tag{9.5}$$

where Γ is the decay rate or inverse coherence time, that is related to the translational diffusion coefficient D by the equation

$$\Gamma = DK^2 \tag{9.6}$$

where K is the scattering vector,

$$K = \frac{4\pi n}{\lambda_0} \sin\left(\frac{\theta}{2}\right) \tag{9.7}$$

λ is the wavelength of light in vacuo, n is the refractive index of the solution and θ is the scattering angle.

The droplet radius R can be calculated from D using the Stokes–Einstein equation,

$$D = \frac{kT}{6\pi\eta_0 R} \tag{9.8}$$

where η_0 is the viscosity of the medium.

The above analysis is valid for dilute monodisperse droplets. With many nano-emulsions, the droplets are not perfectly mono-disperse (usually with a narrow size distribution) and the light scattering results are analysed for polydispersity – the data are expressed as an average size and a polydispersity index that gives information on the deviation from the average size.

9.3 Methods of Emulsification and the Role of Surfactants

As with macroemulsions (see Chapter 6), several procedures may be applied for emulsion preparation: simple pipe flow (low agitation energy L), static mixers and general stirrers (low to medium energy, L-M), high speed mixers such as the Ultraturrex (M), colloid mills and high-pressure homogenizers (high energy, H), and ultrasound generators (M-H). Preparation can be continuous (C) or batch-wise (B). With nano-emulsions, however, a higher power density is required and this restricts their preparation to the use of high-pressure homogenisers and ultrasonics.

An important parameter that describes droplet deformation is the Weber number, W_e, which gives the ratio of the external stress $G\eta$ (where G is the velocity gradient and η is the viscosity) to the Laplace pressure (see Chapter 6),

$$W_e = \frac{G\eta r}{2\gamma} \tag{9.9}$$

Droplet deformation increases with increasing W_e, which means that to produce small droplets one requires high stresses (high shear rates). In other words, nano-emulsions cost more energy to produce than do macroemulsions [4].

The role of surfactants on emulsion formation is detailed in Chapter 6 and the same principles apply to the formation of nano-emulsions. Thus, one must consider the effect of surfactants on the interfacial tension, interfacial elasticity, and interfacial tension gradients.

9.4 Preparation of Nano-Emulsions

Two methods may be applied for the preparation of nano-emulsions (covering the droplet radius size range 50–200 nm). Use of high-pressure homogenisers (aided by appropriate choice of surfactants and cosurfactants) or application of the phase inversion temperature (PIT) concept.

9.4.1 Use of High Pressure Homogenizers

The production of small droplets (submicron) requires application of high energy–emulsification is generally inefficient, as illustrated below.

Simple calculations show that the mechanical energy required for emulsification exceeds the interfacial energy by several orders of magnitude. For example to produce an emulsion at $\phi = 0.1$ with a $d_{32} = 0.6$ μm, using a surfactant that gives an interfacial tension $\gamma = 10$ mN m^{-1}, the net increase in surface free energy is $A\gamma = 6\phi\gamma/d_{32} = 10^4$ J m^{-3}. The mechanical energy required in a homogenizer is 10^7 J m^{-3}, i.e. an efficiency of 0.1% – the rest of the energy (99.9%) is dissipated as heat [10].

The intensity of the process or the effectiveness in making small droplets is often governed by the net power density [$\varepsilon(t)$],

$$p = \varepsilon(t)\,\mathrm{d}t \tag{9.10}$$

where t is the time during which emulsification occurs.

Break up of droplets will only occur at high ε, which means that the energy dissipated at low ε levels is wasted. Batch processes are generally less efficient than continuous processes. This shows why, with a stirrer in a large vessel, most of the energy applied at low intensity is dissipated as heat. In a homogenizer, p is simply equal to the homogenizer pressure.

Several procedures may be applied to enhance the efficiency of emulsification when producing nano-emulsions: One should optimise the efficiency of agitation by increasing ε and decreasing the dissipation time. The emulsion is preferably prepared at high volume faction of the disperse phase and diluted afterwards. However, very high ϕ may result in coalescence during emulsification. Addition of more surfactant creates a smaller γ_{eff} and possibly diminishes recoalescence. A surfactant mixture that shows a reduction in γ compared with the individual components can be used. If possible, the surfactant is dissolved in the disperse phase rather than the continuous phase; this often leads to smaller droplets.

It may be useful to emulsify in steps of increasing intensity, particularly with emulsions having a highly viscous disperse phase.

9.4.2 Phase Inversion Temperature (PIT) Principle

Phase inversion in emulsions can be one of two types: Transitional inversion induced by changing factors that affect the HLB of the system, e.g. temperature and/or electrolyte concentration, and catastrophic inversion, which is induced by increasing the volume fraction of the disperse phase.

Transitional inversion can also be induced by changing the HLB number of the surfactant at constant temperature using surfactant mixtures. This is illustrated in Figure 9.1, which shows the average droplet diameter and rate constant for attaining constant droplet size as a function of the HLB number.

The diameter decreases and the rate constant increases as inversion is approached. To apply the phase inversion principle one uses the transitional inversion method demonstrated by Shinoda and co-workers [11, 12] when using ethoxylate-type nonionic surfactants. These surfactants are highly dependent on temperature, becoming lipophilic with increasing temperature due to the dehydration of the poly(ethylene oxide) chain. When an O/W emulsion prepared using a nonionic surfactant of the ethoxylate type is heated, then, at a critical temperature (the PIT), the emulsion inverts to a W/O emulsion. At the PIT the droplet size reaches a minimum and the interfacial tension also reaches a minimum. However, the small droplets are unstable and they coalesce very rapidly. By rapid cooling of the emulsion that is prepared at a temperature near the PIT, very stable, small emulsion droplets can be produced.

The phase inversion that occurs on heating an emulsion is clearly demonstrated in a study of the phase behaviour of emulsions as a function of temperature. This is illustrated schematically in Figure 9.2 by what happens when the temperature is increased [13, 14].

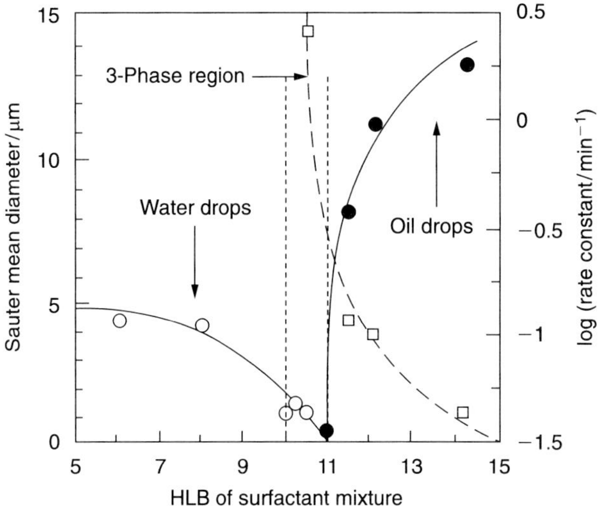

Fig. 9.1. Emulsion droplet diameters (○, ●) and rate constant for attaining steady size (□) as function of HLB, cyclohexane–nonylphenol ethoxylate.

At low temperature, over the Winsor I region, O/W macroemulsions can be formed and are quite stable. On increasing the temperature, the O/W emulsion stability decreases and the macroemulsion finally resolves when the system reaches the Winsor III phase region (both O/W and W/O emulsions are unstable). At higher temperature, over the Winsor II region, W/O emulsions become stable.

Figure 9.3 shows the most clear-cut image of the macroemulsion inversion as a function of temperature; equal volumes of oil and water are emulsified at various temperatures. Five hours after preparation, the macroemulsions sediment completely. Below the balanced temperature (HLB temperature), a stable O/W macroemulsion is formed, whereas above the balanced temperature a stable W/O emulsion is formed. Close to the balanced point (60–68 °C), a three–phase equilibrium is observed and neither O/W or W/O emulsions are stable.

Near the HLB temperature, the interfacial tension reaches a minimum (Figure 9.4). Thus, by preparing the emulsion 2–4 °C below the PIT (near the minimum in γ), followed by rapid cooling, nano-emulsions may be produced.

The minimum in γ can be explained in terms of the change in curvature H of the interfacial region, as the system changes from O/W to W/O.

For O/W systems and normal micelles, the monolayer curves towards the oil and H is given a positive value. For a W/O emulsions and inverse micelles, the monolayer curves towards the water and H is assigned a negative value. At the inversion point (HLB temperature) H becomes zero and γ reaches a minimum.

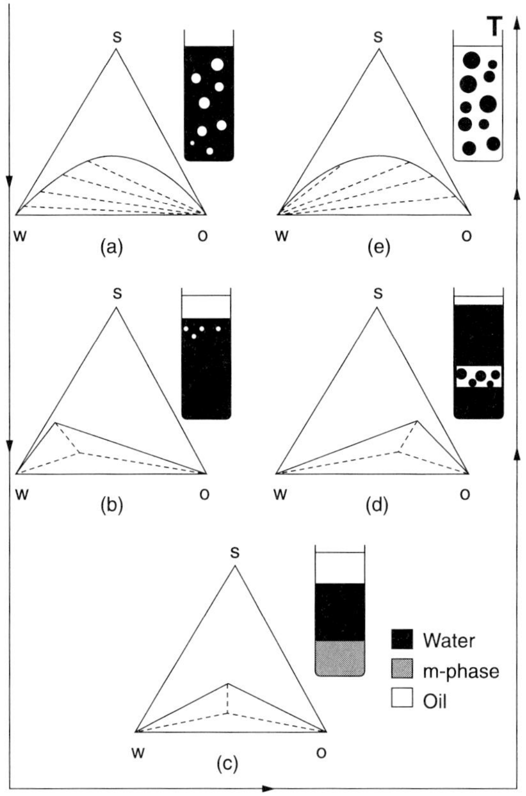

Fig. 9.2. PIT concept.

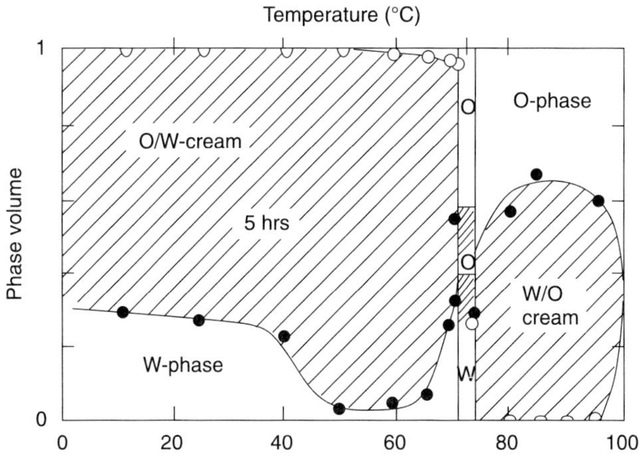

Fig. 9.3. Macroemulsion stability diagram of cyclohexane–water–polyoxyethylene (9.7) nonyl phenol ether system.

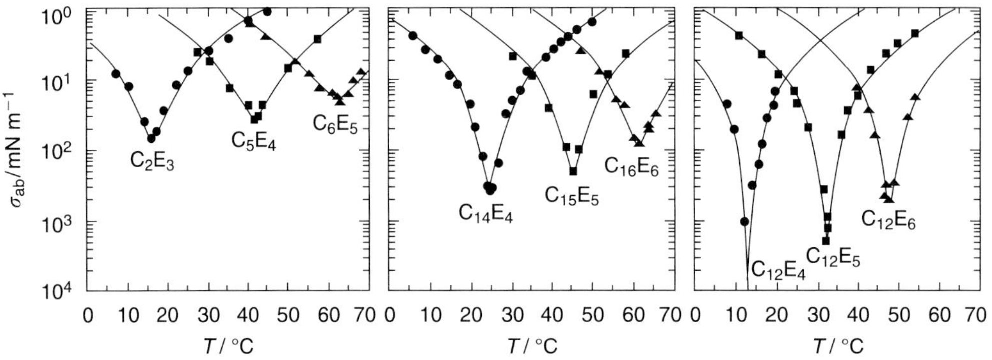

Fig. 9.4. Interfacial tensions of n-octane against water in the presence of various C_nE_m surfactants above the c.m.c. as a function of temperature.

9.5 Steric Stabilization and the Role of the Adsorbed Layer Thickness

Since most nano-emulsions are prepared using nonionic and/or polymeric surfactants, it is necessary to consider the interaction forces between droplets containing adsorbed layers (Steric stabilization). As this is detailed in Chapter 6, only a summary is given here [15, 16].

When two droplets that each contain an adsorbed layer of thickness δ approach to a separation h, whereby h becomes less than 2δ, repulsion occurs as result of two main effects:

(1) Unfavourable mixing of the stabilizing chains A of the adsorbed layers, when these are in good solvent conditions. This is referred to as the mixing (osmotic interaction, G_{mix}) and is given by

$$\frac{G_{mix}}{kT} = \frac{4\pi}{3V_1}\phi_2^2\left(\frac{1}{2} - \chi\right)\left(3a + 2\delta + \frac{h}{2}\right)\left(\delta - \frac{h}{2}\right)^2 \tag{9.11}$$

where k is the Boltzmann constant, T is the absolute temperature, V_1 is the molar volume of the solvent, ϕ_2 is the volume fraction of the polymer (the A chains) in the adsorbed layer and χ is the Flory–Huggins (polymer–solvent interaction) parameter. G_{mix} depends on three main parameters: The volume fraction of the A chains in the adsorbed layer (the denser the layer is the higher G_{mix}); the Flory–Huggins interaction parameter χ (for G_{mix} to remain positive, i.e. repulsive, χ should be lower than $\frac{1}{2}$); and the adsorbed layer thickness δ.

(2) Reduction in configurational entropy of the chains on significant overlap – referred to as elastic (entropic) interaction and is given by the expression

$$G_{el} = 2\nu_2 \ln\left[\frac{\Omega(h)}{\Omega(\infty)}\right] \tag{9.12}$$

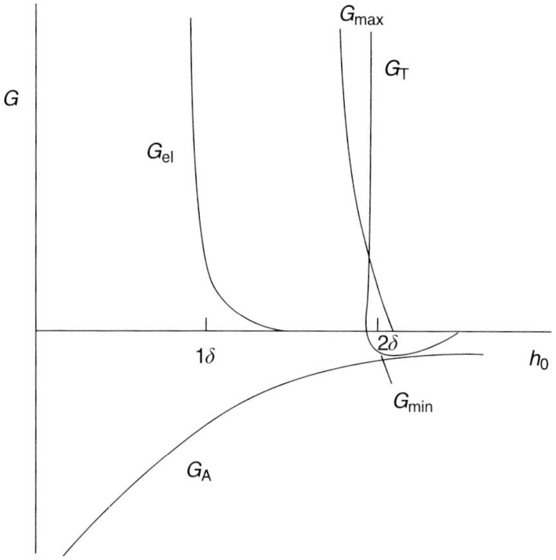

Fig. 9.5. Variation of G_{mix}, G_{el}, G_A and G_T with h.

where ν_2 is the number of chains per unit area, $\Omega(h)$ is the configurational entropy of the chains at a separation distance h and $\Omega(\infty)$ is the configurational entropy at infinite separation.

Combining G_{mix} and G_{el} with the van der Waals attraction G_A gives the total energy of interaction G_T,

$$G_T = G_{mix} + G_{el} + G_A \tag{9.13}$$

Figure 9.5 illustrates the variation of G_{mix}, G_{el}, G_A and G_T with h. As can be seen, G_{mix} increases very rapidly with decreasing h as soon as $h < 2\delta$, G_{el} increase very rapidly with decrease of h when $h < \delta$. G_T shows one minimum, G_{min}, and increases very rapidly with decreasing h when $h < 2\delta$.

The magnitude of G_{min} depends on the particle radius R, the Hamaker constant A and the adsorbed layer thickness δ.

As an illustration, Figure 9.6 shows the variation of G_T with h at various ratios of δ/R. The depth of the minimum clearly decreases with increasing δ/R. This is the basis of the high kinetic stability of nano-emulsions. With nano-emulsions having a radius in the region of 50 nm and an adsorbed layer thickness of say 10 nm, δ/R is 0.2. This relatively high value (for macroemulsions δ/R is at least an order of magnitude lower) results in a very shallow minimum (which could be less than kT).

The above situation results in very high stability with no flocculation (weak or strong). In addition, the very small size of the droplets and the dense adsorbed

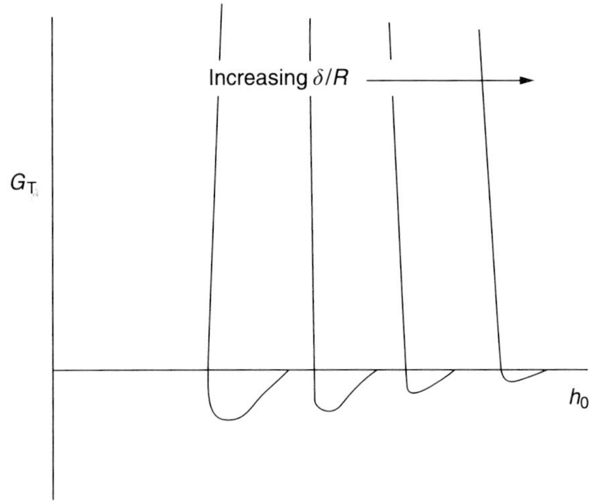

Fig. 9.6. Variation of G_T with h with increasing δ/R.

layers ensures lack of deformation of the interface, lack of thinning and disruption of the liquid film between the droplets and, hence, coalescence is also prevented.

The only instability problem with nano-emulsions is Ostwald ripening (see below).

9.6 Ostwald Ripening

One of the main problems with nano-emulsions is Ostwald ripening, which results from the difference in solubility between small and large droplets. The difference in chemical potential of dispersed phase droplets between different sized droplets was given by Lord Kelvin [17],

$$c(r) = c(\infty) \exp\left(\frac{2\gamma V_m}{rRT}\right) \qquad (9.14)$$

where $c(r)$ is the solubility surrounding a particle of radius r, $c(\infty)$ is the bulk phase solubility and V_m is the molar volume of the dispersed phase.

The quantity $(2\gamma V_m/rRT)$ is termed the characteristic length. It has an order of ~1 nm or less, indicating that the difference in solubility of a 1 μm droplet is of the order of 0.1% or less.

Theoretically, Ostwald ripening should lead to the condensation of all droplets into a single drop (i.e. phase separation). This does not occur in practice since the rate of growth decreases with increasing droplet size.

For two droplets of radii r_1 and r_2 (where $r_1 < r_2$),

$$\left(\frac{RT}{V_m}\right) \ln\left[\frac{c(r_1)}{c(r_2)}\right] = 2\gamma\left(\frac{1}{r_1} - \frac{1}{r_2}\right) \tag{9.15}$$

Equation (9.15) shows that the larger the difference between r_1 and r_2 the higher the rate of Ostwald ripening.

Ostwald ripening can be quantitatively assessed from plots of the cube of the radius versus time t (the Lifshitz–Slesov–Wagner, LSW, theory) [18, 19],

$$r^3 = \frac{8}{9}\left[\frac{c(\infty)\gamma V_m D}{\rho RT}\right] t \tag{9.16}$$

where D is the diffusion coefficient of the disperse phase in the continuous phase and ρ is the density of the disperse phase.

Several methods may be applied to reduce Ostwald ripening [20–22]:

(1) Addition of a second disperse phase component that is insoluble in the continuous phase (e.g. squalene). Here, significant partitioning between different droplets occurs; the component that has low solubility in the continuous phase is expected to be concentrated in the smaller droplets. During Ostwald ripening in a two-component disperse phase system, equilibrium is established when the difference in chemical potential between different size droplets (which results from curvature effects) is balanced by the difference in chemical potential resulting from partitioning of the two components. If the secondary component has zero solubility in the continuous phase, the size distribution will not deviate from the initial one (the growth rate is equal to zero). For limited solubility of the secondary component, the distribution is the same as governed by Eq. (9.16), i.e. a mixture growth rate is obtained that is still lower than that of the more soluble component. The above method is of limited application since one requires a highly insoluble oil, as the second phase, which is miscible with the primary phase.

(2) Modification of the interfacial film at the O/W interface: According to Eq. (9.15) reduction in γ results in a reduction of Ostwald ripening. However, this alone is not sufficient since one has to reduce γ by several orders of magnitude. Walstra [23] suggested that by using surfactants that are strongly adsorbed at the O/W interface (i.e. polymeric surfactants) and which do not desorb during ripening, the rate could be significantly reduced. An increase in the surface dilational modulus and decrease in γ would be observed for the shrinking drops. The difference in γ between the droplets would balance the difference in capillary pressure (i.e. curvature effects).

A-B-A block copolymers that are soluble in the oil phase and insoluble in the continuous phase are useful in achieving the above effect. The polymeric surfactant should enhance the lowering of γ by the emulsifier. In other words, the emulsifier and the polymeric surfactant should show synergy in lowering γ.

9.7 Practical Examples of Nano-Emulsions

Several experiments have been conducted recently to investigate the methods of preparation of nano-emulsions and their stability [24]. The first method applied the PIT principle to prepare nano-emulsions. Experiments were carried out using hexadecane and isohexadecane (Arlamol HD) as the oil phase and Brij 30 ($C_{12}EO_4$) as the nonionic emulsifier. Phase diagrams of the ternary systems water–$C_{12}EO_4$–hexadecane and water–$C_{12}EO_4$–isohexadecane are shown in Figures 9.7 and 9.8, respectively. The main features of the pseudoternary system are (1) a O_m isotropic liquid transparent phase, which extends along the hexadecane–$C_{12}EO_4$ or isohexadecane–$C_{12}EO_4$ axis, corresponding to inverse micelles or W/O microemulsions; (2) a L_α lamellar liquid crystalline phase extending from the water–$C_{12}EO_4$ axis towards the oil vertex; (3) the rest of the phase diagram consists of two- or three-phase regions: a (W_m + O) two-liquid phase region, which appears along the water–oil axis; a ($W_m + L_\alpha$ + O) three-phase region, consisting of a bluish liquid phase (O/W microemulsion), a lamellar liquid crystalline phase (L_α) and a trans-

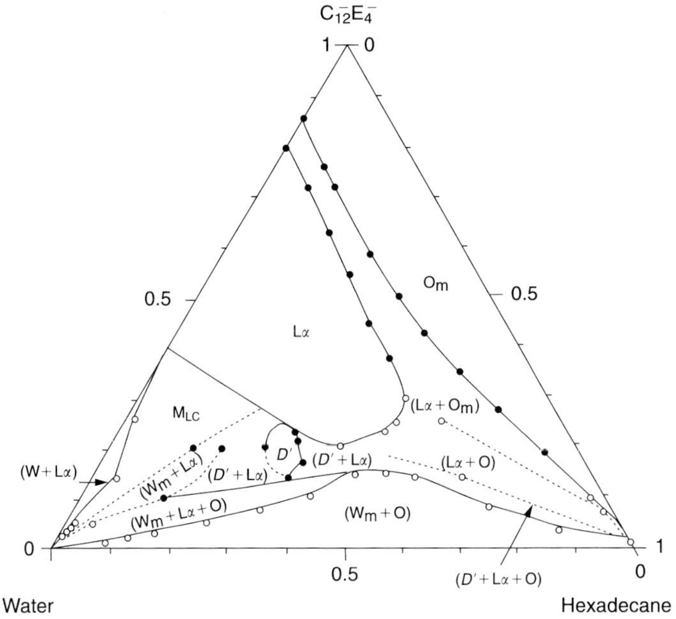

Fig. 9.7. Pseudoternary phase diagram at 25 °C of water–$C_{12}EO_4$–hexadecane.

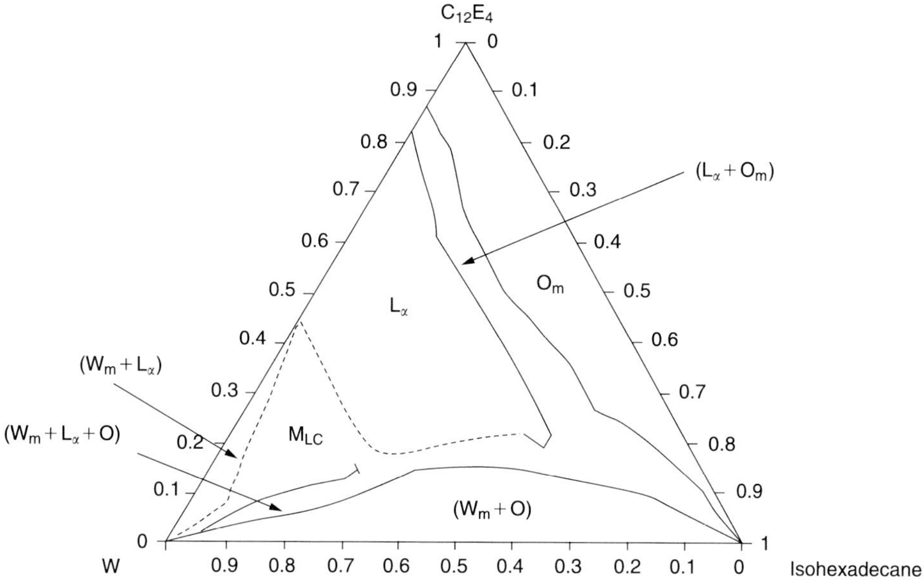

Fig. 9.8. Pseudoternary phase diagram at 25 °C of water–$C_{12}EO_4$–isohexadecane.

parent oil phase; a ($L_\alpha + O_m$) two-phase region, consisting of an oil and liquid crystalline region. M_{LC} is a multiphase region containing a lamellar liquid crystalline phase (L_α).

The HLB temperature was determined using conductivity measurements, whereby 10^{-2} mol dm^{-3} NaCl was added to the aqueous phase (to increase the sensitivity of the conductivity measurements). The concentration of NaCl was low and hence it had little effect on the phase behaviour.

Figure 9.9 shows the variation of conductivity versus temperature for 20% O/W emulsions at different surfactant concentrations. There is a sharp decrease in conductivity at the PIT or HLB temperature.

The HLB temperature decreases with increasing surfactant concentration – this could be due to the excess nonionic surfactant remaining in the continuous phase.

However, at surfactant concentrations higher than 5%, the conductivity plots show a second maximum (Figure 9.9). This was attributed to the presence of an L_α phase and bicontinuous L_3 or D′ phases [25].

Nano-emulsions were prepared by rapidly cooling the system to 25 °C. The droplet diameter was determined using photon correlation spectroscopy (PCS). The results are summarised in Table 9.1, which shows the exact composition of the emulsions, HLB temperature, *z*-average radius and polydispersity index.

O/W nano-emulsions with droplet radii in the range 26–66 nm could be obtained at surfactant concentrations between 4 and 8%. The nano-emulsion droplet size and polydispersity index decreases with increasing surfactant concentration.

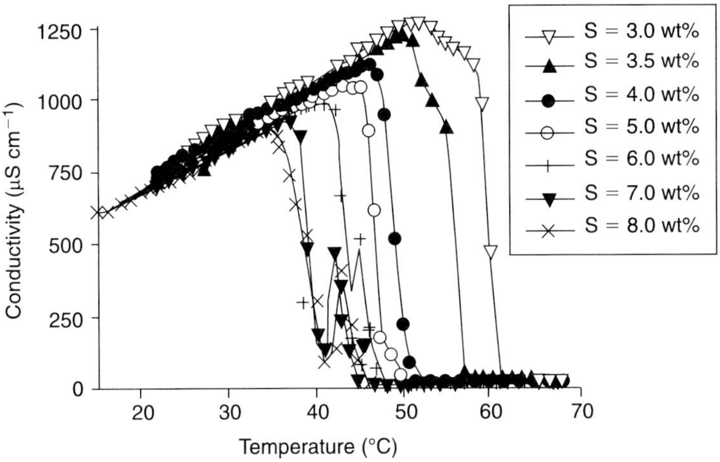

Fig. 9.9. Conductivity versus temperature for a 20:80 hexadecane–water emulsion at various $C_{12}EO_4$ concentrations.

The decrease in droplet size with increase in surfactant concentration is due to the increasing surfactant interfacial area and decrease in interfacial tension, γ. As mentioned above, γ reaches a minimum at the HLB temperature. Therefore, the minimum in interfacial tension occurs at lower temperature as the surfactant concentration increases. This temperature becomes closer to the cooling temperature as the surfactant concentration increases, resulting in smaller droplet sizes.

All nano-emulsions showed an increase in droplet size with time, as a result of Ostwald ripening. Figure 9.10 shows plots of r^3 versus time for all the nano-emulsions studied. The slope of the lines gives the rate of Ostwald ripening ω ($m^3\ s^{-1}$), which showed an increase from 2×10^{-27} to $39.7 \times 10^{-27}\ m^3\ s^{-1}$ as the surfactant concentration is increased from 4 to 8 wt%. This increase could be due to several factors: (1) A decrease in droplet size increases the Brownian diffusion

Tab. 9.1. Composition, HLB temperature (T_{HLB}), droplet radius r and polydispersity index (pol.) for the system water–$C_{12}EO_4$–hexadecane at 25 °C.

Surfactant (wt%)	*Water (wt%)*	*Oil/water*	T_{HLB} *(°C)*	*r (nm)*	*Poly. index*
2.0	78.0	20.4/79.6	–	320	1.00
3.0	77.0	20.6/79.4	57.0	82	0.41
3.5	76.5	20.7/79.3	54.0	69	0.30
4.0	76.0	20.8/79.2	49.0	66	0.17
5.0	75.0	21.2/78.9	46.8	48	0.09
6.0	74.0	21.3/78.7	45.6	34	0.12
7.0	73.0	21.5/78.5	40.9	30	0.07
8.0	72.0	21.7/78.3	40.8	26	0.08

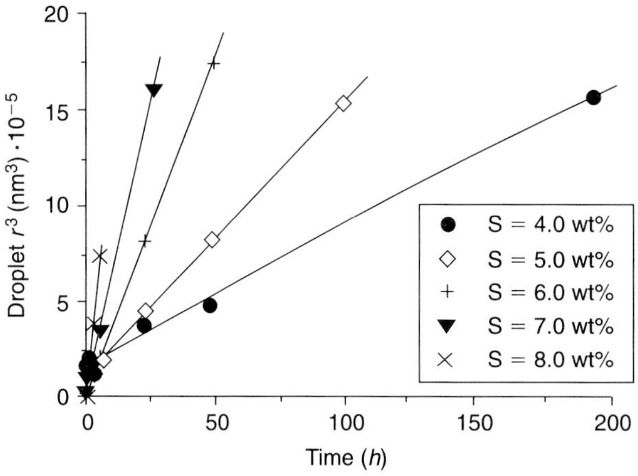

Fig. 9.10. r^3 versus time at 25 °C for nano-emulsions prepared using the system water–$C_{12}EO_4$–hexadecane.

and this enhances the rate. (2) The presence of micelles, which increases with increasing surfactant concentration. This has the effect of increasing the solubilisation of the oil into the core of the micelles, resulting in an increase of the flux J of diffusion of oil molecules from different size droplets. Although the diffusion of micelles is slower than the diffusion of oil molecules, the concentration gradient ($\delta C/\delta X$) can be increased by orders of magnitude as a result of solubilisation. The overall effect will be an increase in J, which may enhance Ostwald ripening. (3) Partition of surfactant molecules between the oil and aqueous phases. With higher surfactant concentrations, the molecules with shorter EO chains (lower HLB number) may preferentially accumulate at the O/W interface and this may result in reduction of the Gibbs elasticity, which in turn results in an increase in the Ostwald ripening rate.

The results with isohexadecane are summarised in Table 9.2. As with the hexadecane system, the droplet size and polydispersity index decreased with increasing

Tab. 9.2. Composition, HLB temperature (T_{HLB}), droplet radius r and polydispersity index (pol.) at 25 °C for emulsions in the system water–$C_{12}EO_4$–isohexadecane.

Surfactant (wt%)	*Water (wt%)*	*O/W*	*T_{HLB} (°C)*	*r (nm)*	*Poly. index*
2.0	78.0	20.4/79.6	–	97	0.50
3.0	77.0	20.6/79.4	51.3	80	0.13
4.0	76.0	20.8/79.2	43.0	65	0.06
5.0	75.0	21.1/78.9	38.8	43	0.07
6.0	74.0	21.3/78.7	36.7	33	0.05
7.0	73.0	21.3/78.7	33.4	29	0.06
8.0	72.0	21.7/78.3	32.7	27	0.12

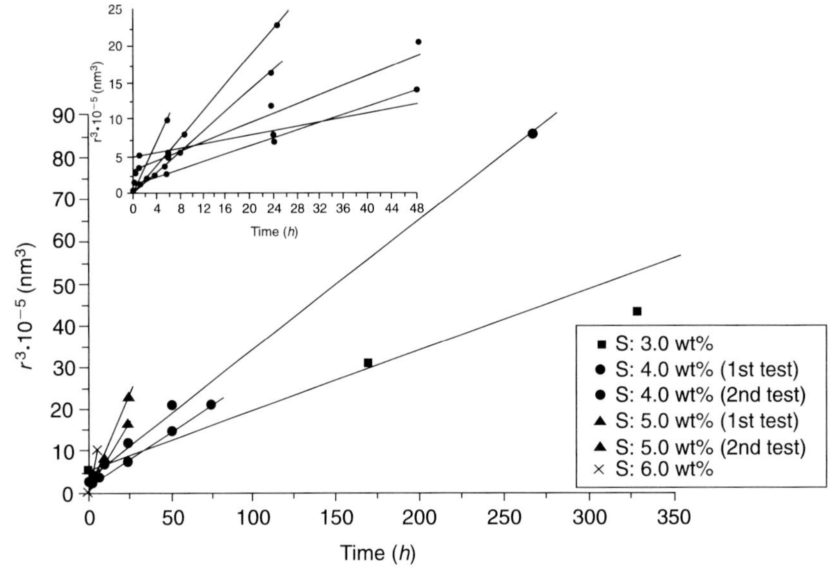

Fig. 9.11. r^3 versus time at 25 °C for the system water–$C_{12}EO_4$–isohexadecane at various surfactant concentrations; O/W ratio 20:80.

surfactant concentration. Nano-emulsions with droplet radii of 25–80 nm were obtained at 3–8 wt% surfactant concentration. Notably, nano-emulsions could be produced at lower surfactant concentrations when using isohexadecane than with hexadecane. This could be attributed to the higher solubility of isohexadecane (a branched hydrocarbon), the lower HLB temperature and the lower interfacial tension.

The stability of the nano-emulsions prepared using isohexadecane was assessed by following the droplet size as a function of time. Plots of r^3 versus time for four surfactant concentrations (3, 4, 5 and 6 wt%) are shown in Figure 9.11. The results show an increase in Ostwald ripening rate as the surfactant concentration is increased from 3 to 6% (the rate increased from 4.1×10^{-27} to 50.7×10^{-27} $m^3\ s^{-1}$). Nano-emulsions prepared using 7 wt% surfactant were so unstable that they showed significant creaming after 8 hours. However, when the surfactant concentration was increased to 8 wt%, a very stable nano-emulsion could be produced with no apparent increase in droplet size over several months. This unexpected stability was attributed to the phase behaviour at such surfactant concentrations. The sample containing 8 wt% surfactant showed birefringence to shear when observed under polarised light. The ratio between the phases ($W_m + L_\alpha + O$) may be a key factor in nano-emulsion stability.

Attempts were made to prepare nano-emulsions at higher O/W ratios (with hexadecane as the oil phase), while keeping the surfactant concentration constant at 4 wt%. When the oil content was increased to 40 and 50% the droplet radius

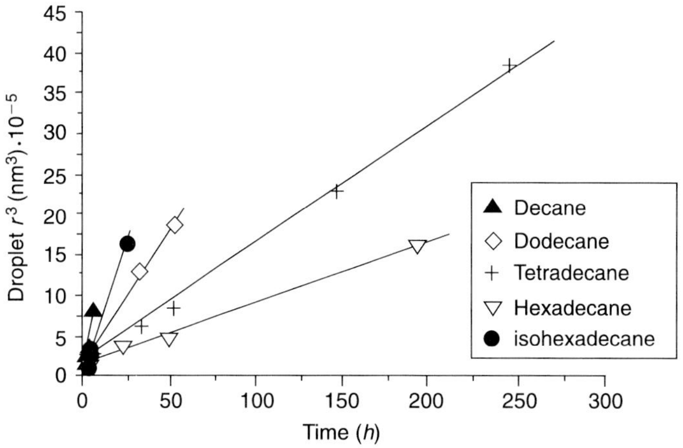

Fig. 9.12. r^3 versus time at 25 °C for nano-emulsions (O/W ratio 20/80) with hydrocarbons of various alkyl chain lengths. System: water–$C_{12}EO_4$–hydrocarbon (4 wt% surfactant).

increased to 188 and 297 nm, respectively. In addition, the polydispersity index also increased to 0.95. These systems become so unstable that they showed creaming within a few hours. This is not surprising, since the surfactant concentration is not sufficient to produce nano-emulsion droplets with high surface area. Similar results were obtained with isohexadecane. However, nano-emulsions could be produced using a 30/70 O/W ratio (droplet size being 81 nm), but with high polydispersity index (0.28). The nano-emulsions showed significant Ostwald ripening.

The effect of changing the alkyl chain length and branching was investigated using decane, dodecane, tetradecane, hexadecane and isohexadecane. Figure 9.12 shows plots of r^3 versus time for a 20/80 O/W ratio and surfactant concentration of 4 wt%. As expected, by reducing the oil solubility from decane to hexadecane, the rate of Ostwald ripening decreases. The branched oil isohexadecane also shows a higher Ostwald ripening rate than with hexadecane. Table 9.3 summarizes the results and also shows the solubility of the oil $C(\infty)$.

Tab. 9.3. HLB temperature (T_{HLB}), droplet radius r, Ostwald ripening rate (ω) and oil solubility for nano-emulsions prepared using hydrocarbons with different alkyl chain length.

Oil	T_{HLB} (°C)	r (nm)	$\omega \times 10^{27}$ ($m^3\ s^{-1}$)	C_∞ ($ml\ ml^{-1}$)
Decane	38.5	59	20.9	710.0
Dodecane	45.5	62	9.3	52.0
Tetradecane	49.5	64	4.0	3.7
Hexadecane	49.8	66	2.3	0.3
Isohexadecane	43.0	60	8.0	–

As expected from the Ostwald ripening theory (LSW theory, Eq. (9.16)), the rate of Ostwald ripening decreases as the oil solubility decreases. Isohexadecane has a rate of Ostwald ripening similar to that of dodecane.

As discussed before, one would anticipate that the Ostwald ripening of any given oil should decrease on incorporation of a second oil with much lower solubility. To test this hypothesis, nano-emulsions were made using hexadecane or isohexadecane to which various proportions of a less-soluble oil, namely squalene, were added. The results using hexadecane did show significantly decreased stability on addition of 10% squalene, which was attributed to coalescence rather than to an increase in Ostwald ripening rate. In some cases addition of a hydrocarbon with a long alkyl chain can induce instability as a result of changes in the adsorption and conformation of the surfactant at the O/W interface.

In contrast to the results obtained with hexadecane, addition of squalene to the O/W nano-emulsion system based on isohexadecane showed a systematic decrease in Ostwald ripening rate as the squalene content was increased. Figure 9.13 shows the results as plots of r^3 versus time for nano-emulsions containing varying amounts of squalene. Addition of squalene up to 20% based on the oil phase showed a systematic reduction in the rate (from 8.0×10^{27} to 4.1×10^{27} $m^3\ s^{-1}$). Notably, when squalene alone was used as the oil phase the system was very unstable and showed creaming within 1 hour. This indicates that the surfactant used is not suitable for the emulsification of squalene.

The effect of HLB number on nano-emulsion formation and stability was investigated by using mixtures of $C_{12}EO_4$ (HLB = 9.7) and $C_{12}EO_4$ (HLB = 11.7). Two surfactant concentrations (4 and 8 wt%) were used and the O/W ratio was kept at 20/80. Figure 9.14 shows the variation of droplet radius with HLB number. This figure shows that the droplet radius remain virtually constant in the HLB range

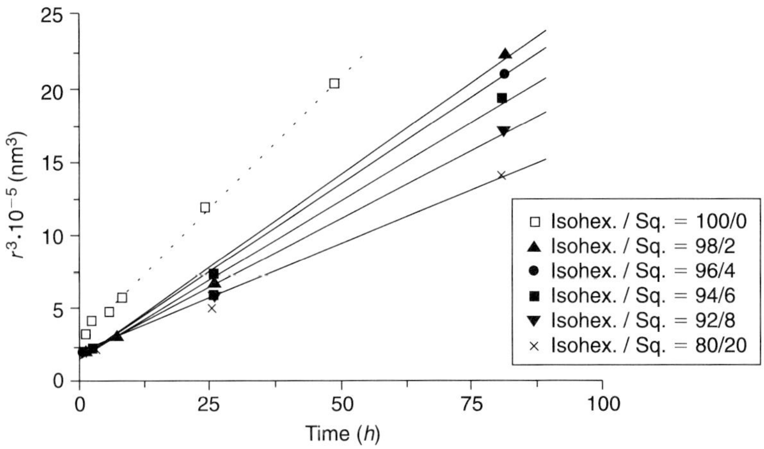

Fig. 9.13. r^3 versus time at 25 °C for the system water–$C_{12}EO_4$–isohexadecane–squalane (20:80 O/W and 4 wt% surfactant).

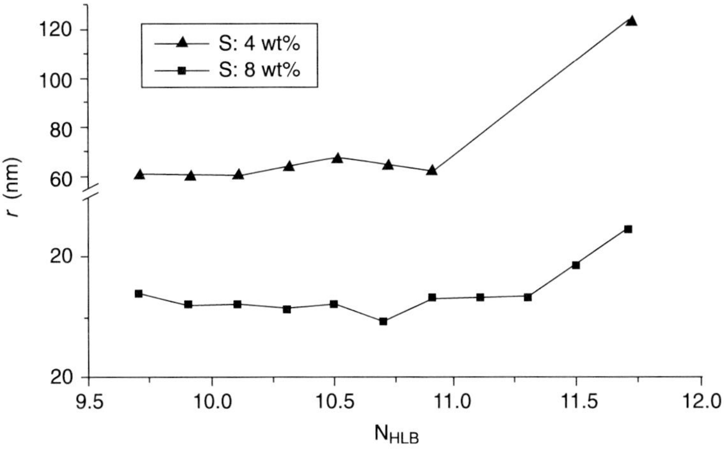

Fig. 9.14. *r* versus HLB number at two different surfactant concentrations (O/W ratio 20:80).

9.7–11.0, after which there is a gradual increase in droplet radius with HLB number of the surfactant mixture. All nano-emulsions showed an increase in droplet radius with time, except for the sample prepared at 8 wt% surfactant with an HLB number of 9.7 (100% $C_{12}EO_4$). Figure 9.15 shows the variation of Ostwald ripening rate constant (ω) with HLB number of surfactant. The rate seems to decrease with increasing surfactant HLB number and, when the latter is >10.5, the rate reaches a low value ($< 4 \times 10^{-27}$ m^3 s^{-1}).

As discussed above, on incorporating an oil-soluble polymeric surfactant that adsorbs strongly at the O/W interface, one would expect the Ostwald ripening rate to

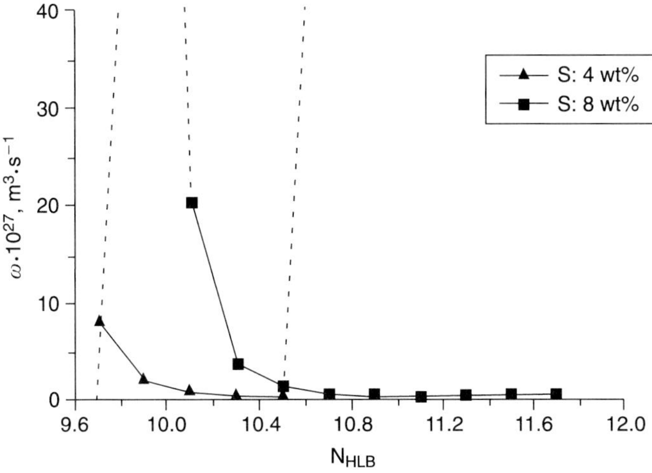

Fig. 9.15. ω versus HLB number in the systems water–$C_{12}EO_4$–$C_{12}EO_6$–isohexadecane at two surfactant concentrations.

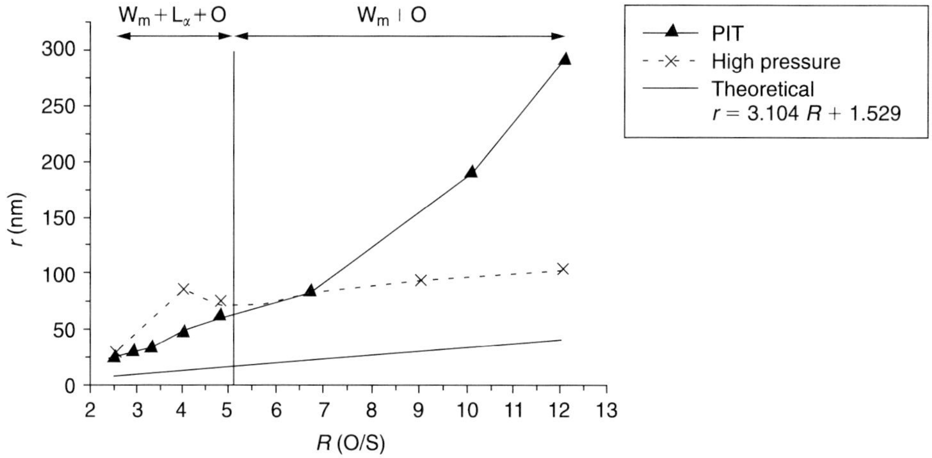

Fig. 9.16. r versus R(O/S) at 25 °C for the system water–$C_{12}EO_4$–hexadecane. W_m = micellar solution or O/W microemulsion, L_α = lamellar liquid crystalline phase; O = oil phase.

reduce. To test this hypothesis, an A-B-A block copolymer of poly(hydroxystearic acid) (PHS, the A chains) and poly(ethylene oxide) (PEO, the B chain), PHS-PEO-PHS (Arlacel P135), was incorporated in the oil phase at low concentrations (the ratio of surfactant to Arlacel was varied between 99:1 and 92:8). For the hexadecane system, the Ostwald ripening rate showed a decrease with the addition of Arlacel P135 surfactant at ratios lower than 94:6. Similar results were obtained using isohexadecane. However, at higher polymeric surfactant concentrations, the nano-emulsion became unstable.

As mentioned above, nano-emulsions prepared using the PIT method are relatively polydisperse and they generally give higher Ostwald ripening rates than nano-emulsions prepared using high-pressure homogenisation techniques. To test this hypothesis, several nano-emulsions were prepared using a Microfluidiser (that can apply pressures in the range 5000–15 000 psi (350–1000 bar)). Using an oil:surfactant ratio of 4:8 and O/W ratios of 20:80 and 50:50, emulsions were prepared first using the Ultturrax followed by high-pressure homogenisation (ranging from 1500 to 15 000 psi). The best results were obtained using a pressure of 15 000 psi (one cycle of homogenisation). The droplet radius was plotted versus the oil:surfactant ratio, R(O/S) (Figure 9.16).

For comparison, the theoretical radii values calculated by assuming that all surfactant molecules are at the interface were calculated using Nakajima's equation [1, 2],

$$r = \left(\frac{3M_b}{AN\rho_a}\right)R + \left(\frac{3\alpha M_b}{AN\rho_b}\right) + d \tag{9.17}$$

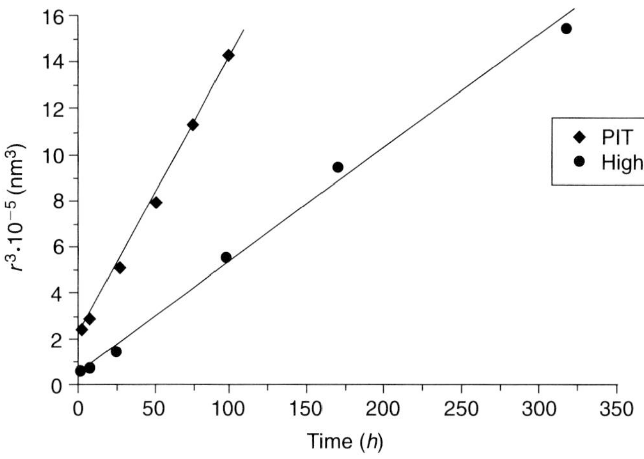

Fig. 9.17. r^3 versus time for nano-emulsion systems prepared using the PIT and Microfluidiser; 20:80 O/W and 4 wt% surfactant.

where M_b is the molecular weight of the surfactant, A is the area occupied by a single molecule, N is Avogadro's number, ρ_a is the oil density, ρ_b is the density of the surfactant alkyl chain, α is the alkyl chain weight fraction and d is the thickness of the hydrated layer of PEO.

In all cases, there is an increase in nano-emulsion radius with increasing $R(O/S)$. However, when using the high-pressure homogeniser, the droplet size can be maintained to below 100 nm at high $R(O/S)$. With the PIT method, there is a rapid increase in r with increase in $R(O/S)$ when the latter exceeds 7.

As expected, nano-emulsions prepared using high-pressure homogenisation showed a lower Ostwald ripening rate than systems prepared using the PIT method. This is illustrated in Figure 9.17, which shows plots of r^3 versus time for the two systems.

References

1 H. Nakajima, S. Tomomossa, M. Okabe: *Proceedings of the First Emulsion Conference, Paris*, 1993.
2 H. Nakajima: *Industrial Applications of Microemulsions*, C. Solans, H. Konieda (ed.): Marcel Dekker, New York, 1997.
3 J. Ugelstadt, M. S. El-Aassar, J. W. Vanderhoff, *J. Polym. Sci.*, **1973**, *11*, 503.
4 M. El-Aasser: *Polymeric Dispersions*, J. M. Asua (ed.): Kluwer Academic, The Netherlands 1997.
5 S. Benita, M. Y. Levy, *J. Pharm. Sci.*, **1993**, *82*, 1069.
6 A. Forgiarini, J. Esquena, J. Gonzalez, C. Solans, *Prog. Colloid Polym. Sci.*, **2000**, *115*, 36.
7 K. Shinoda, H. Kunieda: *Encyclopedia of Emulsion Technology*, P. Becher (ed.): Marcel Dekker, New York, 1983.
8 P. Walstra, *Encyclopedia of Emulsion Technology*, P. Becher (ed.): Marcel Dekker, New York, 1983.

9 P. N. Pusey: *Industrial Polymers: Characterisation by Molecular Weights*, J. H. S. Green, R. Dietz (ed.): Transcripta Books, London, 1973.
10 P. Walstra, P. E. A. Smoulders: *Modern Aspects of Emulsion Science*, B. P. Binks (ed.): Royal Society of Chemistry, Cambridge, 1998, p. 56.
11 K. Shinoda, H. Saito, *J. Colloid Interface Sci.*, **1969**, *30*, 258.
12 K. Shinoda, H. Saito, *J. Colloid Interface Sci.*, **1968**, *26*, 70.
13 B. W. Brooks, H. N. Richmond, M. Zerfa: *Modern Aspects of Emulsion Science*, B. P. Binks (ed.): Royal Society of Chemistry, Cambridge, 1998, p. 175.
14 T. Sottman, R. Strey, *J. Chem. Phys.*, **1997**, *108*, 8606.
15 D. H. Napper: *Polymeric Stabilisation of Colloidal Dispersions*, Academic Press, London, 1983.
16 T. F. Tadros: *The Effect of Polymers on Dispersion Properties*, Th. F. Tadros (ed.): Academic Press, London, 1982.
17 W. Thompson (Lord Kelvin), *Phil. Mag.*, **1871**, *42*, 448.
18 I. M. Lifshitz, V. V. Slesov, *Sov. Phys. JETP*, **1959**, *35*, 331.
19 C. Wagner, *Z. Electrochem.*, **1961**, *35*, 581.
20 A. S. Kabalnov, E. D. Shchukin, *Adv. Colloid Interface Sci.*, **1992**, *38*, 69.
21 A. S. Kabalnov, *Langmuir*, **1994**, *10*, 680.
22 J. G. Weers: *Modern Aspects of Emulsion Science*, B. P. Binks (ed.): Royal Society of Chemistry, Cambridge, 1998, p. 292.
23 P. Walstra, *Chem. Eng. Sci.*, **1993**, *48*, 333.
24 P. Izquierdo, *Thesis Studies on Nabo-Emulsion Formation and Stability*, University of Barcelona, Spain, 2002.
25 H. Kuneida, Y. Fukuhi, H. Uchiyama, C. Solans, *Langmuir*, **1996**, *12*, 2136.

10
Microemulsions

10.1
Introduction

Microemulsions are a special class of "dispersions" (transparent or translucent) that actually have little in common with emulsions. They are better described as "swollen micelles". The term microemulsion was first introduced by Hoar and Schulman [1, 2], who discovered that by titration of a milky emulsion (stabilised by soap such as potassium oleate) with a medium-chain alcohol, such as pentanol or hexanol, a transparent or translucent system was produced. A schematic representation of the titration method adopted by Schulman and co-workers is given below. The final transparent or translucent system is a W/O microemulsion (Scheme 10.1).

O/W emulsion stabilised by soap $\xrightarrow[\text{e.g. } C_5H_{11}OH \;\; C_6H_{13}OH]{\text{Add cosurfactant}}$ Transparent or translucent

Scheme 10.1

A convenient way to describe microemulsions is to compare them with micelles – the latter, which are thermodynamically stable, may consist of spherical units with a radius that is usually less than 5 nm. Two types of micelles may be considered: normal micelles with the hydrocarbon tails forming the core and the polar head groups in contact with the aqueous medium, and reverse micelles (formed in nonpolar media) with a water core containing the polar head groups and the hydrocarbon tails now in contact with the oil. Normal micelles can solubilise oil in the hydrocarbon core, forming O/W microemulsions, whereas reverse micelles can solubilise water to form a W/O microemulsion. Figure 10.1 gives a schematic representation of these systems.

Roughly, the dimensions of micelles, micellar solutions and macroemulsions are: micelles, $R < 5$ nm (they scatter little light and are transparent), macroemulsions, $R > 50$ nm (opaque and milky), and micellar solutions or microemulsions, 5–50 nm (transparent, 5–10 nm, translucent 10–50 nm).

Applied Surfactants: Principles and Applications. Tharwat F. Tadros

ISBN: 3-527-30629-3

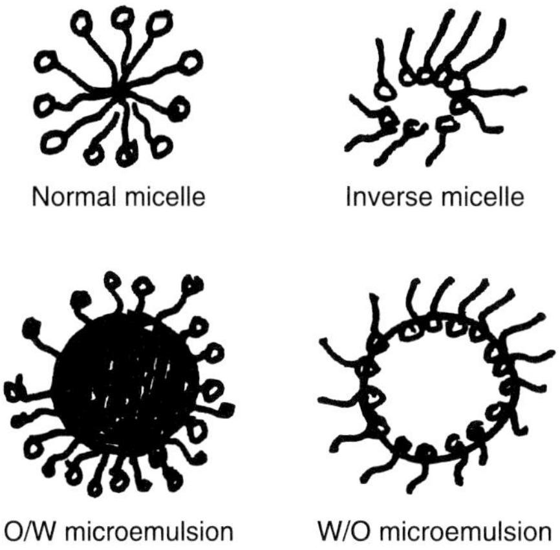

Fig. 10.1. Representation of microemulsions.

The classification of microemulsions based on size is not adequate. Whether a system is transparent or translucent depends not only on the size but also on the difference in refractive index between the oil and the water phases. A microemulsion with small size (in the region of 10 nm) may appear translucent if the difference in refractive index between the oil and the water is large (note that the intensity of light scattered depends on the size and an optical constant that is given by the difference in refractive index between oil and water). Relatively large microemulsion droplets (in the region of 50 nm) may appear transparent if the refractive index difference is very small. The best definition of microemulsions is based on the application of thermodynamics, as discussed below.

10.2
Thermodynamic Definition of Microemulsions

A thermodynamic definition of microemulsions can be obtained from a consideration of the energy and entropy terms for formation of microemulsions, which is schematically represented in Figure 10.2 for the formation of a microemulsion from a bulk oil phase (for O/W microemulsion) or bulk water phase (for a W/O microemulsion).

A_1 is the surface area of the bulk oil phase and A_2 is the total surface area of all the microemulsion droplets. γ_{12} is the O/W interfacial tension.

The increase in surface area when going from state I to state II is ΔA $(= A_2 - A_1)$ and the surface energy increase is equal to $\Delta A \gamma_{12}$. The increase in entropy when going from state I to sate II is $T\Delta S^{\mathrm{conf}}$ (note that state II has higher entropy since a large number of droplets can arrange themselves in many ways, whereas state I with one oil drop has a much lower entropy).

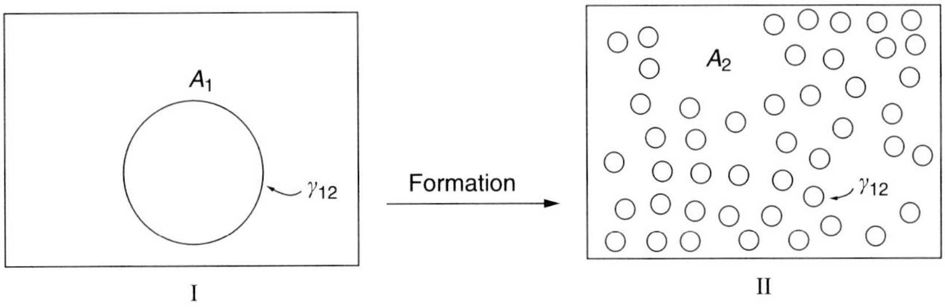

Fig. 10.2. Scheme of microemulsion formation.

According to the second law of thermodynamics, the free energy of formation of microemulsions ΔG_m is given by

$$\Delta G_m = \Delta A \gamma_{12} - T\Delta S^{conf} \tag{10.1}$$

With macroemulsions $\Delta A \gamma_{12} \gg -T\Delta S^{conf}$ and $\Delta G_m > 0$. The system is non-spontaneous (it requires energy to form the emulsion drops) and it is thermodynamically unstable.

With microemulsions $\Delta A \gamma_{12} < -T\Delta S^{conf}$ (due to the ultralow interfacial tension accompanied with microemulsion formation) and $\Delta G_m < 0$. The system is produced spontaneously and it is thermodynamically stable.

The above analysis shows the contrast between emulsions and microemulsions: With emulsions, an increase in either the mechanical energy or surfactant concentration usually results in the formation of smaller droplets, which become kinetically more stable. With microemulsions, neither mechanical energy nor an increase in surfactant concentration can result in its formation. The latter is based on a specific combination of surfactants and specific interaction with the oil and the water phases and the system is produced at optimum composition.

Thus, microemulsions have nothing in common with macroemulsions and it is often better to describe the systems as "swollen micelles". The best definition of microemulsions is as follows [3]: "System of Water + Oil + Amphiphile that is a single Optically Isotropic and Thermodynamically Stable Liquid Solution". Amphiphiles are any molecules that consist of hydrophobic and hydrophilic portions, e.g. surfactants, alcohols, etc.

The driving force for microemulsion formation is the low interfacial energy, which is overcompensated by the negative entropy of dispersion term. The low (ultralow) interfacial tension is produced in most cases by a combination of two molecules, referred to as the surfactant and cosurfactant (e.g. medium-chain alcohol).

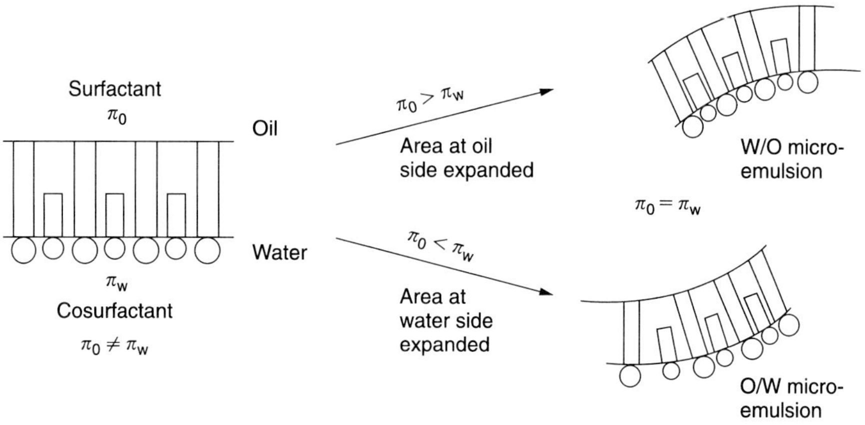

Fig. 10.3. Film bending.

10.3 Mixed Film and Solubilisation Theories of Microemulsions

10.3.1 Mixed Film Theories [4]

The film, which may consist of surfactant and cosurfactant molecules, is considered as a liquid "two-dimensional" third phase in equilibrium with both oil and water. Such a monolayer could be a duplex film, i.e. giving different properties on the water side and oil side. The initial "flat" duplex film (Figure 10.3) has different tensions at the oil and water sides. This is due to the different packing of the hydrophobic and hydrophilic groups (these groups have different sizes and cross sectional areas).

It is convenient to define a two-dimensional surface pressure π,

$$\pi = \gamma_o - \gamma \tag{10.2}$$

γ_o is the interfacial tension of the clean interface, whereas γ is the interfacial tension with adsorbed surfactant.

One can define two values for π at the oil and water phases, π_o and π_w, which for a flat film are not equal, i.e. $\pi'_o = \pi'_w$.

As a result of the difference in tensions, the film will bend until $\pi_o = \pi_w$. If $\pi'_o > \pi'_w$, the area at the oil side has to expand (resulting in reduction of π'_o) until $\pi_o = \pi_w$ – in this case a W/O microemulsion is produced. If $\pi'_w > \pi'_o$, the area at the water side expands until $\pi_w = \pi_o$. In this case an O/W microemulsion is produced. Figure 10.3 gives a schematic representation of film bending for production of W/O or W/O microemulsions.

According to the duplex film theory, the interfacial tension γ_T is given by the following expression [5],

$$\gamma_T = \gamma_{(O/W)} - \pi \tag{10.3}$$

where $(\gamma_{o/w})_a$ is the interfacial tension that is reduced by the presence of the alcohol. $(\gamma_{o/w})_a$ is significantly lower than $\gamma_{o/w}$ in the absence of the alcohol; for example, for hydrocarbon/water $\gamma_{o/w}$ is reduced from 50 to 15–20 mN m^{-1} on the addition of a significant amount of a medium-chain alcohol such as pentanol or hexanol.

Contributions to π are considered to be due to crowding of the surfactant and cosurfactant molecules and penetration of the oil phase into the hydrocarbon chains of the interface.

According to Eq. (10.3) if $\pi > (\gamma_{o/w})_a$, γ_T becomes negative and this leads to expansion of the interface until γ_T reaches a small positive value. Since $(\gamma_{o/w})_a$ is of the order of 15–20 mN m^{-1}, surface pressures of this order are required for γ_T to approach zero.

The above duplex film theory can explain the nature of the microemulsion: The surface pressures at the oil and water sides of the interface depend on the interactions of the hydrophobic and hydrophilic portions of the surfactant molecule at both sides respectively. If the hydrophobic groups are bulky relative to the hydrophilic groups, then for a flat film such hydrophobic groups tend to crowd, forming a higher surface pressure at the oil side of the interface; this results in bending and expansion at the oil side, forming a W/O microemulsion. An example for a surfactant with bulky hydrophobic groups is Aerosol OT (sodium diethyl dioctyl sulphosuccinate). If the hydrophilic groups are bulky, such as with ethoxylated surfactants containing more than five ethylene oxide units, crowding occurs at the water side of the interface. This produces an O/W microemulsion.

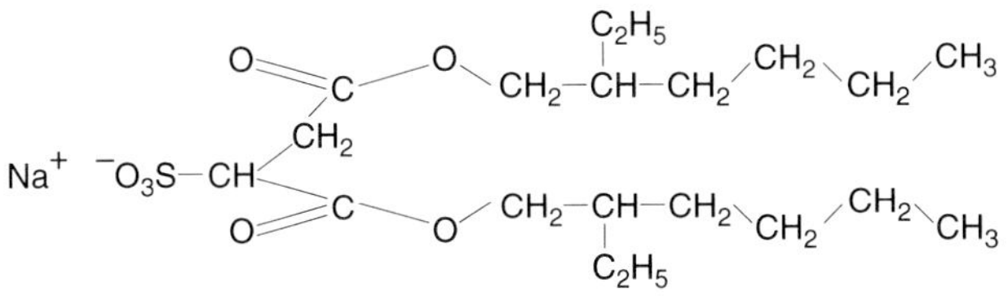

10.1

10.3.2
Solubilisation Theories

These concepts were introduced by Shinoda and co-workers [6] who considered microemulsions to be swollen micelles that are directly related to the phase diagram of their components.

Consider the phase diagram of a three-component system of water, ionic surfactant and medium-chain alcohol (Figure 10.4).

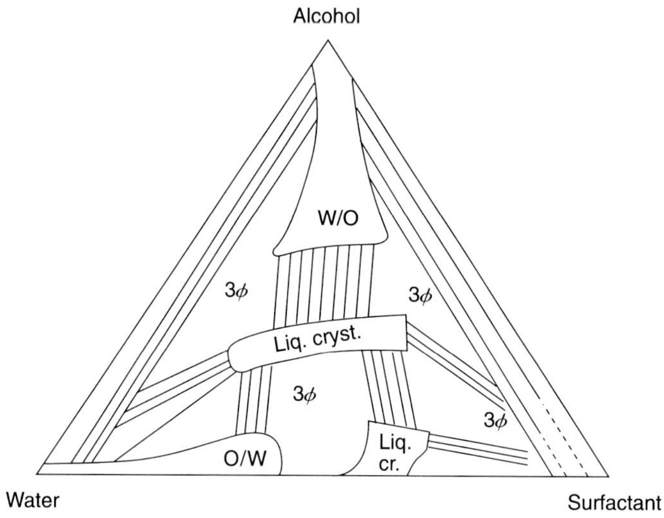

Fig. 10.4. Schematic three-component phase diagram.

At the water corner and at low alcohol concentration, normal micelles (L_1) are formed since in this case there are more surfactant than alcohol molecules. At the alcohol (cosurfactant corner), inverse micelles (L_2) are formed, since in this region there are more alcohol than surfactant molecules.

These L_1 and L_2 are not in equilibrium but are separated by a liquid crystalline region (lamellar structure with an equal number of surfactant and alcohol molecules). The L_1 region may be considered as an O/W microemulsion, whereas L_2 may be considered as a W/O microemulsion.

Addition of a small amount of oil miscible with the cosurfactant, but not with the surfactant and water, changes the phase diagram only slightly. The oil may be simply solubilised in the hydrocarbon core of the micelles. Addition of more oil leads to fundamental changes in the phase diagram, as is illustrated in Figure 10.5, whereby 50:50 of W:O are used. To simplify the phase diagram, the ^{50}W/^{50}O is presented on one corner of the phase diagram.

Near the cosurfactant (co) corner the changes are small compared with the three-phase diagram (Figure 10.4). The O/W microemulsion near the water–surfactant (sa) axis is not in equilibrium with the lamellar phase, but with a non-colloidal oil + cosurfactant phase.

If co is added to such a two-phase equilibrium at fairly high surfactant concentration all oil is taken up and a one-phase microemulsion appears.

Addition of co at low Sa concentration may lead to separation of an excess aqueous phase before all oil is taken up in the microemulsion. A three-phase system is formed, containing a microemulsion that cannot be clearly identified as W/O or W/O and that, presumably, is similar to the lamellar phase swollen with oil or to a more irregular intertwining of aqueous and oily regions (bicontinuous or middle phase microemulsion).

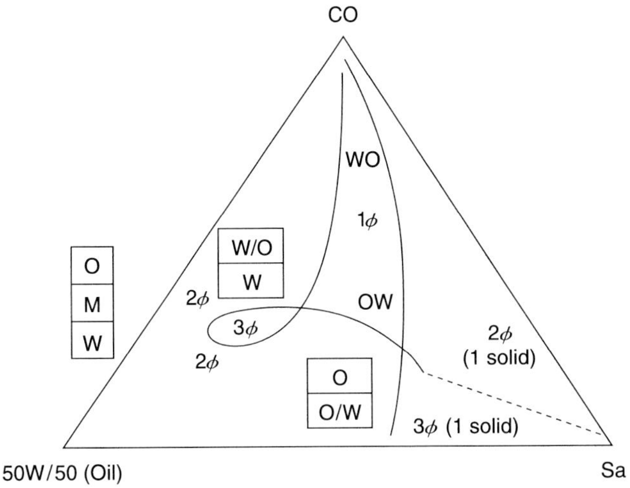

Fig. 10.5. Pseudoternary phase diagram of oil–water–surfactant–cosurfactant.

Interfacial tensions between the three phases are very low (0.1–10^{-4} mN m^{-1}). Further addition of co to the three-phase system makes the oil phase disappear and leaves a W/O microemulsion in equilibrium with a dilute aqueous sa solution.

In the large one-phase region continuous transitions from O/W to middle phase to W/O microemulsions are found.

Solubilization can also be illustrated by considering the phase diagrams of nonionic surfactants containing poly(ethylene oxide) (PEO) head groups. Such surfactants do not generally need a cosurfactant for microemulsion formation. Oil and water solubilisation by nonionic surfactants is represented in Figure 10.6.

At low temperatures, the ethoxylated surfactant is soluble in water and at a given concentration is can solubilise a given amount of oil. The oil solubilisation increases rapidly with rising temperature near the cloud point of the surfactant – this is illustrated in Figure 10.6, which shows the solubilisation and cloud point curves of the surfactant. Between these two curves, an isotropic region of O/W solubilised system exists.

At any given temperature, any increase in the oil weight fraction above the solubilisation limit results in oil separation (oil solubilised + oil). At any given surfactant concentration, any increase in temperature above the cloud point results in separation into oil, water and surfactant.

If one starts from the oil phase with dissolved surfactant and adds water, solubilisation of the latter takes place and solubilisation increases with reduction of temperature near the haze point. Between the solubilisation and haze point curves, an isotropic region of W/O solubilised system exists. At any given temperature, any increase in water weight fraction above the solubilisation limit results in water separation (W/O solubilised + water). At any given surfactant concentration, any decrease in temperature below the haze point results in separation to water, oil and surfactant.

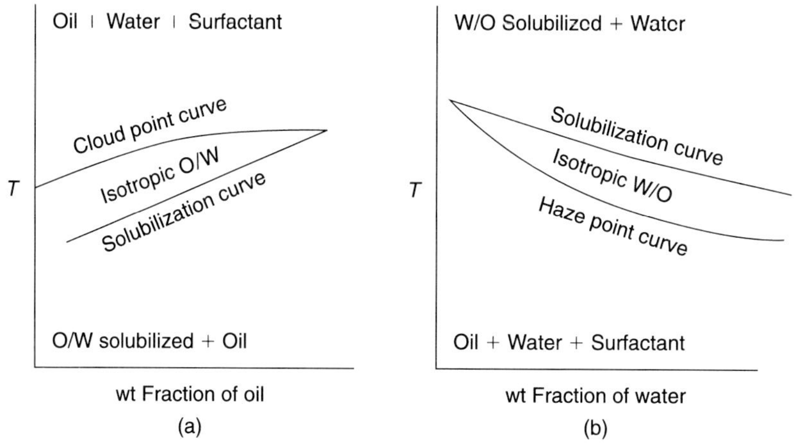

Fig. 10.6. Representation of solubilisation: (a) oil solubilised in a nonionic surfactant solution; (b) water solubilised in an oil solution of a nonionic surfactant.

With nonionic surfactants, both types of microemulsions can be formed, depending on the conditions. With such systems, temperature is the most crucial factor since the solubility of surfactant in water or oil depends on temperature. Microemulsions prepared using nonionic surfactants have a limited temperature range.

10.4
Thermodynamic Theory of Microemulsion Formation

The spontaneous formation of the microemulsion with decreasing free energy can only be expected if the interfacial tension is so low that the remaining free energy of the interface is overcompensated for by the entropy of dispersion of the droplets in the medium [7, 8]. This concept forms the basis of the thermodynamic theory proposed by Ruckenstein and Chi and Overbeek [7, 8].

10.4.1
Reason for Combining Two Surfactants

Single surfactants do lower the interfacial tension γ, but in most cases the critical micelle concentration (c.m.c.) is reached before γ is close to zero. Addition of a second surfactant of a completely different nature (i.e. predominantly oil soluble such as an alcohol) then lowers γ further and very small, even transiently negative, values may be reached [9]. Figure 10.7 illustrates this, showing the effect of addition of the cosurfactant on the $\gamma - \log c_{sa}$ curve. The addition of cosurfactant clearly shifts the whole curve to low γ and the c.m.c. is shifted to lower values.

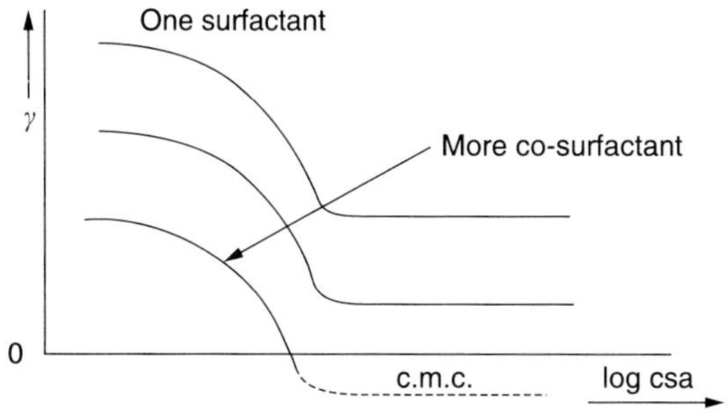

Fig. 10.7. γ versus log c_{sa} for surfactant + cosurfactant.

Why γ is lowered when using two surfactant molecules can be understood from consideration of the Gibbs adsorption equation for multicomponent systems [9]. For a multicomponent system i, each with an adsorption Γ_i (mol m^{-2}, referred to as the surface excess), the reduction in γ, i.e. dγ, is given by

$$d\gamma = -\sum \Gamma_i \, d\mu_i = -\sum \Gamma_i RT \, d \ln C_i \tag{10.4}$$

where μ_i is the chemical potential of component i, R is the gas constant, T is the absolute temperature and C_i is the concentration (mol dm^{-3}) of each surfactant component.

For two components, sa (surfactant) and co (cosurfactant), Eq. (10.4) becomes

$$d\gamma = -\Gamma_{sa} RT \, d \ln C_{sa} - \Gamma_{co} RT \, d \ln C_{co} \tag{10.5}$$

Integration of Eq. (10.5) gives

$$\gamma = \gamma_o - \int_0^{C_{sa}} \Gamma_{sa} RT \, d \ln C_{sa} - \int_0^{C_{co}} \Gamma_{co} RT \, d \ln C_{co} \tag{10.6}$$

which clearly shows that γ_o is lowered by two terms, both from surfactant and cosurfactant.

The two surfactant molecules should adsorb simultaneously and they should not interact with each other, otherwise they lower their respective activities. Thus, the surfactant and cosurfactant molecules should vary in nature, one predominantly water soluble (such as an anionic surfactant) and the other predominantly oil soluble (such as a medium-chain alcohol)

In some cases a single surfactant may be sufficient to lower γ far enough for microemulsion formation to become possible, e.g. Aerosol OT (sodium diethyl hexyl sulphosuccinate) and many non ionic surfactants.

10.5
Free Energy of Formation of Microemulsion

A simple model was used by Overbeek [9] to calculate the free energy of formation of a model W/O microemulsion: The droplets were assumed to be of equal size. The droplets are large enough to consider the adsorbed surfactant layer to have constant composition.

The microemulsion is prepared in several steps and for each step one calculates the Helmholtz free energy F [this was chosen since the pressure inside the drop is higher by the Laplace pressure $2\gamma/a$ (where a is the droplet radius) than the pressure in the medium].

The four steps involved in the preparation of a model W/O microemulsion can be summarised as:

(1) Prepare the oil phase in its final concentration,

$$F_1 = \sum n_i' \mu_i' - p_1 V_1 \tag{10.7}$$

where n_i' and μ_i' are the amount and chemical potential of oil and cosurfactant in the continuous phase, without droplets being mixed in; p_1 is the atmospheric pressure and V_1 is the volume of the oil phase.

(2) Prepare the aqueous phase in its final concentration,

$$F_2 = \sum n_i' \mu_i' - p_1 V_2 \tag{10.8}$$

where i are now water, surfactant and salt and V_2 is the volume of the water phase.

(3) Form the water phase into droplets close packed in the oil phase (i.e. with a packing fraction $\phi = 0.74$) and add all the adsorbed material,

$$F_3 = \gamma A + \Gamma_{sa} A \left[\mu_{sa}' + \left(\frac{2\gamma}{a} \right) \bar{V}_{sa} \right] + \Gamma_i A \mu_i \tag{10.9}$$

where i refers to cosurfactant and oil.

The oil must be negatively adsorbed to keep the volume of the adsorption layer zero (in accordance with the Gibbs dividing surface). Equation (10.9) assumes that the Gibbs plane (the surface of tension in this case) lies close to the surface (where $\Gamma_{water} = 0$).

(4) Allow the close-packed emulsion to expand to its final concentration (volume fraction ϕ),

$$F_4 = n_{dr} RT f(\phi) \tag{10.10}$$

where n_{dr} is the amount of drops (in moles) and $f(\phi)$ is a function of ϕ; $f(\phi)$ may be simply written as

$$f(\phi) = \ln \phi - \ln 0.74 \tag{10.11}$$

More accurately $f(\phi)$ may be calculated using a hard-sphere model [10],

$$f(\phi) = \ln \phi + \phi\left[\frac{4 - 3\phi}{(1 - \phi)^2}\right] - 19.25 \tag{10.12}$$

Combining Eqs. (10.7) to (10.12) gives the Helmholtz free energy of the complete emulsion. The free energy is minimised with respect to a change in the interfacial area A. This involves transfer of adsorbed components to or from the interface, thereby changing the bulk concentration and thus γ; the result is,

$$\gamma = -\text{const.} \times \frac{1}{a^2} \times g(\phi) \tag{10.13}$$

where $g(\phi)$ is similar but not identical to $f(\phi)$.

The droplet radius a can be calculated from a knowledge of the total interfacial area A,

$$A = \frac{n_{sa}}{\Gamma_{sa}} \simeq n_{sa} N_{av} \times (\text{area/molecule}) \tag{10.14}$$

The area per molecule of an anionic surfactant such as sodium dodecyl sulphate (SDS) varies from 0.7 to 1.1 nm^2, depending on the concentration of cosurfactant (pentanol) and salt concentration. The area per pentanol molecule is about 0.3 nm^2. This means that the average area per surfactant molecule is about 0.9 nm^2.

The radius of the droplet can be calculated from the ratio of the volume of the drop to its area,

$$a = \frac{3 \times \frac{4}{3}\pi a^3}{4\pi a^2} = \frac{3V}{A} \tag{10.15}$$

where V is the total volume of the droplets and A is the total interfacial area.

The radius a of the microemulsion droplet has to fit both Eqs. (10.13) and (10.15); γ is the most easily varied quantity in these equations. The correct γ is obtained by adaptation of C_{sa}.

According to Eq. (10.13), any value of a is allowed in the accessible range of γ. If γ is close to zero, very large radii can be obtained, i.e. very large water/sa ratios are allowed. However, the phase diagram shows that, at such high ratios, demixing occurs. This analysis shows the inadequacy of the above simple model and it is necessary to add an explicit influence of the radius of curvature on the interfacial tension. The curvature effect is manifested in the packing of the tails and head groups at the O/W interface. With W/O microemulsions the packing of the short chains and the packing of the head groups will favour W/O curvature with a ratio of 3 or more for co/sa. With O/W microemulsions, a ratio of sa/co of 2 or less is required. Thus O/W microemulsions need less cosurfactant than W/O microemulsions.

10.6 Factors Determining W/O versus O/W Microemulsions

The duplex film theory predicts that the nature of the microemulsion formed depends on the relative packing of the hydrophobic and hydrophilic portions of the surfactant molecule, which determines the bending of the interface. For example, a surfactant molecule such as Aerosol OT (**10.1**) favours the formation of W/O microemulsion, without the need of a cosurfactant. As a result of the presence of a stumpy head group and large volume to length ratio (V/l) of the nonpolar group, the interface tends to bend with the head groups facing onwards, thus forming a W/O microemulsion.

The molecule has $V/l > 0.7$, which is considered necessary for formation of a W/O microemulsion. For ionic surfactants such as SDS, for which $V/l < 0.7$, microemulsion formation needs the presence of a cosurfactant (the latter has the effect of increasing V without changing l).

The importance of geometric packing was considered in detail by Mitchell and Ninham [11], who introduced the concept of the packing ratio P,

$$P = \frac{V}{l_c a_o} \tag{10.16}$$

where a_o is the head group area and l_c is the maximum chain length.

P gives a measure of the hydrophilic–lipophilic balance. For $P < 1$ (usually $P \sim \frac{1}{3}$), normal or convex aggregates are produced (normal micelles). For $P > 1$, inverse micelles are produced. P is influenced by many factors: hydrophilicity of the head group, ionic strength and pH of the medium and temperature.

P also explains the nature of the microemulsion produced using nonionic surfactants of the ethoxylate type: P increases with increasing temperature (as a result of the dehydration of the PEO chain). A critical temperature (PIT) is reached at which P reaches 1, and above this temperature inversion occurs to a W/O system.

The influence of the surfactant structure on the nature of the microemulsion can also be predicted from the thermodynamic theory. The most stable microemulsion would be that in which the phase with the smaller volume fraction forms the droplets (the osmotic pressure increases with increasing ϕ). For a W/O microemulsion prepared using an ionic surfactant such as Aerosol OT, the effective volume (hard-sphere volume) is only slightly larger than the water core volume, since the hydro-

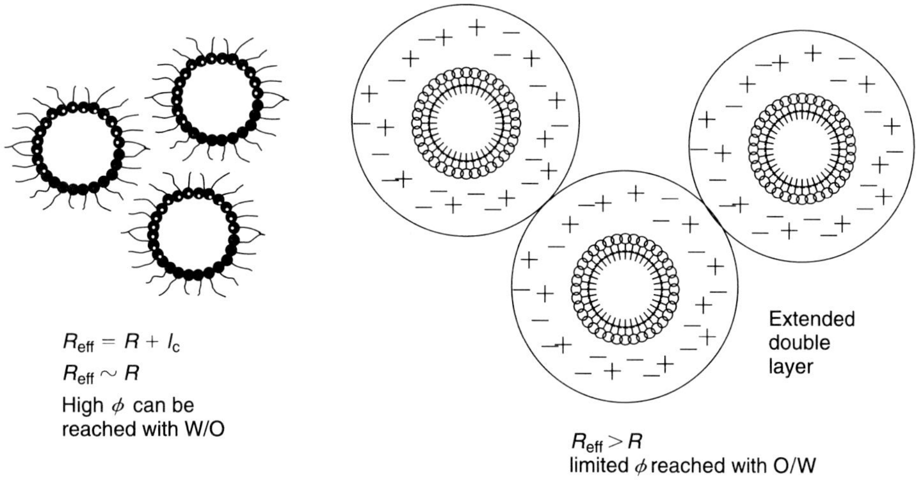

Fig. 10.8. Scheme of W/O and O/W microemulsion droplets.

carbon tails may penetrate to a certain extent when two droplets come together. For an O/W microemulsion, the double layers may expand to a considerable extent, depending on the electrolyte concentration (the double layer thickness is of the order of 100 nm in 10^{-5} mol dm^{-3} 1:1 electrolyte and 10 nm in 10^{-3} mol dm^{-3} electrolyte). Thus the effective volume of O/W microemulsion droplets can be significantly higher than the core oil droplet volume and this explains the difficulty of preparation of O/W microemulsions at high ϕ when using ionic surfactants.

Figure 10.8 gives a schematic representation of the effective volume for W/O and O/W microemulsions.

10.7
Characterisation of Microemulsions Using Scattering Techniques

Scattering techniques provide the most obvious methods for obtaining information on the size, shape and structure of microemulsions. The scattering of radiation, e.g. light, neutrons, X-ray, etc. by particles has been successfully applied to investigate many systems such as polymer solutions, micelles and colloidal particles.

In all the above methods, measurements can be made at sufficiently low concentration to avoid complications arising from particle–particle interactions. The results obtained are extrapolated to infinite dilution to obtain the desirable property such as the molecular weight and radius of gyration of a polymer coil, the size and shape of micelles, etc.

Unfortunately, the above dilution method cannot be applied to microemulsions, which depend on a specific composition of oil, water and surfactants. The microemulsions cannot be diluted by the continuous phase since this dilution results in breakdown of the microemulsion. Thus, when applying the scattering techniques to microemulsions measurements have to be made at finite concentrations and the results obtained have to be analysed using theoretical treatments to take into account the droplet–droplet interactions.

Below, three scattering methods will be discussed: Time-average (static) light scattering, dynamic (quasi-elastic) light scattering, referred to as photon correlation spectroscopy, and neutron scattering.

10.7.1 Time Average (Static) Light Scattering

The intensity of scattered light $I(Q)$ is measured as a function of scattering vector Q [11],

$$Q = \left(\frac{4\pi n}{\lambda}\right) \sin\left(\frac{\theta}{2}\right) \tag{10.17}$$

where n is the refractive index of the medium, γ is the wavelength of light and θ is the angle at which the scattered light is measured.

For a fairly dilute system, $I(Q)$ is proportional to the number of particles N, the square of the individual scattering units V_p and some property of the system (material constant) such as its refractive index,

$$I(Q) - [(\text{Material const.})(\text{Instrument const.})]NV_p^2 \tag{10.18}$$

The instrument constant depends on the geometry of the apparatus (the light path-length and the scattering cell constant).

For more concentrated systems, $I(Q)$ also depends on the interference effects arising from particle–particle interaction,

$$I(Q) = [(\text{Instrument const.})(\text{Material const.})]NV_p^2 P(Q)S(Q) \tag{10.19}$$

where $P(Q)$ is the particle form factor that allows the scattering from a single particle of known size and shape to be predicted as a function of Q. For a spherical particle of radius R,

$$P(Q) = \left[\frac{(3 \sin QR - QR \cos QR)}{(QR)^3}\right]^2 \tag{10.20}$$

$S(Q)$ is the so-called "structure factor", which takes into account the particle–particle interaction. $S(Q)$ is related to the radial distribution function $g(r)$ (which gives the number of particles in shells surrounding a central particle) [12],

$$S(Q) = 1 - \frac{4\pi N}{Q} \int_0^{\infty} [g(r) - 1]r \sin QR \, dr \tag{10.21}$$

For a hard-sphere dispersion with radius R_{HS} (which is equal to $R + t$, where t is the thickness of the adsorbed layer),

$$S(Q) = \frac{1}{[1 - NC(2QR_{HS})]} \tag{10.22}$$

where C is a constant.

One usually measures $I(Q)$ at various scattering angles θ and then plot the intensity at some chosen angle (usually 90°), i_{90}, as a function of the volume fraction ϕ of the dispersion. Alternatively, the results may be expressed in terms of the Rayleigh ratio R_{90},

$$R_{90} = \left(\frac{i_{90}}{I_0}\right) r_s^2 \tag{10.23}$$

I_0 is the intensity of the incident beam and r_s is the distance from the detector.

$$R_{90} = K_0 MCP(90) S(90) \tag{10.24}$$

K_0 is an optical constant (related to the refractive index difference between the particles and the medium); M is the molecular mass of scattering units with weight fraction C.

For small particles (as is the case with microemulsions) $P(90) \sim 1$ and,

$$M = \tfrac{4}{3}\pi R_c^3 N_A \tag{10.25}$$

where N_A is Avogadro's constant.

$$C = \phi_c \rho_c \tag{10.26}$$

where ϕ_c is the volume fraction of the particle core and ρ_c is their density.

Equation (10.24) can be written simply as

$$R_{90} = K_1 \phi_c R_c^3 S(90) \tag{10.27}$$

where $K_1 = K_0(4/3) N_A \rho_c^2$.

Equation (10.27) shows that to calculate R_c from R_{90} one needs to know $S(90)$. The latter can be calculated using Eqs. (10.20) to (10.22).

The above calculations were obtained using a W/O microemulsion of water/xylene/sodium dodecyl benzene sulphonate (NaDBS)/hexanol [11]. The microemulsion region was established using the quaternary phase diagram. The W/O microemulsions were produced at various water volume fractions, using increasing amounts of NaDBS (5, 10.9, 15 and 20%).

Figure 10.9 gives the results for the variation of R_{90} with the volume fraction of the water core droplets at various NaDBS concentrations. With the exception of the 5% NaDBS results, all the others showed an initial increase in R_{90} with increasing

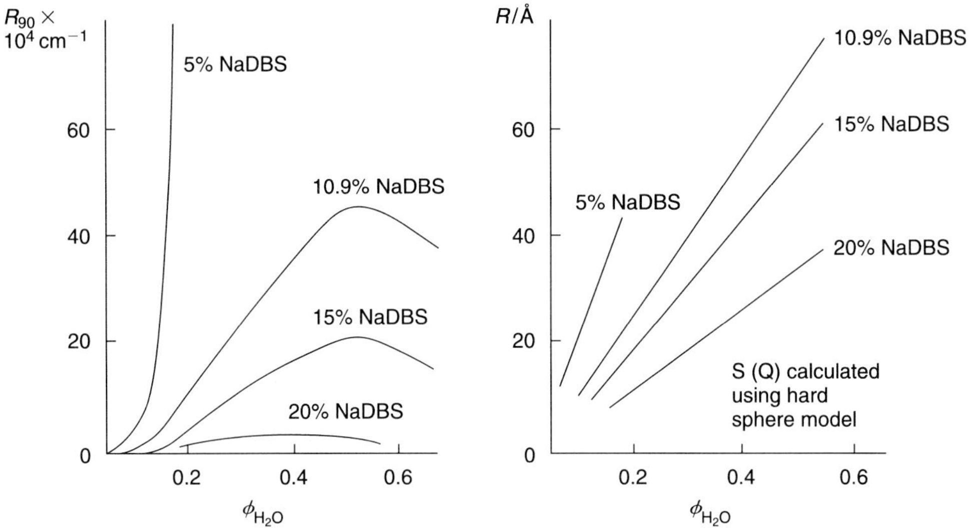

Fig. 10.9. Variation of R_{90} and R with the volume fraction of water for a W/O microemulsion based on xylene–water–NaDBS–hexanol.

ϕ, reaching a maximum at a given ϕ, after which R_{90} decreases with further increase in ϕ.

The above results were used to calculate R as a function of ϕ using the hard-sphere model discussed above (Eq. 10.22). This is also shown in Figure 10.9.

With increasing ϕ, at constant surfactant concentration, R clearly increases (the ratio of surfactant to water decreases with increasing ϕ). At any volume fraction of water, an increase in surfactant concentration results in a decrease in the microemulsion droplet size (the ratio of surfactant to water increases).

10.7.2
Calculation of Droplet Size from Interfacial Area

Assuming all surfactant and cosurfactant molecules are adsorbed at the interface, it is possible to calculate the total interfacial area of the microemulsion from a knowledge of the area occupied by surfactant and cosurfactant molecules.

Total interfacial area = Total number of surfactant molecule$_s$ × area per surfactant molecule A_s + total number of cosurfactant molecules × area per cosurfactant molecule A_{co}.

The total interfacial area A per kg of microemulsion is given by

$$A = \frac{(n_s N_A A_s + n_{co} N_A A_{co})}{\phi} \tag{10.28}$$

where n_s and n_{co} are the number of moles of surfactant and cosurfactant.

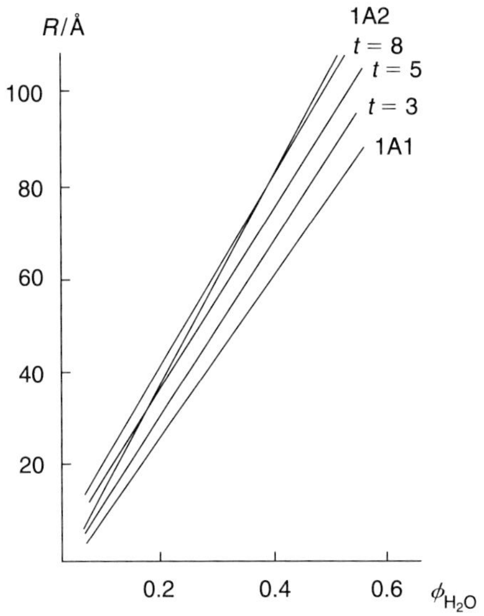

Fig. 10.10. Comparison of the radius obtained using interfacial area calculations with light-scattering results.

A is related to the droplet radius R (assuming all the droplets are of the same size) by

$$A = \frac{3}{R\rho} \tag{10.29}$$

Using reasonable values for A_s and A_{co} (30 $Å^2$ for NaDBS and 20 $Å^2$ for hexanol) R was calculated and the results compared with those obtained using light scattering results. Two conditions were considered: (a) All hexanol molecules were adsorbed 1A1; (b) Part of the hexanol adsorbed to give a molar ratio of hexanol to NaDBS of 2:1 (1A2).

The light scattering results were analysed using different thicknesses of the adsorbed layer ($t = 3, 5$ or 8 Å). All the results and calculations are given in Figure 10.10.

Good agreement is obtained between the light scattering data and R calculated from interfacial area particularly for 1A2 and $t = 8$ Å.

10.7.3
Dynamic Light Scattering (Photon Correlation Spectroscopy, PCS)

In this technique one measures the intensity fluctuation of scattered light by the droplets as they undergo Brownian motion [13]. When a light beam passes through a colloidal dispersion, an oscillating dipole movement is induced in the

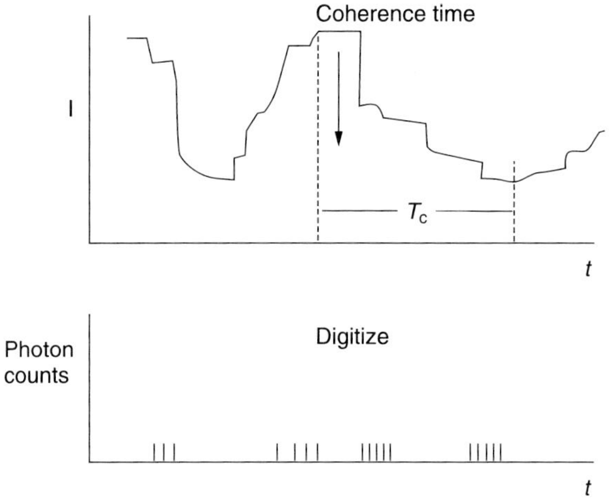

Fig. 10.11. Representation of intensity fluctuation of scattered light.

particles, thereby radiating the light. Due to the random position of the particles, the intensity of scattered light, at any instant, appears as random diffraction ("Speckle" pattern).

As the particles undergo Brownian motion, the random configuration of the pattern will fluctuate, such that the time taken for an intensity maximum to become a minimum (the coherence time) corresponds approximately to the time required for a particle to move one wavelength λ. Using a photomultiplier of active area about the diffraction maximum (i.e. one coherent area), this intensity fluctuation can be measured. The analogue output is digitised (using a digital correlator) that measures the photocount (or intensity) correlation function of scattered light. This fluctuation is schematically illustrated in Figure 10.11.

The photocount correlation function $g^{(2)}(\tau)$ is given by Eq. (10.30), where τ is the correlation delay time.

$$g^{(2)} = B[1 + \gamma^2 g^{(1)}(\tau)]^2 \tag{10.30}$$

The correlator compares $g^{(2)}(\tau)$ for many values of τ. B is the background value to which $g^{(2)}(\tau)$ decays at long delay times. $g^{(1)}(\tau)$ is the normalised correlation function of the scattered electric field and γ is a constant (~ 1).

For monodispersed non-interacting particles,

$$g^{(1)}(\tau) = \exp(-\Gamma\gamma) \tag{10.31}$$

Γ is the decay rate or inverse coherence time, which is related to the translational diffusion coefficient D,

$$\Gamma = DK^2 \tag{10.32}$$

where K is the scattering vector,

$$K = \left(\frac{4\pi n}{\lambda_0}\right) \sin\left(\frac{\theta}{2}\right) \tag{10.33}$$

The particle radius R can be calculated from D using the Stokes–Einstein equation,

$$D = \frac{kT}{6\pi\eta_0 R} \tag{10.34}$$

where η_0 is the viscosity of the medium.

The above analysis only applies for very dilute dispersions. With microemulsions which are concentrated dispersions, corrections are needed to account for the interdroplet interaction. This is reflected in plots of $\ln g^{(1)}(\tau)$ versus τ, which become nonlinear, implying that the observed correlation functions are not single exponentials.

As with time-averaged light scattering, one needs to introduce a structure factor in calculating the average diffusion coefficient. For comparative purposes, one calculates the collective diffusion coefficient D, which can be related to its value at infinite dilution D_0 by [14]

$$D = D_0(1 + \alpha\phi) \tag{10.35}$$

where α is a constant that is equal to 1.5 for hard spheres with repulsive interaction.

10.7.4 Neutron Scattering

Neutron scattering offers a valuable technique for determining the dimensions and structure of microemulsion droplets. The scattering intensity $I(Q)$ is given by

$$I(Q) = (\text{Instrument const.})(\rho - \rho_0)NV_p^2 P(Q)S(Q) \tag{10.36}$$

where ρ is the mean scattering length density of the particles and ρ_0 is the corresponding value for the solvent.

One of the main advantages of neutron scattering over light scattering is the Q range at which one operates: With light scattering the range of Q is small ($\sim$0.0005–0.0015 Å^{-1} while for small-angle neutron scattering the Q range is large (0.02–0.18 Å^{-1}). In addition, neutron scattering can give information on the structure of the droplets.

Figure 10.12 shows illustrative plots of $I(Q)$ versus Q for W/O microemulsions (xylene/water/NaDBS/hexanol) [15].

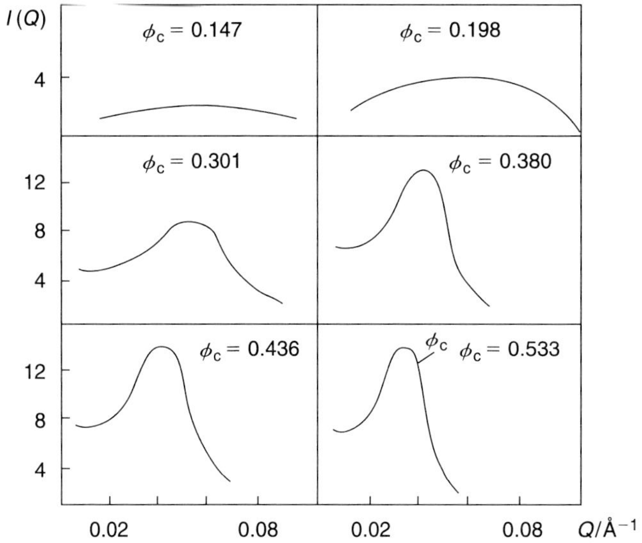

Fig. 10.12. $I(Q)$ versus Q for W/O microemulsions at various water volume fractions.

The Q values at the maximum can be used to calculate the lattice spacing using Bragg's equation. Alternatively, one can use a hard-sphere model to calculate $S(Q)$ and then fit the data of $I(Q)$ versus Q to obtain the droplet radius R.

10.7.5
Contrast Matching for Determination of the Structure of Microemulsions

By changing the isotopic composition of the components (e.g. using deuterated oil and H_2O–D_2O) one can match the scattering length density of the various components: By matching the scattering length density of the water core with that of the oil, one can investigate the scattering from the surfactant "shell". By matching the scattering length density of the surfactant "shell" and the oil, one can investigate the scattering from the water core.

10.7.6
Characterisation of Microemulsions Using Conductivity, Viscosity and NMR

10.7.6.1 Conductivity Measurements

Conductivity measurements may provide valuable information on the structural behaviour of microemulsions. In the early applications of conductivity measurements, the technique was used to determine the nature of the continuous phase. O/W microemulsions should give fairly high conductivity (which is determined by that of the continuous aqueous phase) whereas W/O microemulsions should give fairly low conductivity (determined by that of the continuous oil phase).

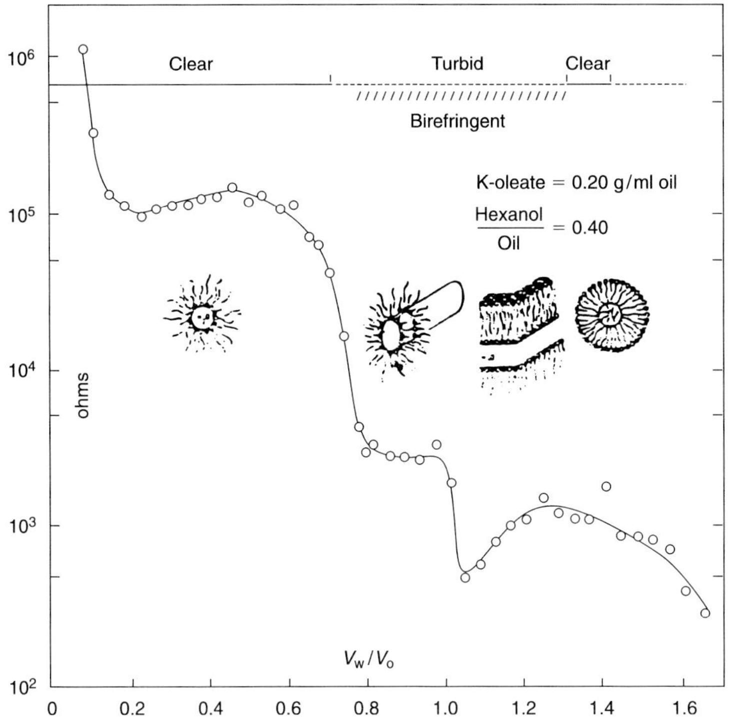

Fig. 10.13. Electrical resistance versus V_w/V_o.

As an illustration Figure 10.13 shows the change in electrical resistance (reciprocal of conductivity) with the ratio of water to oil (V_w/V_o) for a microemulsion system prepared using the inversion method [16]. Figure 10.13 indicates the change in optical clarity and birefringence with the ratio of water to oil.

At low V_w/V_o, a clear W/O microemulsion is produced with a high resistance (oil continuous). As V_w/V_o increases, the resistance decreases, and, in the turbid region, hexanol and lamellar micelles are produced. Above a critical ratio, inversion occurs and the resistance decreases, producing O/W microemulsion.

Conductivity measurements were also used to study the structure of the microemulsion, which is influenced by the nature of the cosurfactant. This can be shown (Figure 10.14) for two systems, one based on water–toluene–potassium oleate–butanol and the other on water–hexadecane–potassium oleate–hexanol [17]. The two systems differ in the nature of the cosurfactant, namely butanol (C_4 alcohol) and hexanol (C_6 alcohol).

The system based on butanol shows a rapid increase in κ above a critical water volume fraction value, whereas the second system based on hexanol shows much lower conductivity, with a maximum and minimum at two water volume fractions, ϕ'_w and ϕ''_w.

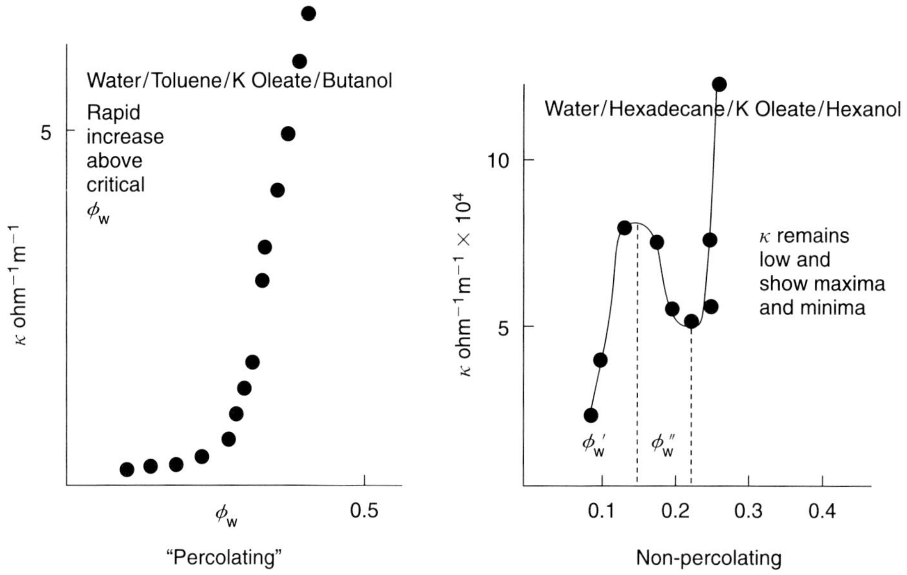

Fig. 10.14. Conductivity versus water volume fraction for two W/O microemulsion systems.

In the first case (when using butanol), the κ–ϕ_w curve can be analysed using the percolation theory of conductivity [18]. In this model, the effective conductivity is practically zero as long as the volume fraction of the conductor (water) is below a critical value ϕ_w^p (the percolation threshold). Beyond this value, κ suddenly takes a non-zero value and increases rapidly with further increase in ϕ_w.

In the above case (percolating microemulsions), the following equations were derived theoretically.

$$\kappa \sim (\phi_w - \phi_w^p)^{8/5} \quad \text{when } \phi_w > \phi_w^p \tag{10.37}$$

$$\kappa \sim (\phi_w^p - \phi_w)^{-0.7} \quad \text{when } \phi_w < \phi_w^p \tag{10.38}$$

By fitting the conductivity data to Eqs. (10.37) and (10.38), ϕ_w^p was found to be 0.176 ± 0.005, in agreement with the theoretical value.

The second system, based on hexanol, does not fit the percolation theory (non-percolating microemulsion). The variation of κ with water volume fraction is due to more subtle changes in the system on changing ϕ_w. The initial increase in κ with increasing ϕ_w can be ascribed to enhanced surfactant solubilisation with added water. Alternatively, it could be due to increasing surfactant dissociation on the addition of water. Beyond the maximum, addition of water mainly causes micelle swelling, i.e. a definite water core (microemulsion droplets) begins to be formed, which may be considered as a dilution process, leading to a decrease in conductivity (the decrease in κ beyond the maximum may be due to the replacement of the hydrated surfactant–cosurfactant aggregates with microemulsion

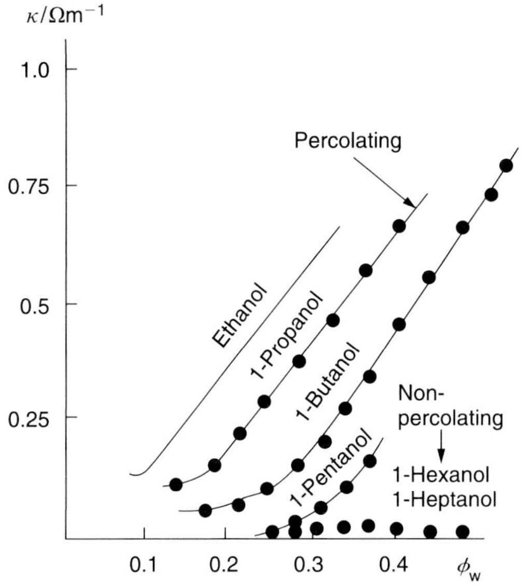

Fig. 10.15. Variation of conductivity with water volume fractions for various cosurfactants.

droplets). The sharp increase in κ beyond the minimum must be associated with a facilitated path for ion transport (formation of non-spherical droplets resulting from swollen micelle clustering and subsequent cluster interlinking).

Clausse and his co-workers carried out a systematic study of the effect of cosurfactant chain length on the conductive behaviour of W/O microemulsions [19]. The cosurfactant chain length was gradually increased from C_2 (ethanol) to C_7 (heptanol). The results for the variation of κ with ϕ_w are shown in Figure 10.15.

With the short-chain alcohols ($C < 5$), the conductivity shows a rapid increase above a critical ϕ. With longer chain alcohols, namely hexanol and heptanol, the conductivity remains very low up to a high water volume fraction.

With the short-chain alcohols, the system shows percolation above a critical water volume fraction. Under these conditions the microemulsion is "bicontinuous". With the longer chain alcohols, the system is non-percolating and one can define definite water cores. This is sometimes referred to as a "true" microemulsion.

10.7.6.2 Viscosity Measurements

Viscosity measurements may be applied in a qualitative manner to give information on the hydrodynmic radius of microemulsions. Figure 10.16 shows illustrative plots of the relative viscosity η_r as a function of the water core volume fraction for W/O microemulsions based on xylene/water/hexanol/sodium dodecyl benzene sulphonate (NaDBS) at various NaDBS concentrations. The results show the typical behaviour obtained with concentrated dispersions, namely a rapid increase in η_r above a critical range of ϕ due to droplet–droplet interaction.

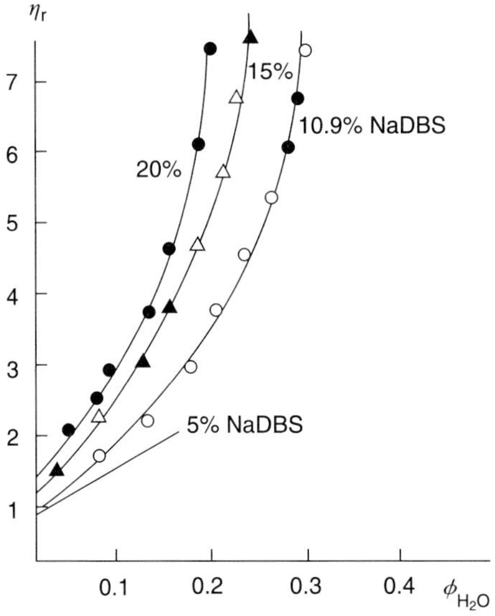

Fig. 10.16. Variation of η_r with ϕ.

The results may be fitted to the Mooney equation [20],

$$\eta_r = \exp\left(\frac{a\phi}{1 - k\phi}\right) \tag{10.39}$$

where a is the intrinsic viscosity (theoretically equal to 2.5 for hard-spheres) and k is the so-called crowding factor (theoretically equal to 1.35–1.90).

The results of Figure 10.16 were fitted to Eq. (10.39) and the values of a and k were obtained as shown in Table 10.1, which also shows the results for the ratio of the effective volume fraction to the core volume fraction ϕ_{eff}/ϕ (denoted V_δ) as well as the ratio of the layer thickness to particle radius δ/a.

Table 10.1 shows reasonable values for the crowding factor k. However, a is much larger than the theoretical value of 2.5. This discrepancy is due to the presence of the surfactant layer, which causes an increase in the core volume fraction,

$$V_\delta = \frac{\phi_{eff}}{\phi} = \left(1 - \frac{\delta}{a}\right)^3 \tag{10.40}$$

Tab. 10.1. Data obtained from viscosity measurements.

wt% NaDBS	*a*	*k*	V_δ	δ/a
5	3.3	1.9	1.3	0.10
10.9	3.0	1.4	1.2	0.07
15	4.0	1.6	1.6	0.17
20	6.0	1.4	2.4	0.35

In addition, Table 10.1 clearly shows that V_δs are larger than 1, resulting from the presence of the surfactant layer. The smaller the core radius a, the larger the effect and the larger the δ/a.

10.7.6.3 NMR Measurements

Lindman and co-workers [21–23] demonstrated that the organisation and structure of microemulsions can be elucidated from self-diffusion measurements of all the components (using pulse-gradient or spin-echo NMR techniques). Within a micelle, the molecular motion of the hydrocarbon tails (translational, reorientation and chain flexibility) is almost as rapid as in a liquid hydrocarbon. In a reverse micelle, water molecules and counter ions are also highly mobile.

For many surfactant–water systems, there is a distinct spatial separation between hydrophobic and hydrophilic domains. The passage of species between different regions is an improbable event and occurs very slowly.

Thus, self-diffusion, if studied over macroscopic distances, should reveal whether the process is rapid or slow, depending on the geometrical properties of the inner structure. For example, a phase that is water continuous and oil discontinuous should exhibit rapid diffusion of hydrophilic components, while the hydrophobic components should diffuse slowly. An oil continuous but water discontinuous system should exhibit rapid diffusion of the hydrophobic components. One would expect a bicontinuous structure to give rapid diffusion of all components.

Using the above principle, Lindman and co-workers [21–23] measured the self-diffusion coefficients of all components, consisting of various components, with particular emphasis on the role of the cosurfactant. For microemulsions consisting of water, hydrocarbon, an anionic surfactant and a short chain alcohol (C_4 and C_5), the self-diffusion coefficients of water, hydrocarbon and cosurfactant were quite high, of the order of 10^{-9} m^2 s^{-1}, i.e. two orders of magnitude higher than the value expected for a discontinuous medium (10^{-11} m^2 s^{-1}). This high diffusion coefficient was attributed to three main effects: Bicontinuous solutions, an easily deformable and flexible interface, and the absence of large aggregates.

With microemulsions based on long-chain alcohols (e.g. decanol), the self-diffusion coefficient for water was low, indicating the presence of definite (closed) water droplets surrounded by surfactant anions in the hydrocarbon medium. Thus, NMR measurements can clearly distinguish between the two types of microemulsion systems.

References

1 T. P. Hoar, J. H. Schulman, *Nature (London)*, **1943**, *152*, 102.
2 L. M. Prince: *Microemulsion Theory and Practice*, Academic Press, New York, 1977.
3 I. Danielsson, B. Lindman, *Colloids Surf.*, **1983**, *3*, 391.
4 J. H. Schulman, W. Stoeckenius, L. M. Prince, *J. Phys. Chem.*, **1959**, *63*, 1677.
5 L. M. Prince, *Adv. Cosmet. Chem.*, **1970**, *27*, 193.
6 K. Shinoda, S. Friberg, *Adv. Colloid Interface Sci.*, **1975**, *4*, 281.

7 E. Ruckenstein, J. C. Chi, *J. Chem. Soc., Faraday Trans. II*, **1975**, *71*, 1690.
8 J. T. G. Overbeek, *Faraday Discuss. Chem. Soc.*, **1978**, *65*, 7.
9 J. T. G. Overbeek, P. L. Bruyn de, F. Verhoeckx: *Surfactants*, Th. F. Tadros (ed.): Academic Press, London, 1984, 111–132.
10 N. F. Carnahan, K. E. Starling, *J. Chem. Phys.*, **1969**, *51*, 635.
11 D. J. Mitchell, B. W. Ninham, *J. Chem. Soc., Faraday Trans. II*, **1981**, *77*, 601.
12 R. C. Baker, A. T. Florence, R. H. Ottewill, T. F. Tadros, *J. Colloid Interface Sci.*, **1984**, *100*, 332.
13 N. W. Ashcroft, J. Lekner, *Phys. Rev.*, **1966**, *45*, 33.
14 P. N. Pusey: *Industrial Polymers: Characterisation by Molecular Weights*, J. H. S. Green, R. Dietz (ed.): Transcripta Books, London, 1973.
15 A. N. Cazabat, D. Langevin, *J. Chem. Phys.*, **1981**, *74*, 3148.
16 D. J. Cebula, R. H. Ottewill, J. Ralston, P. Pusey, *J. Chem. Soc., Faraday Trans. I*, **1981**, *77*, 2585.
17 L. M. Prince: *Microemulsions*, Academic Press, London, 1977.
18 B. Lagourette, J. Peyerlasse, C. Boned, M. Clausse, *Nature*, **1969**, *281*, 60.
19 S. Kilpatrick, *Mod. Phys.*, **1973**, *45*, 574.
20 M. Clausse, J. Peyerlasse, C. Boned, J. Heil, L. Nicolas-Margantine, A. Zrabda: *Solution Properties of Surfactants*, K. L. Mittal, B. Lindman, eds., Plenum Press, New York, 1984, 1583, Volume 3.
21 M. Mooney, *J. Colloid Sci.*, **1950**, *6*, 162.
22 B. Lindman, H. Winnerstrom: *Topics in Current Chemistry*, F. L. Borschke, ed., Springer-Verlag, Heidelberg, 1980, 1–83.
23 H. Winnerstrom, B. Lindman, *Phys. Rep.*, **1970**, *52*, 1.
24 B. Lindman, P. Stilbs, M. E. Moseley, *J. Colloid Interface Sci.*, **1981**, *83*, 569.

11
Role of Surfactants in Wetting, Spreading and Adhesion

11.1
General Introduction

Wetting is important in many processes, both industrial and natural. In many cases, wetting is a prerequisite for application:

(1) In paint films a paint has to wet the substrate completely to form a uniform paint film.
(2) In coatings such as in photographic films, which are coated with a film at very high speed, the dynamics of the wetting process is very important.
(3) In crop sprays applied to plants or weeds it is essential that the spray solution wets the substrate completely and in many cases rapid spreading may be required. Again the dynamics of wetting becomes a very important factor.
(4) Personal care formulations such as creams and lotions require good wetting of the substrate (skin). In many other applications such as hair sprays, droplet impaction and adhesion become important and this may have to be followed by wetting and spreading on the hair surface.
(5) In pharmaceutical applications, e.g. wetting of tablets, which is essential for its disintegration and dispersion.

Wetting of powders is an important prerequisite for dispersion of powders in liquids, i.e. preparation of suspensions. It is essential to wet both the external and internal surfaces of the powder aggregated and agglomerates. Suspensions are applied in many industries such as paints, dyestuffs, printing inks, agrochemicals, pharmaceuticals, paper coatings, detergents, etc.

In all the above processes one has to consider both the equilibrium and dynamic aspects of the wetting process. The equilibrium aspects of wetting can be studied at a fundamental level using interfacial thermodynamics. Under equilibrium, a drop of a liquid on a substrate produces a contact angle θ, which is the angle formed between planes tangent to the surfaces of solid and liquid at the wetting perimeter. This is illustrated in Figure 11.1, which shows the profile of a liquid drop on a flat solid substrate. An equilibrium between vapour, liquid and solid is established with a contact angle θ (which is lower than 90°).

Applied Surfactants: Principles and Applications. Tharwat F. Tadros

ISBN: 3-527-30629-3

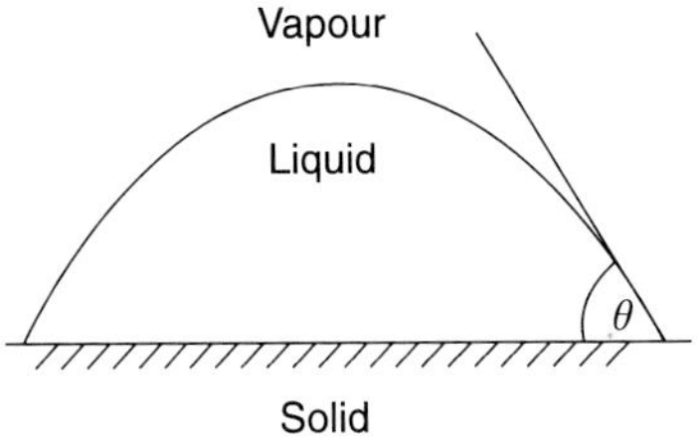

Fig. 11.1. Representation of a liquid drop on a flat substrate.

The wetting perimeter is frequently referred to as the three-phase line (solid/liquid/vapour); The most common name is the wetting line.

Most equilibrium wetting studies centre around measurements of the contact angle – the smaller the angle the better the liquid is said to wet the solid. Typical examples are given in Table 11.1 for water, with a surface tension of 72 mN m^{-1}, on various substrates.

The above values can be roughly used as a measure of wetting of the substrate by water (glass being completely wetted and PTFE very difficult to wet).

The dynamic process of wetting is usually described in terms of a moving wetting line that results in contact angles that change with the wetting velocity. The same name is sometimes given to contact angles that change with time.

Wetting of a porous substrate may also be considered as a dynamic phenomenon. The liquid penetrates through the pores and gives different contact angles depending on the complexity of the porous structure. Study of the wetting of porous substrates is very difficult. However, even measurements of apparent contact angles can be very useful for comparing one porous substrate with another.

Despite increasing attention on the dynamics of wetting, understanding the kinetics of the process at a fundamental level has not been achieved.

The spreading of liquids on substrates is also an important industrial phenomenon, e.g. with crop sprays, which need to spread spontaneously on leaf surfaces to maximise the biological effect. A useful concept introduced by Harkens [1] is the spreading coefficient, which is simply the work done in destroying a unit area of solid/liquid and liquid/vapour interface to produce an area of solid/air interface. The spreading coefficient is simply determined from the contact angle θ and the liquid/vapour surface tension γ_{LV},

Tab. 11.1. Typical contact angle values for a water drop on various substrates.

Substrate	*Contact angle θ (°)*
PTFE (Teflon)	112
Paraffin wax	110
Polyethylene	103
Human skin	75–90
Glass	0

$$S = \gamma_{LV}(\cos\theta - 1) \tag{11.1}$$

For spontaneous spreading S has to be zero or positive. If S is negative only limited spreading is obtained.

Adhesion is defined as "the state in which two surfaces are held together by interfacial forces which may consist of valence forces or interlocking action or both" [2]. The material that can hold the materials together is referred to as the adhesive.

Several industrial applications of adhesives can be recognised: (1) Medical applications, e.g. strips used to cover wounds to prevent infection. (2) Patches for the controlled release of drugs. (3) Various types of hygiene products, e.g. diapers, feminine towels, etc. (4) Adhesives that are used to stick two substrates together – this usually requires very strong adhesive bonds.

Several intermolecular forces are considered to describe the adhesion of surfaces, ranging from physical forces such as van der Waals to more strong chemical bonds obtained by interaction at the interfacial region.

Early adhesives were based on natural products, such as starch, animal glue, natural rubber, etc. Modern adhesives are synthetic materials based on resins, synthetic polymers, epoxides, urethanes, etc.

Several concepts have been introduced to discuss the process of adhesion at a fundamental level: (1) Interfacial concepts based on intermolecular forces; (2) chemical bonds, e.g. ionic and covalent.

The above concepts led to the development of quantitative theories that can be applied to many practical systems.

The process of particle–surface adhesion has many applications in both industrial and biological processes [3]. In industrial processes, particle–surface adhesion is important in detergency, water purification, flotation, fibre manufacture, paper coatings, etc. In the biology, particle–surface adhesion is important in cell contacts, cell adhesion and attachment, phagocytosis, marine fouling, etc.

There is no single quantitative theory that can describe all adhesion phenomena. Both chemical and non-chemical bonds are involved in particle–surface adhesion. Experimental techniques of studying the process are far from adequate. Many processes in industry require strong adhesion of particles to surface, e.g. in paints, dyestuffs, agrochemical particles that need to be strongly attached to leaf surfaces for biological control.

There are many other processes where adhesion of particles to surfaces is undesirable, e.g. in detergency, non-stick surfaces, adhesion of bacteria to tooth surface (dental plaque), adhesion of cells to blood vessels (which can cause thrombosis), etc.

Theories of particle–surface adhesion are available for idealised particles and surfaces. These theories need to be extended to real particles and surfaces. They need to take into account the complex nature of the particle surface as well as the topography of the substrate to which they become attached. Advances in experimental techniques for measurement of particle–surface adhesion with real particles and surfaces have been very slow and this hampered the development of adequate fundamental principles.

Problems encountered in industrial applications are easier to tackle than biological ones. However, advances in understanding the processes encountered in the biology are required since these can be used in prevention of many undesirable effects such as dental plaque, thrombosis, cell attachment, etc.

11.2 Concept of Contact Angle

Wetting is a fundamental interfacial phenomenon in which one fluid phase is displaced completely or partially by another fluid phase from the surface of a solid or a liquid. The most useful parameter that may describe wetting is the contact angle of a liquid on the substrate and this is discussed below.

11.2.1 Contact Angle

When a drop of a liquid is placed on a solid, the liquid either spreads to form a thin (uniform) film or remains as a discrete drop (Figure 11.2).

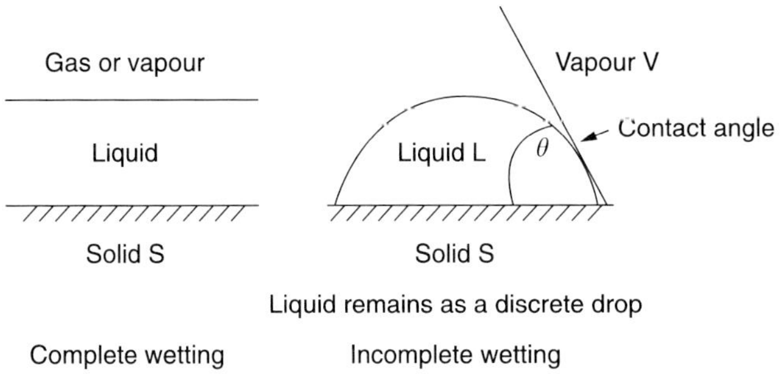

Fig. 11.2. Schematic of complete and partial wetting.

11.2.2 Wetting Line – Three-phase Line (Solid/Liquid/Vapour)

The contact angle θ is the angle formed between planes tangent to the surfaces of the solid and liquid at the wetting perimeter. The wetting perimeter is referred to as the three-phase line (solid/liquid/vapour) or simply the wetting line. The utility of contact angle measurements depends on equilibrium thermodynamic arguments (static measurements).

In practical systems such as in spray applications, one has to displace one fluid (air) with another (liquid) as quickly and as efficiently as possible. Dynamic contact angle measurements (associated with moving wetting line) are more relevant in many practical applications.

Even under static conditions, contact angle measurements are far from simple since they are mostly accompanied by hysteresis. The value of θ depends on the history of the system and whether the liquid is tending to advance across or recede from the solid surface. The limiting angles achieved just prior to movement of the wetting line (or just after movement ceases) are known as the advancing and receding contact angles, θ_A and θ_R, respectively. For a given system $\theta_A > \theta_R$ and θ can usually take any value between these two limits without discernible movement of the wetting line.

11.2.3
Thermodynamic Treatment – Young's Equation [4]

The liquid drop takes the shape that minimises the free energy of the system. Consider a simple system of a liquid drop (L) on a solid surface (S) in equilibrium with the vapour of the liquid (V) (Figure 11.3).

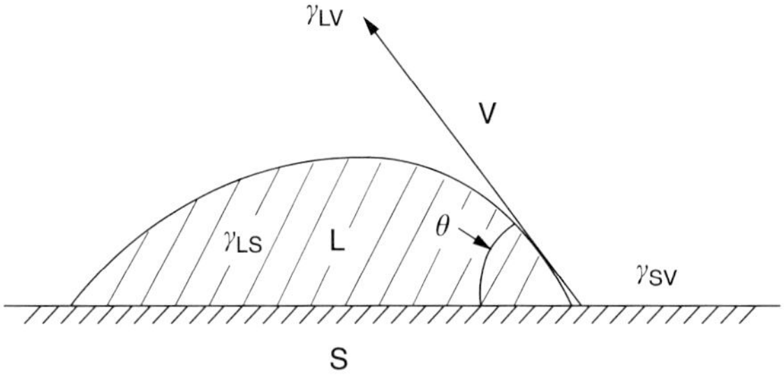

Fig. 11.3. Representation of a liquid drop on a flat substrate and the balance of tensions.

The sum $(\gamma_{SV}A_{SV} + \gamma_{SL}A_{SL} + \gamma_{LV}A_{LV})$ should be a minimum at equilibrium and this leads to Young's equation,

$$\gamma_{SV} = \gamma_{SL} + \gamma_{LV}\cos\theta \tag{11.2}$$

In the above equation θ is the equilibrium contact angle. The angle that a drop assumes on a solid surface is the result of the balance between the cohesion force in the liquid and the adhesion force between the liquid and solid, i.e.

$$\gamma_{LV}\cos\theta = \gamma_{SV} - \gamma_{SL} \tag{11.3}$$

or

$$\cos\theta = \frac{\gamma_{SV} - \gamma_{SL}}{\gamma_{LV}} \tag{11.4}$$

If there is no interaction between solid and liquid, then

$$\gamma_{SL} = \gamma_{SV} + \gamma_{LV} \tag{11.5}$$

i.e. $\theta = 180°$ ($\cos\theta = -1$).

If there is strong interaction between solid and liquid (maximum wetting), the latter spreads until Young's equation is satisfied ($\theta = 0$) and

$$\gamma_{LV} = \gamma_{SV} - \gamma_{SL} \tag{11.6}$$

The liquid spreads spontaneously on the solid surface.

When the surface of the solid is in equilibrium with the liquid vapour, one must consider the spreading pressure, π_e. As a result of the adsorption of the vapour on the solid surface its surface tension γ_s is reduced by π_e, i.e.,

$$\gamma_{SV} = \gamma_s - \pi_e \tag{11.7}$$

and Young's equation can be written as

$$\gamma_{LV}\cos\theta = \gamma_s - \gamma_{SL} - \pi_e \tag{11.8}$$

In general, Young's equation provides a precise thermodynamic definition of the contact angle. However, it suffers from the lack of direct experimental verification since both γ_{SV} and γ_{SL} cannot be directly measured. An important criterion for application of Young's equation is to have a common tangent at the wetting line between the two interfaces.

11.3 Adhesion Tension

There is no direct way by which γ_{SV} or γ_{SL} can be measured. The difference between γ_{SV} and γ_{SL} can be obtained from contact angle measurements. This difference is referred to as the "Wetting Tension" or "Adhesion Tension",

$$\text{Adhesion tension} = \gamma_{SV} - \gamma_{SL} = \gamma_{LV}\cos\theta \tag{11.9}$$

Consider the immersion of a solid in a liquid as is illustrated in Figure 11.4. When the plate is immersed in the liquid, an area $dA\gamma_{SV}$ is lost and an area $dA\gamma_{SL}$ is formed.

The (Helmholtz) free energy change dF is given by

$$dF = dA(\gamma_{SV} - \gamma_{SL}) \tag{11.10}$$

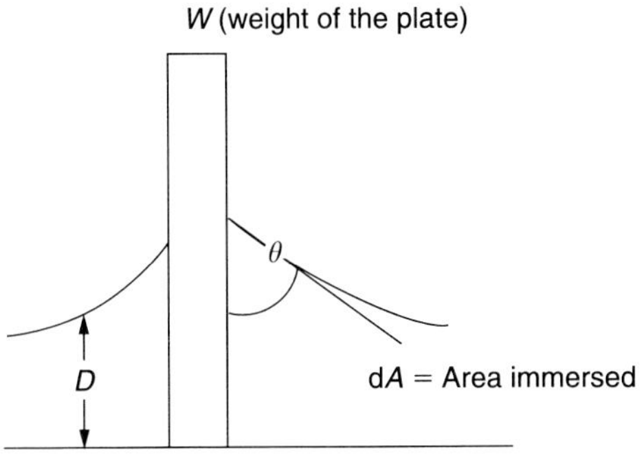

Fig. 11.4. Schematic of immersion of a solid plate in a liquid.

This is balanced by the force on the plate $W\,\mathrm{d}D$,

$$W\,\mathrm{d}D = \mathrm{d}A(\gamma_{SV} - \gamma_{SL}) \tag{11.11}$$

or

$$W\left(\frac{\mathrm{d}D}{\mathrm{d}A}\right) = (\gamma_{SV} - \gamma_{SL}) = \gamma_{LV}\cos\theta \tag{11.12}$$

$\mathrm{d}A/\mathrm{d}D = p =$ plate perimeter, or

$$\frac{W}{p} = \gamma_{LV}\cos\theta \tag{11.13}$$

Equation (11.13) forms the basis of measuring the contact angle θ using an immersed plate (Wilhelmy plate). Equation (11.13) is also the basis of measuring the surface tension of a liquid using the Wilhelmy plate technique. If the plate is made to wet the liquid completely, i.e. $\theta = 0$ or $\cos\theta = 1$, then $W/p = \gamma_{LV}$. By measuring the weight of the plate as it touches the liquid one obtains γ_{LV}.

Gibbs [5] defined the adhesion tension τ as the difference between the surface pressure at the solid/liquid interface π_{SL} and that at the solid/vapour interface π_{SV},

$$\tau = \pi_{SL} - \pi_{SV} \tag{11.14}$$

$$\pi_{SL} = \gamma_s - \gamma_{SL} \tag{11.15}$$

$$\pi_{SV} = \gamma_s - \gamma_{SV} \tag{11.16}$$

Combining Eqs. (11.14) to (11.16) with Young's equation, Gibbs arrived at the following equation for the adhesion tension,

$$\tau = \gamma_{SV} - \gamma_{SL} = \gamma_{LV}\cos\theta \tag{11.17}$$

which is identical to Eq. (11.9).

Thus, the adhesion tension depends on the measurable quantities γ_{LV} and θ. As long as θ is $<90°$, the adhesion tension is positive.

11.4
Work of Adhesion W_a

Consider a liquid drop with surface tension γ_{LV} and a solid surface with surface tension γ_{SV}. When the liquid drop adheres to the solid surface it forms a surface tension γ_{SL} (Figure 11.5).

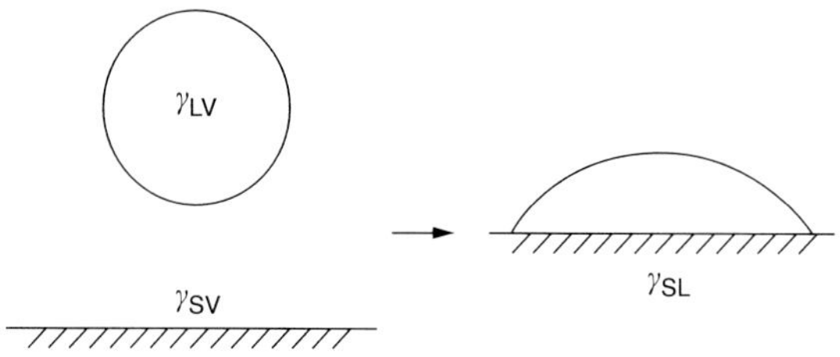

Fig. 11.5. Representation of adhesion of a drop on a solid substrate.

The work of adhesion [6, 7] is simply the difference between the surface tensions of the liquid/vapour and solid/vapour and that of the solid/liquid,

$$W_a = \gamma_{SV} + \gamma_{LV} - \gamma_{SL} \tag{11.18}$$

Using Young's equation,

$$W_a = \gamma_{LV}(\cos\theta + 1) \tag{11.19}$$

11.5
Work of Cohesion

The work of cohesion W_c is the work of adhesion when the two phases are the same. Consider a liquid cylinder with unit cross sectional area. When this liquid is subdivided into two cylinders (Figure 11.6) two new surfaces are formed.

The two new areas will have a surface tension of $2\gamma_{LV}$ and the work of cohesion is simply

$$W_c = 2\gamma_{LV} \tag{11.20}$$

Thus, the work of cohesion is simply equal to twice the liquid surface tension. An important conclusion may be drawn if one considers the work of adhesion given by

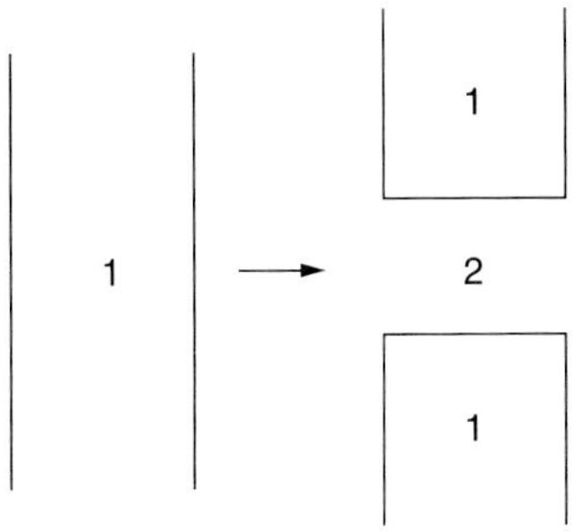

Fig. 11.6. Scheme of subdivision of a liquid cylinder.

Eq. (11.19) and the work of cohesion given by Eq. (11.20): When $W_c = W_a$, $\theta = 0°$. This is the condition for complete wetting. When $W_c = 2W_a$, $\theta = 90°$ and the liquid forms a discrete drop on the substrate surface.

Thus, the competition between cohesion of the liquid to itself and its adhesion to a solid gives an angle of contact that is constant and specific to a given system at equilibrium. This shows the importance of Young's equation in defining wetting.

11.6 Calculation of Surface Tension and Contact Angle

Fowler [8] was the first to calculate the surface tension of a simple liquid. The basic idea is to use the intermolecular forces that operate between atoms and molecules. For this purpose it is sufficient to consider the various van der Waals forces: Dipole–dipole (Keesom), Dipole-induced–dipole (Debye), and Dispersion (London).

Dispersion forces are the most important since they occur between all atoms and molecules and they are additive. The London expression for the dispersion interaction u between two molecules separated by a distance r is

$$u = -\frac{\beta}{r^6} \tag{11.21}$$

where β is the London dispersion constant (which depends on the electric polarizability of the molecules).

Hamaker [9] calculated the attractive forces between macroscopic bodies using a simple additivity principle. For two-semi infinite flat plates separated by a distance d, the attractive force F is given by

$$F = \frac{A}{6\pi d^3} \tag{11.22}$$

where A is the Hamaker constant, which is given by

$$A = \pi^2 n^2 \beta \tag{11.23}$$

where n is the number of interacting dispersion centres per unit volume.

Fowler [8] used the above intermolecular theory to calculate the energy required to break a column of liquid of unit cross section and remove the two halves to infinite separation. Using statistical thermodynamics he calculated the work of cohesion and found it to be equal to twice the surface tension.

11.6.1 Good and Girifalco Approach [10, 11]

Good and Girifalco [10, 11] proposed a more empirical approach to the problem of calculating the surface and interfacial tension. The interaction constant for two different particles was assumed to be equal to the geometric mean of the interaction constants for the individual particles. This is referred to as the Berthelot principle.

For two atoms i and j with London constants β_i and β_j, the interaction constant β_{ij} is given by

$$\beta_{ij} = (\beta_i \beta_j)^{1/2} \tag{11.24}$$

Similarly the Hamaker constant A_{ij} is given by the geometric mean of the individual Hamaker constants,

$$A_{ij} = (A_i A_j)^{1/2} \tag{11.25}$$

By analogy Good and Girifalco [10] represented the work of adhesion between two different liquids W_{a12} as the geometric mean of their respective works of cohesion,

$$W_{a12} = \phi(W_{c1} W_{c2})^{1/2} \tag{11.26}$$

where ϕ is a constant that depends on the relative molecular size and polar content of the interacting media.

The interfacial tension γ_{12} is then related to the surface tension of the individual liquids γ_1 and γ_2 by

$$\gamma_{12} = \gamma_1 + \gamma_2 - 2\phi(\gamma_1 \gamma_2)^{1/2} \tag{11.27}$$

For non-polar media, Eq. (11.27) was found to work well with $\phi \sim 1$. For dissimilar substances such as water and alkanes, ϕ ranges from 0.35 to 1.15.

Good and Grifalco [10] and Good [11] extended the above treatment to the solid/liquid interface and they obtained the following expression for the contact angle θ,

$$\cos\theta = -1 + 2\phi\left(\frac{\gamma_s}{\gamma_1}\right)^{1/2} - \frac{\pi_s}{\gamma_1} \tag{11.28}$$

where π_s is the surface pressure of fluid 1 adsorbed at the solid/gas interface. Equation (11.28) gave reasonable values for π_s for non-polar substrates.

Although the above analysis is semi-empirical it can be usefully applied to predict the interfacial tension between two immiscible liquids. The analysis is also useful for predicting the surface tension of a solid substrate from measurements of the contact angle of the liquid.

11.6.2 **Fowkes Treatment** [12]

Fowkes [12] proposed that the surface and interfacial tensions can be subdivided into independent additive terms arising from different types of intermolecular interactions. For water, in which both hydrogen bonding and dispersion forces operate, the surface tension can be assumed to be the sum of two contributions,

$$\gamma = \gamma^h + \gamma^d \tag{11.29}$$

For non-polar liquids such as alkanes, γ is simply equal to γ^d. By applying the geometric mean relationship to γ^d, Fowkes [12] obtained the following expression for the work of adhesion W_{a12},

$$W_{a12} = 2(\gamma_1^d \gamma_2^2)^{1/2} \tag{11.30}$$

Thus, the interfacial tension γ_{12} is given by

$$\gamma_{12} = \gamma_1 + \gamma_2 - 2(\gamma_1^d \gamma_2^d)^{1/2} \tag{11.31}$$

Fowkes [12] assumed the non-dispersive contributions to γ_1 and γ_2 are unaltered at the 1/2 interface.

Similar equations can be written for the solid/liquid interface. Fowkes derived an expression for the contact angle [12],

$$\cos\theta = -1 + \frac{2(\gamma_s^d \gamma_1^d)^{1/2}}{\gamma_1} - \frac{\pi_s}{\gamma_1} \tag{11.32}$$

Various studies showed that Eqs. (11.31) and (11.32) are quite effective for materials that interact only through dispersion forces and gave reasonable predictions for γ^d for liquids and solids in which other forces are active.

The γ^d for water is ~21.8 mN m^{-1} leaving $\gamma^h \sim 51$ mN m^{-1}. Fowkes [12] showed that for non-polar liquids Equation (11.31) can be obtained by summation of pair-wise interactions.

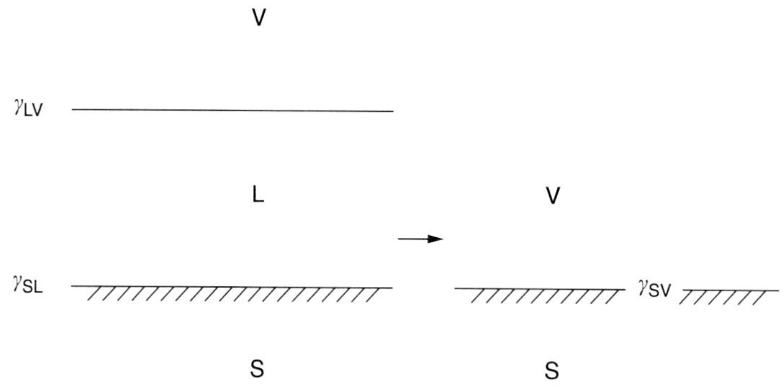

Fig. 11.7. Scheme of spreading coefficient.

11.7
Spreading of Liquids on Surfaces

11.7.1
Spreading Coefficient *S*

Harkins [13, 14] defined the initial spreading coefficient as the work required to destroy a unit area of solid/liquid (SL) and liquid/vapour (LV) and leave a unit area of bare solid (SV) (Figure 11.7).

$$S = \gamma_{SV} - (\gamma_{SL} + \gamma_{LV}) \tag{11.33}$$

Using Young's equation,

$$S = \gamma_{LV}(\cos\theta - 1) \tag{11.34}$$

If S is positive, the liquid will spread until it completely wets the solid so that $\theta = 0°$. If S is negative ($\theta > 0°$) only partial wetting occurs. Alternatively, one can use the equilibrium or final spreading coefficient.

11.8
Contact Angle Hysteresis

For a liquid spreading on a uniform, non-deformable solid (idealised case), there is only one contact angle (the equilibrium value). With real surfaces (practical systems) several stable angles can be measured. Two relatively reproducible angles can be measured: the largest, advancing angle θ_A, and smallest, the receding angle θ_R (Figure 11.8).

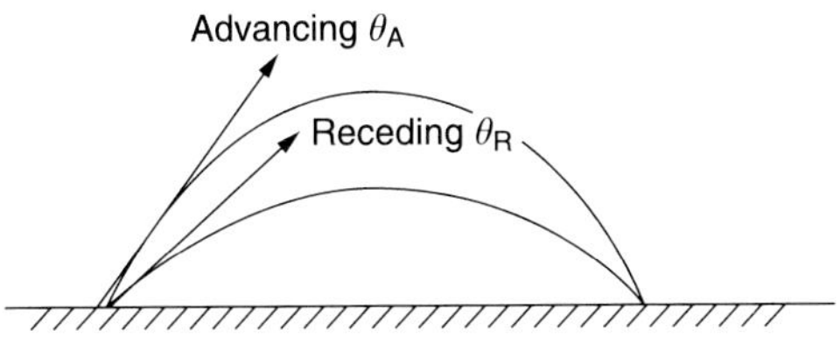

Fig. 11.8. Scheme of advancing and receding angles.

θ_A is measured by advancing the periphery of the drop over the surface (e.g. by adding more liquid to the drop); θ_R is measured by pulling the liquid back (e.g. by removing some liquid from the drop). The difference between θ_A and θ_{3R} is termed "contact angle hysteresis". The contact angle hysteresis can be illustrated by placing a drop on a tilted surface with an angle θ from the horizontal (Figure 11.9).

Advancing and receding angles are clearly shown at the front and the back of the drop on the tilted surface.

Due to the gravity field ($mg \sin \alpha \, dl$, m = mass of the drop, g = acceleration due to gravity), the drop will slide until the difference between the work of dewetting and wetting balances the gravity force.

$$\text{Work of dewetting} = \gamma_{LV}(\cos \theta_R + 1)\omega \, dl \tag{11.35}$$

$$\text{Work of wetting} = \gamma_{LV}(\cos \theta_A + 1)\omega \, dl \tag{11.36}$$

$$mg \sin \alpha \, dl = \gamma_{LV}(\cos \theta_R - \cos \theta_A)\omega \, dl \tag{11.37}$$

$$\frac{mg \sin \alpha}{\omega} = \gamma_{LV}(\cos \theta_R - \cos \theta_A) \tag{11.38}$$

Hysteresis can be demonstrated by measuring the force on a plate that is continuously immersed in the liquid. When the plate is immersed, the force will decrease due to buoyancy. When there is no contact angle hysteresis, the relationship be-

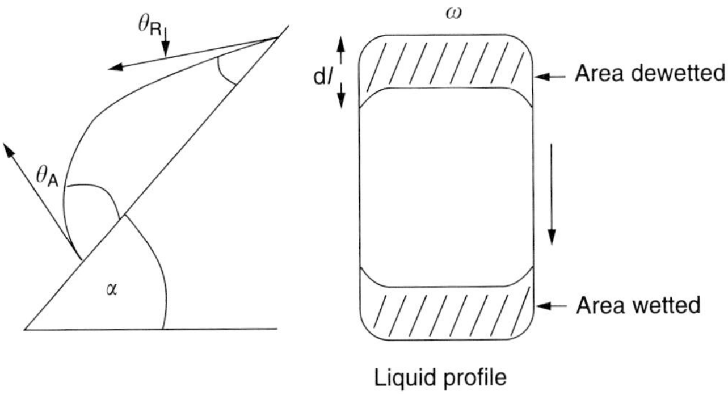

Fig. 11.9. Representation of drop profile on a tilted surface.

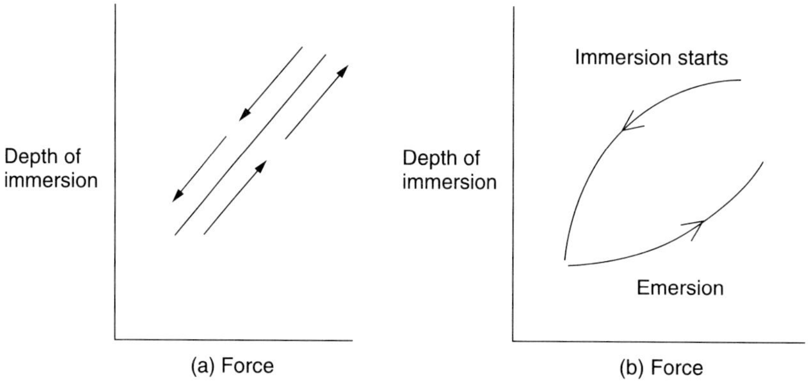

Fig. 11.10. Relationship between depth of immersion and force: (a) no hysteresis; (b) hysteresis present.

tween depth of immersion and force will be as shown in Figure 11.10a. With hysteresis, the relationship between depth of immersion and force will be as shown in Figure 11.10b.

11.8.1 Reasons for Hysteresis

(1) Penetration of wetting liquid into pores during advancing contact angle measurements.
(2) Surface roughness: The first and rear edges meet the solid with the same intrinsic angle θ_0. The macroscopic angles θ_A and θ_R vary significantly at the front and the rear of the drop. This is illustrated in Figure 11.11.

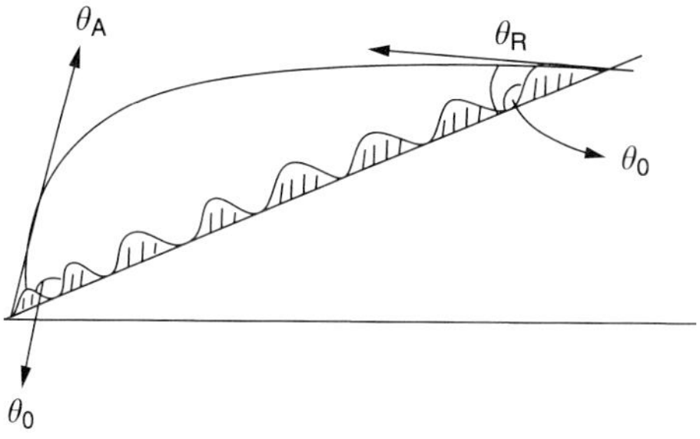

Fig. 11.11. Representation of a drop profile on a rough surface.

11.8.1.1 Wenzel's Equation [15]

Wenzel [15] considered the true area of a rough surface A (which takes into account all the surface topography, peaks and valleys) and the projected area A' (the macroscopic or apparent area). A roughness factor r can be defined as

$$r = \frac{A}{A'} \tag{11.39}$$

r is >1; the higher the value of r the higher the roughness of the surface.

The measured contact angle θ (the macroscopic angle) can be related to the intrinsic contact angle θ_0 through r,

$$\cos\theta = r\cos\theta_0 \tag{11.40}$$

Using Young's equation,

$$\cos\theta = r\left(\frac{\gamma_{SV} - \gamma_{SL}}{\gamma_{LV}}\right) \tag{11.41}$$

If $\cos\theta$ is negative on a smooth surface ($\theta > 90°$) it becomes more negative on a rough surface; θ becomes larger and surface roughness reduces wetting. If $\cos\theta$ is positive on a smooth surface (θ, 90°), it becomes more positive on a rough surface; θ is smaller and surface roughness enhances wetting.

11.8.1.2 Surface Heterogeneity

Most real surfaces are heterogeneous, consisting of patches (islands) that vary in their degrees of hydrophilicity/hydrophobicity. As the drop advances on such a heterogeneous surface, the edge of the drop tends to stop at the boundary of the island. The advancing angle will be associated with the intrinsic angle of the high contact angle region (the more hydrophobic patches or islands). The receding angle will be associated with the low contact angle region, i.e. the more hydrophilic patches or islands.

If the heterogeneities are small compared with the dimensions of the liquid drop, one can define a composite contact angle. Cassie [16, 17] considered the maximum and minimum values of the contact angles and used the following simple expression,

$$\cos\theta = Q_1\cos\theta_1 + Q_2\cos\theta_2 \tag{11.42}$$

Q_1 is the fraction of surface having contact angle θ_1 and Q_2 is the fraction of surface having contact angle θ_2; θ_1 and θ_2 are the maximum and minimum contact angles respectively.

11.9
Critical Surface Tension of Wetting and the Role of Surfactants

A systematic way of characterizing "wettability" of a surface was introduced by Fox and Zisman [18]. The contact angle exhibited by a liquid on a low-energy surface is largely dependent on the surface tension of the liquid γ_{LV}. For a given substrate

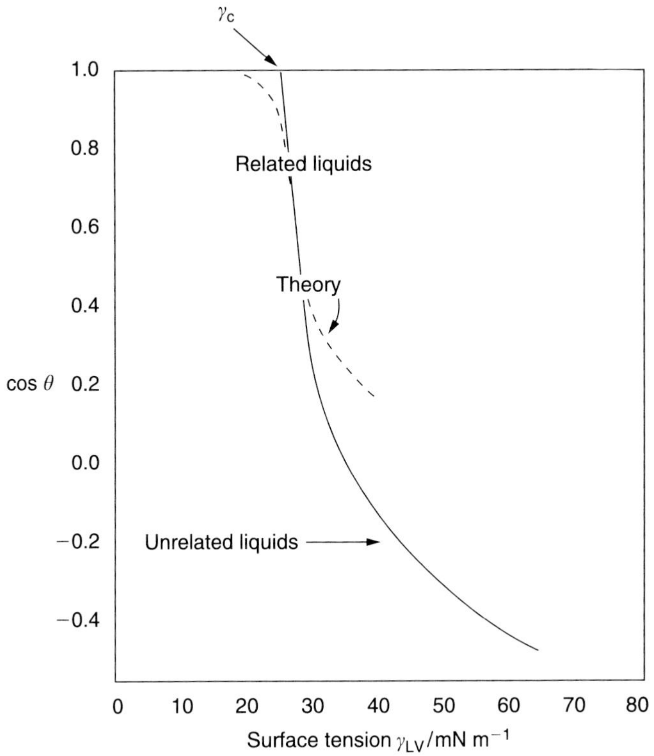

Fig. 11.12. Variation of cos θ with γ_{LV} for related and unrelated liquids on PTFE.

and a series of related liquids (such as n-alkanes, siloxanes or dialkyl ethers) cos θ is a linear function of the liquid surface tension γ_{LV}. Figure 11.12 illustrates this for several related liquids on polytetrafluoroethylene (PTFE). The figure also shows the results for unrelated liquids with widely ranging surface tensions; the line broadens into a band which tends to be curved for high surface tension polar liquids.

The surface tension at the point where the line cuts the cos $\theta = 1$ axis is known as the critical surface tension of wetting. γ_c is the surface tension of a liquid that would just spread on the substrate to give complete wetting.

The above linear relationship can be represented by the following empirical equation,

$$\cos\theta = 1 + b(\gamma_{LV} - \gamma_c) \tag{11.43}$$

High-energy solids such as glass and poly(ethylene terphthalate) have high critical surface tensions ($\gamma_c > 40$ mN m^{-1}). Lower energy solids such as polyethylene have lower γ_c ($\sim$31 mN m^{-1}). The same applies to hydrocarbon surfaces such as paraffin wax. Very low energy solids such as PTFE have lower γ_c, of the order of

18 mN m^{-1}. The lowest known value is ~6 mN m^{-1}, obtained using condensed monolayers of perfluorolauric acid.

11.9.1
Theoretical Basis of the Critical Surface Tension

The value of γ_c depends to some extent on the set of liquids used to measure it. Zisman [18] described γ_c as "a useful empirical parameter" whose relative values act as one would expect of the specific surface free energy of the solid, γ_{os}.

Several authors were tempted to identify γ_c with γ_s or γ_1^d. Good and Girifalco [10, 11] suggested the following expression for the contact angle,

$$\cos\theta = -1 + 2\phi\left(\frac{\gamma_S}{\gamma_{LV}}\right)^{1/2} - \frac{\pi_{SV}}{\gamma_{LV}} \tag{11.44}$$

where π_{SV} is the surface pressure of the liquid vapour adsorbed at the solid/liquid interface.

With $\pi_{SV} = 0$ and $\cos\theta = 0$,

$$\gamma_{SL} = \gamma_{LV} = \phi^2\gamma_S = \gamma_c \tag{11.45}$$

For non-polar liquids and solids $\phi \sim 1$ and $\gamma_S \sim \gamma_c$.

Fowkes [12] obtained the following equation for the contact angle of a liquid on a solid substrate,

$$\cos\theta = -1 + \frac{2(\gamma_S^d\gamma_{LV}^d)^{1/2}}{\gamma_{LV}} - \frac{\pi_{SV}}{\gamma_{LV}} \tag{11.46}$$

Again putting $\cos\theta = 1$ and $\pi_{SV} = 0$,

$$\gamma_{SL} = \gamma_{LV} = (\gamma_{LV}^d\gamma_S^d)^{1/2} = \gamma_c \tag{11.47}$$

Equations (11.44) and (11.46) predict that if $\pi_{SV} = 0$, a plot of $\cos\theta$ versus γ_{LV} should give a straight line with intercept $(\gamma_c)^{-1/2}$ on the $\cos\theta = 1$ axis. The experimental results seem to support this prediction. Thus, for non-polar solids, $\gamma_c = \gamma_S$, provided $\pi_{SV} = 0$, i.e. there is no adsorption of liquid vapour on the substrate. The above condition is unlikely to be satisfied when $\theta = 0$.

11.10
Effect of Surfactant Adsorption

Surfactants lower the surface tension of the liquid, γ_{LV}, and they also adsorb at the solid/liquid interface, lowering γ_{SL}. The adsorption of surfactants at the liquid/air interface can be easily described by the Gibbs adsorption equation [5],

$$\frac{d\gamma_{LV}}{dC} = -2.303\Gamma RT \tag{11.48}$$

where C is the surfactant concentration (mol dm^{-3}) and Γ is the surface excess (amount of adsorption in mol m^{-2}).

Γ can be obtained from surface tension measurements using solutions with various molar concentrations (C). From a plot of γ_{LV} versus log C one can obtain Γ from the slope of the linear portion of the curve just below the critical micelle concentration (c.m.c.).

The adsorption of surfactant at the solid/liquid interface also lowers γ_{SL}. From Young's equation,

$$\cos\theta = \frac{\gamma_{SV} - \gamma_{SL}}{\gamma_{LV}} \tag{11.49}$$

Surfactants reduce θ if either γ_{SL} or γ_{LV} or both are reduced (when γ_{SV} remains constant). Smolders [19] obtained an equation for the change of contact angle with surfactant concentration by differentiating Young's equation with respect to ln C at constant temperature,

$$\frac{d(\gamma_{LV}\cos\theta)}{d\ln C} = \frac{d\gamma_{SV}}{d\ln C} - \frac{d\gamma_{SL}}{d\ln C} \tag{11.50}$$

Using the Gibbs equation,

$$\sin\theta\left(\frac{d\theta}{dlC}\right) = RT(\Gamma_{SV} - \Gamma_{SL} - \Gamma_{LV}\cos\theta) \tag{11.51}$$

Since $\gamma_{LV}\sin\theta$ is always positive, $d\theta/d\ln C$ will always have the same sign as the right-hand side of Eq. (11.51) and three cases may be distinguished:

(1) $(d\theta/d\ln C) < 0$, $\Gamma_{SV} < \Gamma_{SL} + \Gamma_{LV}\cos\theta$; addition of surfactant improves wetting.
(2) $(d\theta/d\ln C) = 0$, $\Gamma_{SV} = \Gamma_{SL} + \Gamma_{LV}\cos\theta$; no effect.
(3) $(d\theta d\ln C) > 0$, $\Gamma_{SV} > \Gamma_{SL} + \Gamma_{LV}\cos\theta$; addition of surfactant causes dewetting.

11.11 Measurement of Contact Angles

11.11.1 Sessile Drop or Adhering Gas Bubble Method

Figure 11.13 gives a schematic representation of a sessile drop on a flat surface and an air bubble resting on a solid surface.

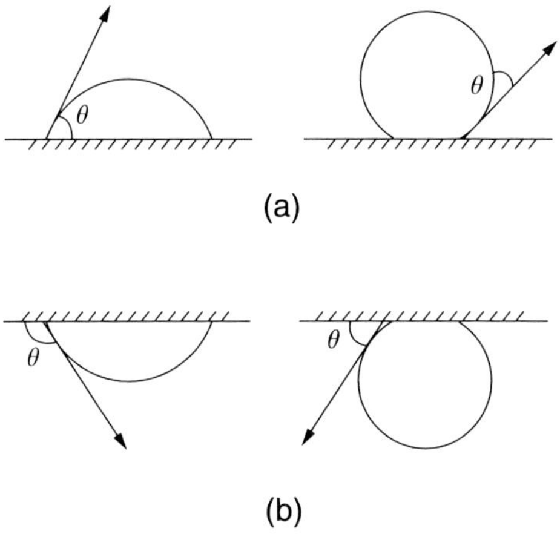

Fig. 11.13. Scheme of the sessile drop (a) and air bubble (b) resting on a surface.

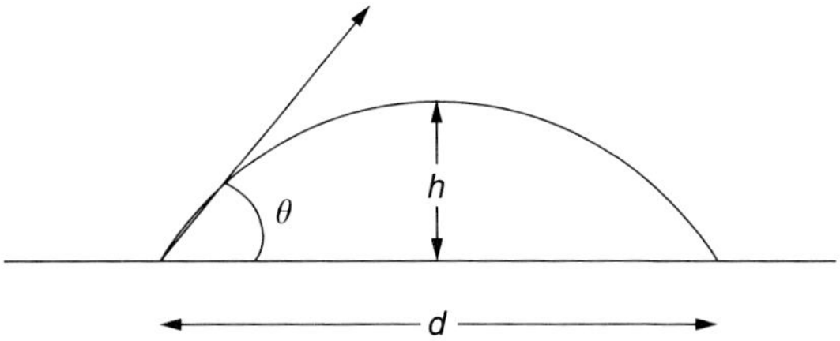

Fig. 11.14. Drop profile for calculation of contact angle.

The contact angle can be measured using a telescope fitted with a goniometer eye piece. Alternatively, it can be measured by taking a photograph or using image analysis.

The accuracy of measurement is $\pm 2°$ for θ between 10° and 160°. For $\theta < 10°$ or $> 160°$, uncertainty is higher and θ can be calculated from the drop profile (applicable to drops $< 10^{-4}$ ml). This is schematically shown in Figure 11.14.

$$\tan\left(\frac{\theta}{2}\right) = \frac{2h}{d} \tag{11.52}$$

$$\frac{d^3}{V} = \frac{24 \sin^3 \theta}{\pi(2 - 3\cos\theta + \cos^3\theta)} \tag{11.53}$$

Care must be taken for kinetic effects and evaporation.

11.11.2
Wilhelmy Plate Method

The substrate in the form of a thin plate is attached to an electrobalance to measure the force. Two procedures may be applied: The plate is allowed to touch the

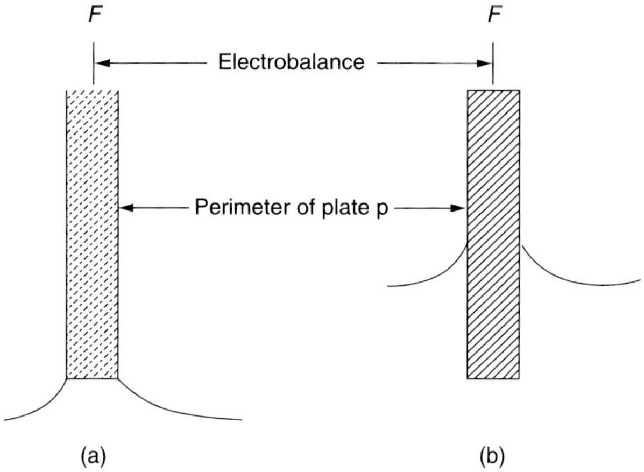

Fig. 11.15. Scheme of the Wilhelmy plate technique for measuring the contact angle. (a) Zero net depth, (b) finite depth.

surface of the liquid in question (i.e. with zero net immersion) or it is allowed to penetrate through the liquid with a finite depth of immersion. This is schematically illustrated in Figure 11.15.

In the case of Figure 11.15(a), the force on the plate F is given by

$$F = (\gamma_{LV} \cos \theta) p \tag{11.54}$$

where p is the plate perimeter.

For Figure 11.15(b), the force is given by

$$F = (\gamma_{LV} \cos \theta) p - \Delta\rho g V \tag{11.55}$$

$\Delta\rho$ is the density difference between the plate and the liquid and V is the volume of liquid displaced.

The above method is convenient and allows one to measure θ as a function of time. Also, θ_A and θ_R can be determined by raising and lowering the liquid in the vessel (using a lab jack).

11.11.3
Capillary Rise at a Vertical Plate

Instead of measuring the capillary pull (the Wilhelmy plate method) one can measure the capillary rise h at a vertical plate,

$$\sin \theta = 1 - \frac{\Delta\rho g h^2}{2\gamma_{LV}} \tag{11.56}$$

Both the Wilhelmy plate and capillary rise methods require a knowledge of γ_{LV}. This may cause uncertainty with surfactant solutions (adsorption alters both γ_{LV} and θ). By combining Eqs. (11.54) and (11.56), one can eliminate γ_{LV} to obtain θ,

$$\cos\theta = \frac{4\Delta m \Delta\rho h^2 p}{4(\Delta m)^2 + p^2(\Delta\rho)^2 h^4} \tag{11.57}$$

where Δm is the weight of the measuring plate.

Alternatively, θ can be eliminated to obtain γ_{LV},

$$\gamma_{LV} = \left(\frac{\Delta mg}{p}\right)^2 \frac{1}{\Delta\rho g h^2} + \frac{\Delta\rho g h^2}{4} \tag{11.58}$$

Thus, by combining the Wilhelmy plate with the capillary rise methods one can obtain θ and γ_{LV} simultaneously.

11.11.4
Tilting Plate Method

Figure 11.16 shows this schematically. The plate can be rotated around an axis normal to the plane of the page, until the liquid meniscus on one side becomes flat. The angle between the plate and the liquid meniscus is the contact angle.

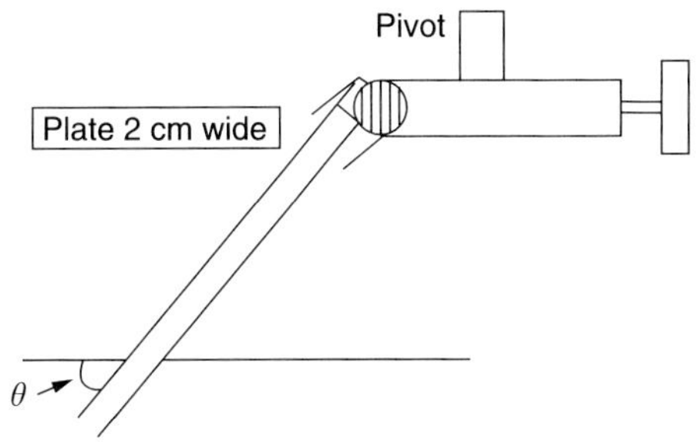

Fig. 11.16. Representation of the tilting plate method.

11.11.5
Capillary Rise or Depression Method

The rise (or depression) h of a liquid inside a partially wetted capillary ($\theta < 90°$) with radius r is related to the liquid surface tension and contact angle by the following equation, which gives the capillary pressure, Δp,

$$\Delta p = \Delta\rho h g = \frac{2\gamma_{LV}\cos\theta}{r} \tag{11.59}$$

11.12
Dynamic Processes of Adsorption and Wetting

Most technological processes (spraying, coating, etc.) work under dynamic conditions and improvement of their efficiency requires the use of surfactants that lower the liquid surface tension γ_{LV} under these dynamic conditions. The interfaces involved (e.g. droplets formed in a spray or impacting on a surface) are freshly formed and have only a small effective age of some seconds or even less than a millisecond.

The most frequently used parameter to characterise the dynamic properties of liquid adsorption layers is the dynamic surface tension (which is time dependent quantity). Techniques should be available to measure γ_{LV} as a function of time (ranging from a fraction of a millisecond to minutes and hours or even days).

To optimise the use of surfactants, polymers, and mixtures of them, specific knowledge of their dynamic adsorption behaviour rather than equilibrium properties is of great interest [20]. It is, therefore, necessary to describe the dynamics of surfactant adsorption at a fundamental level.

11.12.1
General Theory of Adsorption Kinetics

The first physically sound model for adsorption kinetics was derived by Ward and Tordai [21]. It is based on the assumption that the time dependence of surface or interfacial tension, which is directly proportional to the surface excess Γ (mol m^{-2}), is caused by diffusion and transport of surfactant molecules to the interface. This is referred to as "the diffusion-controlled adsorption kinetics model". The interfacial surfactant concentration at any time t, $\Gamma(t)$, is given by Eq. (11.60),

$$\Gamma(t) = 2\left(\frac{D}{\pi}\right)^{1/2} (c_0 t^{1/2} - \int_0^{t^{1/2}} c(0, t-\tau)\, \mathrm{d}(\tau)^{1/2} \tag{11.60}$$

where D is the diffusion coefficient, c_0 is the bulk concentration and τ is the thickness of the diffusion layer.

The above diffusion-controlled model assumes transport by diffusion of the surface active molecules to be the rate-controlled step. The so-called "kinetic controlled model" is based on the transfer mechanism of molecules from solution to the adsorbed state and vice versa [20].

Figure 11.17 gives a schematic picture of the interfacial region, showing three main states: (1) adsorption when the surface concentration Γ is lower than the equilibrium value Γ_0; (2) the equilibrium state when $\Gamma = \Gamma_0$ and (3) desorption when $\Gamma > \Gamma_0$.

The transport of surfactant molecules from the liquid layer adjacent to the interface (subsurface) is simply determined by molecular movements (in the absence of forced liquid flow). At equilibrium, i.e. when $\Gamma = \Gamma_0$, the flux of adsorption is

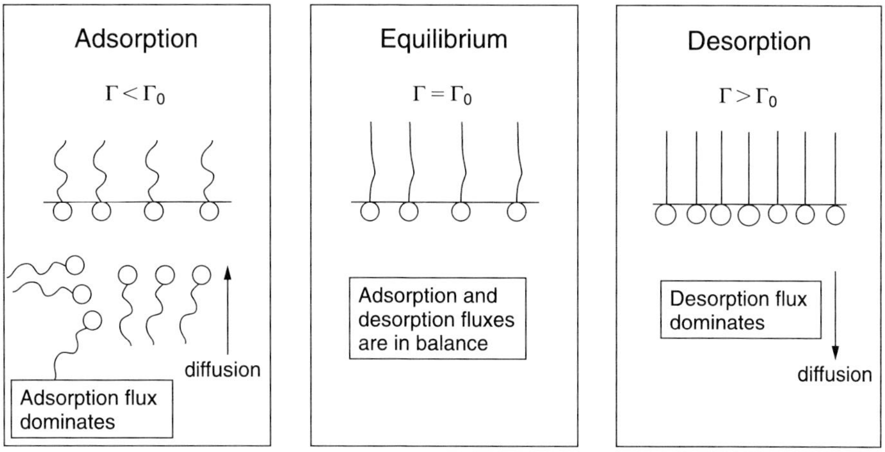

Fig. 11.17. Representation of the fluxes of adsorbing surfactant molecules near a liquid interface in the absence of forced liquid flow [20].

equal to the flux of desorption. Clearly when $\Gamma < \Gamma_0$, the flux of adsorption predominates, whereas when $\Gamma > \Gamma_0$, the flux of desorption predominates [20].

In the presence of liquid flow, the situation becomes more complicated due to the creation of a surface concentration gradients [20]. These gradients, described by the Gibbs dilational elasticity [5], initiate a flow of mass along the interface in the direction of the higher surface or interfacial tension (Marangoni effect). This situation can happen, for example, if an adsorption layer is compressed or stretched (Figure 11.18).

A qualitative model that can describe adsorption kinetics is given by Eq. (61), which affords a rough estimate and results from Eq. (11.60) when the second term on the right-hand side is neglected.

$$\Gamma(t) = c_0 \left(\frac{Dt}{\pi} \right)^{1/2} \tag{11.61}$$

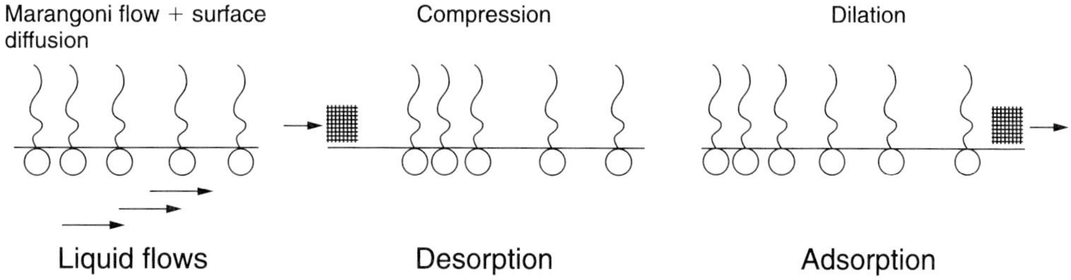

Fig. 11.18. Scheme of surfactant transport at the surface and in the bulk of a liquid [20].

An equivalent equation to Eq. (11.61) has been derived by Paniotov and Petrov [22],

$$c(0,t) = c_0 - \frac{2}{(D\pi)^{1/2}} \int_0^{t^{1/2}} \frac{\mathrm{d}\Gamma(t-\tau)}{\mathrm{d}t} \mathrm{d}\tau^{1/2} \tag{11.62}$$

Hansen [23] as well as Miller and Lukenheimer [24] gave numerical solutions to the integrals of Eqs. (11.60) and (11.62) and obtained a simple expression using a Langmuir isotherm,

$$\Gamma(t) = \Gamma_\infty \frac{c(0,t)}{a_\mathrm{L} + c(0,t)} \tag{11.63}$$

where a_L is the constant in the Langmuir isotherm (mol m^{-3})

The corresponding equation for the variation of surface tension γ with time (the Langmuir–Szyszowski equation) is as follows,

$$\gamma = \gamma_0 + RT\Gamma_\infty \ln\left(1 - \frac{\Gamma(t)}{\Gamma_\infty}\right) \tag{11.64}$$

Figure 11.19 gives a calculation based on Eqs. (11.62–11.64), with different values of c_0/a_L [20].

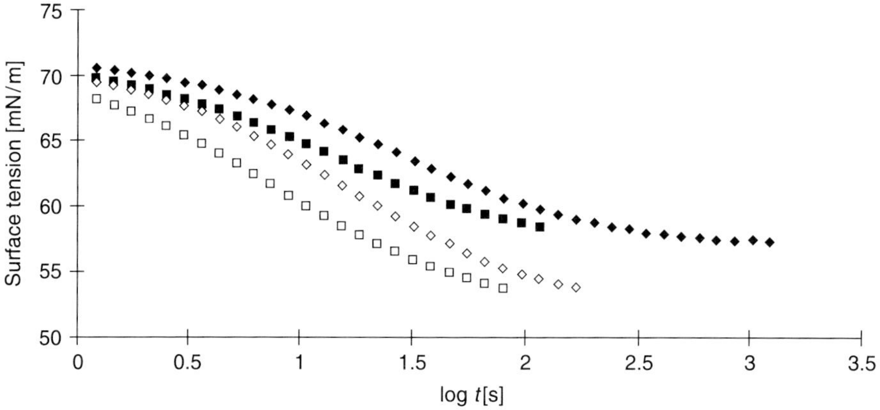

Fig. 11.19. Surface tension–log t curves calculated on the basis of Eqs. (62) to (64) with different values of c_0/a_L – values of a Langmuir isotherm $\Gamma_\infty = 4 \times 10^{-10}$ mol cm^{-2}; $a_\mathrm{L} = 5 \times 10^{-9}$ mol cm^{-3}; $c_0 = 2 \times 10^{-8}$ (◆,■), 3×10^{-3} (□,◇) mol cm^{-3}; $D = 1 \times 10^{-5}$ (◆,◇), 2×10^{-5} (■,□) cm^2 s^{-1} [20].

11.12.2
Adsorption Kinetics from Micellar Solutions

Surfactants form micelles above the critical micelle concentration (c.m.c.) of different sizes and shapes, depending on the nature of the molecule, temperature, electrolyte concentration, etc. (see Chapter 2). The dynamic nature of micellisation can be described by two main relaxation processes, τ_1 (the life time of a monomer in a micelle) and τ_2 (the life time of the micelle, i.e. complete dissolution into monomers).

The presence of micelles in equilibrium with monomers influences the adsorption kinetics remarkably. After a fresh surface has been formed surfactant monomers are adsorbed, resulting in a concentration gradient of these monomers. This gradient will be equalised by diffusion to re-establish a homogeneous distribution. Simultaneously, the micelles are no longer in equilibrium with monomers within the range of the concentration gradient. This leads to a net process of micelle dissolution or rearrangement to re-establish the local equilibrium. Consequently, a concentration gradient of micelles results, which is equalised by diffusion of micelles [20].

Based on the above concepts, one would expect that the ratio of monomers c_1 to micelles c_m, the aggregation number n, the rates of micelle formation (k_f) and micelle dissolution (k_d) will influence the rate of the adsorption process. Figure 11.20 gives a schematic picture of the kinetic process in the presence of micelles.

The above picture shows that to describe the kinetics of adsorption, one must take into account the diffusion of monomers and micelles as well as the kinetics of micelle formation and dissolution. Several processes may take place (Figure 11.21). Three main mechanisms may be considered, namely formation–

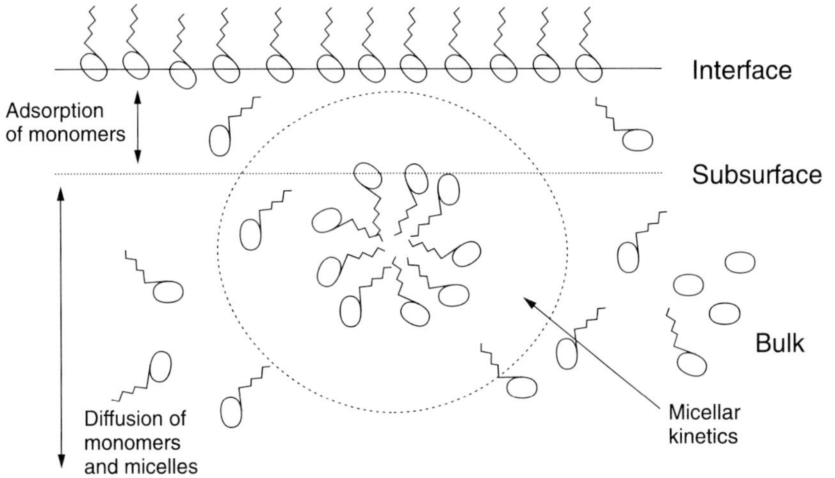

Fig. 11.20. Representation of the adsorption process from a micellar solution.

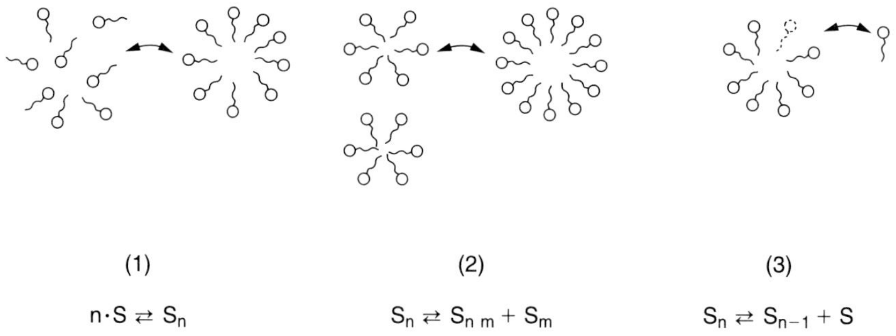

Fig. 11.21. Scheme of micelle kinetics.

dissolution (Figure 11.21(1)), rearrangement (Figure 11.21(2)) and stepwise aggregation-dissolution (Figure 11.21(3)). To describe the effect of micelles on adsorption kinetics, one should know several parameters such as the micelle aggregation number and the rate constants of micelle kinetics [25].

11.12.3
Experimental Techniques for Studying Adsorption Kinetics

The two most suitable techniques for studying adsorption kinetics are the drop volume method and the maximum bubble pressure method. The first method can obtain information on adsorption kinetics in the range of seconds to some minutes. It has the advantage of measurement both at the air/liquid and liquid/liquid interfaces. The maximum bubble pressure method allows one to obtain measurements in the millisecond range, but it is restricted to the air/liquid interface. Both techniques are described below.

11.12.3.1 Drop Volume Technique

A schematic representation of the drop volume apparatus [26] is given in Figure 11.22. A metering system in the form of a motor-driven syringe allows the formation of the liquid drop at the tip of a capillary, which is positioned in a sealed cuvette. The cuvette is either filled with a small amount of the measuring liquid, to saturate the atmosphere, or with a second liquid in the case of interfacial studies. A light barrier arranged below the forming drop enables the detection of drop-detachment from the capillary. Both the syringe and the light barriers are computer-controlled, allowing fully automatic operation of the set-up. The syringe and the cuvette are temperature controlled by a water jacket, which makes interfacial tension measurements possible in the temperature range 10–90 °C.

As mentioned above, the drop volume method is of dynamic character and it can be used for adsorption processes in the time interval of seconds up to some minutes. At small drop time, the so-called hydrodynamic effect has to be considered [27]. This gives rise to apparently higher surface tension. Kloubek et al. [28] used an empirical equation to account for this effect,

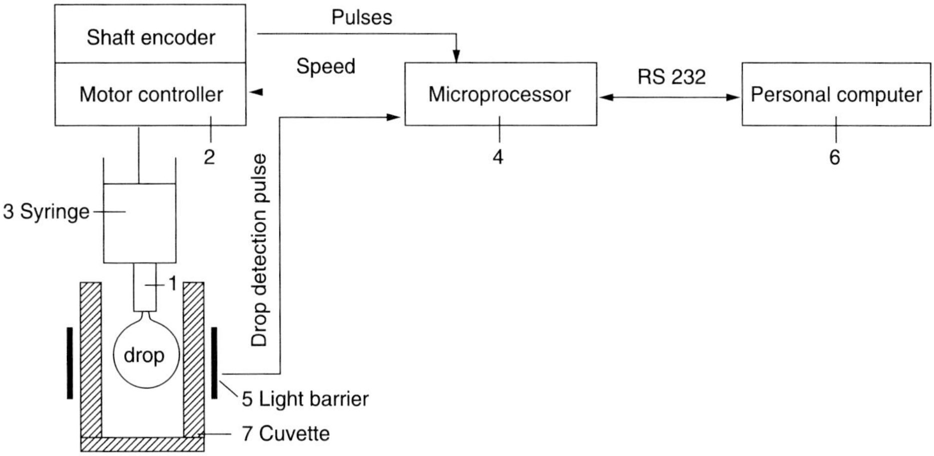

Fig. 11.22. Representation of the drop volume apparatus [26].

$$V_e = V(t) - \frac{K_v}{t} \tag{11.65}$$

V_e is the unaffected drop volume and $V(t)$ is the measured drop volume. K_v is a proportionality factor that depends on surface tension γ, density difference $\Delta\rho$ and tip radius r_{cap}.

Miller [20] obtained the following equation for the variation of $V(t)$ with time,

$$V(t) = V_e + t_0 F = V_e\left(1 + \frac{t_0}{t - t_0}\right) \tag{11.66}$$

where F is the liquid flow per unit time that is given by

$$F = \frac{V(t)}{t} = \frac{V_e}{t - t_0} \tag{11.67}$$

The drop volume technique is limited in its application. Under conditions of fast drop formation and larger tip radii, drop formation shows irregular behaviour.

11.12.3.2 **Maximum Bubble Pressure Technique**

This is the most useful technique for measuring adsorption kinetics at short times, particularly if correction for the so-called "dead time", τ_d, is made. The dead time is simply the time required to detach the bubble after it has reached its hemispherical shape. Figure 11.23 gives a scheme of the principle of maximum bubble pressure, showing the evolution of a bubble at the tip of a capillary as well as the variation of pressure p in the bubble with time.

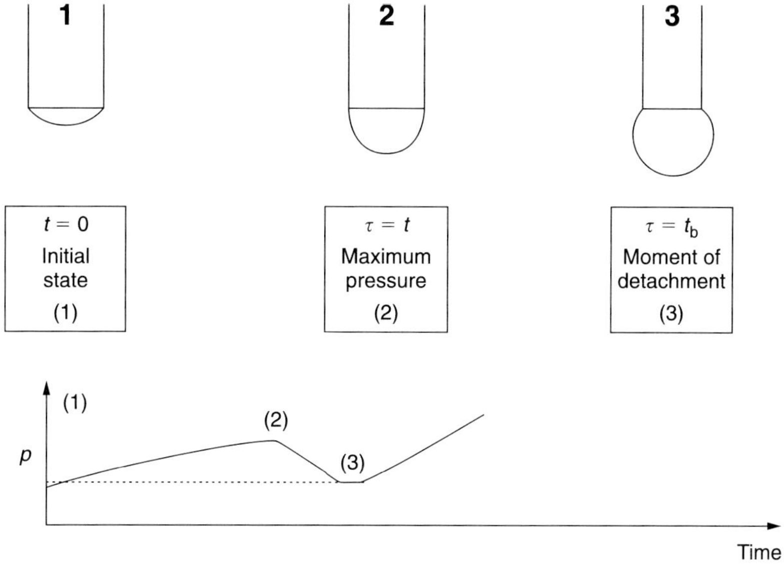

Fig. 11.23. Scheme of bubble evolution and pressure change with time.

At $t = 0$ (initial state), the pressure is low (note that the pressure is equal to $2\gamma/r$; since r of the bubble is large when p is small). At $t = \tau$ (smallest bubble radius, which is equal to the tube radius) p reaches a maximum. At $t = \tau_b$ (detachment time) p decreases since the bubble radius increases. The design of a maximum bubble pressure method for high bubble formation frequencies (short surface age) requires the following: (1) Measurement of bubble pressure; (2) measurement of bubble formation frequency; and (3) estimation of surface lifetime and effective surface age. The first problem can be solved easily if the system volume (which is connected to the bubble) is large enough in comparison with the bubble separating from the capillary. In this case, the system pressure is equal to the maximum bubble pressure. The use of an electric pressure transducer to measure bubble formation frequency presumes that pressure oscillations in the measuring system are distinct enough and this satisfies (2). Estimation of the surface life time and effective surface age, i.e. (3), requires estimation of the dead time τ_d. The set-up for measuring the maximum bubble pressure and surface age is illustrated in Figure 11.24. The air coming from the micro-compressor flows first through the flow capillary. The air flow rate is determined by measuring the pressure difference at both ends of the flow capillary with the electric transducer PS_1. Thereafter, the air enters the measuring cell and the excess air pressure in the system is measured by a second electric sensor (PS_2). In the tube, which leads the air to the measuring cell, a sensitive microphone is placed.

The measuring cell is equipped with a water jacket for temperature control, which simultaneously holds the measuring capillary and two platinum electrodes,

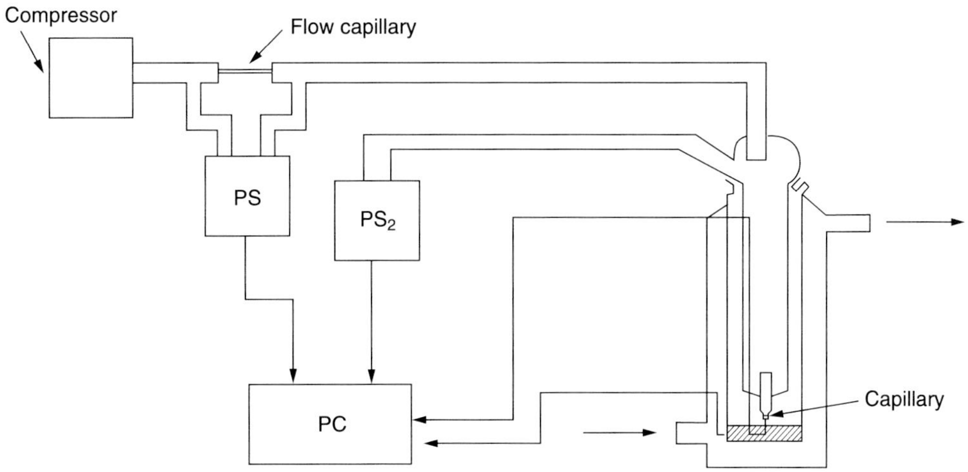

Fig. 11.24. Maximum bubble pressure apparatus.

one of which is immersed in the liquid under study and the second is situated exactly opposite the capillary and controls the size of the bubble. Electric signals from the gas flow sensor PS_1 and pressure transducer PS_2, the microphone and the electrodes, as well as the compressor are connected to a personal computer, which operates the apparatus and acquires the data.

The value of τ_d, equivalent to the time interval necessary to form a bubble of radius R, can be calculated using Poiseuille's law,

$$\tau_d = \frac{\tau_b L}{Kp}\left(1 + \frac{3r_{ca}}{2R}\right) \tag{11.68}$$

K is given by Poiseuille's law,

$$K = \frac{\pi r^4}{8\eta l} \tag{11.69}$$

η is the gas viscosity, l is the length, L is the gas flow rate and r_{ca} is the radius of the capillary.

The calculation of dead time τ_d can be simplified when taking into account the existence of two gas flow regimes for the gas flow leaving the capillary: bubble flow regime when $\tau > 0$ and jet regime when $\tau = 0$ and hence $\tau_b = \tau_d$. Figure 11.25 shows a typical dependence of p on L.

On the right-hand side of the critical point the dependence of p on L is linear, in accordance with Poiseuille law. Under these conditions,

$$\tau_d = \tau_b \frac{L p_c}{L_c p} \tag{11.70}$$

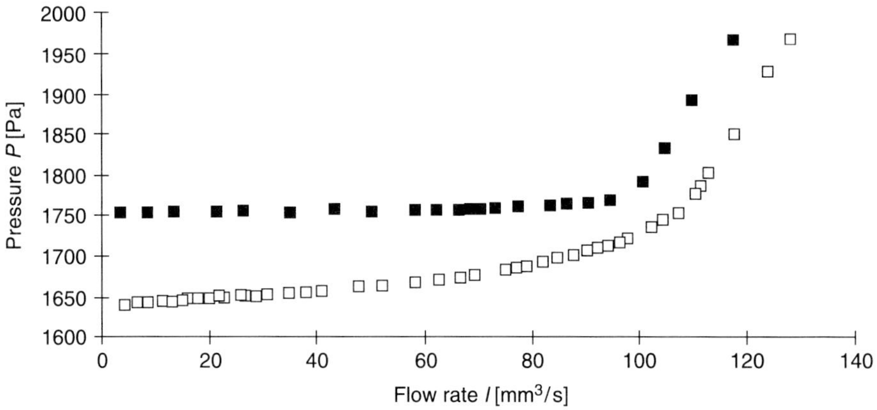

Fig. 11.25. Dependence of p on the gas flow rate L for water (■) and water–glycerine mixture (2:3) (□) at 30 °C, $r = 0.0824$ mm.

where L_c and p_c are related to the critical point, and L and p are the actual values of the dependence left of the critical point.

The surface life-time can be calculated from

$$\tau = \tau_b - \tau_d = \tau_b \left(1 - \frac{Lp_c}{L_c p}\right) \tag{11.71}$$

The critical point dependence on p and L can be easily located and is included in the software of the computer program.

The surface tension in the maximum bubble pressure method is calculated using the Laplace equation,

$$p = \frac{2\gamma}{r} + \rho hg + \Delta p \tag{11.72}$$

where ρ is the density of the liquid, g is the acceleration due to gravity, h is the depth the capillary is immersed in the liquid and Δp is a correction factor to allow for hydrodynamic effects.

11.13 Wetting Kinetics

In many experimental situations, a contact angle that changes with time leads inevitably to the movement of the wetting line [29]. The result is a dynamic contact angle. This situation arises from non-equilibrium conditions and should be distinct from contact angle hysteresis. In dynamic contact angle measurements, some movement of the wetting line is unavoidable and this must cause a temporary disturbance from equilibrium. The state of adsorption at the solid/vapour (SV) and solid/liquid (SL) interfaces will not be the same. Displacement of fluid 2 by fluid 1 (liquid on the solid by the vapour of the liquid) will result in destruction of the SL interface and creation of a fresh SV interface.

If the sorption–desorption processes or the accompanying transport processes to and from the various interfaces are slow compared with the rate of displacement, a non-equilibrium contact angle will result. If the external driving force is now removed, the contact angle will relax to its equilibrium value. If the relaxation processes are sufficiently slow relative to the experimental time scale, a non-equilibrium angle may persist in an apparently stable system.

It is difficult to distinguish the above situation from contact angle hysteresis. This behaviour may arise from the slow molecular orientation following the movement of the wetting line. If contact angle measurements are to be given equilibrium significance, great care must be taken to ensure that equilibrium has been reached. This is particularly the case with surfactant solutions, where there may be no ready mechanism by which a non-volatile solute (the surfactant molecules) can reach the solid/vapour interface, except by prior contact with the solution.

Kinetic factors that may cause contact angle variation can also arise from penetration of the liquid into the solid surface. The penetration of non-polar surfaces by water has been commonly cited in the literature [29]. Zisman [18] reported a relationship between molecular size and the extent of contact angle hysteresis. The observed behaviour simply reflects the non-attainment of a uniformly penetrated surface during the observation time. In extreme cases, penetration may cause swelling of the substrate. The wetting line will rest along a labile ridge and contact angle variation should be similar to that observed when the wetting line is pinned at an edge.

Contact angle variability can be attributed to several factors and proper attention should be paid to attainment of equilibrium. One should be cautious in ascribing the variability to permanent features of the system such as surface roughness or intrinsic heterogeneities. With practical systems all these factors have to be considered.

11.13.1
Dynamic Contact Angle

This is usually ascribed to contact angles that change with time or those associated with moving wetting lines. Contact angles usually depend on the speed and direction of the wetting line displacement. Contact angles are velocity dependent. The advancing contact angle increases and the receding contact angle decreases with increasing rate of displacement. Figure 11.26 illustrates the variation of the contact angle θ with the velocity of wetting v for a system of a poorly wetting liquid with a moderate degree of contact angle hysteresis. θ_A and θ_R are the advancing and receding angles when $v = 0$.

The above behaviour is clearly related to contact angle hysteresis in that both imply thermodynamic irreversibility. In the static case, this is attributed to spontane-

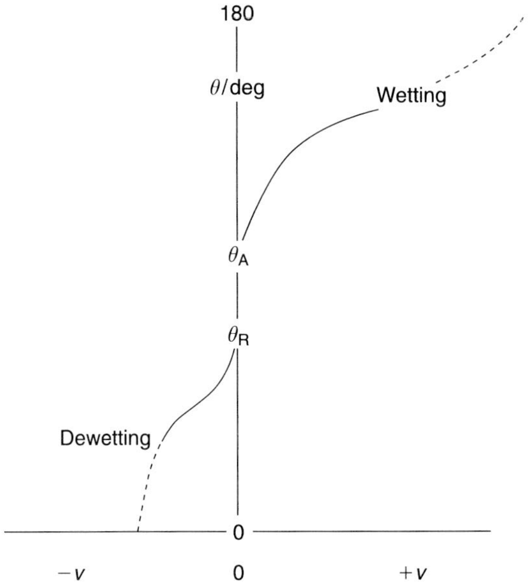

Fig. 11.26. Variation of contact angle (θ) with the velocity of wetting v.

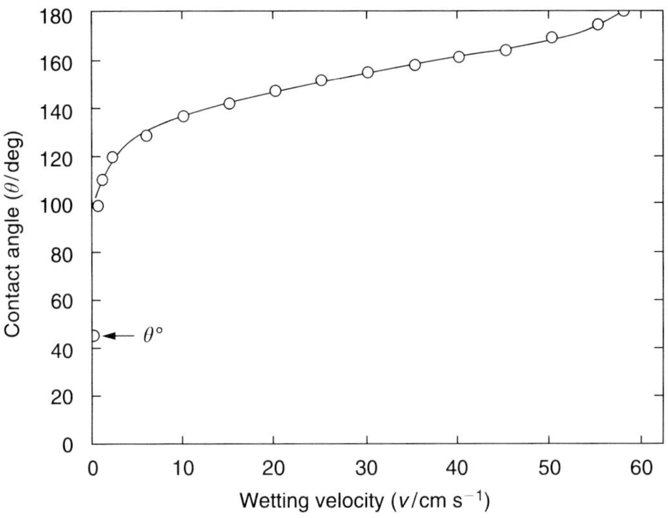

Fig. 11.27. Velocity dependence of the contact angle [30].

ous transitions between metastable equilibrium states. In the dynamic case, the interface may fail to attain any kind of equilibrium in the time available.

Figure 11.27 shows the velocity dependence of the contact angle [30] for aqueous glycerine solutions (0.0456 Pa s) on mylar polyester tape.

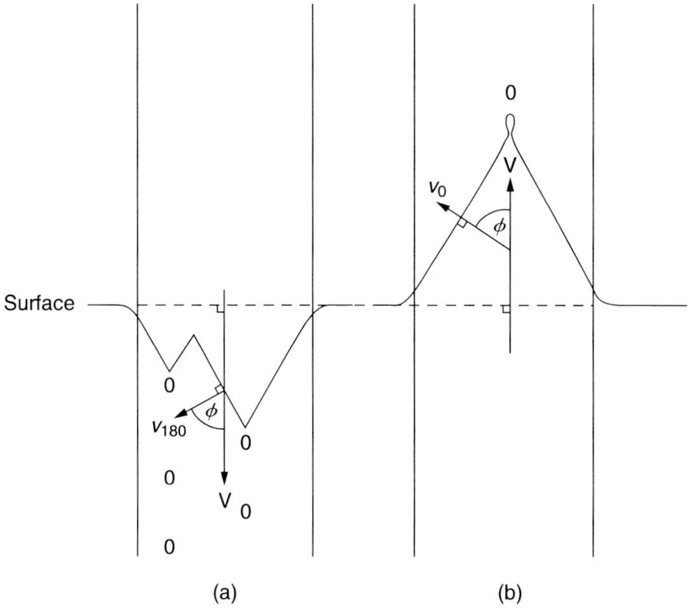

Fig. 11.28. Formation of sawtooth wetting lines, showing entrainment from trailing vertices: (a) wetting, (b) dewetting.

At low rates of wetting, the advancing contact angle appears to be a steep function of the wetting velocity. As v increases, the slope first decreases and then increases again as θ approaches 180° at v_{180}. At still higher velocities, the wetting line develops a sawtooth shape (Figure 11.28a) and air is entrained from the trailing vertices.

Much of the interest in dynamic contact angles lies in maximising the wetting velocity at the onset of entrainment. In liquid coating operations, entrainment of air leads to patchy or uneven coatings. In petroleum recovery, entrainment of crude oil by gas or water flood may reduce the efficiency of recovery. Blake and Ruschak [31] argued that the occurrence of the sawtooth wetting line shows that, for a given system, v_{180} is the maximum velocity at which a wetting line can advance normal to itself. If an attempt is made to wet a solid at some velocity $v > v_{180}$, the wetting line must lengthen and slant at some angle ϕ relative to its orientation at velocities below v_{180} such that

$$\cos\phi = \frac{v_{180}}{v} \tag{11.73}$$

Similarly for dewetting, if an attempt is made to dewet a solid (for which $\theta_R 2 > 0$) at a velocity greater than that at which θ becomes zero, v_0, then the wetting line again consists of at least two straight segments slanted at an angle ϕ relative to the normal orientation. In this case,

$$\cos \phi = \frac{v_0}{v} \tag{11.74}$$

and drops of liquid or a continuous rivulet may be entrained from the trailing vertices.

Blake and Ruschak [31] reported data in good agreement with these simple relationships. The very steep velocity dependence of the contact angle as $v \rightarrow 0$ strongly suggests the kinetic origin of apparent contact angle hysteresis.

11.13.2
Effect of Viscosity and Surface Tension

The velocity-dependence of the contact angle increases with increase in the liquid viscosity η and decrease in the surface tension γ. Several authors found an increase in θ with increasing the capillary number Ca $(= \eta v/\gamma)$. The higher the viscosity, the higher is the velocity-dependence of θ and the lower the value of v_{180}. Viscous forces tend to oppose wetting. The lower the surface tension, the higher is the velocity-dependence of the contact angle and the lower v_{180} is. Experimental results showed that surfactants can improve the rate of wetting.

11.14
Adhesion

Adhesion is defined as "the state in which two surfaces are held together by interfacial forces which may consist of valence forces or interlocking action or both" [32]. An adhesive is a material capable of holding materials together by surface attachment. Valence forces are not required in order that excellent adhesion be obtained since the van der Waals forces are in themselves sufficient to cause excellent adhesion.

The early adhesives were natural products (e.g. glues, starch, natural resins) but most modern adhesives are based on synthetic polymers (e.g. polyacrylates). In adhesion, two materials come sufficiently close for strong interaction to occur. The interface is considered as the zone between the interacting substances, which is somctimes referred to as the interphase.

The main forces responsible for adhesion are van der Waals, which for convenience are considered to be made of three main contributions: Dipole–dipole interaction (Keesom force), dipole-induced–dipole interaction (Debye force) and London dispersion force. A hydrogen-bonding force can also be included in the interaction.

For an adhesive joint to be formed, the adhesive must move into the bond area and remain there until the bond is completely established. The rheology of the polymer systems used as adhesives plays a significant part in adhesion. For adhesion to occur, intimate interaction of the adhesive and substrate must occur and this requires adequate wetting and spreading of the adhesive. Provided intimate intermolecular contact is achieved at the interface, London or dispersion forces are sufficiently strong that good adhesive performance is observed. Poor adhesive performance must be associated with limited interfacial contact.

Electrostatic forces arising from contact or junction potentials between the adhesive layer and the substrate may contribute significantly to the forces required to rupture the bonds. Poor performance results from non-uniform contact or low contact density.

In this section, I will briefly discuss (1) intermolecular forces responsible for adhesion; (2) mechanisms of adhesion: molecular contact at the interface, molecular configuration and conformation; (3) wetting and thermodynamic equilibrium; (4) bond character and adhesive performance; (5) the role of diffusion; (6) the electrostatic contribution; and (7) the locus of adhesive failures.

11.14.1
Intermolecular Forces Responsible for Adhesion

Understanding the intermolecular forces responsible for adhesion and cohesion is quite important. The elastic constants, the plastic deformation, the presence of flaws are functions of inter- and intramolecular forces and steric hindrances to rotation in the various molecules. It is important to relate the magnitude of intermolecular forces between molecular species and a particular substrate to the relative adhesive strengths.

Predictions based solely on intermolecular forces do not always agree with experiment, and other factors such as deformation behaviour and wetting must be considered. The amorphous fraction of polymers used for adhesion is free to conform to the structure of the substrate [33].

11.14.2
Interaction Energy Between Two Molecules

A typical energy–distance curve between two molecules is schematically represented in Figure 11.29, usually referred to as the 6–12 Lennard-Jones potential [34], where A is a constant and r_0 is the intermolecular distance at equilibrium (Eq. 11.75).

$$U = -\frac{A}{2}\left(\frac{2}{r^6} - \frac{r_0^6}{r^{12}}\right) \tag{11.75}$$

The net force F (note that $F = -\mathrm{d}U/\mathrm{d}l$, where l is the distance) is given by

$$F = -\frac{\mathrm{d}U}{\mathrm{d}r} = 6A\left(\frac{1}{r^3} - \frac{r_0^6}{r^{13}}\right) \tag{11.76}$$

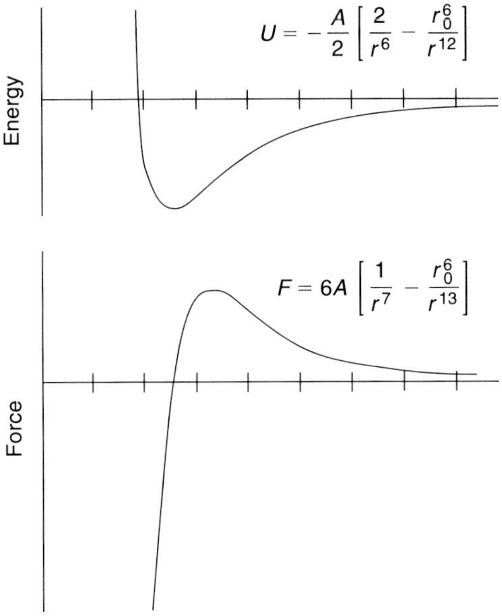

Fig. 11.29. Energy–distance and force–distance curves between two molecules according to the Lennard-Jones potential [34].

Tab. 11.2. Energies of various bonds.

Type of bond	*Energy (kcal mol^{-1})*
Chemical bonds	
Ionic	140–250
Covalent	15–170
Metallic	27–83
Intermolecular force	
Hydrogen bonds	up to 12
Dipole–dipole	up to 5
Dispersion	up to 10
Dipole-induced–dipole	up to 0.5

The force–distance curve is also shown in Figure 11.28. Typical values of the interaction energies of various bonds are summarised in Table 11.2.

11.14.2.1 **Ionic Bonds**

A pure ionic bond is one in which a positive and a negative ion attract each other, and the energy of interaction U_{ionic} is

$$U_{\text{ionic}} = \frac{q^+ q^-}{r} \tag{11.77}$$

where q^+ and q^- are the charges on the ions.

In a medium of relative permittivity ε, U_{ionic} is given by

$$U_{\text{ionic}} = \frac{q^+q^-}{\varepsilon\varepsilon_0 r} \tag{11.78}$$

where ε_0 is the permittivity of free space.

11.14.2.2 **Covalent Bonds**

In covalent bonds between like atoms, the electrons are shared between the two atoms and there is accumulation of electrons in the space between the two atoms. The potential energy function for covalent bonds is often quite well represented by a Lennard-Jones function. Covalent bonds can bridge an interface, and in this case one may consider the interfacial region as a distinct phase. When two atoms have different degrees of electronegativity, the bond between them will have partial ionic character. If the atomic orbitals of the two atoms overlap, the bond will also have partial covalent character.

11.14.2.3 **Metallic Bonds**

An ideal metal crystal consists of a regular array of "ion cores" with the valence electrons nearly free to move throughout the whole mass, as the conduction electrons. In addition to the interaction of conduction electrons, there is mutual interaction of ion cores for each other: Repulsive, ion cores have a net positive charge; attractive, dispersion force of electrons in the ion cores.

11.14.2.4 **Dipole–Dipole Forces**

Covalent bonds are often partially ionic (with an electric dipole moment μ),

$$\mu = ql \tag{11.79}$$

q is the magnitude of the charge separated (as q^+ and q^-) by the distance l.

The unit of μ is Debye (1×10^{-18} e.s.u.), if the dipoles are composed of charges equivalent to 1e separated by 1 Å; μ is determined from the dielectric constant. Examples of dipole moments for some liquids are given in Table 11.3.

The interaction energy U between two dipoles depends on their orientation (vectors),

$$U = -\left(\frac{\mu_1\mu_2}{r^3}\right) f \text{ (angles of rotation)} \tag{11.80}$$

Tab. 11.3. Dipole moments of several liquids.

Liquid	*Water*	*Ethanol*	*n-hexane*	*Acetonitrile*
μ	1.85	1.70	0.00	3.40

If thc dipoles are freely rotating,

$$U = -\left(\frac{\mu_1\mu_2}{r^3}\right) \tag{11.81}$$

11.14.2.5 Hydrogen Bonds

The conditions necessary for hydrogen bonds are: (1) A highly electronegative atom such as O, Cl, F, N or a strongly electronegative group as $-CCl_3$ or $-CN$, with a hydrogen atom attached. (2) Another highly electronegative atom, which may or may not be in a molecule of the same species as the first atom or group (e.g. with a lone pair of electrons). Examples include liquid HCl, which consists of chains of H–Cl$\cdots$H–Cl. Water forms a three-dimensional network of hydrogen bonds (the reason for the high dielectric constant of water). Table 11.4 gives the hydrogen bond energies for some systems.

Tab. 11.4. Hydrogen bond energies for some systems.

System	*Energy (kcal mol^{-1})*
HF–HF	6.3–7.0
HCN–HCN	3.3–4.4
H_2O–H_2O	3.4–5.0
ROH–ROH	3.2–6.2

11.14.2.6 Lewis Acid–Lewis Base Bonding

A Lewis acid is an electron acceptor; a Lewis base is an electron donor. Lewis acid–Lewis base bonding may be strong, with a bond energy of 13–15 kcal mol^{-1}, e.g. anhydrous $AlCl_3$. Lewis acid–Lewis base bonding may also be weak, bond energy less than ca. 7 kcal mol^{-1}, e.g. $I_2 + C_6H_6$, $E = 1.72$ kcal mol^{-1}; I_2 + mesitylene, $E = 7.2$ kcal mol^{-1}.

11.14.2.7 Dipole-Induced–Dipole Forces (Debye)

A dipole in a neighbouring molecule may provide enough electric field to polarize a previously symmetrical non-polar molecule (Scheme 11.1).

The energy of interaction between a dipolar molecule and a non-polar molecule is given by

$$U = -\frac{\mu_1^2\alpha_2}{r^6} \tag{11.82}$$

where α_2 is the polarizability of the non-polar molecule.

$-+ \xrightarrow[\mu=0]{} -+ \quad -+$

$\mu_1 \qquad \mu_2 \qquad \mu_i > 0$

Scheme 11.1

11.14.2.8 London Dispersion Force

This is a universal interaction force that exists between polar and non-polar molecules. It arises from charge fluctuations between the atoms or molecules. Consider two molecules, a and b. At any instant the electrons in molecule a have a definite configuration, so that a has an instantaneous dipole moment. This instantaneous dipole in molecule a induces a dipole in molecule b. The interaction between the two dipoles results in a force of attraction between the two molecules. The dispersion force is the instantaneous force of attraction averaged over all instantaneous configurations of the electrons in molecule a. The magnitude of the instantaneous dipole is proportional to the polarizability (α).

For two molecules of type 1, the London dispersion interaction energy U is given by

$$U = -\frac{3}{4}\left(\frac{\alpha_1^2 C_1}{r_{11}^6}\right) \tag{11.83}$$

where C_1 is a constant; $C_1 = h\nu_0$, where h is the Planck's constant and ν_0 is the characteristic frequency; $h\nu_0$ is approximately equal to the ionisation potential I.

Table 11.5 gives a comparison of the relative magnitude of the forces.

Tab. 11.5. Comparison of the relative magnitudes of the forces.

	μ (D)	*α (Å³)*	*I (eV)*	*Dipole–dipole*	*Dipole/induced dipole*	*London dispersion*
Ar	0	1.63	15.8	–	–	50
CH_4	0	2.58	13.1	–	–	97
CO_2	0	2.86	13.9	–	–	136
Na	0	29.7	5.15	–	–	5340
HCl	1.03	2.63	12.8	18.6	5.4	106
H_2O	1.85	1.48	12.7	190	10	33

11.14.2.9 Forces Across an Interface

If the potential energy functions are known for all the atoms or molecules in a system, and the spatial distribution of all the atoms is also known, it is possible in principle to sum all these forces across an interface. This means that the "ideal" force or energy of cohesion of a single phase or the force or energy of adhesion across the interface can be calculated. If the deformational behaviour of the separate phases is known, one can predict the practical adhesive strength of the system.

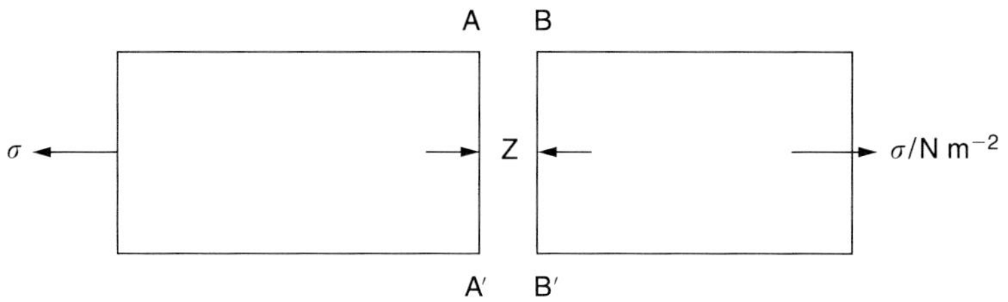

Fig. 11.30. Separation of two bodies.

In practice the above calculations are not possible: Intermolecular potentials are not known (particularly for asymmetric molecules). The microscopic structure of a "real" surface is not known.

Approximate theories exist to calculate the interfacial tension from a knowledge of the cohesive energy of the two pure phases. The specific free energy of cohesion of a pure phase is given by

$$\Delta F^{c} = -2\gamma \tag{11.84}$$

where γ is the specific free energy (J m^{-2}) of the pure phase. Note that J m^{-2} = N m^{-1} so that γ is the surface tension of the pure phase.

ΔF^{c} is the energy required to separate two bodies from a distance Z_0 to infinity. The two bodies attract each other by a force σ in N m^{-2}. This is illustrated in Figure 11.30.

$$-\Delta F_1^{c} = \int_{Z_0}^{\infty} \sigma_1 \, dZ \tag{11.85}$$

Z_0 is the value of Z at equilibrium and σ is the surface energy (N m^{-2}).

The free energy of adhesion ΔF_{12}^{σ} for an interface between phases 1 and 2 is given by

$$\Delta F_{12}^{\sigma} = \gamma_1 + \gamma_2 - \gamma_{12} \tag{11.86}$$

γ_1 and γ_2 are the surface tensions of the pure phases and γ_{12} is the interfacial tension.

Similarly,

$$-\Delta F_{12}^{\sigma} = \int_{Z_0}^{\infty} \sigma_{12} \, dZ \tag{11.87}$$

Good and co-workers [35] derived an expression for the interfacial tension γ_{12} by relating the ratio of free energy of adhesion to the geometric mean of the free energies of cohesion,

$$-\frac{\Delta F^{a}}{(\Delta F_1^{c}\Delta F_2^{c})^{1/2}} = \phi = \frac{(\gamma_1 + \gamma_2 - \gamma_{12})}{2(\gamma_1\gamma_2)^{1/2}} \tag{11.88}$$

$$\gamma_{12} = \gamma_1 + \gamma_2 - 2\phi(\gamma_a\gamma_2)^{1/2} \tag{11.89}$$

Since

$$\frac{-\Delta F^{a}}{2(\gamma_1\gamma_2)^{1/2}} = \phi \tag{11.90}$$

then

$$\phi(\gamma_1\gamma_2)^{1/2} = \int_{Z_0}^{\infty} \sigma_{12}\,dZ = \frac{3C_{12}}{16Z_0} \tag{11.91}$$

If $\phi = 1$, then the ideal adhesion strength will be intermediate between the cohesive strengths of the two separate phases,

$$\sigma_1^{c} < \sigma_{12}^{a} < \sigma_2^{c} \tag{11.92}$$

Adhesion failure occurs if

$$\frac{\sigma_{12}^{a}}{\sigma_1^{c}} < 1 \tag{11.93}$$

or

$$\phi < \left(\frac{\gamma_1}{\gamma_2}\right)^{1/2}\left[\frac{(Z_{0,1} + Z_{0,2})}{2Z_{0,1}}\right] \tag{11.94}$$

11.14.3 Mechanism of Adhesion

A distinction should be made between adhesive performance and adhesion [36]. Adhesive performance comprises experimentally determined values in terms of behaviour under specified conditions: Gross sample geometry, topography of the interface, chemical nature of the materials, mechanical responses of the solid and viscoelastic phases, strain rates, strain geometry and temperature.

Adhesion is the "thermodynamic work of adhesion", i.e. intrinsic interaction across the interface. Several theories of adhesion have been suggested and these may be classified into three categories: (1) Adsorption theories, (2) diffusion theories and (3) electrostatic theories.

11.14.3.1 Molecular Contact at the Interface

Provided sufficiently intermolecular contact is achieved at the interface, the London or dispersion forces are sufficiently strong that good adhesion performance should be observed. Strong interfacial attractions should exist at small intermolecular separations. For two parallel plates, the force of attraction is $\sim 10^9$–10^{11} dyne cm^{-2} at $Z_0 \sim$ 4–5 Å, $\sim 10^8$–10^{10} dyne cm^{-2} at $Z_0 \sim 10$ Å, and $\sim 10^5$–10^7 dyne cm^{-2} at $Z_0 \sim 100$ Å. Thus, poor adhesion performance must be associated with limited interfacial contact.

Evidence for the importance of interfacial contact was obtained from studying the temperature dependence of adhesive performance, which is influenced by the way in which the adhesive bonds are formed. The force required to peel a polymer film of poly(n-butyl methacrylate) from steel was found to depend on the temperature at which the polymer was bonded: Polymer bonded at 100 °C did not achieve interfacial equilibrium; polymer bonded at 150 °C reached interfacial equilibrium. Plasticizers added to poly(methyl methacrylate) bonded to steel also influenced adhesion.

The three main reasons given for limited intermolecular contact are given below.

Molecular Configuration and Conformation Mismatch between atoms in the substrate and adhesive phases has important implications on the magnitude of interfacial interactions. This mismatch may lead to diminished interaction, as suggested by Good and co-workers [35] for two phases 1 and 2,

$$\gamma_{12} = \gamma_1 + \gamma_2 - 2\phi(\gamma_1\gamma_2)^{1/2} \tag{11.95}$$

ϕ depends on the nature of the interaction across the interface and within each phase and upon the configuration of the molecules at the interface. It is a measure of the misfit of molecules of unequal size.

High molecular weight organic polymers are used in most adhesive applications. The conformation of such molecules has an appreciable effect on the number of effective interfacial contacts.

Wetting and Thermodynamic Equilibrium Wetting may be considered as the process of achieving interfacial contact. In practice, coatings or adhesives are mechanically spread over the solid substrate; partial or complete wetting of the substrate must be considered. The free energy accompanying wetting of a substrate is

$$\Delta F = \gamma_{SL}A_{SL} - (\gamma_{SV}A_{SV} + \gamma_{LV}A_{LV}) \tag{11.96}$$

where γ represents interfacial tensions and A the area of each interface. Using Young's equation,

$$\gamma_{SV} = \gamma_{SL} + \gamma_{LV} \cos\theta \tag{11.97}$$

Combining Eqs. (11.96) and (11.97),

$$\Delta F^* = -\gamma_{LV}\left[1 + \left(\frac{A_{SV}}{A_{LV}}\right) \cos\theta\right] \tag{11.98}$$

Unless $(A_{SV}/A_{LV})\cos\theta$ is negative, i.e. $\theta > 90°$, then wetting is spontaneous (ΔF^* is negative). Good adhesives are not necessarily those with a zero or low contact angle. As long as $\theta < 90°$, wetting is spontaneous.

Dynamic Effects The driving force leading to wetting is the capillary pressure, which results from γ_{LV} of the liquid adhesive in the interstices of the substrate. To achieve high capillary pressures, a high γ_{LV} is required for substrates with high surface energy.

For low energy solids, Zisman [18] defined an optimum γ_{LV} for each substrate,

$$\gamma_{LV}(\text{optimum}) = \tfrac{1}{2}\left(\gamma_c + \frac{1}{b}\right) \tag{11.99}$$

where γ_c is the critical surface tension of wetting and b is function of the interaction between liquid and solid.

The rate of wetting is determined largely by the viscosity of the liquid adhesive. Interfacial topography plays a secondary role through its influence on the resistance to flow. The flow rate in the interfacial interstices is directly proportional to the dimensions of these interstices.

11.14.3.2 Adhesives with More Than One Component

All practical adhesives are composed of mixtures of materials (solutions, dispersions, etc.). Selective adsorption of one of the components will lead to changes in the interfacial energy that can produce dramatic changes in wetting rates. Phase separation may also occur in adhesive mixtures. If two separate phases exhibit appreciably different viscosities, the more fluid phase would wet the substrate and fill the interstices even though wetting by the more viscous phase must obtain at equilibrium. If a component of the adhesive composition is volatile, the wetting problem is more complex. The wetting rate will decrease as the solvent evaporates.

Many polymer–solvent mixtures become viscoelastic solids while as much as 15% solvent remains. If wetting by the polymer is not achieved by the time this state is reached, further wetting cannot be achieved within a reasonable time.

11.14.3.3 Role of Diffusion

Adhesion between dissimilar polymers as well as "autoadhesion" is best explained on the basis of diffusion. The adsorption theory cannot account for the high adhesion between non-polar polymers or explain the decreasing adhesion to a polar substrate with increasing polarity of the adhesive. Evidence for the role of diffusion came from studies on the effect of contact time and temperature on the bonding rate, and the influence of polymer molecular weight and structure. Diffusion occurs in cases involving self-adhesion or "autoadhesion". The diffusion mechanism occurs only in regions in which the phases are in contact.

11.14.3.4 Electrostatic Contributions to Adhesive Performance

Electrostatic forces arising from contact or junction potentials between the adhesive layer and the substrate may contribute significantly to the forces required to rupture bonds. Deryaguin and co-workers [37] calculated the energy required to peel polymer films from metal and glass substrates by assuming that, over the small distances involved, the stripping is tantamount to separating the charged plates of an infinite parallel capacitor. The high work of peeling cannot be attributed to the action of van der Waals forces or chemical bonds.

11.14.3.5 Locus of Adhesive Failures

When complete wetting between the adhesive and substrate is achieved, interfacial separation is impossible. This argument follows if one considers the attraction constant A_{12} between two phases,

$$A_{12} = (A_{11}A_{22})^{1/2} \tag{11.100}$$

A_{11} and A_{22} are the attractive constants for the interaction between like molecules of each phase.

Since $(A_{11}A_{22})^{1/2}$ must lie between A_{11} and A_{22}, for a completely wetted interface separation is impossible. However, several factors may reduce the interaction between dissimilar molecules: (1) Disparities between the sizes of atoms or molecular groups. (2) Non-random distribution; This is especially the case with polymers. (3) Difference in polarity.

In the above cases,

$$A_{12} < (A_{11}A_{22})^{1/2} \tag{11.101}$$

Also, when interactions, other than London dispersion forces, are involved the departure from the geometric-mean assumption becomes greater.

Good's parameter ϕ may be used as a criterion for interfacial separation,

$$\phi\left(\frac{\Delta F_a}{\Delta F_b}\right)^{1/2} < 1 \tag{11.102}$$

where F_a is the surface free energy of the substrate and F_b is the surface free energy of the adhesive.

The above discussion explains the possible separation at the interface that leads to adhesive failure. The most satisfactory criterion for selecting an adhesive is its surface tension γ, which should be less than the critical surface tension of wetting γ_c. Measurement of the surface tension of an adhesive can be carried out using the pendent or sessile drop method (due to the high viscosity of most adhesives).

Poor performance results from non-uniform contact or low contact density. Low contact density may be obtained as a result of: (1) Molecular configuration limiting the number of effective interfacial contacts, e.g. with polymers. Control of the polymer structure and molecular weight is essential in developing good adhesives. (2)

Non-equilibrium conditions, i.e. incomplete wetting. This is particularly the case under dynamic conditions. (3) Energy effects associated with polymer morphology.

11.15
Deposition of Particles on Surfaces

The deposition of particles on surfaces is a process that is determined by long-range forces: Van der Waals attraction, electrostatic repulsion or attraction and the presence of adsorbed or grafted surfactants, polymers or polyelectrolytes (referred to as steric interaction).

In this section, I will discuss the role of van der Waals attraction and electrostatic repulsion (or attraction) on particle deposition.

The above discussion is very important for many fields in personal care applications: hair sprays and hair conditioners, foundation, creams and lotions (skin care).

The essentially keratinous bulk makeup of hair and skin (the stratum corneum, SC) is generally accepted. However, the surfaces of skin and hair are less well defined – it is generally recognised that a sebum layer is frequently present on both of these substrates. A lipid layer can also be intrinsic to the surface – the presence of hydrophobic layers affects the deposition of additives through their influence on van der Waals forces, hydrophobic forces and the like.

The complex structure of hair, which consists of four components of different functionality (cortex, medula, cuticle cells and cell membranes), requires careful investigation to study deposition. The hair surface is characterised by surface energy measurements on single hairs before and after treatment with polymer solutions or dispersions of polymer-containing cosmetic formulations. The surface energy of the intact human hair is determined by the outermost layer of the epicuticle, a low-energy, highly hydrophobic surface with low surface energy (not easily wetted by water). Deposition of hydrophilic polymers on the hair fibre surface causes a significant increase in surface energy and it becomes easily wettable.

11.15.1
Van der Waals Attraction

This is detailed in previous chapters, but for the sake of completion of this section a summary is given here. The attraction between atoms or molecules is of three types: Dipole–dipole (Keesom force), dipole-induced–dipole (Debye force) and London dispersion force.

The most important is the London dispersion attraction, which operates for polar and non-polar atoms or molecules. This attractive energy is of short range and it inversely proportional to the sixth power of the distance between the atoms or molecules.

For an assembly of atoms or molecules, the London attraction may be summed, resulting in significant attraction that operates over a large distance of separation h

between the particles. For two identical particles with radius a the attractive energy G_A is given by the simple expression

$$G_A = -\frac{Aa}{12h} \tag{11.103}$$

where A is the effective Hamaker constant that is given by

$$A = (A_{11}^{1/2} - A_{22}^{1/2})^2 \tag{11.104}$$

where A_{11} is the Hamaker constant of the particles and A_{22} is the effective Hamaker constant of the medium.

The Hamaker constant of any substance is given by

$$A = \pi q^2 \beta \tag{11.105}$$

where q is the number of atoms or molecules per unit volume and β is the London dispersion constant.

For two different particles or particle-and-surface with Hamaker constants A_{11} and A_{22} separated by a medium with a Hamaker constant A_{33}, the effective Hamaker constant A is given by

$$A = (A_{11}^{1/2} - A_{33}^{1/2})(A_{22}^{1/2} - A_{33}^{1/2}) \tag{11.106}$$

A schematic representation of the variation of G_A with h is given in Figure 11.31. G_A increases with decreasing h and reaches very high values at very small h. At extremely short h, G_A increases (Born repulsion).

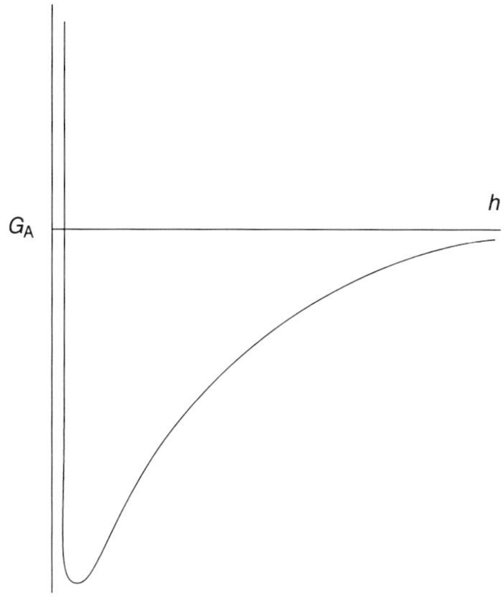

Fig. 11.31. Variation of G_A with h.

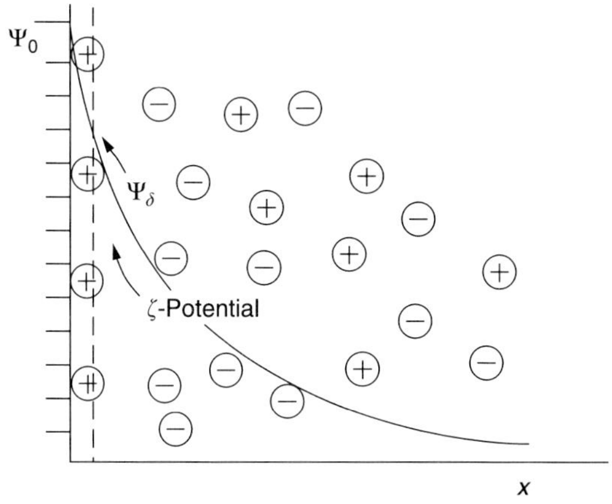

Fig. 11.32. Scheme of the double layer.

11.15.2
Electrostatic Repulsion

Electrostatic repulsion occurs as a result of the presence of an electric double layer consisting of a surface charge that is compensated by an unequal distribution of counter and co-ions. Figure 11.32 shows a schematic representation of the double layer for a negatively charged surface.

The surface potential Ψ_0 decreases linearly with Ψ_δ (the Stern Potential), which is nearly equal to the measurable zeta (ζ) potential.

The double layer extension depends on the electrolyte concentration and valency of the ions as given by $(1/\kappa)$, the "thickness of the double layer". The lower the electrolyte concentration and the lower the valency of the ions, the larger $1/\kappa$ is; for example, for 1:1 electrolyte (e.g. NaCl), $(1/\kappa) = 100$ nm at 10^{-5} mol dm^{-3}, 10 nm at 10^{-3} mol dm^{-3} and 1 nm at 10^{-1} mol dm^{-3}.

When two particles of the same double layer sign approach to a separation h that is less than twice the double layer thickness repulsion occurs, since the double layers begin to overlap. Repulsion between particles or between a particle and a surface decreases with increasing electrolyte concentration. Figure 11.33 shows this schematically, where G_{el} is plotted versus h at both low and high electrolyte concentrations.

Combination of G_A and G_{el} at various h results in the energy–distance curve is illustrated in Figure 11.34, which forms the basis of the Deryaguin–Landau–Verwey–Overbeek theory colloid stability (DLVO theory) [38].

The energy–distance curve is characterised by two minima, a shallow secondary minimum (weak and reversible attraction) and a primary deep minimum (strong and irreversible attraction).

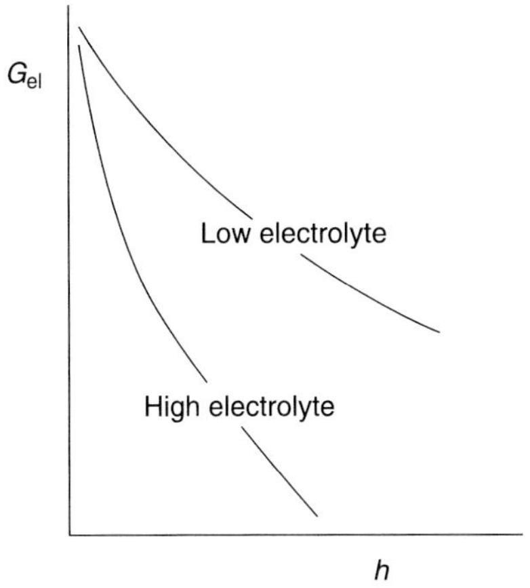

Fig. 11.33. Variation of G_{el} with h at low and high electrolyte concentrations.

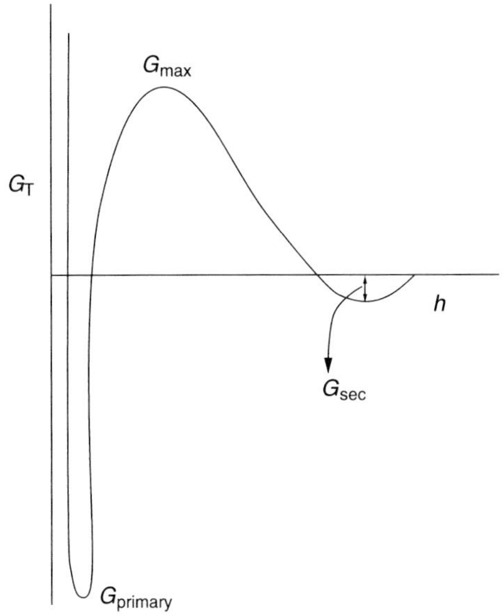

Fig. 11.34. Energy–distance curve (DLVO theory).

Particles deposited under conditions of secondary minimum will be weakly attached, whereas particles deposited under conditions of primary minimum will be strongly attached. At intermediate distances of separation, an energy maximum is obtained whose height depends on the surface or zeta potential, electrolyte concentration and valency of the ions. This maximum prevents particle deposition.

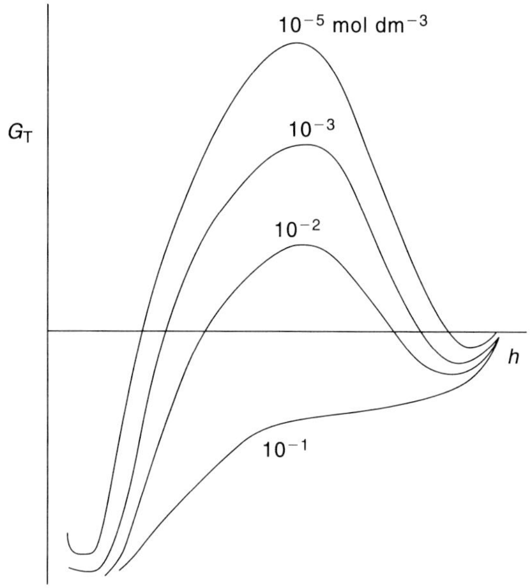

Fig. 11.35. G_T–h curves at various NaCl concentrations.

The magnitude of the energy minima and the energy maximum depends on electrolyte concentration and valency. This is illustrated in Figure 11.35 for a 1:1 electrolyte (e.g. NaCl) at various concentrations.

G_{max} clearly decreases with increasing NaCl concentration and eventually disappears at 10^{-1} mol dm^{-3}. Thus, particle deposition for particles with the same sign as the surface will increase with increasing electrolyte concentrations.

The above trend was confirmed by Hull and Kitchener [39] using a rotating disc coated with a negative film and negative polystyrene latex particles. The number of polystyrene particles deposited was found to increase with increasing NaCl concentration, reaching a maximum at $C_{NaCl} > 10^{-1}$ mol dm^{-3}. The ratio of maximum number of particles deposited N_{max} to the number deposited at any other NaCl concentration N_d (the so-called stability ratio W) was calculated and plotted versus NaCl concentration,

$$W = \frac{N_{max}}{N_d} \qquad (11.107)$$

Figure 11.36 gives such plots, which clearly show that W decreases with increasing NaCl concentration, reaching a minimum above 10^{-1} mol dm^{-3}, whereby maximum deposition occurs. Similar results were obtained by Tadros et al. [45] using a rotating cylinder apparatus (Figure 11.37).

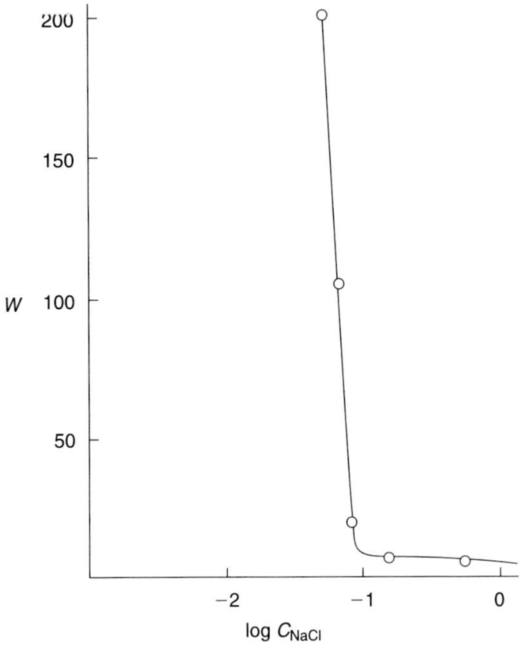

Fig. 11.36. *W* versus log[NaCl] using a rotating disc.

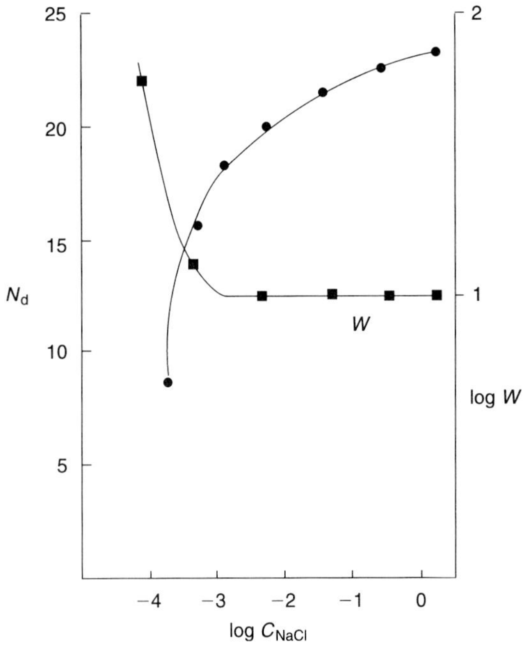

Fig. 11.37. N_d and *W* versus C_{NaCl} using a rotating cylinder.

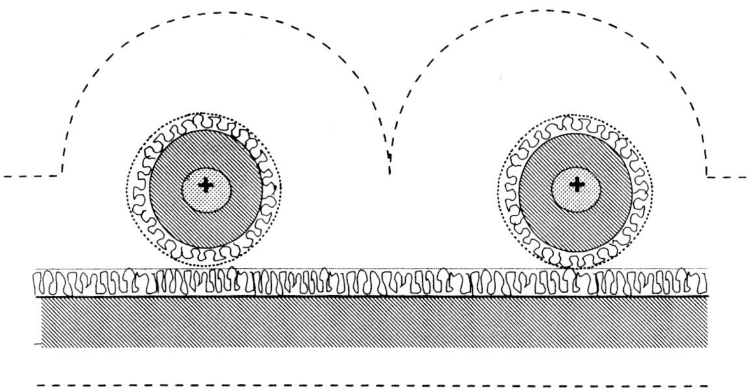

Fig. 11.38. Deposition of positively charged particles on a negatively charged surface.

The above results show that the deposition of particles on substrates of the same sign will increase with increasing electrolyte concentration. However, the situation with an oppositely charged surface to the particles being deposited is very different. In this case, attraction to the oppositely charged surface will occur, a phenomenon referred to as heteroflocculation. This is schematically illustrated in Figure 11.38 for positively charged polystyrene latex particles on a negative surface – both surfaces were covered by a nonionic polymer layer.

The effect of addition of electrolyte in this case will be opposite to that observed with surfaces of the same charge. Attraction between oppositely charged double layers will be higher at lower electrolyte concentrations. In other words, addition of electrolyte in this case will decrease deposition.

11.15.3
Effect of Polymers and Polyelectrolytes on Particle Deposition

Polymers and polyelectrolytes, both of the natural and synthetic, are commonly used in most personal care and cosmetic formulations. These materials are used as thickening agents, film formers, resinous powder and humectants. For example, thickening agents, sometimes referred to as rheology modifiers, are used in many hand creams, lotions, liquid foundations and hair sprays to maintain the product stability.

In many formulations, polymers and surfactants are present and interaction between them can produce remarkable effects. Several structures can be identified, and the aggregates produced can have profound effects on particle deposition. With many hair shampoos, conditioning agents are added that are mostly polyelectrolytes with cationic charges, which are essential for strong attachment to the negatively charged keratin surface.

For convenience, I will consider the effect of three classes of polymers on particle deposition separately: Nonionic polymers, anionic polyelectrolytes and cationic polyelectrolytes.

11.15.4
Effect of Nonionic Polymers on Particle Deposition

Nonionic polymers can be of the synthetic type such as polyvinylpyrrolidone or natural such as many polysaccharides. The role of nonionic polymers in particle deposition depends on the manner in which they interact with the surface and particle to be deposited. With many high molecular weight polymers, the chains adopt a conformation forming loops and tails that may extends several nm from the surface. If there is not sufficient polymer to fully cover the surfaces, bridging may occur, resulting in enhancement of deposition. In contrast, if there is sufficient polymer to cover both surfaces, the loops and tails provide steric repulsion, thereby reducing deposition.

One may be able to correlate particle deposition to the adsorption isotherm. Under conditions of incomplete coverage, i.e. well before the plateau value is reached, particle deposition is enhanced. Under conditions of complete coverage, one observes a reduction in deposition and at sufficient coverage deposition may be prevented altogether. Figure 11.39 shows this schematically, exhibiting the correlation of the adsorption isotherm to particle surface deposition.

The most commonly used nonionic polymers in personal care and cosmetics are polysaccharide based. Polysaccharides perform several functions in cosmetics: Rheology modifiers, suspending agents, hair conditioners, wound healing agents, moisturizing agents, emulsifying agents and emollients.

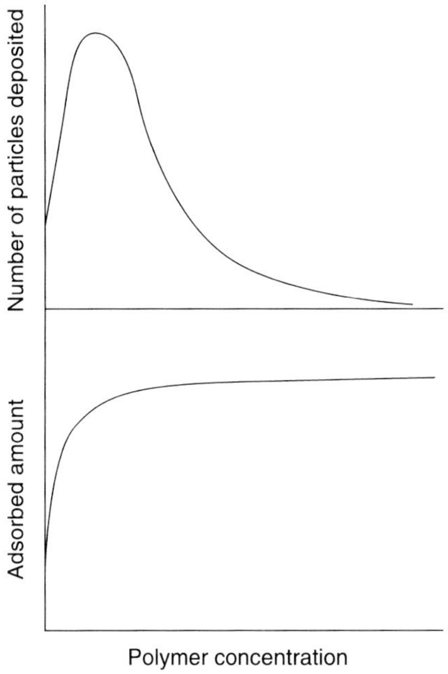

Fig. 11.39. Correlation of particle deposition with polymer adsorption.

Polysaccharides are sometimes referred to as "polyglycans" or "hydrocolloids". Most cosmetically interesting polysaccharides are primarily composed of six-membered cyclic structures known as pyranose ring (five carbon atoms and one oxygen atom). Many polysaccharides form helices, which is a tertiary spatial configuration, arranged to minimize the total energy of the polysaccharide (e.g. xanthan gum).

The behaviour of polysaccharides is critically influenced by the nature of the substituent groups bound to the individual monosaccharides (natural or synthetic). Anionic charges may also occur in natural polysaccharides and this will have a big influence on the adsorption and conformation of the polymer chain.

The effect of polysaccharides on particle deposition is rather complex, and depends on the structure of the molecule and interaction with other ingredients in the formulation.

11.15.5
Effect of Anionic Polymers on Particle Deposition

Many cosmetic formulations contain anionic polymers, mostly of the polyacrylate and polysaccharide type. The role of anionic polymers in particle deposition is complex since these polyelectrolytes interact with ions in the formulation, e.g. Ca^{2+}, as well as with the surfactants used. Two of the most commonly used anionic polysaccharides are carboxymethylcellulose and carboxymethylchitin, obtained by carboxymethylation of cellulose and chitin, respectively.

Several naturally occurring anionic polysaccharides exist: alginic acid, pectin, carrageenans, xanthan gum, hyaluronic acidic, gum exudates (gum arabic, karaya, traganth, etc.). Cross-linking sites that occur when a polyvalent cation (e.g. Ca^{2+}) causes interpolysaccharide binding are called "junction zones".

The above complexes, which may produce colloidal particles, will greatly influence the deposition of other particles in the formulation. They may enhance binding, simply by a cooperative effect in which the polysaccharide complex interacts with the particles and increases the attraction to the surface. The pH of the whole system plays a major role since it affects the dissociation of the carboxylic groups.

Many of the anionic polysaccharides and their complexes affect the rheology of the system and this has a pronounced effect on particle deposition. Any increase in the viscosity of the system will reduce the flux of the particles to the surface and this may reduce particle deposition. This reduction may be offset by specific interaction between the particles and the polyanion or its complex.

11.15.6
Effect of Cationic Polymers on Particle Deposition

Cationic polymers are the traditional conditioners for keratinous substrates, especially hair. This is because these substrates are normally negatively charged (low isoelectric point), and polycations employed, either alone or together with a conditioner, are the most predominant type. These polycationic polyelectrolytes have a pronounced effect on particle deposition due to their interaction with the substrate and the particles.

One of the earliest polycationic polymers was polyethyleneimine (PEI), which was used as a model for studying polycation uptake by hair. This polymer was withdrawn from hair products in the 1970s (for safety reasons). Later, an important class of cellulosic polycationic polymers was introduced with the trade name "Polymer JR" (Amerchol corporation) and this is widely used for hair conditioning. Other synthetic polycationic polymers from Calgon corporation are Merquat 100 (based on dimethyldiallyl amine chloride) and Merquat 550 (based on acrylamide/dimthyldiallylamine chloride).

Several naturally occurring polycationic polymers exist: Chitosan (polyglycan with cationic charges), which is positively charged at pH < 7, cationic hydroxyethyl cellulose and cationic guar gum. These polycationic polymers interact with anionic surfactants present in the formulation and at a specific surfactant concentration a rapid increase in the viscosity of the solution is observed. At higher surfactant concentrations precipitation of the polymer–surfactant complex occurs and at even higher surfactant concentration repeptisation may occur.

Figure 11.40 illustrates this for the interaction of Polymer JR-400 with varying sodium dodecyl sulphate concentration.

Clearly the above interactions will have a pronounced effect on particle deposition. In the absence of any other effects, addition of cationic polyelectrolytes can enhance particle deposition either by simple charge neutralisation or "bridging" between the particle and the surface. At high polyelectrolyte concentrations, when

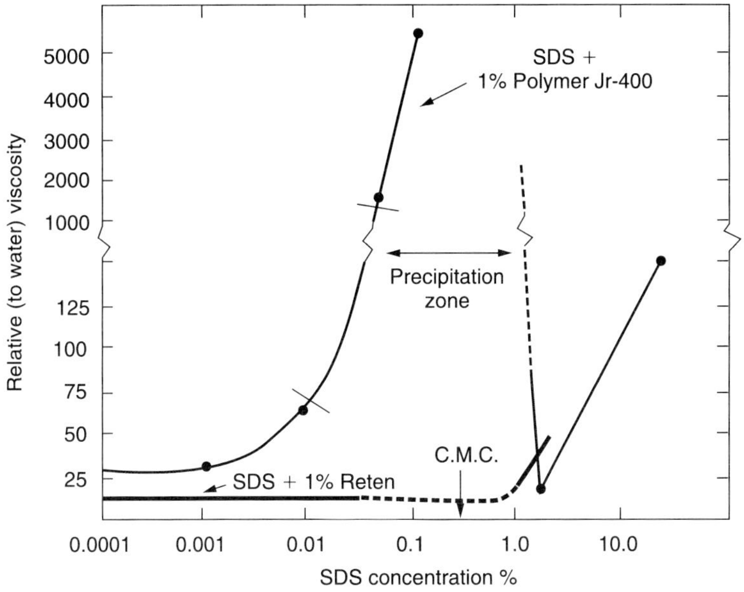

Fig. 11.40. Relative viscosity versus SDS concentration for 1% Polymer JR-400.

there are sufficient molecules to coat both particle and surface, repulsion may occur, resulting in reduction in deposition.

However, the above effects are complicated by the interaction of the polycationic polymer with surfactants in the formulation and this complicates the prediction of particle deposition. Investigations of the interactions that take place between the polycation, surfactants and other ingredients in the formulation are essential before a complete picture on particle deposition is possible.

11.16 Particle–Surface Adhesion

Adhesion is the force necessary to separate adherents; it is governed by short-range forces [38]. Adhesion is more complex than deposition and more difficult to measure. No quantitative theory is available that can describe all adhesion phenomena: Chemical and non-chemical bonds operate. Adequate experimental techniques for measuring adhesion strength are still lacking.

When considering adhesion one must consider elastic and non-elastic deformation that may take place at the point of attachment. The short-range forces could be strong, e.g. primary bonds, or intermediate, e.g. hydrogen and charge-transfer bonds.

In 1934, Deryaguin [40] considered the force of adhesion F in terms of the free energy of separation of two surfaces $[G(h_\infty) - G(h_0)]$ from a distance h_0 to infinite separation (h_∞). For the simple case of parallel plates,

$$-\int_{h_0}^{\infty} F\,\mathrm{d}h = G(h_\infty) - G(h_0) \tag{11.108}$$

F is made up of three contributions,

$$F = F_\mathrm{m} + F_\mathrm{c} + F_\mathrm{e} \tag{11.109}$$

F_m is the molecular component and consists of two parts, an elastic deformation component F_S and a surface energy component F_H,

$$F_\mathrm{m} = F_\mathrm{S} + F_\mathrm{H} \tag{11.110}$$

F_c is the component that depends on prior electrification. F_e is the electrical double layer contribution.

When a sphere adheres to a plane surface, elastic deformation occurs and one can distinguish the radius of the adhesive area r_0.

Figure 11.41 gives a schematic representation of elastic deformation. Usually $r_0/R \ll 1$, where R is the particle radius. The adhesive area can be calculated from a knowledge of the time dependence of the modulus of the sphere and the time dependence of the hardness of the plate.

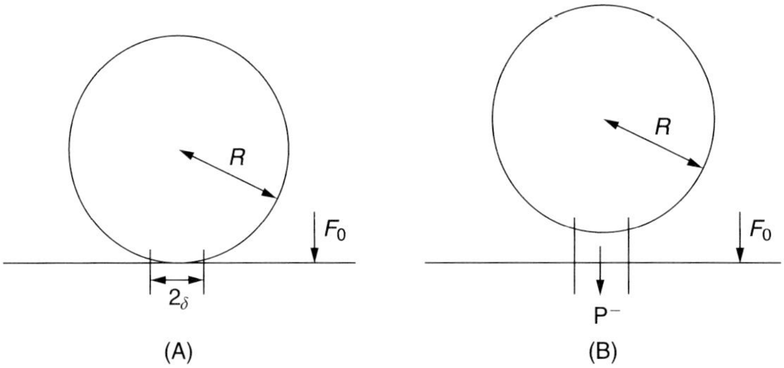

Fig. 11.41. Elastic deformation on adhesion of a sphere to a plane surface.

11.16.1 Surface Energy Approach to Adhesion

Two approaches may be applied to consider the process of adhesion.

11.16.1.1 Fox and Zisman [41] Critical Surface Tension Approach

This approach was initially used to obtain the critical surface tension of wetting of a liquid on solid substrates. Fox and Zisman [40] found that a plot of cos θ (where θ is the contact angle of a liquid drop on the substrate) versus γ_{LV} (the liquid surface tension) for several related liquids gives a straight line, which when extrapolated to cos $\theta = 1$ gives the critical surface tension of wetting γ_c. This is shown in Figure 11.42 for several solids.

A liquid with $\gamma_{LV} < \gamma_c$ will give complete wetting of the substrate. Surfaces with high γ_c (> 40 mN m^{-1}) and small slopes are high energy surfaces (e.g. glass and cellulose). Surfaces with low very γ_c (< 22 mN m^{-1}) and high slope are low energy surfaces, e.g. Teflon. Hydrocarbon surfaces, such as Vaseline, produce intermediate values ($\gamma_c \sim 30$ mN m^{-1}).

The above approach could also be applied for adhesion of "soft" particles to solid substrates. One can define three surface tensions, γ_{PL} (particle/liquid), γ_{PS} (particle/surface) and γ_{SL} (solid/liquid).

The Helmholtz free energy of adhesion ΔF is given by

$$\Delta F = (\gamma_{PS} - \gamma_{PL} - \gamma_{SL})\pi r_0^2 \tag{11.111}$$

For adhesion to occur ΔF should be negative. If ΔF is positive, no adhesion occurs.

11.16.1.2 Neuman's Equation of State Approach

Neuman [42] simplified the analysis by using a simple equation of state approach. He showed that a plot of γ_{LV} cos θ versus γ_{LV} gives a smooth curve (Figure 11.43). This analysis allows one to obtain γ_S from a single contact angle measurement.

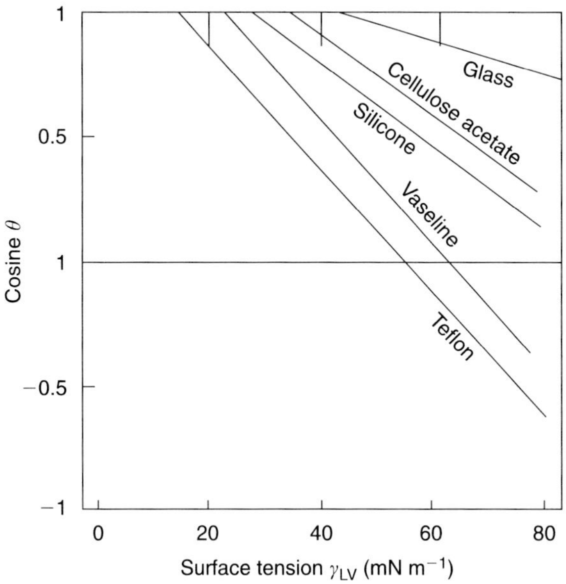

Fig. 11.42. Cos θ versus γ_{LV} for several substrates.

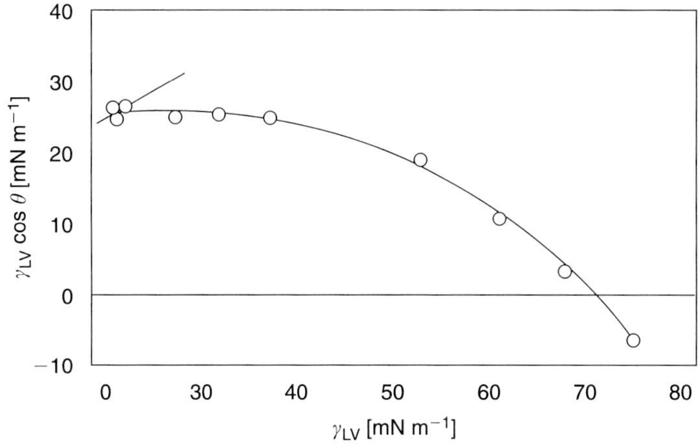

Fig. 11.43. Plot of γ_{LV} cos θ versus γ_{LV}.

11.16.2
Experimental Methods for Measurement of Particle–Surface Adhesion

One has to measure the force required to remove the particle from the substrate:

11.16.2.1 **Centrifugal Method** (Krupp, 1967)

The centrifugal force required to remove a particle is given by the expression [43]

$$F_c = \tfrac{4}{3}\pi a^3(\rho_s - \rho_w)(\omega^2 x + g) \qquad (11.112)$$

where a is the particle radius, ρ_p is the particle density, ρ_w is the density of water, ω is the centrifuge speed, x is the distance from the rotor and g is the acceleration due to gravity.

To remove small particles from substrates one needs to apply very high g values (as high as 10^7). Thus this method has little practical application.

11.16.2.2 **Hydrodynamic Method** (Visser, 1970)

A rotating cylinder apparatus is used and after the particles are deposited on the inner cylinder, the speed is increased [44]. The % detachment is measured (microscopically) as a function of the speed of rotation. The hydrodynamic force required to remove 50% of the particles is taken as a measure of the force of adhesion.

This method was applied by Tadros et al. [45] in 1980 to measure the force of adhesion of polystyrene latex on polyethylene. Figure 11.44 shows the results, giving the % of particle removed versus the speed of rotation of the cylinder.

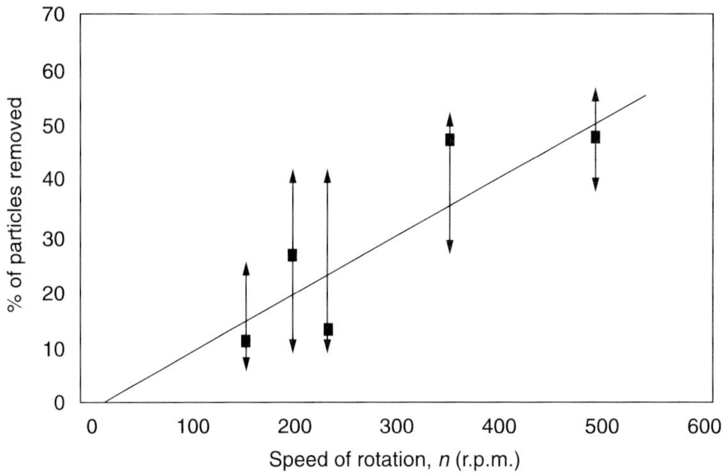

Fig. 11.44. % of particles removed versus speed of rotation.

From knowledge of the hydrodynamic force required for particle removal one could calculate the force of adhesion. The force of adhesion could be compared with the attractive force calculated from Hamaker's equation. The results showed that the force of adhesion was about two orders of magnitude lower than the theoretical value calculated from the van der Waals attraction.

It was concluded from the above results that the latex particles are not perfectly smooth ("hairy" surface) and hence not in intimate contact with the surface.

11.17 Role of Particle Deposition and Adhesion in Detergency

Detergency is defined as the process of removal of liquid or solid dirt from the substrate (usually a solid) with the aid of a liquid cleaning bath. Removal of dirt (liquid or solid) can be from "smooth" surfaces, e.g. in dish washers or from porous or fibrous materials, e.g. from fabrics. A good cleaning agent or detergent must have three main functions: (1) Good wetting power; (2) ability to remove the dirt into the bulk of the liquid or to assist this process; (3) ability to solubilise or disperse the dirt once removed and to prevent its redeposition on the clean surface.

To formulate a good detergent, one has to understand the various processes involved: Wetting, removal of dirt, liquid soiling, prevention of redeposition of dirt.

Below a brief description of the above processes is given, followed by the main topic of the section, namely particle deposition and adhesion and the role of polymers.

11.17.1 Wetting

The best wetting agents are not necessarily the best detergents. For best wetting one needs to lower the dynamic surface tension (which is the value at very short periods of time since the process occurs over very short time scales). This requires molecules with shorter chain alkyl chains (C_8) and surfactants with short relaxation times for the micelles (usually high HLB molecules).

For best detergency one requires molecules that give high surface activity (maximum lowering of the surface tension) and this requires molecules with C_{12}–C_{14} chains. Higher alkyl chain surfactants are not desirable since they have high Krafft temperatures.

In practice most detergents consist of a wide range of molecules with various alkyl chain lengths and different head groups (anionic or nonionic with a range of ethylene oxide units). A detergent formulation will also contain other ingredients such as foam inhibitors, builders to remove multivalent ions, polymers for prevention of redeposition, bleaching agents, enzymes, corrosion inhibitors, perfumes, colours, etc.

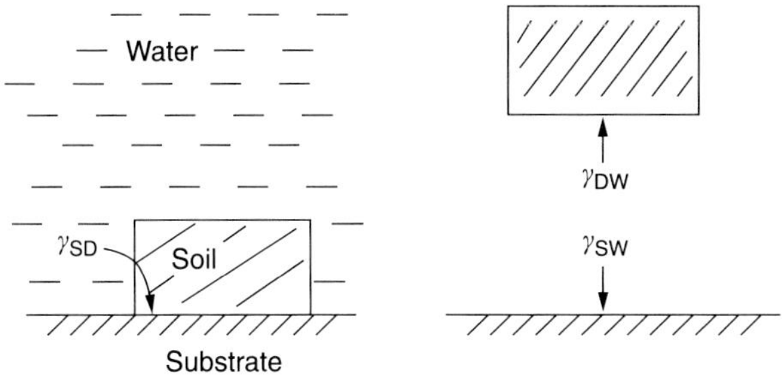

Fig. 11.45. Scheme of dirt removal.

11.17.2
Removal of Dirt

Dirt is generally oily in nature and contains particles of dust, soot and so on. Its removal requires replacement of the soil/surface interface (characterised by a tension γ_{SD}) with a solid/water interface (characterised by a tension γ_{SW}) and a dirt/water interface (characterised by a tension γ_{DW}).

The work of adhesion between a particle of dirt and a solid surface, W_d, is given by

$$W_{SD} = \gamma_{DW} + \gamma_{SW} - \gamma_{SD} \tag{11.113}$$

Figure 11.45 gives a schematic representation of dirt removal.

The task of the detergent is to lower γ_{DW} and γ_{SW} which decreases γ_{SD} and facilitates the removal of dirt by mechanical agitation.

Nonionic surfactants are generally less effective in removal of dirt than anionic surfactants. In practice a mixture of anionic and nonionic surfactants are used.

Liquid soiling: If the dirt is a liquid (oil or fat) its removal depends on the balance of contact angles. The oil or fat forms a low contact angle with the substrate (illustrated in Figure 11.46). To increase the contact angle between the oil and the

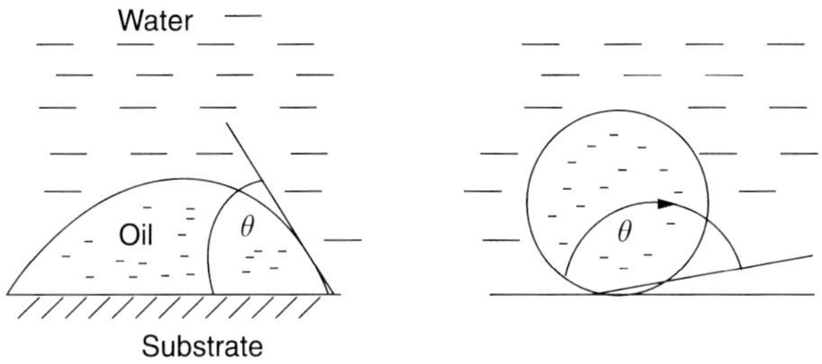

Fig. 11.46. Scheme of oil removal.

substrate (with its subsequent removal), one has to increase the substrate/water interfacial tension, γ_{SW}.

The addition of detergent increases the contact angle at the dirt/substrate/water interface so that the dirt "rolls up" and off the substrate. Surfactants that adsorb both at the substrate/water and the dirt/water interfaces are the most effective. If the surfactant adsorbs only at the dirt/water interface and lowers the interfacial tension between the oil and substrate (γ_{SD}) dirt removal is more difficult. Nonionic surfactants are the most effective in liquid dirt removal since they reduce the oil/water interfacial tension without reducing the oil/substrate tension.

11.17.3 Prevention of Redeposition of Dirt

To prevent dirt particles from redepositing on the substrate once they have been removed, they must be stabilised in the cleaning bath by colloid-chemical means. Prevention can be effected by electrical charge and/or steric barriers (see below) resulting from adsorption of the surfactant molecules from the cleaning bath both by the dirt particles and substrate. The most effective detergents for this purpose are nonionic surfactants of the poly(ethylene oxide) type. In some formulations, nonionic polymers or polyelectrolytes are added to prevent the redeposition of dirt particle (e.g. sodium carboxymethyl cellulose or other nonionic polymers).

11.17.4 Particle Deposition in Detergency

The deposition of particles to fabric or other hard surfaces is a process that is determined by long-range forces, as discussed in the previous section. These long-range forces are of two main types, namely van der Waals attraction and double layer repulsion or attraction. A shorter range force may also operate in particle deposition, when the system contains nonionic surfactants or polymers or polyelectrolytes. The role of these forces in particle deposition has been discussed in detail in the previous sections. Of particular importance is the effect of polymers and polyelectrolytes, which are commonly used in most detergent formulations. As noted, these materials are used as thickening agents (to prevent sedimentation of particles that are present in the detergent formulation, e.g. builders of polyphosphates, zeolites, etc.). Interaction of these polymers and polyelectrolytes with the surfactants used in the detergent formulations plays an important role in particle deposition and its prevention. Many anionic polyelectrolytes also interact with multivalent ions, e.g. Ca^{2+}, that are present in the formulations.

Cationic polymers are also sometimes added to certain detergent formulations, as conditioners for some substrates, e.g. wool. The polycations will adsorb on the negatively charged substrate and they will have a pronounced effect on particle deposition. In addition, polycationic polymers will interact strongly with any anionic surfactants in the formulation, producing polymer–surfactant complexes that will have a pronounced effect on particle deposition.

In the absence of any other effects, addition of cationic polyelectrolytes can enhance particle deposition either by simple charge neutralisation or by "bridging" between the particle and the surface. At high polyelectrolyte concentrations, when there is sufficient molecules to coat both particle and surface, repulsion may occur, resulting in a reduction in deposition. However, these effects are complicated by the interaction of the polycationic polymer with surfactants in the formulation and this complicates the prediction on particle deposition.

11.17.5 Particle–Surface Adhesion in Detergency

As mentioned above, adhesion is the force necessary to separate adherents; it is governed by short-range forces. Adhesion is more complex than deposition and more difficult to measure. The adhesion of "dirt" to substrates is determined by the same short-range forces described above. As already noted, there is no quantitative theory that can describe all adhesion phenomena: Chemical and non-chemical bonds operate. Adequate experimental techniques for measuring adhesion strength in detergency are still lacking.

When considering "dirt" adhesion one must consider elastic and non-elastic deformation that may take place at the point of attachment. The short-range forces could be strong, e.g. primary bonds, or intermediate, e.g. hydrogen and charge-transfer bonds. The force of adhesion can be described by the same theory described above due to Deryaguin [40]. However, this theory is only applicable to idealised systems of spherical particles on rigid non-deformable substrates. Modification is required to consider the case of irregular "dirt" particles on "soft" non-uniform and rough substrates. To date, such modification has not been attempted.

References

1 T. B. Blake: *Surfactants*, Th. F. Tadros (ed.): Academic Press, London, 1984.
2 R. L. Patrick: *Treatise on Adhesion and Adhesives*, Edward Arnold Publishers, London, 1967.
3 T. F. Tadros: *Microbial Adhesion to Surfaces*, R. C. W. Berkeley, et al. (ed.): Ellis Horwood, Chichester, 1980.
4 T. Young, *Phil. Trans. R. Soc. (London)*, **1805**, *95*, 65.
5 J. W. Gibbs: *The Collected Work of J. Willard Gibbs, Vol. 1*, Longman, Harlow, 1928.
6 D. H. Everett, *Pure Appl. Chem.*, **1980**, *52*, 1279.
7 R. E. Johnson, *J. Phys. Chem.*, **1959**, *63*, 1655.
8 R. H. Fowler, *Proc. Royal Soc. Ser. A*, **1937**, *159*, 229.
9 H. C. Hamaker, *Physica (Utrecht)*, **1937**, *4*, 1058.
10 R. J. Good, L. A. Girifalco, *J. Phys. Chem.*, **1960**, *64*, 561.
11 R. J. Good, *Adv. Chem. Ser.*, **1964**, *43*, 74.
12 F. M. Fowkes, *Adv. Chem. Ser.*, **1964**, *43*, 99.
13 W. D. Harkins, *J. Phys. Chem.*, **1937**, *5*, 135.
14 W. D. Harkins: *The Physical Chemistry of Surface Films*, Reinhold, New York, 1952.
15 R. N. Wenzel, *Ind. Eng. Chem.*, **1936**, *28*, 988.

16 A. B. D. Cassie, S. Dexter, *Trans. Faraday Soc.*, **1944**, *40*, 546.

17 A. B. D. Cassie, *Discuss. Faraday Soc.*, **1948**, *3*, 361.

18 H. W. Fox, W. A. Zisman, *J. Colloid Sci.*, **1950**, *5*, 514; W. A. Zisman, *Adv. Chem. Ser.*, **1964**, *43*, 1.

19 C. A. Smolders, *Rec. Trav. Chim.*, **1960**, *80*, 650.

20 S. S. Dukhin, G. Kretzscmar, R. Miller: *Dynamics of Adsorption at Liquid Interfaces*, Elsevier, Amsterdam, 1995.

21 A. F. H. Ward, L. Tordai, *J. Phys. Chem.*, **1946**, *14*, 453.

22 I. Paniotov, J. G. Petrov, *Ann. Univ. Sofia Fac. Chem.*, **1968/69**, *64*, 385.

23 R. S. Hansen, *J. Phys. Chem.*, **1960**, *64*, 637.

24 R. Miller, K. Lunkenheimer, *Z. Phys. Chem.*, **1978**, *259*, 863.

25 R. Zana: *Chemistry and Biology Applied Relaxation Spectroscopy*, 133, Proc. NATO Adv. Study Inst., Ser. C, Volume 18, 1974.

26 R. Miller, A. Hoffmann, R. Hartmann, K. H. Schano, A. Halbig, *Adv. Mater.*, **1992**, *4*, 370.

27 J. T. Davies, E. K. Rideal: *Interfacial Phenomena*, Academic Press, New York, 1969.

28 J. Kloubek, K. Friml, F. Krejci, *Czech. Chem. Commun.*, **1976**, *41*, 1845.

29 T. Blake: in: *Surfactants*, Th. F. Tadros (ed.): Academic Press, London, 1984.

30 R. Burley, B. S. Kennedy, *Chem. Eng. Sci.*, **1976**, *31*, 901.

31 T. Blake, K. J. Ruschak, *Nature*, **1979**, *282*, 489.

32 *Treatise on Adhesion and Adhesives*, R. L. Kilpatrik (ed.): Marcel Dekker, New York, 1967.

33 R. J. Good: *Intermolecular and Intramolecular Forces*, R. L. Kilpatrik (ed.): Marcel Dekker, New York, 1967, Chapter 2.

34 J. H. Hildebrand, R. L. Scott: *Regular Solutions*, Prentice-Hall, Englewood Cliffs, NJ, 1962.

35 R. J. Good, L. A. Girifalco, G. Kraus, *J. Phys. Chem.*, **1958**, *62*, 1418.

36 J. R. Huntsberger: *Treatise on Adhesion and Adhesives*, R. L. Kilpatrik (ed.): Marcel Dekker, New York, 1967, Chapter 4.

37 B. V. Deyaguin, V. P. Smilga, *Proc. Int. Congr. Surface Activity 3rd, Cologne, II, December B*, **1960**, 349.

38 T. F. Tadros: *Microbial Adhesion to Surfaces*, R. C. W. Berkeley, et al. (ed.): Elis Horwood, Chichester, 1980, Chapter 5.

39 M. Hull, S.-A. Kitchener, *Trans. Faraday Soc.*, **1969**, *65*, 3093.

40 B. V. Deryaguin, *Z. Kolloid*, **1934**, *69*, 155.

41 H. W. Fox, W. A. Zisman, *J. Colloid Sci.*, **1952**, *7*, 109.

42 A. W. Neuman, *Adv. Colloid Interface Sci.*, **1974**, *4*, 105.

43 H. Krupp, *Adv. Colloid Interface Sci.*, **1967**, *1*, 111.

44 J. Visser, *J. Colloid Interface Sci.*, **1970**, *34*, 26.

45 T. F. Tadros, et al., unpublished results.

12
Surfactants in Personal Care and Cosmetics

12.1
Introduction

Cosmetic and toiletry products are generally designed to deliver a function benefit and to enhance the psychological well-being of consumers by increasing their aesthetic appeal. Thus, many cosmetic formulations are used to clean hair, skin, etc. and impart a pleasant odour, make the skin feel smooth and provide moisturizing agents, provide protection against sunburn etc. In many cases, cosmetic formulations are designed to provide a protective, occlusive surface layer, which either prevents the penetration of unwanted foreign matter or moderates the loss of water from the skin [1, 2]. To have consumer appeal, cosmetic formulations must meet stringent aesthetic standards such as texture, consistency, pleasing colour and fragrance, convenience of application, etc. In most cases, this results in complex systems that consist of several components of oil, water, surfactants, colouring agents, fragrants, preservatives, vitamins, etc. There has been considerable effort recently to introduce novel cosmetic formulations that provide great beneficial effects to the customer, such as sunscreens, liposomes and other ingredients that may keep skin healthy and provide protection against drying, irritation, etc.

Since cosmetic products come in thorough contact with various organs and tissues of the human body, a most important consideration for choosing ingredients to be used in these formulations is their medical safety. Many cosmetic preparations are left on the skin after application for indefinite periods. Therefore, the ingredients used must not cause any allergy, sensitization or irritation, and they must be free of any impurities that have toxic effects.

One of the main areas of interest of cosmetic formulations is their interaction with the skin [3]. The top layer of the skin, which is the main barrier to water loss, is the stratum corneum, which protects the body from chemical and biological attack [4]. This layer is very thin, approximately 30 μm, and it consists of ~10% by weight of lipids that are organized in bilayer structures (liquid crystalline), which at high water content are soft and transparent. Figure 12.1 gives a schematic of the layered structure of the stratum corneum, suggested by Elias et al. [5]. In this picture, ceramides were considered as the structure-forming elements, but later work by Friberg and Osborne [6] showed the fatty acids to be the essential compounds for the layered structure and that a considerable part of the lipids is

Applied Surfactants: Principles and Applications. Tharwat F. Tadros

ISBN: 3-527-30629-3

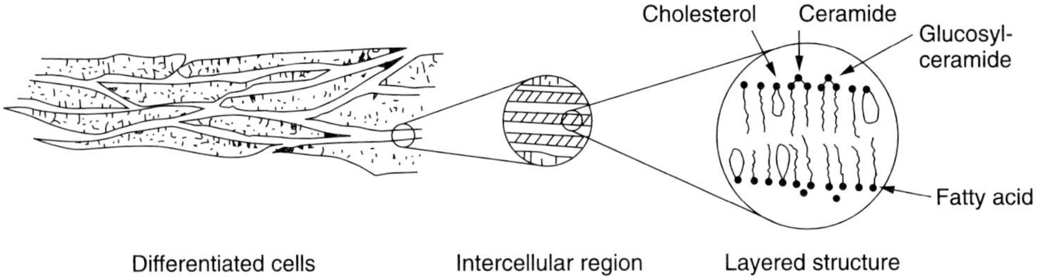

Fig. 12.1. Representation of the stratum corneum structure.

located in the space between the methyl groups. When a cosmetic formulation is applied to the skin, it will interact with the stratum corneum and it is essential to maintain the "liquid-like" nature of the bilayers and prevent any crystallization of the lipids. This happens when the water content is reduced below a certain level. This crystallization has a drastic effect on the appearance and smoothness of the skin ("dry" skin feeling).

The following subsections summarise some of the most commonly used formulations in cosmetics.

12.1.1 Lotions

There are usually oil-in-water (O/W) emulsions that are formulated in such a way (see section on cosmetic emulsions) to give a shear thinning system. The emulsion will have a high viscosity at low shear rates (0.1 s^{-1}) in the region of few hundred Pa s, but the viscosity decreases very rapidly with increasing shear rate, reaching values of few Pa s at shear rates greater than 1 s^{-1}.

12.1.2 Hand Creams

These are formulated as O/W or W/O emulsions with special surfactant systems and/or thickeners to give a viscosity profile similar to that of lotions, but with orders of magnitude greater viscosities. The viscosity at low shear rates (< 0.1 s^{-1}) can reach thousands of Pa s and they retain a relatively high viscosity at high shear rates (of the order of few hundred Pa s at shear rates > 1 s^{-1}). These systems are sometimes said to have a "body", mostly in the form of a gel-network structure that may be achieved by the use of surfactant mixtures to form liquid crystalline structures. In some case, thickeners (hydrocolloids) are added to enhance the gel-network structure.

12.1.3 Lipsticks

These are suspensions of pigments in a molten vehicle. Surfactants are also used in their formulation. The product should show good thermal stability during storage and rheologically it behaves as a viscoelastic solid. In other words, the lipstick should show small deformation at low stresses and this deformation should recover on removal of the stress. Such information could be obtained using creep measurements, which has been described in previous chapters on emulsions and suspensions.

12.1.4
Nail Polish

These are pigment suspensions in a volatile non-aqueous solvent. The system should be thixotropic (see Chapter 7). On application by the brush it should show proper flow for even coating but should have enough viscosity to avoid "dripping". After application, "gelling" should occur in a controlled time scale. If "gelling" is too fast, the coating may leave the "brush marks" (uneven coating). If gelling is too slow, the nail polish may drip. The relaxation time of the thixotropic system should be accurately controlled to ensure good levelling and this requires the use of surfactants.

12.1.5
Shampoos

These are normally a "gelled" surfactant solution of well-defined associated structures, e.g. rod-shaped micelles (see Chapter 2). A thickener such as a polysaccharide may be added to increase the relaxation time of the system. Interaction between the surfactants and polymers is of great importance.

12.1.6
Antiperspirants

These are suspensions of solid actives in a surfactant vehicle. Other ingredients such as polymers that provide good skin feel are added. The rheology of the system should be controlled to avoid particle sedimentation (see Chapter 7). This is achieved by addition of thickeners. Shear thinning of the final product is essential to ensure good spreadability. In stick application, a "semi-solid" system is produced.

12.1.7
Foundations

These are complex systems consisting of a suspension–emulsion system (sometimes referred to as suspoemulsions). Pigment particles are usually dispersed in the continuous phase of an O/W or W/O emulsion. Volatile oils such as cyclomethicone are usually used. The system should be thixotropic to ensure uniformity of the film and good levelling.

Below, a summary, which is by no means exhaustive, is given of the various classes of surfactants commonly used in cosmetics, and personal care, formulations. This is followed by a section on cosmetic emulsions, which are by far the most widely employed systems. Subsequent sections deal with specialized subjects that have been introduced recently in cosmetic systems, namely nano-emulsions, microemulsions, liposomes and multiple emulsions. The last sections will be devoted to some applications of personal care and cosmetic products, illustrating the role of surfactants. The latter are essential ingredients in all personal care and cosmetic formulations.

12.2 Surfactants Used in Cosmetic Formulations

As noted above, surfactants used in cosmetic formulations must be completely free of allergences, sensitisers and irritants. To minimise medical risks, cosmetic formulators tend to use polymeric surfactants, which are less likely to penetrate beyond the stratum corneum and, hence, they are less likely to cause damage.

Conventional anionic, cationic, amphoteric and nonionic surfactants are also used in cosmetic systems. Besides the synthetic surfactants that are used in preparing cosmetic systems such as emulsions, creams, suspensions, etc., several other naturally occurring materials have been introduced and there has been a trend in recent years to use such natural products more widely, in the belief that they are safer for application.

Several synthetic surfactants are applied in cosmetics, such as carboxylates, ether sulphates, sulphate, sulphonates, quaternary amines, betaines, sarcosinates, etc. Ethoxylated surfactants are perhaps the most widely used emulsifiers in cosmetics. Being uncharged, these molecules have a low skin sensitization potential. This is due to their low binding to proteins. Unfortunately, one of the problems of nonionic surfactants is the formation of dioxane, which even in small quantities is unacceptable due to its carcinogeneity. It is, therefore, important when using ethoxylated surfactants to ensure that the level of dioxane is kept at a very low concentration to avoid any side effects. Another drawback of ethoxylated surfactants is their degradation by oxidation or photo-oxidation processes [1]. These problems are reduced by using sucrose esters obtained by esterification of the sugar hydroxyl groups with fatty acids such as lauric and stearic acid. In this case, the danger of dioxane contamination is absent and they are still mild to the skin, since they do not interact to any appreciable extent with proteins.

Phosphoric acid esters are another class of surfactants that are used in cosmetic formulations. These molecules are similar to the phospholipids that constitute the natural building blocks of the stratum corneum. Glycerine esters, in particular the triglycerides, are also used in many cosmetic formulations. These surfactants are important ingredients of sebum, the natural lubricant of the skin. Being naturally occurring, they are claimed to be very safe, causing practically no medical hazard. In addition, these triglycerides can be prepared with a large variety of substituents and hence their HLB values can be varied over a wide range.

Macromolecular surfactants possess considerable advantages for use in cosmetic ingredients. The most commonly used materials are the ABA block copolymers, with A being poly(ethylene oxide) and B poly(propylene oxide) (Pluronics). On the whole, polymeric surfactants have much lower toxicity, sensitization and irritation potentials, provided they are not contaminated with traces of the parent monomers. As will be discussed in the section on emulsions, these molecules provide greater stability and, in some cases, they can be used to adjust the viscosity of the cosmetic formulation.

Several natural surfactants are used in cosmetic formulations, such as those produced from lanolin (wool fat), phytosteroids extracted from various plants and surfactants extracted from beeswax. Unfortunately, these naturally occurring surfactants are not widely used in cosmetics due to their relatively poor physicochemical performance when compared with the synthetic molecules.

Another important class of natural surfactants is proteins, e.g. casein in milk. As with macromolecular surfactants, proteins adsorb strongly and irreversibly at the oil/water interface and hence they can stabilize emulsions effectively. However, the high molecular weight of proteins and their compact structures make them unsuitable for preparation of emulsions with small droplet sizes. For this reason, many proteins are modified by hydrolysis to produce lower molecular weight protein fragments, e.g. polypeptides, or by chemical alteration of the reactive protein side chains. Protein–sugar condensates are sometimes used in skin care formulations. In addition, these proteins impart to the skin a lubricous feel and can be used as moisturizing agents.

Recent years have seen a great trend towards using silicone oils for many cosmetic formulations. In particular, volatile silicone oils have found application in many cosmetic products, owing to the pleasant dry sensation they impart to the skin. These volatile silicones evaporate without unpleasant cooling effects or without leaving a residue. Due to their low surface energy, silicones help spread the various active ingredients over the surface of hair and skin. The chemical structure of silicone compounds used in cosmetic preparations varies according to the application. Figure 12.2 illustrates some typical structures of cyclic and linear silicones. The backbones can carry various attached "functional" groups, e.g. carboxyl, amine, sulfhydryl, etc. [7]. While most silicone oils can be emulsified using conventional hydrocarbon surfactants, there has been a trend in to use silicone surfactants for producing the emulsion [8]. Figure 12.2 shows typical structures of siloxane-poly(ethylene oxide) and siloxane poly(ethylene amine) copolymers. The surface activity of these block copolymers depends on the relative length of the hydrophobic silicone backbone and the hydrophilic (e.g. PEO) chains. The attraction of using silicone oils and silicone copolymers is their relatively small medical and environmental hazards, when compared with their hydrocarbon counterparts [1].

12.3 Cosmetic Emulsions

Cosmetic emulsions need to provide several benefits. For example, such systems should deliver a functional benefit such as cleaning (e.g. hair, skin, etc.), provide a protective barrier against water loss from the skin and in some cases they should

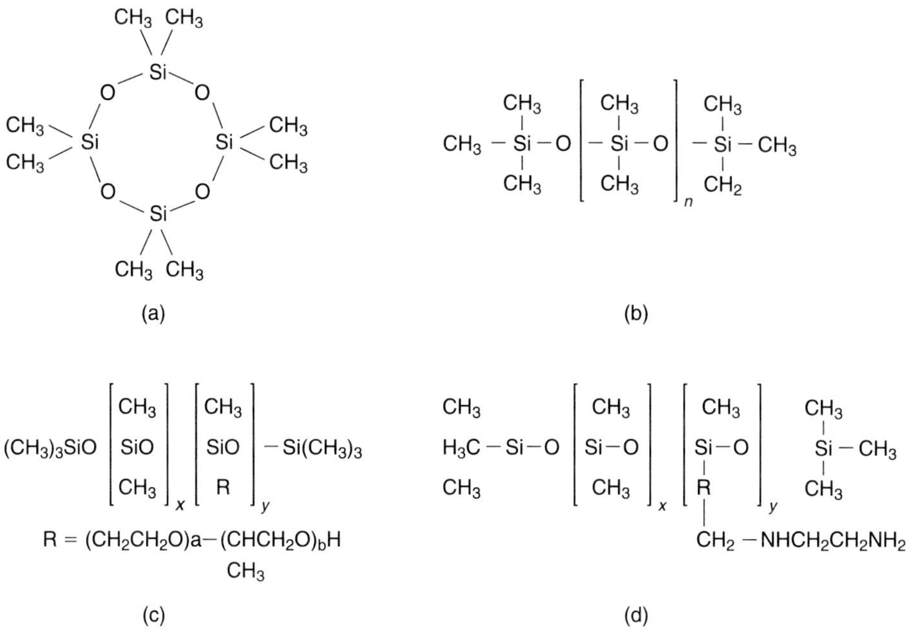

Fig. 12.2. Structural formulae of typical silicone compounds used in cosmetic formulations: (a) cyclic siloxane; (b) linear siloxane; (c) siloxane-polyethylene oxide copolymer; (d) siloxane-polyethylene amine copolymer.

screen out damaging UV light (in which case a sunscreen agent such as titania is incorporated in the emulsion). As mentioned in the introduction, these systems should also impart a pleasant odour and make the skin feel smooth. Emulsions both oil-in-water (O/W) and water-in-oil (W/O) are used in cosmetic applications. As discussed later, more complex systems such as multiple emulsions have been applied in recent years.

The main physico-chemical characteristics that need to be controlled in cosmetic emulsions are their formation and stability on storage as well as their rheology, which controls spreadability and skin feel. Most cosmetic and toiletry brands have a relatively short life span (3–5 years) and hence development of the product should be rapid. Consequently, accelerated storage testing is needed to predict stability and change of rheology with time. These accelerated tests represent a challenge to the formulation chemist.

As noted, the main criterion for any cosmetic ingredient should be medical safety (free of allergances, sensitizers and irritants and impurities that have systemic toxic effects). These ingredients should be suitable for producing stable emulsions that can deliver the functional benefit and the aesthetic characteristics. The main components of an emulsion are the water and oil phases and the emulsifier. Several water-soluble ingredients may be incorporated in the aqueous phase and oil-soluble ingredients in the oil phase. Thus, the water phase may contain functional materials such as proteins, vitamins, minerals and many natural or synthetic water-soluble polymers. The oil phase may contain perfumes and/or pigments (e.g. in make-up). The oil phase may be a mixture of several mineral or vegetable oils. Examples of oils used in cosmetic emulsions are linolin and its derivatives, paraffin and silicone oils. The oil phase provides a barrier against water loss from the skin.

Several emulsifiers, mostly nonionic or polymeric, are used for preparation of O/W or W/O emulsions and their subsequent stabilization (see Chapter 6). For W/O emulsion, the HLB of the emulsifier is in the range 3–6, whereas for O/W emulsions this range is 8–18. Clearly, the exact HLB number depends on the nature of the oil. As mentioned in the previous section, sorbitan esters, sorbitan glyceryl ester, silicone copolymers, sucrose esters, orthophosphoric esters, polyglycerol esters, polymeric surfactants, proteins and amine oxides may be used as emulsifiers.

Cosmetic emulsions are usually referred to as skin creams, which may be classified according to their functional application. The functional and physico-chemical characteristics of these skin creams are summarised in Table 12.1, which also contains some subjective description of the various formulations [1].

To manufacture of cosmetic emulsions, it is necessary to control the process that determines the droplet size distribution, since this controls the rheology of the resulting emulsion. Usually, one starts to make the emulsion on a lab scale (of the order of 1–2 L), which has to be scaled-up to a pilot plant and manufacturing scale. At each stage, it is necessary to control the various process parameters that need to be optimized to produce the desirable effect. It is necessary to relate the process variable from the lab to the pilot plant to the manufacturing scale, and this requires a great deal of understanding of emulsion formation. Two main factors should be considered, namely the mixing conditions and selection of production equipment. For proper mixing, sufficient agitation that produces turbulent flow is necessary to break up the liquid (disperse phase) into small droplets. Various parameters should be controlled, such as flow rate and turbulence, type of impellers, viscosity of the internal and external phases and interfacial properties such as surface tension, surface elasticity and viscosity. The selection of production equipment depends on the characteristics of the emulsion to be produced. Propeller and turbine agitators are normally used for low and medium viscosity emulsions. Agitators that can scrape the walls of the vessel are essential for high viscosity emulsions. Very high shear rates can be produced by using ultrasonics, colloid mills and homogenizers.

Too much heating must be avoided during emulsion preparation, as this may produce undesirable effects such as flocculation and coalescence.

The rheological properties of a cosmetic emulsion that need to be achieved depend on the consumer perspective, which is very subjective. However, the efficacy and aesthetic qualities of a cosmetic emulsion are affected by their rheology. For example, with moisturizing creams one requires fast dispersion and deposition of a continuous protective oil film over the skin surface. This requires a shear thinning system (see below).

Tab. 12.1. Characteristics of skin creams.

Functional	*Physicochemical*	*Subjective*
Cleansing creams Cold creams Massage creams Night creams	Medium to high oil content O/W or W/O Low slip point oil phase Natural pH May contain surfactants to improve penetration and suspension properties	Oily Difficult to rub in May be stiff and rich Also popular as lotions
Moisturizing creams	Low oil content	Easily spreadable and rub in quickly
Foundation creams Vanishing creams	Usually O/W Low slip-point oil phase Natural to slightly acidic pH. May contain emollients and special moisturizing ingredients	Available as creams or lotions
Hand and body protective	Low to medium oil content Usually O/W Medium slip-point oil phase May have slightly alkaline or acidic pH May contain "protective factors" especially silicones and lanolin and lanolin	Easily spreadable but do not rub in with the ease of vanishing creams
All-purpose creams	Medium oil content O/W or W/O	Very often slightly oily but should be easy to spread

To characterize the rheology of a cosmetic emulsion, one needs to combine several techniques, namely steady state, dynamic (oscillatory) and constant stress (creep) measurements. A brief description of these techniques is given below.

With steady-state techniques one measures the shear stress (τ)–shear rate ($\dot{\gamma}$) relationship using a rotational viscometer. A concentric cylinder or cone and plate geometry may be used depending on the emulsion consistency. Most cosmetic emulsions are non-Newtonian, usually pseudo-plastic (Figure 12.3). In this case the viscosity decreases with applied shear rate (shear thinning behaviour; Figure 12.3), but at very low shear rates the viscosity reaches a high limiting value, usually referred to as the residual or zero shear viscosity.

For the above pseudo-plastic flow, one may apply a power law fluid model, a Bingham model [9] or a Casson model [10]. These models are represented by the following equations respectively,

$$\tau = \eta_{\text{app}}\dot{\gamma}^n \tag{12.1}$$

$$\tau = \tau_\beta + \eta_{\text{app}}\dot{\gamma} \tag{12.2}$$

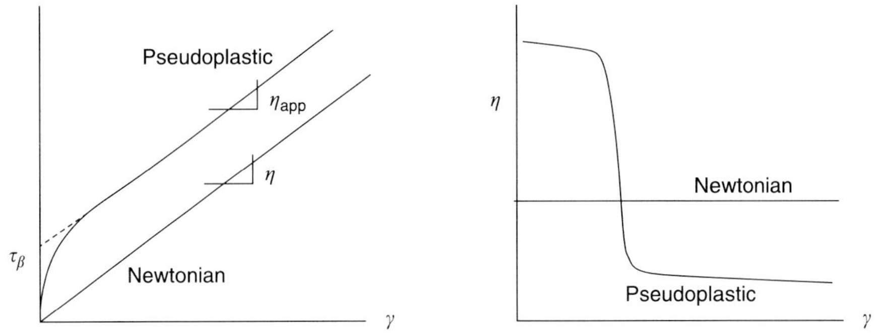

Fig. 12.3. Scheme of Newtonian and non-Newtonian (pseudo-plastic) flow.

$$\tau^{1/2} = \tau_c^{1/2} + \eta_c^{1/2}\dot{\gamma}^{1/2} \tag{12.3}$$

where n is the power in shear rate, which is less than 1 for a shear thinning system (n is sometimes referred to as the consistency index), τ_β is the Bingham (extrapolated) yield value, η is the slope of the linear portion of the τ–$\dot{\gamma}$ curve, usually referred to as the plastic or apparent viscosity, τ_c is Casson's yield value and η_c is Casson's viscosity.

In dynamic (oscillator) measurements, a sinusoidal strain, with frequency ν in Hz or ω in rad s^{-1} ($\omega = 2\pi\nu$) is applied to the cup (of a concentric cylinder) or plate (of a cone and plate) and the stress is measured simultaneously on the bob or the cone, which are connected to a torque bar. The angular displacement of the cup or the plate is measured using a transducer. For a viscoelastic system, such as the case with a cosmetic emulsion, the stress oscillates with the same frequency as the strain, but out-of-phase [11]. Figure 12.4 illustrates the stress and strain sine waves for a viscoelastic system.

From the time shift between the sine waves of the stress and strain, Δt, the phase angle shift δ is calculated as

$$\delta = \Delta t\omega \tag{12.4}$$

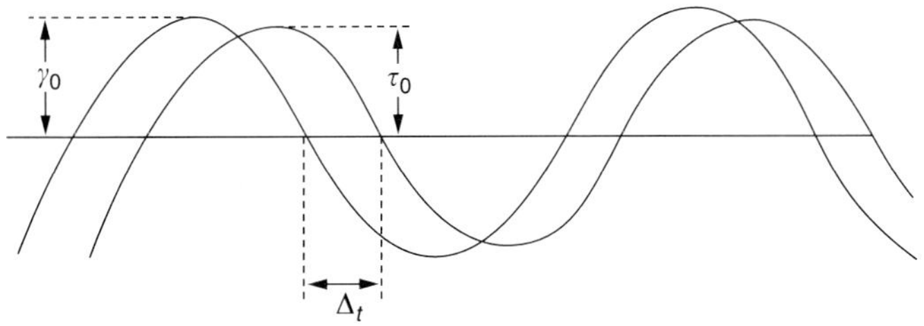

Fig. 12.4. Stress and strain sine waves for a viscoelastic system.

The complex modulus, G^*, is calculated from the stress and strain amplitudes (τ_0 and γ_0 respectively), i.e.

$$G^* = \frac{\tau_0}{\gamma_0} \tag{12.5}$$

The storage modulus, G', which is a measure of the elastic component, is given by Eq. (12.6).

$$G' = |G^*| \cos \delta \tag{12.6}$$

The loss modulus, G'', which is a measure of the viscous component, is given by

$$G'' = |G^*| \sin \delta \tag{12.7}$$

and,

$$|G^*| = G' + iG'' \tag{12.8}$$

where i is $\sqrt{-1}$.

The dynamic viscosity, η', is given by

$$\eta' = \frac{G''}{\omega} \tag{12.9}$$

In dynamic measurements one carries two separate experiments. Firstly, the visco-elastic parameters are measured as a function of strain amplitude, at constant frequency, to establish the linear viscoelastic region, where G^*, G' and G'' are independent of the strain amplitude. Figure 12.5 illustrates this, showing the variation

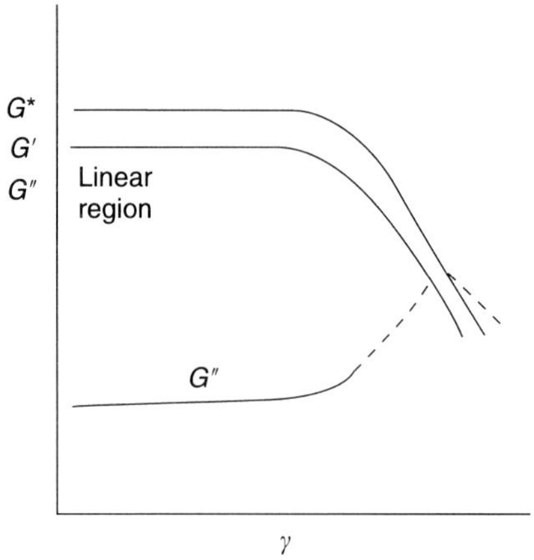

Fig. 12.5. Schematic of the variation of G^*, G' and G'' with strain amplitude (at a fixed frequency).

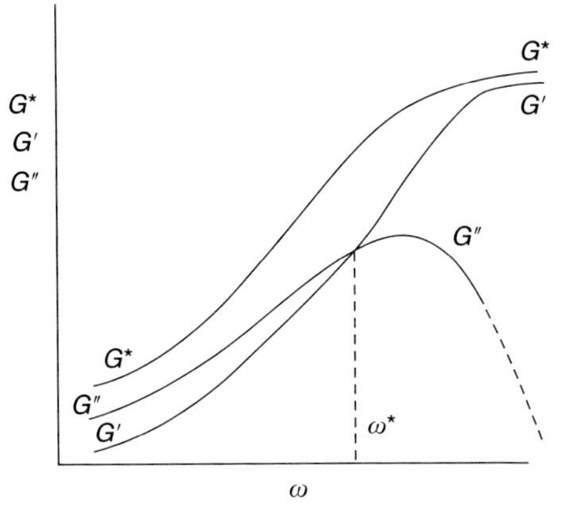

Fig. 12.6. Variation of G^*, G' and G'' with ω for a viscoelastic system.

of G^*, G' and G'' with γ_0. Clearly, the viscoelastic parameters remain constant up to a critical strain value, γ_{cr}, above which G^* and G' start to decrease and G'' starts to increase with further increase in the strain amplitude. Most cosmetic emulsions produce a linear viscoelastic response up to appreciable strains (>10%), indicative of structure build-up in the system ("gel" formation). A short linear region (i.e., a low γ_{cr}) indicates lack of a "coherent" gel structure (in many cases this is indicative of strong flocculation in the system).

Once the linear viscoelastic region is established, measurements are then made of the viscoelastic parameters, at strain amplitudes within the linear region, as a function of frequency. Figure 12.6 shows schematically the variation of G^*, G' and G'' with ν or ω. Below a characteristic frequency, ν^* or ω^*, $G'' > G'$. In this low frequency regime (long time scale), the system can dissipate energy as viscous flow. Above ν^* or ω^*, $G' > G''$, since in this high frequency regime (short time scale) the system can store energy elastically. Indeed, at sufficiently high frequency G'' tends to zero and G' approaches G^* closely, showing little dependency on frequency. The relaxation time of the system can be calculated from the characteristic frequency (the cross over point) at which $G' = G''$, i.e.

$$t^* = \frac{1}{\omega^*} \tag{12.10}$$

Many cosmetic emulsions behave as semi-solids with long t^*. They show only elastic response within the practical range of the instrument, i.e. $G' \gg G''$, and show a small dependence on frequency. Thus, many emulsions creams behave similarly to many elastic gels. This is not surprising, since, in most of cosmetic emulsions systems, the volume fraction of the disperse phase of most cosmetic emulsions is fairly high (usually >0.5) and in many systems a polymeric thickener is added to the continuous phase to stabilize the emulsion against creaming (or sedimentation) and to produce the right consistency for application.

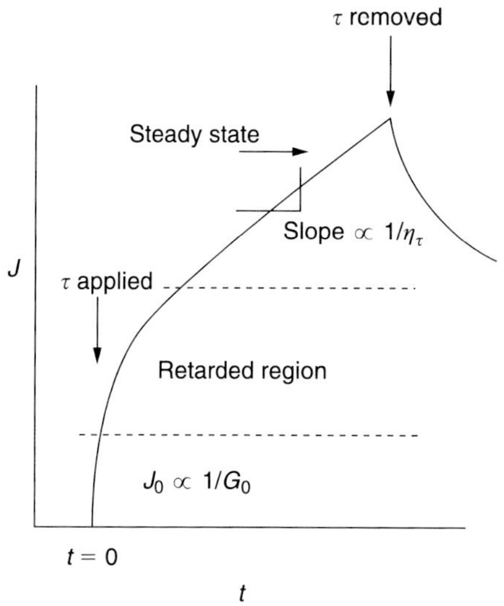

Fig. 12.7. Typical creep curve for a viscoelastic system.

In creep (constant stress) measurements [11], a stress τ is applied on the system and the deformation γ or the compliance $J = \gamma/\tau$ is followed as a function of time. A typical example of a creep curve is shown in Figure 12.7. At $t = 0$, i.e. just after the application of the stress, the system shows a rapid elastic response characterized by an instantaneous compliance J_0, which is proportional to the instantaneous modulus G_0. Clearly, at $t = 0$, all the energy is stored elastically in the system. At $t > 0$, the compliance shows a slow increase, since bonds are broken and reformed, but at different rates. This retarded response is the mixed viscoelastic region. At sufficiently large time scales, which depend on the system, a steady state may be reached with a constant shear rate. In this region J shows linear increase with time and the slope of the straight line gives the viscosity, η_τ, at the applied stress. If the stress is removed after the steady state is reached, J decreases and the deformation reverses sign, but only the elastic part is recovered. By carrying out creep curves at various stresses (starting from very low values, depending on the instrument sensitivity) one can obtain the viscosity of the emulsion at various stresses. A plot of η_τ versus τ typically behaves as shown in Figure 12.8. Below a critical stress, τ_β, the system shows a Newtonian region with a very high viscosity, usually referred to as the residual (or zero shear) viscosity. Above τ_β, the emulsion shows a shear thinning region and, ultimately, another Newtonian region with a viscosity that is much lower than $\eta_{(0)}$. The residual viscosity gives information on the stability of the emulsion on storage. The higher the $\eta_{(0)}$ the lower the creaming or sedimentation of the emulsion. The high stress viscosity gives information on the applicability of the emulsion, such as its spreading and film formation. The critical stress τ_β gives a measure of the true yield value of the system, which is an important parameter both for application purposes and the long-term physical stability of the cosmetic emulsion.

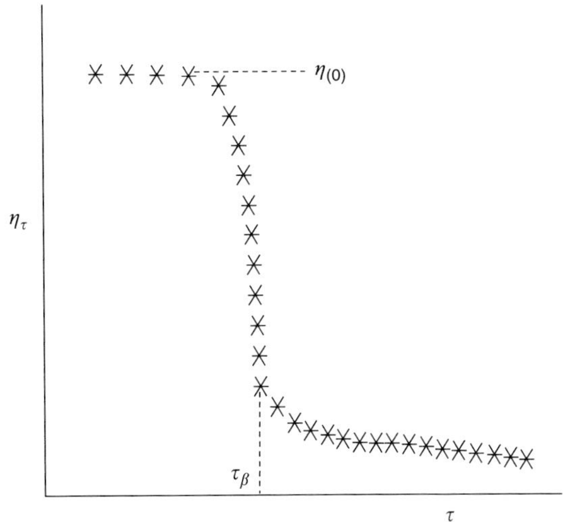

Fig. 12.8. Variation of viscosity with applied stress for a cosmetic emulsion.

The above discussion clearly shows that rheological measurements of cosmetic emulsions are very valuable in determining the long-term physical stability of the system as well as its application. Many cosmetic manufacturers have shown considerable recent interest in this subject. Apart from its value in the above-mentioned assessment, one of the most important considerations is to relate the rheological parameters to the consumer perception of the product. This requires careful measurement of the various rheological parameters for several cosmetic products and relating these parameters to the perception of expert panels that assess the consistency of the product, its skin feel, spreading, adhesion, etc. The rheological properties of an emulsion cream are claimed to determine the final thickness of the oil layer, the moisturizing efficiency and its aesthetic properties such as stickiness, stiffness and oiliness (texture profile). Psychophysical models may be applied to correlate rheology with consumer perception, and the new branch of psychorheology may be introduced.

12.3.1
Manufacture of Cosmetic Emulsions

The process of manufacturing cosmetic emulsions plays an important role in the quality of the final emulsion (such as its droplet size distribution), its long-term stability and its rheological characteristics. This poses difficult problems for the formulation scientist. In the early stages of the development of a cosmetic emulsion, the formulation chemist produces the system on a laboratory scale (usually in the region of 1–2 L). The stability of the system is then followed by storing the emulsion at various temperatures and temperature cycles and by investigations of creaming or sedimentation, flocculation, Ostwald ripening, coalescence and phase inversion, using the methods described in the chapter on emulsions. In many cases, accelerated tests are used to investigate the stability over shorter periods of time. Once the formula has been established, the emulsion is then prepared on a semi-technical scale (pilot plant). This may require some adjustment for the formula to achieve the same physical stability and consistency produced on a lab scale. Finally, the system is scaled up to manufacturing, which may also require further adjustment to the composition. Unfortunately, it is difficult to relate exactly the laboratory scale to the pilot plane and manufacturing scales. However, some chemical engineers may have enough experience to predict the necessary change required in scale-up of the process. Fox has proposed some guidelines [12], discussing the relationships between the chosen processing conditions and the properties of the final emulsion.

As discussed in Chapter 6, the preparation of an emulsion involves the dispersion of an immiscible liquid in a second liquid phase. This requires application of energy that should be strong enough to produce small droplets (typically in the region of few μm). In most cases, turbulent flow is required to produce small droplets. One should avoid foam formation during this process of mixing, since the presence of air bubbles causes an increase in the viscosity of the whole system and this could inhibit emulsification into small droplets.

By choosing suitable impellers and rotational speed, one can regulate the mass flow rate and the pressure head that develop in the fluid during mixing. The processing parameters can change drastically during scale-up and, hence, proper choice of production equipment is very important. Propeller and turbine agitators are used for the preparation of low- and medium-viscosity emulsions [1]. For high-viscosity emulsions, agitators capable of scraping the walls of the container are desirable. When very high shear rates are desired (e.g. for creams and lotions with small droplet sizes), ultrasonic mixers, colloid mills or homogenisers are used.

One of the most useful techniques for preparing cosmetic emulsions is to apply the principle of phase inversion (described in detail in Chapters 6 and 9). For example, to prepare an O/W emulsion one could start with a W/O emulsion, which could be obtained at high temperature (above the HLB temperature of the emulsion). This W/O emulsion is then rapidly cooled to produce the final O/W emulsion. Alternatively, one may start with a W/O emulsion, by dissolving the surfactant in the oil phase and gradually adding water while mixing. When the water content reaches a certain level, inversion to O/W emulsion will occur. This emulsion will have a smaller droplet size distribution than the system produced by directly emulsifying the oil into an aqueous solution of surfactant.

12.4
Nano-Emulsions in Cosmetics

Nano-emulsions are dealt with in detail in Chapter 9; they are transparent or translucent systems having the size range 50–200 nm. They can be prepared using either the phase inversion technique or, more appropriately, by using high-pressure homogenisers. Due to their small droplet sizes, nano-emulsions are stable against creaming or sedimentation, flocculation and coalescence. The only instability problem is Ostwald ripening, which could be significantly reduced by incorporation of a small amount of highly insoluble oil (e.g. squalane) and addition of a polymeric surfactant that adsorbs very strongly at the O/W interface. This polymeric surfactant should chosen to be insoluble in water and have limited solubility in the oil phase. As a result, the polymer molecule remains at the interface, producing a high Gibbs elasticity and this causes a significant reduction in Ostwald ripening.

One of the main advantages of nano-emulsions is the high occlusive film that may be formed on application to the skin. The small size droplets can enter the rough surface of the skin and the droplets may form a close packed structure on the skin surface. This is particularly the case when the droplets have high viscosity or are "solid-like". Another useful application of nano-emulsions is the ability to enhance penetration of actives (e.g. vitamins, antioxidants, etc.) into the skin. This is due to their much higher surface area when compared with coarser emulsions.

12.5
Microemulsions in Cosmetics

Microemulsions are dealt with in detail in Chapter 10. They are thermodynamically stable systems, consisting of oil, water and surfactants, that cover the size range 5–50 nm. They may appear transparent or translucent, depending on the droplet size and the difference in refractive index between the disperse phase and disperse medium. Clear microemulsion gels have been formulated based on nonionic emulsifiers and phosphate esters (15–20%), mineral oil (10–20%), glycols (5–10%) and water (40–60%). Sometimes, long-chain alcohols are also incorporated in small proportions (1–3%). These clear microemulsion gels, sometimes referred to as "lipogels", have specific rheological characteristics (viscoelastic or elastic behaviour) and they form "ringing" gels, due to their characteristic hum when tapped.

Due to their transparency, microemulsions represent a very attractive type of cosmetic formulation, e.g. hair styling gels, perfume gels, bath preparations, sunscreen gels, etc. Their main problem is the relatively high surfactant concentration required for their formulation compared with nano- and macroemulsions. Proper choice of the surfactant system used for their formulation is required to avoid any side-effects, e.g. skin irritation. To arrive at the optimum composition of microemulsion systems, one needs to the phase diagram for these multicomponent formulations.

12.6
Liposomes (Vesicles)

Liposomes (multilamellar bilayers) are produced by dispersion of phospholipids, e.g. lecithin, in water by simple agitation. When these multilamellar phases are

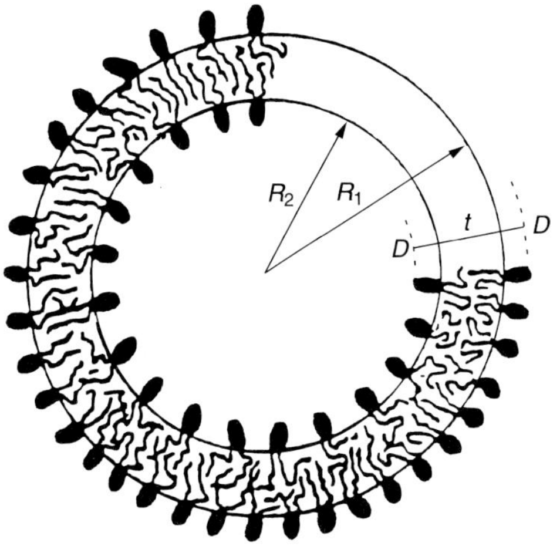

Fig. 12.9. Representation of a phospholipid vesicle.

sonified, they produce "singular" bilayers or vesicles. Figure 12.9 illustrates the latter, showing the bilayer of the phospholipid of thickness t, outer radius R_1 and inner radius R_2 [13]. Vesicles are sometimes referred to as water-in-water dispersions that are separated by a membrane, namely the phospholipid bilayer. The phospholipids employed in cosmetic formulations are usually from a natural source such as egg or soybean lecithin. These molecules are derivatives of glycerol with two alkyl groups and a zwitterionic group, with the general formula given by structure **12.1**, where R usually has a C_{16}–C_{18} chain and may contain unsaturation to ensure that the hydrocarbon chains are above their "melting temperature" (fluid-like above 0 °C).

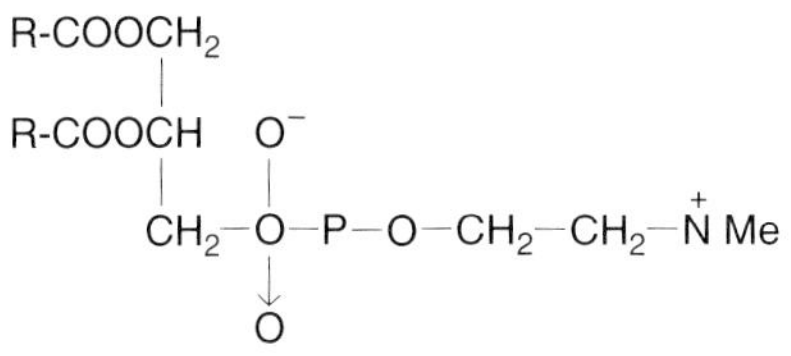

12.1

The packing ratio for the vesicles ($P = v/al$, where v is the volume of the hydrocarbon chain, l its length and a is the cross sectional area of the head group) is greater than $\frac{2}{3}$, implying that globular and cylindrical micelles are prohibited, for which $P < \frac{2}{3}$. The free energy for the amphiphile in a spherical vesicle, with outer and inner radii R_1 and R_2, respectively, depends on the interfacial tension between hydrocarbon and water (γ), the number of molecules in the outer and inner layers (n_1 and n_2), the charge for the polar head group (e), the thickness of the head group (D), and the hydrocarbon volume per amphiphile (v, taken to be constant). The minimum free energy configuration per amphiphile, for a particular aggregation number N, is given by

$$\mu_N^0(\text{min}) \approx 2a_0\gamma\left[1 - \frac{2\pi Dt}{Na_0}\right] \qquad (12.11)$$

where a_0 is the surface area per amphiphile in a planar bilayer, i.e. when $N = \infty$. Several principles may be drawn from the analysis of Israelachvili et al. [13]: (1) μ_N^0 is slightly lower than $\mu_N^0(\text{min})$ for a bilayer ($= 2a_0\gamma$); (2) since a spherical vesicle has much lower N than a planar bilayer, then spherical vesicles are more favoured than planar bilayers; (3) due to packing constraints, the vesicle cannot go below a critical size R_1^c; (4) $a_1 < a_0 < a_2$; (5) for vesicles with radius greater than R_1^c, there are no packing constraints. These vesicles are not thermodynamically favoured over smaller vesicles that have lower N; (6) the vesicle size distribution is nearly Gaussian with a narrow range, e.g. egg phosphatidylcholine vesicles have $R_1 100 \pm 4$ Å. The maximum hydrocarbon chain length is ~17.5 Å; (7) once formed, vesicles are thermodynamically stable and are not affected by the time and strength of sonication. The latter is necessary in most cases to break up the lipid bilayers that are first produced when the phospholipid is dispersed in water. Figure 12.10 gives a schematic representation of how a vesicle might form spontaneously from a bilayer [13].

Vesicles are ideal systems for cosmetic applications. They offer a convenient method for solubilizing active substances in the hydrocarbon core of the bilayer. They will always form a lamellar liquid crystalline structure on the skin and, therefore, they do not disrupt the structure of the stratum corneum. No facilitated transdermal transport is possible, thus eliminating skin irritation (unless the surfactant molecules used for making the vesicles are themselves skin irritants). Indeed, phospholipid liposomes may be used as in vitro indicators for studying skin irritation by surfactants [14].

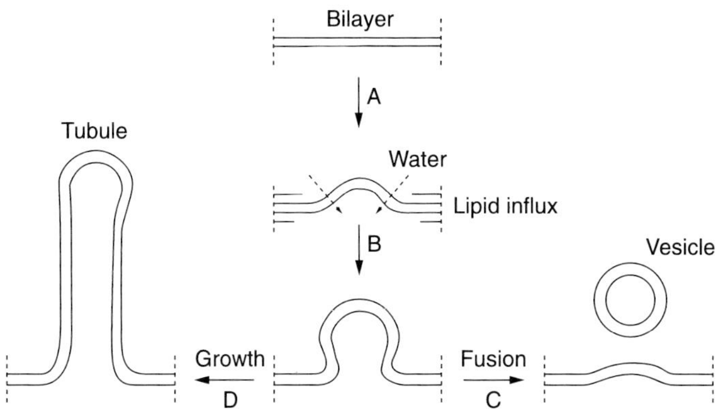

Fig. 12.10. Mechanism of the spontaneous formation of a vesicle from a bilayer.

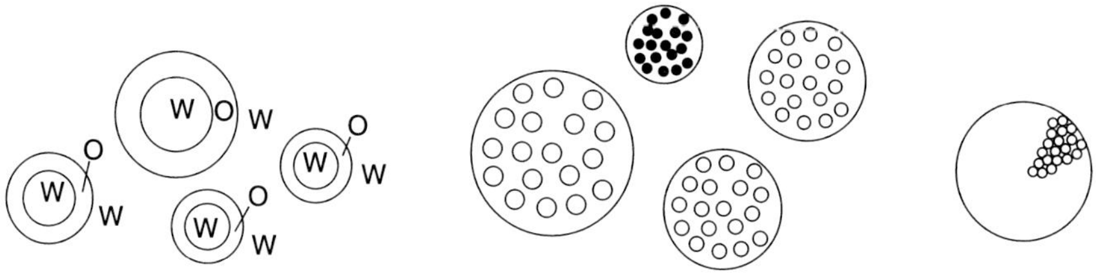

(a) (b) (c)

Fig. 12.11. Three different types of multiple emulsion droplets.

12.7 Multiple Emulsions

Multiple emulsions (W/O/W or O/W/O) are ideal systems for application in cosmetics for the following reasons [15]: (1) one can dissolve additives in three different compartments, e.g. with W/O/W multiple emulsions, one can incorporate two different water-soluble additives (proteins, enzymes and vitamins) and an oil-soluble additive (perfume); (2) they can be usefully applied for sustained release by control of the breakdown process on application; (3) they allow one to produce the same cream consistency as produced by emulsions, e.g. by incorporating thickeners in the outer phase.

Three types of multiple emulsions may be distinguished [16] (Figure 12.11). This classification is based on the predominance of the multiple emulsion droplet type. Using isopropyl myristate as the oil phase, 5% Span 80 to prepare the primary W/O emulsion, and various surfactants to prepare the secondary emulsion, three main types of multiple emulsions were observed [16]: Type "A" droplets contained on a large internal droplet, similar to that observed by Matsumoto et al. [17]. This type was produced when polyoxyethylene oxide (4) lauryl ether (Brij 30) was used as secondary emulsifier at 2%. Type "B" droplets contained several small internal droplets. These were prepared using 2% polyoxyethylene (16.5) nonylphenyl ether (Triton X-165). Type "C" drops entrapped a large number of small internal droplets. These were prepared using a 3:1 Span 80–Tween 80 mixture.

The main criteria for preparation of stable multiple emulsions are: two emulsifiers, one with low and one with high HLB number. Emulsifier 1 should ideally produce a viscoelastic film to reduce transport during storage. A very stable primary emulsion is required; coalescence should be minimised to reduce leakage. Optimum osmotic balance: the osmotic pressure of the electrolyte in the external phase should be slightly lower than that of the external phase. The secondary emulsifier should produce an effective barrier to prevent flocculation and coalescence of the multiple emulsion drops. Figure 12.12 shows a schematic representation of the optimum procedure for preparing a W/O/W multiple emulsion. Several formulation variables must be considered. (1) Primary W/O emulsifier: various low HLB number surfactants are available, including decaglycerol decaoleate; mixed

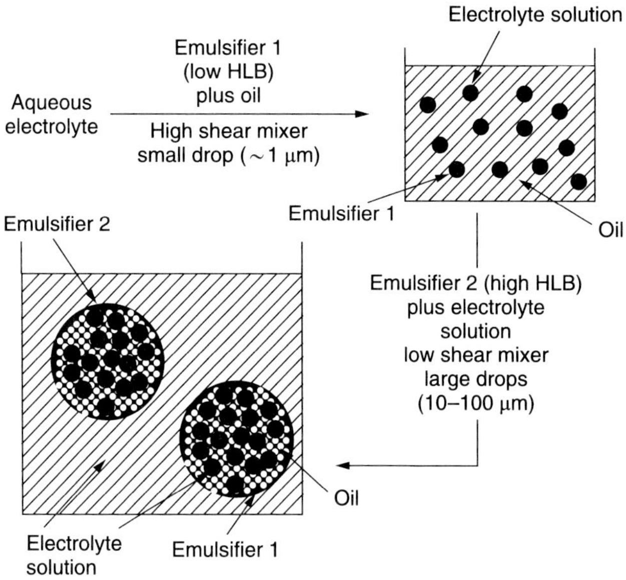

Fig. 12.12. Scheme for preparation of a W/O/W multiple emulsion.

triglycerol trioleate and sorbitan trioleate; ABA block copolymers of PEO and polyhydroxystearic acid. (2) Primary volume fraction of the W/O or O/W emulsion: usually volume fractions between 0.4 and 0.6 are produced, depending on the requirements. (3) Nature of the oil phase: various paraffinic oils (e.g. heptamethyl nonane), silicone oil, soybean and other vegetable oils may be used. (4) Secondary O/W emulsifier: high HLB number surfactants or polymers may be used, e.g. Tween 20, poly(ethylene oxide)-poly(propylene oxide) block copolymers (Pluronics). (5) Secondary volume fraction: this may be varied between 0.4 and 0.8, depending on the consistency required. (6) Electrolyte nature and concentration: e.g. NaCl, $CaCl_2$, $MgCl_2$ or $MgSO_4$. (7) Thickeners and other additives: in some cases a gel coating for multiple emulsion drops may be beneficial, e.g. poly(methacrylic acid) or carboxymethyl cellulose. Gels in the outside continuous phase for a W/O/W multiple emulsion may be produced using xanthan gum (Keltrol or Rhodopol), Carbopol or alginates. (8) Process: to prepare the primary emulsion, high-speed mixers such as Elado (Ystral), Ultraturrax or Silverson may be used. For the secondary emulsion preparation, a low shear mixing regime is required, in which case paddle stirrers are probably the most convenient. The mixing times, speed and order of addition need to be optimized.

Figure 12.13 gives a schematic representation of the multiple emulsion drop. Multiple emulsions require several physical measurements, of which the following are worth mentioning. The droplet size distribution of the primary emulsion can be obtained using photon correlation spectroscopy (provided the droplets are submicron) or diffraction methods. The size distribution of the multiple emulsion

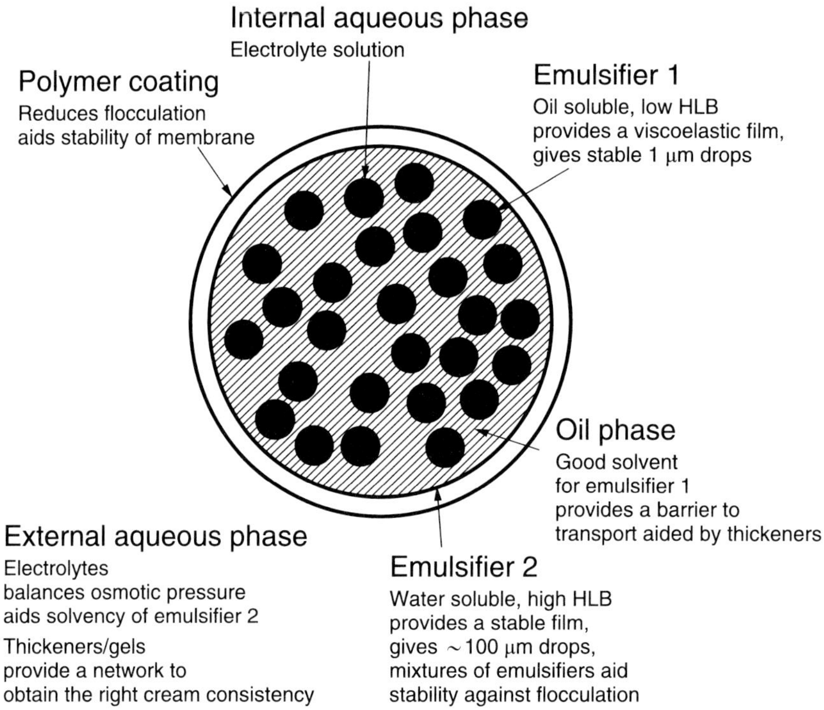

Fig. 12.13. Structure of a multiple emulsion drop showing the role of the various components.

drops can be determined using the Coulter counter or diffraction techniques (e.g. Malvern Master sizer). Dialysis methods can be applied to measure the concentration of free electrolyte in the outer continuous phase and any leakage from the internal droplets. Centrifugation methods may be used qualitatively to assess the physical stability, e.g. creaming or sedimentation. Interference contrast optical microscopy and electron microscopy (in conjunction with freeze–fracture) offer the most informative techniques on the structure of multiple emulsions.

The consistency of the multiple emulsion, which is very important for cosmetic applications, can be evaluated using rheological methods, in the same manner as used for emulsions (see above). These methods can be applied for the primary as well as the final multiple emulsion, which may also contain a gel phase.

12.8
Polymeric Surfactants and Polymers in Personal Care and Cosmetic Formulations

This subject has attracted special attention in recent years [18]. As mentioned above, the use of polymeric surfactants as emulsifiers and dispersants is desirable since these high molecular weight substances cannot penetrate the skin and, hence, they cause no damage on application. In addition, high molecular weight materials such as hydroxyethyl cellulose and xanthan gum are used in many formulations as rheology modifiers (to control the consistency of the product) and they are essential components for the stabilisation of emulsions and suspensions (e.g. prevention of creaming or sedimentation).

A-B, A-B-A block and BA_n graft type polymeric surfactants are used to stabilise emulsions and suspensions [18]. B is the "anchor" chain that adsorbs very strongly at the O/W or S/L interface, whereas the A chains are the "stabilising" chains that provide steric stabilisation. These polymeric surfactants exhibit surface activity at the O/W or S/L interface. The adsorption and conformation of these polymeric surfactant at the interface has been described in detail in reference 18.

Silicone-based materials are an important class of polymeric surfactants that are commonly used in the cosmetic industry. They consist of poly(dimethyl siloxane) (PDMS) that is modified by incorporation of specific groups for special applications. For example, dimethicone copolyol (used as emulsifier or dispersant) is typically a copolymer of PDMS and polyoxyalkylene ether. Aminofunctional silicones provide excellent hair-conditioning benefit. Polyether-modified silicones, including terpolymers containing an alkyl or polyglucoside moiety, are very effective emulsifiers for water–silicone emulsions. These silicone surfactants act as defoamers, depending on the amount and type of glycol modification. They are also used to reduce skin irritation.

Most of the synthetic polymers used as rheology modifiers in the cosmetic industry are based on monomers that can produce carbon–carbon bonds on polymerisation, e.g. acrylic, vinyl, allyl and ethylene oxide. The resulting polymers are used as thickening agents. Chain entanglement is the simplest and most straightforward mechanism of polymeric thickening (see chapters on emulsions and suspensions). The polymers will also interact with the surfactants in the formulation, producing some synergistic effect [18]. An alternative thickening agent commonly used in the personal care industry is the lightly cross linked poly(acrylic acid) (carbomer), which, when neutralised by alkali (such as triethanolamine), produces a thickening effect as a result of the production of charged carboxylate groups that extend in solution and form a "gel" by double layer repulsion.

Lipid thickeners, e.g. waxes, are also used in cosmetic formulations as rheology modifiers. They also form water-repellent films and improve the smoothness and texture of emulsions. Silica or polyethylene are used to formulate anhydrous lipogels.

12.9 Industrial Examples of Personal Care Formulations and the Role of Surfactants

A useful text that gives many examples of commercial cosmetic formulations has been written by Polo [19] to which the reader should refer for detailed information. Below only a summary of some personal care and cosmetic formulations is given, illustrating the use of surfactants. As far as possible, a qualitative description of the role of the surfactants is given. For more fundamental information, the reader should refer to the chapters on emulsions and suspensions.

12.9.1 Shaving Formulations

Three main types of shaving preparations may be distinguished: (1) Wet shaving formulations; (2) dry shaving formulations and (3) after shave preparations.

The main requirements for wet shaving preparations are to soften the beard, to lubricate the passage of the razor over the face and to support the beard hair. The hair of a typical beard is very coarse and difficult to cut and hence it is important to soften the hair for easier shaving and this requires the application of soap and water. The soap makes the hair hydrophilic and hence it becomes easy to wet by water which also may cause swelling of the hair. Most soaps used in shaving preparations are sodium or potassium salts of long-chain fatty acids (sodium or potassium stearate or palmitate). Sometimes, the fatty acid is neutralised with triethanolamine. Other surfactants such as ether sulphates and sodium lauryl sulphate are included in the formulation to produce stable foam. Humectants such as glycerol may also be included to hold the moisture and prevent drying of the lather during shaving.

The most commonly used shaving formulations are those of the aerosol type, whereby hydrocarbon propellants (e.g. butane) are used to dispense the foam. The amount of propellant is critical for foam characteristics. More recently, several companies have introduced the concept of post-foaming gel, whereby the product is discharged in the form of a clear gel which can be easily spread on the face and the foam is then produced by vaporisation of low-boiling hydrocarbons such as isopentene. Due to the high viscosity of the gel, the latter is packed in a bag separated from the propellant used to expel the gel.

The above aerosol type formulations are complex, consisting of an O/W emulsion (whereby the propellant forms most of the oil phase) with the continuous phase consisting of soap/surfactant mixtures. The aerosol shaving foam that was introduced first is relatively simple, whereby a pressurised can is used to release the soap/surfactant mixture in the form of a foam. The sudden release of pressure results in the formation of fine foam bubbles throughout the emerging liquid phase. Two main factors should be considered, the first of which is the foam stability that should be maintained during shaving. In this case, one has to consider the intermolecular forces that operate in a foam film, which are discussed in detail in Chapter 8. The life-time of a foam film is determined by the disjoining pressure that operates across the liquid lamellae. By using the right combination of soap and surfactants, one can optimise the foam characteristics. The second important property of the foam is its feel on the skin. This is determined by the amount of propellant used in the formulation. If the propellant level is too low, the foam will appear "watery". In contrast, a high amount of propellant will produce a "rubbery" dry foam. The humectant added also plays an important role in the skin feel of the foam. Again, an optimum concentration is required to prevent drying out of the foam during shaving. However, if the humectant level is too high it may cause problems by pulling out moisture from the hair, thus making it more difficult to shave.

Clearly, to formulate a shave foam, the chemist has to consider many physicochemical factors, such as the interaction between the soap and surfactant, the quality of the emulsion produced and the bulk properties of the foam produced. Unsurprisingly, most shave foams consist of complex recipes and the role of each component at the molecular level is far from being understood.

As mentioned above, the aerosol shave foam has been replaced with the more popular aerosol post-forming gel. The latter is more difficult to produce, since one has to produce a clear gel with the right rheological characteristics for discharge from the aerosol container, with good spreading on the surface of the skin. The foam should then be produced by vaporisation of a low-boiling liquid such as isobutene or isopentene.

The first problem that must be addressed is the gel characteristics, which are produced by a combination of soap/surfactant mixtures and some polymer (that acts as a "thickener"), e.g. poly(vinylpyrrolidone). Interaction between the surfactants and polymer should be considered to arrive at the optimum composition. The heat of the skin causes the isopentene to evaporate, forming a rich thick gel. One can incorporate skin conditioners and lubricating agent in the gel to obtain good skin feel. Again, most aerosol post-forming gels consist of complex recipes and the interactions between the various components is difficult to understand at a molecular level. A fundamental colloid and interface science investigation is essential to arrive at the optimum composition. In addition, the rheology of the gel, in particular its viscoelastic properties, must be considered in detail. Measurements of the viscoelasticity of these gels are difficult, since the foam is produced during such measurements.

One of the main properties to be considered in these shave foams and post-forming gels is the lubricity of the formulation. Skin friction can be reduced by incorporation of some oils, e.g. silicone, and gums. When shaving, the first stroke by the razor causes no problem since the shave foam or gel is present in sufficient quantities to ensure lubricity of the skin. However, the second stroke in shaving will produce a very high frictional force and hence one should ensure that a residual amount of a lubricant is present on the skin after the first stroke.

Another type of wet-shaving preparation is the non-aerosol type, which is now much less popular than the aerosol type. Two types may be distinguished, namely the brushless and lather shave creams. These formulations are still marketed, although they are much less popular than the aerosol type systems. The brushless shaving cream is an O/W emulsion with high concentrations of oil and soap. The thick film of lubricant oil provides emolliency and protection to the skin surface. This reduces razor drag during shaving. The main disadvantages of these creams are the difficulty of rinsing them from the razor and the formulation may leave a "greasy" feeling on the skin. Due to the high oil content of the formulation, the hair softening action is less effective than for the aerosol type.

The lather shave cream is a concentrated dispersion of alkali metal soap in a glycerol–water mixture. This formulation has adequate physical stability, particularly if the manufacturing process is carefully optimised. Phase separation of the formulation may occur at elevated temperatures.

Dry shaving is a process using electric shavers. In contrast to wet shaving, when using an electric razor the hair should remain dry and stiff. This requires removal of the moisture film and sebum from the face. This may be achieved by using a lotion based on an alcohol solution. A lubricant such as fatty acid ester or isopropyl myristate may be added to the lotion. Alternatively a dry talc stick may be used that can absorb the moisture and sebum from the face.

Another important formulation that is used after shaving is that used to reduce skin irritation and provide a pleasant feel. This can be achieved by providing emolliency accompanied by a cooling effect. In some cases an antiseptic agent is added to keep the skin free from bacterial infection. Most of these after-shave formulations are aqueous-based gels, which should be non-greasy and easy to rub into the skin.

12.9.2
Bar Soaps

These are one of the oldest toiletries products, having been used for over centuries. The earliest formulations were based on simply fatty acid salts, such as sodium or potassium palmitate. However, these simple soaps suffer from the problem of calcium soap precipitation in hard water. For that reason, most soap bars contain other surfactants such as cocomonoglyceride sulphate or sodium cocoglyceryl ether sulphonate that prevent precipitation with calcium ions. Other surfactants used in soap bars include sodium cocyl isethinate, sodium dodecyl benzene sulphonate and sodium stearyl sulphate.

Several other functional ingredients are included in soap bar formulations, e.g. antibacterials, deodorants, lather enhancers, anti-irritancy materials, vitamins, etc. Other soap bar additives include antioxidants, chelating agents, opacifying agents (e.g. titanium dioxide), optical brighteners, binders, plasticisers (for ease of manufacture), anticracking agents, pearlescent pigments, etc. Fragrants are also added to impart pleasant smell to the soap bar.

12.9.3
Liquid Hand Soaps

Liquid hand soaps are concentrated surfactant solutions that can be simply applied from a plastic squeeze bottle or a simple pump container. The formulation consists of a mixture of various surfactants such as alpha olefin sulphonates, lauryl sulphates or lauryl ether sulphates. Foam boosters such as cocoamides are added to the formulation. A moisturizing agent such as glycerine is also added. A polymer such as polyquaternium-7 is added to hold the moisturizers and to impart a good skin feel. More recently, some manufacturers used alkyl polyglucosides in their formulations. The formulation may also contain other ingredients such as proteins, mineral oil, silicones, lanolin, etc. In many cases a fragrant is added to impart a pleasant smell to the liquid soap.

One of the major properties of liquid soaps that needs to be addressed is its rheology, which affects its dispensing properties and spreading on the skin. Most liquid soap formulations have high viscosities to give them a "rich" feel, but some shear thinning properties are required for ease of dispensation and spreading on the surface of the skin.

12.9.4 Bath Oils

Three types of bath oils may be distinguished: floating or spreading oil, dispersible, emulsifying or blooming oil and milky oil. The floating or spreading bath oils (usually mineral or vegetable oils or cosmetic esters such as isopropyl myristate are the most effective for lubricating the dry skin as well as carrying the fragrant. However, they suffer from "greasiness" and deposit formation around the bath tub. These problems are overcome by using self-emulsifying oils that are formulated with surfactant mixtures. When added to water they spontaneously emulsify, forming small oil droplets that deposit on the skin surface. However, these self-emulsifying oils produce less emolliency than the floating oils. These bath oils usually contain a high level of fragrance since they are used in a large amount of water.

12.9.5 Foam (or Bubble) Baths

These can be produced in the form of liquids, creams, gels, powders, granules (beads). Their main function is to produce maximum foam into running water. The basic surfactants used in bubble bath formulations are anionic, nonionic or amphoteric together with some foam stabilisers, fragrants and suitable solublisers. These formulations should be compatible with soap and they may contain other ingredients for enhancing skin care properties.

12.9.6 After-Bath Preparations

These are formulations designed to counteract the damaging effects caused after bathing, e.g. skin drying caused by removal of natural fats and oils from the skin. Several formulations may be used, e.g. lotions and creams, liquid splashes, dry oil spray, dusting powders or talc, etc. The lotions and creams that are the most commonly used formulations are simply O/W emulsions with skin conditioners and emollients. The liquid splashes are hydroalcoholic products that contain some oil to provide skin conditioning. They can be applied as a liquid spread on the skin by hand or by spraying.

12.9.7
Skin Care Products

The skin forms an efficient permeability barrier with the following essential functions: (1) Protection against physical injury, wear and tear and it may also protect against ultraviolet (UV) radiation. (2) Protection against penetration of noxious foreign materials, including water and micro-organisms. (3) It controls loss of fluids, salts, hormones and other endogenous materials from within. (4) It provides thermoregulation of the body by water evaporation (through sweat gland).

For the above reasons skin care products are essential materials for protection against skin damage. A skin care product should have two main ingredients, a moisturiser (humectant) that prevents water loss from the skin and an emollient (the oil phase in the formulation) that provides smoothing, spreading, degree of occlusion and moisturizing effect. The term emollient is sometimes used to encompass both humectant and oils.

The moisturizer should keep the skin humid and it should bind moisture in the formulation (reducing water activity) and protect it from drying out. The term water content implies the total amount of water in the formulation (both free and bound), whereas water activity is a measure of the free (available) water only. The water content of the deeper, living epidermic layers is of the order of 70% (same as the water content in living cells). Several factors can be considered to account for drying of the skin. One should distinguish between the water content of the dermis, viable epidermis and the horny layer (stratum corneum). During dermis ageing, the amount of mucopolysaccharides decreases, leading to a decrease in the water content. This ageing process is accelerated by UV radiation (in particular the deep penetrating UVA, see section on sunscreens). Chemical or physical changes during ageing of the epidermis also lead to dry skin. As discussed in the introduction, the structured lipid/water bilayer system in the stratum corneum forms a barrier towards water loss and protects the viable epidermis from the penetration of exogenous irritants. The skin barrier may be damaged by extraction of lipids by solvents or surfactants and the water loss can also be caused by low relative humidity.

Dry skin, caused by a loss of horny layer, can be cured by formulations containing extracts of lipids from horny layers of humans or animals. Due to loss of water from the lamellar liquid crystalline lipid bilayers of the horny layer, phase transition to crystalline structures may occur, causing contraction of the intercellular regions. The dry skin becomes inflexible and inelastic and it may also crack.

For the above reasons, it is essential to use skin care formulations that contain moisturisers (e.g. glycerine) that draw and strongly bind water, thus trapping water on the skin surface. Formulations prepared with non-polar oils (e.g. paraffin oil) also help in water retention. Occlusion of oil droplets on the skin surface reduces the rate of trans-epidermal water loss. Several emollients can be applied, e.g. petrolatum, mineral oils, vegetable oils, lanolin and its substitutes and silicone fluids. Apart from glycerine, which is the most widely used humectant, several other moisturisers can be used, e.g. sorbitol, propylene glycol. poly(ethylene glycol)s (with molecular weights in the range 200–600). As noted above for liposomes or vesicles, neosomes can also be used as skin moisturisers.

Emollients may be described as products that have softening and smoothing properties. They could be hydrophilic substances such as glycerine, sorbitol, etc. (mentioned above) and lipophilic oils such as paraffin oil, castor oil, triglycerides, etc. For the formulation of stable O/W or W/O emulsions for skin care application, the emulsifier system has to be chosen according to the polarity of the emollient. The polarity of an organic molecule may be described by its dielectric constant or dipole moment. Oil polarity can also be related to the interfacial tension of oil against water γ_{OW}. For example, a non-polar substance such as isoparaffinic oil will give an interfacial tension in the region of 50 mN m^{-1}, whereas a polar oil such as cyclomethicone gives γ_{OW} in the region of 20 mN m^{-1}. The physicochemical nature of the oil phase determines its ability to spread on the skin, the degree of occlusivity and skin protection. The optimum emulsifier system also depends on the properties of the oil (its HLB number) as detailed in the chapter on emulsions.

The choice of an emollient for a skin care formulation is mostly based on sensorial evaluation using well-trained panels. These sensorial attributes are classified into several categories: ease of spreading, skin feeling directly after application and 10 minutes later, softness, etc. A lubricity test is also conducted to establish a friction factor. Spreading of an emollient may also be evaluated by measurement of the spreading coefficient (see Chapter 11).

12.9.8
Hair Care Formulations

Hair care consists of two main operations: (1) Care and stimulation of the metabolically active scalp tissue and its appendages the pilosibaceous units. This is normally carried out by dermatologists or specialised hair saloons. (2) Protection and care of the lifeless hair shaft as it passes beyond the surface of the skin. The latter is the subject of cosmetic preparations, which should acquire one or more of the following functions: (1) Hair conditioning for ease of combing. This could also include formulations that can easily manage styling by combing and brushing and the hairs' capacity to stay in place for a while. The difficulty in managing hair is due to the static electric charge, which may be eliminated by hair conditioning. (2) Hair "body", i.e. the apparent volume of a hair assembly as judged by sight and touch.

Another important type of cosmetic formulation is that used for hair dyeing, i.e. changing the natural colour of the hair. This will also be briefly discussed in this section.

Hair is complex multicomponent fibre with both hydrophilic and hydrophobic properties. It consists of 65–95% by weight of protein and up to 32% water, lipids, pigments and trace elements. The proteins are made of structured hard α-keratin embedded in an amorphous, proteinaceous matrix. Human hair is a modified epidermal structure, originating from small sacs called follicles that are located at the border line of dermis and hypodermis. A cross section of human hair shows three morphological regions, the medulla (inner core), the cortex that consists of fibrous proteins (α-keratin and amorphous protein), and an outer layer namely the cuticle. The major constituents of the cortex and cuticle of hair are protein or polypeptides (with several amino acid units). Keratin has an α-helix structure (molecular weight in the region of 40 000–70 000 Da, i.e. 363–636 amino acid units).

The surface of hair has both acidic and basic groups (i.e. amphoteric in nature). For unaltered human hair, the maximum acid combining capacity is approximately 0.75 mmol per g hydrochloric, phosphoric or ethyl sulphuric acid. This corresponds to the number of dibasic amino acid residues, i.e. arginine, lysine or histidine. The maximum alkali combining capacity for unaltered hair is 0.44 mmol per g of potassium hydroxide, which corresponds to the number of acidic residues, i.e. aspartic and glutamic side-chains. The isoelectric point (i.e.p.) of hair keratin (i.e. the pH at which there is an equal number of positive, $-NH^+$ and negative, $-COO^-$ groups) is ~pH 6.0. However, for unaltered hair, the i.e.p. is at pH 3.67.

The above charges on human hair play an important role in the reaction of hair to cosmetic ingredients in a hair-care formulation. Electrostatic interaction between anionic or cationic surfactants in any hair-care formulation will occur with these charged groups. Another important factor in the application of hair care products is the water content of the hair, which depends on the relative humidity (RH). At low RH ($< 25\%$), water is strongly bound to hydrophilic sites by hydrogen bonds (sometimes this is referred to as "immobile" water). At high RH ($> 80\%$), the binding energy for water molecules is lower because of the multimolecular water–water interactions (this is sometimes referred to as "mobile" or "free" water). With increasing RH, the hair swells; on increasing relative humidity from 0 to 100% the hair diameter increases by ~14%. When water-soaked hair is put into a certain shape while drying, it will temporarily retain its shape. However, any change in RH may lead to the loss of setting.

Both surface and internal lipids exist in hair. The surface lipids are easily removed by shampooing with a formulation based on an anionic surfactant. Two successive steps are sufficient to remove the surface lipids. However, the internal lipids are difficult to remove by shampooing due to the slow penetration of surfactants.

Analysis of hair lipid reveals that they are very complex, consisting of saturated and unsaturated, straight and branched fatty acids with chain lengths of from 5 to 22 carbon atoms. The difference in composition of lipids between persons with "dry" and "oily" hair is only qualitative. Fine straight hair is more prone to "oiliness" than curly coarse hair.

From the above discussion, hair treatment clearly requires formulations for cleansing and conditioning of hair, and this is mostly achieved by using shampoos. The latter are now widely used by most people and various commercial products are available with different claimed attributes. The primary function of a shampoo is to clean both hair and scalp of soils and dirt. Modern shampoos fulfill other purposes, such as conditioning, dandruff control and sun protection. The main requirements for a hair shampoo are: (1) Safe ingredients (low toxicity, low sensitisation and low eye irritation); (2) low substantivity of the surfactants; (3) absence of ingredients that can damage the hair.

The main interactions of the surfactants and conditioners in the shampoo occur in the first few μm of the hair surface. Conditioning shampoos (sometimes referred to as 2-in-1 shampoos) deposit the conditioning agent onto the hair surface. These conditioners neutralise the charge on the surface of the hair, thus decreasing hair friction, making the hair easier to comb. The adsorption of the ingredients in a hair shampoo (surfactants and polymers) occurs both by electrostatic and hydrophobic forces. The hair surface has a negative charge at the pH at which a shampoo is formulated. Any positively charged species such as a cationic surfactant or cationic polyelectrolyte will adsorb by electrostatic interaction between the negative groups on the hair surface and the positive head group of the surfactant. The adsorption of hydrophobic materials such as silicone or mineral oils occurs by hydrophobic interaction (hydrophobic bonding is discussed in detail in Chapter 2).

Several hair conditioners are used in shampoo formulations, e.g. cationic surfactants such as stearyl benzyl dimethyl ammonium chloride, cetyltrimethylammonium chloride, distearyl dimethyl ammonium chloride or stearamidopropyldimethyl amine. As mentioned above, these cationic surfactants cause dissipation of static charges on the hair surface, thus allowing ease of combing by decreasing the hair friction. Sometimes, long-chain alcohols such as cetyl alcohol, stearyl alcohol and cetostearyl alcohol are added, which is claimed to have a synergistic effect on hair conditioning. Thickening agents, such as hydroxyethyl cellulose or xanthan gum are added, which act as rheology modifiers for the shampoo and may also enhance deposition to the hair surface. Most shampoos also contain lipophilic oils such as dimethicone or mineral oils, which are emulsified into the aqueous surfactant solution. Several other ingredients, such as fragrants, preservatives and proteins are also incorporated in the formulation. Thus, a formula of shampoo contains several ingredients and the interaction between the various components should be considered both for the long-term physical stability of the formulation and its efficiency in cleaning and conditioning the hair.

Another hair care formulation is that used for permanent-waving, straightening and depilation. The steps in hair waving involve reduction, shaping and hardening of the hair fibres. Reduction of cysteine bonds (disulphide bonds) is the primary reaction in permanent waving, straightening and depilation of human hair. The most commonly used depilatory ingredient is calcium thioglycollate, which is applied at pH 11–12. Urea is added to increase the swelling of the hair fibres. In permanent waving, this reduction is followed by molecular shifting through stressing the hair on rollers and ended by neutralisation with an oxidising agent where cysteine bonds are reformed. Recently, superior "cold waves" have replaced the "hot waves" by using thioglycollic acid at pH 9 to 9.5. Glycerylmonothio-glycolate is also used in hair waving. An alternative reducing agent is sulphite, which could be applied at pH 6 and is followed by hydrogen peroxide as neutraliser.

Another process also applied in the cosmetic industry is hair bleaching, which has the main purpose of lightening the hair. Hydrogen peroxide is used as the primary oxidising agent and salts of persulphate are added as "accelerators". The system is applied at pH 9–11. The alkaline hydrogen peroxide disintegrates the melanin granules, which are the main source of hair colour, with subsequent destruction of the chromophore. Heavy metal complexants are added to reduce the rate of decomposition of the hydrogen peroxide. Notably, during hair bleaching, an attack of the hair keratin occurs, producing cystic acid.

Another important formulation in the cosmetic industry is that used for hair dyeing. Three main steps may be involved in this process: bleaching, bleaching and colouring combined as well as dyeing with artificial colours. Hair dyes can be classified into several categories: permanent or oxidative dyes, semipermanent dyes and temporary dyes or colour rinses. The colouring agent for hair dyes may consist of an oxidative dye, an ionic dye, a metallic dye or a reactive dye. Permanent or oxidative dyes are the most commercially important systems and they consist of dye precursors such as *p*-phenylenediamine which is oxidised by hydrogen peroxide to a diimminium ion. The active intermediate condenses in the hair fibre with an electron-rich dye coupler such as resorcinol and possibly with electron-rich side chain groups of the hair, forming di-, tri- or polynuclear products that are oxidised into an indo dye.

Semipermanent dyes are formulations that dye the hair without the use of hydrogen peroxide, to a colour that only persists for 4–6 shampooings. Temporary hair dyes or colour rinses aim to provide colour that is removed after the first shampooing process.

12.9.9
Sunscreens

The damaging effect of sunlight (in particular ultraviolet light) has been recognised for several decades and has led to a significant demand for improved photoprotection by topical application of suncreening agents. Three main wavelengths of ultraviolet (UV) radiation may be distinguished, referred to as UV-A [wavelength range 320–400 nm, sometimes subdivided into UV-A1 (340–360 nm) and UV-A2 (320–340 nm)], UV-B (covering 290–320 nm) and UV-C (covering 200–290 nm). UV-C is of little practical importance since it is absorbed by the ozone layer of the stratosphere. UV-B is energy rich and produces intense short- and long-range pathophysiological damage to the skin (sun burn). About 70% is reflected by the horny layer (stratum corneum), 20% penetrates into the deeper layers of the epidermis and 10% reaches the dermis. UV-A is of lower energy, but its photobiological effects are cumulative, causing long-term effects. UV-A penetrates deeply into the dermis and beyond, i.e. 20–30% reaches the dermis. As it has a photoaugmenting effect on UV-B, it contributes about 8% to UV-B erythema.

Several studies have shown that sunscreens are able not only to protect against UV-induced erythema in humans and animal skin but also to inhibit photocarcinogenesis in animal skin. The increasing harmful effect of UV-A on UV-B has led to a quest for sunscreens that absorb in the UV-A with the aim of reducing the direct dermal effects of UV-A, which causes skin ageing and several other photosensitivity reactions. Sunscreens are given a sun protection factor (SPF), which is a measure of the ability of a sunscreen to protect against sunburns with in the UV-B wavelength (290–320 nm). Formulation of sunscreen with a high SPF (>50) has been the object of many cosmetic industries.

An ideal sunscreen formulation should protect against both UV-B and UV-A. Repeated exposure to UV-B accelerates skin ageing and can lead to skin cancer. UV-B can cause thickening of the horny layer (producing "thick" skin). UV-B can also damage DNA and RNA. Individuals with fair skin cannot develop a protective tan and they must protect themselves from UV-B.

UV-A can also cause several effects: (1) Large amounts of UV-A radiation penetrate deep into the skin and reach the dermis, causing damage blood vessels, collagen and elastic fibres. (2) Prolonged exposure to UV-A can cause skin inflammation and erythema. (3) UV-A contributes to photoageing and skin cancer. It augments the biological effect off UV-B. (4) UV-A can cause phytotoxicity and photoallergy and it may cause immediate pigment darkening (immediate tanning), which may be undesirable for some ethnic populations.

From the above discussion, the formulation of effective sunscreen agents that meet the following requirements is clearly necessary: (1) Maximum absorption in the UV-B and/or UV-A. (2) High effectiveness at low dosage. (3) Non-volatile agents with chemical and physical stability. (4) Compatibility with other ingredients in the formulation. (5) Sufficient solubility or dispersibility in cosmetic oils, emollients or in the water phase. (6) Absence of dermato-toxological effects with minimum skin penetration. (7) Resistance to removal by perspiration.

Sunscreen agents may be classified into organic light filters of synthetic or natural origin and barrier substances or physical sunscreen agents. Examples of UV-B filters are cinnamates, benzophenones, *p*-aminobenzoic acid, salicylates, camphor derivatives and phenyl benzimidazosulphonates. Examples of UV-A filters are dibenzoyl methanes, anthranilates and camphor derivatives. Several natural sunscreen agents are available, e.g. camomile or aleo extracts, caffeic acid, unsaturated vegetable or animal oils. However, these natural sunscreen agents are less effective and they are seldom used in practice.

Barrier substances or physical sunscreens are essentially micronized insoluble organic molecules such as gaunine or micronised inorganic pigments such as titanium dioxide and zinc oxide. Micropigments act by reflection, diffraction and/or absorption of UV radiation. Maximum reflection occurs when the particle size of the pigment is about half the wavelength of the radiation. Thus, for maximum reflection of UV radiation, the particle radius should be in the region of 140 to 200 nm. Uncoated materials such as titanium and zinc oxide can catalyse the photodecomposition of cosmetic ingredients such as sunscreens, vitamins, antioxidants and fragrances. These problems can be overcome by special coating or surface treatment of the oxide particles, e.g. using aluminium stearate, lecithins, fatty acids, silicones and other inorganic pigments. Most of these pigments are supplied as dispersions ready to mix in the cosmetic formulation. However, one must avoid any flocculation of the pigment particles or interaction with other ingredients in the formulation, which causes severe reduction in their sunscreening effect.

A topical sunscreen product is formulated by the incorporation of one or more sunscreen agents (referred to as UV filters) in an appropriate vehicle, mostly an

O/W or W/O emulsion. Several other formulations are also produced, e.g. gels, sticks, mousse (foam), spray formulation or an anhydrous ointment. In addition to the usual requirements for a cosmetic formulation, e.g. ease of application, pleasant aspect, colour or touch, sunscreen formulations should also be (1) Effective in thin films, strongly absorbing both in UV-B and UV-A. (2) Non-penetrating and easily spreading on application. (3) Possess a moisturising action and be waterproof and sweat resistant. (4) Free from any phototoxic and allergic effects. Most sunscreens on the market are creams or lotions (milks) and progress has been made in recent years to provide high SPF at low levels of sunscreen agents.

12.9.10 Make-up Products

Make-up products include many systems such as lipstick, lipcolour, foundations, nail polish, mascara, etc. All these products contain a colouring agent, which could be a soluble dye or a pigment (organic or inorganic). Examples of organic pigments are red, yellow, orange and blue lakes. Examples of inorganic pigments are titanium dioxide, mica, zinc oxide, talc, iron oxide (red, yellow and black), ultramarines, chromium oxide, etc. Most pigments are modified by surface treatment using amino acids, chitin, lecithin, metal soaps, natural wax, polyacrylates, polyethylene, silicones, etc.

The main function of colour cosmetics, such as foundation, blushers, mascara, eyeliner, eyeshadow, lip colour and nail enamel, is to improve appearance, impart colour, even out skin tones, hide imperfections and produce some protection. Several types of formulations are produced, ranging from aqueous and non-aqueous suspensions to oil-in-water and water-in-oil emulsions and powders (pressed or loose).

Make-up products have to satisfy several criteria for acceptance by the consumer: (1) Improved, wetting spreading and adhesion of the colour components. (2) Excellent skin feel. (3) Skin and UV protection and absence of skin irritation.

For these purposes, the formulation has to be optimised to achieve the desirable property. This is achieved by using surfactants and polymers as well as using modified pigments (by surface treatment). The particle size and shape of the pigments should also be optimised for proper skin feel and adhesion.

Pressed powders require special attention to achieve good skin feel and adhesion. The fillers and pigments have to be surface treated to achieve these objectives. Binders and compression aids are also added to obtain a suitable pressed powder. These binders can be dry powders, liquids or waxes. Other ingredients that may be added are sunscreens and preservatives. These pressed powders are readily applied by simple "pick-up", deposition and even coverage. The appearance of the pressed powder film is very important and great care should be taken to achieve uniformity on application. A typical pressed powder may contain 40–80% fillers, 10–40% specialised fillers, 0–5% binders, 5–10% colourants, 0–10% pearls and 3–8% wet binders.

As an alternative to pressed powders, liquid foundations have attracted special attention in recent years. Most foundation make-ups are made of O/W or W/O emulsions in which the pigments are dispersed either in the aqueous or the oil phase. These are complex systems consisting of a suspension/emulsion (suspoemulsion) formulation. Special attention should be paid to the stability of the emulsion (absence of flocculation or coalescence) and suspension (absence of flocculation). This is achieved by using specialised surfactant systems such silicone polyols or block copolymers of poly(ethylene oxide) and poly(propylene oxide). Some thickeners may be also added to control the consistency (rheology) of the formulation.

The main purpose of a foundation make-up is to provide colour in an even way, even out any skin tones and minimise the appearance of any imperfections. Humectants are also added to provide a moisturising effect. The oil used should be chosen to be a good emollient. Wetting agents are also added to achieve good spreading and even coverage. The oil phase could be a mineral oil, an ester such as isopropyl myristate or volatile silicone oil (e.g. cyclomethicone). An emulsifier system of fatty acid/nonionic surfactant mixture may be used. The aqueous phase contains a humectant of glycerine, propylene glycol or poly(ethylene glycol). Wetting agents such as lecithin, low HLB surfactant or phosphate esters may also be added. A high HLB surfactant may also be included in the aqueous phase to provide better stability when combined with the oil emulsifier system. Several suspending agents (thickeners) may be used, such as magnesium aluminium silicate, cellulose gum, xanthan gum, hydroxyethyl cellulose or hydrophobically modified polyethylene oxide. A preservative such as methyl paraben is also included. Surface-treated pigments are dispersed either in the oil or aqueous phase. Other additives such as fragrances, vitamins and light diffusers may also be incorporated.

Liquid foundations are, quite clearly, a challenge to the formulation chemist due to both the numerous components used and the interaction between them. Particular attention should be made to the interaction between the emulsion droplets and pigment particles (a phenomenon referred to as heteroflocculation), which may have adverse effects of the final property of the deposited film on the skin. Even coverage is the most desirable property and the optical properties of the film, e.g. its light reflection, adsorption and scattering, play important roles in the final appearance of the foundation film.

Several anhydrous liquid (or "semi-solid") foundations are also marketed by cosmetic companies. These may be described as cream powders that consist of a high content of pigment/fillers (40–50%), a low HLB wetting agent (such as polysorbate 85), an emollient such as dimethicone combined with liquid fatty alcohols and some esters (e.g. octyl palmitate). Some waxes, such as stearyl dimethiicone or microcrystalline or carnuba wax, are also included in the formulation.

Lipsticks, one of the most important make-up systems, may be simply formulated with a pure fat base having a high gloss and excellent hiding power. However, these simple lipsticks tend to come off the skin too easily. In recent years, there has been a great tendency to produce more "permanent" lipsticks that contain hydrophilic solvents such as glycols or tetrahydrofurfuryl alcohol. The raw materials for a lipstick base include ozocerite (good oil absorbent that also prevents crystallisation), microcrystalline ceresin wax (which also is a good oil absorbent), Vaseline (that forms an impermeable film), bees wax (that increases resistance to fracture), myristyl myristate (which improves transfer to the skin), cetyl and meristyl lactate (which forms an emulsion with moisture on the lip and is non-sticky), carnuba wax (an oil binder that increases the melting point of the base and gives some surface luster), lanolin derivatives, olyl alcohol and isopropyl myristate. This shows the complex nature of a lipstick base and several modifications of the base can produce some desirable effects that help good marketing of the product.

Mascara and eyeliners are also complex formulations, and need to be carefully applied to the eye lashes and edges. Some of the preferred criteria for mascara are good deposition, ease of separation and lash curling. The appearance of the mascara should be as natural as possible. Lash lengthening and thickening are also desirable. The product should also remain for an adequate time and it should also be easily removed. Three types of formulations may be distinguished: anhydrous solvent-based suspension, water-in-oil emulsion and oil-in-water emulsion. Water resistance can be achieved by addition of emulsion polymers, e.g. poly(vinyl acetate).

References

1 M. M. Breuer: *Encyclopedia of Emulsion Technology*, P. Becher (ed.): Marcel Dekker, New York, 1985, Volume 2, Chapter 7.
2 S. Harry: *Cosmeticology*, J. B. Wilkinson, R. J. Moore (ed.): Chemical Publish-ing, New York, 1981.
3 S. E. Friberg, *J. Soc. Cosmet. Chem.*, **1990**, *41*, 155.
4 A. M. Kligman: *Biology of the Stratum Corneum in Epidermis*, W. Montagna (ed.): Academic Press, London, 1964, 421–446.
5 P. M. Elias, B. E. Brown, P. T. Fritsch, R. J. Gorke, G. M. Goay, R. J. White, *J. Invest. Derm.*, **1979**, *73*, 339.
6 S. E. Friberg, D. W. Osborne, *J. Disp. Sci. Technol.*, **1985**, 6, 485.
7 S. C. Vick, *Soaps Cosmet. Chem. Spec.*, **1984**, *36*.
8 M. S. Starch, *Drug Cosmet. Ind.*, **1984**, *134*, 38.
9 R. W. Wahrlow: *Rheological Techniques*, John Wiley & Sons, New York, 1980.
10 N. Casson: *Rheology of Disperse Systems*, C. C. Hill (ed.): Pergamon Press, Oxford, 1959, pp. 84.
11 J. W. Goodwin: *Solid/Liquid Dispersions*, Th. F. Tadros (ed.): Academic Press, London, 1987, 199–224.
12 C. Fox: *Cosmetic Science*, M. M. Breuer (ed.): Academic Press, London, 1980, Volume 2.
13 J. N. Israelachvili, D. J. Mitchell, B. W. Ninham, *J. Chem. Soc., Faraday Trans. II*, **1976**, *72*, 1525.
14 U. K. Charaf, G. L. Hart, *J. Soc. Cosm. Chem.*, **1991**, *42*, 71.
15 T. F. Tadros, *Int. J. Cosmet. Sci.*, **1992**, *14*, 93.
16 A. T. Florence, D. J. Whitehill, *J. Colloid Interface Sci.*, **1981**, *79*, 243.
17 S. Matsumoto, Y. Kita, D. Yonezawa, *J. Colloid Interface Sci.*, **1976**, *57*, 353.
18 E. D. Goddard, J. V. Gruber: *Principles of Polymer Science and Technology in Cosmetics and Personal Care*, Marcel Dekker, New York, 1999.
19 K. F. Polo De: *A Short Textbook of Cosmetology*, Verlag für Chemishe Industrie, H. Ziolkowsky, Augsburg, Germany, 1998.

13
Surfactants in Pharmaceutical Formulations

13.1
General Introduction

Surfactants are used in all disperse systems employed in pharmaceutical formulations. Several types of disperse systems can be identified (Table 13.1).

This chapter deals with the use of surfactants in the first three classes, namely suspensions, emulsions and gels. Such disperse systems cover a wide size range: Colloidal (1 nm–1 μm) to non-colloidal (> 1 μm). Generally speaking they are prepared by two main processes [1]:

(1) *Condensation methods*, i.e. nucleation and growth. Here, one starts with molecular units that are condensed to form nuclei that grow further to produce the particles. An example of such a process is the formation of particles by a precipitation technique, e.g. production of colloidal AgI particles by reaction of $AgNO_3$ with KI. Many suspensions are produced by addition of a solution of the chemical in a suitable solvent, which is then added to another miscible solvent in which the drug is insoluble. A third example of condensation methods is the production of polymer particles from their monomers by a suitable polymerisation technique.

(2) *Dispersion methods.* In this case one starts with preformed particles of the bulk chemical, which are then subdivided into smaller particles by a suitable dispersion/comminution process. An example of this method is the production of suspensions by wet milling (using grinding equipments). Emulsification of oils (using high speed stirrers or homogenisers) is also a kind of dispersion method.

Tab. 13.1. Types of disperse systems in pharmaceutical formulations.

Disperse phase	*Dispersion medium*	*Class*
Solid	Liquid	Suspensions
Liquid	Liquid	Emulsions
Liquid	Solid	Gels
Liquid	Air	Aerosols
Gas	Liquid	Foams
Solid	Gas	Smokes
Solid	Solid	Composites

Applied Surfactants: Principles and Applications. Tharwat F. Tadros

ISBN: 3-527-30629-3

All disperse systems are thermodynamically unstable, i.e. the free energy of their formation is positive, as illustrated below.

13.1.1 Thermodynamic Consideration of the Formation of Disperse Systems

For thermodynamic analysis, let us consider the formation of suspensions from the large bulk phase (with surface area A_1) to a many much smaller particles (with a total surface area A_2 that is much larger than A_1). The interfacial tension between the solid and the liquid γ_{SL} is considered to be the same for the large and small particles. This is schematically shown in Figure 13.1a. The same process can be considered for the formation of emulsions (Figure 13.1b) [2]. The bulk oil has an area A_1 whereas the numerous smaller oil droplets have a total area A_2. The interfacial tension between oil and water γ_{ow} is considered to be the same for the large and small droplets.

In the above processes, the surface energy of the system increases by $\Delta A \gamma_{SL}$ for suspensions and $\Delta A \gamma_{OW}$ for emulsions (ΔA is the increase in surface area in the dispersion process). In the dispersion process, numerous particles or droplets are produced and this is accompanied by an increase in entropy ΔS.

According to the second law of thermodynamics, the free energy of formation of the system ΔG is given by the following two expressions for suspensions and emulsions,

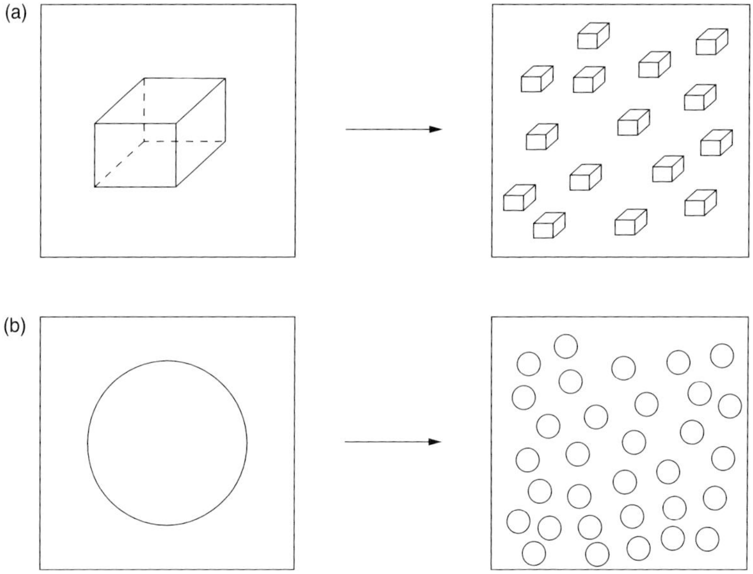

Fig. 13.1. (a) Formation of small suspension particles from much larger particles; (b) formation of small emulsion droplets from much larger oil drops.

$$\Delta G = \Delta A\gamma_{SL} - T\Delta S \tag{13.1}$$

$$\Delta G = \Delta A\gamma_{OW} - T\Delta S \tag{13.2}$$

In the above systems $\Delta A\gamma_{SL} \gg -T\Delta S$ and $\Delta A\gamma_{OW} \gg -T\Delta S$ and hence $\Delta G > 0$. This implies thermodynamic instability and the production of suspension or emulsions by the dispersion process is non-spontaneous, i.e. energy is required to produce the smaller particles or droplets from the larger ones. In the absence of any stabilisation mechanism (which will be discussed below), the smaller particles or droplets tend to aggregate and/or coalesce to reduce the total interfacial area, hence reducing the total surface energy of the system.

Prevention of aggregation and/or coalescence of suspensions or emulsions requires fundamental understanding of the various interaction forces between the particles or droplets [3] and these will be discussed in subsequent sections. For full details one can refer to the chapters on emulsions and suspensions.

13.1.2
Kinetic Stability of Disperse Systems and General Stabilisation Mechanisms

As mentioned above disperse systems lack thermodynamic stability and they tend to reduce their surface energy by aggregation and/or coalescence of the particles or droplets. The main driving force for the aggregation process is the universal van der Waals attraction, which will be discussed in subsequent sections.

To overcome the aggregation and/or coalescence processes, one must overcome the van der Waals attraction by some repulsive mechanism that will give the system kinetic stability with an adequate shelf-life. Normally one requires a shelf-life of 2–3 years under various storage conditions (e.g. temperature variation).

Several stabilisation mechanisms are encountered with disperse systems, and these are summarised below.

13.1.2.1 Electrostatic Stabilisation

This is achieved by some sort of charge separation at the solid/liquid or liquid/liquid interface and the creation of an electrical double layer.

When two particles or droplets containing electrical double layers with the same charge sign approach each other to a distance of separation whereby the double layers begin to overlap, repulsion occurs since the double layers cannot be fully developed in the confined space between the particles or droplets. This repulsive mechanism will be discussed in detail in subsequent sections.

Double layer repulsion counteracts the van der Waals attraction, particularly at intermediate separations. As a result an energy barrier is created between the particles or droplets, thus preventing their close approach. The magnitude of the barrier depends on several parameters: the surface charge or potential at the particle or droplet surface, the electrolyte concentration and valency of the counter ions, particle size and magnitude of the van der Waals attraction. By adjusting these parameters, one can give the system enough kinetic stability and adequate shelf-life.

13.1.2.2 Steric Stabilisation

This is produced by the presence of adsorbed surfactant and/or polymer layers. These layers will extend from the particle or droplet surface to some distance in the bulk solution. Provided these layers are strongly solvated by the molecules of the medium, they produce repulsion as a result of the unfavourable mixing of these chains. When two particles or droplets approach to a distance of separation h that is smaller than twice the adsorbed layer thickness 2δ, overlap and/or compression of these layers may take place, resulting in strong repulsion. In addition, when these adsorbed layers begin to overlap, the chains lose configurational entropy, resulting in an additional repulsion.

Combination of the mixing and entropic repulsion produces a strong steric repulsion that counteracts the van der Waals attraction, particularly at distances smaller than 2δ.

13.1.2.3 Electrosteric Repulsion

This is produced by a combination of electrostatic and steric repulsion, as with polyelectrolytes that can be used as dispersants or emulsifiers.

The above repulsive mechanisms ensure the colloid stability of suspensions or emulsions. This colloid stability should be distinct from the overall physical stability of the system, which implies complete homogeneity of the system with no separation on storage.

13.1.3 Physical Stability of Suspensions and Emulsions

The main driving force for separation of suspensions or emulsions is gravity. Most suspensions or emulsions have particle or droplet size ranges whereby the Brownian diffusion kT (where k is the Boltzmann constant and T is the absolute temperature) of the particles or droplets is insufficient to overcome the gravity force (that is given by the mass of each particle $\times$ acceleration due to gravity g $\times$ height of the container L), i.e.

$$\tfrac{4}{3}\pi R^3 \Delta\rho g L \gg kT \tag{13.3}$$

R is the particle or droplet radius, $\Delta\rho$ is the density difference between particles or droplets and the medium.

To prevent sedimentation or creaming, one needs to add suspending agents in the continuous phase. Such agents are usually high molecular weight polymers (such as hydroxyethyl cellulose or xanthan gum), usually referred to as thickeners.

These thickeners produce very high viscosities at low shear rates or shear stresses and hence they overcome the stresses exerted by the sedimenting or creaming particles of droplets. Generally, these high molecular weight polymers produce non-Newtonian flow with a "yield stress" that prevents separation on storage.

13.2
Surfactants in Disperse Systems

Surface active agents (usually referred to as surfactants) are amphipathic molecules that consist of a non-polar hydrophobic portion, usually a straight or branched hydrocarbon or fluorocarbon chain containing 8–18 carbon atoms, which is attached to a polar or ionic portion (hydrophilic) (see Chapter 1). The hydrophilic portion can be nonionic, ionic or zwitterionic, accompanied by counter ions in the last two cases.

The hydrocarbon chain interacts weakly with water molecules, whereas the polar or ionic head group interacts strongly with water molecules (ion–dipole or dipole–dipole interaction). The strong interaction of the head group with water molecules renders the surfactant molecule soluble in water. Cooperative action of dispersion and hydrogen bonding between the water molecules tends to "squeeze" the hydrophobic group out of the water (hydrophobic chains).

The balance between hydrophilic and hydrophobic part of the chain (sometimes referred to as the hydrophilic lipophilic balance, HLB) gives these molecules their special properties: Accumulation at various interfaces (adsorption); association in solution to form micelles.

13.2.1
General Classification of Surfactants

The most common surfactant classification is based on the nature of the head group: anionic, cationic, zwitterionic, nonionic (various examples are given in Chapter 1). A special class of surfactants with high molecular weights (polymeric surfactants) is given below.

13.2.2
Surfactants of Pharmaceutical Interest

13.2.2.1 Anionic Surfactants

a) Soaps
The most commonly used soaps are the alkali metal soaps, RCOOX, where X is sodium, potassium or ammonium, and R is generally between C_{10} and C_{20}.

b) Sulphated fatty alcohols
These are esters of sulphuric acid – the most commonly used compound is sodium lauryl sulphate, which is a mixture of sodium alkyl sulphates. The main component is sodium dodecyl sulphate, $C_{12}H_{25}$-O-$SO_3^-Na^+$. It is used pharmaceutically as a preoperative skin cleanser having bacteriostatic action against Gram positive bacteria. It is also used in medicated shampoos and tooth paste (as foam producer).

c) Ether sulphates (sulphated polyoxyethylated alcohols)
$R\text{-}(OCH_2\text{-}CH_2)_n\text{-}O\text{-}SO_3^-M^+$ ($n < 6$). This has better water solubility than the alkyl sulphates, better resistance to electrolyte and less irritation to the eye and the skin.

d) Sulphated oils
These include, for example, sulphated castor oil (triglyceride of the fatty acid 12-hydroxyoleic acid). This is used as an emulsifying agent for oil-in-water creams and ointments (non-irritant).

13.2.2.2 **Cationic Surfactants**

a) Cetrimide B.P.
This is a mixture consisting of tetradecyl (~68%), dodecyl (~22%) and hexadecyl (~7%) trimethylammonium bromide. Solutions containing 0.1–1% cetrimide are used to clean skin, wounds and burns, in shampoos to remove scales of seborrhoea, and also in Cetavlon cream.

b) Benzalkonium chloride
This consists of a mixture of alkyl benzylammonium chlorides. In dilute solutions (0.1–0.2%) it is used as pre-operative disinfection of the skin and mucous membranes, and as a preservative for eye-drops.

13.2.2.3 **Zwitterionic Surfactants**

The most commonly used zwitterionic surfactant in pharmacy is lecithin (phosphatidylcholine) which is applied as an oil-in-water emulsifier. It has the general structure shown by **13.1**.

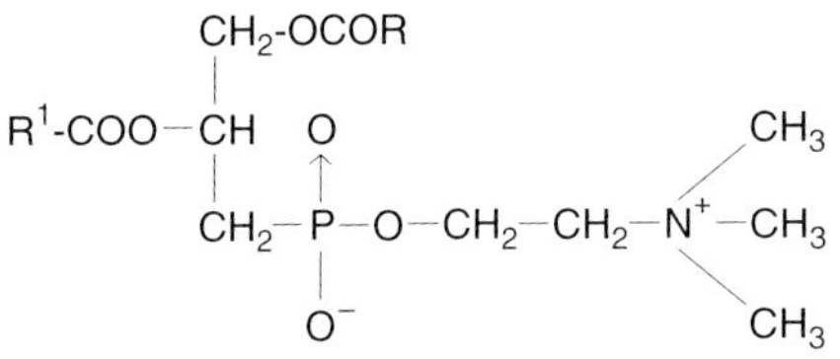

13.1

13.2.2.4 **Nonionic Surfactants**

These have the advantage over ionic surfactants in their compatibility with most other types of surfactants, they are little affected by moderate pH changes and moderate electrolyte concentrations. A useful scale for describing nonionic surfactants is the hydrophilic–lipophilic balance (HLB), which simply gives the relative proportion of hydrophilic to lipophilic components. For a simple nonionic surfactant such as an alcohol ethoxylate, the HLB is simply given by the percentage of hydrophilic components (PEO) divided by 5.

a) Sorbitan esters

Commercial products are mixtures of the partial esters of sorbitol and its mono- and di-anhydrides (**13.2**). Several sorbitan esters can be identified (Table 13.2). Such esters are water insoluble (low HLB) and oil soluble, and are used as water-in-oil emulsifiers.

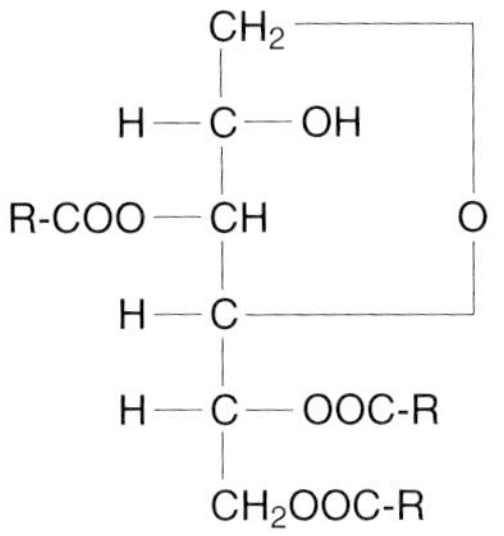

13.2

Tab. 13.2. Sorbitan esters.

Chemical name	*Commercial name*	*HLB*
Sorbitan monolaurate	Span 20	8.6
Sorbitan monopalmitate	Span 40	6.7
Sorbitan monostearate	Span 60	4.7
Sorbitan tristearate	Span 65	2.1
Sorbitan monooleate	Span 80	4.3
Sorbitan trioleate	Span 85	1.8

b) Polysorbates

These are the ethoxylated derivatives of sorbitan esters. Commercial products are complex mixtures of partial esters of sorbitol and its mono- and di-anhydrides condensed with an approximate number of moles of ethylene oxide. They have high HLB numbers, are water soluble and are used as oil-in-water emulsifiers. A list of polysorbates is given in Table 13.3.

Tab. 13.3. Polysorbates.

Chemical name	*Commercial name*	*HLB*
Polyoxyethylene (20) sorbitan monolaurate	Tween 20	16.7
Polyoxyethylene (20) sorbitan monopalmitate	Tween 40	15.6
Polyoxyethylene (20) sorbitan monostearate	Tween 60	14.9
Polyoxyethylene (20) sorbitan tristearate	Tween 65	10.5
Polyoxyethylene (20)	Tween 80	15.0

c) Polyoxyethylated glycol monoethers

These have the general structure C_xE_y, where x and y denote the alkyl and ethylene oxide chain length, e.g. $C_{12}E_6$ represents hexaoxyethylene glycol monododecyl ether.

One of the most widely used compounds is Cetromacrogel 1000 B.P.C., which is water soluble, with an alkyl chain length of 15 or 17 and an ethylene oxide chain length between 20 and 24. It is used in the form of cetomacrogel emulsifying wax in the preparation of oil-in-water emulsions and also as a solubilising agent for volatile oils.

Several other polyoxyethylated monoethers are commercially available, such as the Brij series from ICI: $C_{12}E_4$ (Brij 30)–$C_{12}E_{23}$ (Brij 35).

13.2.2.5 Polymeric Surfactants

The most commonly employed polymeric surfactants used in pharmacy are the A-B-A block copolymers, with A being the hydrophilic chain [poly(ethylene oxide), PEO] and B being the hydrophobic chain [poly(propylene oxide), PPO]. The general structure is PEO-PPO-PEO and is commercially available with different proportions of PEO and PPO (Pluronics, BASF or Synperonic PE, Poloxamers ICI). The commercial name is followed by a letter L (Liquid), P (Paste) and F (Flake). This is followed by two numbers that represent the composition – the first digit represents the PPO molecular mass and the second digit represents the % of PEO, e.g. Pluronic F68 (PPO Mol Wt 1501–1800) + 140 mol EO, and Pluronic L62 (PPO Mol Wt 1501–1800) + 15 mol EO.

13.2.3 Physical Properties of Surfactants and the Process of Micellisation

This subject is discussed in detail in Chapter 2, and only a summary is given here. The physical properties of surface active agents differ from those of smaller molecules in one major aspect, namely the abrupt changes in their properties above a critical concentration. The abrupt changes within a critical concentration range are consistent with the fact that, at and above this concentration, surface active ions or molecules in solution associate to form larger units (micelles). The concentration at which this occurs is known as the critical micelle concentration (c.m.c.). The c.m.c. depends on the structure of the surfactant molecule, the composition of the aqueous phase and the temperature. For a homologous series with the same hydrophilic group, the c.m.c. decreases with increasing alkyl chain length. The c.m.c. for nonionics is about two orders of magnitude lower than that for ionics with the same alkyl chain length. For a given alkyl chain length, the c.m.c. for nonionic ethoxylates increases with increasing number of EO units. For ionic surfactants, the c.m.c. increases with rising temperature, but decreases on addition of electrolytes. For nonionic surfactants, the c.m.c. decreases with rising temperature and/or addition of electrolytes.

The above trends are schematically illustrated in Figures 13.2 to 13.4, where the c.m.c. (log scale) is plotted as a function of alkyl chain length (at a given hydro-

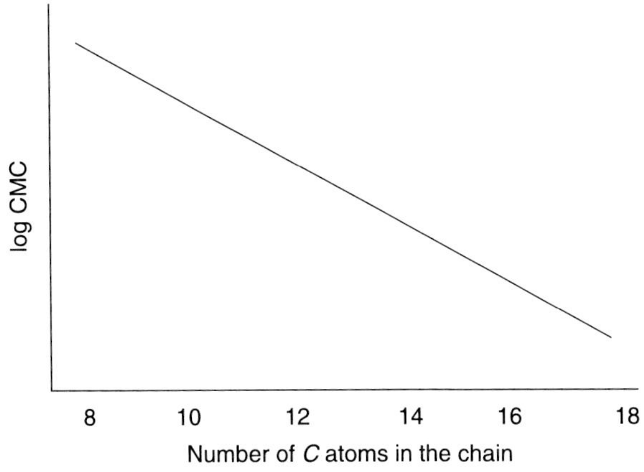

Fig. 13.2. Variation of c.m.c. with alkyl (R) chain length at a given hydrophilic group (ionic or PEO).

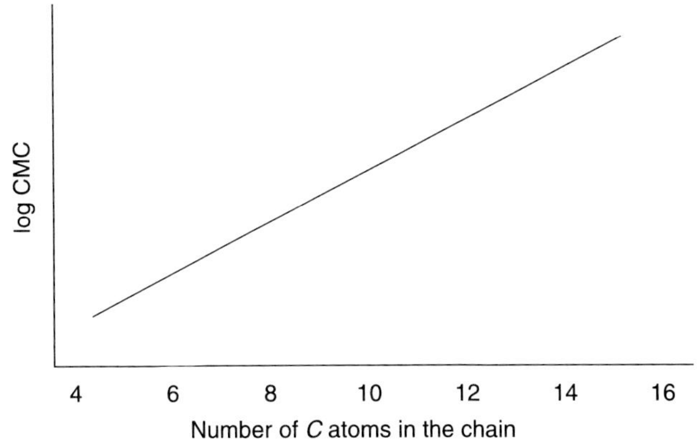

Fig. 13.3. Variation of c.m.c. with EO units at a given R.

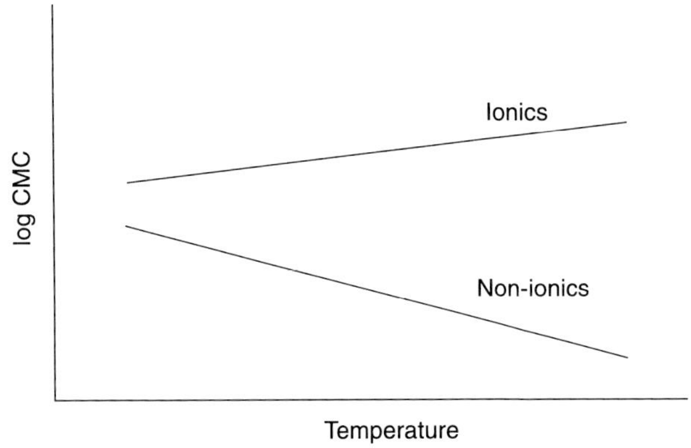

Fig. 13.4. Variation of c.m.c. with temperature.

philic group), as a function of number of EO units (at a given alkyl chain length), and as a function of temperature, respectively.

The presence of micelles can account for many of the unusual properties of surfactants found above the c.m.c.: Constant surface tension, reduction in molar conductance, rapid increase in light scattering, and increase in solubilisation.

13.2.4
Size and Shape of Micelles

McBain, Hartley and Adam [4–7] suggested that micelles are spherical, with a radius approximately equal to the chain length of R (for ionic surfactants), and an aggregation number of 50–100. The micelle interior is "liquid-like". Debye and Anacker suggested the presence of rod-shaped micelles to explain the light scattering results for hexadecyltrimethylammonium bromide. McBain also suggested the presence of lamellar micelles to account for the drop in molar conductance, and this were confirmed by X-ray scattering. Figure 2.2 (see Chapter 2) gives a schematic representation of the shape of micelles.

The solubility of ionic surfactants increases gradually with increasing temperature, but at a critical temperature there is a rapid increase of solubility with further increase in temperature. This critical temperature is termed the Krafft Temperature (it increases with increasing alkyl chain length).

Solutions of nonionic surfactants of the ethoxylate type show special behaviour with increasing temperature, namely "clouding" above a certain critical temperature. This is defined as the cloud point (CP), which depends on the surfactant concentration and its composition. CP decreases with increasing alkyl chain length (at a given EO number) and increases with increasing EO number (at a given R). In addition, CP decreases with increasing electrolyte concentration (for most electrolytes). With rising temperature, the PEO chain becomes dehydrated (breaking of hydrogen bonds) and, at the CP, the dehydrated micelles aggregate, which is probably the origin of the clouding phenomenon.

13.2.5
Surface Activity and Adsorption at the Air/Liquid and Liquid/Liquid Interfaces

This is dealt with in detail in Chapter 3 and only a summary is given here. Adsorption of surfactants at the air/liquid or liquid/liquid interface lowers the surface or interfacial tension γ. Just before the c.m.c., the $\gamma - \log[C_{SAA}]$ curve is linear and above the c.m.c. γ becomes virtually constant. From the slope of the linear portion of the $\gamma - \log C$ curve one can obtain the amount of surfactant adsorption Γ (mol m^{-2}), usually referred to as the surface excess using the Gibbs equation [8], where R is the gas constant and T is the absolute temperature,

$$\frac{d\gamma}{d \log C} = -2.303\Gamma RT \qquad (13.4)$$

From Γ one can calculate the area per molecule,

$$\text{Area per molecule} = \frac{10^{18}}{\Gamma \times N_{\text{av}}}\ \text{nm}^2 \tag{13.5}$$

The area per molecule gives information on the orientation of the surfactant molecules at the interface: For a flat orientation, the area per molecules is given by the area of the hydrocarbon chain and the head group (it increases with increasing R chain length and/or number of EO units). For vertical orientation (as is mostly the case near the c.m.c.), the area per molecule is determined by the cross sectional area of the head group (it does not significantly depend on the R chain length, but it increases with rising number of EO units).

From the c.m.c., one can determine the free energy of micellisation, $\Delta G^\circ_{\text{m}}$,

$$\Delta G^\circ_{\text{m}} = RT \ln[\text{c.m.c.}] \tag{13.6}$$

The free energy of micellisation is large and negative, indicating that micelle formation is spontaneous and that micelles are thermodynamically stable.

13.2.6
Adsorption at the Solid/Liquid Interface

The adsorption of surfactants at the solid/liquid interface may be described by the Langmuir equation [9],

$$\Gamma = \frac{\Gamma_\infty bC}{1 + bC} \tag{13.7}$$

where Γ_{∞} is the plateau value at $C >$ c.m.c. and C_2 is the equilibrium surfactant concentration.

The adsorption of many surfactants at the solid/liquid interface may show several steps that are accompanied by various structures on the surface (bilayers, hemi-micelles and micelles). This is represented in detail in Chapter 4.

13.2.7
Phase Behaviour and Liquid Crystalline Structures

This is discussed in detail in Chapter 3, which includes typical phase diagrams for nonionic surfactants such as dodecyl hexaoxyethylene glycol monoether–water mixture. Three main liquid crystalline phases could be distinguished: M (hexagonal) phase, an anisotropic phase consisting of hexagonally packed rod-shaped units that appears like a transparent gel; N (Lamellar L_α) phase, an anisotropic phase consisting of sheets of molecules in a bimolecular packing. This is less viscous than the middle phase. In some cases, a cubic phase that is viscous and isotropic (consisting of hexagonally packed spherical units) may be produced between the hexagonal and the lamellar phases.

A schematic representation of the liquid crystalline structures is given in Chapter 3, such structures can be identified using polarising microscopy, which shows particular textures for each phase. They can also be studied using low angle X-rays.

13.3
Electrostatic Stabilisation of Disperse Systems

As mentioned in the general introduction, disperse systems are thermodynamically unstable, since the surface energy of a large number of small particles or droplets is large and positive and the system tends to aggregate and/or coalesce to reduce such surface energy. The main driving force for aggregation of particles or droplets is the universal van de Waals attraction, which increases very sharply at small separations. To overcome the van der Waals attraction, one needs to have a repulsive force (energy) between the particles, particularly at intermediate and small separations.

Such a repulsive energy can be produced by charge separation and the creation of electrical double layers, as discussed in detail in Chapters 6 and 7. Combination of the van der Waals attraction and double layer repulsion at various separation distances between the particles produce an energy–distance curve, which will have an energy barrier at intermediate separations [10] and this is the origin of electrostatic stabilisation. The energy–distance curve is controlled by the following parameters.

13.3.1
Van der Waals Attraction

Equation (13.8) gives the van der Waals attractive energy, G_A, for two particles or droplets with equal radius R, and surface-to-surface separation h (when $h \ll R$),

$$G_A = -\frac{AR}{12h} \tag{13.8}$$

where A is the effective Hamaker constant, which is given by

$$A = (A_{11}^{1/2} - A_{22}^{1/2})^2 \tag{13.9}$$

where A_{11} is the Hamaker constant of the particles or droplets and A_{22} is the Hamaker constant of the medium.

The Hamaker constant of any material depends on the number of atoms per unit volume q and the London dispersion constant β,

$$A = \pi q^2 \beta \tag{13.10}$$

G_A is seen to increase very sharply with decreasing h when the latter reaches small values. In the absence of repulsion between the particles or droplets, the latter will aggregate (flocculate) by simple diffusion through the medium. This leads to fast flocculation kinetics and the rate constant for the process k_0 has been calculated using the Smolulokowski equation,

$$k_0 = \frac{4kT}{3\eta} = 5.5 \times 10^{-18}\ \mathrm{m^3\ s^{-1}} \tag{13.11}$$

k is the Boltzmann constant, T is the absolute temperature and η is the viscosity of the medium.

13.3.2
Double Layer Repulsion

An electrical double layer can be created at the solid/liquid or liquid/liquid interface by charge separation due to the presence of ionogenic groups (e.g. –OH, –COOH) or by adsorption of ionic surfactants at the interface.

The double layer is characterised by the surface charge (σ_0), the charge in the Stern layer (σ_s), the charge of the diffuse layer σ_d (note that $\sigma_0 = \sigma_s + \sigma_d$) the surface potential $(\Psi)_0$ and the Stern potential Ψ_d ($\sim$zeta potential).

The double layer extension is determined by the electrolyte concentration and the valency of the counter ions, as given by the reciprocal of the Debye–Hückel parameter $(1/\kappa)$ – referred to as the thickness of the double layer,

$$\left(\frac{1}{\kappa}\right) = \left(\frac{\varepsilon_r \varepsilon_0 kT}{2n_0 Z_i^2 e^2}\right) \tag{13.12}$$

where ε_r is the permittivity (dielectric constant) of the medium, ε_0 is the permittivity of free space, n_0 is the number of ions per unit volume of each type present in bulk solution, Z_i is the valency of the ions and e is the electronic charge.

The double layer thickness increases with decreasing electrolyte concentration, 10^{-5} mol dm^{-3} NaCl $(1/\kappa) = 100$ nm and 10^{-3} mol dm^{-3} NaCl $(1/\kappa) = 10$ nm.

When two particles or droplets with double layers of the same sign approach to a separation h that is smaller than twice the double layer thickness, double layer repulsion occurs, since the two double layers cannot be fully extended in the confined space. This leads to repulsion energy G_{el}, which is given by

$$G_{el} = \frac{[4\pi\varepsilon_r\varepsilon_0 R^2 \Psi_0^2 \exp(-\kappa h)]}{[2R + h]} \tag{13.13}$$

The above expression shows that G_{el} decreases exponentially with increasing h and the rate of this decrease depends on electrolyte concentration.

13.3.3 Total Energy of Interaction

Combination of G_A with G_{el} at various h results an the total energy G_T–distance curve, as illustrated in Figure 13.5.

This presentation forms the basis of the theory of colloid stability due to Deryaguin–Landau–Verwey–Overbeek (DLVO theory) [10]. The G_T–h curve shows two minima and one maximum: A shallow minimum (of the order of few kT) units at large separations, which may result in weak and reversible flocculation. A deep primary minimum (several $100kT$ units) is seen at short separations – this results in strong flocculation (coagulation). An energy maximum, G_{max}, at intermediate distances prevents flocculation into the primary minimum.

To ensure adequate colloid stability, G_{max} has to be greater than $25kT$. The height of the maximum depends on the surface (or zeta) potential, electrolyte concentration, particle radius and Hamaker constant. When the zeta potential is higher than 40 mV, and the electrolyte concentration is $<10^{-2}$ mol dm^{-3} for 1:1 electrolyte, G_{max} is $>25kT$.

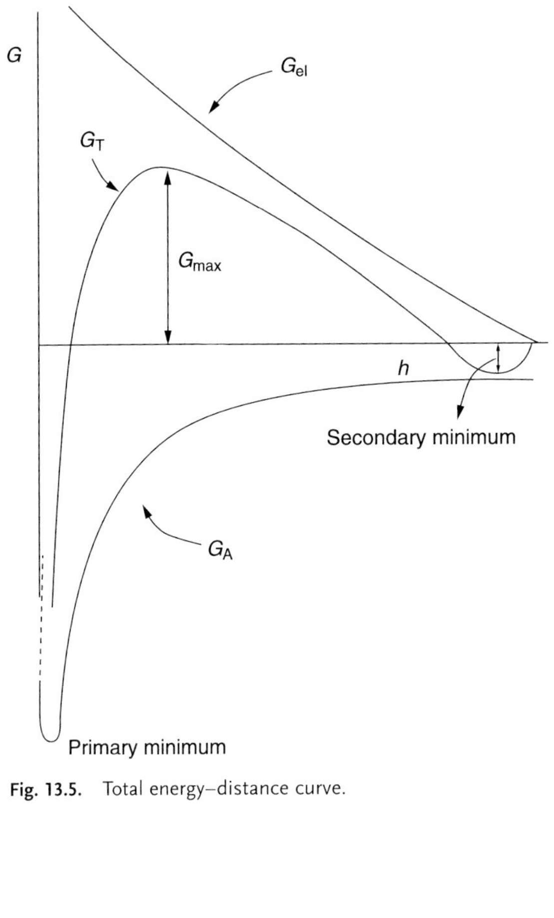

Fig. 13.5. Total energy–distance curve.

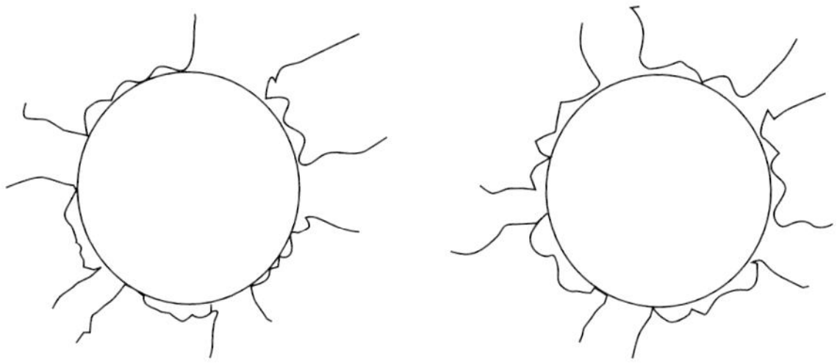

Fig. 13.6. Scheme of the adsorption of Poloxamers on suspension particles and oil droplets.

To maintain colloid stability over a long period of time (i.e. 2–3 years), one needs to ensure the following conditions: High zeta potential by ensuring adequate coverage of the particles or droplets by ionic surfactant; low electrolyte concentration; low valency of the electrolyte (multivalent ions should be avoided).

13.4 Steric Stabilization of Disperse Systems

Many polymers are used for the preparation of disperse systems (suspensions and emulsions) in pharmaceutical formulations. An example of such systems is the A-B-A block copolymer of poly(ethylene oxide)-poly(propylene oxide)-poly(ethylene oxide), PEO-PPO-PEO, commercially available as Poloxamers (ICI), Pluronics (BASF) and Synperonic PE (ICI). On hydrophobic drug particles or oil droplets, the polymer adsorbs with the B hydrophobic chain (PPO) close to the surface, leaving the two hydrophilic A chains dangling in solution.

Figure 13.6 gives a schematic picture of the adsorption and conformation of the Poloxamers.

These nonionic polymers provide stabilisation against flocculation and/or coalescence by a mechanism usually referred to as steric stabilisation (discussed in detail in Chapters 6 and 7).

To understand the principles of steric stabilisation, one must first consider the adsorption and conformation of the polymer at the solid/liquid or liquid/liquid interface. This is discussed in detail in Chapter 5, and only a summary is given here.

13.4.1 Adsorption and Conformation of Polymers at Interfaces

Consider the case of the PEO-PPO-PEO block copolymer at the interface (represented by a simple flat surface). The PPO chain adsorbs on the surface with many attachment points, forming small "loops", whereas the A chains (sometimes referred to as "tails") extend to some distance (few nm) from the surface [11–13]. The chain segments in direct contact with the surface are termed "trains". A

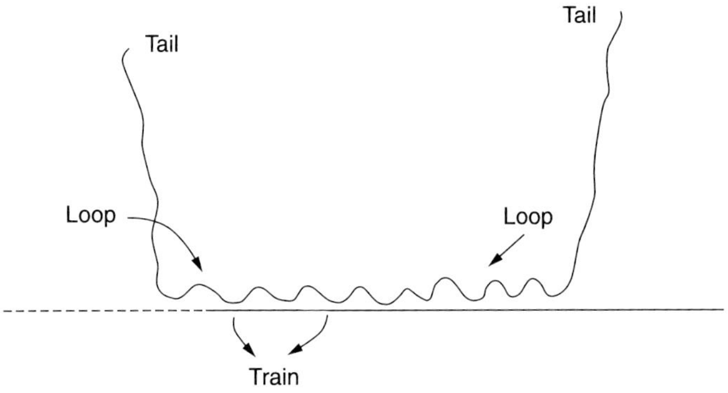

Fig. 13.7. Representation of the conformation of an ABA block copolymer at the interface.

scheme of the configuration of the block copolymer at the interface is shown in Figure 13.7.

Adsorption of polymers at interfaces differs significantly from that of simple surfactant molecules (described in Chapter 5): The adsorption isotherm is of the high affinity type, i.e. the first added molecules are virtually completely adsorbed and the plateau adsorption is reached at low equilibrium concentration. Adsorption is practically "irreversible" because the molecule is attached with several segments to the surface.

Figure 13.8 shows a schematic representation of the adsorption isotherm, whereby the amount of adsorption Γ (mg m^{-1} or mol m^{-2}) is plotted versus the equilibrium concentration C_2.

The isotherm depends on the structure, molecular weight and environment (temperature, electrolyte) of the chains. To fully characterise polymer adsorption, one needs to obtain information on the following parameters: The amount of ad-

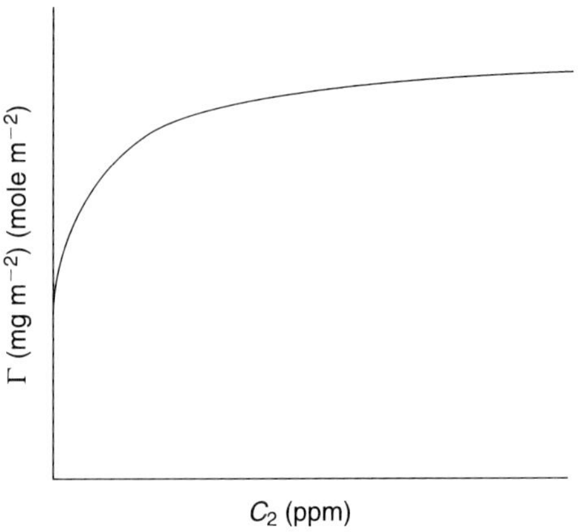

Fig. 13.8. Typical adsorption isotherm for a polymer.

sorption (Γ), the fraction of segments in "trains" and the adsorption energy per segment, the extension of the A chains in bulk solution, usually described as "segment density distribution", $\rho(z)$, or hydrodynamic thickness δ_h. It is essential to know how these parameters vary with the system parameters such as proportion of hydrophobic to hydrophilic chains, molecular weight, flexibility, temperature, and addition of electrolyte.

The most important parameter for steric stabilisation is the strong "anchoring" of the B chain to the surface and the extension of the A chains (adsorbed layer thickness, δ_h) and its solvation by the molecules of the medium.

13.4.2
Interaction Forces (Energies) Between Particles or Droplets Containing Adsorbed Non-ionic Surfactants and Polymers

When two particles or droplets each with a radius R and containing an adsorbed surfactant or polymer layer with a hydrodynamic thickness δ_h, approach each other to a surface–surface separation h that is smaller than $2\delta_h$, the surfactant or polymer layers interact, resulting in two main conditions: polymer chains may overlap and the polymer layers may undergo compression. In both cases, the local segment density of the chains in the interaction region will increase (Figure 13.9).

The real situation lies, perhaps, between the above two cases, i.e. the polymer chains may undergo some interpenetration and some compression. Providing the dangling chains (A chains) are in a good solvent (i.e. strongly solvated by the solvent molecules), this local increase in segment density will result in strong repulsion due to two main effects, given below.

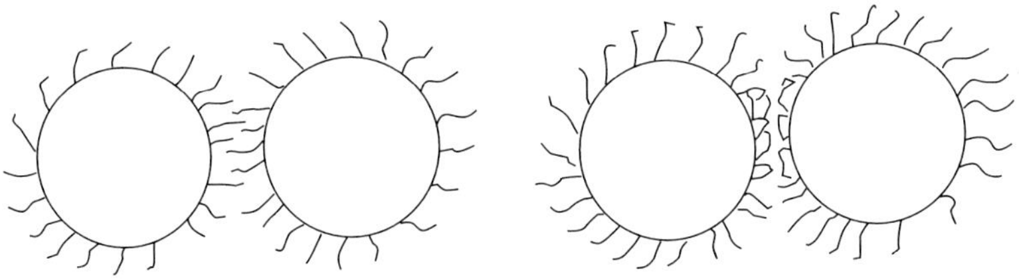

Interpenetration without compression   Compression without interpenetration

Fig. 13.9. Scheme of interaction of adsorbed layers.

13.4.2.1 Osmotic or Mixing Interaction

As a result of the unfavourable mixing of the chains, when these are in good solvent, the osmotic pressure in the interaction zone increases – this is described as a free energy of interaction, G_{mix}, that is given by

$$\frac{G_{mix}}{kT} = \left(\frac{2V_2^2}{V_1}\right)\nu^2\left(\frac{1}{2}-\chi\right)\left(\delta-\frac{h}{2}\right)^2\left(3R+2\delta+\frac{h}{2}\right) \qquad (13.14)$$

where k is the Boltzmann constant, T is the absolute temperature, V_2 is the molar volume of polymer, V_1 is the molar volume of solvent, ν_2 is the number of polymer chains per unit area, and χ is the Flory–Huggins interaction parameter.

The sign of G_{mix} depends on the value of the Flory–Huggins interaction parameter. When the chains are in good solvent conditions (strongly solvated by the molecules of the medium), $\chi < \frac{1}{2}$ and G_{mix} is positive, i.e. the mixing interaction free energy is positive and this leads to strong repulsion as soon as $h < 2\delta$. Clearly, to main stability of a suspension or emulsion, one must ensure that χ is less than $\frac{1}{2}$ under all conditions of storage (e.g. temperature variation, addition of electrolyte, etc.).

13.4.2.2 Entropic, Volume Restriction or Elastic Interaction

This results in significant overlap of the chains, which lose configurational entropy. The entropy loss leads to a positive free energy of interaction, G_{el}, which increases very sharply when $h < \delta$.

G_{el} may be given by the following simple expression (assuming the chains to be represented by simple rods that rotate in a circle with a radius δ),

$$\frac{G_{el}}{kT} = 2kT\nu^2 \ln\left[\frac{\Omega(h)}{\Omega(\infty)}\right] \tag{13.15}$$

where $\Omega(h)$ is the number of chain configurations after overlap ($h < \delta$), and $\Omega(\infty)$ is the number of chain configurations before overlap ($h > 2\delta$).

G_{el} is always positive and could play a major role in steric stabilisation. The steric free energy of interaction G_s is given by the sum of G_{mix} and G_{el}, and when this is added to the van der Waals attraction gives the total interaction energy G_T,

$$G_T = G_s + G_A = G_{mix} + G_{el} + G_A \tag{13.16}$$

Figure 13.10 gives a schematic representation of the variation of G_{mix}, G_{el}, G_A and G_T with h .

G_{mix} increases very sharply with decrease of h when $h < 2\delta$. G_{el} increases very sharply with decreasing h when $h < \delta$. G_T versus h shows a minimum, G_{min}, at separation distances comparable to 2δ – G_T shows a rapid increase with further decrease in h [11–13].

Unlike the G_T–h predicted by the DLVO theory (which shows two minima and one energy maximum), the G_T–h curve for systems that are sterically stabilised shows only one minimum, G_{min}, followed by a sharp increase in G_T when $h < 2\delta$. The depth of the minimum depends on the Hamaker constant A, particle radius R and adsorbed layer thickness. At a given A and R, G_{min} increases with decreasing δ. When δ is small (say less than 5 nm), G_{min} may reach sufficient depth (few kT units) for weak flocculation to occur. However, this flocculation is reversible, and by gentle shaking of the container the suspension or emulsion can de easily redispersed.

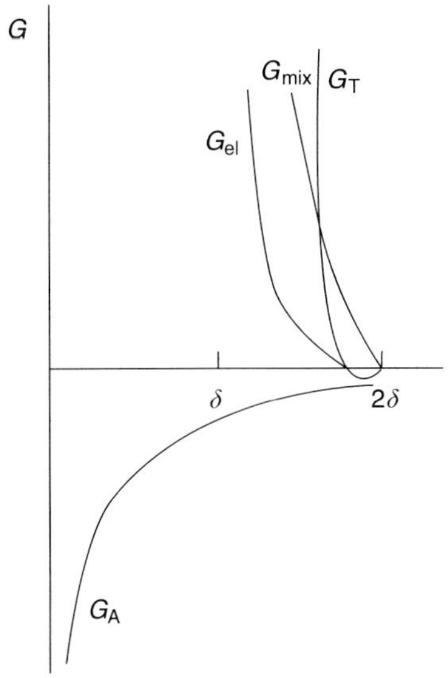

Fig. 13.10. Variation of G_{mix}, G_{el}, G_A and G_T with surface–surface distance between the particles.

When δ is sufficiently large (say > 10 nm), G_{min} may become so small that no flocculation occurs. In this case the suspension or emulsion approaches thermodynamic stability and no aggregation occurs over very long periods (more than two years). Clearly, the particle radius also plays a major role – the larger the particles, the deeper G_{min} becomes (at a given δ).

13.4.3
Criteria for Effective Steric Stabilisation

(1) The particles or droplets should be completely covered by the surfactant or polymer (the amount should correspond to the plateau value). Any bare patches may cause flocculation either by van der Waals attraction (between the bare patches) or by bridging flocculation (where a polymer molecule will become simultaneously adsorbed on two or more particles or droplets).

(2) The polymer should be strongly "anchored" to the particles or droplets surface, to prevent any displacement during particle approach – this is particularly important for concentrated suspensions or emulsions. With an A-B-A block copolymer, the B chain is chosen to be highly insoluble in the medium and has a strong affinity to the surface.

(3) The stabilising chain(s) A should be highly soluble in the medium and strongly solvated by its molecules. In other words the Flory–Huggins interaction parameter for the A chains should always be less than 1/2 – the most commonly used A chains are those based on PEO (e.g. with Poloxamers).

(4) The adsorbed layer thickness δ should be sufficiently large (> 5–10 nm) to prevent weak flocculation. This is particularly the case with concentrated suspensions and emulsions, since such flocculation may cause an increase in the viscosity of the system, making it difficult to redisperse on shaking.

13.5
Surface Activity and Colloidal Properties of Drugs

Many drugs are surface active, e.g. chlorpromazine, diphenylmethane derivatives (such as diphenhydramine) and tricyclic antidepressants (such as amitriptyline) [14]. As an illustration is the structure of chlorpromazine (**13.3**).

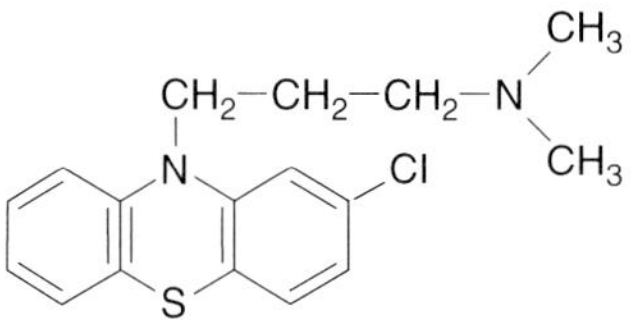

13.3

The biological and pharmaceutical consequences of the surface activity will be discussed here. The solution properties of these surface active drugs and their mode of association play an important role in their biological efficacy.

Surfactants are also used in many pharmaceutical formulations, e.g. to prepare suspensions or emulsions of insoluble drugs, or as solubilizers (in the micelles) for many compounds for application as injectables or enhancement of the drug efficacy. Many surfactants are also used as germicides or antibacterials (e.g. the cationic quaternary ammonium salts).

Another important class of surfactants is the bile salts (that are synthesized in the liver), phospholipids and cholesterol, which are the main constituents of membranes. These naturally occurring surfactants will also be discussed, briefly, in this section.

13.5.1
Association of Drug Molecules

As mentioned before, many drugs exhibit surface active properties that are similar to surfactants, e.g. they accumulate at interfaces and produce aggregates (micelles) at critical concentrations. However, micellization of drugs represents only one pattern of association, since, with many drug molecules, rigid aromatic or heterocycles replace the flexible hydrophobic chains present in most surfactant systems. This will have a pronounced effect on the mode of association, to an extent that the process may not be regarded as micellization. A self-association structure may be produced by hydrophobic interaction (charge repulsion plays an insignificant role in this case) and the process is generally continuous, i.e. with no abrupt change in the properties. However, many drug molecules may contain aromatic groups with

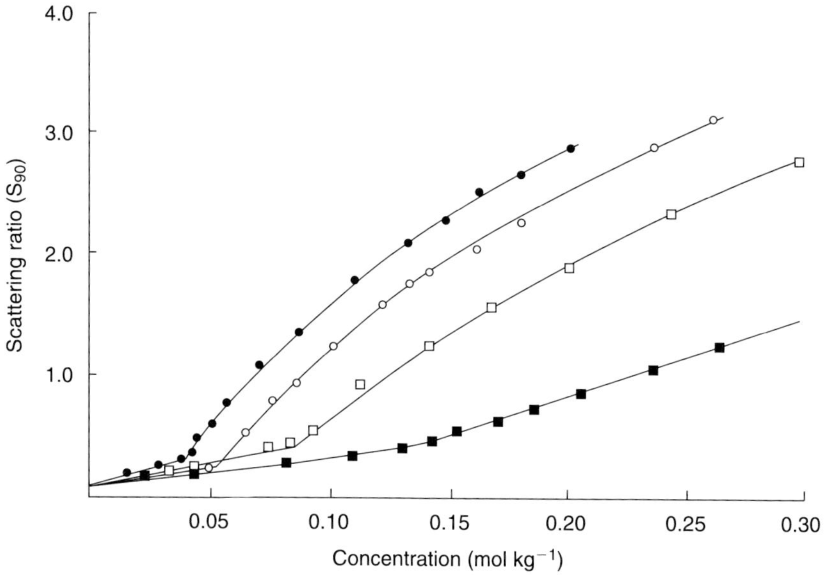

Fig. 13.11. Variation of the scattering ratio, S_{90}, with concentration for aqueous solutions of diphenylmethane antihistamines: (●) chlorocyclizine hydrochloride; (○) bromodiphenhydramine hydrochloride; (□) diphenylpyraline hydrochloride; (■) diphenylhydramine hydrochloride; (– –) calculated from mass action theory.

a high degree of flexibility. In this case, the association structures resemble surfactant micelles. Figure 13.11 illustrates light scattering results for several diphenylmethane antihistamines [15, 16].

The results of Figure 13.11 clearly show distinct inflection points, which may be identified with the c.m.c. However, the aggregation numbers of these association units are much lower (in the region of 9–12) than those encountered with micellar surfactants (which show aggregation numbers of 50 or more depending on the alkyl chain length, see Chapter 2). These lower aggregation numbers cast some doubt on micelle formation and a continuous association process may be envisaged instead. The light scattering results could be fitted by Attwood and Udeala [16] using the mass action model for micellization (see Chapter 2).

Considering the ionic micelle M^{p+} to be formed by association of n drug ions, D^+, and $(n-p)$ firmly bound counter ions, X^-,

$$nD^+ + (n-p)X^- \rightleftharpoons M^{p+} \tag{13.17}$$

The equilibrium constant for micelle formation assuming ideality is given by

$$K_m = \frac{x_m}{[x_s]^n[x_x]^{n-p}} \tag{13.18}$$

where x_x is the mole fraction of counter ion.

The standard free energy of micellization per mole of monomeric drug is given by

$$\Delta G^\circ_m = -\frac{RT}{n} \ln K_m = -\frac{RT}{n} \ln \left[\frac{x_m}{[x_s]^n [x_x]^{n-p}} \right] \quad (13.19)$$

which on rearrangement gives Eq. (13.20),

$$\log x_s = -\left(1 - \frac{p}{n}\right) \log x_x + \frac{\Delta G^\circ_m}{2.303RT} + \frac{1}{n} \log x_m \quad (13.20)$$

Assuming the monomeric drug concentration x_s, in the presence of micelles, to be equal to the c.m.c., Eq. (13.20) may be written in a simple form,

$$\log \text{c.m.c.} = -a \log x_x + b \quad (13.21)$$

where a is equal to $(1 - p/n)$, i.e. $(1 - \alpha)$, where α is the degree of dissociation and b is equal to $(\Delta G^\circ_m/2.303RT) + (1/n) \log x_m$.

The solid line in Figure 13.11 is based on calculations using Eq. (13.20). Addition of electrolyte to solutions of these diphenylmethane antihistamines produces an increase in the aggregation number and a decrease in the c.m.c., as commonly found with simple surfactants. Figure 13.12 shows plots of log-c.m.c. versus counter-ion concentration [17]. These plots are linear, as predicted from Eq.

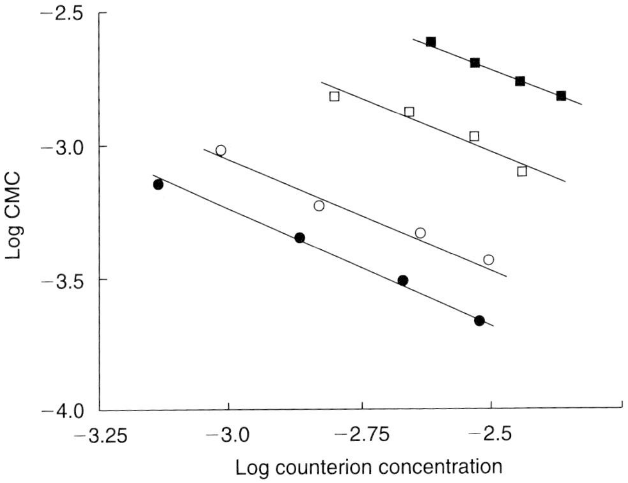

Fig. 13.12. Log-c.m.c. against counter ion concentration: (●) bromodiphenylhydramine hydrochloride; (○) chlorocyclizine hydrochloride; (□) diphenylpyraline hydrochloride; (■) diphenhydramine hydrochloride.

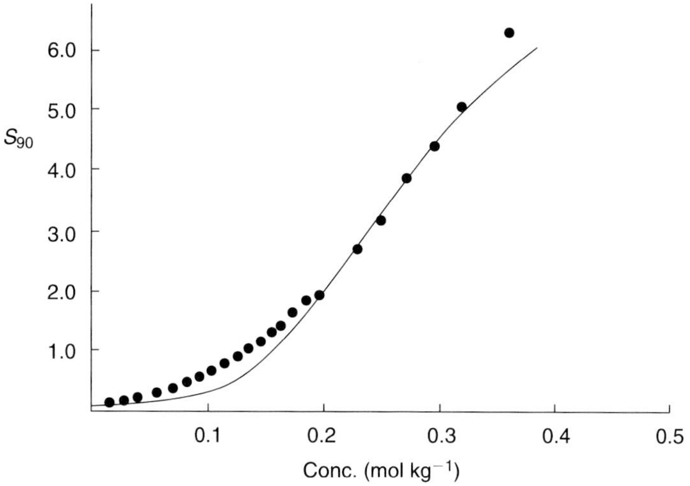

Fig. 13.13. Concentration dependence of the scattering ratio, S_{90}, for mepyramine maleate.

(13.20). Values of α derived from the slopes of these lines are in agreement with those obtained from the light-scattering data. In addition, the standard free energy of micellization, ΔG_m°, determined from the intercept of the lines is in reasonable agreement with the expected value derived from consideration of the free energy associated with the transfer of two phenyl rings from an aqueous to a non-aqueous environment. The micellar charge and hydration of the diphenylmethane antihistamines have been examined in detail by Attwood and Udeala in reference [18], to which the reader should refer for further information.

The above results indicate that the diphenylmethane derivatives of histamines behave as normal surfactants with a clear c.m.c. However, this is not general since other derivatives such as mepyramine maleate (a pyridine derivative) did not show a clear break point. This is illustrated in Figure 13.13, which shows the light-scattering results that indicate a continuous association process with no apparent c.m.c.

The solid line in Figure 13.13 was obtained using Eq. (13.20) with $n = 10$, $K_m = 10^{42}$ and $\alpha = 0.2$. Using such values, an inflection point is obtained, which is not present in the experimental data. However, surface tension results showed, in many cases. a break point in the $\gamma - \log C$ curves (Figure 13.14) [18].

Later studies on other drugs with non-micellar association patterns showed that the apparent c.m.c. detected by surface tension techniques arose because of the very limited change of monomer concentration with total solution concentration at high concentrations.

Several other examples may be quoted to illustrate the association of drug molecules producing either micelles or other association structures. For more detail the reader should refer to the text by Attwood and Florence [14].

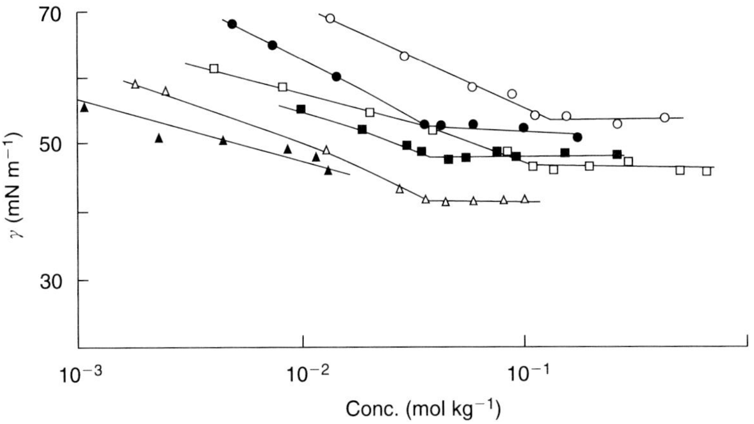

Fig. 13.14. γ *versus* log C for several antihistamines: (○) pheniramine; (●) bromopheniramine maleate; (□) diphenhydramine; (■) bromdiphenhydramine hydrochloride; (△) cyclizine; (▲) chlorocyclizine hydrochloride in water at 30 °C.

13.5.2
Role of Surface Activity and Association in Biological Efficacy

Both the surface activity and micellization have implications on the biological efficacy of many drugs. Surface active drugs tend to bind hydrophobically to proteins and other biological macromolecules. They also tend to associate with other amphipathic molecules such as other drugs, bile salts and, of course, with receptors. Guth and Spirtes [19] attributed the activity of phenothiazines to their interaction with membranes, which may be correlated with their surface activity. These compounds act by altering the conformation and activity of enzymes and by altering membrane permeability and function.

Several other examples may be quoted to illustrate the importance of surface activity of many drugs. Many drugs produce intralysosomal accumulation of phospholipids, which are observable as multilamellar objects within the cell. Drugs implicated in phospholipidosis induction are often amphipathic compounds [20]. Interaction between the surfactant drug molecules and phospholipid renders the phospholipid resistant to degradation by lysosomal enzymes, resulting in their accumulation in cells.

Many local anaesthetics have significant surface activity and it is tempting to correlate their surface activity to their action. However, one should not forget other important factors such as partitioning of the drug into the nerve membrane (a factor that depends on the pK_a) and the distribution of hydrophobic and cationic groups, which must be important for the appropriate disruption of nerve membrane function.

The biological relevance of micelle formation by drug molecules is not as clear as their surface activity, since the drug is usually applied at concentrations well below that at which micelles are formed. However, accumulation of drug molecules in certain sites may allow them to reach concentrations whereby micelles are produced. Such aggregate units may cause significant biological effects. For example, the concentration of monomeric species may increase only slowly or may decrease with increasing total concentration and the transport and colligative properties of the system are changed. In other words, the aggregation of the compounds will affect their thermodynamic activity and hence their biological efficacy in vivo.

13.5.3 Naturally Occurring Micelle Forming Systems

Several naturally occurring amphipathic molecules (in the body) exist, such as bile salts, phospholipids, cholesterol, which play an important role in various biological processes. Their interactions with other solutes, such as drug molecules, and with membranes are also very important. Below a brief summary of some of these biological surfactants is given, illustrating their interactions.

Bile salts are synthesized in the liver and consist of alicyclic compounds possessing hydroxyl and carboxyl groups; an illustration is cholic acid (**13.4**).

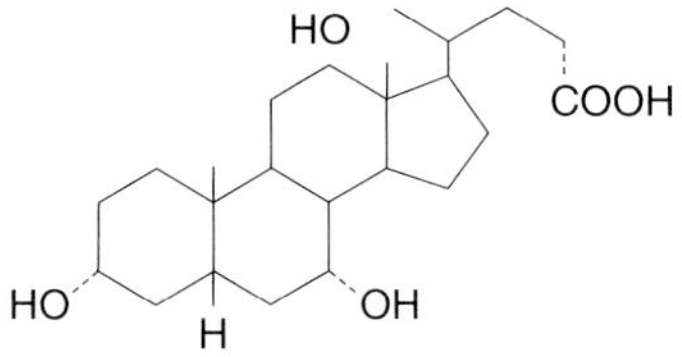

13.4

It is the positioning of the hydrophilic groups in relation to the hydrophobic steroidal nucleus that gives the bile salts their surface activity, and determines the ability to aggregate.

Figure 13.15 shows the possible orientation of cholic acid at the air/water interface, the hydrophilic groups being oriented towards the aqueous phase [21, 22]. The steroid portion of the molecule is shaped like a "saucer" as the A ring is cis with respect to the B ring.

Small [23] suggested that small or primary aggregates with up to 10 monomers form above the c.m.c. by hydrophobic interactions between the non-polar sides of the monomers. These primary aggregates form larger units by hydrogen bonding between the primary micelles. This is schematically illustrated in Figure 13.16. Oakenfull and Fisher [22, 24] stressed the role of hydrogen bonding rather than hydrophobic bonding in the association of bile salts. However, Zana [25] regarded the association as a continuous process with hydrophobic interaction as the main driving force.

The c.m.c. of bile salts is strongly influenced by their structure; the trihydroxycholanic acids have higher c.m.c.s than the less hydrophilic dihydroxy derivatives. As expected, the pH of solutions of these carboxylic acid salts influences micelle

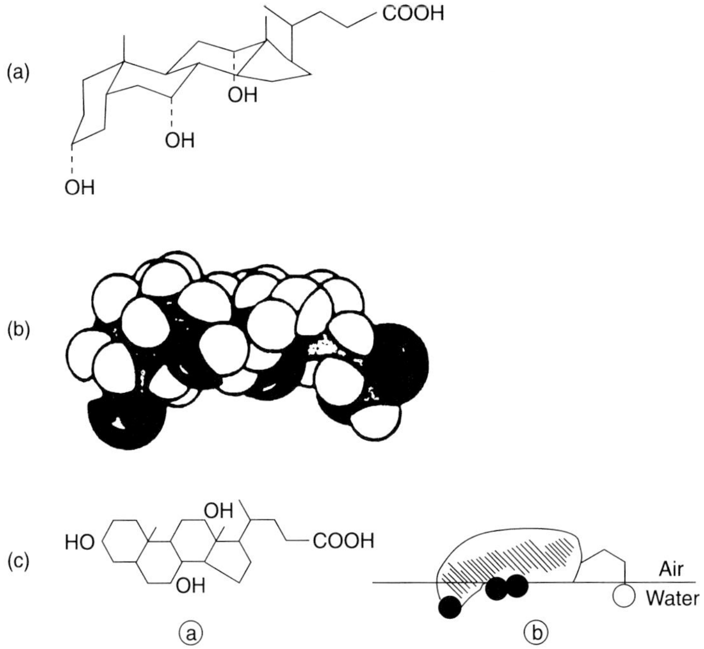

Fig. 13.15. (a) Structural formula of cholic acid, showing the cis position of the A ring; (b) Courtauld space-filling model of cholic acid; (c) orientation of cholic acid molecules at the air–water interface (hydroxyl groups represented by filled circles and carboxylic acid groups by open circles).

formation. At sufficiently low pH, bile acids, which are sparingly soluble, will be precipitated from solution, initially being incorporated or solubilized in the existing micelles. The pH at which precipitation occurs, on saturation of the micellar system, is generally about one pH unit higher than the pK_a of the bile acid.

Bile salts play important roles in physiological functions and drug absorption. It is generally agreed that bile salts aid fat absorption. Mixed micelles of bile salts, fatty acids and monogylycerides can act as vehicles for fat transport. The role of bile salts in drug transport is not well understood, although several suggestions have been made, such as facilitation of transport from liver to bile by direct effect on canicular membranes, stimulation of micelle formation inside the liver cells, binding of drug anions to micelles, etc. The enhanced absorption of medicinals on administration with deoxycholic acid may be due to reduction in interfacial tension or micelle formation. The administration of quinine and other alkaloids in combination with bile salts has been claimed to enhance their parasiticidal action. Quinine, taken orally is considered to be absorbed mainly from the intestine, and a considerable amount of bile salts is required to maintain a colloidal dispersion of quinine. Bile salts may also influence drug absorption either by affecting membrane permeability or by altering normal gastric emptying rates. For example, so-

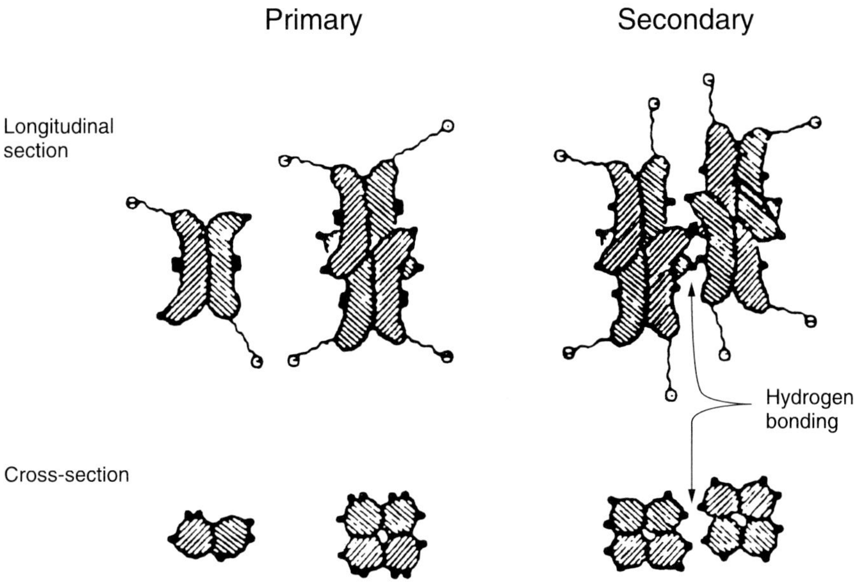

Fig. 13.16. Representation of the structure of bile acid salt micelles.

dium taurcholate increases the absorption of sulphaguanidine from the stomach, jejunum and ileum. This is due to an increase in membrane permeability induced by calcium depletion and interference with the bonding between phospholipids in the membrane.

Another important naturally occurring class of surfactants that are widely found in biological membranes are the lipids, which include phosphatidylcholine (lecithin), lysolecithin, phosphatidylethanolamine and phosphatidyl inositol (Figure 13.17). These lipids are also used as emulsifiers for intravenous fat emulsions, anaesthetic emulsions as well as for production of liposomes or vesicles for drug delivery. The lipids form coarse turbid dispersions of large aggregates (liposomes), which on ultrasonic irradiation form smaller units or vesicles. Liposomes are smectic mesophases of phospholipids organised into bilayers that assume a multilamellar or unilamellar structure. Multilamellar species are heterogeneous aggregates, most commonly prepared by dispersal of a thin film of phospholipid (alone or with cholesterol) into water. Sonication of the multilamellar units can produce the unilamellar liposomes, sometimes referred to as vesicles. The net charge of liposomes can be varied by incorporation of a long-chain amine, such as stearyl amine (to give a positively charged vesicle) or dicetyl phosphate (giving negatively charged species). Both lipid-soluble and water-soluble drugs can be entrapped in liposomes. Liposoluble drugs are solubilized in the hydrocarbon interiors of the lipid bilayers, whereas the water-soluble drugs are intercalated in the aqueous layers. The reader is referred to the review of Fendler and Romero [26] for details of liposomes as drug carriers. Liposomes, like micelles, may provide a special me-

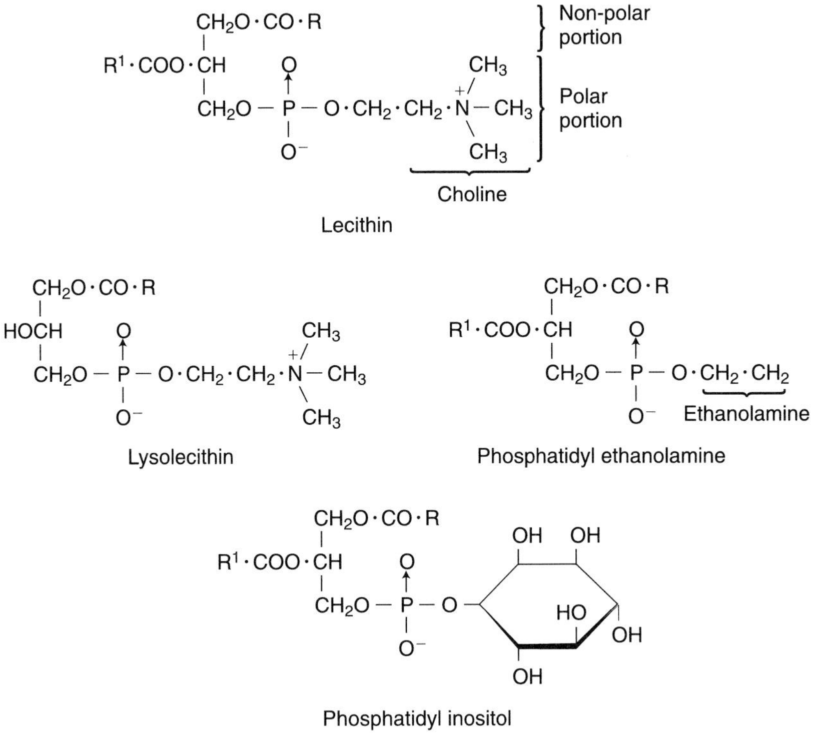

Fig. 13.17. Structure of lipids.

dium for reactions to occur between the molecules intercalated in the lipid bilayers or between the molecules entrapped in the vesicle and free solute molecules.

Phospholipids play an important role in lung functions. The surface active material to be found in the alveolar lining of the lung is a mixture of phospholipids, neutral lipids and proteins. Lowering of surface tension by the lung surfactant system and the surface elasticity of the surface layers assists alveolar expansion and contraction. Deficiency of lung surfactants in newborns leads to a respiratory distress syndrome and this led to the suggestion that instillation of phospholipid surfactants could cure the problem.

13.6 Biological Implications of the Presence of Surfactants in Pharmaceutical Formulations

The use of surfactants as emulsifying agents, solubilizers, dispersants for suspensions and as wetting agents in the formulation can lead to significant changes in the biological activity of the drug in the formulation. Surfactant molecules incorpo-

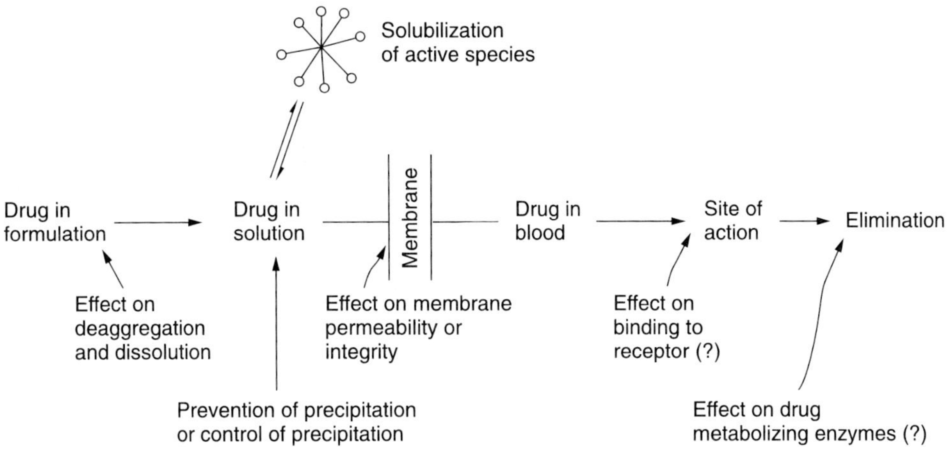

Fig. 13.18. Effect of surfactants on drug absorption and its activity.

rated in the formulation can affect drug availability and its interaction with various sites in several ways (Figure 13.18).

The surfactant may influence the desegregation and dissolution of solid dosage forms, by controlling the rate of precipitation of drugs administered in solution form, by increasing membrane permeability and affecting membrane integrity. Release of poorly soluble drugs from tablets and capsules for oral use may be increased by the presence of surfactants, which may decrease the aggregation of drug particles and, therefore, increase the area of the particles available for dissolution. The lowering of surface tension may also be a factor in aiding the penetration of water into the drug mass. This wetting effect operates at low surfactant concentration. Above the c.m.c., the increase in saturation solubility of the drug substance by solubilization in the surfactant micelles can result in more rapid rates of drug dissolution. This will increase the rate of drug entry into the blood and may affect peak blood levels. However, very high concentrations of surfactant can decrease drug absorption by decreasing the chemical potential of the drug. This results when the surfactant concentration exceeds that required to solubilize the drug. Complex interactions between the surfactants and protein may take place, and this will result in alteration of drug-metabolizing enzyme activity. There have also been some suggestions that the surfactant may influence the binding of the drug to the receptor site. Some surfactants have direct physiological activity of their own, and in the whole body these molecules can affect the physiological environment, e.g. by altering gastric residence time.

Numerous studies on the influence of surfactants on drug absorption have shown that they can increase, decrease or exert no effect on the transfer of drugs through membranes [14]. As discussed above, the presence of surfactant affects the dissolution rate of the drug, although the effect is less than predicted by the

Noyes–Whitney equation [27], which shows that the rate of dissolution dc/dt is related to the surface area A and the saturation solubility c_s,

$$\frac{dc}{dt} = kA(C_s - c) \tag{13.22}$$

Higuchi [28] assumed that an equilibrium exists between the solute and solution at the solid/liquid interface and that the rate of movement of the drug into the bulk is governed by the diffusion of free solute and solubilized drug across a stagnant diffusion layer. Drug solubilized in micelles will have a lower diffusion coefficient than the free solute molecules. This means that the effect of surfactant on the dissolution rate will be related to the dependence of dissolution rate on the diffusion coefficients of the species and not on their solubilities, as suggested by Eq. (13.22). Thus, the rate of dissolution will be given by

$$\frac{dc}{dt} = \left[\frac{D_f c_f}{h} + \frac{D_m c_m}{h}\right] \tag{13.23}$$

where the subscripts f and m refer to free and micellar drug, and c_m is thus the increase in solubility due to the micellar phase; h is the thickness of the diffusion layer.

Predictions of dissolution rate may be made using diffusion coefficients of the solutes in their solubilized state by applying the Stokes–Einstein equation,

$$D = \frac{RT}{6\pi\eta N_A}\left(\frac{4\pi N_A}{3Mv}\right)^{1/3} \tag{13.24}$$

where R is the gas constant, T is the absolute temperature, η is the viscosity of the solvent, N_A is the Avogadro's constant, M is the micellar molecular weight and v is the partial specific volume of the micelles.

13.7 Aspects of Surfactant Toxicity

The presence of surfactants in drug formulations may produce unwanted side or toxic effects because of their interaction with proteins, lipids, membranes and enzymes. To fully understand these interactions, it is essential to have information on the metabolic fate of the ingested surfactant. Membrane disruption by surfactants involves binding of the surfactant monomers to the membrane components, followed by the formation of co-micelles of the surfactant with segments of the membrane. The interaction between surfactants and proteins can lead to solubilization of the insoluble-bound protein or to changes in the biological activity of enzyme

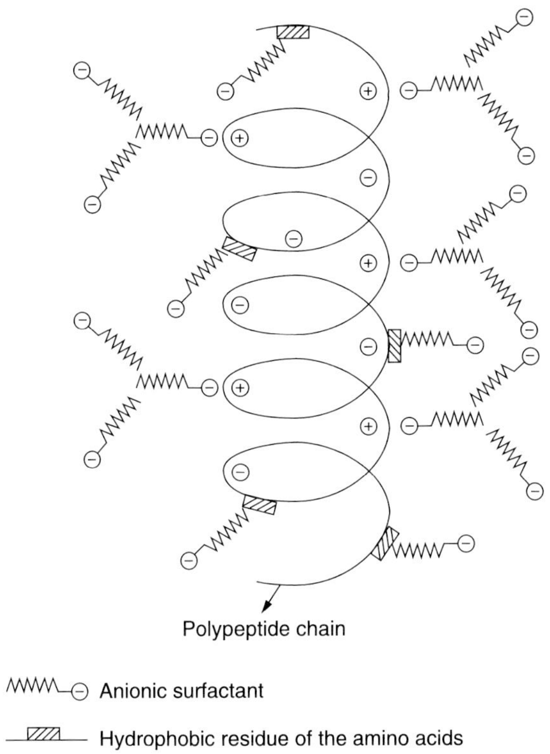

Fig. 13.19. Representation of binding of an anionic surfactant to a protein.

systems. It has long been known that surfactants could precipitate, form complexes with or denature proteins at low concentrations. Figure 13.19 illustrates this diagrammatically, showing the modes of binding of an anionic surfactant to a protein. The hydrophobic interactions with the amino acid hydrophobic residues would be equally appropriate for cationic and nonionic surfactants. However, cationic surfactants could attach themselves electrostatically to anionic sites. Surfactants produce conformational changes in proteins at low concentrations.

Solubilization of toxic substances may result in their enhanced absorption and, thus, the presence of surfactants in river and tap water could increase the absorption of the carcinogenic polycyclic compounds, which are generally insoluble in body fluids, as a result of their solubilization. The hazards of exposure to household surfactants, in washing-up liquids, in toothpaste and in water supplies should not be underestimated. In addition, the irritant effects of surfactants (cutaneous toxicity) present in many cosmetic products should be addressed, and this led to the introduction of surfactants that are milder to the skin. For more detail on surfactant toxicity, the reader should refer to the text by Attwood and Florence.

13.8 Solubilised Systems

Solubilisation is the process of preparation of a thermodynamically stable isotropic solution of a substance (normally insoluble or sparingly soluble in a given solvent) by incorporation of an additional amphiphilic component(s) [29]. It is the incorporation of the compound (referred to as solubilisate or substrate) within a micellar (L_1 phase) or reverse micellar (L_2 phase) system.

L_1 and L_2 can be described by considering the phase diagram of a ternary system of water–surfactant–cosurfactant system, as shown in Figure 13.20 for water–ionic sulphate–long-chain alcohol system.

The aqueous micellar solution A solubilises some alcohol to form normal micelles (L_1), whereas the alcohol solution B dissolves large amounts of water, forming inverse micelles (L_2). These two phases are not in equilibrium, but are separated by a third region, namely the lamellar liquid crystalline phase (L_- phase).

Lipophilic (water-insoluble) substances become incorporated in the L_1 (normal micelle) phase.

Hydrophilic (water-soluble) substances are incorporated in the L_2 phase. The lamellar liquid crystalline phase can also incorporate solubilisates. This also applies to the hexagonal (middle phase) that may be present in concentrated nonionic surfactant systems.

Since the above-mentioned liquid crystals are anisotropic, the above definition of solubilisation does not strictly apply. The site of incorporation of the solubilisate

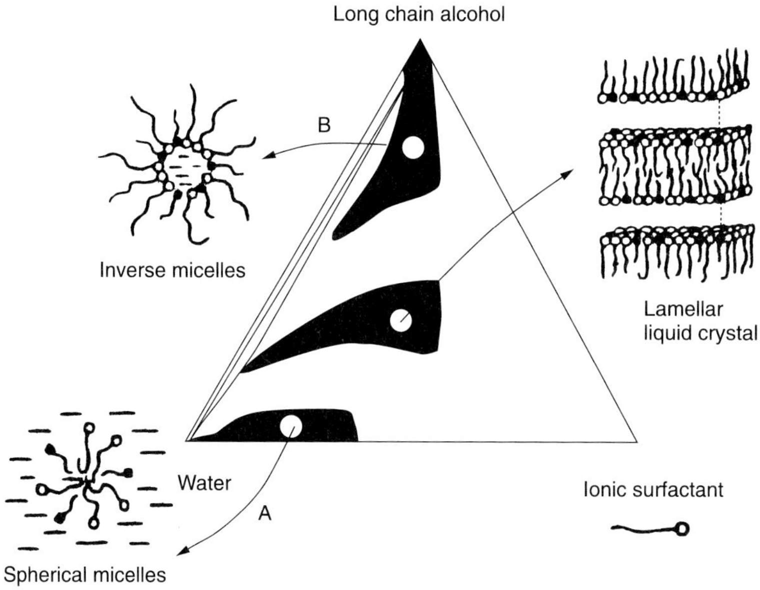

Fig. 13.20. Ternary phase diagram of the water–surfactant–alcohol system.

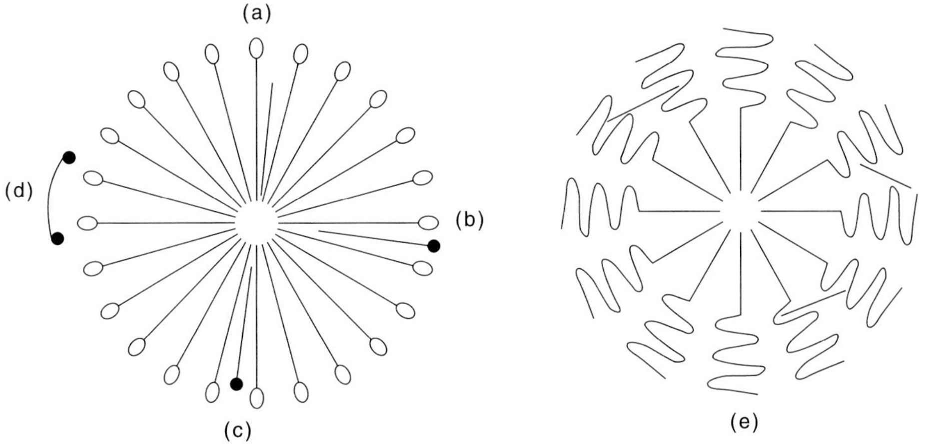

Fig. 13.21. Site of incorporation of solubilisate: (a) in the hydrocarbon core; (b) short penetration; (c) deep penetration; (d) adsorption; (e) in the polyoxyethylene chain.

is closely related to its structure (Figure 13.21): Non-polar solubilisate in the hydrocarbon core; semi-polar or polar solubilisate oriented within the micelle (short or deep) [30].

13.8.1 Experimental Methods of Studying Solubilisation

13.8.1.1 Maximum Additive Concentration

The concentration of solubilisate that can be incorporated into a given system with the maintenance of a single isotropic solution (saturation concentration or maximum additive concentration, MAC) is obtained using the same procedures for measurement of solubility of any compound in a given solvent [30]. Since solubilisation is temperature sensitive, adequate temperature control is essential.

If the refractive indices of the solubilizing system and solubilisate are sufficiently different, saturation is detected by the presence of supra-colloidal aggregates with a concomitant increase on the opacity. A long time may be required to reach equilibrium saturation, particularly with highly insoluble drugs. An excess of solubilisate is shaken up with the surfactant solution until equilibrium is reached and the two phases could be separated by centrifugation or using millipore filters. Data are best expressed as concentration of solubilisate versus concentration of surfactant or as ratio of solubilisate dissolved per gram of surfactant versus surfactant concentration. The results can also be expressed using a ternary phase diagram of solubilisate–solvent–surfactant.

13.8.1.2 Micelle–Water Distribution Equilibria

With solubilisates having significant water solubility, it is of interest to know both the distribution ratio of solubilisate between micelles and water under saturation and unsaturation conditions. To measure the distribution ratio under unsaturation conditions, a dialysis technique can be employed, using membranes that are permeable to solubilisate but not to micelles. Ultrafiltration and gel filtration techniques can be applied to obtain the above information. The data are treated using the phase-separation model of micellisation (micelles are considered to be a separate phase in equilibrium with monomers).

The partition coefficient, P_{m}, between micelles and solution is given by

$$P_{\mathrm{m}} = \frac{C_3^{\mathrm{m}}}{C_3^{\mathrm{a}}} \tag{13.25}$$

where C_3^{m} is the moles of solubilisate per mole of micellar surfactant and C_3^{a} is the number of moles of free solubilisate per mole of water.

Equation (13.25) does not include the volumes of the micellar or aqueous phase, which can be obtained from the partial molar values of the surfactant. A better expression is

$$P_{\mathrm{m}} = \frac{D_{\mathrm{b}}/V}{D_{\mathrm{f}}/(1-V)} \tag{13.26}$$

where D_{b} and D_{f} are the amount of solute in the micellar and aqueous phases, respectively, V is the volume of micellar phase and $(1 - V)$ is the volume fraction of the aqueous phase.

An alternative method of expressing solubilisation data is

$$\frac{D_{\mathrm{t}}}{D_{\mathrm{f}}} = 1 + k[C] \tag{13.27}$$

where D_{t} is the total solute concentration and $[C]$ is the surfactant concentration; k is a measure of the binding capacity of the surfactant – it is given by the slope of plots of $D_{\mathrm{t}}/D_{\mathrm{f}}$ versus $[C]$.

An alterative expression that treats solubilisation as a process of "binding" of solute molecules to binding sites on the surfactant is

$$r = \frac{nK[D_{\mathrm{f}}]}{1 + K[D_{\mathrm{f}}]} \tag{13.28}$$

where r is the molar ratio of bound solute to total surfactant,

$$r = \frac{[D_{\mathrm{b}}]}{[C]} \tag{13.29}$$

n is the total number of independent binding sites on the surfactant micelle; K is an intrinsic dissociation constant for the binding of solute molecules to one of the sites. Eq. (13.28) is a form of a Langmuir isotherm.

In some cases, plots of r/D_f are curved, indicating more than one adsorption site, and this requires modification of Eq. (13.28). For example, for two adsorption sites with dissociation constant K_1 and K_2,

$$r = \frac{n_1 K_1 [D_f]}{1 + K_1 [D_f]} + \frac{n_2 K_2 [D_f]}{1 + K_2 [D_f]} \tag{13.30}$$

Analysis of the curves allows one to obtain n_1, n_2, K_1 and K_2.

13.8.1.3 Determination of Location of Solubilisate

The site of incorporation of solubilisate is closely related to its chemical structure (Figure 13.21). Although in many cases a particular location is preferred, the lifetime of a solubilisate within the micelle is long enough for a rapid interchange between different locations.

For a nonionic surfactant, consisting of an alkyl group R and PEO chain, one may determine the number of equivalents of alkyl chain moiety, C_R, and that of the PEO chain, C_{PEO}. The solute may be considered to be distributed between the R PEO chains. The total amount solubilised S' is given by

$$S' = a_{PEO} + bC_R \tag{13.31}$$

where a and b are proportionality constants.

Rearrangement of Eq. (13.31) gives,

$$\frac{S'}{C_{PEO}} = a + b\frac{C_R}{C_{PEO}} \tag{13.32}$$

Plots of (S'/C_{PEO}) versus (C_R/C_{PEO}) gives straight lines, from which a and b can be determined. This allows one to obtain the relative incorporation of solubilisate in the R and PEO chains. Several quantitative methods have been applied to obtain the exact location of the solubilisate.

X-ray Diffraction This is based on application of the Bragg's equation,

$$n\lambda = 2d \sin\theta \tag{13.33}$$

where d is the distance between two parallel plates, n is an integer, and θ is the angle of incidence to the plane of the X-ray beam with wavelength λ.

In addition to the diffraction caused by the solvent, three diffraction bands appear: an "S" or short spacing band, giving a repeating distance of 0.4–0.5 nm, corresponding to the thickness of the hydrocarbon chain; "M", a micelle thickness band that varies with the length of the R chain (value slightly less than twice the extended length of the R chain); and an "I" or long spacing band (greater than "M" or "S" bands) that is sensitive to surfactant concentration. Both "M" and "I" bands show an increase in length of spacing with addition of apolar solubilisates, but show little or slight increase with the addition of polar solubilisates. Assuming the micelles are spherical, their radius could be obtained from the long spacing,

$$d = \left(\frac{8\pi}{3 \times 2^{1/2}}\right)^{1/2} \phi^{-1/3} r \tag{13.34}$$

where r is the radius of a sphere occupying a fraction ϕ of the total volume.

An alternative X-ray technique is to plot the scattering intensity (I_s) versus s [$= 2\lambda \sin(\theta/2)$]. The diffuse maximum in the small angle hit shows a shift and increase in intensity on solubilisation. These changes are attributed to the change in radii and electron density of the core and polar regions of the micelle.

Absorption Spectrometry The amount of vibrational fine structure in the UV absorption spectrum of a compound in solution is a function of the interaction between solvent and solute. The extent of interaction between solvent and solute increases with increasing solvent polarity, leading to decreasing fine structure. As the micelle is characterised by regions of different polarity, UV spectra have been used to obtain information on the environment of the solubilisate in the micelle.

NMR Methods NMR can be used to obtain information on solubilisation, by measuring the shift in the peak positions on addition of the solubilisate. For example, by measuring the ^1H NMR shift for a compound with an aromatic ring versus the concentration of a surfactant that contains no aromatic ring (e.g. SDS) one can determine the location of the solubilisate. This leads to an upfield shift of the ^1H peak, indicating a more hydrophobic environment.

Fluorescence Depolarisation This is based on the use of fluorescence probes such as pyrene, which has been used to study the interior of micelles. The fluorescence spectrum of pyrene shows a significant change on solubilisation in the core of the micelle.

Electron-Spin Resonance (ESR) This is based on the introduction of free-radical probe such as nitroxide. The ESR spectrum reflects the microenvironment of the micelle and, hence, on solubilisation this spectrum shows significant changes.

13.8.1.4 Mobility of Solubilisate Molecules

As with surfactant monomers, the solubilisate molecules are not rigidly fixed in the micelle, but have a freedom of motion that depends on the solubilisation site. The lifetime of a solubilisate in the micelle is very short, usually less than 1 ms. These short relaxation times have been determined using NMR and ultrasonic techniques.

13.8.1.5 Factors Affecting Solubilisation

Several factors affect solubilisation:

Solubilisate Structure Generalisation about the manner in which structure affects solubilisation is complicated by the existence of different solubilisation sites. The main parameters that may be considered when investigating solubilisates are: Polarity, polarisability, chain length and branching, molecular size and shape. The most significant effect is, perhaps, the polarity of the solublisate and, sometimes, they are classified into polar and apolar; however, difficulty exists with intermediate compounds.

Some correlation exists between hydrophilicity/lipophilicity of solubilisate and partition coefficient between octanol and water (the log *P* number concept – the higher the value the more lipophilic the compound is).

Surfactant Structure For solubilisates incorporated in the hydrocarbon core, the extent of solubilisation increases with increasing the alkyl chain length. For the same R, solubilisation increases in the order: anionics < cationics < nonionics.

The solubilisation power, normally described by the ratio of moles solubilisate to mole surfactant, increases with increasing PEO chain length – this is due to the decrease in micelle size. With increasing PEO chain length, the aggregation number decreases and hence the number of micelles per mole surfactant increases.

Temperature Mostly, solublisation increases with increase of temperature as a result of the increase in solubility of the compound and decrease in the c.m.c. (for nonionic surfactants) with rising temperature.

Addition of Electrolytes and Non-electrolytes Most electrolytes cause a reduction in the c.m.c. and they may increase the aggregation number (and size) of the micelle. This may lead to an increase in solublisation. Addition of non-electrolytes, e.g. alcohols, can lead to an increase in solubilisation.

The above discussion clearly demonstrates that solubilisation above the c.m.c. offers an approach to formulation of poorly soluble drugs. This approach has several limitations: Finite capacity of micelles for the drug; short- or long-term adverse effects; solubilisation of other ingredients such as preservatives, flavours and colouring agents, which may cause alteration in stability and effectiveness.

Future research is required for: Solubilising agents that increase bio-availability; use of co-solvents; effect of surfactants on properties of solubilised systems and interaction with components of the body; and mixed micelle formation between surface active drugs and surfactants.

13.8.2 Pharmaceutical Aspects of Solubilisation

The presence of micelles and surfactant monomers in a drug formulation can have pronounced effects on the biological efficacy. Surfactants (both micelles and monomers) can influence the disintegration and dissolution of solid dosage forms by controlling the rate of precipitation (drug administration in solution), increasing membrane permeability and by affecting membrane integrity. The release of poorly soluble drugs from tablets and capsules (oral use) may be increased in the presence of surfactants. The reduction of aggregation on disintegration of tablets and capsules increases the surface area. Lowering of the surface tension aids penetration of water in the drug mass. Above the c.m.c., an increase in flux by solublisation can lead to a rapid increase in the rate of dissolution.

The above effect has been analysed by Noyes and Whitney [14]; the dissolution rate dC/dt depends on the surface area of the drug and its saturation solubility C_s,

$$\frac{dC}{dt} = KA(C_s - C) \tag{13.35}$$

Higuchi [31, 32] assumed an equilibrium between solute and solution at the solid–solution interface. The rate of drug movement into the bulk is governed by the diffusion of the free solute (with a diffusion coefficient D_t) and the solubilised drug (with a diffusion coefficient D_m) across a stagnant diffusion layer of thickness h,

$$\frac{dC}{dt} = \left(\frac{D_t C_t}{h} + \frac{D_m C_m}{h}\right) \tag{13.36}$$

Prediction of dissolution rates may be made using the Stokes–Einstein equation for D,

$$D = \frac{RT}{6\pi\eta N_A}\left(\frac{4\pi N_A}{3MV}\right)^{1/3} \tag{13.37}$$

where R is the gas constant, T is the absolute temperature, η is the viscosity of the solvent, N_A is Avogadro's constant, M is the micellar molecular weigh and V is the partial specific volume of the micelles.

However, very high surfactant concentrations (above that required for solubilisation) may decrease drug absorption by decreasing the chemical potential of the drug. The complex interaction between surfactant micelles, monomers and proteins may alter the drug metabolising activity. Surfactants may also alter the binding of the drug to the receptor site.

13.8.2.1 Solubilisation by Block Copolymers

Block copolymers, particularly those of the PEO-PPO-PEO type (sold under the trade name Pluronics, BASF, Synperonic PE or Poloxamers, ICI), have also shown significant ability to solubilise drugs. At low concentrations, approximating to those at which conventional nonionic surfactants form micelles, these block copolymer may produce monomolecular micelles by a change in configuration in solution. At higher surfactant concentrations, these monomolecular micelles aggregate to form aggregates, of varying size, that can solubilise drugs and increase the stability of the solubilising agent.

13.8.2.2 **Hydrotropes in Pharmaceutical Systems**

Hydrotropes are substances that increase the solubility of a solute, without having any significant surface activity. The mechanism of action of hydrotropes is complex and depends on different effects. Some hydrotropes act simply by complexation with the drug, e.g. piperazine, sodium salicylate, adenosine and diethanolamine, which have been applied to solubilise theophylline. Apart from the possible prevention of unwanted physiological effects, hydrotropes can have a direct effect on efficacy. Complexation may occur by donor–acceptor interaction (hydrophobic and hydrogen bonding are thought to play a less important role). Several other hydrotropes have been suggested, e.g. *p*-toluene sulphonate and cumine sulphonate.

13.9 Pharmaceutical Suspensions

Pharmaceutical suspensions are dispersions of solid particles of an insoluble or sparingly soluble drug in a liquid vehicle, usually water [33, 34]. Several examples of pharmaceutical suspensions are used: Oral Suspensions, antibiotic preparations, antiacid and clay suspensions, radioopaque suspensions, barium sulphate suspensions. The vehicle is syrup, sorbitol solution or gum thickener with added artificial sweetener. Topical suspensions (externally applied "shake lotion") such as calamine lotion USP are also formulated as suspensions. Several dermatological preparations are also used in pharmacy.

Safety to skin is very important and protective action and cosmetic properties are also essential. These systems require the use of high concentrations of disperse phase. The vehicle may be oil-in-water (O/W) or water-in-oil (W/O) emulsion, dermatological paste and clay suspensions. Parenteral suspensions are examples with low solid content (usually 0.5–5%), except penicillin (antibiotic content > 35%).

Sterile preparations are designed for intramuscular, intradermal, intralesional, intraarticular or subcutaneous administration. The most important property of parenteral suspensions is its viscosity, which should be low enough to facilitate injection. Syringeability (the ability of a parenteral suspension to pass easily through a hypodermic needle) is controlled by the viscosity of the suspension during transfer from the vial and during flow through the needle. A shear thinning system (rapid reduction of viscosity with applied shear rate) is vital. Common vehicles for parenteral suspensions are preserved sodium chloride or parenterally acceptable vegetable oil. Ophthalmic suspensions that are instilled into the eye must be sterile and the vehicle employed is aqueous isotonic solution.

13.9.1 Main Requirements for a Pharmaceutical Suspension

The main criteria for pharmaceutical suspensions are [35–37]: Colloid and physical stability, with acceptable shelf-life under storage conditions; colloid stability indicates the lack of any strong irreversible flocculation. On application, any aggregates should be broken down to single particles (weak flocks are broken under shear).

One of the main requirements of pharmaceutical suspension is the lack of Ostwald ripening (crystal growth), i.e. the growth of particles on storage that results in shift of particle size distribution to larger values. This may affect bioavailability and results in physical instability.

Physical stability implies uniformity of the suspension whereby the particles are equally distributed across the whole vehicle. Uniformity of the suspension on application is vital. This means that the viscosity of the suspension should be low enough to ensure uniformity.

Another important criterion for pharmaceutical is the absence of settling, caking or claying. The particles (which have density greater than the medium) tend to settle under gravity, resulting in a concentration gradient of the particles across the container. The particles at the bottom may move across each other, forming a hard sediment (referred to as "cake" or "clay"). Such hard sediments are very difficult to redisperse by gentle shaking.

13.9.2 Basic Principles for Formulation of Pharmaceutical Suspensions

The insoluble drug, usually in a powder form, is formulated using the following four steps:

(1) Dispersion of the powder in the vehicle. This requires adequate wetting of the powder (external and internal surfaces) by the liquid vehicle, necessitating the use of a wetting/dispersing agent, mostly a nonionic surfactant (such as polysorbates) or polymeric surfactant (such as Poloxamer). This process is described in detail in Chapter 7.

(2) Maintenance of the particles as individual units in the dispersed state. This is achieved by the strong repulsive forces between the particles as a result of the presence of adsorbed surfactant or polymer layers.

(3) Comminution (grinding) of the dispersed particles into smaller units (usually in the region of 1–10 μm). Wet grinding is achieved by the use of mills (Dyno mill) or using a Microfluidiser.

(4) Addition of a suspending (antisettling) agent to reduce settling caking or claying. This is achieved by addition of "thickeners" (such as xanthan gum, hydroxyethyl cellulose, silica or swellable clays).

After preparation of the suspension, its colloid and physical stability should be assessed over periods of time under various conditions (temperature changes and temperature cycling) to ensure the shelf-life of the product.

13.9.3 Maintenance of Colloid Stability

After preparation of the dispersion, one should ensure the presence of enough repulsive energy to prevent aggregation of the particles and their rejoining. The origin of the repulsive energy has been discussed in Chapters 6 and 7 and only a summary is given here. Two main repulsive energies may be applied: (1) Electrostatic (by creation of electrical double layers) when using ionic surfactants. This is seldom used since most ionic surfactants are unacceptable in drug formulations (due to toxic effects). (2) Steric repulsion when using nonionic surfactants (such as polysorbates) or polymeric surfactants (such as Poloxamers). The energy–distance curve for sterically stabilised dispersions (shown in Chapter 7) exhibits a shallow minimum at a particle–particle separation distance h comparable to twice the adsorbed layer thickness 2δ, followed by steep rise in repulsion when $h < 2\delta$. As discussed before, weak (reversible) flocculation may occur when the depth of the attractive minimum is large ($>5kT$). This weak flocculation is not a problem, since on gently shaking the suspension can be easily redispersed. This weak flocculation may be beneficial in reducing settling and the prevention of caking.

13.9.4
Ostwald Ripening (Crystal Growth)

Many drugs have some solubility in the vehicle (usually water) in which they are dispersed. Since the suspension has a wide size distribution (that may range from 0.1 to 10 μm), the solubility of the various size particles differ due to curvature effects (the higher the curvature, i.e. the smaller the particle size, the larger the solubility). Lord Kelvin related the solubility of a particle with radius r, $S(r)$, to that of an infinitely large particle ($r = \infty$), $S(\infty)$, by the following equation,

$$S(r) = S(\infty) \exp\left(\frac{2\gamma_{SL} V_m}{rRT}\right) \tag{13.38}$$

where V_m is the molar volume of the drug.

Theoretically, Ostwald ripening should lead to condensation of all particles into a single one. However, this does not occur in practice since the rate of Ostwald ripening decreases with time. For two particles with radii r_1 and r_2 ($r_1 < r_2$),

$$\left(\frac{RT}{V_m}\right) \ln\left[\frac{S(r_1)}{S(r_2)}\right] = 2\gamma\left(\frac{1}{r_1} - \frac{1}{r_2}\right) \tag{13.39}$$

Eq. (13.39) shows that the higher the difference between r_1 and r_2 the higher the rate of Ostwald ripening.

The Ostwald ripening rate can be obtained by plotting the cube of the average radius versus time [39],

$$r^3 = \tfrac{8}{9}\left[\frac{S(\infty)\gamma V_m D}{\rho RT}\right] t \tag{13.40}$$

where D is the diffusion coefficient of the drug and ρ is its density.

Eq. (13.40) shows that a plot of r^3 versus t should give a straight line, from which the rate of crystal growth can be obtained. Eq. (13.38) shows that the rate of crystal growth is directly proportional to $S(\infty)$, i.e. the bulk solubility of the drug and it is also proportional to the solid/liquid interfacial tension, γ_{SL}.

Materials that reduce γ_{SL} (e.g. surfactants) will cause a reduction in the rate of crystal growth.

A second driving force for crystal growth is due to polymorphic changes. If the drug has two polymorphs A and B, the more soluble polymorph, say A (which may be more amorphous) will have higher solubility than the less soluble (more stable) polymorph B. During storage, polymorph A will dissolve and recrystallise as polymorph B. This can have a detrimental effect on bioefficacy, since the more soluble polymorph may be more active.

Crystal growth may be inhibited by the addition of additives, usually referred to as crystal growth inhibitors. Trace concentrations of certain additives (one part in 10^4) can have profound effects on crystal growth and habit modification – it is generally accepted that the additive must adsorb on the surface of the crystals. Surfactants and polymers (interfacially active) are expected to affect crystal size and habit. They may influence the diffusion process (transport to and from the boundary layer) and they may also affect the surface-controlled process (by adsorption at the surface, edge or specific sites). A good example of a polymer that inhibits crystal growth of sulphathiazole is poly(vinyl pyrrolidone) (PVP). The molecular weight of the polymer can be important. Many block ABA and graft BA_n copolymers (with B being the "anchor" part and A the stabilising chain) are very effective in inhibiting crystal growth. The B chain adsorbs very strongly on the surface of the crystal and sites become unavailable for deposition. This has the effect of reducing the rate of crystal growth. Apart from their influence on crystal growth, the above copolymers also provide excellent steric stabilisation, providing the A chain is chosen to be strongly solvated by the molecules of the medium. This is discussed in detail in Chapter 7.

13.9.5
Control of Settling and Prevention of Caking of Suspensions

Gravity is the driving force for settling of suspensions. When the gravity force exceeds the Brownian diffusion (kT) sedimentation occurs,

$$\tfrac{4}{3}\pi R^3 \Delta\rho g L \gg kT \tag{13.41}$$

R is the particle radius, $\Delta\rho$ is the buoyancy (difference between density of particles and medium), g is the acceleration due to gravity and L is the height of the container.

For a very dilute suspension (volume fraction $\phi < 0.01$), the rate of settling v_0 is given by Stokes' law,

$$v_0 = \frac{2R^2 \Delta\rho g}{9\eta} \tag{13.42}$$

where η is the viscosity of the medium (~1 mPa s for water at 25 °C).

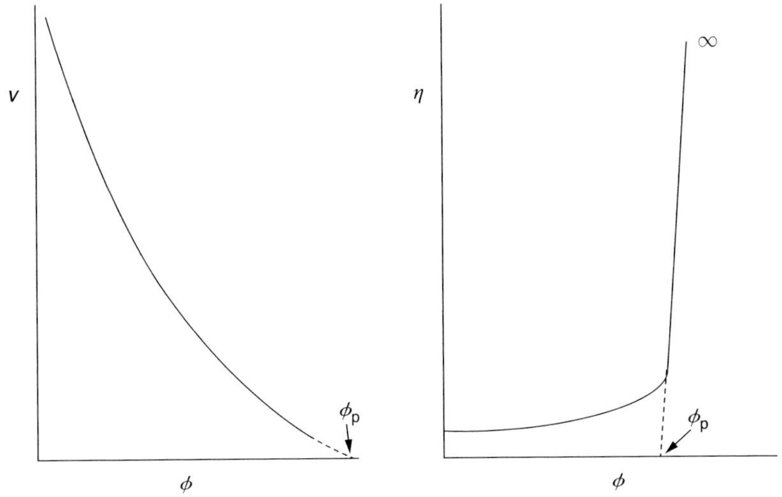

Fig. 13.22. Variation of ν and η with ϕ.

For a suspension with intermediate concentration, $\phi \sim 0.1$, the rate v is reduced below v_0 as a result of hydrodynamic interaction between the particles,

$$v = V_0(1 - k\phi) \tag{13.43}$$

k is a constant equal to 6.5.

For a more concentrated suspension, $\phi > 0.2$, the rate of sedimentation becomes a complex function of the volume fraction. Figure 13.22 illustrates this, showing the variation of v with ϕ, as well as the variation of η with ϕ.

Clearly, v decreases with increasing ϕ and reaches ~ 0 at a critical volume fraction ϕ_p (the so-called maximum packing fraction); η reaches infinity as ϕ approaches ϕ_p. The maximum packing fraction is ~ 0.6 for random packing of equal sized particles. It is >0.6 for polydisperse suspensions. Most practical suspensions are prepared well below the maximum packing and hence settling is the rule rather than the exception.

Several methods may be applied to reduce settling and the formation of cakes or clays:

13.9.5.1 **Balance of Density of Disperse Phase and Medium**

This can only be applied if the density difference between the particles and medium is small (<0.1), whereby the density of the medium can be increased by dissolving an "inert" material. This procedure suffers from the disadvantage that density matching can only be achieved at one temperature.

13.9.5.2 **Reduction of Particle Size**

As can be seen from Eq. (13.42), the rate of sedimentation is proportional to R^2. If R is reduced say below 0.1 μm, the rate becomes very small and the Brownian

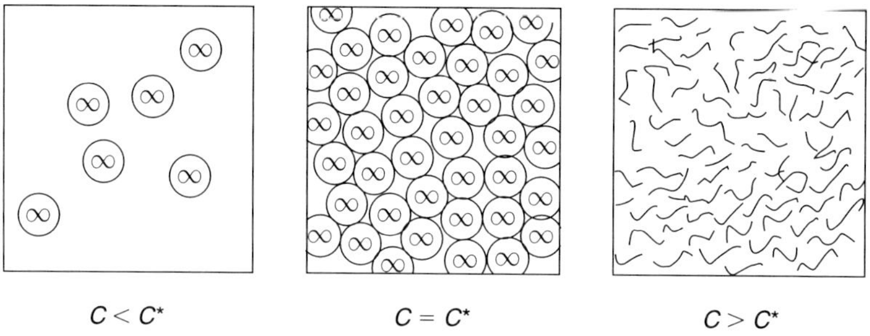

Fig. 13.23. Scheme of polymer chain overlap.

diffusion of the particles can overcome the gravity force. This is the principle of preparation of nano-suspensions, which could be achieved by proper choice of dispersant and application of a Microfluidiser. The major problem with nano-suspensions is Ostwald ripening, since these small particles will have significant solubility. Since the particles are of different sizes, the smaller particles will dissolve and become deposited on the larger ones – with storage, the nano-suspensions may grow sufficiently for sedimentation to occur.

13.9.5.3 **Use of Thickeners**

The most effective thickeners are high molecular weight materials, natural or synthetic, such as hydroxyethyl cellulose, xanthan gum, alginates, carrageenans, etc. Above a critical concentration C^* (referred to as the semi-dilute region), the polymer coils or extended chains begin to overlap, showing a rapid increase in viscosity with further increasing concentration above C^*. Figure 13.23 shows a schematic representation of chain overlap.

Figure 13.24 shows the variation of log η with log C for a polymer solution.

C^* is inversely proportional to the molecular weight of the polymer. Above C^*, the polymer solution shows non-Newtonian flow, pseudo-plastic or shear thinning

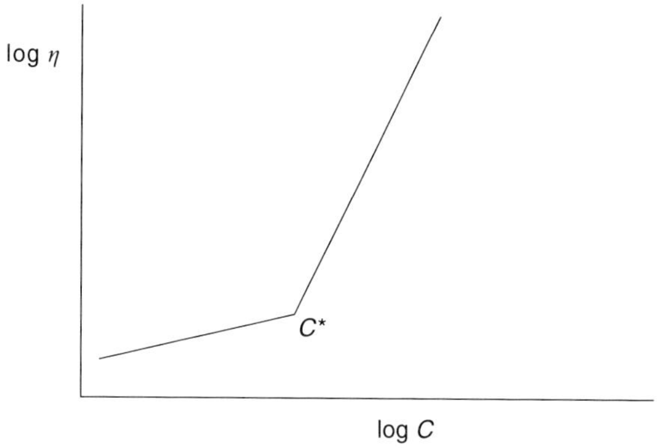

Fig. 13.24. Variation of log η with log C.

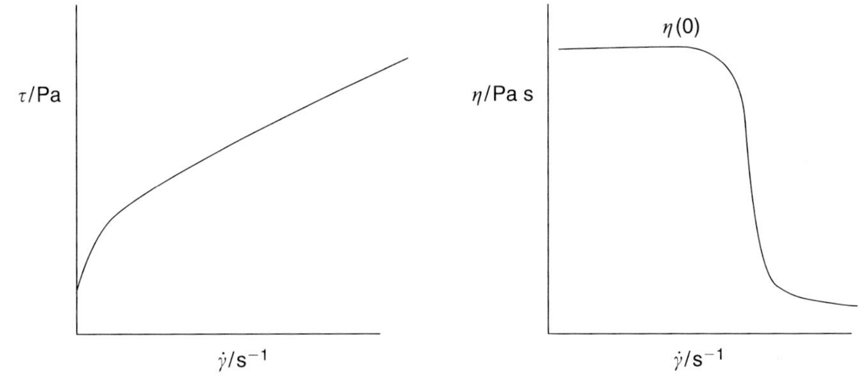

Fig. 13.25. Flow curves for a polymer solution at $C > C^*$.

behaviour. Figure 13.25 illustrates this, showing the variation of shear stress τ (Pa) and viscosity (Pa s) with applied shear rate $\dot{\gamma}(s^{-1})$.

The above systems show a "yield value", τ_β, and a high viscosity at low shear rates [residual or zero shear viscosity $\eta(o)$]. Providing τ_β is higher than a certain value ($>$ 0.1 Pa) and $\eta(o) > 1000$ Pa s, no sedimentation occurs with most suspensions.

Several other thickeners may be used: Swellable clays, such as sodium montmorillonite (commercially known as "Bentopharm"). These clay particles consist of very thin plates and they can produce a "gel" in the continuous phase by a mechanism known as "House-of-Card" structure. The plates have negative surfaces and positive edges and they produce the "House-of-Card" structure by edge-to-face association. Fumed silica, such as Aerosil 200 (manufactured by DeGussa), can also be used. These finely divided particles produce "gels" in the continuous phase by association between the particles forming chains and cross-chains.

13.10 Pharmaceutical Emulsions

As mentioned in Chapter 6, emulsions are dispersions of one immiscible liquid into another, stabilized by a third component (the emulsifier). Two main classes of emulsions can be distinguished: oil-in-water (O/W) and water-in-oil (W/O).

Emulsions of a polar liquid in a non-polar liquid or vice versa are also possible. These are sometimes referred to as oil-in-oil (O/O) emulsions.

Several pharmaceutical products are formulated as emulsions: (1) Parenteral emulsion systems, e.g. parenteral nutritional emulsions, lipid emulsions as drug carriers; (2) Perfluorochemical emulsions as artificial blood substitute; (3) Emulsions as vehicles for vaccines; (4) topical formulations, e.g. for treatment of some skin diseases (dermatitis).

To prepare any of the above systems, the formulation chemist must choose an optimum emulsifier system that is suitable for the preparation of the emulsion and maintenance of its long-term physical stability. The oil that may be used as drug carrier has to be nontoxic, e.g. vegetable oils (soybean and safflower), synthetic glycerides (including simulated human fats), and acetoglycerides.

The emulsifier system chosen should be safe (having no undesirable toxic effects) and it should be approved by the FDA (Food and Drug Administration). Below are some examples of approved emulsifiers:

- Anionic: sodium cholate – bile salts.
- Zwitterionic: lecithin (mainly phosphatidylcholine, phosphatidylethanolamine).
- Non-ionic: Polyethylene glycol stearate – polyoxyethylene monostearate (Myrj), Sorbitan esters (Spans) and their ethoxylates (Tweens), Poloxamers (polyoxyethylene–polyoxypropylene block copolymers).

For nonionic surfactants, particularly those of the ethoxylate type, a selection can be made based on the hydrophilic–lipophilic balance (HLB) concept (Chapter 6). A closely related system developed by Shinoda and his collaborators is based on the phase inversion temperature (PIT) concept. This is also described in detail in Chapter 6.

13.10.1
Emulsion Preparation

For the preparation of macroemulsions (with droplet size in the range 1–5 μm), such as in many topical application creams, high speed stirrers such as Silverson or Ultraturrax can be used. However, for parenteral emulsions (such as fat emulsions and anaesthetics), a much smaller particle size range is required, usually in the range 200–500 nm (sometimes referred to as nano-emulsions). These systems can be prepared using high temperature and/or high-pressure techniques (using homogenisers such as the Microfluidiser).

The role of surfactant in emulsion formation is crucial and is described in detail in Chapter 6. It reduces the oil–water interfacial tension, γ_{OW}, by adsorption at the interface. The droplet size R is directly proportional to γ_{OW}.

It enhances deformation and break-up of the droplets by reducing the Laplace pressure p,

$$p = \frac{2\gamma_{OW}}{R} \tag{13.41}$$

It prevents coalescence during emulsification by creating a tangential stress at the interface (Gibbs–Marangoni effect). During emulsification, deformation of the droplets results in an increase in the interfacial area A and hence the surfactant molecules will show a concentration gradient at the interface that gives an interfacial dilational (Gibbs) elasticity ε,

$$\varepsilon = \frac{d\gamma}{d \ln A} \tag{13.45}$$

As a result of this interfacial tension gradient, surfactant molecules will diffuse to the regions with higher γ, and they carry liquid with them (i.e. they force liquid in between the droplets) and this prevents coalescence (Marangoni effect).

13.10.2 Emulsion Stability

Several breakdown processes of emulsions can be distinguished on standing, and these are described in detail in Chapter 6. Apart from creaming or sedimentation that is governed by gravity, all other breakdown processes are determined by the interaction forces (energies) between the droplets. Below a brief description of the various instability processes is given, along with methods to overcome such instability.

13.10.2.1 Creaming or Sedimentation

As discussed in the suspensions section, the driving force for creaming (whereby the droplets have a density lower than the medium) or sedimentation is gravity. When the gravity force exceeds the Brownian diffusion, creaming or sedimentation will occur. With macroemulsions (droplets > 1 μm), the creaming or sedimentation rate is very fast and it may be completed in a matter of hours or days.

The most common procedure for eliminating creaming or sedimentation is to use a "thickener" in the continuous phase, e.g. a high molecular weight polymers such as hydroxyethyl cellulose or xanthan gum. These thickeners produce a "gel" in the continuous phase that has a yield value (> 0.1 Pa) and a high zero shear viscosity (>1000 Pa s), thus preventing any creaming or sedimentation.

13.10.2.2 Flocculation

Flocculation of emulsions resembles that of suspensions and the driving force is the van der Waals attraction, which becomes significant at short distances of separation. To prevent flocculation, one needs a strong repulsive force that operates at intermediate separations, thus preventing the close approach of the droplets. Two main repulsive forces can be distinguished: Electrostatic and steric (discussed in detail in Chapters 6 and 7).

13.10.2.3 Ostwald Ripening

As with suspensions, the driving force for Ostwald ripening is the difference in solubility between small and large droplets (as a result of curvature effect). The smaller droplets have a higher solubility than the larger ones and with time diffusion of molecules occurs from the small to the large droplets, shifting the size distribution to larger values.

Several methods may be applied to reduce Ostwald ripening in emulsions:

(a) Addition of a second, much less soluble oil (e.g. squalane). With time, diffusion of molecules of the more soluble oil (the major component) occurs from the small to the large droplets, with the result of concentrating the less soluble oil in the smaller droplets. Equilibrium is established when the difference in chemical potential between small and large droplets (due to the difference in composition) balances the difference resulting from curvature effect. This reduces Ostwald ripening.

(b) Modification of the interfacial film: a reduction in interfacial tension by surfactant adsorption causes a reduction in the rate of Ostwald ripening. The most important factor is the Gibbs elasticity – by addition of polymeric surfactants that are oil soluble and adsorb strongly at the O/W interface, the rate of Ostwald ripening is greatly reduced.

13.10.2.4 Coalescence

The driving force for coalescence is the thinning and disruption of the liquid film between the droplets. This can occur in a "floc", in a cream or sedimented layer or during Brownian collision. This is described in detail in Chapter 6.

When the film thickness reaches a critical value, film rupture occurs as a result of the strong van der Waals attraction in such thin films. To prevent film instability, one needs a strong repulsive force between the surfactant layers.

A useful concept for stability is the concept of disjoining pressure π (introduced by Deryaguin) (see Chapter 6); π has three contributions, due to van der Waals attraction, π_A, electrostatic repulsion, π_{el}, and steric repulsion, π_s,

$$\pi = \pi_A + \pi_{el} + \pi_s \tag{13.46}$$

π_{el} and π_s are positive, whereas π_A is negative.

To ensure film stability $\pi_{el} + \pi_s \gg \pi_A$, which can be achieved by several mechanisms: (a) Using mixed surfactant films (ionic + nonionic), which provide both electrostatic and steric repulsion. These films are condensed and they give high interfacial elasticity and viscosity, thus preventing any film fluctuation. (b) Using lamellar liquid crystalline phases. These lamellar liquid crystals consist of several surfactant bilayers and they "wrap" around the droplets, providing a barrier against coalescence. (c) Use of macromolecular surfactants of the A-B, A-B-A or BA_n type that strongly adsorb at the O/W interface and provide a strong repulsive energy between the droplets.

13.10.2.5 Phase Inversion

Two types of phase inversion may be distinguished: (a) Catastrophic, e.g. when the disperse phase volume exceeds its maximum packing and (b) transitional, which occurs over a long period of time due to change in the conditions. An example of transitional phase inversion is when the emulsion is subjected to a temperature change. For example, when an emulsion, prepared using an ethoxylated nonionic surfactant, is subjected to a temperature increase, the poly(ethylene oxide) (PEO) chains become dehydrated and, above a critical temperature, the surfactant may become "oil soluble" and more suitable for producing a W/O emulsion.

13.10.3 Lipid Emulsions

Lipid emulsions are used for parenteral nutrition, e.g. Intralipid that consists of 10–20% soybean, 1.2% egg lecithin and 2.5% glycerol [38]. The advantage of a fat emulsion is that a large amount of energy can be delivered in a small volume of isotonic fluid via a peripheral vein. The main problem with these fat emulsions is their long-term stability. A wide range of nonionic emulsifying agent has been investigated as potential emulsifying agents for intravenous fat.

Commercial fat emulsions employed in parenteral nutrition are stabilised by egg lecithin, which is a complex mixture of phospholipids with the following composition: phosphatidylcholine (PC) 7.3%, lysophosphatidylcholine (LPC) 5.8%, phosphatidylethanolamine 15.0%, lysophosphatidylethanolamine (LPE) 2.1%, phosphatidylinositol (PI) 0.6%, and sphingomyelin (SP) 2.5%.

In the development of suitable fat emulsions, pure PC and PE were employed but the emulsion had poor stability. This is due to the lack of formation of an electrical or mechanical barrier against coalescence. Introduction of ionic lipids such as phosphatidic acid (PA) and phosphatidylserine was essential to improve the stability of the emulsion.

Fat emulsions must have a small particle size (200–500 nm), which requires the use of high-pressure homogenisers. It is essential to store the emulsion at various temperatures and investigate any increase in fatty acid composition that causes lipoprotein lipase reactions. Also, an increase in droplet size increased the toxicity of the emulsion. Addition of drugs and nutrients to fat emulsions can also cause instability and/or cracking of the emulsion. Following the administration of fat emulsions to the body, it will be distributed rapidly throughout the circulation and then cleared.

13.10.4 Perfluorochemical Emulsions as Artificial Blood Substitutes

Perfluorochemicals can dissolve large quantities of oxygen and hence can be used as red blood substitutes [38]. Several emulsifying agents have been examined and the best stability was obtained using the block copolymers of poly(ethylene oxide) poly(propylene oxide) (Poloxamers or Pluronics, e.g. Pluronic F68). Several advantages of fluorochemical emulsions can be quoted: Good shelf-life, good stability in surgical procedures, no blood-group incompatibility problems, ready accessibility, and no problem with hepatitis.

The final emulsion should have (a) low toxicity; (b) no adverse interaction with normal blood; (c) little effect on blood clotting; (d) satisfactory oxygen and carbon dioxide exchange; (e) satisfactory rheological characteristics; (f) satisfactory clearance from the body.

To date the use of perfluorocarbon emulsions as blood substitutes have been investigated using animal studies, although a product from Japan (Fluosol-DA) containing perfluorodecalin has been studied in humans. The formulation of a suitable perfluorocarbon emulsion is still in its infancy since oils that produce stable emulsions are not cleared from the body and the choice of a suitable emulsifier is still difficult. The most suitable emulsifier system was based on a mixture of lecithin and poloxamer. The preferred oil, perfluorodecalin, gave initially fine droplet size, but on storage the droplets grew in size by an Ostwald ripening mechanism. The size of the emulsion droplets can have a pronounced effect on the biological results. Fluosol-DA has a mean particle size in the region 100–200 nm. Large particles were shown to have toxic effects.

13.11 Multiple Emulsions in Pharmacy

As mentioned in Chapter 12 multiple emulsions are complex systems of "Emulsions of Emulsions". Two main types can be distinguished: water-in-oil-in-water (W/O/W) multiple emulsions, whereby the dispersed oil droplets contain emulsified water droplets, and oil-in-water-in-oil (O/W/O) multiple emulsions, whereby the dispersed water droplets contain emulsified oil droplets. The most commonly used multiple emulsions in pharmacy are the W/O/W, which may be considered as water/water emulsions, whereby the internal water droplets are separated from the outer continuous phase by an "oily layer" (membrane).

Application of multiple emulsions in pharmacy for control and sustained release of drugs has been investigated over several decades using animal studies. The only successful application of multiple emulsions in industry is in the field of personal care and cosmetics, as discussed in Chapter 12. Products based on W/O/W systems have been introduced by several cosmetic companies. The potential application of multiple emulsions in drug delivery has attracted particular attention: W/O and W/O/W systems have been compared for delivery of aqueous solutions of medicine, e.g. it was shown that in giving ovalbumin to mice the W/O/W system was easier to inject than W/O and it gave a slightly better response. The potential of prolonged release of water-soluble drugs can be understood since the drug has to diffuse out through the oil layer of the droplet. Any biologically active material in the innermost phase will be protected from the external environment.

Despite the above-mentioned advantages of multiple emulsions, progress on commercial application in pharmacy has been slow for the following reasons: The complex nature of the system, which may require a carefully designed two step preparation. Limited availability of surfactants that can be used in preparation of the essential two emulsions, e.g. for a W/O/W systems one has to start by preparing a W/O emulsion, which is then further emulsified into water to produce the W/O/W drops (two surfactants with low and high HLB numbers are required).

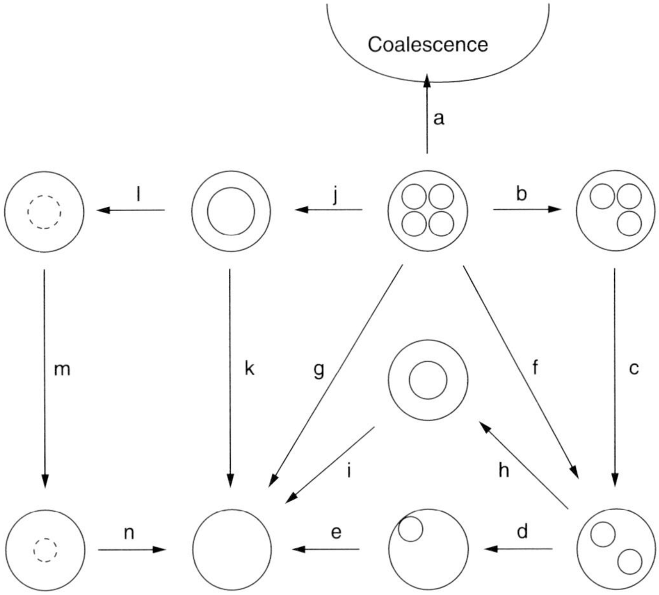

Fig. 13.26. Schematic representation of the possible breakdown pathways in W/O/W multiple emulsions: (a) coalescence; (b) to (e) expulsion of one or more internal aqueous droplets; (f) and (g) less frequent expulsion; (h) and (i) coalescence of water droplets before expulsion; (j) and (k) diffusion of water through the oil phase; (l) to (n) shrinking of internal droplets.

Lack of understanding of the factors that affect the long-term stability and that could limit the shelf-life of the product.

Florence and Whitehill [14] distinguished between three types of multiple emulsions, (W/O/W) that were prepared using isopropyl myristate as the oil phase, 5% Span 80 to prepare the primary W/O emulsion and various surfactants to prepare the secondary emulsion (see Chapter 12 for details). A schematic representation of some breakdown pathways that may occur in W/O/W multiple emulsions is shown in Figure 13.26.

One of the main instabilities of multiple emulsions is the osmotic flow of water from the internal to the external phase or vice versa. This leads to shrinkage or swelling of the internal water droplets, respectively. This process assumes the oil layer to act as a semi-permeable membrane (permeable to water but not to solute).

The volume flow of water, J_W, may be equated with the change of droplet volume with time dv/dt,

$$J_W = \frac{dv}{dt} = -L_p ART(g_2 c_2 - g_1 c_1) \tag{13.47}$$

L_p is the hydrodynamic coefficient of the oil "membrane", A is the cross sectional area, R is the gas constant and T is the absolute temperature.

The flux of water ϕ_W is,

$$\phi_W = \frac{J_W}{V_m} \tag{13.48}$$

where V_m is the partial molar volume of water.

An osmotic permeability coefficient P_0 can be defined,

$$P_0 = \frac{L_p RT}{V_m} \tag{13.49}$$

Combining Eqs. (13.47) to (13.49),

$$\phi_W = -P_0 A(g_2 c_2 - g_1 c_1) \tag{13.50}$$

The diffusion coefficient of water D_W can be obtained from P_0 and the thickness of the diffusion layer Δx,

$$-P_0 = \frac{D_W}{\Delta x} \tag{13.51}$$

For isopropyl myristate W/O/W emulsions, Δx is ~8.2 μm and $D_W \sim 5.15 \times 10^{-8}$ cm^2 s^{-1}, the value expected for diffusion of water in reverse micelles.

13.11.1
Criteria for Preparation of Stable Multiple Emulsions

These were discussed in detail in Chapter 12 and a summary is given below:

(1) Two emulsifiers: with low and high HLB numbers. Emulsifier 1 should prevent coalescence of the internal water droplets, preferably producing a viscoelastic film that also reduces water transport.

(2) The secondary emulsifier should also produce an effective steric barrier at the O/W interface to prevent any coalescence of the multiple emulsion droplets.

(3) Optimum osmotic balance: This is essential to reduce water transport. Osmotic balance can be achieved by addition of electrolytes or non-electrolytes. The osmotic pressure in the external phase should be slightly lower than that of the internal phase to compensate for curvature effects.

13.11.2
Preparation of Multiple Emulsions

The most convenient method of preparation is a two stage process as is illustrated in Chapter 12.

The yield of the multiple emulsion can be determined using dialysis for W/O/W multiple emulsions. A water-soluble marker is used and its concentration in the outside phase is determined.

$$\% \text{ Multiple emulsion} = \frac{C_i}{C_i + C_e} \times 100 \tag{13.52}$$

where C_i is the amount of marker in the internal phase and C_e is the amount of marker in the external phase.

It has been suggested that if a yield of more than 90% is required, the lipophilic (low HLB) surfactant used to prepare the primary emulsion must be $\sim 10 \times$ higher in concentration than the hydrophilic (high HLB) surfactant.

13.11.3
Formulation Composition

Oils to be used for the preparation of multiple emulsions must be pharmaceutically acceptable (no toxicity). The most convenient oils are vegetable oils such as soybean or safflower oil. Paraffinic oils with no toxic effect may be used. Also, some polar oils such as isopropyl myristate can be applied.

The low HLB emulsifiers (for the primary W/O emulsion) are mostly the sorbitan esters (Spans), but these may be mixed with other polymeric emulsifiers such as silicone emulsifiers. The high HLB surfactant can be chosen from the Tween series, although the block copolymers PEO-PPO-PEO (Poloxamers or Pluronics) may give much better stability.

To control the osmotic pressure of the internal and external phases, electrolytes such as NaCl or non-electrolytes such as sorbitol may be used.

In most cases, a "gelling agent" is required both for the oil and the outside external phase. For the oil phase, fatty alcohols may be employed. For the aqueous continuous phase one can use the same "thickeners" that are used in emulsions, e.g. hydroxyethyl cellulose, xanthan gum, alginates, carrageenans, etc. Sometimes, liquid crystalline phases are applied to stabilise the multiple emulsion droplets. These can be generated using a nonionic surfactant and long-chain alcohol. "Gel" coating around the multiple emulsion droplets may also be formed to enhance stability.

13.11.4
Characterisation of Multiple Emulsions

(1) Droplet size analysis: The droplet size of the primary emulsion (internal droplets of the multiple emulsion, are usually in the region 0.5–2 μm, with an average of ~0.5–1.0 μm. The multiple emulsion droplets cover a wide range of sizes, usually 5–100 μm, with an average in the region of 5–20 μm. Optical microscopy (differential interference contrast) can be applied to assess the droplets of the multiple emulsion. Optical micrographs may be taken at various storage times to assess the

stability. Freeze–fracture and electron microscopy can give a quantitative assessment of the structure of the multiple emulsion droplets. Photon correlation spectroscopy (PCS) can be used to measure the droplet size of the primary emulsion. This depends on measuring the intensity fluctuation of scattered light by the droplets as they undergo Brownian motion. Light diffraction techniques can be applied to measure the droplet size of the multiple emulsion. Since the particle size is >5 μm (i.e. the diameter is ≫ than the wavelength of light), they show light diffraction (Fraunhofer's diffraction). For this purpose, a master sizer could be used.

(2) Dialysis: As mentioned above, this could be used to measure the yield of the multiple emulsion; it can also be applied to follow any solute transfer from the inner droplets to the outer continuous phase.

(3) Rheological techniques [40]: Three rheological techniques may be applied:

(a) Steady-state shear stress (τ)–shear rate ($\dot{\gamma}$) measurements. A pseudo-plastic flow curve is obtained that can be analysed using, for example, the Herschel–Buckley equation,

$$\tau = \tau_\beta + k\dot{\gamma}^n \tag{13.53}$$

where τ_β is the "yield value", k is the consistency index and n is the shear thinning index.

The above equation can be used to determine the viscosity η as a function of shear rate. By following the change in viscosity with time, one can obtain information on multiple emulsion stability. For example, if there is water flow from the external phase to the internal water droplets ("swelling"), the viscosity will increase with time. If, after sometime, the multiple emulsion droplets begin to disintegrate, forming O/W emulsion, the viscosity will drop.

(b) Constant stress (creep) measurements. In this case, a constant stress is applied and the strain γ (or compliance $J = \gamma/\tau$) is followed as a function of time. If the applied stress is below the yield stress, the strain will initially show a small increase and then remain virtually constant. Once the stress exceeds the yield value, the strain shows a rapid increase with time and eventually it reaches a steady state (with constant slope). From the slopes of the creep tests one can obtain the viscosity at any applied stress. A high plateau value is obtained below the yield stress (residual or zero shear viscosity) followed by rapid decrease when the yield stress is exceeded. By following the creep curves as a function of storage time one can assess the stability of the multiple emulsion. Apart from swelling or shrinking of the droplets, which cause a reduction in zero shear viscosity and yield value, any separation will also show a change in the rheological parameters.

(c) Dynamic (oscillatory) technique. Here a sinusoidal strain (or stress) is applied and the stress (or strain) is simultaneously measured. For a viscoelastic system, such with multiple emulsions, the strain and stress sine waves will be shifted by a phase angle δ ($90° > \delta > 0°$). This allows one to obtain the elastic component of the complex modulus, G', and the viscous component of the complex modulus, G''. Both G' and G'' are measured as a function of strain amplitude (at constant frequency) and as a function of frequency (at constant strain amplitude in the linear viscoelastic region). Any change is the structure of the multiple emulsion will be accompanied by a change in G' and G''. For example, if the multiple emulsion droplets undergo "swelling" by flow of water from the external to the internal phase, G' will increase with time. Once the multiple emulsion droplets disintegrate to form an O/W emulsion, a drop in G' is observed. Alternatively, if the multiple emulsion droplets shrink, G' decreases with time.

13.12
Liposomes and Vesicles in Pharmacy

Liposomes are multilamellar structures that consist of several bilayers of lipids (several μm) (see Chapter 12). They are produced by simply shaking an aqueous solution of phospholipids, e.g. egg lecithin. When sonicated, these multilayer structures produce unilamellar structures (size range 25–50 nm) that are referred to as liposomes. Figure 13.27 gives a schematic picture of liposomes and vesicles.

Glycerol-containing phospholipids are used for the preparation of liposomes and vesicles: phosphatidylcholine (**13.1**), phosphatidylserine, phosphatidylethanolamine, phosphatidylanisitol, phosphatidylglycerol, phosphatidic acid and cholesterol. In most preparations, a mixture of lipids is used to obtain the optimum structure.

Liposomes and vesicles are ideal systems for drug delivery, due to their high degree of biocompatability, in particular for intravenous delivery [41, 42]. For effective application, larger unilamellar vesicles are preferred (diameter 100–500 nm). In addition to drug delivery, liposomes have been used as model membranes, as carriers of drugs, DNA, ATP, enzymes and diagnostic agents. Both water-soluble and insoluble drugs can be incorporated by encapsulation in the aqueous space or intercalation into the lipid bilayer.

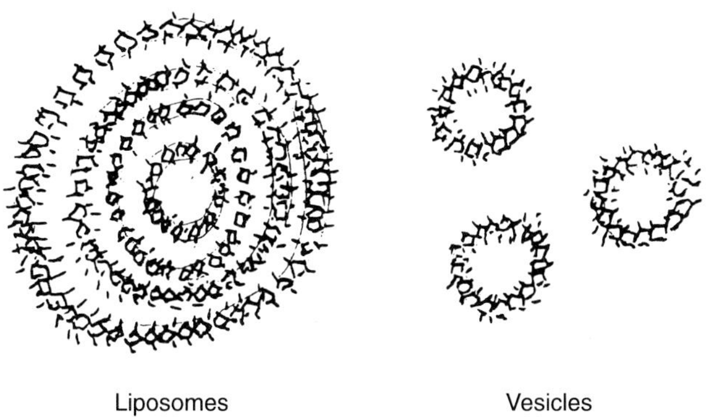

Fig. 13.27. Representation of liposomes and vesicles.

13.12.1
Factors Responsible for Formation of Liposomes and Vesicles – The Critical Packing Parameter Concept

The critical packing parameter (CPP) is a simple geometrical concept that determines the shape of any aggregation unit (see Chapter 6). It is simply the ratio of the cross sectional area of the hydrocarbon chain, a_0, to the cross sectional area of the hydrophilic head group, a,

$$\mathrm{CPP} = \frac{a_0}{a} = \frac{v}{a l_c} \tag{13.54}$$

where v is the volume of the hydrocarbon chain and l_c is the extended length of the chain.

For spherical normal micelles $a \gg a_0$ and packing constraints require $\mathrm{CPP} < \frac{1}{3}$. For cylindrical micelles, the head group area becomes smaller (for example by addition of electrolyte to an ionic surfactant) and packing constraints require $\mathrm{CPP} < \frac{1}{2}$. For vesicles, with two hydrocarbon chains, a_0 increases and packing constraints require $\mathrm{CPP} < \frac{2}{3}$. For lamellar micelles $\mathrm{CPP} \sim 1$; for inverse micelles $\mathrm{CPP} > 1$.

As discussed in Chapter 12, the free energy for an amphiphile with a spherical micelle of outer and inner radii R_1 and R_2, respectively, depends on: (1) γ the interfacial tension between hydrocarbon and water; (2) n_1, n_2 the number of molecules in the outer an inner layers; (3) e the charge on the polar head group; (4) D the thickness of the head group; and (5) v the hydrocarbon volume per amphiphile (taken to be constant).

The minimum free energy, μ_N°, for a particular aggregation number N is given by

$$\nu_N^\circ \approx 2a_0\gamma\left(\frac{1 - 2\pi Dt}{Na_0}\right) \tag{13.55}$$

where a_0 is the surface area per amphiphile in a planar bilayer ($N = \infty$).

As discussed in Chapter 12, several conclusions can be drawn from the thermodynamic analysis of vesicle formation: (1) μ_N° is slightly lower than $\mu_N^\circ(\mathrm{min}) = 2a_0\gamma$. (2) Since a spherical vesicle has a much lower aggregation number N than a planar bilayer, then spherical vesicles are favoured over planar bilayers. (3) $a_1 < a_0 < a_2$. (4) For vesicles with a radius $> R_1^c$, there are no packing constraints. These vesicles are not favoured over smaller vesicles which have lower N. (5) Vesicle size distribution is nearly Gaussian, with a narrow range. For example, vesicles produced from phosphatidylcholine (egg lecithin) have $R_1 \sim 10.5 \pm 0.4$ nm. The maximum hydrocarbon chain length is $\sim$1.75 nm. (6) Once formed, vesicles are homogeneous and stable and they are not affected by the length of time and strength of sonication. (7) Sonication is necessary in most cases to break-up the lipid bilayers that are first produced when the phospholipid is dispersed into water.

Figure 12.10 gives a schematic representation of the formation of bilayers and their break-up into vesicles (see Chapter 12). Spherical bilayer vesicles with smaller aggregation numbers are thermodynamically favoured over bilayer sheets. The main problem with any picture is the need for an activation energy to separate the vesicle from the extended bilayer.

13.12.2
Solubilisation of Drugs in Liposomes and Vesicles and their Effect on Biological Enhancement

Liposomes can solubilise both water-soluble and lipid-soluble drugs. The amount and location of a drug within a liposome depends on several factors: (1) The location of a drug within a liposome is based on the partition coefficient of the drug between aqueous compartments and lipid bilayers. (2) The maximum amount of drug that can be entrapped within a liposome depends on its total solubility in each phase. (3) Drugs with limited solubility in polar and non-polar solvents cannot be encapsulated in liposomes. (4) Efficient capture depends on the use of drugs at concentrations that do not exceed the saturation limit in the aqueous compartment (for polar drugs) or the lipid bilayers (for non-polar drugs).

If liposomes are prepared by mixing the drug with the lipids, the drug will eventually partition to an extent depending on the partition coefficient of the drug and the phase volume ratio of water to bilayer. Release rates are highest when the drug has an intermediate partition coefficient.

The bilayer/aqueous compartment partition coefficient is usually estimated by determining the partition coefficient between octanol and water (the log P number). Drug solubilisation in liposomes has important biological effects: (1) The ultimate efficacy of a liposomal dosage depends on the control of the amount of free drug that can reach the exact "site of action". (2) Generally, the exact "site of action" is not known and one relies on attaining reproducible blood levels of the drug. (3) With non-parenteral dosage forms, only the free drug is absorbed and hence one can measure the amount of drug that enters the blood as a function of time.

Parenteral, especially intravenous, administration of drugs encapsulated in liposomes requires control of the pharmacokinetics of the drug and this necessitates control of (1) concentration of the free drug in the blood. (2) Concentration of liposomes and their entrapped drug in blood. (3) Leakage rate of drug from liposome in the blood. (4) The disposition of the intact drug-carrying liposomes in the blood.

To control the pharmacokinetics of these complex systems, one must separate out the leakage rate of the drug from the liposome in the blood and the disposition of the intact carrying liposomes in the blood.

One of the major problems with application of liposomes for drug delivery is their interaction with high molecular weight substances such as inulin albumin.

The instability of liposomes in plasma appears to be the result of transfer of bilayer lipids to albumin and high-density lipoproteins (HDL). Some protein is also transferred from the lipoprotein to the liposome. Both lecithin and cholesterol can exchange with membranes of red blood cells. The susceptibility of liposomal phospholipid and phospholipase is strongly dependent on liposome size and type. Generally MLVs (multilamellar vesicles) are most stable and SUVs (singular vesicles) are least stable. Liposomes prepared with higher chain length phospholipids are more stable in buffer and in plasma. Cholesterol and sphingomyelin are very effective in reducing instability. As we will see later, incorporation of block copolymers such as poloxamers can enhance the stability of liposomes.

Despite the above limitations, the therapeutic promise of liposomes as a drug delivery system is becoming a reality in the following applications: (1) Parenteral administration; (2) inhalation treatment; (3) percutaneous administration; (4) ophthalmics; (5) cancer treatment; and (6) controlled-release formulations.

13.12.3
Stabilisation of Liposomes by Incorporation of Block Copolymers

Sterically stabilised vesicles prepared with the addition of triblock copolymers of the poly(ethylene oxide) (PEO)-poly(propylene oxide) (PPO) type, namely Poloxamers or Pluronics (PEO-PPO-PEO), have shown enhanced stabilisation [43]. Steric stabilisation of phospholipid vesicles by the copolymer molecules has been attempted by two different techniques: (1) Addition of the block copolymer to preformed vesicles (method A) and (2) addition of the block copolymer to the lipid before formation of the vesicles (method I). In the latter, both the lipid and copolymer participate in the construction of the vesicle. A schematic picture of the resulting vesicle structure for the two methods is given in Figure 13.28.

Vesicles prepared according to method I are more stable than those prepared according to method A for the following reasons: Association of the block copolymer as an integral part of the bilayer (method I) gives a better "anchor" to the bilayer than vesicles prepared by simple adsorption (method A). Increased rigidity of the lipid–polymer bilayer structure (for method I); the increased rigidity decreases the interaction with HDL. The (I) vesicles do not exhibit osmotic swelling.

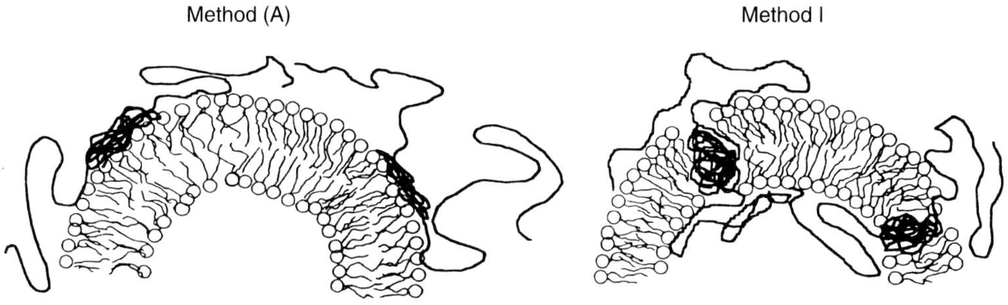

Fig. 13.28. Representation of vesicle structure in the presence of block copolymers.

13.13
Nano-particles, Drug Delivery and Drug Targeting

The concept of delivering a drug to its pharmacologically site of action in a controlled manner has many advantages [44, 45]: (1) Protection of the drug against metabolism or recognition by the immune system. (2) Reduction of toxic side-effects, especially for potent chemotherapeutic drugs and poor tissue specificity. (3) Improved patient compliance by avoiding repetitive administration.

As discussed in the previous section, liposomes have been used as drug delivery systems, due to the natural origin of their principal components (phospholipids and cholesterol). However, liposomes suffer from the problem of the long-term stability, although attempts have been made to improve their stability, e.g. by incorporation of block copolymers.

This section will deal with the possibility of using nano-particles for drug delivery and drug targeting. Polymeric nano-particles have some advantages in terms of their long-term physical stability and also their stability in vivo. Both model non-degradable and biodegradable particulate drug carriers have been investigated. The main problem to be overcome is their removal by phagocytic cells (macrophages) of the reticuloendothelial system (RES) and in particular the Kupffer cells of the liver. The main target of any research on nano-particles is to modify the surface of the particles in such a way to avoid RES recognition.

The above approach has been investigated both for non-biodegradable polymer particles (such as polystyrene or cyanoacrylate) and biodegradable particles, such as poly(lactic acid)/poly(lactic acid-*co*-glycolic acid) [46, 47].

13.13.1
Reticuloendothelial System (RES)

Phagocytic cells (macrophages of the liver and spleen) of the RES remove particulate systems (considered as foreign bodies). This process is facilitated by adsorption of proteins at the solid/liquid interface, a process that is referred to as opsonisation. Suppression of phagocytosis by other components of the blood, such as immunoglobin IgA and secretory IgA is referred to as dysopsonosis and is sometimes attributed to the hydrophilicity of IgA. However, coating polystyrene nano-particles with IgA had little effect on liver uptake.

13.13.2
Influence of Particle Characteristics

(1) Particle size: Particles greater than 7 μm are larger than blood capillaries (~6 μm) and become entrapped in the capillary beds of the lungs (which may have fatal effects). Most particles that pass the lung capillary bed accumulate in the elements of the RES (spleen, liver and bone marrow). The degree of splenic uptake increases with particle size. Removal of particles > 200 nm is due to a non-phagocytic process (physical filtration) in the spleen and phagocytosis (by

Kupffer cells) by the liver. Particles < 200 nm decrease splenic uptake and the particles are cleared by the liver and bone marrow. Colloidal particles not cleared by the RES can potentially exit the blood circulation via the sinusoidal fenestration of the liver and bone marrow.

(2) Surface charge: Surface charges only influence the particle–protein or particle–macrophage interactions at very short distances. The surface charge may affect the surface hydrophobicity. which can affect protein adsorption.

(3) Surface hydrophobicity: Serum components adsorb on the surface of colloidal particles via their hydrophobic sites. Increasing surface hydrophobicity increases opsonisation. To reduce opsonisation, a predominantly hydrophilic surface is required. This led to the conclusion that adsorption of poly(ethylene glycol) (PEG) type block copolymers on the surface of the particles should reduce opsonisation. This will be discussed in the next section.

13.13.3
Surface-modified Polystyrene Particles as Model Carriers

Polystyrene nano-particles have been used as model systems for investigation of the effect of surface modification in the various processes of phagocytosis, opsonisation and dysopsonisation. The surface of polystyrene particles can be modified by either adsorption of block copolymers containing PEG or by grafting PEG chains on the surface of the particles. PEG has the advantage of being non-toxic and has been approved by the FDA. Earlier work using liposomes containing PEG-phospholipid derivatives showed prolonged circulation times and prevention of phagocytic clearance. The PEG chains act as a barrier towards adsorption of proteins, thus preventing phagocytic clearance.

Two methods for surface modification could be applied:

(1) Adsorption of block copolymers of PEO-PPO-PEO, namely Poloxamers, or Poloxamines that are made of poly(ethylene diamine) with four branches of PEG chains. The molecular weight of the PEG chain and hence the adsorbed layer thickness is crucial in preventing phagocytosis. For example, Poloxamer 338 (with PEG chains of $M_w = 5600$) is more efficient than Poloxamer 108 (with PEG $M_w = 1800$) in preventing phagocytosis. A long PPO chain is also important to ensure anchoring of the block copolymer to the surface.
   The particle size of the polystyrene particles is also important. Both 60 and 150 nm particles coated with Poloxamer 407 were not sequestered by the macrophages in the bone marrow (they avoided capture by the Kupffer cells), whereas 250 nm particles (also coated with Poloxamer 207) were sequestered by the spleen and liver and only a small portion reached the bone marrow.

(2) Chemically grafting the PEG chains: Particles with different surface densities of PEG were prepared by copolymerisation of styrene with methoxy(PEG) acrylate macromonomer. Particle uptake by the Kupffer cells (in rat studies) decreased with increasing graft density. Only particles with a very low PEG den-

sity resulted in considerable liver deposition. However, the higher PEG density achieved with grafting did not improve the blood circulation time when compared with particles containing adsorbed block copolymers.

13.13.4 Biodegradable Polymeric Carriers

Several biodegradable polymers have been investigated as drug carriers [47, 48]: (1) poly(lactic acid)/poly(lactic acid-*co*-glycolic acid), (2) poly(anhydrides), (3) poly (ortho esters), (4) poly(β-malic acid-*co*-benzyl malate), and (5) poly(alkylcyanoacrylates).

The most widely used biodegradable polymer is poly(lactic acid) (PLA)/poly (lactic acid-*co*-glycolic acid) (PLGA), which has been used to produce a wide range of drug delivery formulations (microparticles, implants and fibres).

To avoid contamination, the nano-particles were produced by precipitation by mixing acetone solution with water. PLGA nano-particles less than 150 nm in diameter were produced. The surface of the PLGA particles was modified by adsorption of water-soluble PLA-PEG block copolymer or Poloxamine 908 (forming an adsorbed layer thickness $\sim$ 10 nm). Block copolymer micelles were also used as drug carriers – A-B or A-B-A block copolymers can produce micelles with aggregation numbers of several tens or hundreds of molecules (10–30 nm diameter). The hydrophobic core can be used to solubilise insoluble drugs (lipophilic molecules), whereas the hydrophilic chains provide the steric barrier, preventing protein adsorption and phagocytosis.

The most systematically studied micelle forming block copolymers are those base on PLA-PEG assemblies. The structure of the PLA-PEG micelles depend on the copolymer composition: the PLA to PEG ratio. With a fixed PEG block ($M = 5000$), the size of the PLA-PEG micelle increases with increasing PLA molecular weight. Characteristics of the corona, in a good solvent, are determined by the aggregation number and surface curvature. PLA-PEG nano-particles prepared from a copolymer with PLA molecular weight of 15 000 or less are colloidally stable with a reasonable high surface coverage of stabilising PEG chain. The grafting density of PEG chains appears to increase as the molecular weight of the PLA block is increased from 2000 to 15 000, which results in increasing thickness of the steric layer. As the PEG chains become radially more extended, they become less compressible. As the PLA molecular weight increases above 15 000, the colloid stability of the nano-particles is impaired, suggesting a reduction in PEG surface coverage. For PLA-PEG copolymer with a PLA block $M > 45\,000$, it is possible to increase the aggregation number and particle size. To produce long-circulating particulate carriers (i.e. to reduce opsonisation), one has to enhance the surface coverage of the brush-like PEG chains – copolymers with intermediate PLA-to-PEG ratio (e.g. PLA-PEG 15:5) appear to give optimal protein-resistant surface properties.

To study the effect of nano-particle structure on blood circulation, a hydrophobic radiolabelled γ-emitter ^{111}In-oxine (8-hydroxyquinoline) was incorporated within PLA and PLA-PEG nano-particles [49, 50]. The PLA nano-particles ($\sim$125 nm) were rapidly cleared from blood circulation with only 13% of the injected dose still circulating after 5 minutes. After 3 hours 70% of the nano-particles were removed by the liver. The rate and extent of release of the radiolabelled compound (using in vitro studies with rat serum) was higher for nano-particles produced from the PLA-PEG copolymers with a lower M PLA. After 3 hours, 77% and 88% radiolabelled compound remained associated with the PLA-PEG 6:5 and 30:5.

In vivo studies showed that the free radiolabelled compound remained in the blood at moderate levels after 3 hours and there was low liver accumulation. Contrary to expectation, the smaller size micelles of PLA-PEG copolymers did accumulate in the liver. It was necessary to have micelles with size >100 nm to evade phagocytosis. By optimising the size of the micelles and controlling the surface characteristics, it is possible to produce nano-particles that can be applied as drug carriers.

13.14 Topical Formulations and Semi-solid Systems

Topical formulations that are applied externally can be ointments, creams, pastes and gels. Ointments are semi-solid drug formulations that are intended for application to healthy, diseased or injured skin. Ointments for injuries and corticosteroid ointments, which penetrate the upper layers of the skin, have a local curative effect. Some ointments are designed for deeper penetration into the skin [51].

Ointments are single-phase, spreadable formulations that can be hydrophobic (based on Vaseline, oils, fats and waxes) or hydrophilic (based on macrogels that are miscible or soluble in water). Creams are based on oil-in-water (O/W) or water-in-oil (W/O) emulsions that are "structured" with a "gel network", mostly consisting of lamellar liquid crystalline phases (produced by a mixture of long-chain alcohol such as cetyl or stearyl alcohol and a surfactant that can be ionic, e.g. cetrimide or nonionic, e.g. cetomacrogel). Creams are also semi-solid systems that have specific rheological properties. Gels are also semi-solid systems, being either suspensions of small inorganic particles (such as clays or silica) or large organic molecules (polymers such xanthan gum, alginates, carrageenans, etc.) interpenetrated with liquid.

13.14.1 Basic Characteristics of Semi-Solids

Perhaps the best way to define semi-solids is based on their rheological characteristics. They show flow behaviour intermediate between liquids and solids, i.e. they are viscoelastic systems. To understand the flow behaviour of semi-solids, let us first consider the characteristics of elastic solids and viscous liquids. Elastic solids follow Hooke's law, which states that the relative strain (γ, dimensionless) increases linearly with the applied stress (τ/Pa),

$$\tau = G'\gamma \tag{13.56}$$

where G' is the shear (elastic modulus) in Pa.

Viscous liquids follow Newton's law, which states that the shear rate (γ/s^{-1}) varies linearly with the applied stress τ,

$$\tau = \eta\dot{\gamma} \tag{13.57}$$

where η is the viscosity.

Systems that obey Newton's law (such as simple liquids) are defined as Newtonian.

With semi-solids that are viscoelastic, they neither obey Hooke's law or Newton's law. This can be illustrated if one plots the shear stress τ or viscosity η versus shear rate $\dot{\gamma}$. The flow curve (sometimes referred to as "pseudo-plastic" or shear thinning) can be analysed using, for example, the Herschel–Buckley equation as discussed above.

Another technique that can be applied to characterise the viscoelastic behaviour of semi-solids is constant stress or creep measurements, which was also discussed above.

A third technique to investigate the viscoelastic properties of semi-solids is to apply dynamic (oscillatory) measurements. A sinusoidal strain or stress (with amplitudes γ_0 or τ_0) is applied on the system at a frequency $\varepsilon/\mathrm{rad\ s}^{-1}$ and the resulting stress or strain is measured simultaneously. For a viscoelastic system, the strain and stress will oscillate with the same frequency but out of phase (showing a time shift Δt or phase angle shift $\delta = \Delta t\varepsilon$). From γ_0, τ_0 and δ, one can calculate the various viscoelastic parameters: The complex modulus G^*, the storage modulus G' (the elastic component of the complex modulus) and the loss modulus G'' (the viscous component of the complex modulus).

For a viscoelastic system $\delta < 90°$ and $G' > G''$. The lower δ is the more "solid-like" (elastic) the system is.

A useful parameter is tan δ,

$$\tan\delta = \frac{G''}{G'} \tag{13.58}$$

Semi-solids and gels have tan $\delta \ll l$ (usually in the region of 0.2 or even lower for ointments and 0.2–0.4 for creams and gels).

13.14.2 Ointments

Several ointments are used in topical formulations: For skin applications and for ophthalmic use. As mentioned before, ointments used for skin applications are single-phase systems containing paraffin, Vaseline, vegetable oils and fats and waxes. The drug can be simply incorporated in these systems.

Ophthalmic ointments should be non-irritating, homogeneous, relatively non-greasy and should not cause blurred vision – they must also be sterilised by autoclaving and sometimes preservatives are added to avoid microbial growth. Absorption of the drug by ophthalmic ointments depends on several factors: nature of the eye, its limited capacity to hold the administered dosage form, tear fluid and aqueous humor dynamics (secretion and drainage rates), absorption by tissues, penetration through the cornea, spillage, blinking rate, reflex tearing, etc. These complex factors necessitate a great deal of research to produce the optimum ophthalmic ointment.

13.14.3
Semi-Solid Emulsions

Semi-solid emulsions or creams should have both physical and chemical stability, acceptable cosmetic effect and optimum environment for the active ingredient to reach the skin. The most commonly used creams are those based on a mixture of an ionic surfactant, such as cetrimide, with cetostearyl alcohol (e.g. in Cetavlon cream) or a mixture of nonionic surfactant (such as Cetomacrogel, an alcohol ethoxylate) with cetyl or stearyl alcohol [52].

The above mixtures produce a viscoelastic structure that gives a "bodying effect" to the formulation through the formation of a gel network of lamellar liquid crystalline phases [53]. Sometimes, glycerol is also added and this is thought to affect the hydration of the ethoxylate chain, thus affecting the final gel network structure. White soft paraffin may also be added to enhance the gel structure.

The structure of the final cream is rather complex and it can be investigated using various techniques: polarizing optical microscopy, freeze–fracture electron microscopy, differential scanning calorimetry (DSC), low-angle X-ray and NMR. Figure 13.29 gives a schematic picture of the structure of the cream; (a) and (b) illustrate an example of a cavity in the continuous gel phase whereas part (c) shows the interfacial area between dispersed phase and continuous phase on a molecular scale.

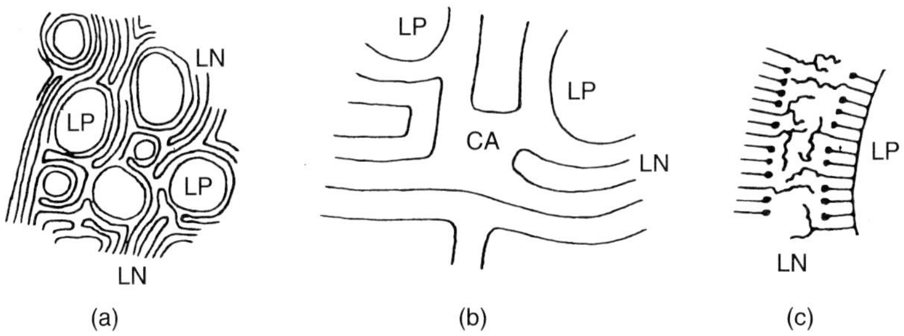

Fig. 13.29. Structure of a typical cream formulation.
LP = liquid lipophilic dispersed phase; LN = lamellar gel network (continuous gel phase); CA = cavity.

The lamellar gel phase (LN) surrounds the liquid lipophilic dispersed phase (LP) and this stabilises the emulsion. The main route for the release of, say, a hydrophilic drug in the cream is through the aqueous layers of the gel phase. Binding of the hydrophilic drug to the PEO chain of the nonionic surfactant can play an important role. For a lipophilic drug, the LP dispersed phase acts as a depot for the drug. Liberation of the drug from the LP phase is restricted by the crystalline character of the hydrocarbon sheets of the gel structure. Influencing the fluidity of the hydrocarbon sheet can substantially affect the drug release. In addition, partition of the drug between vehicle and skin will also change. Addition of a hygroscopic component will lead to uptake of water from the skin and this can affect the release rate.

An important study for emulsion creams is the change on ageing, which can affect the physical stability of the system and the release rate. Ageing of creams can be prevented by proper choice of surfactant structure and ratio of surfactant to alcohol. Polymerisable surfactants and alcohols can also prevent ageing. The most stable systems have been obtained using commercial cetostearyl alcohol (a mixture of cetyl and stearyl alcohol) and anionic (sodium dodecyl sulphate), cationic (cetrimide, a mixture of alkyltrimethylammonium bromide) and nonionic (Cetomacrogel, an alcohol ethoxylate with an alkyl chain length in the region 15–17 and EO chain of 20–40). The ratio of alcohol to surfactant was kept constant at 9:1. The alcohol was melted and dispersed (as droplets) in the surfactant solution. The water initially penetrates to form an isotropic L_2 phase and mixed micelles of alcohol and surfactant were also formed. On further dilution with water, the above phase rapidly transforms into a highly viscous liquid crystalline phase (LC) with a lamellar structure. Individual globules of molten cetyl alcohol stream through the system, forming elongated threads of liquid crystalline structures (Figure 13.30).

Oil/water emulsions can be prepared using the above ternary system. This is illustrated in Figure 13.31 for an emulsion prepared using liquid paraffin as the oil phase.

13.14.4 Gels

Gels encompass semi-solids of a wide range of characteristics, from fairly rigid gelatin slabs to suspensions of colloidal clays and polymer networks. They may be considered as being composed of two interpenetrating phases. For convenience, gels may be classified into two main categories: (1) gels based on macromolecules (polymer gels) and (2) gels based on solid particulate materials.

Numerous examples of gels based on polymers may be identified: Gels produced by overlap or "entanglement" of polymer chains (physical gels), gels produced by association of polymer chains ("associative thickeners") and gels produced by physical or chemical cross-linking of polymer chains.

The most common particulate gels are those based on "swelling clays" (both aqueous and non-aqueous) and finely divided oxides (such as silica gels).

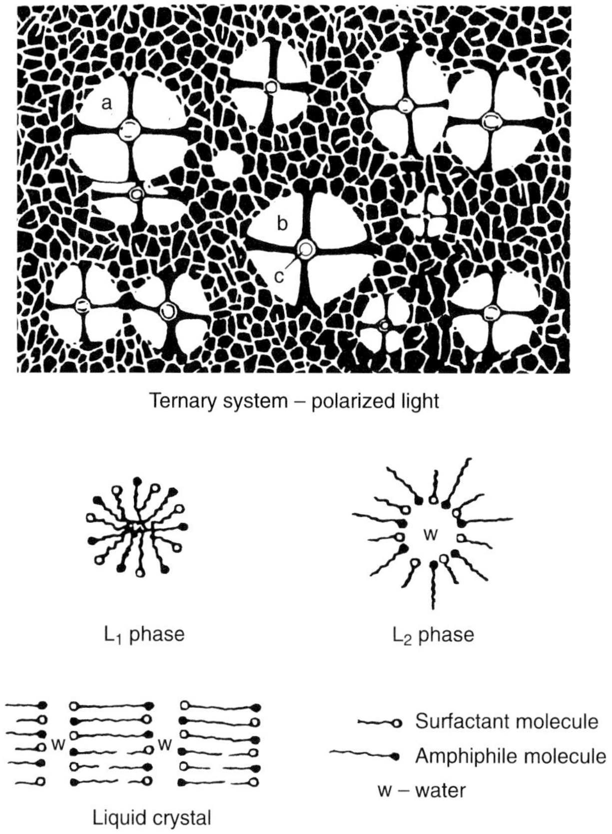

Fig. 13.30. Various structures in surfactant–alcohol mixtures.

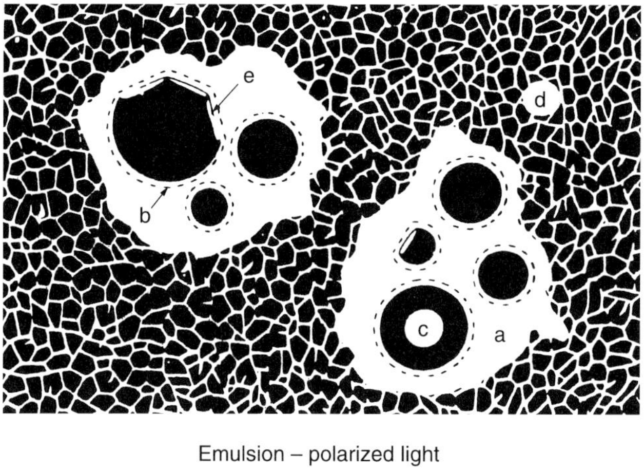

Fig. 13.31. Structures formed in emulsions containing alcohol–surfactant mixtures.

Apart from the above two main classes, gels can also be produced from surfactant liquid crystalline phases, mostly lamellar phases, as discussed above using mixtures of long-chain alcohols and surfactants.

13.14.4.1 Polymer Gels

Physical gels can be produced by simple "coil" (such as hydroxyethyl cellulose) or "extended chain" overlap (such as xanthan gum). Above a critical polymer concentration, C^*, overlap of the chains or "rods" occurs, forming a gel.

Hydrophobically modified polymer chains, referred to as associative thickeners, can produce gels at low concentrations, e.g. hydrophobically modified hydroxyethyl cellulose (Natrosol Plus, Hercules) or hydrophobically modified poly(ethylene oxide) (Rhom and Haas).

The most commonly used "cross-linked" polymer gels are those based on poly(acrylic acid) (Carbopol, Goodric). On neutralisation with alkali (such as KOH or ethanolamine), ionisation of the COOH groups occurs, resulting in swelling (by double layer expansion), and a gel can be produced at low concentrations.

13.14.4.2 Particulate Gels

Two main interactions can cause gel formation with particulate materials: (1) Long-range repulsion between the particles, e.g. using extended double layers or steric repulsion, using adsorbed surfactant or polymer layers. (2) Van der Waals attraction between the particles (flocculation), which can produce "three-dimensional" gel networks in the continuous phase.

The above systems produce non-Newtonian systems that show a "yield value" and high viscosity at low stresses or shear rates. Several examples may be quoted for particulate gels. Swellable clays, e.g. sodium montmorillonite (technically referred to as bentonite, e.g. Bentopharm). At low electrolyte concentrations, these clays produce gels as a result of double layer repulsion. At moderate electrolyte concentrations, gels are produced as result of face-to-edge association ("house of card structures"), by attraction between the negative surfaces and the positive edges. Silica gels – these are produced by association of finely divided silica particles (e.g. using Aerosil 200, De Gussa) to form three-dimensional gel network structures. Sometimes mixtures of particulate solids and polymers are used to produce gels.

References

1 *Solid/Liquid Dispersions*, T. F. Tadros (ed.): Academic Press, London, 1987.

2 T. F. Tadros, B. Vincent: *Encyclopedia of Emulsion Technology*, P. Becher (ed.): Marcel Dekker, New York, 1983, Chapter 3.

3 *Modern Aspects of Emulsion Science*, B. P. Binks (ed.): Royal Society Chemistry, Cambridge, 1998.

4 *Surfactants*, T. F. Tadros (ed.): Academic Press, London, 1984.

5 T. F. Tadros: *Surfactants in Agrochemicals*, Marcel Dekker, New York, 1994.

6 D. Attwood, A. T. Florence: *Surfactant Systems*, Chapman & Hall, London, 1983.

7 K. Holmberg, B. Jönsson, B. Kronberg, B. Lindman, *Surfactants and Polymers in Aqueous Solution*, Wiley, New York, 2002.

8 J. W. Gibbs: *Collected Papers*, Longman, Harlow, 1928.
9 G. D. Parfitt, C. H. Rochester: *Adsorption from Solution at the Solid/liquid Interface*, G. D. Parfitt, C. H. Rochester (ed.): Academic Press, London, 1983.
10 E. J. W. Verwey, J. T. G. Overbeek: *Theory of Stability of Lyophobic Colloid*, Elsevier Science, Amsterdam, 1948.
11 *Effect of Polymers on Dispersion Properties*, T. F. Tadros (ed.): Academic Press, London, 1982.
12 D. H. Napper: *Polymeric Stabilisation of Colloidal Dispersions*, Academic Press, London, 1983.
13 T. F. Tadros: *Polymer Colloids*, R. Buscall, T. Corner, J. Stageman (ed.): Elsevier, London. Applied Sciences, 1985.
14 D. Attwood, A. T. Florence: *Surfactant Systems, Their Chemistry, Pharmacy and Biology*, Chapman & Hall, New York, 1983.
15 D. Attwood, *J. Pharm. Pharmacol.*, **1972**, *24*, 751.
16 D. Attwood, O. K. Udeala, *J. Phys. Chem.*, **1975**, *79*, 889.
17 D. Attwood, O. K. Udeala, *J. Pharm. Pharmacol.*, **1975**, *27*, 395; **1974**, *26*, 854.
18 D. Attwood, O. K. Udeala, *J. Pharm. Pharmacol.*, **1975**, *27*, 754.
19 P. S. Guth, M. A. Spirtes, *Int. Rev. Neurobiol.*, **1964**, *7*, 231.
20 T. R. Blohm, *Pharmacol. Rev.*, **1979**, *30*, 593.
21 B. W. Barry, G. M. T. Gray, *J. Colloid Interface Sci.*, **1975**, *52*, 314.
22 D. G. Oakenfull, L. R. Fisher, *J. Phys. Chem.*, **1977**, *81*, 1838.
23 D. M. Small, *Adv. Chem. Ser.*, **1968**, *84*, 31.
24 D. G. Oakenfull, L. R. Fisher, *J. Phys. Chem.*, **1978**, *82*, 2443.
25 R. Zana, *J. Phys. Chem.*, **1978**, *82*, 2440.
26 J. H. Fendler, A. Romero, *Life Sci.*, **1977**, *20*, 1109.
27 A. T. Florence: *Techniques of Solubilization of Drugs*, S. Yalkowsky (ed.): Marcel Dekker, New York, 1982, Chapter 2.
28 T. Higuchi, K. Kuramoto, *J. Amer. Pharm. Assoc.*, **1954**, *43*, 398.
29 E. L. McBain, E. Hutchinson: *Solubilisation and Related Phenomena*, Academic Press, New York, 1955.
30 A. T. Florence: *Techniques of Solubilisation of Drugs*, S. Yalkowsky (ed.): Marcel Dekker, New York, 1982.
31 W. I. Higuchi, *J. Pharm. Sci.*, **1964**, *53*, 532.
32 W. I. Higuchi, *J. Pharm. Sci.*, **1967**, *56*, 315.
33 *Pharmaceutical Dosage Forms: Disperse Systems*, H. A. Liberman, M. M. Rieger, G. S. Banker (ed.): Marcel Dekker, New York, 1988, Volumes 1 and 2.
34 J. T. Cartensen: *Theory of Pharmaceutical Systems*, Academic Press, London, 1973.
35 *Solid/Liquid Dispersions*, T. F. Tadros (ed.): Academic Press, London, 1987.
36 H. Mollet, A. Grubenmann: *Formulation Technology*, Wiley-VCH, Weinheim, 2001.
37 C. L. Foy, D. W. Pritchard, G. B. Beestman: *Formulation Science*, The Association of Formulation Chemists, USA, 1998.
38 S. S. Davis, J. Hodgraft, K. L. Palin: *Encyclopedia of Emulsion Technology*, P. Becher (ed.) Volume 2. Marcel Dekker, New York, 1985.
39 *Modern Aspects of Emulsion Science*, B. P. Binks (ed.): Royal Society of Chemistry, Cambridge, 1998.
40 *Multiple Emulsions: Structure Properties and Application*, J. L. Grossierd, M. Seiller (ed.): Editions de Sante, France, 1997.
41 H. A. Liberman, M. M. Rieger, G. S. Banker, *Pharmaceutical Dosage Forms: Disperse Systems*, Volume 2. Marcel Dekker, New York, 1989.
42 H. Sasaki, T. Kakutani, M. Hashida et al., *J. Pharm. Pharmacol.*, **1985**, *37*, 46.
43 K. Kostarelos, T. F. Tadros, P. F. Luckham, Physical conjugation of (tri-) block copolymers to liposomes toward the construction of sterically stabilised vesicle systems, *Langmuir*, **1999**, *15*, 369–376.
44 J. Kreuter: *Colloidal Drug Delivery Sytems*, Marcel Dekker, New York, 1994.
45 S. M. Moghimi, S. S. Davis, Innovations in avoiding particle clearance from blood by Kupffer cells: cause for reflection, *Crit Rev. Ther. Drug. Carriers Syst.*, **1994**, *11*, 31.

46 S. J. Douglas, S. S. Davis, L. Illum, Nanoparticles in drug delivery, *Crit. Rev. Ther. Drug. Carriers Syst.*, **1987**, *3*, 233–261.

47 S. Stonik, S. E. Dunn, M. C. Davies, A. G. A. Coombes, D. C. Taylor, M. P. Irving, S. C. Purkiss, T. F. Tadros, S. S. Davis, L. Illum, Surface Modification of Poly(lactide-co-glycolide) Nano-spheres by biodegradable poly(lactide)-poly(ethylene oxide) copolymers, *Pharm. Res.*, **1994**, *11*, 1800–1808.

48 S. Hagen, A. G. A. Coombes, M. C. Garnett, S. E. Dunn, M. C. Davies, L. Illum, S. S. Davis, Poly(lactide)-poly(ethylene glycol) Copolymers as Drug Delivery Systems, 1. Characterisation of water dispersible micelle forming systems, *Langmuir*, **1996**, *12*, 2153–2161.

49 S. M. Moghimi, C. J. Porter, I. S. Muir, L. Illum, S. S. Davis, Non-phagocytic uptake of i. v. injected microspheres in the rat spleen: influence of particle size and hydrophilic coating, *Biochem. Biophys. Res. Commun.*, **1991**, *177*, 861–866.

50 S. Stolnik, L. Illum, S. S. Davis, Long circulating drug carriers, *Adv. Drug. Rev.*, **1995**, *16*, 195–214.

51 *Pharmaceutical Dosage Forms: Disperse Systems*, H. A. Liebermann, M. M. Rieger, G. S. Banker (ed.) Volumes 1 and 2, Marcel Dekker, New York, 1989.

52 B. W. Barry: *Dermatological Formulations, Precutaneous Absorption*, Marcel Dekker, New York, 1983.

53 T. Vringer de, J. G. H. Joosten, H. E. Juninger, *Colloid Polym. Sci.*, **1984**, *262*, 56.

14
Applications of Surfactants in Agrochemicals

14.1
Introduction

Agrochemical formulations cover a wide range of systems that are prepared to suit a specific application. All these formulations require the use of a surfactant, which is not only essential for its preparation and maintenance of long-term physical stability, but also to enhance biological performance of the agrochemical. In some cases, an agrochemical is a water-soluble compound, of which paraquat and glyphosate (both are herbicides) are probably the most familiar. Paraquat is a 2,2′-bipyridium salt and the counter ions are normally chloride. It is formulated as a 20% aqueous solution, which on application is simply diluted into water at various ratios (1:50 up to 1:200 depending on the application). To such an aqueous solution, surface active agents (sometimes referred to as wetters) are added, which are essential for several reasons. The most obvious reason for adding surfactants is to enable the spray solution to adhere to the target surface, and spread over it to cover a large area. In addition, the surface active agent plays a very important role in the optimisation of the biological efficacy. The role of surfactants in biological control is discussed in detail in the last part of this chapter. Thus, the choice of the surfactant system in an agrochemical formulation is crucial since it has to perform several functions. To date, such a choice is made by a trial and error procedure, owing to the complex nature of application and, due to the lack of understanding of the complex processes that occur during application of the agrochemical and the mode of action of the agrochemical, the choice of a particular surfactant system presents a challenge to the formulation chemist and the biologist. This chapter aims to rationalise the role of surfactants in formulations, subsequent application and optimisation of biological efficacy.

Most agrochemicals are water-insoluble compounds with various physical properties, which have first to be determined to decide on the type of formulation. Accurate physical data on the compound is essential, such as its solubility, its lipophilic character and its chemical stability. One of the earliest types of formulations is wettable powders (WP), which are suitable for formulating solid water-insoluble compounds that can be produced in a powder form. The chemical (which may be micronised) is mixed with a filler, such as china clay, and a solid surfactant, such as sodium alkyl or alkyl aryl sulphate or sulphonate, is added. When the

Applied Surfactants: Principles and Applications. Tharwat F. Tadros

ISBN: 3-527-30629-3

powder is added to water, the particles are spontaneously wetted by the medium, and, on agitation, the particles are dispersed. Clearly, the particles should remain suspended in the continuous medium for a period of time that depends on the application. This requires maintenance of the particles as individual units. Some physical testing methods are available to evaluate the suspensibility of the WP. The surfactant system obviously plays a crucial role in wettable powders. Firstly, it enables spontaneous wetting and dispersion of the particles. Secondly, by adsorption on the particle surface, it provides a repulsive force that prevents aggregation of the particles – aggregation enhances the settling of particles and may also cause problems on application, such as nozzle blockage.

The second and most familiar type of agrochemical formulation is the emulsifiable concentrates (ECs). These are produced by mixing an agrochemical oil with another one such as xylene or trimethylbenzene or a mixture of various hydrocarbon solvents. Alternatively, a solid pesticide could be dissolved in a specific oil to produce a concentrated solution. In some cases, the pesticide oil may be used without any extra addition of oils. In all cases, a surfactant system (usually a mixture of two or three components) is added for several purposes. Firstly, the surfactant enables self-emulsification of the oil on addition to water. This occurs by a complex mechanism that involves several physical changes such as lowering of the interfacial tension at the oil/water interface, enhancement of turbulence at that interface with the result of spontaneous production of droplets. Secondly, the surfactant film that adsorbs at the oil/water interface stabilises the produced emulsion against flocculation and/or coalescence. As we will see in later sections, emulsion breakdown must be prevented, otherwise excessive creaming, sedimentation or oil separation may take place during application. This results in an inhomogeneous application of the agrochemical, on the one hand, and possible losses on the other. The third role of the surfactant system in agrochemicals is in enhancement of biological efficacy. As shown in later sections, it is essential to arrive at optimum conditions for effective use of the agrochemicals. In this case, the surfactant system will help in spreading the agrochemical at the target surface and may enhance its penetration.

Recent years have seen a great demand to replace ECs with concentrated aqueous oil-in-water (O/W) emulsions, technically referred to as EWs. Several advantages may be envisaged for such replacements. In the first place, one can replace the added oil with water, which is of course much cheaper and environmentally acceptable. Secondly, removal of the oil could help in reducing undesirable effects such as phytotoxicity, skin irritation, etc. Thirdly, by formulating the pesticide as an O/W emulsion, one can control the droplet size to an optimum value, which may be crucial for biological efficacy. Fourthly, water-soluble surfactants, which may be desirable for biological optimisation, can be added to the aqueous continuous phase. As shown in later sections, the choice of a surfactant, or a mixed surfactant system, is crucial for the preparation of a stable O/W emulsion. In recent years, macromolecular surfactants have been designed to produce very stable O/W emulsions, which can be easily diluted into water and applied without detrimental effects to the emulsion droplets.

A similar concept has been applied to replace wettable powders, namely with aqueous suspension concentrates (SCs). These systems are more familiar than ECs and they have been introduced for several decades. Indeed, SCs are probably the most widely used systems in agrochemical formulations. Again, SCs are much more convenient to apply than WP's. Dust hazards are absent, and the formulation can be simply diluted in the spray tanks, without the need for any vigorous agitation. As we will see later, SCs are produced by a two- or three-stage process. The agrochemical powder is first dispersed in an aqueous solution of a surfactant or a macromolecule (usually referred to as the dispersing agent) using a high-speed mixer. The resulting suspension is then subjected to a wet milling process (usually bead milling) to break any remaining aggregates or agglomerates and reduce the particle size. One usually aims at a particle size distribution ranging from 0.1 to 5 μm, with an average of 1–2 μm. The surface or polymer added adsorbs on the particle surfaces, resulting in their colloidal stability. The particles need to be maintained stable over a long period of time, since any strong aggregation in the system may cause various problems. Firstly, the aggregates, being larger than the primary particles, tend to settle faster. Secondly, any gross aggregation may result in a lack of dispersion on dilution. Large aggregates can block spray nozzles and may reduce biological efficacy as a result of the inhomogeneous distribution of the particles on the target surface. Apart from their role in ensuring the colloidal stability of the suspension, surfactants are added to many SCs to enhance their biological efficacy. This is usually produced by solubilisation of the insoluble compound in the surfactant micelles. This will be discussed in later sections. Another role, a surfactant may play in SCs, is the reduction of crystal growth (Ostwald ripening). The latter process may occur when the solubility of the agrochemical is appreciable (say greater than 100 ppm) and when the SC is polydisperse. Smaller particles will have higher solubility than larger ones. With time, the small particles dissolve and become deposited on the larger ones. Surfactants may reduce this Ostwald ripening by adsorption on the crystal surfaces, thus preventing deposition of the molecules at the surface. This will be described in detail in the section on SCs.

Mixtures of suspensions and emulsions, referred to as suspoemulsions, have been formulated to allow application of two active ingredients, one being solid and the other an immiscible liquid. Such multiphase systems are difficult to formulate due to the complex interaction between the suspension particles and emulsion droplets. Such complex formulations will be briefly described below.

Very recently, microemulsions have been considered as potential systems for formulating agrochemicals. Microemulsions are isotropic, thermodynamically stable, systems consisting of oil, water and surfactant(s) whereby the free energy of formation of the system is zero or negative. Obviously such systems, if can be formulated, are very attractive since they will have an indefinite shelf-life (within a certain temperature range). Since the droplet size of microemulsions is very small (usually less than 50 nm), they appear transparent. As we will see in later sections, microemulsion droplets may be considered as swollen micelles and, hence, they will solubilise the agrochemical. This may afford considerable enhancement in biological efficacy. Thus, microemulsions may offer several advantages over the commonly used macroemulsions. Unfortunately, formulating the agrochemical as a microemulsion is not straightforward since one usually uses two or more surfactants, an oil and the agrochemical. These tertiary systems produce various complex phases and it is essential to investigate the phase diagram before arriving at the optimum composition of microemulsion formation. As shown in Chapter 10, a high concentration of surfactant (10–20%) is needed to produce such a formulation. This makes such systems more expensive to produce than macroemulsions. However, the extra cost incurred could be offset by an enhancement of biological efficacy, which means that a lower agrochemical application rate could be achieved.

The above introduction illustrates the role of surfactants in various agrochemical formulations. It is necessary to know the physical properties of surfactants, their adsorption at various interfaces and their phase behaviour. These topics are dealt with in Chapters 2 to 5. The role of surfactants in stabilisation of emulsions and suspensions is been dealt with in detail in Chapters 6 and 7.

In this chapter, I will concentrate on the application of surfactants in the various agrochemical formulations. The role of surfactants in enhancing biological efficacy will be dealt with in some detail.

14.2 Emulsifiable Concentrates

Many agrochemicals are formulated as emulsifiable concentrates (ECs), which when added to water produce oil-in-water emulsions either spontaneously or by gentle agitation. Such formulations are produced by addition of surfactants to the agrochemical if the latter is an oil with reasonably low viscosity or to an oil solution of the agrochemical if the latter is a solid or a liquid with high viscosity. Spontaneous emulsification requires several criteria to be met, which may be achieved by control of the properties of the interfacial region. This requires the use of two emulsifiers, which have to be optimised to achieve the required effect. The most commonly used mixture is based on calcium or magnesium salts of alkyl aryl sulphonates and a nonionic surfactant of the ethoxylate type [1]. With such blends, a 5% emulsifier concentration in the formulation may be sufficient for spontaneous emulsification and production of an emulsion with adequate stability within the time of application. Unfortunately, little fundamental work has been carried out to explain the good performance of such a blend. Indeed, most ECs are based on such mixture of anionic/nonionic surfactants and only slight modifications have been made to such recipes. However, with the advent of agrochemical compounds, such simple blends were found, in some cases, to give inferior ECs. In addition, specific surfactants have to be developed to overcome some of the problems encountered with certain agrochemicals that may interact chemically with one or the two of the above-mentioned blends. Another problem that may be encountered with ECs is their sensitivity to variations in the batch of the chemical or the surfactants, which may result in lack of spontaneity of emulsification and/or the stability of the resulting emulsion. Consequently, most manufacturers of ECs adopt rigorous quality control tests to ensure the adequacy of the resulting formulation under the practical conditions encountered in the field. This requires laborious testing of the effect of temperature, water hardness, agitation in the spray tank and batch-to-batch variation of the ingredients of the formulation. As we will see later, the tests used by the formulation chemists are often too simple to provide adequate quantitative evaluation of the EC. Thus, despite the wide use of ECs in agrochemicals, relatively little effort has been devoted to establish quantitative tests for assessment of the quality of the EC. A fundamental surface and colloid chemistry approach to the formulation of ECs is also lacking.

Due to the above shortcomings, a review on emulsifiable concentrates will be at best qualitative and will only help the reader in identifying the areas that require further research, rather than providing any principles or guide lines of how one can best formulate ECs. This contrasts with the situation with concentrated emulsions and suspension concentrates, whereby the basic principles are relatively more established. A useful review on ECs has been published by Becher [1]. The first part of the present section will deal with the common practice of formulation of ECs. This is followed by a section on spontaneous emulsification, which is an important criterion for ECs. The question of stability of the resulting emulsion on standing will not be dealt in this chapter, since this topic was adequately covered in the chapter on emulsions. The third sub-section will deal with a specific example of an emulsifiable concentrate that has been investigated using fundamental studies.

14.2.1
Formulation of Emulsifiable Concentrates

As mentioned in the introduction, ECs are formulated by trial and error, whereby a pair of emulsifiers is selected for a specific agrochemical formulation. As stated by Becher [1], the hydrophilic–lipophilic balance (HLB) method, which is normally used to select surfactants in emulsions (see Chapter 7), is inadequate for the formulation of ECs [2, 3]. This is not surprising since with ECs one requires, in the first place, spontaneity of dispersion on dilution, which as mentioned above is governed by the properties of the interfacial region. Other indices such as the cohesive energy ratio concept suggested by Beerbower and Hill [4] may provide a better option. This concept was discussed in some detail in Chapter 6. Essentially, the method involves selecting suitable emulsifiers by balancing the interactions of their hydrophobic parts with the oil phase and the hydrophilic parts with the aqueous phase. This involves knowledge of the solubility and hydrogen-bonding parameters of the various components. However, as explained in Chapter 6, these parameters are not always available. Owing to the lack of this information, it is not possible to check whether this approach could be applied to emulsifier selection.

From the above discussion, it is clear why the choice of surfactants for ECs is still made on a trial and error basis. Once this choice is made, an extensive work is required to optimise the composition to produce an acceptable product that satisfies the criteria of spontaneous emulsification and stability under practical conditions. In many cases, it is also essential to add other components such as dyes, defoamers, crystal growth inhibitors and various other stabilizers. The amount of work required to select emulsifiers may be illustrated from the publication of Kaertkemeyer and Ahmed [5] who investigated the emulsification of non-phytotoxic pesticidal oils using a set of eight nonionic surfactants. Four different hydrophobic groups, namely linear and branched alcohols (with an average of about 13 carbons) and linear and branched nonylphenol, were each ethoxylated to two different degrees to form four pairs of surfactants investigated. In each case the oleophilic emulsifier was made by adding about one mole of ethylene oxide, while the hydrophilic one contained about 6 moles. The authors investigated the four combinations of two pairs possible when one pair had an alkyl hydrophobe and the other an alkyl aryl one. The results were presented in the form of planes cutting the tetrahedral phase diagrams. Sections of the phase diagrams where good ECs existed were found, but they were only a small part of the total volume. No correlation was found between the HLB number of the surfactant blend and emulsion spontaneity or stability. The authors also noted that branching of the alkyl chains affected the results significantly. For example, the straight-chain alkyl phenols were found to be more effective than their branched counterparts. This shows the great sensitivity of the quality of the EC to minor changes in surfactant structure, rendering formulation of ECs somewhat tedious (at least in some cases).

As mentioned above, testing of ECs is carried out using fairly simple procedures. The most common procedure is that based on the recommendation of the World Health Organization (W.H.O.), which was originally designed for DDT emulsifiable concentrates. The W.H.O. specification states,

> "any creaming of the emulsion at the top, or separation of sediment at the bottom, of a 100-ml cylinder shall not exceed 2 ml when the concentrate is tested as described in Annex 12 in *Specification for Pesticides* [6, 7]."

This test is described as follows:

> "Into a 250-ml beaker having an internal diameter of 6–6.5 cm and 100-ml calibration mark and containing 75–80 ml of standard hard water, 5 ml of EC is added, using a Mohr-type pipette, while stirring using a glass rod, 4–6 mm in diameter, at about four revolutions per second. The standard hard water is designed to contain 342 ppm, calculated as calcium carbonate. This is prepared by adding 0.304 g of anhydrous $CaCl_2$ and 0.139 g $MgCl_2 \cdot 6H_2O$ to make one liter using distilled water. The concentrate should be added at a rate of 25–30 ml per minute, with the point of the pipette 2 cm inside the beaker, the flow of the concentrate being directed towards the centre, and not against the side, of the beaker. The final emulsion is made to 100 ml with standard hard water, stirring continuously, and then immediately poured into a clean, dry 100-ml graduated cylinder. The emulsion is kept at 29–31 °C for one hour and examined for any creaming or separation."

The reason for the above procedure is because both temperature and water hardness have a major effect on the performance of ECs. This was illustrated previously [8], showing the effect of water hardness on the amount of cream that separates from a typical formulation. The best performance appears to be observed at a water hardness of 300 ppm, but this may not be general with all other formulations. The rate that the amount of cream approaches equilibrium is fairly independent of water hardness, which means that taking an arbitrary time of 1 hour to measure the separated cream or sediment is adequate for relative comparison of various formulations. Increasing the time to 3 to 4 hours for full creaming or sedimentation does not, in general, change the rating of various systems [9]. The stability of the produced emulsion first improves as the water hardness is increased, but above a critical value of ~500 ppm it rapidly decreases with further increase in hardness of water [9]. The decrease in the volume of cream at high water hardness is caused by the appearance of macroscopic oil droplets. At still higher water hardness the oil will separate as a distinct layer. The effect of temperature on the stability of an emulsion produced from an EC [9] has also been investigated. Generally speaking, raising the temperature shifts the optimum performance to softer water and lowering it has the opposite effect.

The above dependence of the performance of ECs on water hardness and temperature may be related to the dependence of emulsifier properties on these parameters. This is discussed in more detail in Chapter 6. Nonionic surfactants are particularly sensitive to these parameters. For example, the solubility in water and, therefore, the effective HLB number of a typical nonionic surfactant decreases as the temperature or salt content of the solution increases. This is evident from the decrease in c.m.c. and cloud point with increasing salt concentration and/or temperature [10, 11].

Carino and Nagy [12] investigated the properties of ECs of toxaphene and diazinon dissolved in kerosene and xylene using surfactant blends of calcium dodecyl benzene sulphonate, CaDBS, (70%) and a series of ethoxylated nonylphenols. The results were compared in soft (water hardness of ~11.5 ppm) and hard (~290 ppm) water. The number of ethylene oxide (EO) groups on the nonionic surfactant that gave the most stable emulsions and the number of days before any cream separated from them was obtained as a function of CaDBS concentration.

As the amount of CaDBS was increased, the number of EO units in the surfactant required to maintain stability also increased. The highest stability overall was found at a ratio of anionic to nonionic surfactant that was a function of water hardness. The authors also found a strong dependence of stability on the amount of surfactant used.

14.2.2
Spontaneity of Emulsification

Spontaneous emulsification was first demonstrated by Gad [13], who observed that when a solution of lauric acid in oil is carefully placed onto an aqueous alkaline solution, an emulsion is spontaneously formed at the interface. As explained in Chapter 10, such spontaneous emulsification could be due to the very low (or transient negative) interfacial tension produced by the surfactant. Using an aqueous alkaline solution causes partial neutralisation of lauric acid. A mixture of lauric acid

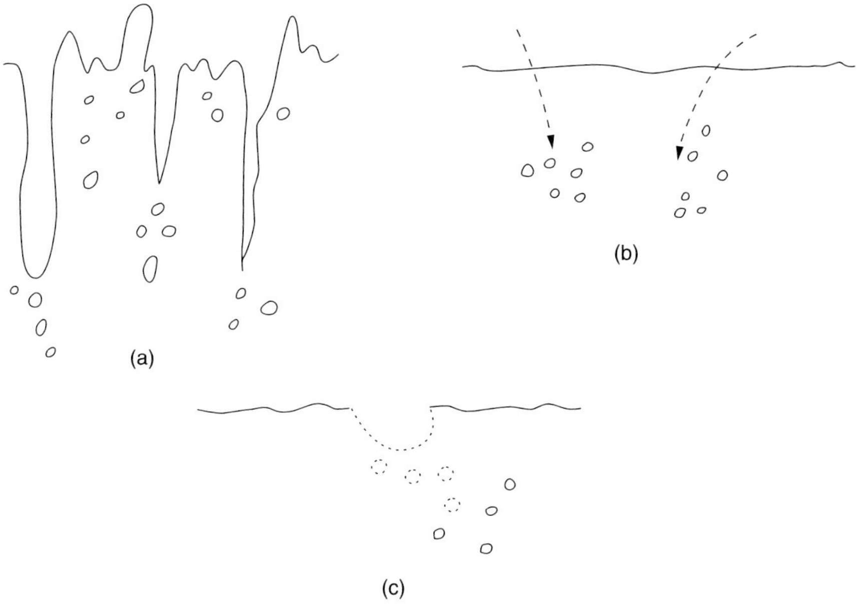

Fig. 14.1. Schematic representation of spontaneous emulsification: (a) interfacial turbulence; (b) diffusion and stranding; (c) ultralow interfacial tension.

and laurate can produce an ultralow interfacial tension. The process of spontaneous emulsification described by Gad [13] appears to occur with minimum external agitation, thus supporting the view that disruption of the interface may occur as a result of the combined surfactant film. Three main mechanisms may be responsible for spontaneous emulsification and these are briefly summarised below.

The first mechanism is due to interfacial turbulence, which may occur as a result of mass transfer. In many cases the interface shows unsteady motions; streams of one phase are ejected and penetrate into the second phase, shredding small droplets (Figure 14.1). Localised reductions in interfacial tension are caused by the non-uniform adsorption of the surfactant at the oil/water interface [14] or by mass transfer of surfactant molecules across the interface [15, 16]. With two phases that are not in chemical equilibrium, convection currents may form, conveying liquid rich in surfactants towards areas of liquid deficient of surfactant [17, 18]. These convection currents may give rise to local fluctuations in interfacial tension, causing oscillation of the interface. Such disturbances may amplify themselves, leading to violent interfacial perturbations and eventual disintegration of the interface, when liquid droplets of one phase are "thrown" into the other [19].

The second mechanism that may account for spontaneous emulsification is based on diffusion and stranding. This is best illustrated by carefully placing an ethanol–toluene mixture (containing say 10% alcohol) onto water. The aqueous layer eventually becomes turbid due to the presence of toluene droplets [20]. In this case, interfacial turbulence does not occur although spontaneous emulsification apparently takes place. Alcohol molecules are suggested to diffuse into the aqueous phase, carrying some toluene in a saturated three-component sub-phase [20, 21]. At some distance from the interface the alcohol becomes sufficiently diluted in water to cause the toluene to precipitate as droplets in the aqueous phase. Such a phase transition might be expected to occur when the third component increases the mutual solubility of the two previously immiscible phases.

The third mechanism of spontaneous emulsification may be due to the production of an ultralow (or transiently negative) interfacial tension. This mechanism is thought to be the cause of formation of microemulsions when two surfactants, one essentially water soluble and one essentially oil soluble, are used [22, 23]. This mechanism is described in detail in Chapter 10 on microemulsions.

14.2.3
Fundamental Investigation on a Model Emulsifiable Concentrate

Lee and Tadros [24–27] carried out some fundamental studies on a model EC of xylene containing a nonionic and a cationic surfactant. The objective was to study the effect of stability of the resulting emulsion on herbicidal activity of a model compound, namely 2,4-dichlorophenoxyacetic acid ester. The nonionic surfactant used was Synperonic NPE 1800 (supplied by ICI), ethoxylated-propoxylated nonylphenol (**14.1**). The cationic surfactant was Ethoduomeen T20 (ET 20) (Supplied by Armour Hess) (**14.2**).

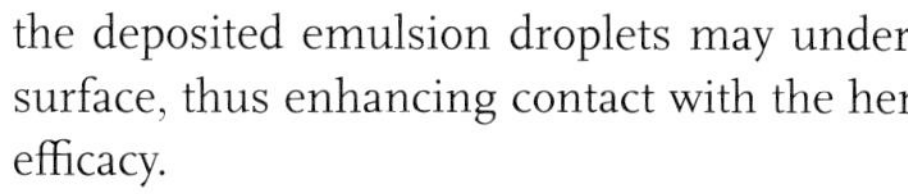

14.1

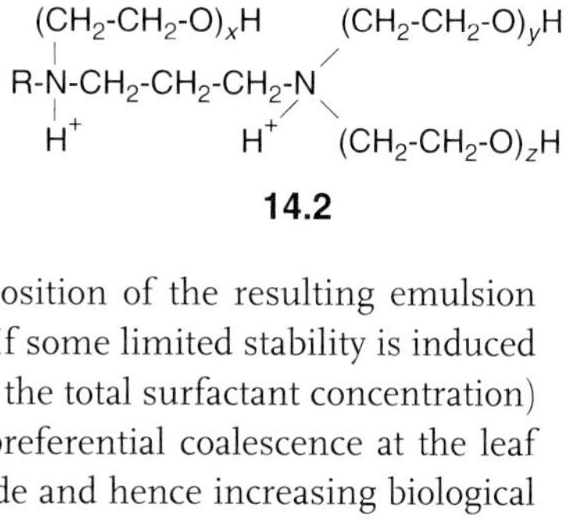

14.2

The cationic surfactant was used to ensure deposition of the resulting emulsion droplets on the negatively charged leaf surfaces. If some limited stability is induced in the resulting emulsion produced (by reducing the total surfactant concentration) the deposited emulsion droplets may undergo preferential coalescence at the leaf surface, thus enhancing contact with the herbicide and hence increasing biological efficacy.

The effect of surfactant concentration of SNPE + ET20 (in a 1:1 ratio by weight) on the spontaneity of dispersion on dilution of the xylene EC was studied using the CIPAC test and measuring the dispersed phase mean droplet diameter. In the CIPAC test 1 ml of an EC was added by free fall from a 1 ml pipette held 1 cm above the surface, to 100 ml of water contained in a 100 ml measuring cylinder. The apparent ease of dispersion was termed either good, moderate or poor. The measuring cylinder was then immediately inverted three times and the fineness and uniformity of the emulsion expressed on a scale of 0 (the sample does not emulsify) to 6 (the sample disperses to form a clear or opalescent solution, i.e. L_1, indicated on the phase diagram; see below). The average droplet diameter of each emulsion was measured immediately after dispersal of the EC, using the Coulter Nanosizer, which measures the time-dependent fluctuations in the intensity of scattered light by a dispersion, and calculates the average diffusion coefficient of the droplets and hence their average droplet diameter. Before the measurement, the emulsions were diluted in water to avoid multiple scattering. The results showed that a minimum of about 1% total surfactant is necessary to produce spontaneous emulsification, after which there is a gradual improvement in spontaneity and a reduction in droplet size with increasing surfactant concentration up to 5%. Beyond this concentration the average droplet size of the emulsions increases with increase in surfactant concentration, although the spontaneity of emulsification is well maintained until a concentration of 20% surfactant is reached. Any further increase in surfactant concentration results in a deterioration of spontaneity, accompanied by further increase in droplet size, until, with 40% surfactant, the dispersal of the oil becomes relatively poor. When the concentration of surfactant reaches 60%, a very viscous EC is formed, which disperses very slowly to form a "solubilized" system with a small mean droplet diameter.

Figure 14.2 shows the change in viscosity of the ECs with increasing concentration of surfactant. This figure shows a rapid increase in viscosity above ~30% surfactant, which may be due to the formation of surfactant aggregates, e.g. inverse micelles produced in the presence of small amounts of water that is present in the hydrated surfactants.

Figure 14.3 shows the interfacial tension (γ) at the xylene–water interface versus log concentration (expressed as %) for SNPE, ET20 and a 1:1 mixture of SNPE and ET20. SNPE is seen to be more surface active at the xylene–water interface than ET20. The γ values for the 1:1 mixture are closer to those for SNPE, indicating preferential adsorption of SNPE at the oil–water interface. At sufficiently high concen-

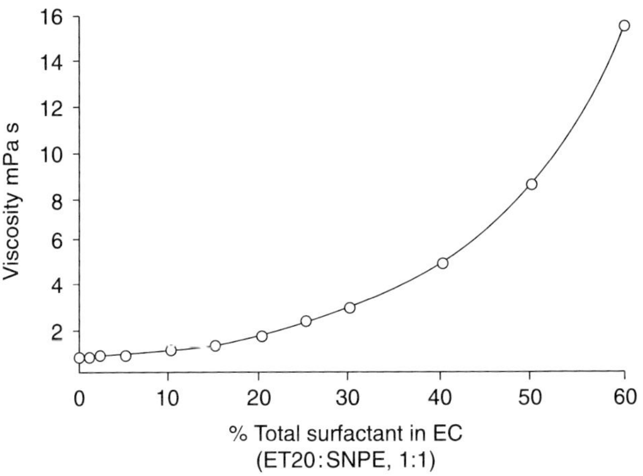

Fig. 14.2. Variation of viscosity with surfactant concentration (1:1 mixture of SNPE and ET20) for a xylene EC.

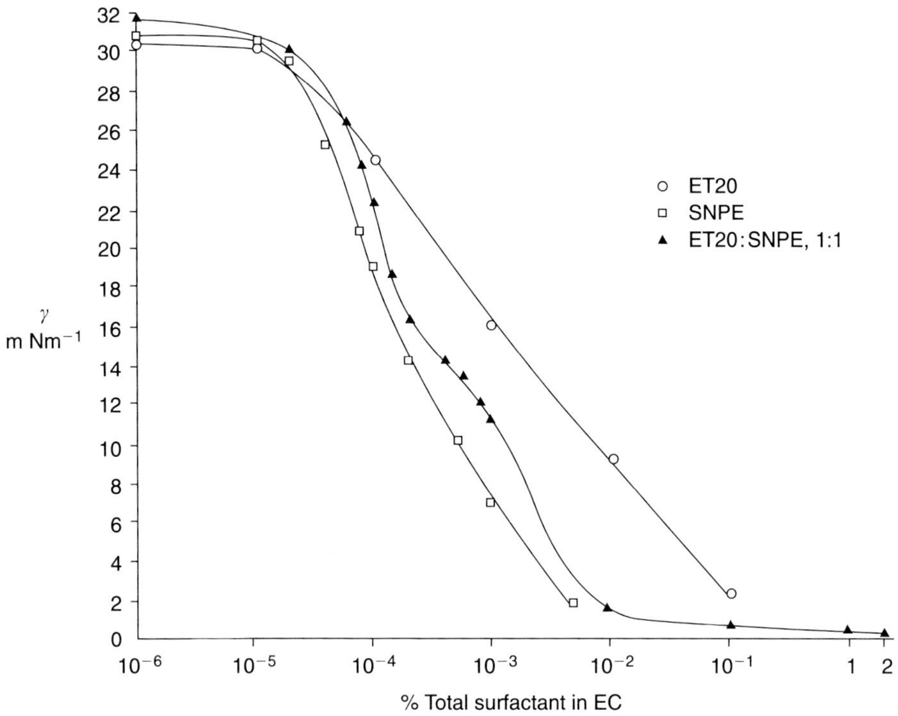

Fig. 14.3. Interfacial tension versus log concentration of surfactants.

tration of surfactant, γ becomes quite low (<1 mN m^{-1}). This is clearly illustrated for the 1:1 mixture for which the concentration was extended to 2%. These low interfacial tensions were measured using the spinning drop technique [28]. No measurements could be made above 2% surfactant since the droplets disintegrated as soon as spinning of the tube started.

The above interfacial tension results may throw some light on the mechanism of spontaneous emulsification in the present model EC. As mentioned before, there are basically two main mechanisms of spontaneous emulsification, namely creation of local supersaturation (i.e. diffusion and stranding) or by mechanical breakup of the droplets as a result of interfacial turbulence and/or the creation of an ultralow (or transiently negative) interfacial tension. Diffusion and stranding is not the likely mechanism in the present system since no water-soluble co-solvent was added. To check whether the low interfacial tension produced is sufficient to cause spontaneous emulsification, a rough estimate may be made from consideration of the balance between the entropy of dispersion and the interfacial energy, i.e.

$$\Delta G^{\mathrm{form}} = \gamma \, \mathrm{d}A - T\Delta S^{\mathrm{config}} \tag{14.1}$$

where ΔG^{form} is the free energy of formation of the emulsion from the EC, dA is the increase in interfacial area when a bulk oil phase is dispersed into droplets, T is the absolute temperature and ΔS^{config} is the configurational entropy of the droplets in the resulting dispersion. To a first approximation [29],

$$\Delta S^{\text{config}} = -nk\left[\ln \phi_2 + \left(\frac{1-\phi_2}{\phi_2}\right)\ln(1-\phi_2)\right] \tag{14.2}$$

where k is the Boltzmann constant and ϕ_2 is the volume fraction of the dispersed phase. Clearly, $\gamma\,\mathrm{d}A$ must be $< T\Delta S^{\text{config}}$ for spontaneous emulsification to occur. The limiting value of γ where this occurs is obtained by equating ΔG^{form} to zero. Replacing dA by $n4\pi r^2$, where n is the number of droplets and r their radius, one obtains

$$\gamma = -kT\frac{[\ln \phi_2 + (1-\phi_2)/\phi_2 \ln(1-\phi_2)]}{4\pi r^2} \tag{14.3}$$

Taking an example from the present investigation, e.g. with 5% surfactant $\phi_2 = 0.01$ (the dilution used) and $r = 0.27$ μm, the value of γ required for spontaneous emulsification to occur is found from Eq. (14.3) to be $\sim 2 \times 10^{-5}$ mN m^{-1}. Values of this order have not yet been reached, thus ruling out the possibility of an ultralow interfacial tension as responsible for spontaneous emulsification. The most likely mechanism in the present system is interfacial turbulence that may be caused by mass-transfer of surfactant molecules across the interface, which will also lead to interfacial tension gradients.

Another useful fundamental study for ECs is to establish the phase diagram of the various components. Figure 14.4 illustrates this for the present model EC of xylene/SNPE/ET20/water. The phase diagrams show the effect of addition of water on the three-phase system of xylene/SNPE/ET20. The anhydrous ECs are all isotropic liquids (Figure 14.4a), with the exception of those containing SNPE concentrations in excess of its solubility in the other components, in which case solid SNPE is also present. At very high concentrations of SNPE, a gel is formed. ET20 is, apparently, miscible with the other components at all concentrations. Addition of 5% water to the ECs (Figure 14.4b) produces a large area of L_2 phase, consisting of water solubilised with the inverse micelles of surfactant in xylene [30]. The phase diagram also shows small areas where an emulsion or an emulsion in equilibrium with liquid crystalline phase is formed. This is particularly the case at low surfactant concentrations, where the water is not completely solubilized. With increasing concentrations of water (10, 50, 90 and 99% in Figure 14.4c–f respectively) the area of the emulsion phase increases, and inversion from water-in-oil to oil-in-water takes place at some unidentified concentration. With 10 and 50% water, a significant region of liquid crystalline phase is observed. With 50% water, a gel and L_1 phase (oil solubilized in an aqueous micellar solution) appear. With a further increase in water concentration, the area of the L_1 phase extends at the ex-

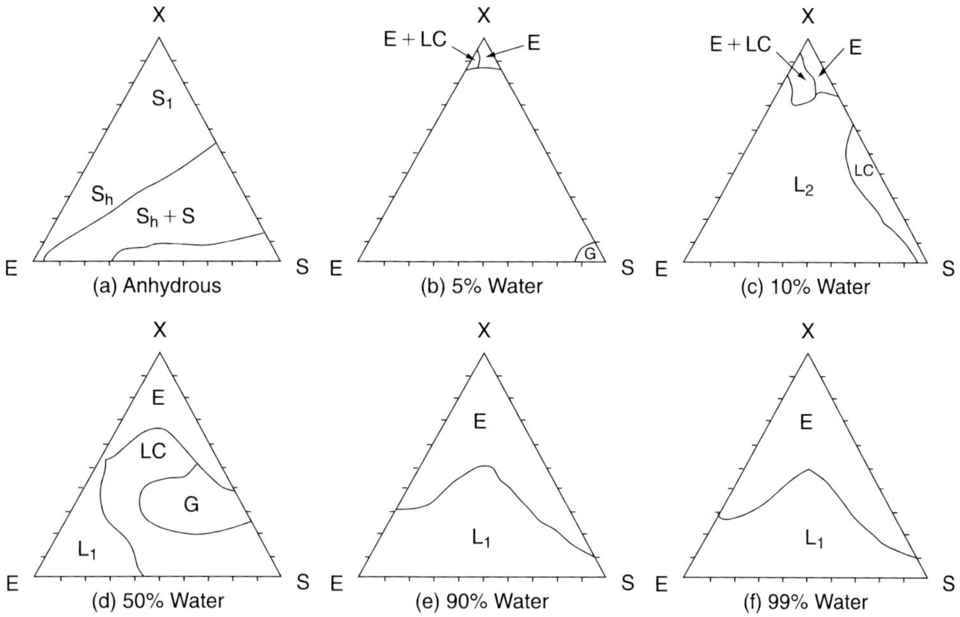

Fig. 14.4. Phase diagrams for the system SNPE–ET20–xylene at various dilutions in water. Key: X = xylene; E = ET20, S = SNPE; S_1 = low viscosity solution; S_h = high viscosity solution; S = solid; G = gel; E (inside the triangular axes) = emulsion; LC = liquid crystal; L_2 = organic isotropic solution; L_1 = aqueous isotropic liquid.

pense of the liquid crystalline and gel phases. The latter phase disappears completely with 90% water leaving only emulsion and L_1 phases.

For quantitative assessment of emulsion stability after dilution of the EC, it is necessary to measure the coalescence rate. This could be done by measuring the droplet number as a function of time, using, for example, a Coulter counter. As an illustration, Figure 14.5 shows the results obtained using the model xylene EC.

The number of droplets shown in Figure 14.5 is those greater than 1 μm, since the Coulter counter can not count submicron droplets. Assuming that coalescence occurs between aggregated oil droplets and that, on average, each droplet is in contact with two other droplets, then the rate of coalescence could be described by a first-order kinetics that is governed by the rupture of the aqueous film (lamella) separating neighbouring droplets [31].

Thus if N_0 is the number of oil droplets at $t = 0$ and N_t is that at time t, then

$$N_t = N_0 \exp(-Kt) \tag{14.4}$$

where K is the rate of coalescence. Eq. (14.4) predicts that a plot of log N_t versus t should be a straight line. However, straight lines were only obtained in very few

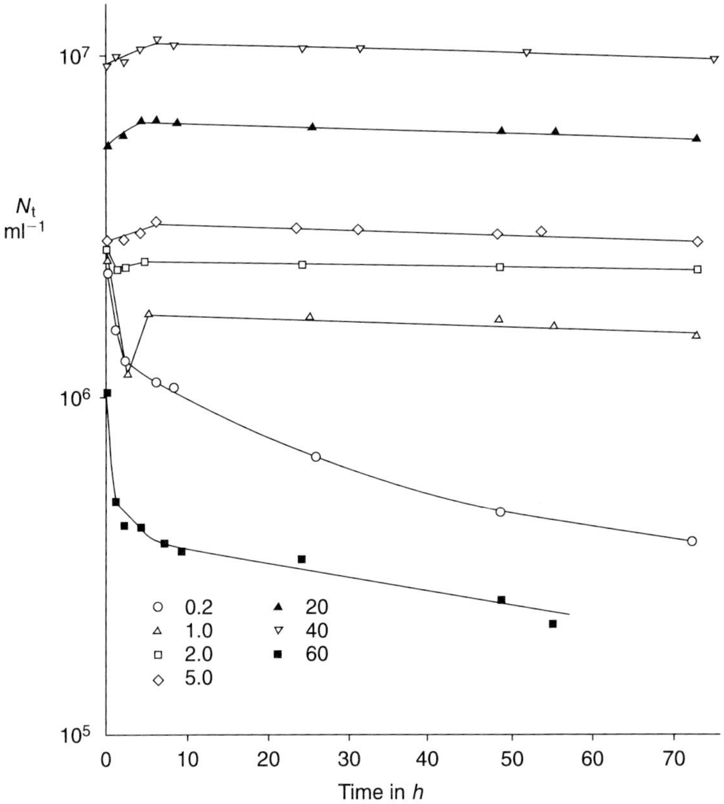

Fig. 14.5. Log(droplet number > 1 μm) versus time for spontaneously formed emulsions produced at various surfactant (1:1 ratio of SNPE–ET20 mixture) concentrations.

cases (Figure 14.8 below) and in most cases the log N_t versus t plots were curved, showing some fluctuations in log N_t during the initial period (< 5 h). However, for comparison, the apparent coalescence rates were calculated from the slopes of the straight lines (when these were obtained) or of the tangents to the initial portion of each curve (within the first few hours). The coalescence rates of all emulsion were then plotted versus total surfactant concentration (Figure 14.6). The later also shows the variation of the initial number of droplets >1 μm versus surfactant concentration.

Figure 14.6 shows that the coalescence rate K decreases very rapidly with increasing surfactant concentration from 0.2 to 1%, after which there is only a slight reduction in K with a further increase in surfactant concentration up to 40%. However, when the surfactant concentration is increased from 40 to 60%, the coalescence rate apparently increased to 2.14×10^{-4} s^{-1}, a value that is higher than that obtained with the lowest surfactant concentration (0.2%). There was also a simul-

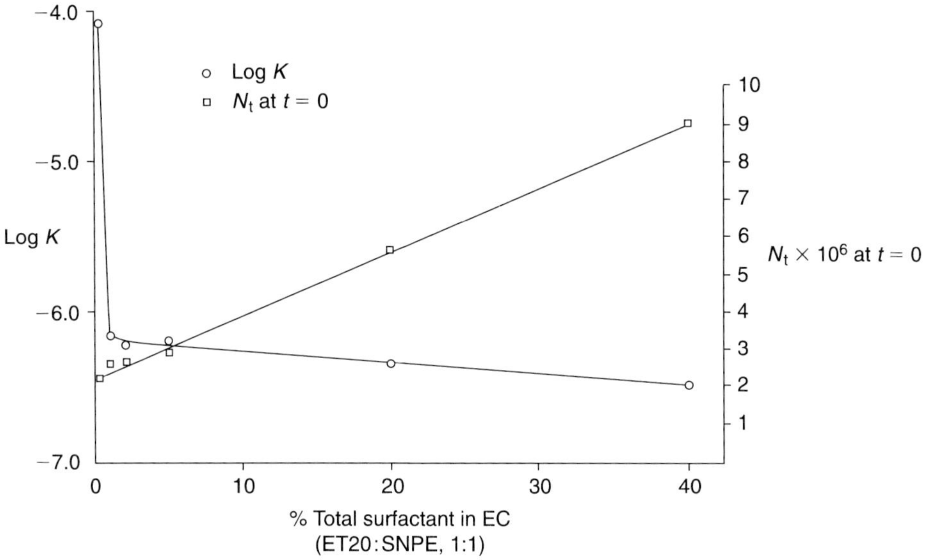

Fig. 14.6. Coalescence rate and droplet number > 1 μm (N_t) at $t = 0$ as a function of surfactant concentration.

taneous reduction in the initial number of droplets > 1 μm in diameter to 1.08×10^6 and the average droplet, diameter as measured by the Coulter Nanosizer, was 1.06 μm.

The increase in emulsion stability with increasing surfactant concentration up to 40% is what one would expect. This enhanced stability is due to an increase in the viscoelastic properties of the film, e.g. surface elasticity and/or viscosity and possible formation of liquid crystalline structures at the oil/water interface (for details see Chapter 6). However, the increase in coalescence rate above 40% surfactant concentration is probably due to Ostwald ripening, which may be enhanced by solubilization by the surfactant micelles. The latter in particular is known to enhance crystal growth of solid/liquid dispersions [32]. The driving force for this process is the difference in solubility between small and large droplets. The smaller droplets with their greater solubility are thermodynamically unstable with respect to the larger ones. This can be expressed using the Ostwald–Freundlich equation [33],

$$S_r = S_\infty \exp \frac{2\gamma M}{r\rho RT} \tag{14.5}$$

where S_r is the solubility of a droplet with radius r, S_∞ is the solubility of a droplet with infinite size, γ is the interfacial tension, M and ρ are the molecular weight and density of the droplet material, R is the gas constant and T the absolute temperature. Substitution of reasonable values of γ in Eq. (14.5) shows that the difference between S_r and S_∞ becomes significant as the droplet radius becomes very small. For example, taking an γ of ~1 mN m^{-1} (the value obtained at relatively high surfactant concentrations) S_r/S_∞ is 1.01 for a 0.01 µm droplet of xylene at 25 °C. Such solubility difference does not in itself provide sufficient driving force to account for the instability of the emulsion at surfactant concentrations above 40%. Moreover, if the difference in solubility between large and small droplets is the only driving force accounting for instability, there is no reason why instability continues to increase with increasing surfactant concentration in the range where the droplet size does not change significantly. Thus, some other process must be responsible for the transfer of the oil molecules from the smaller to the larger droplets. This is probably due to the solubilization of the oil by the surfactant micelles, an effect that is appreciable at high surfactant concentrations. The effect of this on droplet stability can be explained if one considers the diffusion of the oil molecules to the droplet–continuum interface. The diffusion flux, J, of the oil molecules in the aqueous continuous phase, in mol cm^{-2} s^{-1}, is given by Fick's first law,

$$J = -D\left(\frac{\delta c}{\delta x}\right) \tag{14.6}$$

where D is the diffusion coefficient and $(\delta c/\delta x)$ is the concentration gradient. As a result of solubilization, the oil molecules become incorporated into the micelles. Since the diffusion coefficient is roughly proportional to the radius of the diffusing particle [34, 35], D is reduced by a factor of about 10, which would correspond to a micelle having a volume 1000× larger than that of the solubilizate. However, as a result of solubilization, the concentration gradient will increase greatly (in direct proportion to the extent of solubilization). This is because Fick's law involves the absolute gradient of concentration, which is small so long as the solubility is small, rather than its relative value. If S represents the saturation value, then

$$J = -DS\left(\frac{\delta \ln S}{\delta x}\right) \tag{14.7}$$

Equation (14.7) shows that, for the same gradient of relative saturation, the flux caused by diffusion is directly proportional to saturation. Hence solubilization will, in general, increase transport by diffusion, since it can increase the saturation value by many orders of magnitude, even though it decreases the diffusion coefficient. Thus, as a result of the large extent of solubilization at high surfactant concentrations, the diffusional flux increases and enhances the extent of Ostwald ripening. The greater the difference in size between the droplets, the greater the rate of growth of the larger droplets, particularly when there is a significant proportion of droplets in the submicron region. This was confirmed using homogenization after dilution of the emulsions, to create smaller droplets. In this case Ostwald ripening was detected at much lower surfactant concentration, namely 15% [24].

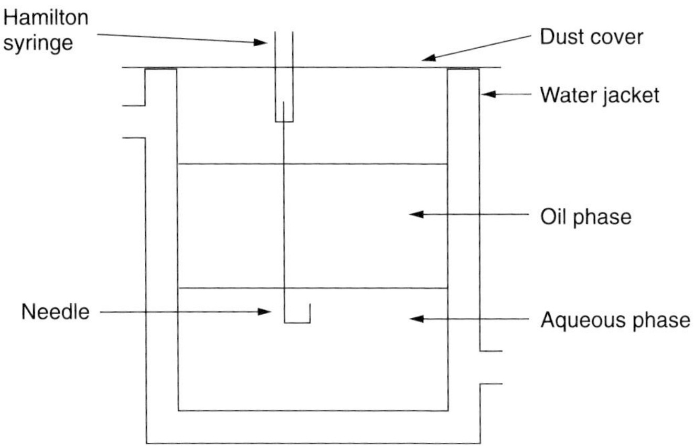

Fig. 14.7. Setup for measuring coalescence at a planar oil–water interface.

Another method of investigating the stability of emulsions, produced by dilution of ECs, is to study the coalescence of droplets at a planar oil–water interface at various surfactant concentrations. This may be carried out using the set-up shown in Figure 14.7, which was suggested by Cockbain and McRoberts [36]. Equal volumes of water and EC are placed in the cell, which is kept at constant temperature by circulating water from a thermostat bath through the double-walled vessel. Equal volumes of water and EC are left for several hours in the cell to equilibrate. Subsequently, droplets of equilibrated EC, of the same volume, are individually formed from a Hamilton syringe in the aqueous phase near the interface, and their rest times before coalescence with the bulk organic phase are measured. Not less than 80 droplets should be measured when the rest times are long and more than 120 droplets when the rest times are short. Many investigators [36–38] have noted that the rest times of several oil droplets produced from the same EC at constant surfactant concentration are not constant but show considerable variation. Two methods may be used to treat the data. In the first method, the rest times are assumed to be symmetrically distributed around a mean value (a Gaussian distribution) and so an arithmetic mean (t_{mean}) and standard deviation (σ) are calculated for each system. The results obtained using this method are summarised in Table 14.1 for xylene ECs containing various concentrations of a 1:1 W/W mixture of SNPE and ET20.

The second method used by Cockbain and McRoberts [36] involves plotting the results as a distribution curve. The number N_t of droplets that have not yet coalesced within a time t is plotted versus time. This distribution curve consists of two fairly well defined regions, one in which N_t is nearly constant with t, followed by a region in which N_t decreases with time in an exponential fashion. The first region corresponds to the process of drainage of the thin liquid film of the continuous phase from between the droplets and the planar interface, whereas the second region is that where rupture of the thin film and coalescence take place. Since

Tab. 14.1. Mean rest times, half-life for film rapture and drainage time at various surfactant concentrations.

% Total surfactant in EC	*Gaussian distribution*		*Cockbain–McRoberts*	
	t_{mean} (s)	σ (s)	$T_{1/2}$ (s)	t_D (s)
0	5.1	3.2	4.5	0.6
1×10^{-4}	3.0	1.0	1.3	1.7
1×10^{-3}	1.3	0.3	0.5	0.8
2×10^{-3}	1.8	1.0	1.6	0.2
3×10^{-3}	7.0	3.8	7.4	(−0.4)
4×10^{-3}	14.0	10.6	19.0	(−5.0)
5×10^{-3}	98	40	41	57
6×10^{-3}	229	33	39	190
7×10^{-3}	240	19	38	202
1×10^{-2}	249	41	62	187
2×10^{-2}	272	27	48	224
3×10^{-2}	265	53	54	211
4×10^{-2}	259	36	47	212
5×10^{-2}	262	40	43	219
7.5×10^{-2}	267	31	37	230
1×10^{-1}	>500	–	–	–
1.0	>500	–	–	–

film rupture and coalescence usually follow a first order process [36], the rate constant k can be calculated from the slope of the line of log N_t versus t. As an illustration, the results obtained for xylene ECs containing 1:1 W/W mixture of SNPE:ET20 are shown in Figures 14.8 and 14.9 for various total surfactant concentrations. In these figures, log N_t/N_0 is plotted against t, where N_0 is the total number of droplets measured at $t = 0$. All these plots show the two regions mentioned above. A first-order half-life for film rupture ($T_{1/2}$) was calculated from the slope of the second section of each plot, using the relationship [36],

$$\log(N_t/N_0) = -kt \tag{14.8}$$

where k is the rate constant of film rupture that is equal to $\ln(2/T_{1/2})$. The drainage time t_D was also calculated from

$$t_D = t_{mean} - T_{1/2} \tag{14.9}$$

where t_{mean} is the geometric mean rest time, which is numerically equal to the experimental half-life $t_{1/2}$. t_D is a measure of the rate of drainage of the liquid film antecedent to its rupture.

The results for $T_{1/2}$ and t_D summarised in Table 14.1 show that the addition of surfactant to xylene affords an initial decrease in mean rest time and $T_{1/2}$, both of which reach a minimum at 1×10^{-3}% total surfactant. With further increase in surfactant concentration, the mean rest time and $T_{1/2}$ increase, reaching their orig-

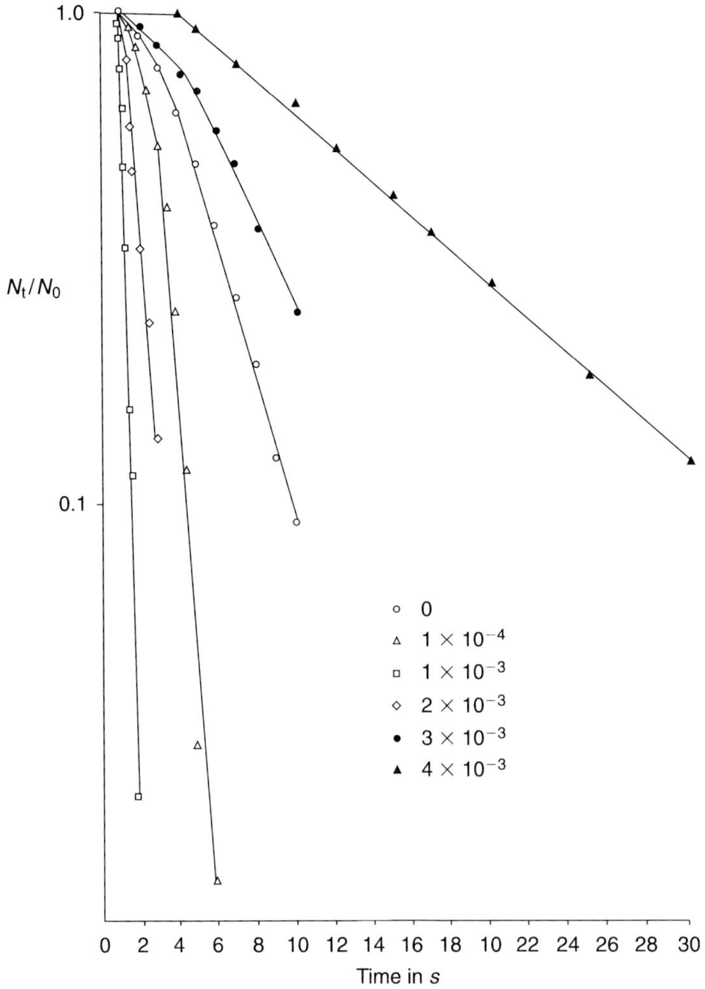

Fig. 14.8. Cockbain–McRoberts plots for the ECs (0–0.004% surfactant).

inal values for xylene at ca. 3×10^{-3}% total surfactant. Above 3–4 $\times 10^{-3}$% there is a sharp increase in drainage time and $T_{1/2}$ and an increase in the mean rest time.

The reduction in droplet rest time in the presence of a small concentration of surfactant ($< 4 \times 10^{-3}$%, Table 14.1) may be attributed to interfacial turbulence caused by the diffusion of surfactant molecules across the interface [36]. Such disturbances should increase the possibility of film rupture. However, this should not occur extensively in a chemically equilibrated system. The decrease in rest time at these low surfactant concentrations, therefore, indicates that either the systems were not fully equilibrated after having stood overnight, or that interfacial turbulence was caused by changes in interfacial tension gradients caused by the distur-

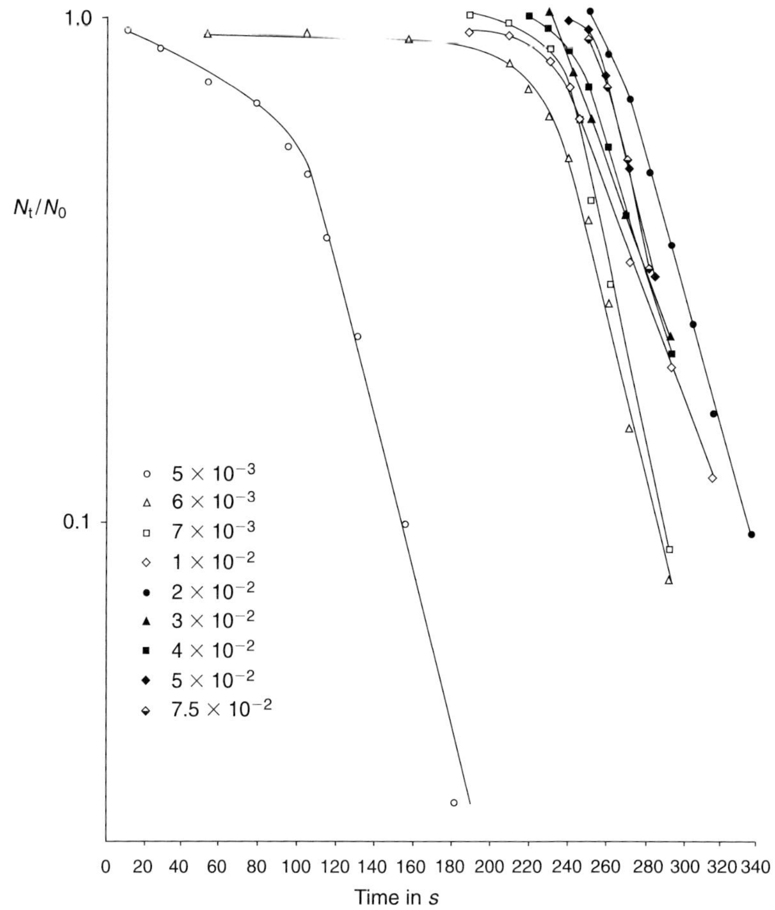

Fig. 14.9. Cockbain–McRoberts plots for ECs (0.005–0.075% surfactant).

bance of the surfactant film during drainage. Thus, one has to be careful in applying this technique for studying the stability of emulsions produced by dilution of the EC.

The sharp increase in drainage time and $T_{1/2}$ above 3–4 × 10^{-3}% surfactant concentration implies the existence of a considerable resistance to film thinning (i.e. high emulsion stability) above this surfactant concentration. Several factors may account for such film stability and enhancement of the stability of the emulsion above this surfactant concentration: a reduction in interfacial tension, an increase in interfacial electrostatic potential and/or the formation of a highly condensed interfacial film giving strong steric interactions between the droplet and the planar interface. For comparison, the mean rest time, interfacial tension and zeta potential are plotted as a function of surfactant concentration in Figure 14.10. The co-

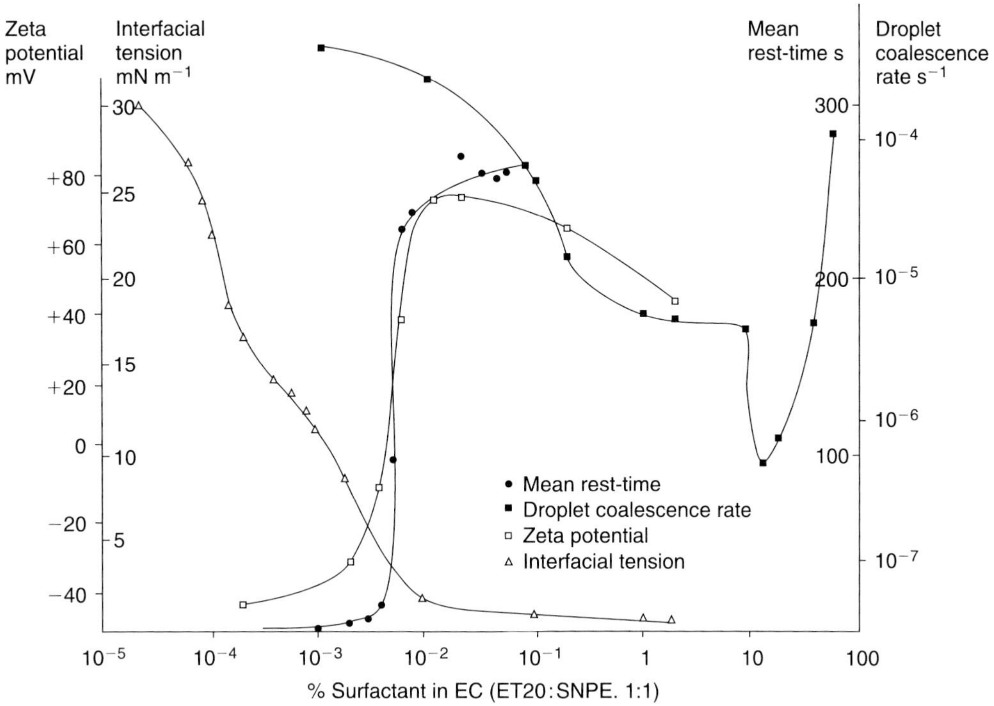

Fig. 14.10. Mean rest times, interfacial tensions, zeta potentials and coalescence rates as a function of surfactant concentration.

alescence rates for the emulsions in bulk solution obtained by dilution of the ECs are also shown on the same diagram. Clearly, the increase in stability against drainage or coalescence is not directly related to a reduction in interfacial tension, as the mean rest time did not significantly increase in the surfactant concentration range where the interfacial tension showed a sharp decrease. Indeed, it was not until the interfacial tension was significantly reduced below 5 mN m^{-1} that the rest time began to increase sharply. The same applies to the stability of the diluted emulsions. This behaviour is not surprising since previous studies using nonionic [37] and anionic [38] surfactants showed little correlation between interfacial tension and resistance to drainage at a planar oil/water interface. However, the sharp increase in rest times seems to correlate with the sharp increase in zeta potential, although the correlation with coalescence rate of the diluted ECs is not as good.

Seemingly, a minimum surfactant concentration is required to ensure the stability of the emulsion produced by dilution of the ECs. Mixed interfacial films with specific rheological properties are required for stabilisation of the emulsions. These films should provide high dilational viscoelasticity and they should prevent film thinning and drainage. This is discussed in more detail in Chapter 6.

14.3
Concentrated Emulsions in Agrochemicals (EWs)

As mentioned in the introduction, concentrated oil-in-water emulsions (EWs) are preferable for emulsifiable concentrates for formulations of agrochemicals. The latter can be an oil with sufficiently low viscosity to be directly emulsified into water using a suitable surfactant system. If the agrochemical oil viscosity is high, it can be mixed with an oil of low viscosity such as xylene or Solvesso (commercially available trimethylbenzene). The mixture of agrochemical and oil can be emulsified into an aqueous solution containing the suitable emulsifier system. In some cases, a solid agrochemical (with low melting point) can be dissolved in a suitable oil, and the oil solution is used to formulate an O/W emulsion.

The major advantages of EWs are their relatively low toxicity when compared with ECs, their high flash points and possibility of incorporation of adjuvants in the oil and aqueous phases. In addition, by controlling the droplet size distribution of the oil, one can enhance deposition and spreading and this may increase biological efficiency.

The main drawback of O/W emulsions is the control of their physical stability, which needs to be controlled at various temperatures with adequate shelf life (usually a shelf life of 1–2 years is required at temperatures that can vary from $-10°$ to 50 °C). This represents a challenge to the formulation chemist.

As discussed in Chapter 6, emulsions are thermodynamically unstable. This can be considered by examining the formation of the emulsion. When the bulk oil phase is subdivided into a large number of oil droplets, a large increase in the interfacial energy is produced. Assuming that the interfacial tension γ_{12} of the bulk oil and the droplets to be the same (this is usually true for droplets that are not too small, i.e. greater than say 0.1 μm), then the free energy of formation of the emulsion from the bulk phase is given by the simple expression

$$\Delta G^{\text{form}} = \Delta A_{12} - T\Delta S^{\text{config}} \tag{14.10}$$

The first term $\Delta A\gamma_{12}$ is the energy required to expand the interface ($\Delta A = A_2 - A_1$ is the increase in interfacial area). This term is positive since γ_{12} is positive. In the absence of an emulsifier, γ_{12} is of the order of 30–50 mN m^{-1} and hence $\Delta A\gamma_{12}$ is large and positive. Therefore, to reduce the energy for emulsification, one has to reduce γ_{12} by at least one order of magnitude. This is achieved by adsorption of the emulsifier at the O/W interface. As discussed in Chapter 6, the emulsifier will also play other roles to prevent breakdown of the emulsion by flocculation and coalescence. This is achieved by creating an energy barrier (due to electrostatic or steric repulsion) that prevents flocculation and an interfacial tension gradient (Gibbs elasticity) that prevents coalescence.

The second term in Eq. (14.10), $-T\Delta S^{\text{config}}$, is the configurational entropy resulting from the increase in the number of possible configurations resulting from the production of numerous droplets. This term, being negative, actually helps in the formation of emulsions. However, with macroemulsions, $\Delta A|\gamma_{12}|$ is much larger than $|-T\Delta S^{\text{config}}|$ and hence ΔG^{form} is positive. In other words, emulsion formation is a non-spontaneous process and an energy barrier must be created to prevent reversal to state I (by flocculation and coalescence). This means that emulsions are only stable in the kinetic sense, and to give them a practical shelf-life one has to maximise the energy barrier against flocculation and coalescence.

Several types of emulsifiers may be used to prepare an emulsion (listed in Chapter 1). The emulsifier plays several roles in formation of the emulsion and its subsequent stabilisation. This is discussed in detail in Chapter 6 and only a summary is given here. Emulsification may be envisaged to start by formation of a film of the future (continuous) phase around the droplets. If no surfactant is present, this film is very unstable, draining rapidly under gravity, until complete drainage occurs. However, in the presence of a surfactant, the film can exist for sometime due to the creation of an interfacial tension gradient $\mathrm{d}\gamma/\mathrm{d}z$. Such a gradient creates a tangential stress on the liquid or, alternatively, if the liquid streams along the interface with the surfactant, an interfacial tension gradient develops. This interfacial tension gradient supports the film, preventing its rupture by drainage (due to the gravitational force) providing $2\,\mathrm{d}\gamma/\mathrm{d}z > \rho_c hg$, where h is the film thickness, ρ_c its density and g the acceleration due to the gravity.

Notably, however, the energy required for emulsification exceeds the thermodynamic energy $\Delta A\gamma_{12}$ by several orders of magnitude [39]. This is because a significant amount of energy is needed to overcome the Laplace pressure, Δp, which results from the production of a highly curved interface (small droplets), i.e.

$$\Delta p = \left(\frac{1}{R_1} + \frac{1}{R_2}\right) = \frac{2\gamma}{R} \tag{14.11}$$

where R_1 and R_2 are the principal radii of curvature. For a spherical droplet with radius r, $\Delta p = 2\gamma/r$ and hence deformation leads to a large Δp and energy is needed to overcome this. This explains why emulsification is an inefficient process and, to produce very small droplets, one needs to apply special methods, e.g. valve homogenisers, ultrasonics, static mixers, etc.

Five general main roles may be identified for the emulsifier. The first and most obvious is to lower γ, as mentioned above. This has a direct effect on droplet size; generally, the lower the interfacial tension, the smaller the droplet size. This is the case when viscous forces are predominant, whereby the droplet diameter is proportional to γ. When turbulence prevails, $d \propto \gamma^{3/5}$. When emulsification continues, and an equilibrium is set up between the amount adsorbed and the concentration in the continuous phase, C, the effective γ depends on the surface dilational modulus, ε, which is given by

$$\varepsilon = \frac{\mathrm{d}\gamma}{\mathrm{d}\ln A} = A\left(\frac{\mathrm{d}\gamma}{\mathrm{d}A}\right) \tag{14.12}$$

where A is the area of the interface (number of moles of surfactant adsorbed per unit area). Clearly, ε depends on the nature of the surfactant. During emulsification, ε decreases as a result of depletion of surfactants and an increase of $\mathrm{d}\ln A/\mathrm{d}t$. Hence the effective γ during breakup will be between the equilibrium value γ and γ_0 (interfacial tension of the bare liquid/liquid interface).

The second role of the surfactant is through its effect on the surface free energy for enlarging the drop surface. Both dilational elasticity and viscosity have an effect on this surface free energy. This surface free energy is now $\gamma\,\mathrm{d}A + A\,\mathrm{d}\gamma$, which implies that more energy is needed, although this energy is lower than $\gamma_0\,\mathrm{d}A$. Moreover, if the surface dilational viscosity $(\mathrm{d}\gamma/\mathrm{d}\ln A)/\mathrm{d}t$ is large, viscous resistance to surface enlargement may cost entropy.

The third role of the surfactant is to create interfacial tension gradients. This has been discussed before. As a result of the tangential stress $\mathrm{d}\gamma/\mathrm{d}z$, which can build up on a pressure of the order of 10^4 Pa (for $\gamma \sim 10$ mN m^{-1} and droplet diameter of 1 μm), the internal circulation in the droplet is impeded or even prevented, thus facilitating droplet formation and breakup.

The fourth role of the surfactant is to reduce coalescence during emulsification. The stabilising mechanism of a surfactant during emulsification is usually ascribed to the Gibbs–Marangoni effect [40]. During emulsification, adsorption of surfactant is usually incomplete, so that the interfacial tension decreases with time and the film becomes rapidly depleted with surfactant as a result of its adsorption. The Gibbs elasticity, E_f, is given by Eq. (14.13) [40–42],

$$E_f = \frac{2\gamma(\mathrm{d}\ln\Gamma)}{1 + \frac{1}{2}h(\mathrm{d}C/\mathrm{d}\Gamma)} \tag{14.13}$$

where Γ is the surface excess (number of moles of surfactant adsorbed per unit area of the interface). As shown in Eq. (14.13), the Gibbs elasticity E_f will be highest in the thinnest part of the film. As a result the surfactant will move in the direction of the highest γ and this motion will drag liquid along with it. The latter effect is the Marangoni effect. The final result is to reduce further thinning and hence coalescence is reduced. The Marangoni effect can be explained as liquid motion caused by the tangential stress $\mathrm{d}\gamma/\mathrm{d}z$. This gradient causes considerable streaming of liquid, which forces its way into the gap between the approaching droplets, thus preventing their approach.

The fifth role of the surfactant is to initiate interfacial instability. Disruption of a plane interface may take place by turbulence, Rayleigh instabilities and Kelvin–Helmholtz instability. Turbulence eddies tend to disrupt the interface [43] since they create local pressures of the order of $(\rho_1 - \rho_2)u_e^2$ (where u_e is the shear stress velocity of the eddy, which may exceed the Laplace pressure $2\gamma/R$. The interface may be disrupted if the eddy size l_e is about twice R. However, disruption turbulent eddies do not take place unless γ is very low. The Kelvin–Helmholtz instability arises when the two phases move with different velocities u_1 and u_2 parallel to the interface [44].

Interfacial instabilities may also occur for cylindrical threads of disperse phase that form during emulsification or when a liquid is injected into another from small orifices. Such cylinders undergo deformation [45–48] and become unstable under certain conditions. With a sinusoidal disruption of the radius of the cylinder, the latter becomes unstable when the wavelength λ of the perturbation exceeds the circumference of the undisturbed cylinder. Under these conditions, the waves are amplified until the thread breaks up into droplets [47]. The presence of surfactants will accelerate this breakup process due to interfacial tension gradients – the curved part will have a higher γ since it receives the smallest amount of surfactant per unit area. Hence, surfactant is transported towards the point of strongest curvature, carrying liquid streaming to that part. This may cause droplet shredding.

14.3.1 Selection of Emulsifiers

The selection of different surfactants in the preparation of EWs emulsion is still made on an empirical basis. This is discussed in detail in Chapter 6, and only a summary is given here. One of the earliest semi-empirical scales for selecting an appropriate surfactant or blend of surfactants was proposed by Griffin [49, 50] and is usually referred to as the hydrophilic–lipophilic balance or HLB number. Another closely related concept, introduced by Shinoda and co-workers [51–53, 58], is the phase inversion temperature (PIT) volume. Both the HLB and PIT concepts are fairly empirical and one should be careful in applying them in emulsifier selection. A more quantitative index that has received little attention is that of the cohesive energy ratio (CER) concept introduced by Beerbower and Hill [54] (see Chapter 6). The HLB system that is commonly used in selecting surfactants in agrochemical emulsions is described briefly below.

The HLB is based on the relative percentage of hydrophilic to lipophilic groups in the surfactant molecule(s). Surfactants with a low HLB number normally form W/O emulsions, whereas those with a high HLB number form O/W emulsions. A summary of the HLB range required for various purposes is given in Chapter 6. For O/W emulsions HLB is in the range 8–18.

The relative importance of hydrophilic and lipophilic groups was first recognised when mixtures of surfactants were used with varying properties of surfactants having low and high HLB numbers. The efficiency of any combination as judged by phase separation passes through a maximum [50] when the blend contains a particular concentration of the surfactant with the high HLB number. The original method for determining HLB numbers, developed by Griffin [50], is quite laborious and requires several trial and error procedures. Later, Griffin [51] developed a simple equation that permits calculation of the HLB number of certain numbers of nonionic surfactants such as fatty acid esters and alcohol ethoxylates of the type $R(CH_2\text{-}CH_2\text{-}O)_n\text{–}OH$. For polyhydroxy fatty acid esters, the HLB number is given by

$$\mathrm{HLB} = 20\left(1 - \frac{S}{A}\right) \qquad (14.14)$$

where S is the saponification number of the ester and A the acid number of the acid. Thus, a glyceryl monostearate, with $S = 161$ and $A = 198$, will have an

HLB number of 3.8, i.e. it is suitable for a W/O emulsifier. However, in many cases, accurate estimation of the saponification number is difficult, e.g. ester of tall oil, resin, beeswax and linolin. For the simpler ethoxylate alcohol surfactants, HLB can be calculated simply from the weight per cent of oxyethylene E and polyhydric alcohol P, i.e.

$$\mathrm{HLB} = (E + P)/5 \tag{14.15}$$

If the surfactant contains poly(ethylene oxide) as the only hydrophilic group, e.g. in the primary alcohol ethoxylates $R(CH_2\text{-}CH_2\text{-}O)_n\text{–}OH$, the HLB number is simply (E/5) (the content from one OH group is simply neglected).

The above equation cannot be used for nonionic surfactants containing propylene oxide or butylene oxide, nor can it be used for ionic surfactants. In the latter case, ionisation of the head groups tends to make them even more hydrophilic in character, so that the HLB number cannot be calculated from the weight per cent of the ionic groups. In that case, the laborious procedure suggested by Griffin [51] must be used.

Davies [55] derived a method for calculating the HLB number of surfactants directly from their chemical formulae, using empirically determined group numbers. Thus, a group number is assigned to various emulsifier component groups. These numbers are tabulated in Chapter 6. The HLB number is then calculated from these numbers using the following empirical relationship [55, 56],

$$\mathrm{HLB} = 7 + \sum(\text{Hydrophilic group nos}) - \sum(\text{Lipophilic group nos}) \tag{14.16}$$

HLB numbers calculated using the empirical Eq. (14.16) show quite satisfactory agreement with those determined experimentally.

Various procedures were later devised to determine the HLB number of different surfactants. For example, Griffin [51] found a good correlation between the cloud point of a 5% solution of various nonionic surfactants and their HLB number. This enables one to obtain the HLB number from a simple measurement of the cloud point. A more accurate method of determining HLB number is based on gas-liquid chromatography [57].

14.3.2
Emulsion Stability

Various emulsion breakdown process may be identified (schematically represented in Chapter 6). These breakdown processes will be briefly summarised below, with particular attention as to how one can stabilise the emulsion against the instability described.

Creaming and sedimentation result from external forces, usually gravitational or centrifugal. When such forces exceed the thermal motion of the droplets (Brownian motion) a concentration gradient builds up in the system, with the larger droplets moving faster to the top (if their density is lower than that of the medium) or to the bottom (if their density is larger than that of the medium) of the container. In limiting cases, the droplets may form a close-packed (random or ordered) array at the top or the bottom of the system, with the remainder of the volume being occupied by the continuous phase liquid. The case where the droplets move to the top is referred to as creaming, whereas that whereby the droplets move to the bottom is referred to as sedimentation. Strictly speaking, when one refers to creaming or sedimentation it is understood that no change in droplet size or its distribution takes place. Creaming or sedimentation is opposed by the thermal (Brownian) motion of the droplets; since this force increases with decreasing droplet size, both processes are significantly reduced with decreasing droplet size. Indeed for no separation to occur, the Brownian diffusion kT (where k is the Boltzmann constant and T the absolute temperature) must exceed the gravitational force, i.e.

$$kT \gg \tfrac{4}{3}\pi R^3 \Delta\rho g L \tag{14.17}$$

where $\Delta\rho$ is the density difference between dispersed phase and medium, and L is the height of the container. Clearly, to reduce creaming or sedimentation, $\Delta\rho$ has to be made close to zero (i.e. the density of the oil should be as close as possible to that of the medium) and R as small as possible. This simply follows from Stokes' law, which gives the sedimentation velocity for a very dilute emulsion consisting of non-interacting droplets, i.e.

$$v_0 = \frac{2R^2 \Delta\rho g}{9\eta_0} \tag{14.18}$$

Equation (14.18) only applies for an infinitely dilute emulsion. For a concentrated emulsion, the creaming or sedimentation rate is v reduced with increasing volume fraction of the emulsion. This can be empirically expressed as

$$v = v_0(1 - k\phi) \tag{14.19}$$

where k is an empirical constant that accounts for droplet–droplet interaction. For the simplest case, where only hydrodynamic interaction is considered, k is in the region of 5–6. This is usually the case at $\phi < 0.1$. However, for more concentrated emulsions k becomes a complex function of ϕ. Usually v decreases with increasing ϕ, approaching zero as ϕ approaches ϕ_p, the so-called maximum packing fraction; ϕ_p is in the region of 0.7 for fairly monodisperse emulsion, but it can reach higher values (>0.8) for polydisperse systems. However, when one approaches the maximum packing fraction, the viscosity of the emulsion becomes very high (at $\phi = \phi_p$, $\eta \sim \infty$). Therefore, most practical emulsions have volume fractions well below ϕ_p and creaming or sedimentation is the rule rather than the exception. In this case, various procedures must be applied to avoid emulsion separation. As mentioned above, one of the simplest methods is to reduce $\Delta\rho$ or R. This may be achieved by matching the density of the oil to that of the medium (by using oil mixture) and/or reducing R by the use of homogenisers. In cases where this is not possible in practice one may use thickeners, or apply concepts of controlled flocculation. Thickeners are perhaps the most widely used materials for reducing creaming or sedimentation. These are usually high molecular weight polymers of the synthetic or natural type, e.g. hydroxy ethyl cellulose, poly(ethylene oxide), xanthan gum, guar gum, alginates, carrageenans, etc. All these materials, when dissolved in the continuous phase, increase the viscosity of the medium and hence reduce creaming or sedimentation. However, their action is not simple, since these materials give non-Newtonian solutions that are viscoelastic. This means that the viscosity of these polymer solutions depend on the applied shear rate $\dot{\gamma}$. Generally, such systems produce pseudo-plastic flow with an apparent yield value τ_β (the stress extrapolated to $\dot{\gamma} = 0$) and an apparent viscosity η_{app} that decreases with increasing γ. Moreover, the viscosity of these polymer solutions increases with increase in their concentration in a peculiar way. Initially, η increases with C, but above a certain concentration, denoted C^*, there is a much more rapid increase in η with further increase in C. The concentration C^* is the point at which polymer coil overlap begins, and, above C^*, the solution shows elastic behaviour that increases with rising C. Usually, thickeners are added at concentrations above C^*, in which case a viscoelastic system is produced. Moreover, C^* decreases with increasing molecular weight of the added polymer and, hence, to reduce the polymer concentration above which a viscoelastic system is produced one uses higher molecular weights.

The above-mentioned viscoelastic polymer solutions reduce (or eliminate) creaming or sedimentation of the emulsion, providing they produce an "elastic" network in the continuous phase that is sufficient to overcome the stresses exerted by the creaming or sedimenting droplets. Such viscoelastic solutions produce a very high zero shear viscosity that is sufficient to eliminate creaming or sedimentation.

Another method of reducing creaming or sedimentation is to induce weak flocculation in the emulsion system. This may be achieved by controlling some parameters of the system, such as electrolyte concentration, adsorbed layer thickness and droplet size. These weakly flocculated emulsions are discussed in the next section. Alternatively, weak flocculation may be produced by addition of a "free" (non-adsorbing) polymer. Above a critical concentration of the added polymer, polymer–polymer interaction becomes favourable as a result of polymer coil overlap and the polymer chains are "squeezed out" from between the droplets. This results in a polymer-free zone between the droplets, and weak attraction occurs as a result of the higher osmotic pressure of the polymer solution outside the droplets. This phenomenon is usually referred to as depletion flocculation [59] and can be applied for "structuring" emulsions and hence reduction of creaming or sedimentation.

Flocculation refers to aggregation of the droplets, without any change in the primary droplet size, into larger units. Flocculation is the result of the van der Waals attraction that is universal with all dispelsed system. For two droplets of equal radii R, the van der Waals attractive energy G_{A} is given by Eq. (14.20) [60] (when the separation between the droplets h is much smaller than the droplet radius),

$$G_A = -\frac{AR}{12h} \tag{14.20}$$

where A is the net Hamaker constant between droplets A_{11} and medium A_{22}, i.e.

$$A = (A_{11}^{1/2} - A_{22}^{1/2})^2 \tag{14.21}$$

The Hamaker constant of any material is given by

$$A_{ii} = \pi q^2 \beta_{ii} \tag{14.22}$$

where q is the number of atoms or molecules per unit volume and β_{ii} is the London dispersion constant (which is related to the polarizability).

From Eq. (14.20), G_A clearly increases rapidly with decreasing separation between the droplets and, in the absence of repulsion between the droplets, flocculation is very fast, producing clusters of droplets. Thus, to stabilise droplets against aggregation, a repulsive force must be created to prevent close-approach of the droplets. Two general stabilising mechanisms may be envisaged. The first is based on the creation of an electrical double layer around the droplets. This may be produced, for example, by adsorption of an ionic surfactant. Here, the surface of the droplets becomes covered with a layer of charged head-groups (negative with anionic and positive with cationic surfactants). This charge becomes compensated by counter ions, some of which approach the surface closely (in the so-called Stern layer) while the rest extend into bulk solution to a distance that is determined by the double layer extension. The extension of the double layer depends on electrolyte concentration. When two droplets with their extended double layers (as is the case at low electrolyte concentration) approach to a separation h such that the double layers begin to overlap, repulsion occurs due to the increase in free energy of the whole system. In the simple case of two large droplets and low potential, Ψ_0, the repulsion interaction free energy G_E is given by the expression [61, 62],

$$G_E = 2\pi R \varepsilon_r \varepsilon_0 \Psi_0^2 \ln[1 + \exp(-\kappa h_0)] \tag{14.23}$$

where ε_r is the permittivity of the medium, ε_0 that of free space, and κ is the Debye–Hückel parameter that is related to electrolyte concentration, C,

$$\kappa = \left(\frac{2Z^2 e^2 C}{\varepsilon_r \varepsilon_0 kT}\right)^{1/2} \tag{14.24}$$

where Z is the valency of the electrolyte and e is the electronic charge.

The combination of van der Waals attraction and double layer repulsion results in the well-known theory of colloid stability due to Deryaguin, Landau, Verwey and Overbeek (DLVO theory) [61, 62]. The energy–distance curve is schematically represented in Figure 14.11.

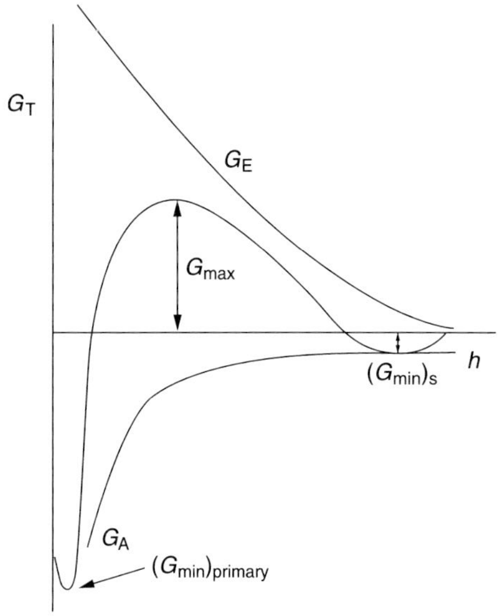

Fig. 14.11. Scheme of the energy–distance curve according to the DLVO theory [61, 62].

It is characterised by two minima and one maximum. At long separations, attraction prevails, resulting in a shallow minimum (secondary minimum) whose depth depends on particle size, Hamaker constant and electrolyte concentration. The attraction energy in this minimum is usually small, of the order of a few kT units. In contrast, at very short separations, the attractive force becomes much larger than the repulsion force, resulting in a deep primary minimum. If the droplets can reach such separation distances (i.e. in the absence of a sufficiently energy barrier) very strong attraction occurs and the droplets form large aggregate units with a small separation between the surfaces. This strong attraction, sometimes referred to as coagulation, is prevented by the presence of an energy maximum at intermediate distances of separation. The height of this maximum is directly proportional to the surface potential Ψ_0 and inversely proportional to the electrolyte concentration. Thus, by controlling Ψ_0 and C, one can make this height sufficient ($>25kT$) to prevent coagulation in the primary minimum. However, in some situations, one may need to create weak attraction in the secondary minimum to reduce creaming or sedimentation. This is achieved by using intermediate electrolyte concentrations and larger emulsion droplets.

The second mechanism by which flocculation may be prevented is that of steric stabilisation. This is produced using nonionic surfactants or polymers that adsorb at the liquid/liquid interface with their hydrophobic portion, leaving a thick layer of hydrophilic chains in bulk solution, e.g. poly(ethylene oxide) (PEO) or poly(vinyl alcohol). These thick hydrophilic chains produce repulsion as a result of two main effects. The first, usually referred to as mixing interaction (osmotic repulsion), results from the unfavourable mixing of the hydrophilic layers on close approach of the droplets. These chains may overlap when the latter approach to a separation h that is smaller than twice the adsorbed layer thickness (2δ) [63]. However, when these chains are in good solvent condition (such as PEO or PVA in water) such overlap becomes unfavourable because of the increase of the osmotic pressure in the overlap region. This results in diffusion of solvent molecules into this overlap region, thus separating the droplets, i.e. resulting in repulsion. The free energy of repulsion due to this overlap effect can be calculated from the free energy of mixing of the two polymer layers. This results in the following expression for G_{mix} [68],

$$\frac{G_{mix}}{kT} = \frac{4\pi}{3V_1}\phi_2^2 N_{av}\left(\frac{1}{2} - \chi\right)\left(3R + 2\delta + \frac{h}{2}\right)\left(\delta - \frac{h}{2}\right)^2 \quad (14.25)$$

where V_1 is the molar volume of the solvent, ϕ_2 is the volume fraction of the polymer or surfactant in the adsorbed layer, N_{av} is Avogadro's constant and χ is the Flory–Huggins chain–solvent interaction parameter.

Equation (14.25) clearly shows that G_{mix} is positive (i.e. repulsive) when $\chi < 0.5$, i.e. when the chains are in good solvent conditions; when $\chi > 0.5$, G_{mix} becomes negative, i.e. attractive. There is one point at which $\chi = 0.5$, referred to as the θ-point for the chain, which determines the onset of attraction.

The second effect that results from the presence of adsorbed layer is the loss in configurational entropy of the chains when significant overlap occurs. This effect, which is always repulsive, is usually referred to as the entropic, elastic or volume restriction effect, G_{el}.

Combination of steric interaction with the van der Waals attraction results in an energy–distance curve as schematically represented in Figure 14.12.

G_{mix} starts to increase rapidly as soon as h becomes smaller than 2δ. Conversely, G_{el} begins to increase with decreasing h when the latter becomes significantly smaller than 2δ. When G_{mix} and G_{el} are combined with G_A, the total energy G_T–distance curve only shows one minimum, whose location depends on 2δ and whose magnitude depends on the Hamaker constant and droplet radius R. Clearly, if 2δ is made sufficiently large and R sufficiently small, the depth of the minimum can become very small and one may approach thermodynamic stability. This explains why nonionic surfactants and polymers are relatively more effective in stabilising emulsions against flocculation. However, one should ensure that the medium for the chains remains a good solvent, otherwise incipient flocculation occurs.

Ostwald ripening results from the finite solubility of the liquid phases. With emulsions that are polydisperse (this is usually the case) the smaller droplets will have a larger chemical potential (larger solubility) than the larger droplets. The higher solubility of the smaller droplets is the result of their higher radii of curvature (note that $S = 2\gamma/r$). With time, the smaller droplets disappear by dissolution and diffusion and become deposited on the larger droplets. This process, usually referred to as Ostwald ripening, is determined by the difference in solubility between small and large droplets, as given by the Ostwald equation,

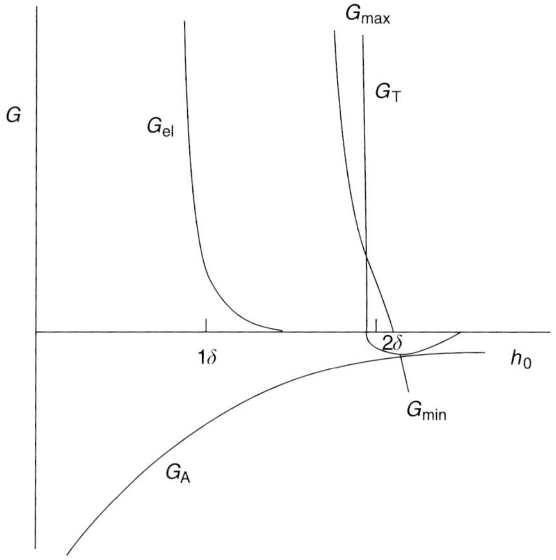

Fig. 14.12. Variation of G_{mix}, G_{el}, G_A and G_T with h (schematic).

$$\frac{RT}{M} \ln \frac{S_1}{S_2} = \frac{2\gamma}{\rho} \left(\frac{1}{R_1} - \frac{1}{R_2} \right) \tag{14.26}$$

where S_1 is the solubility of a droplet of radius R_1 and S_2 that of a droplet with radius R_2 (note that $S_1 > S_2$ when $R_1 < R_2$), M is the molecular weight and ρ is the density of the droplets.

The above process of Ostwald ripening is reduced by the presence of surfactants, which play two main roles. Firstly, by adsorption of surfactants, γ is reduced, thus reducing the driving force for Ostwald ripening. Secondly, surfactants produce a surface tension gradient (Gibbs elasticity) that will also reduce Ostwald ripening. This can be understood from the following argument [64]. A droplet is in mechanical equilibrium if $\mathrm{d}p/\mathrm{d}R > R$, i.e. when $\mathrm{d}\gamma/\mathrm{d}\ln R > \gamma$. Since $A = 4\pi R^2$, then $2\,\mathrm{d}\gamma/\mathrm{d}\ln A > \gamma$ or $2\varepsilon > \gamma$, where ε is the interfacial dilational modulus. Thus, when twice the interfacial elasticity exceeds the interfaced tension, Ostwald ripening is significantly reduced.

Another method of reducing Ostwald ripening, introduced by Davies and Smith [65], is to incorporate a small proportion of a highly insoluble oil within the emulsion droplets. This reduces the molecular diffusion of the oil molecules, which are assumed to be the driving force for Ostwald ripening.

Coalescence is the process of thinning and disruption of the liquid film between the droplets, resulting in their fusion. When two emulsion droplets come into close contact in a floc or during Brownian collision, e.g. in a creamed or sedimented layer, thinning and disruption of the liquid film may occur, resulting in its eventual rupture and, hence, the droplets join each other, i.e. they coalesce. The process of thinning and disruption of liquid lamellae between emulsion droplets is complex. For example, during a Brownian encounter, or in a cream or sediment, emulsion droplets may produce surface or film thickness fluctuations in the region of closest approach. The surface fluctuations produce waves that may grow in amplitude and, during close approach, the apexes of these fluctuations may join, causing coalescence (region of high van der Waals attraction). Alternatively, any film thickness fluctuations may result in regions of small thicknesses for van der Waals attraction to cause even more thinning, with the ultimate disruption of the whole film. Unfortunately, the process of coalescence is far from well understood, although some guidelines may be obtained by considering the balance of surface forces in the liquid lamellae between the droplets. Deryaguin and co-workers [66] introduced the useful concept of the disjoining pressure $\pi(h)$ for thin films adhering to substrates; $\pi(h)$ balances the excess normal pressure $P(h) - P_0$ in the film. $P(h)$ is the normal pressure of a film of thickness h, whereas P_0 is the normal pressure of a sufficiently thick film such that the interaction free energy is zero. Notably, $\pi(h)$ is the net force per unit area acting across the film, i.e. normal to the interfaces. Thus $\pi(h)$ is simply equal to $-\mathrm{d}V_\mathrm{T}/\mathrm{d}h$, where V_T is the net force that results from three main contributions, van der Waals, electrostatic and steric forces, i.e.

$$\pi(h) = \pi_\mathrm{A} + \pi_\mathrm{E} + \pi_\mathrm{S} \tag{14.27}$$

To produce a stable film $\pi(h)$ needs to be positive, i.e. $\pi_\mathrm{E} + \pi_\mathrm{S} > \pi_\mathrm{A}$. Thus, to reduce coalescence one needs to enhance the repulsion between the surfactant layers, e.g. by using either a charged film or surfactants with long hydrophilic chains that produce a strong steric repulsion.

To reduce coalescence, one needs to dampen the fluctuation in surface waves or film thickness. This is produced by enhancement of the Gibbs–Marangoni effect. Several methods can be applied to reduce or eliminate coalescence, and these are summarised below.

One of the earliest methods for reducing coalescence is to use mixed surfactant films. These will increase the Gibbs elasticity and/or interfacial viscosity. Both effects reduce film fluctuations and, hence, reduce coalescence. In addition, mixed surfactant films are usually more condensed and hence diffusion of the surfactant molecules from the interface is greatly hindered. An alternative explanation for enhanced stability using surfactant mixture was introduced by Friberg and co-workers [67] who considered the formation of a three-dimensional association structure (liquid crystals) at the oil/water interface. These liquid crystalline structures prevent coalescence since one has to remove several surfactant layers before droplet–droplet contact may occur.

Another method of reducing coalescence is to use macromolecular surfactants such as gums, proteins and synthetic polymers, e.g. A-B, A-B-A block copolymers. The latter in particular could produce very stable films by strong adsorption of the B groups of the molecules, leaving the A chains dangling in solution and providing a strong steric barrier that prevents coalescence. Examples of such molecules are poly(vinyl alcohol) and poly(ethylene oxide)–poly(propylene oxide) block copolymers.

Phase inversion is the process whereby the internal and external phases of an emulsion suddenly invert, i.e. O/W to W/O and vice versa. Phase inversion can be easily observed if the oil volume fraction of, say, an O/W emulsion is gradually increased. For example, at a given emulsifier concentration, the viscosity of an emulsion often gradually increases with increase in ϕ but, at a certain critical volume fraction ϕ_{cr}, there is a sudden decrease. The same sudden change is observed in the specific conductivity (κ), which initially decreases slowly with increasing ϕ but, above ϕ_{cr}, it decreases much more rapidly. The critical volume fraction corresponds to the point at which the O/W emulsion inverts to a W/O emulsion. The sharp decrease in η observed at the inversion point is due to the sudden reduction in disperse phase volume fraction. The sudden rapid decrease in κ is due to the fact that the emulsion now becomes oil continuous, with a much lower conductivity than the aqueous continuous phase emulsion.

Early theories of phase inversion postulated that the inversion takes place as a result of difficulty in packing emulsion droplets above a certain volume fraction (the maximum packing fraction). For example, if the emulsion is monodisperse $\phi_p = 0.74$ and any attempt to increase ϕ above this value leads to inversion. However, several investigations have clearly indicated the invalidity of this argument, inversion being found to occur at values much greater or smaller than 0.74. At present, there seems to be no quantitative theory that explains phase inversion. However, location of the inversion point is of practical importance, particularly on storage of the emulsion. As mentioned above, the phase inversion temperature (PIT) can be an important criterion in assessing the long-term physical stability of emulsions.

14.3.3
Characterisation of Emulsions and Assessment of their Long-term Stability

To characterize emulsion systems, it is necessary to obtain fundamental information on the liquid/liquid interface (e.g. interfacial tension and interfacial rheology) and properties of the bulk emulsion system, such as droplet size distribution, flocculation, coalescence, phase inversion and rheology. The information obtained, if analyzed carefully, can be used for the assessment and (in some cases) prediction of the long-term physical stability of the emulsion.

Chapter 6 gives details of the methods used to investigate the interfacial properties. These include measurement of the interfacial tension, interfacial viscosity and interfacial elasticity. Correlation between the parameters obtained and the long-term physical stability of emulsions is also described in detail.

Measurement of the droplet size distribution, rates of flocculation, Ostwald ripening and coalescence are also described in Chapter 6. These methods should be applied for agrochemical emulsions to ensure their long-term physical stability. In addition, the bulk rheology of the system should be investigated after storage at various temperatures.

14.4
Suspension Concentrates (SCs)

The formulation of agrochemicals as dispersions of solids in aqueous solution (referred to as suspension concentrates or SCs) has attracted considerable attention in recent years. Such formulations are a natural replacement for wettable powders (WPs). The latter are produced by mixing the active ingredient with a filler (usually a clay material) and a surfactant (dispersing and wetting agent). These powders are dispersed into the spray tank to produce a coarse suspension that is applied to the crop. Although wettable powders are simple to formulate they are not the most convenient for the farmer. Apart from being dusty (and occupying a large volume due to their low bulk density), they tend to settle rapidly in the spray tank and they do not provide optimum biological efficiency due to the large particle size of the system. In addition, one cannot incorporate the necessary adjuvants (mostly surfactants) in the formulation. These problems are overcome by formulating the agrochemical as an aqueous SC.

Several advantages may be quoted for SCs: Firstly, one may control the particle size by controlling the milling conditions and proper choice of the dispersing agent. Secondly, it is possible to incorporate high concentrations of surfactants in the formulation, which is sometimes essential for enhancing wetting, spreading and penetration (see Chapter 11). Stickers may also be added to enhance adhesion and, in some cases, to provide slow release.

Factors that govern the stability of suspension concentrates have been the subject of considerable research [68–70]. Theories of colloid stability could be applied to predict the physical states of these systems on storage. In addition, the problem of sedimentation of SCs has been analyzed at a fundamental level [71]. Since the density of the particles is usually larger than that of the medium (water) SCs tend to separate as a result of sedimentation. The sedimented particles tend to form a compact layer at the bottom of the container (sometimes referred to as clay or cake), which is very difficult to redisperse. It is, therefore, essential to reduce sedimentation and formation of clays by incorporation of an antisettling agent.

In this section, I will attempt to address the above-mentioned phenomena at a fundamental level. The section will start with a sub-section on the preparation of suspension concentrates and the role of surfactants (dispersing agents). This is followed by a sub-section on the control of the physical stability of suspensions. The problem of Ostwald ripening (crystal growth) will also be briefly described and particular attention will be paid to the role of surfactants. The problem of sedimentation and prevention of claying will then be covered. The various methods that may be applied to reduce sedimentation and prevention of the formation of hard clays will be summarised. The last sub-section will deal with methods that may be applied to assess the physical stability of SCs. To assess flocculation and crystal growth, particle size analysis techniques are commonly applied. The bulk properties of the suspension, such as sedimentation and separation, redispersion on dilution, may be assessed using rheological techniques. The latter will be summarised, with particular emphasis on their application in predicting the long-term physical stability of suspension concentrates.

14.4.1
Preparation of Suspension Concentrates and the Role of Surfactants

This is dealt with in detail in Chapter 7 and a summary is given here. Suspension concentrates are usually formulated using a wet milling process, which requires the addition of a surfactant (dispersing agent). The latter should satisfy the following criteria: be a good wetting agent for the agrochemical powder (both external and internal surfaces of the powder aggregates or agglomerates must be spontaneously wetted), be a good dispersing agent to break such aggregates or agglomerates into smaller units and subsequently help in the milling process (one usually aims at a dispersion with a volume mean diameter of 1–2 μm); provide good stability in the colloid sense (essential for maintaining the particles as individual units once formed). Powerful dispersing agents are particularly important for the preparation of highly concentrated suspensions sometimes required for seed dressing). Any flocculation will cause a rapid increase in the viscosity of the suspension and this makes the wet milling of the agrochemical difficult. The next sub-section will briefly discuss the wetting of agrochemical powders (for detailed analysis see Chapter 7), their subsequent dispersions and milling. Colloid stability will be dealt with subsequently.

14.4.2
Wetting of Agrochemical Powders, their Dispersion and Comminution

Dry powders of organic compounds usually consist of particles of various degrees of complexity, depending on the isolation stages and the drying process. Generally, particles in a dry powder form aggregates (in which the particles are joined together with their crystal faces) or agglomerates, in which the particles touch at edges or corners, forming a looser more open structure. It is essential in the dispersion process to wet the external as well as the internal surfaces and displace the air entrapped between the particles. This is usually achieved by the use of surface active agents of the ionic or nonionic type. In some cases, macromolecules or polyelectrolytes may be efficient in this wetting process. This may be the case since these polymers contain a very wide distribution of molecular weights and the low molecular weight fractions may act as efficient wetting agents. For efficient wetting the molecules should lower the surface tension of water and they should diffuse rapidly in solution and quickly become adsorbed at the solid/solution interface.

Wetting of a solid is usually described in terms of the equilibrium contact angle θ and the appropriate interfacial tensions, using the classical Young's equation (see Chapter 11),

$$\gamma_{SV} - \gamma_{SL} = \gamma_{LV} \cos \theta \tag{14.28}$$

or

$$\cos \theta = \frac{(\gamma_{SV} - \gamma_{SL})}{\gamma_{LV}} \tag{14.29}$$

where γ represents the interfacial tension and the symbols S, L and V refer to the solid, liquid and vapour, respectively. Equation (14.29) clearly shows that, if $\theta < 90°$, a reduction in γ_{LV} improves wetting. Hence the use of surfactants that reduce both γ_{LV} and γ_{SL} aid wetting. However, the process of wetting particulate solids is more complex and involves at least three distinct types of wetting [72, 73], namely adhesional wetting, spreading wetting and immersional wetting (see Chapters 7 and 11). For a cube with unit surface area the work of dispersion is given by

$$W_d = W_a + W_i + W_s = -6\gamma_{SL} - 6\gamma_{SV} = -6\gamma_{LV} \cos \theta \tag{14.30}$$

Thus, the wetting of a solid by a liquid depends on two measurable quantities, γ_{LV} and θ and hence Eq. (14.30) may be used to predict whether the process is spontaneous, i.e. W_d is negative. The adhesion process is invariably spontaneous, whereas the other two processes depend on θ. For example, spreading is only spontaneous when $\theta = 0$, whereas immersion and dispersion are spontaneous when $\theta < 90°$.

The next stage to consider is the wetting of the internal surface, which implies penetration of the liquid into channels between and inside the agglomerates. This is more difficult to define precisely. However, one may make use of the equation derived for capillary phenomena. To force a liquid into a capillary tube of radius r, a pressure P is required such that

$$P = \frac{-2\gamma_{LV} \cos \theta}{r} = -\frac{2(\gamma_{SV} - \gamma_{SL})}{r} \tag{14.31}$$

Equation (14.31) shows that to increase penetration θ and γ_{SL} have to be made as small as possible, e.g. by sufficient adsorption of surfactant. However, when $\theta = 0$, P is proportional to γ_{LV}, i.e. a large surface tension is required. These two opposing effects show that the proper choice of a surfactant is not simple. In most cases a compromise has to be made to minimise θ while not having a too small surface tension to aid penetration.

Another important factor in the wetting process is the role of penetration of the liquid into the channels between and inside the agglomerates. This has to be as fast as possible to aid the dispersion process. Penetration of liquids into powders can be qualitatively treated by the Rideal–Washburn equation [74], assuming the powder to be represented by a bundle of capillaries. For horizontal capillaries (where gravity may be neglected), the rate of penetration of a liquid into an air-filled tube is given by

$$\frac{dl}{dt} = \frac{r\gamma \cos \theta}{4\eta l} \tag{14.32}$$

where l is the distance the liquid has travelled along the pore in time t, a is the radius of the capillary tube and γ and η are the surface tension and viscosity of the liquid respectively. Thus, to enhance penetration, γ_{LV} should be made as high as possible and θ as low as possible. Moreover, to increase penetration, the powder has to be made as loose as possible.

Integration of Eq. (14.32) leads to

$$l^2 = \frac{r\gamma_{LV} \cos \theta t}{2\eta} \tag{14.33}$$

which has to be modified for powders by replacing r with a factor K that contains an "effective radius" for the bed and a tortuosity factor to allow for the random shape and size of the capillaries, i.e.

$$l^2 = \frac{K\gamma_{LV} \cos \theta t}{2\eta} \tag{14.34}$$

Equation (14.34) forms the basis of the method commonly used for measuring θ for a powder/liquid system (see Chapter 11). A known weight of the dry powder is packed in a glass tube fitted at one end with a sintered glass disc and the rate of rise of the liquid into the powder bed is measured. A plot of l^2 versus t is usually linear with a slope equal to $K\gamma_{LV} \cos \theta/2\eta$. The value of K may be obtained by using a liquid of known surface tension that gives zero θ.

Thus, in summary, the dispersion of a powder in a liquid depends on three main factors, namely the energy of wetting of the external surface, the pressure involved in the liquid penetrating inside and between the agglomerates, and the rate of penetration of the liquid into the powder. All these factors are related to two main parameters, namely γ_{LV} and θ. In general, the process is likely to be more spontaneous the lower θ and the higher γ_{LV}. Since these two factors tend to operate in opposite senses, the choice of the proper surfactant (dispersing agent) can be difficult.

To disperse aggregates and agglomerates into smaller units one requires high-speed mixing, e.g. a Silverson mixer. In some cases the dispersion process is easy and the capillary pressure may be sufficient to break up the aggregates and agglomerates into primary units. The process is aided by the surfactant, which becomes adsorbed on the particle surface. However, one should be careful during the mixing process not to entrap air (foam), which causes an increase in the viscosity of the suspension and prevents easy dispersion and subsequent grinding. If foam formation becomes a problem, one should add antifoaming agents such as polysiloxane surfactants.

After completion of the dispersion process, the suspension is transferred to a ball or bead mill for size reduction. Milling or comminution (the generic term for size reduction) is a complex process and there is little fundamental information on its mechanism. For the breakdown of single crystals into smaller units, mechanical energy is required. This energy in a bead mill, for example, is supplied by impaction of the glass beads with the particles. As a result, permanent deformation of the crystals and crack initiation result. This will eventually lead to the fracture of the crystals into smaller units. However, since the milling conditions are random, some particles inevitably receive impacts that are far in excess of those required for fracture, whereas others receive impacts that are insufficient to fracture them. This makes the milling operation grossly inefficient and only a small fraction of the applied energy is actually used in comminution. The rest of the energy is dissipated as heat, vibration, sound, interparticulate friction, friction between the particles and beads, and elastic deformation of unfractured particles. For these reasons, milling conditions are usually established by a trial and error procedure. Of particular importance is the effect of various surface active agents and macromolecules on the grinding efficiency. The role played by these agents in the comminution process is far from understood, although Rehbinder and collaborators [75–80] have given this problem particular consideration. As a result of the adsorption of surfactants at the solid/liquid interface, the surface energy at the boundary is reduced and this facilitates the process of deformation or destruction. Adsorption of the surfactant at the solid/solution interface in cracks facilitates their propagation. This is usually referred to as the "Rehbinder effect" [77]. The surface energy manifests itself in destructive processes on solids, since the generation and growth of cracks and separation of one part of a body from another is directly connected with the development of new free surface. Thus, fine grinding is facilitated as a result of adsorption of surface active agents at structural defects in the surface of the crystals. In the extreme case where there is a very great reduction in surface energy at the sold/liquid boundary, spontaneous dispersion may take place, resulting in the formation of colloidal particles (< 1 μm). Rehbinder [77] has developed a theory for such spontaneous dispersion. Unfortunately, there are insufficient experimental data to prove or disprove the "Rehbinder effect".

14.4.3
Control of the Physical Stability of Suspension Concentrates

Powerful dispersing agents, e.g. surfactants of the ionic or nonionic type, nonionic polymers or polyelectrolytes, are used to control stability against irreversible flocculation (where the particles are held together in aggregates that cannot be redispersed by shaking or on dilution). These dispersing agents must be strongly adsorbed onto the particle surfaces and fully cover them. With ionic surfacetants, irreversible flocculation is prevented by the repulsive force generated from the presence of an electrical double layer at the particle solution interface (see Chapter 7). Depending on the conditions, this repulsive force can be made sufficiently large to overcome the ubiquitous van der Waals attraction between the particles, at intermediate distances of separation. With nonionic surfactants and macromolecules, repulsion between the particles is ensured by the steric interaction of the adsorbed layers on the particle surfaces (see Chapter 7). With polyelectrolytes, both electrostatic and steric repulsion exist. The role of surfactants in stabilization of particles against flocculation is summarised below.

Ionic surfactants, such as sodium dodecyl benzene sulphonate (NaDBS) or cetyltrimethylammonium chloride (CTACl), adsorb on hydrophobic particles of agrochemicals, as a result of the hydrophobic interaction between the alkyl group of the surfactant and the particle surface. As a result, the particle surface will acquire a charge that is compensated by counter ions (Na^+ in the case of NaDBS and Cl^- in the case of CTACl), forming an electrical double layer.

Adsorption of ionic surfactants at the solid/solution interface is of vital importance in determining the stability of suspension concentrates. As discussed in Chapter 5, the adsorption of ionic surfactants on solid surfaces can be measured directly by equilibrating a known amount of solid (with known surface area) with surfactant solutions of various concentrations. After reaching equilibrium, the solid particles are removed (for example by centrifugation) and the concentration of surfactant in the supernatant liquid is determined analytically. From the difference between the initial and final surfactant concentrations (C_1 and C_2 respectively) the number of moles of surfactant adsorbed, Γ, per unit area of solid is determined and the results may be fitted to a Langmuir isotherm,

$$\Gamma = \frac{\Delta C}{mA} = \frac{abC_2}{1 + bC_2} \tag{14.35}$$

where $\Delta C = C_1 - C_2$, m is the mass of the solid with surface area A, a is the saturation adsorption and b is a constant that is related to the free energy of adsorption, ΔG [$b \propto \exp(\Delta G/RT)$]. From a, the area per surfactant ion on the surface can be calculated (area per surfactant ion $= 1/aN_{av}$).

Results on the adsorption of ionic surfactants on pesticides are scarce. However, Tadros [81] obtained some results on the adsorption of NaDBS and CTABr on a fungicide, namely ethirimol. For NaDBS, the shape of the isotherm was of a Langmuir type, giving an area/DBS^- at saturation of $\sim$0.14 nm^2. The adsorption of CTA^+ showed a two-step isotherm with areas/CTA^+ of 0.27 and 0.14 nm^2. These results suggest full saturation of the surface with surfactant ions that are vertically oriented.

The above discussion shows that ionic surfactants can be used to stabilise agrochemical suspensions by producing sufficient electrostatic repulsion. When two particles with adsorbed surfactant layers approach each other to a distance where the electrical double layers begin to overlap, strong repulsion occurs, preventing any particle aggregation (see Chapter 7). The energy–distance curve for such electrostatically stabilised dispersions is schematically shown in Figure 7.21. This shows an energy maximum, which, if high enough ($> 25kT$), prevents particle aggregation into the primary minimum. However, ionic surfactants are the least attractive dispersing agents for the following reasons. Adsorption of ionic surfactants is seldom strong enough to prevent some desorption, resulting in production of "bare" patches that may induce particle aggregation. The system is also sensitive to ionic impurities present in the water used for suspension preparation. In particular, the system will be sensitive to bivalent ions (Ca^{2+} or Mg^{2+}), which produce flocculation at relatively low concentrations.

Nonionic surfactants of the ethoxylate type, e.g. $R(CH_2CH_2O)_nOH$ or $RC_6H_5(CH_2CH_2O)_2OH$, provide a better alternative provided the molecule contains sufficient hydrophobic groups to ensure their adsorption and enough ethylene oxide units to provide an adequate energy barrier. As discussed before, the origin of steric repulsion arises from two main effects. The first effect arises from the unfavourable mixing of the poly(ethylene oxide) chains, which are in good solvent conditions (water as the medium). This effect is referred to as the mixing or osmotic repulsion. The second effect arises from the loss in configurational entropy of the chains when these are forced to overlap on approach of the particles. This is referred to as the elastic or volume restriction effect. The energy–distance curve for such systems shown before clearly demonstrates the attraction of steric stabilization. Apart from a small attractive energy minimum (which can be reasonably shallow with sufficiently long poly(ethylene oxide) chains), strong repulsion occurs and there is no barrier to overcome. A better option is to use block and graft copolymers (polymeric surfactants) consisting of A and B units combined together in A-B, A-B-A or BA_n fashion. B represents units that have high affinity for the particle surface and are basically insoluble in the continuous medium, thus providing strong adsorption ("anchoring units"). Conversely, A represents units with high affinity to the medium (high chain–solvent interaction) and little or no affinity to the particle surface. An example of such a powerful dispersant is a graft copolymer of poly(methyl methacrylate–methacrylic acid) (the anchoring portion) and methoxy poly(ethylene oxide) (the stabilising chain) methacrylate [79]. Adsorption measurements of such a polymer on a pesticide, namely ethirimol (a fungicide) showed a high affinity isotherm with no desorption. Using such macromolecular surfactant, a suspension of high volume fractions could be prepared.

The third class of dispersing agents commonly used in SC formulations is that of polyelectrolytes. Of these, sulphonated naphthalene-formaldehyde condensates and lignosulphonates are the most commonly used is agrochemical formulations. These systems show a combined electrostatic and steric repulsion. The energy–distance curve shows a shallow minimum and maximum at intermediate distances (characteristic of electrostatic repulsion) as well as strong repulsion at relatively short distances (characteristic of steric repulsion). The stabilization mechanism of polyelectrolytes is sometimes referred to as electrosteric. These polyelectrolytes offer some versatility in SC formulations. Since the interaction is fairly long-range (due to the double layer effect), one does not obtain the "hard-sphere" type behaviour that may lead to the formation of hard sediments. Steric repulsion ensures the colloid stability and prevents aggregation on storage.

14.4.4
Ostwald Ripening (Crystal Growth)

There are several ways in which crystals can grow in an aqueous suspension. One of the most familiar is the phenomenon of "Ostwald ripening", which occurs as a result of the difference in solubility between the small and large crystals [82–86],

$$\frac{RT}{M}\ln\left(\frac{S_1}{S_2}\right)=\frac{2\sigma}{\rho}\left(\frac{1}{r_1}-\frac{1}{r_2}\right) \tag{14.36}$$

where S_1 and S_2 are the solubilities of crystals of radii r_1 and r_2 respectively, σ is the specific surface energy, ρ is the density and M is the molecular weight of the solute molecules, R is the gas constant and T the absolute temperature. Since r_1 is smaller than r_2 then S_1 is larger than S_2.

Another mechanism for crystal growth is related to polymorphic changes in solutions, and again the driving force is the difference in solubility between the two polymorphs. In other words, the less soluble form grows at the expense of the more soluble phase. This is sometimes also accompanied by changes in the crystal habit. Different faces of the crystal may have different surface energies and deposition may preferentially take place on one of the crystal faces modifying its shape. Other important factors are the presence of crystal dislocations, kinks, surface impurities, etc. Most of these effects have been discussed in detail in monographs on crystal growth [83–85].

The growth of crystals in suspension concentrates may create undesirable changes. As a result of the drastic change in particle size distribution, the settling of the particles may be accelerated leading to caking and cementing together of some particles in the sediment. Moreover, increase in particle size may lead to a reduction in biological efficiency. Thus, prevention of crystal growth or at least reducing it to an acceptable level is essential in most suspension concentrates. Surfactants affect crystal growth in several ways. The surfactant may affect the rate of dissolution by affecting the rate of transport away from the boundary layer at the crystal solution interface. However, if the surfactant form micelles that can solubilize the solute, crystal growth may be enhanced as a result of increasing the concentration gradient. Thus by proper choice of dispersing agent one may reduce crystal growth of suspension concentrates. This has been demonstrated by Tadros [86] for terbacil suspensions. When using Pluronic P75 [poly(ethylene oxide)-poly(propylene oxide) block copolymer)] crystal growth was significant. By replacing the Pluronic surfactant with poly(vinyl alcohol) the rate of crystal growth was greatly reduced and the suspension concentrate was acceptable.

Many surfactants and polymers may act as crystal growth inhibitors if they adsorb strongly on the crystal faces, thus preventing solute deposition. However, the choice of an inhibitor is still an art and there are not many rules that can be used to select crystal growth inhibitors.

14.4.5
Stability Against Claying or Caking

Once a dispersion that is stable in the colloid sense has been prepared, the next task is to eliminate claying or caking. This is the consequence of settling of the colloidally stable suspension particles. The repulsive forces necessary to ensure this colloid stability allows the particles to move past each other forming a dense sediment that is very difficult to redisperse. Such sediments are dilatant (shear thickening, see section on rheology) and hence the SC becomes unusable. Before describing the methods used for controlling settling and prevention of formation of dilatant clays, an account is given on the settling of suspensions and the effect of increasing the volume fraction of the suspension on the settling rate. This was discussed in detail in Chapter 7 and only a summary is given here.

The sedimentation velocity v_0 of a very dilute suspension of rigid non-interacting particles with radius a can be determined by equating the gravitational force with the opposing hydrodynamic force as given by Stokes' law, i.e.

$$v_0 = \frac{2a^2(\rho - \rho_0)g}{9\eta_0} \tag{14.37}$$

where ρ is the density of the particles, ρ_0 that of the medium, η_0 is the viscosity of the medium and g is the acceleration due to gravity. Equation (14.37) predicts a sedimentation rate for particles with radius 1 μm in a medium with a density difference of 0.2 g cm^{-3} and a viscosity of 1 mPa s (i.e. water at 20 °C) of 4.4×10^{-7} m s^{-1}. Such particles will sediment to the bottom of 0.1 m container in about 60 hours. For 10 μm particles, the sedimentation velocity is 4.4×10^{-5} m s^{-1} and such particles will sediment to the bottom of 0.1 m container in about 40 minutes.

The above treatment using Stokes' law applied only to very dilute suspensions (volume fraction $\phi < 0.01$). For more concentrated suspensions, the particles no longer sediment independent of each other and one has to take into account both the hydrodynamic interaction between the particles (which applies for moderately concentrated suspensions) and other higher order interactions at relatively high volume fractions. A theoretical relationship between the sedimentation velocity v of non-flocculated suspensions and particle volume fraction has been derived by Maude and Whitmore [87] and by Batchelor [88]. Such theories apply to relatively low volume fractions (< 0.1) and they show that the sedimentation velocity v at a volume fraction ϕ is related to that at infinite dilution v_0 (the Stokes' velocity) by an equation of the form

$$v = v_0(1 - k\phi) \tag{14.38}$$

where k is a constant in the region of 5–6. Batchelor [88] derived a rigorous theory for sedimentation in a relatively dilute dispersion of spheres. He considered that the reduction in Stokes' velocity with increasing particle number concentration arises from hydrodynamic interactions. k in Eq. (14.38) was calculated and found to be 6.55. This theory applies up to a volume fraction of 0.1. At higher volume fractions, the sedimentation velocity becomes a complex function of ϕ and only empirical equations are available to describe the variation of v with ϕ. For example, Reed and Anderson [89] developed a virial expansion technique to describe the settling rate of concentrated suspensions. They derived the following expression for the average velocity, v_{av},

$$v_{av} = v_0 \frac{1 - 1.83\phi}{1 + 4.70\phi} \tag{14.39}$$

Good agreement between experimental settling rates and those calculated using Eq. (14.39) was obtained up to $\phi = 0.4$.

Buscall et al. [90] measured the rate of settling of polystyrene latex particles with $a = 1.55$ µm in 10^{-3} mol dm^{-3} up to $\phi = 0.5$. They showed that v/v_0 decreases exponentially with increase in ϕ, approaching zero at $\phi > 0.5$, i.e. in the region of close packing. An empirical equation for the relative settling rate has been derived using the Dougherty–Krieger equation [91] for the relative viscosity, η_r $(= \eta/\eta_0)$,

$$\eta_r = \left(1 - \frac{\phi}{\phi_p}\right)^{-[\eta]\phi_p} \tag{14.40}$$

where $[\eta]$ is the intrinsic viscosity (equal to 2.5 for hard spheres) and ϕ_p is the maximum packing fraction (which is close to 0.6). Assuming that $v/v_0 = \alpha(\eta_0/\eta)$, it is easy to derive the following empirical relationship for the relative sedimentation velocity, v_r $(= v/v_0)$,

$$v_r = \left(1 - \frac{\phi}{\phi_p}\right)^{\alpha[\eta]\phi_p} = \left(1 - \frac{\phi}{\phi_p}\right)^{k\phi_p} \tag{14.41}$$

By allowing the latex to settle completely and then determining the volume concentration of the packed bed, a value of 0.58 was obtained for ϕ_p (close to random packing). Using this value and $k = 5.4$, the relative rate of sedimentation was calculated and found to agree very well with experimental results.

It seems from the above discussion that there is a correlation between the reduction in sedimentation rate and the increase in relative viscosity of the suspension as the volume fraction of the suspension is increased. Clearly, when the maximum packing fraction is reached, the sedimentation velocity approaches zero. However, such dense suspensions have extremely high viscosities and are not a practical solution for reduction of settling. In most cases one prepares a suspension concentrate at practical volume fractions (0.2–0.4) and then uses an antisettling agent to reduce settling. As we will discuss below, most antisettling agents used in practice are high molecular weight polymers. These materials show an increase in the viscosity of the medium with increase in their concentration. However, at a critical polymer concentration (which depends on the nature of the polymer and its molecular weight) they show a very rapid increase in viscosity with further increase in their concentration. This critical concentration (sometimes denoted by C^*) represents the situation where the polymer coils or rods begin to overlap. Under these conditions, the solutions become significantly non-Newtonian (viscoelastic, see section on rheology) and they produce stresses that are sufficient to overcome the stress exerted by the particles. The settling of suspensions in these non-Newtonian fluids is not simple since one has to consider the non-Newtonian behaviour of these polymer solutions. This problem has been addressed by Buscall et al. [90]. To adequately describe the settling of particles in non-Newtonian fluids one needs to know how the viscosity of the medium changes with shear rate or shear stress. Most of these viscoelastic fluids show a gradual increase of viscosity with decreasing shear rate or shear stress but, below a critical stress or shear rate, they show a Newtonian region with a limiting high viscosity that is denoted as the residual (or zero shear) viscosity. This is illustrated in Chapter 7, which shows the variation of the viscosity with shear stress for several solutions of ethyl hydroxyethyl cellulose at various concentrations. The viscosity increases with decreasing stress and the limiting value, i.e. the residual viscosity $\eta(o)$, increases rapidly with increase in polymer concentration. The shear thinning behaviour of these polymer solutions was clearly shown, since above a critical stress value the viscosity decreases rapidly with increasing shear stress. The limiting value of the viscosity is reached at low shear stress.

It is now important to calculate the stress exerted by the particles. This stress is equal to $a\Delta\rho g/3$. For polystyrene latex particles with radius 1.55 μm and density 1.05 g cm^{-3}, this stress is equal to 1.6×10^{-4} Pa. Such stress is lower than the critical stress for most EHEC solutions. In this case, one would expect a correlation between the settling velocity and the zero shear viscosity. This is illustrated in Chapter 7, whereby v/a^2 is plotted versus $\eta(o)$. A linear relationship between $\log(v/a^2)$ and $\log \eta(o)$ is obtained, with a slope of -1, over three decades of viscosity. This indicated that the settling rate is proportional to $[\eta(o)]^{-1}$. Thus, the settling rate of isolated spheres in non-Newtonian (pseudo-plastic) polymer solutions is determined by the zero shear viscosity in which the particles are suspended. As discussed in Chapter 7, on rheological measurements, determination of the zero shear viscosity is not straightforward and requires the use of constant stress rheometers.

The above correlation applies to the simple case of relatively dilute suspensions. For more concentrated suspensions, other parameters should be taken into consideration, such as the bulk (elastic) modulus. Clearly, the stress exerted by the particles depends not only on the particle size but on the density difference between the particle and the medium. Many suspension concentrates have particles with radii up to 10 μm and a density difference of more than 1 g cm^{-3}. However, the stress exerted by such particles will seldom exceed 10^{-2} Pa and most polymer solutions will reach their limiting viscosity value at higher stresses than this. Thus, in most cases the correlation between settling velocity and zero shear viscosity is justified, at least for relatively dilute systems. For more concentrated suspensions, an elastic network is produced in the system which encompasses the suspension particles as well as the polymer chains. Here, settling of individual particles may be prevented. However, in this case the elastic network may collapse under its own weight and some liquid be squeezed out from between the particles. This is manifested in a clear liquid layer at the top of the suspension, a phenomenon usually referred to as syneresis. If such separation is not significant, it may not cause any problem on application since by shaking the container the whole system redisperses. However, significant separation is not acceptable since it becomes difficult to homogenise the system. In addition, such extensive separation is cosmetically unacceptable and the formulation rheology should be controlled to minimise such separation.

Several methods are applied to control settling and prevent the formation of dilatant clays (discussed in Chapter 7). As mentioned in Chapter 7, balancing the density of disperse phase and medium is obviously the simplest method for retarding settling, since, as is clear from Eq. (14.37), if $\rho = \rho_0$, then $v_0 = 0$. However, this method can only be applied to systems where the difference in density between the particle and the medium is not too large. For example, with many organic solids with densities between 1.1 and 1.3 g cm^{-3} suspended in water, some soluble substances such as sugar or electrolytes may be added to the continuous phase to increase the density of the medium to a level that is equal to that of the particles. However, one should be careful that the added substance does not cause any flocculation of the particles. This is particularly the case when using electrolytes, whereby one should avoid any "salting out" materials, which causes the medium to be a poor solvent for the stabilizing chains. In addition, density matching can only be achieved at one temperature. Liquids usually have larger thermal expansion coefficients than solids and if, say, the density is matched at room temperature, settling may occur at higher temperatures. Thus, one has to be careful when applying the density matching method, particularly if the formulation is subjected to large temperature changes.

The most practical method for reducing settling is to use high molecular weight materials such as natural gums, hydroxyethyl cellulose or synthetic polymers such as poly(ethylene oxide). The most commonly used material in agrochemical formulations is xanthan gum (produced by converting waste sugar into a high molecular weight material using a micro-organism and sold under the trade names Kelzan or Rhodopol), which is effective at relatively low concentrations (of the order of 0.1–0.2% depending on the formulation). As mentioned above, these high molecular weight materials produce viscoelastic solutions above a critical concentration. This viscoelasticity produces sufficient residual viscosity to stop the settling of individual particles. The solutions also give enough elasticity to overcome separation of the suspension. However, one cannot rule out interaction of these polymers with the suspension particles that may result in "bridging" and, hence, the role by which such molecules reduce settling and prevent the formation of clays may be complex. To arrive at the optimum concentration and molecular weight of the polymer necessary to prevent settling and claying, one should study the rheological characteristics of the formulation as a function of the variables of the system such as its volume fraction, concentration and molecular weight of the polymer and temperature.

Fine inorganic materials such as swellable clays and finely divided oxides (silica or alumina), when added to the dispersion medium of coarser suspensions, can eliminate claying or caking. These fine inorganic materials form a "three-dimensional" network in the continuous medium, which by virtue of its elasticity prevents sedimentation and claying. With swellable clays such as sodium montmorillonite, the gel arises from the interaction of the plate-like particles in the medium – such particles consist of an octahedral alumina sheet sandwiched between two tetrahedral silica sheets [92]. In the tetrahedral sheets, tetravalent Si may be replaced by trivalent Al, whereas in the octahedral sheet there may be replacement of trivalent Al with divalent Mg, Fe, Cr or Zn. This replacement is usually referred to as isomorphic substitution [92], i.e. an atom of higher valency is replaced by one of lower valency. This results in a deficit of positive charges or excess of negative charges. Thus, the faces of the clay platelets become negatively charged and these negative charges are compensated by counter ions such as Na^+ or Ca^{2+}. As a result, a double layer is produced with a constant charge (that is independent of the pH of the solution). However, at the edges of the platelets, some disruption of the bonds occurs, resulting in the formation of an oxide-like layer, e.g. –Al–OH, which undergoes dissociation to give a negative $-Al-O^-$ group or a positive $-Al-OH_2^+$ group, depending on the pH of the solution. An isoelectric point may be identified for the edges (usually between pH 7 and 9). This means that the double layer at the edges differs from that at the faces and the surface charges can be positive or negative depending on the pH of the solution. For that reason, van Olphen [92] suggested an edge-to-face association of clay platelets (which he termed the "house of card" structure) and this was assumed to be the driving force for gelation of swellable clays. However, Norrish [93] suggested that clay gelation is caused simply by the interaction of the expanded double layers. This is particularly the case in dilute electrolyte solutions whereby the double layer thickness can be several orders of magnitude higher than the particle dimensions.

With oxides, such as finely divided silica, gel production is caused by the formation of chain aggregates, which interact to form an elastic three-dimensional network. Clearly, the formation of such networks depends on the nature and particle size of the silica particles. For effective gelation, one should choose silicas with very small particles and highly solvated surfaces.

Mixtures of polymers such as hydroxyethyl cellulose or xanthan gum with finely divided solids such as sodium montmorillonite or silica offer one of the most robust antisettling systems. By optimising the ratio of the polymer to the solid particles, one can arrive at the right viscosity and elasticity to reduce settling and separation. Such systems are more shear thinning than the polymer solutions and hence they are more easily dispersed in water on application. The most likely mechanism by which these mixtures produce viscoelastic network is probably through bridging or depletion flocculation. Polymer–particulate mixtures also show less temperature dependence of viscosity and elasticity than polymer solutions and hence they ensure long-term physical stability at high temperatures.

Another method of curbing sedimentation is controlled flocculation. For systems where the stabilizing mechanism is electrostatic in nature, for example those stabilized by surfactants or polyelectrolytes, the energy–distance curve shows a secondary minimum at larger particle separations. This minimum can be quite deep (few tens of kT units), particularly for large (> 1 μm) and asymmetric particles. The

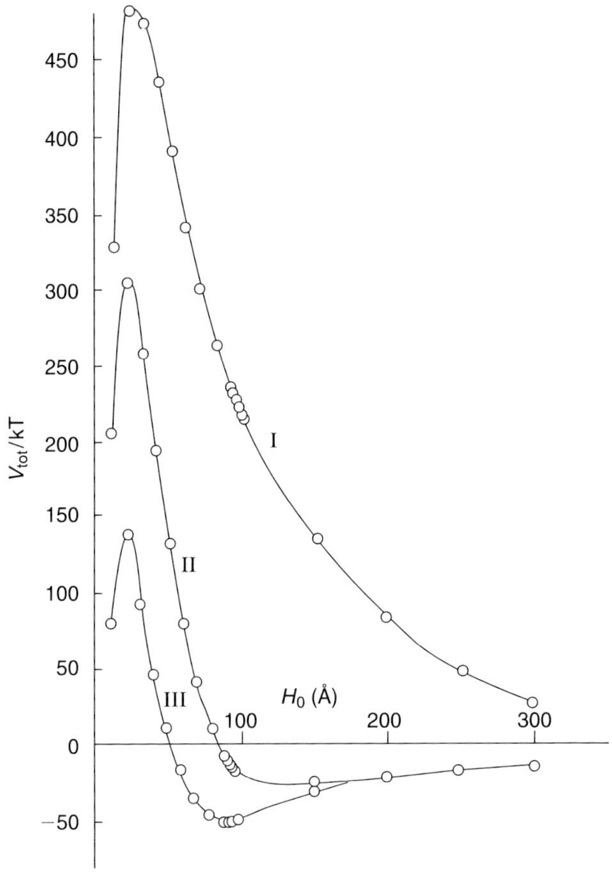

Fig. 14.13. Energy–distance curves for ethirimol suspensions at three NaCl concentrations: I = 10^{-3}; II = 10^{-2}; III = 10^{-1} mol dm^{-3}.

depth of the minimum also depends on electrolyte concentration. Thus, by adding small amounts of electrolyte, weak flocculation may be obtained. Weakly flocculated systems may produce a gel network (self-structured systems) that has sufficient elasticity to reduce settling and eliminate claying. Tadros [94] demonstrated this for ethirimol suspensions stabilised with phenol formaldehyde sulphonated condensate (a polyelectrolyte with modest molecular weight). Energy–distance curves for such suspensions at three NaCl concentrations (10^{-3}, 10^{-2} and 10^{-1} mol dm^{-3}) are shown in Figure 14.13. By increasing NaCl concentration, the depth of the secondary minimum increases, reaching $\sim 50kT$ at the highest electrolyte concentration. By using electrolytes of higher valency such as $CaCl_2$ or $AlCl_3$, such deep minima are produced at much lower electrolyte concentrations. Thus, by controlling electrolyte concentration and valency, one can reach sufficiently deep secondary minimum to produce a gel with enough elasticity to reduce settling and

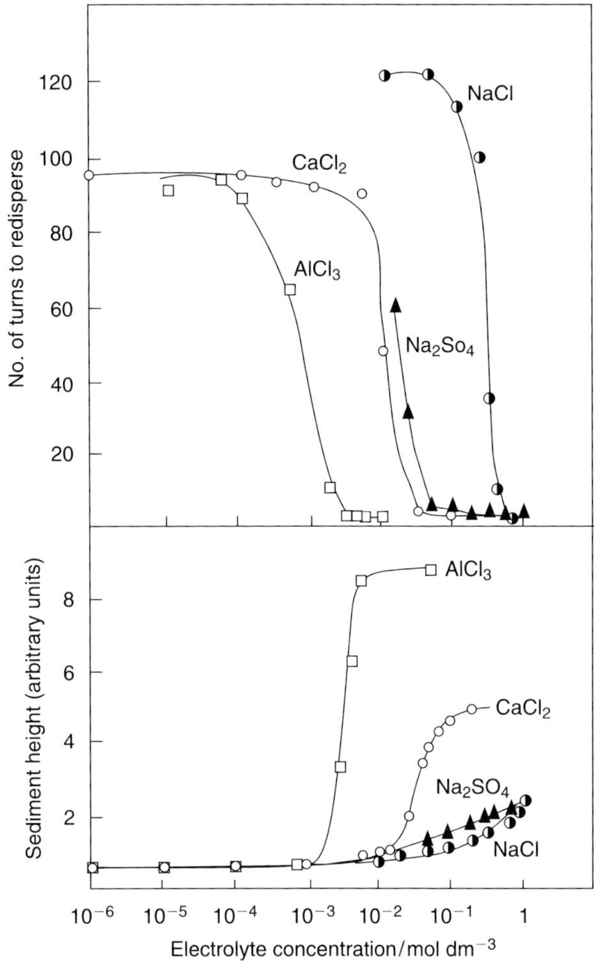

Fig. 14.14. Sediment height and redispersion as a function of electrolyte concentration.

eliminate claying. Figure 14.14 illustrates this, showing the variation of sediment height and redispersion as a function of electrolyte concentration for four electrolytes, namely NaCl, Na_2SO_4, $CaCl_2$ and $AlCl_3$. Clearly, above a critical electrolyte concentration, the sediment height increases and this prevents the formation of clays. Above this critical electrolyte concentration, redispersion of the suspension becomes easier (Figure 14.14).

For systems stabilised by nonionic surfactants or macromolecules, the energy–distance curve also shows a minimum whose depth depends on particle size, the Hamaker constant and the thickness of the adsorbed layer [94, 95]. This is illustrated in Figure 14.15, which shows the energy–distance curves for polystyrene latex particles containing poly(vinyl alcohol) (PVA) layers of various molecular

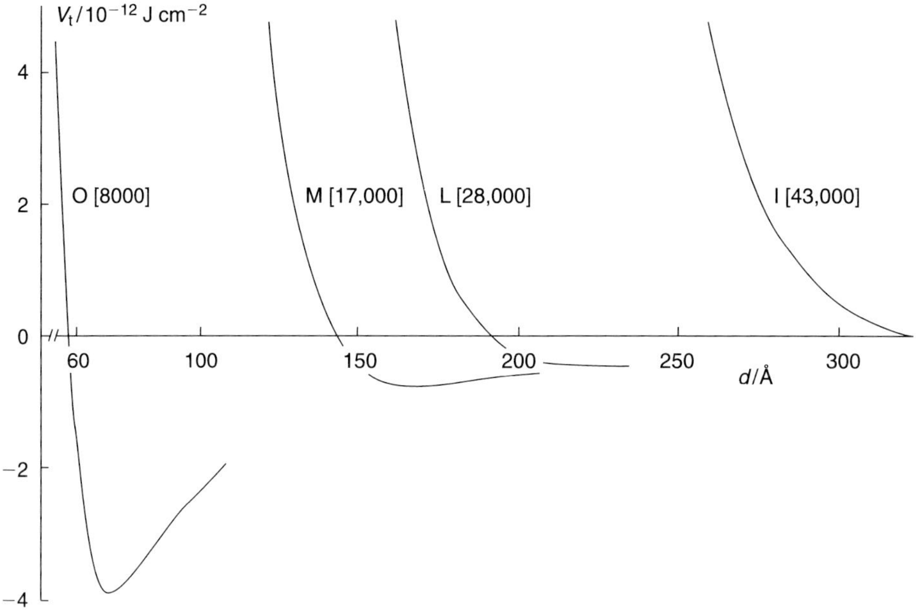

Fig. 14.15. Energy–distance curves for polystyrene latex dispersions with adsorbed PVA layers with various molecular weights.

weights [94, 95]. Clearly, with the high molecular weight polymers ($M > 17\,000$ with an adsorbed layer thickness $\delta > 9.8$ nm), the energy minimum is too small for flocculation to occur. However, as the molecular weight of the polymer is reduced below a certain value, i.e. as the adsorbed layer becomes small ($M = 8000$ and $\delta = 3.3$), the energy minimum becomes deep enough for flocculation to occur. This was demonstrated for the latex containing PVA with $M = 8000$, whereby scanning electron micrographs of a freeze-dried sediment showed flocculation and an open structure. In this case claying was prevented.

Another method of reducing sedimentation is to employ the principle of depletion flocculation (described in Chapter 7). The addition of "free" (non-adsorbing) polymer can induce weak flocculation of the suspension, when the concentration or volume fraction of the free polymer (ϕ_p) exceeds a critical value denoted by ϕ_p^+. Asakura and Oosawa reported the first quantitative analysis of the phenomenon [96]. They showed that when two particles approach to a separation that is smaller than the diameter of the free coil, polymer molecules are excluded from the interstices between the particles, leading to the formation a polymer-free zone (depletion zone). Figure 14.16 shows this for the situation below and above ϕ_p^+.

As a result of this process, an attractive force, associated with the lower osmotic pressure in the region between the particles, is produced. This weak flocculation process can be applied to prevent sedimentation and formation of clays. Heath et al. [97] have illustrated this using ethirimol suspensions stabilized by a graft

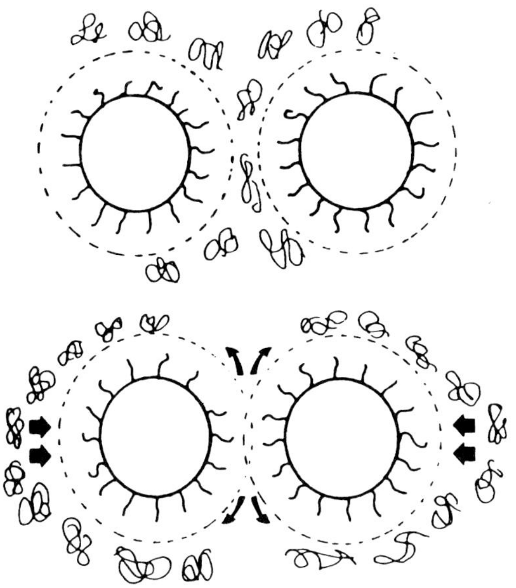

Fig. 14.16. Representation of depletion flocculation (top, below ϕ_p^+; bottom, above ϕ_p^+).

copolymer containing poly(ethylene oxide) (PEO) side chains (with $M = 750$) to which free PEO with various molecular weights (20 000, 35 000 and 90 000) was added. Above a critical volume fraction of the free polymer (which decreased with increasing molecular weight) weak flocculation occurred, and this prevented the formation of dilatant sediments.

14.4.6
Assessment of the Long-term Physical Stability of Suspension Concentrates

To fully assess the properties of suspension concentrates, three main types of measurements are required. Firstly some information is needed on the structure of the solid/solution interface at a molecular level. This requires investigation of the double layer properties (for systems stabilised by ionic surfactants and polyelectrolytes), adsorption of the surfactant or polymer as well as the extension of the layer from the interface (adsorbed layer thickness). Secondly, one needs to obtain information on the state of dispersion on standing, such as its flocculation and crystal growth. This requires measurement of the particle size distribution as a function of time and microscopic investigation of flocculation. The spontaneity of dispersion on dilution, i.e. reversibility of flocculation needs also to be assessed. Finally, information on the bulk properties of the suspension on standing is required, which can be obtained using rheological measurements. The methods that may be applied for suspension concentrates are described briefly below.

Electrokinetic measurements provide the most practical method for investigating the double layer at the solid/solution interface [98]. The most common such measurement is that of microelectrophoresis, which allows one to obtain the particle mobility as a function of system paramcters such as surfactant and electrolyte concentration. The dilute dispersion is investigated microscopically in a capillary tube connected to two containers fitted with electrodes. By applying an electric field with a strength E/l, where E is the applied voltage and l is the distance between the electrodes, one can measure the average velocity v of the particles and hence its mobility u ($u = v/(E/l)$). From the mobility u, the zeta potential, ζ, can be calculated using the Smoluchowski equation, which is valid for most coarse suspensions [98],

$$u = \frac{\varepsilon\varepsilon_0\zeta}{\eta} \tag{14.42}$$

where ε is the relative permittivity, ε_0 is the permittivity of free space and η is the viscosity of the medium. For aqueous dispersions at 25 °C,

$$\zeta = 1.282 \times 10^6 u \tag{14.43}$$

ζ is calculated in volts when u is expressed in $m^2\ V^{-1}\ s^{-1}$.

By measuring the zeta potential as a function of concentration of added ionic surfactant, one obtains information on the limiting zeta potential that can be reached. This value usually coincides with saturation adsorption. Qualitatively, the higher the zeta potential, the stronger the repulsion between the particles, and the higher the colloid stability.

Adsorption of ionic, nonionic and polymeric surfactant on the agrochemical solid gives valuable information on the magnitude and strength of the interaction between the molecules and the substrate as well as the orientation of the molecules. The latter is important in determining colloid stability. Adsorption isotherms are fairly simple to determine, but require careful experimental techniques. A representative sample of the solid with known surface area A per unit mass must be available. The surface area is usually determined using gas adsorption. N_2 is usually used as the adsorbate, but for materials with relatively low surface area, such as those encountered with most agrochemical solids, it is preferable to use Kr as the adsorbate. The surface area is obtained from the amount of gas adsorbed at various relative pressures by application of the BET equation [96]. However, the surface area determined by gas adsorption may not represent the true surface area of the solid in suspension (the so-called "wet" surface). In this case it is preferable to use dye adsorption to measure the surface area [99].

An analytical method that is sensitive enough to determine surfactant or polymer adsorption then needs to be established. Several spectroscopic and colorimetric methods may be applied, which in some cases need to be developed for a particular surfactant or polymer. A reproducible method for dispersing the solid in the surfactant or polymer solution also needs to be established. In this case, high-speed stirrers or ultrasonic radiation may be applied, provided these do not cause any comminution of the particles during dispersion. The time required for adsorption must be established by carrying out experiments as a function of time. Finally, the solid needs to separated from the solution, which may be carried out using centrifugation.

Once the above procedure is established, the adsorption isotherm can be determined, whereby the amount of adsorption in moles per unit area, Γ, is plotted as function of equilibrium concentration, C_2. With many surfactants, where adsorption reversible, the results can usually be fitted with a Langmuir-type equation, i.e. Γ increases gradually with increasing C_2 and, eventually, reaches a plateau value, Γ_∞, which corresponds to saturation adsorption,

$$\Gamma = \frac{\Gamma_\infty bC_2}{1 + bC_2} \tag{14.44}$$

where b is a constant that is related to the adsorption free energy ($b \propto \exp -\Delta G_{ads}/RT$). A linearlised form of the Langmuir equation may be used to obtain Γ_∞ and b, i.e.

$$\frac{1}{\Gamma} = \frac{1}{\Gamma_\infty} + \frac{1}{\Gamma_\infty bC_2} \tag{14.45}$$

A plot of $1/\Gamma$ versus $1/C_2$ gives a straight line with intercept $1/\Gamma_\infty$ and slope $1/\Gamma_\infty bC_2$ from which both Γ_∞ and b can be calculated. As discussed in Chapter 5, from Γ_∞, the area per surfactant molecule can be calculated (area per molecule $= 1/\Gamma_\infty N_{av}$). The area per surfactant molecule gives information on the orientation at the solid/solution interface. For example, for vertically orientated ionic surfactant molecules, such as sodium dodecyl sulphate on a hydrophobic surface, an area per surfactant ion in the region of 0.3–0.4 nm^2 is to be expected. The area per surfactant ion is determined by the cross sectional area of the ionic head group. However, many surfactant ions may undergo association on the solid surface and this is usually accompanied by steps in the adsorption isotherm. In the region of surfactant association, the amount of adsorption increases rapidly with increasing surfactant concentration and, finally, another plateau is reached when the aggregate units (sometimes referred to as hemimicelles) become close packed on the surface.

With nonionic surfactants, the adsorption isotherm may also show steps that are characteristic of various orientation and association of the molecules on the surface. Nonionic surfactants of the ethoxylate type, such as $R\text{-}(CH_2\text{-}CH_2\text{-}O)_n\text{–}OH$, show complex adsorption isotherms that are very sensitive to small changes in concentration, temperature or molecular structure. The main interaction with a hydrophobic surface is usually through hydrophobic bonding (see Chapter 5) with the alkyl group, leaving the poly(ethylene oxide) (PEO) chain dangling in solution. As a result, the area per molecule is usually large since it is determined by the area occupied by the PEO chain, which occupies a large area, particularly when the chain contains several EO units. As with ionic surfactants, the molecules may aggregate on the surface to form hemimicelles or even micelles. Figure 5.4 gives a schematic representation of nonionic surfactant molecules adsorbed on a solid surface (see Chapter 5), showing the possible association structures. Moreover, adsorption increases rapidly with rising temperature near the phase-separation point.

Polymer adsorption is more complex than surfactant adsorption, since one must consider the various interactions (chain–surface, chain–solvent and surface–solvent) as well as the conformation of the polymer on the surface. Various conformations of different macromolecular surfactants are schematically shown in Figure 5.5 (see Chapter 5). Complete information on polymer adsorption may be obtained if one can determine the segment density distribution, i.e. the segment concentration in all layers parallel to the surface where such segments are accommodated. In practice, however, such information is unavailable and, therefore, one determines three main parameters: the amount of adsorption per unit area, Γ, the fraction of segments in trains p and the adsorbed layer thickness δ. The value of Γ can be determined in the same way as for surfactants, although in this case some complications may arise. Firstly, adsorption may be very slow, requiring long equilibration times (sometimes days). Secondly, most commercially available polymers have a wide distribution of molecular weights. Instead of the theoretically predicted high-affinity isotherms, in which the plateau value starts at near zero polymer concentration, the experimental isotherm is rounded. Such rounded isotherms are due to the heterodispersity of the samples. In this case, the amount adsorbed Γ depends on the area to volume ratio A/V, with Γ decreasing as A/V increases (i.e. the amount of adsorption decreases with increasing concentration of the suspension).

The second parameter that needs to be established in polymer adsorption is the fraction p of segments in direct contact with the surface. As mentioned in Chapter 5, direct and indirect methods may be applied. Direct methods are based on spectroscopic techniques such as infrared (IR), electron sin resonance (ESR), and nuclear magnetic resonance (NMR) [95]. Of the indirect methods, microcalorimetry is perhaps the most convenient to apply. By measuring the heat of adsorption of the chain, one may obtain p by referring to the heat of adsorption of a segment.

The most practical methods for measuring the adsorbed layer thickness are based on hydrodynamic techniques (see Chapter 5). Unfortunately, these methods can only be applied to spherical particles with small radii such that the ratio of the adsorbed layer thickness to particle radius δ/a is significant (of the order of 10%). The adsorbed layer thickness is determined from a comparison of the hydrodynamic radius of the particle with the adsorbed layer a_δ with that of the bare particle a. Thus, measurement of the adsorbed layer thickness on agrochemical suspension particles is not possible. One has to use model particles such as polystyrene latex particles to obtain such information and it is assumed that the adsorbed layer thickness obtained on such model particles is comparable to that on the practical agrochemical suspension particles. This assumption is not serious, since in, most cases one only needs a comparison between various polymer samples. Four different hydrodynamic methods may be applied to obtain δ: measurement of the sedimentation coefficient (using an ultracentrifuge), diffusion coefficient (using dynamic light scattering or photon correlation spectroscopy), viscosity and slow-speed centrifugation. These methods are described in some detail in Chapter 5.

To assess the state of the suspension concentrate, one needs to obtain information on flocculation, crystal growth and separation on storage. Two general techniques are widely used to monitor the flocculation rate of suspensions, both of which can be only applied to dilute systems. The first method is based on measurement the turbidity τ (at a given wavelength of light λ) as a function of time during the early stages of flocculation. This method can be only applied if the particles are smaller than $\lambda/20$ and hence it cannot be used for coarse suspensions. In the latter case, direct particle counting as a function of time is the most suitable procedure. This can be carried out manually using a light microscope or automatically using a Coulter counter or an ultramicroscope. Recently, optical microscopy has been combined with image analysis techniques for counting the particles. An alternative procedure is to use light diffraction, e.g. using the commercial Master sizer instrument (Malvern). The rate constant for flocculation (assumed to follow a bimolecular process) is determined by plotting $1/n$ versus t, where n is the particle number at time t, i.e.

$$\frac{1}{n} = \frac{1}{n_0} + kt \tag{14.46}$$

where n_0 is the number of particles at $t = 0$. The rate constant k can be related to the rapid flocculation rate k_0 given by Smoluchowski [62], i.e.

$$k_0 = \frac{8\,kT}{6\,\eta} \tag{14.47}$$

For particles dispersed in an aqueous phase at 25 °C, $k_0 = 5.5 \times 10^{-18}$ m^3 s^{-1}. k is usually related to k_0 by the stability ratio W, i.e. $W = k_0/k$. The higher W is the more stable the dispersion. Thus, by plotting W versus system parameters such as surfactant and/or electrolyte concentration, one can obtain a quantitative assessment of the stability of the suspension under various conditions. Notably, the stability ratio W is related to the energy maximum in the energy–distance curve for electrostatically stabilized suspensions. The higher this energy maximum is, the higher the value of W.

Incipient flocculation of sterically stabilized suspensions, i.e. the condition when the chains are in a poor solvent condition, can be investigated using turbidity measurements. The suspension is placed in a spectrophotometer cell placed in a block that can be heated at a controlled rate. From a plot of turbidity versus temperature one can obtain the critical flocculation temperature, which is the point at which there is rapid increase in turbidity.

Crystal growth (Ostwald ripening) can be investigated by following the particle size distribution as a function of time using a Coulter counter or an optical disc centrifuge. The percentage number cumulative frequency over size is plotted versus time for various particle radii. Curves are produced at various intervals of time. When crystal growth occurs, the cumulative counts are shifted towards coarser particle sizes. Horizontal lines corresponding to various percentage cumulative counts are then made to cut the curves in their steep portions. From the intersections, a plot of equivalent diameter against time is drawn, allowing one to obtain the rate of crystal growth.

Sediment height (volume) measurement provides a qualitative method for assessing the state of dispersion. The suspension is placed in graduated stoppered cylinders and the sediment height (volume) is followed as function of time until equilibrium is reached. Normally in sediment height (volume) measurements one compares the initial height H_0 or volume V_0 with that reached at equilibrium, H and V, respectively. A clayed suspension gives low values for the relative height (H/H_0) or volume (V/V_0), whereas a "structured" suspension containing an anti-settling agent or weakly flocculated gives high values. Clearly, the higher H/H_0 or V/V_0 is the better the suspension. One aims at a relative value of unity, which implies no separation on standing. However, one must be careful since strong flocculation must be avoided. A strongly flocculated system may give a high relative sediment height (volume) but in this case the suspension cannot be redispersed or adequately diluted. Thus, at the end of the sedimentation experiment, one should redisperse the system by rotating the cylinders end-over-end and carry out a dispersion test. The suspension is poured into a beaker containing water and the dispersion is observed visually. In most application methods one requires a spontaneous dispersion on dilution with minimum agitation. For a more quantitative assessment of the state of flocculation and dispersion of the suspension concentrate one should apply the rheological methods discussed below.

Rheological measurements are used to investigate the bulk properties of suspension concentrates (see Chapter 7 for details). Three types of measurements can be applied: (1) Steady-state shear stress–shear rate measurements that allow one to obtain the viscosity of the suspensions and its yield value. (2) Constant stress or creep measurements, which allow one to determine the residual or zero shear viscosity (which can predict sedimentation) and the critical stress above which the structure starts to "break-down" (the true yield stress). (3) Dynamic or oscillatory measurements that allow one to obtain the complex modulus, the storage modulus (the elastic component) and the loss modulus (the viscous component) as a function of applied strain amplitude and frequency. From a knowledge of the storage modulus and the critical strain above which the structure starts to "break-down", one can obtain the cohesive energy density of the structure.

The above rheological parameters can be used to assessment and predict the long-term physical stability of suspension concentrates. They offer valuable tools to the formulation chemist for the development of stable systems. In addition, one can design a simple rheological technique for evaluation of the suspension concentrate during manufacture (quality assurance test).

14.5
Microemulsions in Agrochemicals

As mentioned before, for oil-insoluble agrochemicals one of the most common formulations is emulsifiable concentrates (ECs), which when added to water produce oil-in-water (O/W) emulsions either spontaneously or by gentle agitation. However, there has great concern recently in using ECs in agrochemical formulations for several reasons. The use of aromatic oils is undesirable due to their possible phytotoxic effect and their environmental disadvantages. An alternative and more attractive system to ECs are oil-in-water emulsions, as discussed before. In this case, the pesticide which may be an oil is emulsified into water and a water-based concentrated emulsion (EW) is produced. With very viscous or semi-solid pesticides, a small amount of an oil (which may be aliphatic) may be added before the emulsification process. Unfortunately, EWs suffer from several problems, such as the difficulty of emulsification and their long-term physical stability. A very attractive alternative for the formulation of agrochemicals is to use microemulsion systems. The latter are single optically isotropic and thermodynamically stable dispersions, consisting of oil, water and amphiphile (one or more surfactants) [100]. As discussed in Chapter 10, the origin of the thermodynamic stability arises from the low interfacial energy of the system, which is outweighed by the negative entropy term of dispersion [101]. These systems offer several advantages over O/W emulsions: Once the composition of the microemulsion is identified, the system is prepared by simply mixing all the components without the need of any appreciable shear. Owing to their thermodynamic stability, these formulations undergo no separation or breakdown on storage (within a certain temperature range, depending on the system). The low viscosity of the microemulsion systems ensures their ease of pourability, dispersion on dilution and they leave little residue in the container. Another main attraction of microemulsions is their possible enhancement of biological efficacy of many agrochemicals. This, as we will see later, is due to the solubilization of the pesticide by the microemulsion droplets.

This section will summarise the basic principles involved in the preparation of microemulsions and the origin of their thermodynamic stability (see Chapter 10 for more details). A sub-section is devoted to emulsifier selection for both O/W and W/O microemulsions. Physical methods that may be applied for characterization of microemulsions will be briefly described. Finally a sub-section is devoted to the possible enhancement of biological efficacy using microemulsions. The role of microemulsions in enhancing wetting, spreading and penetration will be discussed. Solubilization is also another factor that may enhance the penetration and uptake of an insoluble agrochemical.

14.5.1 Basic Principles of Microemulsion Formation and their Thermodynamic Stability

As discussed in Chapter 6, the formation of oil droplets from a bulk oil is accompanied by an increase in the interfacial area, ΔA, and hence an interfacial energy, $\Delta A\gamma$. The entropy of dispersion of the droplets is equal to $T\Delta S$. With macroemulsions (EWs), the interfacial energy term is much larger than the entropy term and, hence, emulsification is non-spontaneous. In other words, energy is needed to produce the emulsion, e.g. by the use of high-speed mixers. With microemulsions, the interfacial tension is made sufficiently low such that the interfacial energy becomes comparable to or even lower than the entropy of dispersion [101]. In this case, the free energy of formation of the system becomes zero or negative. This explains the thermodynamic stability of microemulsions. Thus, the main driving force for microemulsion formation is the ultralow interfacial tension that is usually achieved by the use of two or emulsifiers. This is discussed in detail in Chapter 10. The role of surfactants in microemulsion formation was considered by Schulman and co-workers [102, 103] and later by Prince [104] who introduced the concept of a two-dimensional mixed liquid film as a third phase in equilibrium with both oil and water. This implies that the monolayer of the mixed surfactant film may be represented by a duplex film that has different properties on the oil and the water side. A two-dimensional surface pressure π (where π is given by the difference between the interfacial tension of the clean interface and that with the adsorbed surfactant film) describes the property of the film at both sides of the interface. Initially, the flat film will have two different surface pressures at the oil and water sides, namely π'_o and π'_w respectively. This is due to the different "crowding" of the hydrophobic and hydrophilic components at both interfaces. For example, if the hydrophobic part of the chain is "bulkier" than the hydrophilic part, "crowding" will occur at the oil side of the interface and π'_o will be higher than π'_w. This inequality between π'_o and π'_w will result in a stress at the interface that must be relieved by bending. In this case, the film has to be expanded at the oil side of the interface until the surface pressures become equal at both sides of the duplex film, i.e. $\pi_o = \pi_w = \frac{1}{2}(\pi_{O/W})_a$ (the subscript a is used to indicate the alcohol cosurfactant, which reduces the interfacial tension on its own right). This leads to the formation of a W/O microemulsion. Conversely, if $\pi'_w > \pi'_o$, the film has to expand at the water side and an O/W microemulsion is formed. This is described in Chapter 10.

Contributions to π are considered to be the crowding of surfactant and cosurfactant molecules and penetration of the oil phases into the hydrocarbon part of the molecules. If $\pi > (\gamma_{O/W})_a$ then γ_T becomes negative, leading to the expansion of the interface until γ_T becomes zero or a small positive value. Since $(\gamma_T)_a$ is of the order of 15–20 mN m^{-1}, surface pressures of that order have to be reached for γ_T to reach an ultralow value that is required for microemulsion formation. This is best achieved by the use of two surfactant molecules, as discussed above.

The above simple theory can explain the nature of the microemulsion produced when using surfactants with different structures. For example, if the molecules have bulky hydrophobic groups such as Aerosol OT, a W/O microemulsion is produced. Conversely, if the molecule has bulky hydrophilic chains such as alcohol ethoxylates with high ethylene oxide units, an O/W microemulsion is produced. These concepts will be rationalised using the packing ratio concept discussed in detail in Chapter 6.

Microemulsions may also be considered as swollen micellar systems, as suggested by Shinoda and co-workers [105–107]. These authors considered the phase diagrams of the components of the microemulsion systems. As discussed in Chapter 10, the phase diagram of the three-component system water/surfactant/cosurfactant (alcohol) shows two main regions at the water and alcohol corners, namely L_1 (normal micelles) and L_2 (inverse micelles). These regions are separated by liquid crystalline structures. Addition of a small amount of oil, miscible with the cosurfactant, changes the phase diagram only slightly. In the presence of substantial amounts of oil, the phase diagram changes significantly (see Chapter 10). The O/W microemulsion near the water/surfactant axis is now not in equilibrium with the lamellar phase (as is the case with the three-phase system), but with a non-colloidal oil + cosurfactant phase. If cosurfactant is added to such a two-phase equilibrium at sufficiently high surfactant concentration, all oil is taken up and a one-phase microemulsion appears. However, addition of cosurfactant at low surfactant concentration may lead to separation of an excess aqueous phase before all oil is taken up in the microemulsion. A three-phase system (3ϕ) is formed that contains a microemulsion that cannot be identified as W/O or O/W (bicontinuous or Winsor III phase system). This phase, sometimes referred to as middle-phase microemulsion, is probably similar to the lamellar phase swollen with oil or to a still more irregular intertwining aqueous and oil region (bicontinuous structure). This middle microemulsion phase has a very low interfacial tension with both oil and water (10^{-4}–10^{-2} mN m^{-1}). Further addition of cosurfactant to the three-phase system makes the oil phase disappear and leaves a W/O microemulsion in equilibrium with a dilute aqueous surfactant solution. In the large one-phase region continuous transitions from O/W to middle phase (bicontinuous) to W/O microemulsions are found.

Solubilization and formation of swollen micelles can also be illustrated by considering the phase diagrams of nonionic surfactants containing poly(ethylene oxide). Such surfactants do not generally need a cosurfactant for microemulsion formation. At low temperatures, the ethoxylated surfactant is soluble in water, and at a given concentration it can solubilize a given amount of oil. However, by adding more oil to such a solution, separation into two phases occurs: O/W solubilized + oil. If the temperature of such a two-phase system is increased the excess oil may be solubilized. This occurs at the solubilization temperature of the system. Above this temperature an isotropic O/W microemulsion is produced. By further increasing the temperature of this microemulsion, the cloud point of the surfactant is reached and separation into oil + water + surfactant takes place. Thus, an O/W microemulsion is produced between the solubilization temperature and cloud point temperature of the surfactant. The isotropic O/W microemulsion region is located between the solubilization curve and the cloud point curve. This phase diagram shows the temperature range within which a microemulsion is produced. This range decreases as the oil weight fraction is increased and, above a certain weight fraction (which depends on surfactant concentration), there will be no single isotropic region.

Nonionic ethoxylated surfactants can also be used to produce isotropic W/O microemulsions. A low HLB number surfactant may be dissolved in an oil, and such a solution can solubilize water to a certain extent, depending on surfactant concentration. If water is added above the solubilization limit, the system separates into two phases: W/O solubilized + water. If the temperature of such a two-phase system is reduced an isotropic W/O microemulsion is formed below the solubilization temperature. If the temperature is then further reduced below the haze point, separation into water + oil + surfactant occurs. Thus, a W/O microemulsion can be identified between the solubilization and the haze point curves.

From the above discussion, nonionic surfactants of the ethoxylate type clearly can be used to produce O/W or W/O microemulsions. However, such microemulsions have limited temperature stability and are of limited practical application. Their temperature range of stability may be increased by addition of ionic surfactants, which usually increase the cloud point of the surfactant.

Several factors play a role in determining whether a W/O or an O/W microemulsion is formed. They may be considered in the light of the theories described above. For example, the duplex theory predicts that the nature of the microemulsion formed depends on the relative packing of the hydrophobic and hydrophilic portions of the surfactant molecule, which determines the bending of the interface. This can be illustrated by considering an ionic surfactant molecule such as Aerosol OT (diethyl hexyl sulphosuccinate). This molecule has a bulky hydrophobe (two alkyl groups) with a large volume to length (v/l) and a stumpy head group. When this molecule adsorbs at a flat O/W interface, the hydrophobic groups become crowded and the interface tends to bend with the head groups facing inwards, thus forming a W/O microemulsion. This geometric constraint for the Aerosol OT molecule has been considered in detail by Oakenfull [108] who showed that the molecule has a v/l greater than 0.7, which was considered to be necessary for W/O microemulsion formation. For single-chain ionic surfactants, such as sodium dodecyl sulphate, v/l is less than 0.7 and W/O microemulsion formation requires a cosurfactant, which increases v without affecting l (if the chain length of the cosurfactant does not exceed that of the surfactant). These cosurfactant molecules act as "padding", separating the head groups.

The importance of geometric packing on the nature of microemulsion has been considered in detail by Mitchell and Ninham [109]. According to these authors, the nature of the aggregate unit depends on the packing ratio, P, given by

$$P = \frac{v}{a_0 l_c} \tag{14.48}$$

where v is the partial molecular volume of the surfactant, a_0 is the head group area of a surfactant molecule and l_c is the maximum chain length. Thus, this packing ratio provides a quantitative measure of the hydrophilic–lipophilic balance (HLB). For $P < 1$, normal (i.e. convex) aggregates are predicted, whereas inverse drops are expected for $P > 1$. The packing ratio is affected by many factors, including hydrophobicity of the head group, ionic strength of the solution, pH and temperature and the addition of lipophilic compounds such as cosurfactants. With Aerosol OT, $P > 1$ since both a_0 and l_c are small. Thus this molecule favours the formation of W/O microemulsions.

The packing ratio also explains the nature of microemulsions formed by using nonionic surfactants. If $v/a_0 l_c$ increases with rising temperature (as a result of the reduction of a_0 with temperature), one would expect the solubilization of the hydrocarbon to increase with temperature until $v/a_0 l_c$ reaches the value of 1, where phase inversion would be expected. At higher temperatures, $v/a_0 l_c$ is >1 and W/O microemulsion would be expected. Moreover, the solubilization of water would decrease as the temperature rises, as expected.

The influence of surfactant structure on the nature of the microemulsion can be predicted from the thermodynamic theory suggested by Overbeek [110]. According to this theory, the most stable microemulsion would be that in which the phase with the smaller volume fraction, ϕ, forms the droplets, since the osmotic pressure of the system increases with increasing ϕ. For a W/O microemulsion prepared using an ionic surfactant, the hard-sphere volume is only slightly larger than the water-core volume since the hydrocarbon tails may interpenetrate to some extent when two droplets come together. For an O/W microemulsion, however, the double layer may extend considerably, depending on the electrolyte concentration. For example, at 10^{-5} mol dm^{-3} 1:1 electrolyte, the double layer is 100 nm thick. Under these conditions, the effective volume of the microemulsion droplets is much larger than the core oil volume. In 10^{-3} mol dm^{-3}, the double layer thickness is still significant (10 nm) and the hard-sphere radius is increased by 5 nm. Thus, this effect of the double layer extension limits the maximum volume fraction that can be achieved with O/W microemulsions. This explains why W/O microemulsions with higher volume fractions are generally easier to prepare than O/W microemulsions. Furthermore, establishing a curvature of the adsorbed layer at a given adsorption is easier with water as the disperse phase since the hydrocarbon chains will have more freedom than if they were inside the droplets. Thus, to prepare O/W microemulsions at high volume fractions, it is preferable to use nonionic surfactants. As mentioned above, to extend the temperature range, a small proportion of an ionic surfactant must be incorporated and some electrolyte should be added to compress the double layer.

14.5.2
Selection of Surfactants for Microemulsion Formulation

The formulation of microemulsions is still an art, since understanding the interactions, at a molecular level, at the oil and water sides of the interface is far from achieved. However, some rules may be applied for the selection of emulsifiers for formulating O/W and W/O microemulsions. These rules are based on the same principles applied to select emulsifiers for macroemulsions described in Chapter 6. Three main methods may be applied for such selection, namely the hydrophilic–lipophilic balance (HLB), the phase inversion temperature (PIT) and the cohesive energy ratio (CER) concepts. As mentioned before, the HLB concept is based on the relative percentage of hydrophilic to lipophilic (hydrophobic) groups in the surfactant molecule. Surfactants with a low HLB number (3–6) normally form W/O emulsions, whereas those with high HLB number (8–18) form O/W emulsions. Given an oil to be microemulsified, the formulator should first determine its required HLB number. Several procedures may be applied to determine the HLB number, depending on the type of surfactant that needs to be used. These procedures are described in Chapter 6. Once the HLB number of the oil is known one must try to find the chemical type of emulsifier that best matches the oil. Hydrophobic portions of surfactants that are similar to the chemical structure of the oil should be looked at first.

The PIT system provides information on the type of oil, phase volume relationships and concentration of the emulsifier. The PIT system is established on the proposition that the HLB number of a surfactant changes with temperature and that the inversion of the emulsion type occurs when the hydrophilic and lipophilic tendencies of the emulsifier just balance. At this temperature no emulsion is produced. From a microemulsion viewpoint the PIT has an outstanding feature since it can throw some light on the chemical type of the emulsifier needed to match a given oil. Indeed, the required HLBs for various oils estimated from the PIT system compare very favourably with those prepared using the HLB system described above. This shows a direct correlation between the HLB number and the PIT of the emulsion.

As discussed in Chapter 6, the CER concept provides a more quantitative method for selecting emulsifiers. The same procedure can also be applied for microemulsions.

14.5.3
Characterisation of Microemulsions

Several physical methods may be applied to characterize microemulsions (described in detail in Chapter 10). Conductivity [111–113], light scattering [114], viscosity [114] and nuclear magnetic resonance (NMR) [115] are probably the most commonly used. Early applications of conductivity measurements were used to determine the nature of the continuous phase. O/W microemulsions are expected to give high conductivity, whereas W/O ones should be poorly conducting. Later conductivity measurements were employed to give more information on the structure of the microemulsion system (see Chapter 10). Light scattering, both static (time average) and dynamic (quasi-elastic or photon correlation spectroscopy) is the most widely used technique for measuring the average droplet size and its distribution [114]. This is also discussed in detail in Chapter 10. Viscosity measurements can be applied to obtain the hydrodynamic radius of microemulsions if the results of viscosity versus volume fraction can be fitted to some models. NMR can be applied to obtain the self-diffusion coefficient of all components in the microemulsion [115] and this could give information on the structure of the system. This is also discussed in detail in Chapter 10.

14.5.4
Role of Microemulsions in Enhancement of Biological Efficacy

The role of microemulsions in enhancing biological efficiency can be described in terms of the interactions at various interfaces and their effect on transfer and performance of the agrochemical. This will be described in detail below, and only a summary is given here. Application of an agrochemical as a spray involves several interfaces, where interaction with the formulation plays a vital role. The first interface during application is that between the spray solution and the atmosphere (air), which governs the droplet spectrum, rate of evaporation, drift, etc. In this respect the rate of adsorption of the surfactant at the air/liquid interface is of vital importance. Since microemulsions contain high concentrations of surfactant and mostly more than one surfactant molecule is used for their formulation, then on diluting a microemulsion on application, the surfactant concentration in the spray solution will be sufficiently high to cause efficient lowering of the surface tension γ. As discussed above, two surfactant molecules are more efficient in lowering γ than either of the two components. Thus, the net effect is to produce small spray droplets that, as we will see later, adhere better to the leaf surface. In addition, the presence of surfactants in sufficient amounts will ensure that the rate of adsorption (which is the situation under dynamic conditions) is fast enough to ensure coverage of the freshly formed spray by surfactant molecules.

The second interaction is between the spray droplets and the leaf surface, whereby the droplets impinging on the surface undergo several processes that determine their adhesion and retention and further spreading on the target surface. The most important parameters that determine these processes are the volume of the droplets and their velocity, the difference between the surface energy of the droplets in flight, E_0, and their surface energy after impact, E_s. As mentioned above, microemulsions that are effective in lowering the surface tension of the spray solution ensure the formation of small droplets, which do not usually undergo reflection if they are able to reach the leaf surface. Clearly, if the droplets are too small, drift may occur. One usually aims at a droplets spectrum in the region of 100–400 μm. As discussed next, the adhesion of droplets is governed by the relative magnitude of the kinetic energy of the droplet in flight and its surface energy as it lands on the leaf surface. Since the kinetic energy is proportional to the third power of the radius (at constant droplet velocity), whereas the surface energy is proportional to the second power, one would expect that sufficiently small droplets will always adhere. For a droplet to adhere, the difference in surface energy between free and attached drop $(E_0 - E_s)$ should exceed the kinetic energy of the drop, otherwise bouncing will occur. Since E_s depends on the contact angle, θ, of the drop on the leaf surface, low θ are clearly required to ensure adhesion, particularly with large drops that have high velocity. Microemulsions when diluted in the spray solution usually give low contact angles of spray drops on leaf surfaces as a result of lowering the surface tension and their interaction with the leaf surface.

Another factor that can affect biological efficacy of foliar spray application of agrochemicals is the extent to which a liquid wets and covers the foliage surface. This, in turn, governs the final distribution of the agrochemical over the areas to be protected. Several indices may be used to describe the wetting of a surface by the spray liquid, of which the spread factor and spreading coefficient are probably the most useful. The spread factor is simply the ratio between the diameter of the area wetted on the leaf, D, and the diameter of the drop, d. This ratio is determined by the contact angle of the drop on the leaf surface. The lower θ is the higher the spread factor. As noted above, microemulsions usually give a low contact angle for the drops produced from the spray. The spreading coefficient is determined by the surface tension of the spray solution as well as θ. Again, with microemulsions diluted in a spray both γ and θ are sufficiently reduced, resulting in a positive spreading coefficient. This ensures rapid spreading of the spray liquid on the leaf surface.

Another important factor in controlling biological efficacy is the formation of "deposits" after evaporation of the spray droplets, which ensure the tenacity of the particles or droplets of the agrochemical. This will prevent removal of the agrochemical from the leaf surface by falling rain. Many microemulsion systems form liquid crystalline structures after evaporation, which have high viscosity (hexagonal or lamellar liquid crystalline phases). These structures will incorporate the agrochemical particles or droplets and ensure their "stickiness" to the leaf surface.

One of the most important roles of microemulsions in enhancing biological efficacy is their effect on penetration of the agrochemical through the leaf. Two complementary effects may be considered. The first is due to enhanced penetration of the chemical as a result of the low surface tension. For penetration to occur through fine pores, a very low surface tension is required to overcome capillary (surface) forces. These forces produce a high pressure gradient that is proportional to the surface tension of the liquid – the lower the surface tension, the lower the pressure gradient and the higher the rate of penetration. The second effect is due to solubilization of the agrochemical within the microemulsion droplet. Solubilization results in an increase in the concentration gradient, thus enhancing the flux due to diffusion. This can be understood from a consideration of Fick's first law,

$$J_D = D\left(\frac{\partial C}{\partial x}\right) \tag{14.49}$$

where J_D is the flux of the solute (amount of solute crossing a unit cross section in unit time), D is the diffusion coefficient and $(\partial C/\partial x)$ is the concentration gradient. The presence of the chemical in a swollen micellar system will lower the diffusion coefficient. However, the solubilising agent (the microemulsion droplet) increases the concentration gradient in direct proportion to the increase in solubility. This is because Fick's law involves the absolute gradient of concentration, which is necessarily small as long as the solubility is small, but not its relative rate. If saturation denoted by S, Fick's law may be written as

$$J_D = D100S\left(\frac{\partial \% S}{\partial x}\right) \tag{14.50}$$

where $(\partial\%S/\partial x)$ is the gradient in relative value of S. Eq. (14.50) shows that, for the same gradient of relative saturation, the flux caused by diffusion is directly proportional to saturation. Hence, solubilization will in general increase transport by diffusion, since it can increase the saturation value by many orders of magnitude (outweighing any reduction in D). In addition, solubilization enhances the rate of dissolution of insoluble compounds, thereby increasing the availability of molecules for diffusion through membranes.

14.6 Role of Surfactants in Biological Enhancement

The discovery and development of effective agrochemicals that can be used with maximum efficiency and minimum risk to the user requires the optimization of their transfer to the target during application. In this way the agrochemical can be used effectively, thus minimising any waste during application. Optimization of the transfer of the agrochemical to the target requires careful analysis of the steps involved during application. Most agrochemicals are applied as liquid sprays [8], particularly for foliar application. The spray volume applied range from high values of the order of 1000 litres per hectare (whereby the agrochemical concentrate is diluted with water) to ultralow volumes of the order of 1 litre per hectare (when the agrochemical formulation is applied without dilution). Various spray application techniques are used, of which spraying using hydraulic nozzles is probably the most common. In this case, the agrochemical is applied as spray droplets with a wide spectrum of droplet sizes (usually in the range 100–400 μm in diameter). On application, parameters such as droplet size spectrum, their impaction and adhesion, sliding and retention, wetting and spreading are of prime importance in ensuring maximum capture by the target surface as well as adequate coverage of the target surface. These factors will be discussed in some detail below. In addition to these "surface chemical" factors, i.e. the interaction with various interfaces, other parameters that affect biological efficacy are deposit formation, penetration and interaction with the site of action. As we will see later, deposit formation, i.e. the residue left after evaporation of the spray droplets, has a direct effect on the efficacy of the pesticide, since such residues act as "reservoirs" of the agrochemical and hence they control the efficacy of the chemical after application. The penetration of the agrochemical and its interaction with the site of action is very important for systemic compounds. Enhancement of penetration is sometimes crucial to avoid removal of the agrochemical by environmental conditions such as rain and/or wind. All these factors are influenced by surfactants and polymers and this will be discussed in detail below. In addition, some adjuvants that are used in combination with the formulation consist of oils and/or surfactant mixtures. The role of these adjuvants in enhancing biological efficacy is far from understood and, in most cases, they are arrived at by a trial and error procedure. Much research is required in this area, which would involve understanding the surface chemical processes both static and dynamic, e.g. static and dynamic surface tension and contact angles, as well as their effect on penetration and uptake of the chemical. In recent years, some progress has been made in the techniques that can be applied to such a complex problem and these should, hopefully, lead to a better understanding of the role of adjuvants. The role of these complex mixtures of oils and/or surfactants in controlling agrochemical efficiency is important from several points of view. In the first place there is greater demand to reduce the application rate of chemicals and to make better use of the present agrochemicals, for example in greater selectivity. In addition, environmental pressure concerning the hazards to the operator and the long-term effects of such residues and wastage demands better understanding of the role of the adjuvants in the application of agrochemicals. This should lead to optimization of the efficacy of the chemical as well as reduction of hazards to the operators, the crops and the environment.

There are, generally, two main approaches to selecting adjuvants: (1) An interfacial (surface) physico-chemical approach, which is designed to increase the dose of the agrochemical received by the target plant or insect, i.e. enhancement of spray deposition, wetting, spreading, adhesion and retention. (2) Uptake activation that is enhanced by addition of surfactant, which is the result of specific interactions between the surfactant, the agrochemical and the target species. These interactions may not be related to the intrinsic surface active properties of the surfactant/adjuvant.

Both approaches must be considered when selecting an adjuvant for a given agrochemical and the type of formulation that is being used. The most important adjuvants are (1) surface active agents and (2) polymers. In some cases these are used in combination with crop oils (e.g. methyl oleate). Several complex recipes may be used and in many cases the exact composition of an adjuvant is unknown.

Adjuvants are applied in two ways: (1) by incorporation in the formulation – mostly the case with flowables (SCs and EWs). (2) In tank mixtures during application. Such adjuvants can be complex mixtures of several surfactants, oils, polymers, etc.

The choice of an adjuvant depends on the (1) nature of the agrochemical, water soluble or insoluble (lipophilic), whereby its solubility and log P values are important. (2) Mode of action of the agrochemical, i.e. systemic or non-systemic, selective or non-selective. (3) Type of formulation that is used, i.e. flowable, EC, grain, granule, capsule, etc.

As mentioned above, most important adjuvants are surface active agents of the anionic, nonionic or zwitterionic type. In some cases polymers are added as stickers or anti-drift agents. As mentioned in Chapters 4 and 5, surfactant molecules accumulate at various interfaces as a result of their dual nature. Basically, a surfactant molecule consists of a hydrophobic chain (usually a hydrogenated or fluorinated alkyl or alkyl aryl chain with 8 to 18 carbon atoms) and a hydrophilic group or chain [ionic or polar nonionic such as poly(ethylene oxide)]. At the air/water interface (as for spray droplets) and the solid/liquid interface (such as the leaf surface), the hydrophobic group points towards the hydrophobic surface (air or leaf), leaving the hydrophilic group in bulk solution. This results in a lowering of the air/liquid surface tension, γ_{LV}, and the solid/liquid interfacial tension, γ_{SL}. As the surfactant concentration is gradually increased both γ_{LV} and γ_{SL} decrease until the critical micelle concentration (c.m.c.) is reached, after which both values remain virtually constant. This situation represents the conditions under equilibrium whereby the rate of adsorption and desorption are the same. The situation under dynamic conditions, such as during spraying, may be more complicated since the rate of adsorption is not equal to the rate of formation of droplets. Above the c.m.c., micelles are produced, which at low C are essentially spherical (with an aggregation number in the region of 50–100 monomers). Depending on the conditions (e.g. temperature, salt concentration, structure of the surfactant molecules) other shapes may be produced, e.g. rod-shaped and lamellar micelles. Since micelles play a vital role when considering adjuvants, their properties must be understood in some detail. As mentioned in Chapter 2, micelle formation is a dynamic process, i.e. a dynamic equilibrium is set up whereby surface active agent molecules are constantly leaving the micelles while others enter the micelles (the same applies to the counter-ions). The dynamic process of micellization is described by two relaxation processes: (1) A short relaxation time τ_1 (of the order of 10^{-8}–10^{-3} s), which is the lifetime for a surfactant molecule in a micelle. (2) A longer relaxation time τ_2 (of the order of 10^{-3}–1 s), which is a measure of the micellization-dissolution process. Both τ_1 and τ_2 depend on the surfactant structure, its chain length, and these relaxation times determine some of the important factors in selecting adjuvants, such as the dynamic surface tension (discussed below).

The c.m.c. of nonionic surfactants is usually two orders of magnitude lower than the corresponding anionic of the same alkyl chain length. This explains why nonionics are generally preferred when selecting adjuvants. For a given series of nonionics, with the same alkyl chain length, the c.m.c. decreases with decreasing number of ethylene oxide (EO) units in the chain. Under equilibrium, the $\gamma - \log C$ curves shift to lower values as the EO chain length decreases. However, under dynamic conditions, the situation may be reversed, i.e. the dynamic surface tensions could become lower for the surfactant with the longer EO chain. This trend is understandable if one considers the dynamics of micelle formation. The surfactant with the longer EO chain has a higher c.m.c. and it forms smaller micelles than the surfactant containing shorter EO chain. This means that the life time of a micelle with longer EO chain is shorter than that with a longer EO chain. This explains why the dynamic surface tension of a solution of a surfactant containing a longer EO chain can be lower than that of a solution of an analogous surfactant (at the same concentration) with a shorter EO chain.

For a series of anionic surfactants with the same ionic head group, the life time of a micelle decreases with decreasing alkyl chain length of the hydrophobic component. Branching of the alkyl chain could also play a notable role in the life time of a micelle. Consequently, dynamic surface tension measurements need to be performed when selecting a surfactant as an adjuvant as this may have an important influence on spray retention.

However, the above measurements should not be taken in isolation as other factors may also play an important role, e.g. solubilization, which may require larger micelles. Selecting a surfactant as an adjuvant requires knowledge of the factors involved, which will be discussed in some detail below.

At high surfactant concentrations (usually above 10%) several liquid crystalline phases are produced (see Chapters 2 and 3). Three main types of liquid crystals may be distinguished: (1) Hexagonal (middle) phase that consists of cylindrical anisotropic units with high viscosity. (2) Cubic, body-centered isotropic phase with a viscosity that is higher than the hexagonal phase. (3) Lamellar (neat) phase consisting of sheet-like units that are anisotropic, but with a viscosity that is lower than the hexagonal phase.

The above phases may form during evaporation of a spray drop. In some cases a middle phase is first produced that on further evaporation may afford a cubic phase, which, due to its very high viscosity, may entrap the agrochemical. This could be advantageous for the systemic fungicides that require "deposits" to act as reservoirs for the chemical. Viscous cubic phases may also enhance the tenacity of the agrochemical particles (particularly with SCs) and, hence, enhance rain fastness. In some other applications, a lamellar phase is preferred as this provides some mobility (due to its lower viscosity).

The various phases produced by a surfactant can be related to its structure. An important parameter that can be used to predict the phase behaviour of surfactants is the critical packing parameter (CPP) described above. ($CPP = v/la$, where v is the volume of the hydrocarbon chain with a length l and a is the cross sectional area of the hydrophilic head group.) For spherical micelles, $CPP < \frac{1}{3}$, for cylindrical micelles $1 > CPP > \frac{1}{2}$, and for lamellar micelles $CPP \sim 1$.

Study of the phase behaviour of surfactants (which can be performed using polarizing microscopy) is crucial in the selection of adjuvants. Interaction of the above units with the agrochemical is crucial in determining performance (e.g. solubilization). Similar interactions may also occur between the above structural units and the leaf surface (wax solubilization).

Application of an agrochemical, as a spray, involves several interfaces, where the interaction with the formulation plays a vital role. The first interface during application is that between the spray solution and the atmosphere (air), which governs the droplet spectrum, rate of evaporation, drift, etc. In this respect, the rate of adsorption of the surfactant and/or polymer at the air/liquid interface is of prime importance. This requires dynamic measurements of parameters such as surface tension (see Chapter 11), which will give information on the rate of adsorption. This subject is dealt with in the first part of this section. The second interface is that between the impinging droplets and the leaf surface (with insecticides the interaction with the insect surface may be important). Droplets impinging on the surface undergo several processes that determine their adhesion and retention and further spreading on the target surface. The rate of evaporation of the droplet and the concentration gradient of the surfactant across the droplet governs the nature of the deposit formed. These processes of impaction, adhesion, retention, wetting and spreading will be discussed in subsequent parts in this section. Interaction with the leaf surface will be described in terms of the various surface forces involved.

14.6.1 Interactions at the Air/Solution Interface and their Effect on Droplet Formation

In a spraying process, a liquid is forced through an orifice (the spray nozzle) to form droplets by application of a hydrostatic pressure. Before describing what happens during spraying, it is beneficial to consider the processes that occur when a drop is formed at various time intervals. If the time to form a drop is large (greater say than 1 minute), the volume of the drop depends on the properties of the liquid, such as its surface tension and the dimensions of the orifice, but is independent of the time of its formation. However, at shorter times of drop formation (less than 1 minute), the drop volume depends on the time of its formation. The loosening of the drop that occurs when its weight W exceeds the surface force $2\pi r\gamma$ (i.e. $W > 2\pi r\gamma$) progresses at a speed determined by the viscosity and surface tension of the liquid. However, during this loosening, the hydrostatic pressure pumps more liquid into the drop and this is represented by a "hump" in the W–t curve. The height of the "hump" increases with increasing viscosity, perhaps because the rate of contraction diminishes as the viscosity rises.

At short t, W becomes smaller since the liquid in the drop has considerable kinetic energy even before the drop breaks loose. The liquid coming into the drop imparts downward acceleration and this may cause separation before the drop has reached the value given by

$$W = 2\pi r\gamma f\left(\frac{g\rho r^2}{2\gamma}\right) \tag{14.51}$$

where ρ is the density and r is the radius of the orifice. Equation (14.51) is the familiar equation for calculating the surface or interfacial tension from the drop weight or volume (see Chapter 6).

When the hydrostatic pressure is raised further, i.e. when at even shorter t than those described above, no separate drops are formed and a continuous jet issues from the orifice. At even higher hydrostatic pressure, the jet breaks into droplets, the phenomenon usually referred to as spraying. The break-up of jets (or liquid sheets) into droplets is the result of surface forces. The surface area and, consequently, the surface free energy (area × surface tension) of a sphere is smaller than that of a less symmetrical body. Hence, small liquid volumes of other shapes tend to give rise to smaller spheres. For example, a liquid cylinder becomes unstable and divides into two smaller droplets as soon as the length of the liquid cylinder is greater than its circumference. This occurs on accidental contraction of the long liquid cylinder. A prolate spheroid tends to give two spherical drops when the length of the spheroid is greater than 3–9 times its width. A very long cylinder with radius r (as for example a jet) tends to divide into drops with a volume equal to $(9/2)\pi r^3$. Since the surface area of two unequal drops is smaller than that of two equal drops with the same total volume, the formation of a polydisperse spray is more probable.

The effect of surfactants and/or polymers on the droplet size spectrum of a spray can be, to a first approximation, described in terms of their effect on the surface tension. Since surfactants lower the surface tension of water, one would expect that their presence in the spray solution would result in the formation of smaller droplets. This is similar to the process of emulsification described in Chapter 6. Owing to the low surface tension in the presence of surfactants, the total surface energy of the droplets produced on atomization is lower than that in the absence of surfactants. This implies that less mechanical energy is required to form the droplets when a surfactant is present. This leads to smaller droplets at the same energy input. However, the actual situation is not simple since one is dealing with a dynamic situation. In a spraying process a fresh liquid surface is continuously formed. The surface tension of that liquid depends on the relative ratio between the time taken to form the interface and the rate of adsorption of the surfactant from bulk solution to the air/liquid interface. The rate of adsorption of a surfactant molecule depends on its diffusion coefficient and its concentration (see below). Clearly, if a fresh interface is formed at a much faster rate than the rate of adsorption of the surfactant, the surface tension of the spray liquid will not be far from that of pure water. Alternatively, if the rate of formation of the fresh surface is much slower than the rate of adsorption, the surface tension of the spray liquid will be close to that of the equilibrium value of the surface tension. The actual situation is somewhere in between and the rate of formation of a fresh surface is comparable to that of the rate of surfactant adsorption. In this case, the surface tension of the spray liquid will lie between that of a clean surface (pure water) and the equilibrium value of the surface tension that is reached at times larger than that required to produce the jet and the droplets. This shows the importance of measuring dynamic surface tension and the rate of surfactant adsorption.

The rate of surfactant adsorption may be described by application of Fick's first law. When concentration gradients are set up in the system, or when the system is stirred, diffusion to the interface may be expressed in terms of Fick's first law, i.e.

$$\frac{\mathrm{d}\Gamma}{\mathrm{d}t} = \frac{D}{\delta}\frac{N_\mathrm{A}}{100}C(1-\theta) \tag{14.52}$$

where Γ is the surface excess (number of moles of surfactant adsorbed per unit area), t is the time, D is the diffusion coefficient of the surfactant molecule, δ is the thickness of the diffusion layer, N_A is Avogadro's constant and θ is the fraction of the surface already covered by adsorbed molecules. Equation (14.52) shows that the rate of surfactant diffusion increases with increase of D and C. The diffusion coefficient of a surfactant molecule is inversely proportional to its molecular weight. This implies that shorter chain surfactant molecules are more effective in reducing the dynamic surface tension. However, the limiting surface tension reached by a surfactant molecule decreases with increase of its chain length and hence a compromise is usually made when selecting a surfactant molecule. Usually, one chooses a surfactant with a chain length of the order of 12 carbon atoms. In addition, the higher the surfactant chain length, the lower its c.m.c. (see Chapter 2) and, hence, lower concentrations are required when using a longer chain surfactant molecule. Again, a problem with longer chain surfactants is their high Krafft temperatures (becoming only soluble at temperatures higher than ambient). Thus, an optimum chain length is usually necessary to optimise the spray droplet spectrum.

As mentioned above, the faster the rate of adsorption of surfactant molecules, the greater the effect of reducing the droplet size. However, with liquid jets there is an important factor that may enhance surfactant adsorption. Addition of surfactants reduce the surface velocity (which is generally lower than the mean velocity of flow of the jet) below that obtained with pure water. This results from surface tension gradients, which can be explained as follows. Where the velocity profile is relaxing, the surface is expanding, i.e. it is newly formed, and might even approach the composition and surface tension of pure water. A little further downstream, appreciable adsorption of the surfactant will have occurred, giving rise to a back spreading tendency from this part of the surface in the direction back towards the cleaner surface immediately adjacent to the nozzle. Thus, this phenomenon is a form of the Marangoni effect (see Chapter 6), which reduces the surface velocity near the nozzle and induces some liquid circulation, which accelerates the adsorption of the surfactant molecules by as much as ten times. This effect casts doubt on the use of liquid jets to obtain the rate of adsorption. Indeed, under conditions of jet formation, it is likely that the surface tension approaches its equilibrium value very closely. Thus, one should be careful in using dynamic surface tension values, as for example measured using the maximum bubble pressure method (see Chapter 11).

The influence of polymeric surfactants on the droplet size spectrum of spray liquids is relatively more complicated since adsorbed polymers at the air/liquid interface produce other effects than simply reducing the surface tension. In addition, polymeric surfactants diffuse very slowly to the interface and it is doubtful if they have an appreciable effect on the dynamic surface tension. In most agrochemical formulations, polymers are used in combination with surfactants and this makes the situation more complicated. Depending on the ratio of polymer to surfactant in the formulation, various effects may be envisaged. If the concentration of the polymer is appreciably greater than the surfactant and interaction between the two components is strong, the resulting "complex" will behave more like a polymer. Conversely, if the surfactant concentration is appreciably higher than that of the polymer and interaction between the two molecules is still strong, one may end up with polymer–surfactant "complexes" as well as free surfactant molecules. The latter will behave as free molecules and the reduction in the surface tension may be sufficient even under dynamic conditions. However, the role of the polymer–surfactant "complex" could be similar to that of the free polymer molecules. The latter produce a viscoelastic film at the air/water interface, which may modify the droplet spectrum and the adhesion of the droplets to the leaf surface. The situation is far from understood and fundamental studies are required to evaluate the role of polymer in spray formation and droplet adhesion.

The above discussion is related to the case where a polymeric surfactant is used for the formulation of agrochemicals as discussed in Chapters 6 and 7 on emulsions and suspension concentrates. However, in many agrochemical applications high molecular weight materials such as polyacrylamide, poly(ethylene oxide) or guar gum are sometimes added to the spray solution to reduce drift. Incorporation of high molecular weight polymers favours the formation of larger drops. The effect can be reached at very low polymer concentrations when the molecular weight of the polymer is fairly high ($> 10^6$). The most likely explanation of how polymers affect the droplet size spectrum is in terms of their viscoelastic behaviour in solution. High molecular weight polymers adopt spatial conformations in bulk solution, depending on their structure and molecular weight. Many flexible polymer molecules adopt a random coil configuration that is characterized by a root mean square radius of gyration, R_G. The latter depends on the molecular weight and the interaction with the solvent. If the polymer is in good solvent conditions, e.g. poly(ethylene oxide) in water, the polymer coil becomes expanded and R_G can reach high values, of the order of several tens of nms. At relatively low polymer concentrations, the polymer coils are separated and the viscosity of the polymer solution increases gradually with increase of its concentration. However, at a critical polymer concentration, denoted C^*, the polymer coils begin to overlap and the solutions show a rapid increase in the viscosity with further increase above C^*. This concentration C^* is defined as the onset of the semi-dilute region. C^* decreases with increasing molecular weight of the polymer and at very high molecular weights it can be as low as 0.01%. Under this condition of polymer coil overlap, the spray jet opposes deformation, resulting in the production of larger drops. This phenomenon is applied successfully to reduce drift. Some polymers also produce conformations that approach a rod-like or double helix structure. An example is xanthan gum, which is used with many emulsions and suspension concentrates to reduce sedimentation. If the concentration of such a polymer is appreciable in the formulation, then even after extensive dilution on spraying (usually by 100- to 200-fold) the concentration of the polymer in the spray solution may be sufficient to cause production of larger drops. This effect may be beneficial if drift is a problem. However, it may be undesirable if relatively small droplets are required for adequate adhesion and coverage. Again, the ultimate effect required depends on the application methods and the mode of action of the agrochemical. Fundamental studies of the various effects are required to arrive at the optimum conditions. The effect of the various surfactants and polymers should be studied in spray application during the formulation of the agrochemical. In most cases, the formulation chemist concentrates on producing the best system that produces long-term physical stability (shelf life). It is crucial to investigate the effect of the various formulation variables on the droplet spectrum, their adhesion, retention and spreading. Simultaneous investigations should be made on the effect of the various surfactants on the penetration and uptake of the agrochemical.

One of the problems with many anti-drift agents is their shear degradation. At the high shear rates involved in spray nozzles (which may reach several thousand s^{-1}) the polymer chain may degrade into smaller units, resulting in a considerable reduction of the viscosity. This will reduce the anti-drift effect. It is, therefore, essential to choose polymers that are stable to the high shear rates involved in a spraying process.

14.6.2 Spray Impaction and Adhesion

When a drop of a liquid impinges on a solid surface, e.g. a leaf, one of several states may arise, depending on the conditions. The drop may bounce or undergo fragmentation into two or more droplets that, in turn, may bounce back and return to the surface with a lower kinetic energy. Alternatively, the drop may adhere to the leaf surface after passing through several stages, where it flattens, retracts, spreads and finally rests to form a hemispherical cap. In some cases, the droplet may not adhere initially but floats as an individual drop for a fraction of a second or even several seconds and can either adhere to the surface or leave it again.

The most important parameters that determine which of the above stages is reached are the mass (volume) of the droplet, its velocity in flight and the distance between the spray nozzle and the target surface, the difference between the surface energy of the droplet in flight, E_0, and its surface energy after impact, E_s and displacement of air between the droplet and the leaf.

Droplets in the region of 20–50 μm in diameter do not usually undergo reflection if they are able to reach the leaf surface. Such droplets have a low momentum and can only reach the surface if they travel in the direction of the air stream. Conversely, large droplets of the order of few thousand micrometres in diameter undergo fragmentation. Droplets in the range 100–400 μm, which covers the range produced by most spray nozzles, may be reflected or retained depending on several parameters such as the surface tension of the spray solution, surface roughness and elasticity of the drop surface. A study by Brunskill [116] revealed, with drops of 250 μm, 100% adhesion when the surface tension of the liquid was lowered (using methanol) to 39 Nm^{-1}, whereas only 4% adhesion occurred when the surface tension, γ, was 57 Nm^{-1}. For any given spray solution (with a given surface tension), a critical droplet diameter exists, below which adhesion is high and above which adhesion is low. The critical droplet diameter increases as the surface tension of the spray solution decreases. The viscosity of the spray solution has only a small effect on the adhesion of large drops, but with small droplets adhesion increases with increasing viscosity. As expected, the percentage of adhered droplets decreases as the angle of incidence of the target surface increases.

Hartley and Brunskill [117] have formulated a simple theory for bouncing and droplet adhesion, considering an ideal case where there are no adhesion (short range) forces between the liquid and solid substrate and the liquid has zero viscosity. During impaction, the initially spherical droplet will flatten into an oblately spheroid until the increased area has stored the kinetic energy as increased surface energy. This is often followed by an elastic recoil towards the spherical form and later beyond it, with the long axis normal to the surface. During this process, energy will be transformed into upward kinetic energy and the drop may leave the surface in a state of oscillation between the spheroidal forms. This sequence was confirmed using high-speed flash illumination.

When the reflected droplet leaves in an elastically deformed condition, the coefficient of restitution must be less than unity since part of the translational energy is transformed into vibrational energy. Moreover, the distortion of droplets involves loss of energy as heat by operation of viscous forces. The effect of increasing the viscosity of the liquid is rather complex, but at a very high viscosity liquids usually have a form of elasticity operating during deformations of very short duration. Reduction of deformation as a result of an increase in viscosity will affect adhesion.

As mentioned above, the adhesion of droplets is governed by the relative magnitude of the kinetic energy of the droplet in flight and its surface energy as it lands on the leaf surface. Since the kinetic energy is proportional to the third power of the radius (at constant droplet velocity) whereas the surface energy is proportional to the second power of the radius, one would expect that sufficiently small droplets will always adhere. However, this is not always the case since smaller droplets fall with smaller velocities. Indeed, the kinetic energy of sufficiently small drops, in the Stokesian range, falling at their terminal velocity, is proportional to the seventh power of the radius. In the 100–400 μm range, it is nearly proportional to the fourth power of the radius.

Consider a droplet of radius r (sufficiently small for gravity to be neglected) falling onto a solid surface and spreading with an advancing contact angle, θ_A, and having a spherical upper surface of radius R. The surface energy of the droplet in flight, E_0, is given by

$$E_0 = 4\pi r^2 \gamma_{LA} \tag{14.53}$$

where γ_{LA} is the liquid/air surface tension.

The surface energy of the spread drop is given by Eq. (14.54)

$$E_s = A_1\gamma_{LA} + A_2\gamma_{SL} - A_2\gamma_{SA} \tag{14.54}$$

where A_1 is the area of the spherical air/liquid interface, A_2 is that of the plane circle of contact with the solid surface, γ_{SL} is the solid/liquid interfacial tension and γ_{SA} that of the solid/air interface.

From Young's equation,

$$\gamma_{SA} = \gamma_{SL} + \gamma_{LA}\cos\theta \tag{14.55}$$

Therefore, the surface energy of the droplet spreading on the leaf surface is given by

$$E_s = \gamma_{LA}(A_1 - A_2\cos\theta) \tag{14.56}$$

The volume of a free drop is $(4/3)\pi r^3$, whereas that of the spread drop is $\pi R^3[(1 - \cos\theta) + (1/3)(\cos^3\theta)]$ so that

$$\tfrac{4}{3}\pi r^3 = \pi R^3\left[(1 - \cos\theta) + \tfrac{1}{3}(\cos^3\theta - 1)\right] \tag{14.57}$$

and

$$A_1 = 2\pi R^2(1 - \cos\theta) \tag{14.58}$$

$$A_2 = \pi R^2 \sin^2\theta \tag{14.59}$$

Combining Eqs. (14.56) to (14.59) one can obtain the minimum energy barrier between attached and free drop,

$$\frac{E_0 - E_s}{E_0} = 1 - 0.39[2(1 - \cos\theta) - \sin^2\theta\cos\theta][1 - \cos\theta + (1/3)(\cos^3\theta - 1]^{-2/3} \tag{14.60}$$

A plot of $(E_0 - E_s)/E_0$ shows that this ratio decreases rapidly from its value of unity when $\theta = 0$ to a near zero value when $\theta > 160°$. This plot can be used to calculate the critical contact angle required for adhesion of water droplets, with a surface tension $\gamma = 72$ mN m^{-1} at 20 °C, of various sizes and velocities. As an illustration, consider a water droplet 100 μm in diameter falling with its terminal velocity v of ~0.25 m s^{-1}. The kinetic energy of the drop is 1.636×10^{-9} J, whereas its surface energy in flight is 2.26×10^{-9} J. The surface energy of the attached drop at which the kinetic energy is just balanced is 2.244×10^{-9} J. The contact angle at which this occurs can be obtained by calculating the fraction $(E_0 - E_s)/E_0$ and interpolation using the above-mentioned plot. This gives $(E_0 - E_s)/E_0 = 0.00723$ and $\theta \sim 160°$. Thus, providing droplets of this size form an angle that is less than 160°, they will stick to the leaf surface. Unsurprisingly, droplets of this size do not need any surfactant for adhesion. For a 200 μm droplet, with a velocity of 1 m s^{-1}, the critical contact angle is 87°, showing that, in this case, surfactants are required for adhesion. The higher the velocity of the drop, the lower the critical contact angle required for adhesion. With larger drops, this critical contact angle becomes smaller and smaller and this clearly shows the importance of surfactants for ensuring drop adhesion.

Notably, however, the above calculations are based on "idealized" conditions, i.e. droplets falling on a smooth surface. Deviation is expected when dealing with practical surfaces such as leaf surfaces. The latter are rough, containing leaf hairs and wax crystals that are distributed in different ways, depending on the nature of the leaf and climatic conditions. Under such conditions, the adhesion of droplets may occur at critical contact angle values that are either smaller or larger than those predicted from the above calculations. The critical θs will certainly be determined by the topography on the leaf surface. As we will see later, the definition and measurement of the contact angle on a rough surface are not straightforward. Despite these complications, experimental results on droplet adhesion [8] seem to support the predictions from the above simple theory. These experimental results showed little dependence of adhesion of spray droplets on surfactant concentration. Since, with most spray systems, the contact angles obtained were lower than the critical value for adhesion (except for droplets > 400 μm), then in most circumstances surfactant addition had only a marginal effect on droplet adhesion. However, one should not forget that the surfactant in the spray solution determines the droplet size spectrum. Also, addition of surfactants will certainly affect the adhesion of droplets moving at high velocities and on various plant species. The situation is further complicated by the dynamics of the process, which depend on the nature and concentration of the surfactant added. For fundamental investigations, measurements of the dynamic surface tension and contact angle are required both on model and practical surfaces. These measurements are now easy to perform due to advances in instrumentation such as the maximum bubble pressure method for measuring dynamic surface tension and high-speed video equipment for measuring the dynamic contact angle. Such techniques will enable the formulation chemist and the biologist to understand the role and the function of the surfactant in spray solutions.

14.6.3
Droplet Sliding and Spray Retention

Many agrochemical applications involve high volume sprays, whereby with continuous spraying the volume of the drops continue to grow in size by impaction of more spray droplets upon them and by coalescence with neighbouring drops on the surface. During this process, the amount of spray retained steadily increases, provided the liquid drops, which are impacted, are also retained. However, on further spraying the drops continue to grow until they reach a critical size, above which they begin to slide down the surface and "drop off" – the so-called "run-off" condition. At the point of "incipient run-off" the volume of the spray retained is a maximum. Retention at this point is governed by the movement of the liquid drops on the solid surface. Bikerman [118] stated that the percentage of droplets sticking to a plant after having touched it should depend upon the tilt of the leaf, the size of the droplets and the contact angle at the plant leaf/droplet/air interface. However, such a process is complicated and governed by many other factors [119] such as droplet spectrum, velocity of impacting droplets, volatility and viscosity of the spray liquid and ambient conditions.

Several authors [119] have tried to relate the resistance to movement of liquid drops on a tilted surface to the surface tension and the contact angles (advancing and receding) of liquid droplets with the solid surface. The detailed analysis given by Furmidge [119] is summarized below.

Consider a droplet of mass m on a plane surface that is inclined at an angle α from the horizontal (Figure 14.17).

Owing to gravity the droplet will start to slide down with a slow constant velocity. Assuming the droplet will have a rectangular plan view (Figure 14.17), with width w and it has moved a distance $\mathrm{d}l$, then the work done by the droplet in moving such a distance, W_g, is given by

$$W_g = mg \sin \alpha \qquad (14.61)$$

The above force is opposed by the surface force resulting from wetting and dewetting of the leaf surface as the droplet slides downwards. In moving down, an area $w\,\mathrm{d}l$ of the leaf is wetted by the droplet and a similar area is dewetted by the trailing edge. The work of wetting per unit area of the surface is equal to $\gamma_{LA}(\cos \theta_A + 1)$, whereas that of dewetting is given by $\gamma_{LA}(\cos \theta_R + 1)$, where θ_A and θ_R are the advancing and receding contact angles, respectively. Thus the surface force, W_s, is given by

$$W_s = \gamma_{LA} w\,\mathrm{d}l(\cos \theta_R - \cos \theta_A) \qquad (14.62)$$

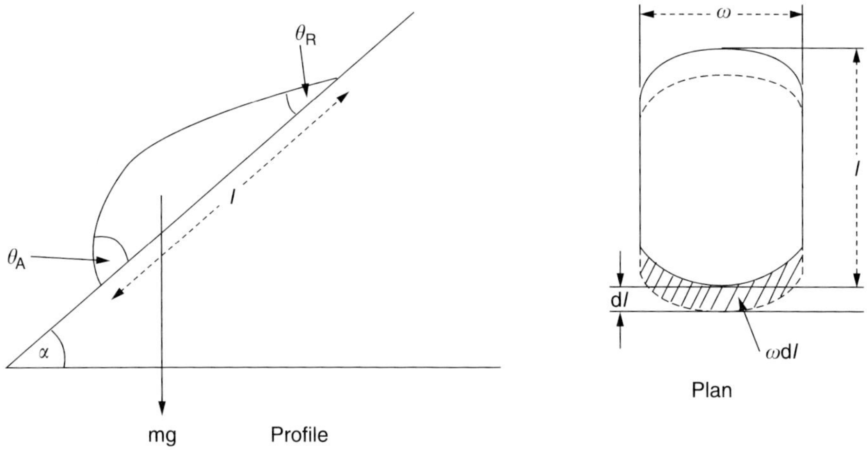

Fig. 14.17. Profile and plan view of a drop during sliding.

At equilibrium, $W_g = W_s$ and

$$\frac{mg \sin \alpha}{w} = \gamma_{LA}(\cos \theta_R - \cos \theta_A) \tag{14.63}$$

If the impaction of the spray is uniform and the spray droplets are reasonably homogeneous in size, the total volume of spray retained in an area L^2 of surface is proportional to the time of spraying until the time when the first droplet runs off the surface. Also, the volume v of spray retained per unit area, R, at the moment of "incipient run-off" is given by

$$R = \frac{kv}{w^2} \tag{14.64}$$

where k is a constant. Eq. (14.63) gives the critical relationship of m/w for the movement of liquid droplets on a solid surface. As the surface is sprayed the adhering drops grow in size until the critical value of m/w is reached, and during this period they remain more or less circular in plan form. Since the droplets are small, the deforming effect of gravity may be ignored and the droplets can be regarded as spherical caps whose volume v is given by

$$v = \frac{\pi(1 - \cos \theta)^2 (2 + \cos \theta) w^3}{24 \sin^3 \theta} = \frac{m}{\rho} \tag{14.65}$$

Combining Eqs. (14.64) and (14.65), it is possible to obtain an expression for w, the diameter of adhering droplet in terms of the surface forces given above, i.e. γ_{LA} and θ,

$$w = \left[\frac{24 \sin^3 \theta_A \gamma_{LA}(\cos \theta_R - \cos \theta_A)}{\pi\rho(1 - \cos \theta_A)^2(2 + \cos \theta) g \sin \alpha}\right]^{1/2} \tag{14.66}$$

Combining Eqs. (14.64)–(14.66), one obtains an expression of spray retention, R, in terms of γ_{LA} and θ, i.e.

$$R = k\left[\frac{\pi\gamma_{LA}(\cos \theta_R - \cos \theta_A)}{24\rho g \sin \alpha}\right]^{1/2}\left[\frac{(1 - \cos \theta_A)^2(2 + \cos \theta_A)}{\sin^3 \theta_A}\right]^{1/2} \tag{14.67}$$

The value of k depends on the droplet spectrum, since it relates to the rate of build-up of critical droplets and their distribution. However, Eq. (14.67) does not take into account the flattening effect of the droplet on impact, which results in reduction of θ and increase of w above the value predicted by Eq. (14.66). Thus, Eq. (14.67) is only likely to be valid under conditions of small impaction velocity. In this case, retention is governed by the surface tension of the spray liquid, the difference between θ_A and θ_R (i.e. the contact angle hysteresis) and the value of θ_A.

Equation (14.67) can be further simplified by removing the constant terms and standardizing $\sin \alpha$ as equal to 1. A further simplification is to replace the second term between square brackets on the right-hand side of Eq. (14.67) by θ_M, the arithmetic mean of θ_A and θ_R. In this way a retention factor, F, may be defined by the simple expression

$$F = \theta_M\left[\frac{\gamma_{LA}(\cos \theta_R - \cos \theta_A)}{\rho}\right]^{1/2} \tag{14.68}$$

Equation (14.68) shows that F depends on γ_{LA} and the difference between θ_R, θ_A and θ_M. At any given θ_A and γ_{LA}, F increases rapidly with increasing $(\theta_A - \theta_R)$, reaches a maximum, and then decreases. At any given $(\theta_A - \theta_R)$ and γ_{LA}, F increases rapidly with increasing θ_A (and also θ_M). With systems having the same contact angles, F increases with increasing γ_{LA} but the effect is not very large since $F \propto \gamma_{LA}^{1/2}$. Obviously, any variation in γ_{LA} is accompanied by a change in contact angles and hence one cannot investigate these parameters in isolation. In general, by increasing the surfactant concentration, γ_{LA}, θ_A and θ_R are reduced. The relative extent to which these three values are affected depends on surfactant nature, its concentration and the surface properties of the leaf. This is a very complex problem and predictions are almost impossible.

Notably, the above treatment does not take into account the effect of surface roughness and presence of hairs, which play a significant role. A difference in the amount of liquid retained of up to an order of magnitude may be encountered, at constant F, between, say, a hairy and a smooth leaf. Besides these large variations in surface properties between leaves of various species, there are also variations within the same species, depending on age, environmental conditions and position. However, contact angle measurements on leaf surfaces are not easy and one has to make several measurements and subject the results to statistical analysis. Thus, at best, the measured *F*s can be used as a guide to compare various surface active agents on leaf surfaces of a particular species that are grown under standard conditions.

Several other factors affect retention, of which droplet size spectrum, droplet velocities and wind speed are probably the most important. Usually, retention increases with reduction of droplet size, but is reduced significantly at high droplet velocities and wind speeds. The impact velocity effect becomes more marked as the receding contact angle decreases. Wind reduces the volume of spray that can be retained, particularly when θ_A and θ_R are fairly large, because little force is required to move the drop along the surface. As θ_A and θ_R become small, the wind effect becomes less significant and it becomes negligible when θ_A and θ_R are close to zero. The leaf structure is also important, since less spray is lost due to wind movements from leaves with a very rough surface when compared with smooth leaves. Thus, care should be taken when results are obtained on plants grown under standard conditions, such as glass houses. These results should not be extrapolated to field conditions, since plants grown under normal environmental conditions may have surfaces that are vastly different from those grown in glass houses. To obtain a realistic picture of spray retention, measurements should be made on field grown plants and the results obtained may be correlated to those obtained on glass house plants. In this case, it is possible to use glass house plants in selecting surfactants, if an allowance is made for the difference between the two sets of results.

14.6.4 Wetting and Spreading

Another factor that can affect the biological efficacy of foliar spray application of agrochemicals is the extent to which the liquid wets, spreads and covers the foliage surface. This, in turn, governs the final distribution of the agrochemical over the area to be protected [8]. The optimum degree of coverage in any spray application depends on the mode of action of the agrochemical and the nature of the pest to be controlled. With non-systemic agrochemicals, the cover required depends on the mobility or location of the pest. The more static the pest, the greater the need for complete coverage on those areas of the plant liable to attack. Under those conditions, good spreading of the liquid spray with maximum coverage is required. Conversely, with systemic agrochemicals, satisfactory cover is ensured provided the spray liquid is brought into contact with those areas of the plant through which the agrochemical is absorbed. Since, as we will see later, high penetration requires high concentration gradients, an optimum situation may be required here, whereby one achieves adequate coverage of those areas where penetration occurs, without too much spreading over the total leaf surface since this usually results in "thin" deposits. These "thin" deposits do not give adequate "reservoirs", which are sometimes essential to maintain a high concentration gradient, thus enhancing penetration. In addition, thick deposits produced from droplets with limited spreading can increase the tenacity of the agrochemical and so ensure the longer term protection. This situation may be required with many systemic fungicides.

Many leaf surfaces represent the most unwettable of most known surfaces. This is due to the predominantly hydrophobic nature of the leaf surface, which is usually covered with crystalline wax of straight chain paraffinic alcohols (24–35 carbon atoms). The crystals may be less than 1 μm thick and only few μms apart, giving the surface "microroughness" – the "real" area of the surface can be several times the "gross" (apparent) area. When a water drop is placed on a leaf surface, it takes the form of a spherical cap that is characterized by the contact angle θ. From the balance of tensions, one obtains the familiar Young's equation, which applies to a liquid drop on a smooth surface,

$$\gamma_{SA} = \gamma_{SL} + \gamma_{LA} \cos \theta \tag{14.69}$$

Wetting is sometimes simply assessed by the value of θ; the smaller the angle the better the liquid is said to wet the solid. Complete wetting implies a contact angle of zero, whereas complete non-wetting dictates an angle of contact of 180°. However, contact angle measurements are not easy on real surfaces since a great variation in the value is obtained at various locations of the surface. In addition, it is very difficult to obtain an equilibrium value. This is due to the heterogeneity of the surface and its roughness. Thus, in most practical systems, such as spray drops on leaf surfaces, the contact angle exhibit hysteresis, i.e. its value depends on the history of the system and varies according to whether the given liquid is tending to advance across or recede from the leaf surface. Limiting angles achieved just prior to movement of the wetting line (or just after movement ceases) are known as the advancing and receding contact angles, θ_A and θ_R, respectively.

For a given system, $\theta_A > \theta_R$ and θ can usually take any value between these two limits without discernable movement of the wetting line. Since smaller angles imply better wetting, it is clear that the contact angle always changes in such a direction as to oppose wetting line movement.

The use of contact angle measurements to assess wetting depends upon equilibrium thermodynamic arguments, which, unfortunately, is not the real situation. In the practical situation of spraying, the liquid has to displace the air or another fluid attached to the leaf surface and hence measurement of dynamic contact angles, i.e. those associated with moving wetting lines is more appropriate. Such measurements require special equipment such as video cameras and image analysis that should enable one to obtain a more accurate assessment of wetting by the spray liquid.

As mentioned above, the contact angle often undergoes hysteresis so that θ cannot be defined unambiguously by experiment. This hysteresis is accounted for by surface roughness, surface heterogeneity and metastable configurations. Surface roughness can be taken into account by introducing into the Young's equation a term r, which is the ratio of real to apparent surface area, i.e.

$$r\gamma_{SA} = r\gamma_{SL} + \gamma_{LA} \cos\theta \tag{14.70}$$

Thus, the contact angle on a rough surface is given by

$$\cos\theta = \frac{r(\gamma_{SA} - \gamma_{SL})}{\gamma_{LA}} \tag{14.71}$$

In other words, the contact angle on a rough surface, θ, is related to that on a smooth surface, θ°, by

$$\cos\theta = r\cos\theta^\circ \tag{14.72}$$

Equation (14.72) shows that surface roughness increases the magnitude of $\cos\theta^\circ$, whether its value is positive or negative. If $\theta^\circ < 90^\circ$, $\cos\theta^\circ$ is positive and it becomes more positive as a result of roughness, i.e. $\theta < \theta^\circ$ or roughness in this case enhances wetting. In contrast, if $\theta^\circ > 90^\circ$, $\cos\theta^\circ$ is negative and roughness increases the negative value of $\cos\theta^\circ$, i.e. roughness results in $\theta > \theta^\circ$. This means that, if $\theta^\circ > 90^\circ$, roughness makes the surface even more difficult to wet.

The influence of surface heterogeneity was analyzed by Cassie and Baxter [127] (as described in Chapter 11). Deryaguin has suggested the possibility of adoption of metastable configurations as a result of surface roughness [120]. He considered the wetting line to move in a series of thermodynamically irreversible jumps from one metastable configuration to the next.

Assuming an idealized rough surface, consisting of concentric patterns of sinusoidal corrugations (Figure 14.18a), one may simply relate the apparent contact angle θ to that on a smooth surface θ° by Eq. (14.120),

$$\theta = \theta^\circ + \alpha \tag{14.73}$$

where α is the slope of the solid surface at the wetting line. The value of θ depends, therefore, on the location of the wetting line and, hence, upon factors such as drop volume and gravitational forces. A model heterogeneous surface may be repre-

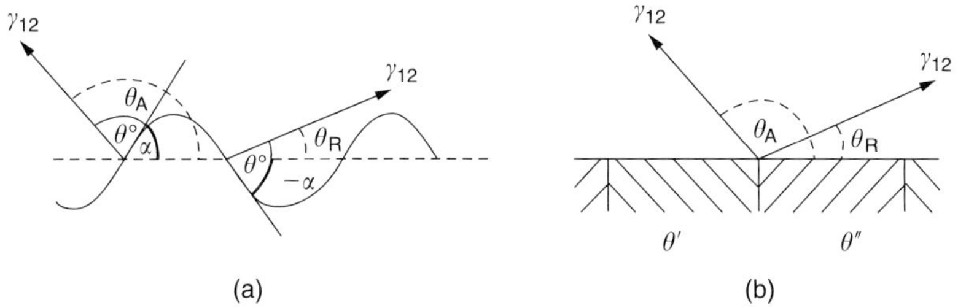

Fig. 14.18. Origin of contact angle hysteresis on model surfaces: (a) rough surface, (b) heterogeneous surface.

sented by a series of concentric bands having alternate characteristic contact angles θ' and θ'', such that $\theta' > \theta''$ (Figure 14.18b). A drop of a liquid placed on this type of a surface will spread or retract until the wetting line assumes some configuration such that $\theta' > \theta > \theta''$.

Despite the above complications, measurement of contact angles of spray liquids on leaf surfaces are still most useful in defining the wetting and spreading of the spray. A very useful index for measurement of spreading of a liquid on a solid surface is Harkin's spreading coefficient, S, which is defined by the change in tension when solid/liquid and liquid/air interfaces are replaced by a solid/air interface. In other words S is the work required to destroy a unit area each of the solid/liquid and liquid/air interfaces while forming a unit area of the solid/air interface, i.e.

$$S = \gamma_{SA} - (\gamma_{SL} + \gamma_{LA}) \tag{14.74}$$

If S is positive, the liquid will usually spread until it completely wets the solid. If S is negative, the liquid will form a non-zero contact angle. This can be clearly shown if Eq. (14.24) is combined with the Young's equation, i.e.

$$S = \gamma_{LV}(\cos\theta - 1) \tag{14.75}$$

Clearly, if $\theta > 0$, S is negative and this implies only partial wetting. In the limit $\theta = 0$, S is zero and this represents the onset of complete wetting. A positive S implies rapid spreading of the liquid on the solid surface. Indeed, by measuring the contact angle only, one can define a spread factor, SF, which is the ratio between the diameter of the area wetted on the leaf, D, and the diameter of the drop d, i.e.

$$\mathrm{SF} = \frac{D}{d} \tag{14.76}$$

Provided θ is not too small ($> 5°$), the spread factor can be calculated from θ, i.e.

$$\mathrm{SF} = \left[\frac{4\sin^3\theta}{(1-\cos\theta)^2(2+\cos\theta)}\right]^{1/3} \tag{14.77}$$

A plot of SF versus θ shows a rapid increase in SF when $\theta < 35°$. The most practical method of measuring the spread factor is to apply drops of known volume using a microapplicator on the leaf surface. By using a tracer material, such as a fluorescent dye, one may be able to measure the spread area directly using for example image analysis. This area can be converted into an equivalent sphere, allowing D to be obtained.

An alternative method of defining wetting and spreading is through measurement of the work of adhesion, W_a, which is the work required to separate a unit area of the solid/liquid interface to leave a unit area each of the liquid/air and solid/air respectively (232), i.e.

$$W_a = (\gamma_{LA} + \gamma_{SA}) - \gamma_{SL} \tag{14.78}$$

Again using the Young's equation, one obtains the following expression for W_a,

$$W_a = \gamma_{LA}(\cos\theta + 1) \tag{14.79}$$

Another useful concept for assessing the wettability of surfaces is that introduced by Zisman and collaborators [121], namely the critical surface tension of wetting, γ_c, that was discussed in detail in Chapter 11. These authors found that, for a given surface and a series of related liquids such as n-alkanes, siloxanes or dialkyl ethers, $\cos\theta$ is a reasonably linear function of γ_{LA}. The surface tension at the point where the line cuts the $\cos\theta = 1$ axis is known as the critical surface tension of wetting, γ_c. It is the surface tension of a liquid that would just spread to give complete wetting.

Several authors have tried to relate the critical surface tension to the solid/liquid interfacial tension, or at least its dispersion component, γ_S^d. This is discussed in detail in Chapter 11. From the above discussion, to enhance the wetting and spreading of liquids on leaf surfaces one clearly needs to lower the contact angle of the droplets. This is usually achieved by the addition of surfactants, which adsorb at various interfaces and modify the local interfacial tension. The general relationship between the change in contact angle with surfactant concentration and adsorption is discussed in Chapter 11. Since most leaf surfaces are non-polar low energy surfaces, an increase in surfactant concentration enhances wetting. This explains why most agrochemical formulations contain high concentrations of surfactants to enhance wetting and spreading. However, as shown below, surfactants play other roles in deposit formation, distribution of the agrochemical on the target surface and enhancement of penetration of the chemical.

Although the role of a surfactant is complex, these materials, sometimes referred to as wetting agents or simply adjuvants, need to be carefully selected to optimise biological efficacy. To date, surfactants are still selected by the formulation chemist on the basis of a trial and error procedure. However, some guidelines may be applied in such selection. As discussed in Chapter 6, the HLB system may be initially applied to choose the most common wetting agents. The latter have HLB numbers between 7 and 9. As discussed in Chapter 2, nonionic surfactants usually have a critical micelle concentration (c.m.c.) that two orders of magnitude lower than that of their ionic counter parts at the same alkyl chain length. Since the limiting value of the surface tension is reached at concentrations above the c.m.c., many nonionic surfactants are clearly more effective as wetting agents since, after dilution of the formulation, the concentration of the nonionic surfactant in the spray solution may be higher than its c.m.c. However, many nonionic surfactants with HLB numbers in the range 7–9 undergo phase separation at high concentrations and/or temperatures. This may limit their incorporation in the formulation at high concentrations. In some cases addition of a small amount of an ionic surfactant may be beneficial in reducing this phase separation and raising the cloud point of the nonionic surfactant. Thus, many agrochemical formulations contain complex mixtures of surfactants that are carefully arrived at by the formulation chemist. The composition of such mixtures is usually kept confidential.

Another important property of the surfactant that is selected for a given agrochemical is its effect on the leaf structure and the cuticle. Surfactants that cause significant damage to the leaf are described as phytotoxic and in many crops such damage must be avoided. This can sometimes limit the choice, since in some cases the best wetter may not be the best from the phytotoxicity point of view and a compromise has to be made. This shows that selecting the surfactant can be difficult and requires careful investigation of many surface chemical properties as well as its interaction with the leaf surface and the cuticle. In addition, its effect on deposit formation and penetration of the agrochemical need to be investigated separately.

14.6.5
Evaporation of Spray Drops and Deposit Formation

The object of spraying is often to leave a long-lasting deposit of particulate fungicide or insecticide or a residue able to penetrate the cuticle in the case of systemic pesticides and herbicides or to be transferred locally within the crop by its own slower evaporation. The form of residue left by evaporation of the carrier liquid depends largely on the rate of evaporation and most importantly on the nature and concentration of surfactant and other ingredients in the formulations. Evaporation from a spray drop tends to occur most rapidly near the edges since these receive the necessary heat most rapidly from the air by conduction through the dry surround of the leaf. This results in a higher concentration of surfactant at the edge, causing surface tension gradients and convection (arising from the associated density difference). Surface tension gradients cause a Marangoni effect (see Chapter 5) with liquid circulation within the drop that causes the particles to be preferentially deposited at the edge. Convection within the drop leads to preferential precipitation near the edge because the particles can first become "wedged" between the solid/liquid and liquid/air interface.

The type and composition of the spray deposit depends to a large extent on the type of formulation as well as the concentration and type of dispersing agent (for suspensions) or emulsifier (for emulsifiable concentrates and emulsions). Additives such as wetters, humectants, stickers also affect the nature of the deposit. Notably, a spray droplet containing dispersed particles or droplets may undergo some physical changes during evaporation. For example, the solid particles of a suspension may undergo recrystallization, forming different shaped particles that will affect the final form of the deposit. Both suspension particles and emulsion droplets may also undergo flocculation, coalescence and Ostwald ripening, all of which affect the nature of the deposit. Following such changes during evaporation is not easy and requires special techniques such as microscopy and differential scanning calorimetry.

Another important factor in deposits is the tenacity of the resulting particles or droplets. Strong adhesion between the particles or droplets and the leaf surface is required to prevent removal of these particles or droplets by the rain. Adhesion forces between a particle or droplet are determined by the van der Waals attraction and the area of contact between the particles and the surface [8]. Several other factors may affect adhesion, namely electrostatic attraction, chemical and hydrogen bonding. The area of contact between the particle and the surface is determined by its size and shape. Obviously, on reducing the particle size of a suspension, one increases the total area of contact between them and the leaf surface, when compared with coarser particles of the same total mass. The shape of the particle also affects the area of contact. For example, flat or cubic-like particles will have larger areas of contact than needle-shaped crystals of equivalent volume. Several other factors may affect adhesion, such as the water solubility of the agrochemical. In general, the lower the solubility the greater the rain fastness.

One of the most important factors that affect deposit formation is the phase separation that occurs during evaporation. As discussed in Chapters 2 and 3, surfactants form liquid crystalline phases when their concentration exceeds a certain value that depends on the nature of the surfactant, its hydrocarbon chain length and the nature and length of the hydrophilic portion of the molecules. During evaporation, liquid crystals of very high viscosity such as hexagonal or cubic phases may be produced at first. Such highly viscous (and elastic structures) will incorporate any particles or droplets and act as reservoirs for the chemical. As a result of solubilization of the chemical, penetration and uptake may be enhanced (see below). With further evaporation, hexagonal and cubic phases may produce lamellar structures with lower viscosity than the former phases. Such structures will affect the distribution of particles or droplets in the deposit. Thus, the choice of a particular surfactant for an agrochemical formulation necessitates study of its phase diagram to identify the nature of the liquid crystalline phases that are produced on increasing its concentration. The effect of temperature on the liquid crystalline structures is also important. Liquid crystalline structures "melt" above a critical temperature, producing liquid phases that contain micelles. These liquid phases have much lower viscosity and hence the particles or droplets of the agrochemical within these liquid phases become mobile. The temperature at which such melting occurs depends on the structure of the surfactant molecule and, hence, the choice depends on the mode of action of the agrochemical and the environmental conditions encountered (such as temperature and humidity). Liquid crystalline structures will be affected by other additives in the formulation, such as the antifreeze and electrolytes. In addition, the particles or droplets of the agrochemical may affect the liquid crystalline structures produced and this requires a detailed study of the phase diagram in the presence of the various additives as well as in the presence of the agrochemical. Various methods may be applied for such investigations, such as polarizing microscopy, differential scanning calorimetry and rheology.

14.6.6 Solubilisation and its Effect on Transport

Solubilization by micelles has been described in detail in Chapter 13, and only a summary is given here. As discussed in Chapter 13, solubilisation is the incorporation of an "insoluble substance (usually referred to as the substrate) into surfactant micelles (the solubilizer). Solubilization may also be referred to as the formation of a thermodynamically stable, isotropic solution of a substance, normally insoluble or slightly soluble in water, by the introduction of an additional amphiphilic component or components. Solubilzation can be determined by measuring the concentration of the chemical that can be incorporated in a surfactant solution while remaining isotropic, as a function of its concentration. At concentrations below the c.m.c., the amount of chemical that can be incorporated in the solution increases slightly above its solubility in water. However, just above the c.m.c., the concentration of the chemical that can be incorporated in the micellar solution increases rapidly with further increasing surfactant concentration. This rapid increase, just above the c.m.c., is usually described as the onset of solubilization. One may differentiate three different locations of the substrate in the micelles (see Chapter 13). The most common location is in the hydrocarbon core of the micelle. This is particularly the case for a lipophilic non-polar molecule, as is the case with most agrochemicals. Alternatively, the substrate may be incorporated between the surfactant chains of the micelle, i.e. by co-micellization. This is sometimes referred to as penetration in the palisade layer, in which one may distinguish between deep and short penetration. The third way of incorporation is by simple adsorption on the surface of the micelle. This is particularly the case with polar compounds.

Several factors affect solubilization, of which the structure of the surfactant and solubilizate, temperature and addition of electrolyte are probably the most important. Generalisation about the manner in which the structural characteristics of the surfactant affect its solubilizing capacity are complicated by the existence of different solubilzation sites within the micelles. For deep penetration within the hydrocarbon core of the micelle, solubilization increases with increasing alkyl chain length of the surfactant. Conversely, if solubilization occurs in the hydrophilic portion of the surfactant molecules, e.g. its poly(ethylene oxide) chain, then the capacity increases with increasing hydrophilic chain length. The solubilizate structure can also play a major role. For example, polarity and polarizability, chain branching, molecular size and shape and structure have various effects. The temperature also has an effect on the extent of micellar solubilization that depends on the structure of the solubilizate and of the surfactant. In most cases, solubilization increases with rising temperature. This is usually due to the increase in solubility of the solubilizate and increase of the micellar size with nonionic ethoxylated surfactants. Addition of electrolytes to ionic surfactants usually causes an increase in the micelle size and a reduction in the c.m.c., and hence an increase in the solubilization capacity. Non-electrolytes that can be incorporated in the micelle, e.g. alcohols, lead to an increase in the micelle size and, hence, to increased solubilization.

As discussed in Chapter 10, microemulsions, which may be considered as swollen micelles, are more effective in solubilization of many agrochemicals. Oil-in-water microemulsions contain a larger hydrocarbon core than surfactant micelles and hence they have a larger capacity for solubilizing lipophilic molecules such as agrochemicals. However, with polar compounds, O/W microemulsions may not be as effective as micelles of ethoxylated surfactants in solubilizing the chemical. Thus, one has to be careful in applying microemulsions without knowledge of the interaction between the agrochemical and the various components of the microemulsion system.

The presence of micelles or microemulsions will have significant effects on the biological efficacy of an insoluble pesticide. In the first instance, surfactants will affect the rate of solution of the chemical. Below the c.m.c., surfactant adsorption can aid wetting of the particles and, consequently, increases the rate of dissolution of the particles or agglomerates [8]. Above the c.m.c., the rate of dissolution is affected as a result of solubilization. According to the Noyes–Whitney relation [122], the rate of dissolution is directly related to the surface area of the particles A and the saturation solubility, C_s, i.e.

$$\frac{dC}{dt} = kA(C_s - C) \tag{14.80}$$

where C is the concentration of the solute.

Higuchi [123] assumed that an equilibrium exists between the solute and solution at the solid/solution interface and that the rate of movement of the solute into the bulk is governed by the diffusion of the free and solubilized solute across a stagnant layer. Thus, the effect of surfactant on the dissolution rate will be related to the dependence of that rate on the diffusion coefficient of the diffusing species and not on their solubilities as suggested by Eq. (14.80). However, experimental results have not confirmed this hypothesis and it was concluded that the effect of solute solubilization involves more steps than a simple effect on the diffusion coefficient. For example, it has been argued that the presence of surfactants may facilitate the transfer of solute molecules from the crystal surface into solution, since the activation energy of this process was found to be lower in the presence of surfactant than its absence in water [8]. Conversely, Chan, Evans and Cussler [124] considered a multi-stage process in which surfactant micelles diffuse to the surface of the crystal, become adsorbed (as hemimicelles) and form mixed micelles with the solubilizate. The latter is dissolved and it diffuses away into bulk solution, removing the solute from the crystal surface. This multi-stage process, which directly involves surfactant micelles, will probably enhance the dissolution rate.

Apart from the above effect on dissolution rate, surfactant micelles also affect the membrane permeability of the solute [8]. Solubilization can, under certain circumstances, help the transport of an insoluble chemical across a membrane. The driving force for transporting the substance through an aqueous system is always the difference in its chemical potential (or to a first approximation the difference in its relative saturation) between the starting point and its destination. The principal steps involved are dissolution, diffusion or convection in bulk liquid and crossing of a membrane. As mentioned above, solubilization will enhance the diffusion rate by affecting transport away from the boundary layer adjacent to the crystal [8]. However, to enhance transport the solution should remain saturated, i.e. excess solid particles must be present since an unsaturated solution has a lower activity.

Diffusion in bulk liquid obeys Fick's first law, i.e.

$$J_D = D\left(\frac{\partial C}{\partial x}\right)$$

where J_D is the flux of solute (amount of solute crossing a unit cross section in unit time), D is the diffusion coefficient and $(\partial C/\partial x)$ is the concentration gradient. The presence of the chemical in a micelle will lower D since the radius of a micelle is obviously greater than that of a single molecule. Since the diffusion coefficient is inversely proportional to the radius of the diffusing particle, D is generally reduced when the molecule is transported by a micelle. Assuming that the volume of the micelle is about 100 times greater than a single molecule, the radius of the micelle will only be about 10 times larger than that of a single molecule. Thus, D will be reduced by about a factor of 10 when the molecule diffuses within a micelle when compared with that of a free molecule. However, the presence of micelles increases the concentration gradient in direct proportion to the increase in incorporation of the chemical by the micelle. This is because Fick's law involves the absolute concentration gradient, which is necessarily small as long as the solubility is small, and not its relative rate. If the saturation is represented by S, Fick's law may be written as

$$J_D = D100S\left(\frac{\partial \%S}{\partial x}\right)$$

where $(\partial\%S/\partial x)$ is the gradient in relative value of S. Eq. (14.50) shows that, for the same gradient of relative saturation, the flux caused by diffusion is directly proportional to saturation. Hence, solubilization will in general increase transport by diffusion, since it can increase the saturation value by many orders of magnitude (outweighing the reduction in D).

Solubilization also increases transport by convection since the flux of this process, J_C, is directly proportional to the velocity of the moving liquid and the concentration of the solute C. Moreover, one would expect that solubilization enhances transport through a membrane [8] by an indirect mechanism. Since solubilization reduces the steps involving diffusion and convection in bulk liquid, it permits application of a greater fraction of the total driving force to transport through the membrane. In this way, solubilization accelerates the transport through the membrane, even if the resistance to this step remains unchanged. However, enhancement of transport as a result of solubilization does not necessarily involve transport of any micelles. The latter are generally too large to pass through membranes.

The above discussion clearly demonstrates the role of surfactant micelles in the transport of agrochemicals. Since the droplets applied to foliage undergo rapid evaporation, the concentration of the surfactant in the spray deposits can reach very high values, which allow considerable solubilization of the agrochemical. This will certainly enhance transport, as discussed above. Since the life time of a micelle is relatively short, usually less than 1 ms (see Chapter 2), such units break up, quickly releasing their contents near the site of action, and produce a large flux by increasing the concentration gradient. However, there have been few systematic investigations to study this effect in more detail and this should certainly be a topic of future research.

14.6.7 Interaction Between Surfactant, Agrochemical and Target Species

In selecting adjuvants that can be used to enhance biological efficacy one has to consider the specific interactions that may take place between the surfactant, agrochemical and target species. This is usually described in terms of an activation process for uptake of the chemical into the plant. This mechanism is particularly important for systemic agrochemicals.

Several key factors may be identified in the uptake activation process: (1) In the spray droplet, (2) in the deposit formed on the leaf surface, (3) in the cuticle before or during penetration and (4) in tissues underlying the site of application. Four main sites were considered by Stock and Holloway [125] for increasing the uptake of the agrochemical into a leaf: (1) Surface of the cuticle' (2) within the cuticle itself, (3) the outer epidermal wall underneath the cuticle and (4) the cell membrane of internal tissues.

The activator surfactant is initially deposited together with the agrochemical and it can penetrate the cuticle, reaching other sites of action and, hence, the role of surfactant in the activation process can be very complex. The net effect of surfactant interactions at any of the sites of action is to enhance the mass transfer of an agrochemical from a solid or liquid phase on the outside of the cuticle to the aqueous phase of the internal tissues of the treated leaf. As discussed above, solubilisation can play a major role in activating the transport of the agrochemical molecules. With many non-polar systemic fungicides, which are mostly applied as suspension concentrates, the presence of micelles can enhance the rate of dissolution of the chemical and this results in increased availability of the molecules. It also leads to an increase in the flux as discussed above.

It has been suggested that cuticular wax can be solubilized by surfactant micelles (by the same mechanism of solubilization of the agrochemical). However, no evidence could be presented (for example using SEM) to show the wax disruption by the micelles. Schönherr [126] suggested that the surfactants interact with the waxes of the cuticle and thus increase the fluidity of this barrier. This hypothesis is sometimes referred to as wax "Plasticization" (similar to the phenomenon of the glass transition temperature reduction of polymers by addition of plasticizers). Some measurements of uptake using surfactants with various molecular weights and HLB numbers offered some support for this hypothesis.

Several other mechanisms have been suggested by Stock and Holloway [125] for uptake activation: (1) Prevention of crystal formation in deposits. It is often assumed that the foliar uptake of an agrochemical from a crystalline deposit will be less than that from an amorphous one. (2) Retention of moisture in deposits by humectant action. The humectant theory has arisen mainly from the observation that the uptake of highly soluble chemicals is promoted by high EO surfactants such as Tween 20. (3) Promotion of uptake of solutions via stomatal infiltration. This hypothesis stemmed from the observation of rapid uptake of agrochemicals (within the first 10 minutes) when using superwetters such as Silwett L-77, which can reduce the surface tension of water to values as low as 20 mN m^{-1}.

References

1 D. Z. Becher: *Encyclopedia of Emulsion Technology*, P. Becher (ed.): Marcel Dekker, New York, 1985, Chapter 4.

2 P. L. Lindner: *Emulsions and Emulsion Technology*, K. J. Lissant (ed.): Marcel Dekker, New York, 1974, 179.

3 K. Meusberger: *Advances in Pesticide Formulation Technology*, H. B. Scher (ed.): 1984, ACS Symp. Ser., 254.

4 C. Hansen, A. Beerbewer: *Kirk-Othmer Encyclopedia of Chemical Technology*, 2nd edition, John Wiley & Sons, New York, 1974.

5 L. Kaertkemeyer, J. Ahmad: *Pesticide Formulation Technology*, Gordon & Breach, New York, 1966, 28, SCI Monograph 21.

6 *Specifications for Pesticides Used in Public Health, 4th edition*, World Health Organization, Geneva, 1973.

7 *CIPAC Handbook*, G. R. Raw (ed.): Heffer, Cambridge, 1970, Volume 1.

8 T. F. Tadros: *Surfactants in Agrochemicals*, Marcel Dekker, New York, 1994.

9 R. W. Behrens, W. C. Griffin, *J. Agric. Food Chem.*, **1953**, *1*, 720.

10 T. Nagakawa: *Nonionic Surfactants*, M. J. Schick (ed.): Marcel Dekker, New York, 1967, 572.

11 F. E. Bailey, J. V. Koleske: *Nonionic Surfactants*, M. J. Schick (ed.): Marcel Dekker, New York, 1967, 794.

12 L. Carino, G. Nagy, *Pest. Sci.*, **1971**, *2*, 23.

13 J. Gad, *Arch. Anat. Physiol.*, **1878**, *181*.

14 G. Quinck, *Weidemanns Ann.*, **1888**, *35*, 593.

15 F. H. Garner, C. W. Nutt, M. F. Mohtadi, *Nature*, **1955**, *175*, 603.

16 J. T. Davies, E. K. Rideal: *Interfacial Phenomena, 2nd edition*, Academic Press, New York, 1963.

17 H. A. Hartung, O. K. Rice, *J. Colloid Interface Sci.*, **1955**, *10*, 436.

18 J. T. Davies, D. A. Haydon: *Proc. Int. Congr. Surface Activity, 2nd edition*. 1961, 417, Volume 1.

19 A. Kaminski, J. W. McBain, *Proc. R. Soc.*, **1949**, *A198*, 447.

20 K. Ogino, M. Ota, *Bull. Chem. Soc. Jpn.*, **1976**, *49*, 1187.

21 K. Ogino, H. Umetsu, *Bull. Chem. Soc. Jpn.*, **1978**, *51*, 1543.

22 J. T. G. Overbeek, *Faraday Discuss. Chem. Soc.*, **1978**, *65*, 7.

23 E. Ruckenstein, J. C. Chi, *J. Chem. Soc. Faraday Trans.*, **1975**, *71*, 1690.

24 G. W. J. Lee, T. F. Tadros, *Colloids Surf.*, **1982**, *5*, 105.

25 G. W. J. Lee, T. F. Tadros, *Colloids Surf.*, **1982**, *5*, 117.

26 G. W. J. Lee, T. F. Tadros, *Colloids Surf.*, **1982**, *5*, 129.

27 L. M. Boize, G. W. J. Lee, T. F. Tadros, *Pesticide Sci.*, **1983**, *14*, 427.

28 J. L. Cayias, R. S. Schechter, W. H. Wade, *ACS Symp. Ser.*, **1975**, *8*, 234.

29 T. F. Tadros, B. Vincent: *Encyclopedoia of Emulsion Technology*, P. Becher (ed.): Marcel Dekker, New York, 1983, 139.

30 B. J. Boffey, R. Collison, A. S. C. Lawrence, *Trans. Faraday Soc.*, **1959**, *55*, 654.

31 M. Tempel van den, *Rec. Trav Chim. Pay. Bas.*, **1953**, *72*, 419.

32 T. F. Tadros, *S.C.I. Monograph Crystal Growth*, A. L. Smith (ed.), *28*, **1973**, 221.

33 W. Ostwald, *Z. Phys. Chem.*, *34*, 493.

34 P. Mukerjee, *Adv. Colloid Interface Sci.*, **1967**, *1*, 214.

35 K. J. Mysels, *Adv. Chem. Ser.*, **1969**, *86*, 24.

36 E. C. Cockbain, J. S. McRoberts, *J. Colloid Sci.*, **1953**, *8*, 440.

37 R. M. Edge, M. Greaves, *J. Chem. Eng. Symp.*, **1967**, *26*, 57.

38 C. Hansen, A. H. Brown, *J. Chem. Eng. Symp.*, **1967**, *26*, 57.

39 T. F. Tadros, B. Vincent: *Encyclopedia of Emulsion Technology*, P. Becher (ed.): Marcel Dekker, New York, 1980, Volume I, Chapter 3.
40 P. Walstra: *Encyclopedia of Emulsion Technology*, P. Becher, eds., Marcel Dekker, New York, 1980, Volume I, Chapter 2.
41 M. van den Tempel, *Proceedings of the 3rd Conf. Surface Activity, Cologne*. **1960**, 573, Volume 2.
42 A. Prins, C. Acuri, M. van der Tempel, *J. Colloid Interface Sci.*, **1967**, *24*, 84.
43 J. Lucassen: *Physical Chemistry of Surfactant Actions*, J. Lucassen, E. H. Reynders (ed.): Marcel Dekker, New York, 1979, Volume 10.
44 J. T. Davies: *Turbulance Phenomenon*, Academic Press, New York, 1972, Chapters 8–10.
45 S. Chandrosekhav: *Hydrodynamics and Hydrodynamic Instability*, Cleeverdon, Oxford, 1961, Chapters 10–12.
46 Lord Rayleigh, *Phil. Mag.*, **1884**, *14*, 184.
47 Lord Rayleigh, *Phil. Mag.*, **1892**, *34*, 145.
48 F. D. Ruscheidt, G. Mason, *J. Colloid Interface Sci.*, **1982**, *17*, 260.
49 W. C. Griffin, *J. Cosmetics Chem.*, **1949**, *1*, 311.
50 W. C. Griffin, *J. Cosmetics Chem.*, **1954**, *5*, 249.
51 K. Shinoda, *J. Colloid Interface Sci.*, **1967**, *25*, 396.
52 K. Shinoda, H. Saito, *J. Colloid Interface Sci.*, **1969**, *34*, 238.
53 K. Shinoda, *Proc. Int. Congr. Surface Activity (5th)*, Butterworths, London, **1968**, 295, Volume 2.
54 A. Beerbower, M. W. Hill, *Am. Cosmet. Pref.*, **1972**, *87(6)*, 85.
55 J. T. Davies: *Proc. Int. Congr. Surface Act. (2nd)*, Butterworths, London, 1959, 426, Volume I.
56 J. T. Davies, E. K. Rideal: *Interfacial Phenomen*, Academic Press, New York, 1963.
57 P. Becher, R. L. Birkmeer, *J. Am. Oil Chem. Soc.*, **1964**, *41*, 169.
58 K. Shinoda, H. Saito, H. Arai, *J. Colloid Interface Sci.*, **1971**, *35*, 624.
59 J. H. M. Scheutjens, G. J. Fleer, B. Vincent, *ACS Symp. Ser.*, **1984**, *240*, 245.
60 H. C. Hamaker, *Physica*, **1937**, *4*, 1058.
61 B. V. Deryaguin, L. Landau, *Acta Physicochem., USSR*, **1949**, *14*, 633.
62 E. J. W. Verwey, J. T. G. Overbeek: *Theory of Stability of Lyophobic Colloids*, Elsevier Science, Amsterdam, 1948.
63 D. H. Napper: *Polymeric Solubilisation of Dispersion*, Academic Press, London, 1983.
64 B. Vincent: *Surfactants*, Th. F. Tadros (ed.): Academic Press, London, 1984.
65 S. S. Davis, A. Smith: *Theory and Practice of Emulsion Technology*, A. L. Smith (ed.): Academic Press, London, 1974, 285.
66 B. V. Deryaguin, E. Obucher, *J. Colloid Chem.*, **1930**, *1*, 385; B. V. Deryaguin, R. C. Scherbaker, *Kolloid Z.*, **1961**, *23*, 33.
67 S. Friberg, P. O. Jansson, E. Cederberg, *J. Colloid Interface Sci.*, **1976**, *55*, 614.
68 T. F. Tadros, *Adv. Colloid Interface Sci.*, **1980**, *12*, 141.
69 T. F. Tadros, *Adv. Colloid Interfce Sci.*, **1990**, *32*, 205.
70 T. F. Tadros, *Pest. Sci.*, **1989**, *26*, 51.
71 Th. F. Tadros: *Solid/Liquid Dispersions*, Th. F. Tadros (ed.): Academic Press, London, 1967.
72 G. D. Parrfitt: *Dispersions of of Powders in Liquids*, Applied Science Publishers, London, 1973.
73 M. Patton: *Paint Flow and Pigment Dispersion*, Interscience, New York, 1964.
74 E. D. Washburn, *Phys. Rev.*, **1921**, *17*, 273, 374; E. K. Rideal, *Phil. Mag.*, **1922**, *44*, 1152.
75 P. A. Rehbinder, *J. Colloid USSR*, **1958**, *20*, 493.
76 P. A. Rehbinder, V. I. Likhtman: *Proc. 2nd Int. Conference Surface Activity*. Butterworths, London, 1957, 157, Volume 3.
77 E. D. Sscukin, P. A. Rehbinder, *Colloid J. USSR*, **1958**, *20*, 601.
78 E. D. Schukin, *Sov. Sci. Rev.*, **1972**, *3*, 157.
79 G. S. Khodakouar, P. A. Rehbinder, *Colloid J. USSR*, **1960**, *223*, 75.
80 G. M. Bartenov, I. V. Iudena, P. A. Rehbinder, *Colloid J. USSR*, **1958**, *20*, 611.
81 D. Heath, R. D. Knott, D. A. Knowles, T. F. Tadros, *ACS Symp. Ser.*, **1984**, *254*, 11.

82 W. Ostwald, *Z. Phys. Chem.*, **1900**, *34*, 493.
83 H. E. Buckley: *Crystal Growth*, John Wiley & Sons, London, 1951.
84 E. V. Khaminski: *Crystallisation from Solution*, Consultants Bureau, New York, 1969.
85 G. H. Nancollas, N. Purdie, *Quarterly Rev.*, **1964**, *18*, 1.
86 Th. F. Tadros: *Particle Growth in Suspensions*, A. L. Smith (ed.): Academic Press, London, 1973.
87 A. D. Maude, R. L. Whitmore, *Br. J. Appl. Phys.*, **1958**, *9*, 477.
88 G. K. Batchelor, *J. Fluid Mech.*, **1972**, *52*, 245.
89 C. C. Reed, J. L. Anderson, *AICHE J.*, **1980**, *26*, 814.
90 R. Buscall, J. W. Goodwin, R. H. Ottewill, T. F. Tadros, *J. Colloid Interface Sci.*, **1982**, *85*, 78.
91 I. M. Krieger, *Adv. Colloid Interface Sci.*, **1971**, *3*, 45.
92 H. Olphen van: *Clay Colloid Chemistry*, John Wiley & Sons, New York, 1963.
93 K. Norrish, *Discuss. Faraday Soc.*, **1954**, *18*, 120.
94 T. F. Tadros, *Colloids Surf.*, **1986**, *18*, 427.
95 T. F. Tadros, *Effect of Polymers on Dispersion Properties*, Th. F. Tadros (ed.): Academic Press, London, 1982.
96 S. Asakura, F. Oosawa, *J. Chem. Phys.*, **1958**, *22*, 1255; *J. Polym. Sci.*, **1958**, *23*, 183.
97 D. Heath, P. K. Thomas, R. W. Warrington, T. F. Tadros, *ACS Symp. Ser.*, **1984**, *254*, 11.
98 R. J. Hunter: *Zeta Potential in Colloid Science*, Academic Press, London, 1981.
99 J. J. Kipling, R. B. Wilson, *J. Appl. Chem.*, **1960**, *10*, 109.
100 I. Danielsson, B. Lindman, *Colloids Surf.*, **1981**, *3*, 391.
101 J. Th. G. Overbeek, P. L. Bruyn de, F. Verhoecks: *Surfactants*, Th. F. Tadros (ed.): Academic Press, London, 1984, 111.
102 J. E. Bowcott, J. H. Schulman, *J. Electrochem.*, **1955**, *54*, 283.
103 J. H. Schulman, W. Stockenius, L. M. Prince, *J. Phys. Chem.*, **1959**, *63*, 1677.
104 L. M. Prince, *J. Colloid Interface Sci.*, **1967**, *23*, 165.
105 K. Shinoda, S. Friberg, *Adv. Colloid Interface Sci.*, **1975**, *4*, 281.
106 H. Saito, K. Shinoda, *J. Colloid Interface Sci.*, **1967**, *24*, 10; **1968**, *26*, 70.
107 K. Shinoda, H. Kunieda, *J. Colloid Interface Sci.*, **1973**, *42*, 381.
108 D. Oakenfull, *J. Chem. Soc., Faraday Trans. I*, **1980**, *76*, 1875.
109 D. J. Mitchell, B. W. Ninham, *J. Chem. Soc., Faraday Trans. II*, **1981**, *77*, 601.
110 J. T. G. Overbeek, *Faraday Discuss. Chem. Soc.*, **1978**, *65*, 7.
111 D. O. Shah, R. M. Hamlin, *Science*, **1971**, *171*, 483.
112 T. P. Hoar, J. H. Schulman, *Nature*, **1943**, *152*, 102.
113 M. Clausse, J. Peyerlesse, C. Boned, J. Heil, L. Nicolas-Margantine, A. Zrabda: *Solution Properties of Surfactants*, K. L. Mittal, B. Lindman, eds., Plenum Press, New York, 1984, 1583, Volume 3.
114 R. C. Baker, A. T. Florence, R. H. Ottewill, T. F. Tadros, *J. Colloid Interface Sci.*, **1984**, *100*, 332.
115 B. Lindman, P. Stilbs, M. E. Moseley, *J. Colloid Interface Sci.*, **1981**, *83*, 569.
116 R. T. Brunskill: *Proceeding of the Third Weed Conference*, Association of British Manufacturers, 1956, 593.
117 G. S. Hartley, R. G. Brunskill: *Surface Phenomena in Chemistry and Biology*, Pergamon Press, New York, 1958, 214–223.
118 J. J. Bikerman, *Ind. Eng. Chem.*, **1941**, *13*, 443.
119 C. G. L. Furmidge, *J. Colloid Interface Sci.*, **1962**, *17*, 309.
120 B. V. Deryaguin, *C. R. Acad. Sci. USSR*, **1946**, *51*, 361.
121 W. A. Zisman, *Adv. Chem. Ser.*, **1964**, *43*.
122 D. Attwood, A. T. Florence: *Surfactant Systems*, Chapman & Hall, London, 1983.
123 W. I. Higuchi, *J. Pharm. Sci.*, **1964**, *53*, 532; **1967**, *56*, 315.
124 A. F. Chan, D. F. Evans, E. L. Cussler, *A. J. Chem.*, **1976**, *22*, 1006.
125 D. Stock, P. J. Holloway et al., *Pestic. Sci.*, **1993**, *37*, 233.
126 J. Schönherr, *Pestic. Sci.*, **1993**, *39*, 213.
127 A. D. B. Cassie, S. Baxter, *Trans. Faraday Soc.*, **1944**, *40*, 546.

15
Surfactants in the Food Industry

15.1
Introduction

Surfactants have been used in the food industry for many centuries. Naturally occurring surfactants such as lecithin from egg yolk and various proteins from milk are used for the preparation of many food products such as mayonnaise, salad creams, dressings, deserts, etc. Later, polar lipids such as monoglycerides were introduced as emulsifiers for food products. More recently, synthetic surfactants such as sorbitan esters and their ethoxylates and sucrose esters have been used in food emulsions. For example, esters of monostearate or mono-oleate with organic carboxylic acids, e.g. citric acid, are used as antispattering agents in margarine for frying.

Many foods are colloidal systems, containing particles of various kinds. The particles may remain as individual units suspended in the medium, but in most cases aggregation takes place to form three-dimensional structures, generally referred to as "gels". These aggregation structures are determined by the interaction forces between the particles that are determined by the relative magnitudes of attractive (van der Waals forces) and repulsive forces. The latter can be electrostatic or steric, depending on the composition of the food formulation. Clearly, the repulsive interactions will be determined by the nature of the surfactant present in the formulation. Such surfactants can be ionic or polar, or they may be polymeric. The latter are sometimes added not only to control the interaction between particles or droplets in the food formulation, but also to control the consistency (rheology) of the system.

Many food formulations contain mixtures of surfactants (emulsifiers) and hydrocolloids. Interaction between the surfactant and polymer molecule plays a major role in the overall interaction between the particles or droplets, as well as the bulk rheology of the whole system. Such interactions are complex and require fundamental studies of their colloidal properties. As discussed later, many food products contain proteins that are used as emulsifiers. Interaction between proteins and hydrocolloids is also very important in determining the interfacial properties and bulk rheology of the system. In addition, the proteins can also interact with the emulsifiers present in the system and this interaction requires particular attention.

Applied Surfactants: Principles and Applications. Tharwat F. Tadros

ISBN: 3-527-30629-3

This chapter, which is by no means exhaustive, focuses on some specific aspects of surfactants in the food industry. Firstly, the interaction between food grade agent surfactants and water will be described, with some highlights on the structure of the liquid crystalline phases. Some examples of the phase diagrams of the monogylceride–water systems are given [1]. This is followed by a section on proteins, which are used in many food emulsions [2]. A brief description of casein micelles and their primary and secondary structures will be given. These systems are widely used in many food products. A sub-section is devoted to interfacial phenomena in food colloids, in particular their dynamic properties and the competitive adsorption of the various components at the interface. The interaction between proteins and polysaccharides in food colloids is described briefly. This is followed by a section on the interaction between polysaccharides and surfactants. Surfactant association structures, microemulsions and emulsions in food are reviewed [3]. Finally, the effect of food surfactants on the interfacial and bulk rheology of food emulsions is described briefly.

The structures of many food emulsions are complex and, often, several phases may exist. Such structures may exist under non-equilibrium conditions and the state of the system may depend largely on the preparation process employed, its prehistory and the conditions to which it is subjected. Unsurprisingly, therefore, fundamental studies on such systems are not easy, and in many cases one is content with some qualitative observations. However, due to the great demand of producing consistent food products and the introduction of new recipes, a great deal of fundamental understanding of the physical chemistry of such complex systems is required.

15.2 Interaction Between Food-grade Surfactants and Water

15.2.1 Liquid Crystalline Structures

Krog et al. [1] have reviewed on this subject, and the reader should refer to this publication for more details. As discussed by these authors, food grade surfactants are, in general, not soluble in water, but they can form association structures in aqueous media that are liquid crystalline in nature. Three main liquid crystalline structures may be distinguished, namely the lamellar, hexagonal and cubic phases. Figure 15.1 shows a model of the crystalline state of a surfactant that forms a lamellar phase (Figure 15.1a). When dispersed in water above its Krafft temperature (T_c) it produces a lamellar mesophase (Figure 15.1b) with a thickness d_a of the bilayer and a thickness d_w of the water layer. The lamellar layer thickness d is simply $d_a + d_w$. These thicknesses can be determined using low-angle X-ray diffraction. The surface area per molecule of surfactant is denoted by S.

The lamellar mesophase can be diluted with water and it has almost infinite swelling capacity, provided the lipid bilayers contain charged molecules and the

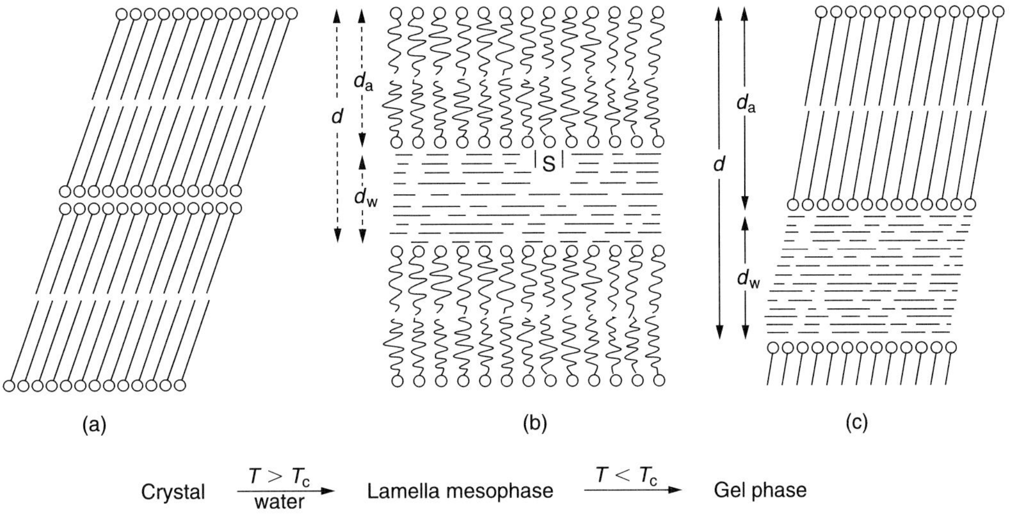

Fig. 15.1. Representation of lamellar liquid crystalline structures.

water phase has a low ion concentration [4]. These diluted lamellar phases may form liposomes (multilamellar vesicles) [5], which are spherical aggregates with internal lamellar structures. Under the polarizing microscope, lamellar structures show an "oil-streaky" texture (see Chapter 3).

When the surfactant solution containing the lamellar phase is cooled below the Krafft temperature of the surfactant, a gel phase is formed (Figure 15.1c). The crystalline structure of the bilayer is now similar to that of the pure surfactant and the aqueous layer with thickness d_w is the continuous phase of the gel.

The hexagonal mesophase structure is periodic in two dimensions and it exists in two modifications, hexagonal I and hexagonal II. The hexagonal I phase consists of cylindrical aggregates of surfactant molecules with the polar head groups oriented towards the outer (continuous) water phase and the surfactant hydrocarbon chains filling out the core of the cylinders. These structures show a fan-like or angular texture under a polarizing microscope (see lecture on concentrated surfactant solutions). The hexagonal II phase consists of cylindrical aggregates of water in a continuous medium of surfactant molecules with the polar head groups oriented towards the water phase and the hydrocarbon chains filling out the exterior between the water cylinders. This phase shows the same angular texture, under the polarizing microscope, as the hexagonal I phase. Whereas the hexagonal I phase can be diluted with water to produce micellar (spherical) solutions, the hexagonal II phase has a limited swelling capacity (usually not more than 40% water in the cylindrical aggregates).

The viscous isotropic cubic phase, which is periodic in three dimensions, is produced with monogylceride–water systems at chain lengths above C_{14}. This isotropic phase has a bicontinuous structure, consisting of a lamellar bilayer, which separates two water channel systems [6, 7]. The cubic phase behaves as a very viscous liquid phase, which can accommodate up to ~40% water.

Of the above liquid crystalline structures, the lamellar phase is the most important for food applications. As we will see later, these lamellar structures are very good stabilizers for food emulsions. In addition, they can be diluted with water, forming liposome dispersions that are easy to handle (pumpable liquids) and they interact with water-soluble components such as amylose in starch particles. Hexagonal and cubic phases, in contrast, when formed give problems in food processing due to their highly viscous nature (viscous particles may block filters).

15.2.2
Binary Phase Diagrams

Figure 15.2 shows typical binary (surfactant + water) phase diagrams of monoglycerides for three molecules with decreasing Krafft temperature (1-monopalmitin, 1-mono-elaidin and 1-mono-olein). With 1-monopalmitin, the dominant mesophase is the lamellar (neat) phase, which swells to a maximum water layer thickness, d_w, of 2.1 nm at 40% water. At higher water content (>60%) a disperse phase is produced in the temperature range 55–68 °C, whereas above 68 °C a cubic phase in equilibrium with water is formed.

With the mono-elaidin–water phase diagram (Figure 15.2b), the lamellar region becomes smaller, whereas the cubic phase region becomes larger, when compared with the monopalmitin–water phase diagram. The temperature at which the lamellar phase is formed (Krafft point) is decreased from 55 to 33 °C. At higher water concentrations (> 40%), the mono-elaidin forms a cubic phase in equilibrium with bulk water.

The mono-olein–water phase diagram (Figure 15c) shows the formation of lamellar liquid crystalline structure at room temperature (20 °C) at water content between 2 and 20%. At higher water concentrations, a cubic phase is formed, which above 40% water exists in equilibrium with water. If the temperature of the cubic phase is increased above 90 °C, a hexagonal II phase is produced.

Distilled monoglycerides from edible fats (lard, tallow or vegetable oils) showed similar mesophase formation to that of the pure monoglycerides. However, the phase regions may differ in size, depending on the purity and fatty acid composition of the commercial monoglycerides. As mentioned above, the continuous swelling of the lamellar phase in the water-rich region of the phase diagram is controlled by the charge of the lipid, which can be obtained by neutralization of the free fatty acid in the monoglyceride (by adding sodium bicarbonate or sodium hydroxide). The formation of charged $RCOO^-$ molecules in the lipid bilayer of the lamellar phase increases swelling by water. This has been confirmed using low-angle X-ray diffraction methods to measure the water thickness. For example, with fully hydrogenised lard (C_{16}/C_{18}), the water layer thickness at 60 °C was found to be 1.6 nm before neutralization, and increased to 1.6 nm after neutralization of the free fatty acids with NaOH. At high water concentration, these neutralized monoglycerides form transparent dispersions (liposomes).

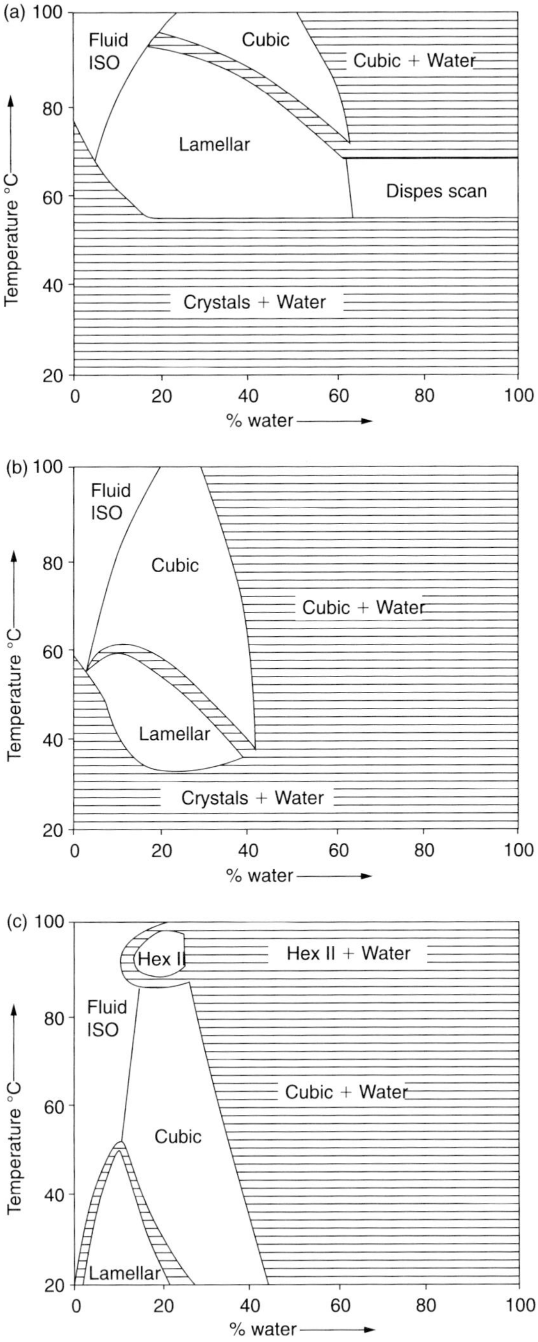

Fig. 15.2. Binary phase diagrams of pure 1-monoglycerides in water: (a) 1-monopalmitin, (b) monoelaidin, (c) mono-olein.

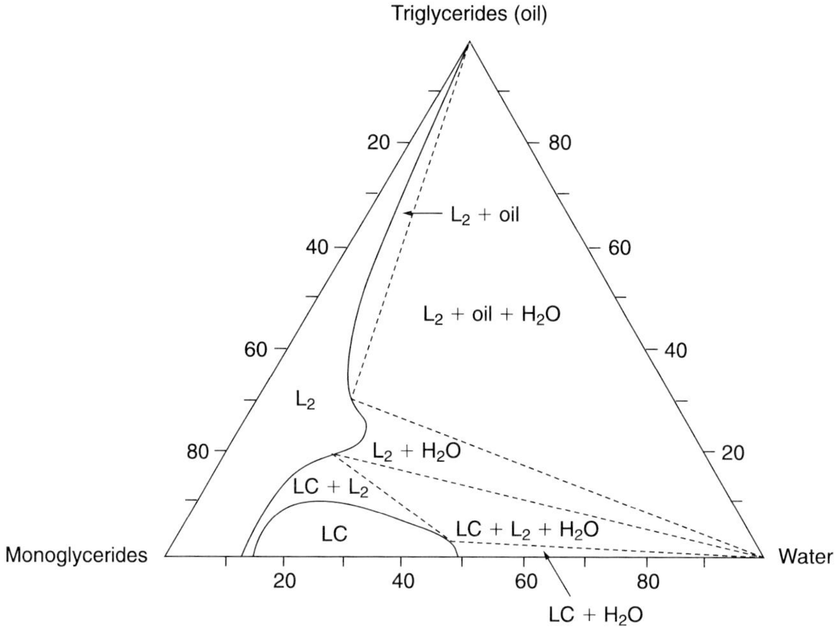

Fig. 15.3. Ternary phase diagram of soybean oil–sunflower oil monoglyceride and water.

15.2.3 Ternary Phase Diagrams

Figure 15.3 shows a typical ternary phase diagram of soybean oil (triglyceride), sunflower oil monoglyceride and water at 25 °C [8]. It clearly shows the LC phase and the inverse micellar (L_2) phase. This inverse micellar phase is relevant to the formation of water-in-oil emulsions. The interfacial tension between the micellar L_2 phase and water is about 1–2 mN m^{-1} and that between the L_2 and oils even lower. The L_2 phase is proposed to form an interfacial film during emulsification, and the droplet size distribution should then be expected to be related mainly to the interfacial and rheological properties of the L_2 phase.

Surface pressure measurements at the air–water interface showed that the lipid molecules begin to associate at low monoglyceride monomer concentration (typically 10^{-6} mol dm^{-3} for mono-olein), forming a cubic structure. Monoglycerides of saturated fatty acids associate to form lamellar liquid crystalline phases at low concentrations. These condensed layers form at the oil–water interface at and above the critical temperature T_c, which is the temperature used for emulsification. These liquid crystalline phases play a major role in emulsion stabilisation (see Chapter 6). As discussed before, lamellar liquid crystalline phases form at the O/W interface, providing a barrier against coalescence. These multilayers significantly reduce the attraction potential and also produce a viscoelastic film with much higher viscosity than that of the oil droplet, i.e. the multilayers produce a form of "mechanical barrier" against coalescence.

15.3 Proteins as Emulsifiers

A protein is a linear chain of amino acids that assumes a three-dimensional shape dictated by the primary sequence of the amino acids in the chain [2]. The side chains of the amino acids play an important role in directing the way in which the protein folds in solution. The hydrophobic (non-polar) side chains avoid interaction with water, while the hydrophilic (polar) side chains seek such interaction. This results in a folded globular structure with the hydrophobic side chains inside and the hydrophilic side chains outside [9]. The final shape of the protein (helix, planar or "random coil") is a product of many interactions, which form a delicate balance [10, 11]. These interactions and structural organisations are briefly discussed below [11].

Three levels of structural organisation have been suggested: (1) Primary structure, referring to the amino acid sequence. (2) Secondary structure, denoting the regular arrangement of the polypeptide back bone. (3) Tertiary structure, as the three-dimensional organisation of globular proteins. A quaternary structure, consisting of the arrangement of aggregates of globular proteins, may also be distinguished. The regular arrangement of the protein polypeptide chain in the secondary structure is determined by the structural restrictions. C–N bonds in the peptide amide groups have a partial double bond character that restricts free rotation about the C–N bond. This influences the formation of secondary structures. The polypeptide backbone forms a linear group, if successive peptide units assume identical relative orientations. The secondary structures are stabilized by hydrogen bonds between peptide amide and carbonyl groups. In the α-helix, the C=O bond is parallel to the helix axis and a straight hydrogen bond is formed with the N–H group, and this is the most stable geometrical arrangement. Interaction of all constituent atoms of the main chain, which are closely packed together, allows the van der Waals attraction to stabilize the helix. This shows that the α-helix is the most abundant secondary structure in proteins. Several other structures may be identified, designated as p-helix, β-sheet, etc.

Proteins are classified according to the secondary structures: α-proteins with α-helix only (e.g. myoglobein), β-proteins mainly with β-sheets (e.g. immunoglobin), $\alpha + \beta$ proteins with α-helix and β-sheet regions that exist apart in the sequence (e.g. lysozome).

The protein structure is stabilized by covalent disulphide bonds and a complexity of non-covalent forces, e.g. electrostatic interactions, hydrogen bonds, hydrophobic interactions and van der Waals forces. Both the average hydrophobicity and the charge frequency (parameter of hydrophobicity) are important in determining the physical properties such as the solubility of the protein. The latter can be expressed as the equilibrium between hydrophilic (protein–solvent) and hydrophobic (protein–protein) interactions.

Protein denaturation can be defined as the change in the native conformation (i.e. in the region of secondary, tertiary and quaternary structure) that takes place without change of the primary structure, i.e. without splitting of the peptide bonds. Complete denaturation may correspond to totally unfolded protein.

When the protein is formed, the structure produced adopts the conformation with the least energy. This structure is referred to as the native or naturated form of the protein. Modification of the amino side chains or their hydrolysis may lead to different conformations. Similarly, addition of molecules that interact with the amino acids may cause conformational changes (denaturation of the protein). Proteins can be denaturated by adsorption at interfaces, as a result of hydrophobic interaction between the internal hydrophobic core and the non-polar surfaces.

Many examples of proteins used in interfacial adsorption studies may be quoted: (1) Small and medium size globular proteins, e.g. those present in milk such as β-lactoglobulin, α-lactoalbumin and serum albumin, and egg white, e.g. lysozyme and ovalalbumin. At pHs below the isoelectric point (4.2–4.5), these proteins associate to form dimers, trimers and higher aggregates. α-Lactoalbumin is stabilized by Ca^{2+} against thermal unfolding. X-ray analysis of lysozyme showed that all charged and polar groups are at the surface, whereas the hydrophobic groups are buried in the interior. Bovine serum albumin (which represents about 5% of whey proteins in bovine milk) forms a triple domain structure that includes three very similar structural domains, each consisting of two large double loops and one small double loop. Below pH 4, the molecule becomes fully uncoiled within the limits of its disulphide bonds. Ovalbumin, the major component of egg white, is a monomeric phosphoglycoprotein with a molecular weight of 43 kDa. During storage of eggs, even at low temperatures, ovalbumin is modified by SH/SS exchange into a variant with greater heat stability, called s-ovalbumin.

Proteins that form micelles, namely caseins, are the major protein fraction in bovine milk (about 80% of the total milk protein). Several components may be identified, namely $\alpha_{s,1}$ and $\alpha_{s,2}$-caseins, β-casein and κ-casein. A proteolytic breakdown product of β-casein is γ-casein. Similar to ovalbumin, caseins are phosphoproteins. Large spherical casein micelles are formed by association of α_s-, β and κ-casein in the presence of free phosphate and calcium ions. Molecules are held together by electrostatic and hydrophobic interactions. The α_s- and β-caseins are surrounded by the flexible hydrophilic κ-casein, which forms the surface layer of the micelle. The high negative charge of the κ-casein prevents collapse of the micelle by electrostatic repulsion. The micelle diameter varies between 50 and 300 nm.

Several oligomeric plant storage proteins can be identified. They are classified according to their sedimentation behaviour in the analytical ultracentrifuge, namely 11 S, 7 S and 2 S proteins. Both 11 S and 7 S proteins are oligomeric globular proteins. The 11 S globulins are composed of six non-covalently linked subunits, each of which contains a disulphide bridged pair of a rather hydrophilic acidic 30–40 kDa α-polypeptide chain and a more hydrophilic basic 20 kDa β-polypeptide chains.

The molar mass and size of the protein as well as its shape depends on the nature of the plant from which it is extracted. These plant proteins can be used as emulsifying and foaming agents.

15.3.1
Interfacial Properties of Proteins at the Liquid/Liquid Interface

Since proteins are used as emulsifying agents for oil-in-water emulsions, it is important to understand their interfacial properties, in particular the structural changes that may occur on adsorption. Protein adsorption layers differ significantly from those of simple surfactant molecules. In the first place, surface denaturation of the protein molecule may take place, resulting in unfolding of the molecule, at least at low surface pressures. Secondly, the partial molar surface area of proteins is large and can vary depending on the conditions for adsorption. The number of configurations of the protein molecule at the interface exceeds that in bulk solution, resulting in a significant increase of the non-ideality of the surface entropy. Thus, one cannot apply thermodynamic analysis, e.g. the Langmuir adsorption isotherm, for protein adsorption. The question of reversibility versus irreversibility of protein adsorption at the liquid interface is still subject to a great deal of controversy. Consequently, protein adsorption is usually described using statistical mechanical models. Scaling theories proposed by de Gennes [12] could also be applied.

One of the most important investigations of protein surface layers is to measure their interfacial rheological properties (e.g. it viscoelastic behaviour). Several techniques can be applied to study such properties of protein layers, e.g. using constant stress (creep) or stress relaxation measurements. At very low protein concentrations, the interfacial layer exhibits Newtonian behaviour, independent of pH and ionic strength. At higher protein concentrations, the extent of surface coverage increases and the interfacial layers exhibit viscoelastic behaviour, revealing features of solid-like phases. Above a critical protein concentration, protein–protein interactions become significant, resulting in the formation of a "two-dimensional" structure. The dynamics of formation of protein layers at the liquid/liquid interface should be considered in detail when one applies protein molecules as stabilizers for emulsions. Several kinetic processes must be considered: solubilisation of non-polar molecules, resulting in the formation of associates in the aqueous phase; diffusion of solutes from bulk solution to the interface; adsorption of the molecules at the interface; orientation of the molecules at the liquid/liquid interface; formation of aggregation structures, etc.

15.3.2
Proteins as Emulsifiers

When a protein is used as an emulsifier it may adopt various conformations, depending on the interaction forces involved. The protein may adopt a folded or unfolded conformation at the oil/water interface. In addition, the protein molecule may interpenetrate in the lipid phase to various degrees. Several layers of proteins may also exist. The protein molecule may bridge one drop interface to another. The actual structure of the protein interfacial layer may be complex, combining any or all of the above possibilities. For these reasons, measurement of protein conformations at various interfaces still remains a difficult task, even when using several techniques such as UV, IR and NMR spectroscopy as well as circular dichroism [13].

At an oil/water interface, it is usually assumed that the protein molecule undergoes some unfolding and that this accounts for the lowering of the interfacial tension on protein adsorption. As mentioned above, multilayers of protein molecules may be produced and one should take into account intermolecular interactions as well as the interaction with the lipid (oil) phase.

Proteins act in a similar way to polymeric stabilizers (steric stabilization). However, molecules with compact structures may precipitate to form small particles that accumulate at the oil/water interface. These particles stabilize the emulsions (sometimes referred to as Pickering emulsions) by a different mechanism. As a result of the partial wetting of the particles by the water and the oil, they remain at the interface. The equilibrium location at the interface provides the stability, since their displacement into the dispersed phase (during coalescence) results in an increase in the wetting energy.

From the above discussion, proteins clearly act as stabilizers for emulsions by different mechanisms depending on their state at the interface. If the protein molecules unfold and form loops and tails they provide stabilization in a similar way to synthetic macromolecules. Conversely, if the protein molecules form globular structures, they may provide a mechanical barrier that prevents coalescence. Finally, precipitated protein particles located at the oil/water interface provide stability as a result of the unfavourable increase in wetting energy on their displacement. Clearly, in all cases, the rheological behaviour of the film plays an important role in the stability of the emulsions (see below).

15.4 Protein–Polysaccharide Interactions in Food Colloids

Proteins and polysaccharides are present in nearly all food colloids [14]. The proteins are used as emulsion and foam stabilizers, whereas the polysaccharide acts as a thickener and also for water-holding. Both proteins and polysaccharides contribute to the structural and textural characteristics of many food colloids through their aggregation and gelation behaviour. Several interactions between proteins and polysaccharides may be distinguished, ranging from repulsive to attractive. Repulsive interactions may arise from excluded volume effects and/or electrostatic interaction. These repulsive interactions tend to be weak, except at very low ionic strength (expanded double layers) or with anionic polysaccharides at pHs above the isoelectric point of the protein (negatively charged molecules). Attractive interaction can be weak or strong and either specific or non-specific. A covalent linkage between protein and polysaccharide represents a specific strong interaction. A non-specific protein–polysaccharide interaction may occur as a result of ionic, dipolar, hydrophobic or hydrogen bonding interaction between groups on the biopolymers. Strong attractive interaction may occur between positively charged protein (at a pH below its isoelectric point) and an anionic polysaccharide. In any particular system, the protein–polysaccharide interaction may change from repulsive to attractive as the temperature or solvent conditions (e.g. pH and ionic strength) change.

Aqueous solutions of proteins and polysaccharides may exhibit phase separation at finite concentrations. Two types of behaviour may be recognised, namely coacervation and incompatibility. Complex coacervation involves spontaneous separation into solvent-rich and solvent-depleted phases. The latter contains the protein–polysaccharide complex, which is caused by non-specific attractive protein–polysaccharide interaction, e.g. opposite charge interaction. Incompatibility is caused by spontaneous separation into two solvent-rich phases, one composed of predominantly protein and the other predominantly polysaccharide. Depending on the interactions, a gel formed from a mixture of two biopolymers may contain a coupled network, an interpenetrating network or a phase-separated network. In food colloids the two most important proteinaceous gelling systems are gelatin and casein micelles. An example of a covalent protein–polysaccharide interaction is that produced when gelatin reacts with propylene glycol alginate under mildly alkaline conditions. Non-covalent, non-specific interaction occurs in mixed gels of gelatin with sodium alginate or low-methoxy pectin. In food emulsions containing protein and polysaccharide, any of the mentioned interactions may take place in the aqueous phase. This results in specific structures with desirable rheological characteristics and enhanced stability. The nature of the protein–polysaccharide interaction affects the surface behaviour of the biopolymers and the aggregation properties of the dispersed droplets.

Weak protein–polysaccharide interactions may be exemplified by a mixture of milk protein (sodium casinate) and a hydrocolloid such as xanthan gum. Sodium casinate acts as the emulsifier and xanthan gum (with a molecular weight in the region of 2×10^6 Da) is widely used as a thickening agent and as a synergistic gelling agent (with locust bean gum). In solution, xanthan gum exhibits pseudoplastic behaviour that is maintained over a wide range of temperature, pH and ionic strength. Xanthan gum at concentrations exceeding 0.1% inhibits creaming of emulsion droplets by producing a gel-like network with a high residual viscosity (see Chapter 6). At lower xanthan gum concentrations ($< 0.1\%$), creaming is enhanced as a result of depletion flocculation. Other hydrocolloids such as carboxymethylcellulose (with a lower molecular weight than xanthan gum) are less effective in reducing the creaming of emulsions.

Covalent protein–polysaccharide conjugates are sometimes used to avoid any flocculation and phase separation that is produced with weak non-specific protein–polysaccharide interactions. An example of such conjugates is that produced with globulin–dextran or bovine serum albumin–dextran. These conjugates produce emulsions with smaller droplets and narrower size distribution and they stabilize the emulsion against creaming and coalescence.

15.5
Polysaccharide–Surfactant Interactions

One of the most important aspects of polymer–surfactant systems is their ability to control stability and rheology over a wide range of composition [14]. Surfactant molecules that bind to a polymer chain generally do so in clusters that closely resemble the micelles formed in the absence of polymer [15]. If the polymer is less polar or contains hydrophobic regions or sites, there is an intimate contact between the micelles and polymer chain. In such a situation, contact between one surfactant micelle and two polymer segments will be favourable. The two segments can be in the same polymer chain or in two different chains, depending on the polymer concentration. For a dilute solution, the two segments can be in the same polymer chain, whereas in more concentrated solutions the two segments can be in two polymer chains with significant chain overlap. The cross-linking of two or more polymer chains can lead to network formation and dramatic rheological effects.

Surfactant–polymer interaction can be treated in different ways, depending on the nature of the polymer. A useful approach is to consider the binding of surfactant to a polymer chain as a co-operative process. The onset of binding is well defined and can be characterised by a critical association concentration (CAC). The latter decreases with increasing alkyl chain length of the surfactant. This implies an effect of polymer on surfactant micellisation. The polymer is considered to stabilize the micelle by short- or long-range (electrostatic) interaction. The main driving force for surfactant self-assembly in polymer–surfactant mixtures is generally the hydrophobic interaction between the alkyl chains of the surfactant molecules. Ionic surfactants often interact significantly with both nonionic and ionic polymers. This can be attributed to the unfavourable contribution to the energetics of micelle formation from the electrostatic effects and their partial elimination due to charge neutralisation or lowering of the charge density. For nonionic surfactants, there is little to gain in forming micelles in the presence of a polymer and, hence, the interaction between nonionic surfactants and polymers is relatively weak. However, if the polymer chain contains hydrophobic segments or groups, e.g. with block copolymers, the hydrophobic polymer–surfactant interaction will be significant.

For hydrophobically modified polymers [such as hydrophobically modified hydroxyethyl cellulose or poly(ethylene oxide)] the interaction between the surfactant micelles and the hydrophobic chains on the polymer can result in the formation of cross-links, i.e. gel formation (Figure 15.4). However, at high surfactant concentrations, there will be more micelles that can interact with the individual polymer chains and the cross-links are broken.

The above interactions are manifested in the variation of viscosity with surfactant concentration. Initially, the viscosity shows an increase with increasing surfactant concentration, reaching a maximum and then decreasing with further increase in surfactant concentration. The maximum is consistent with the for-

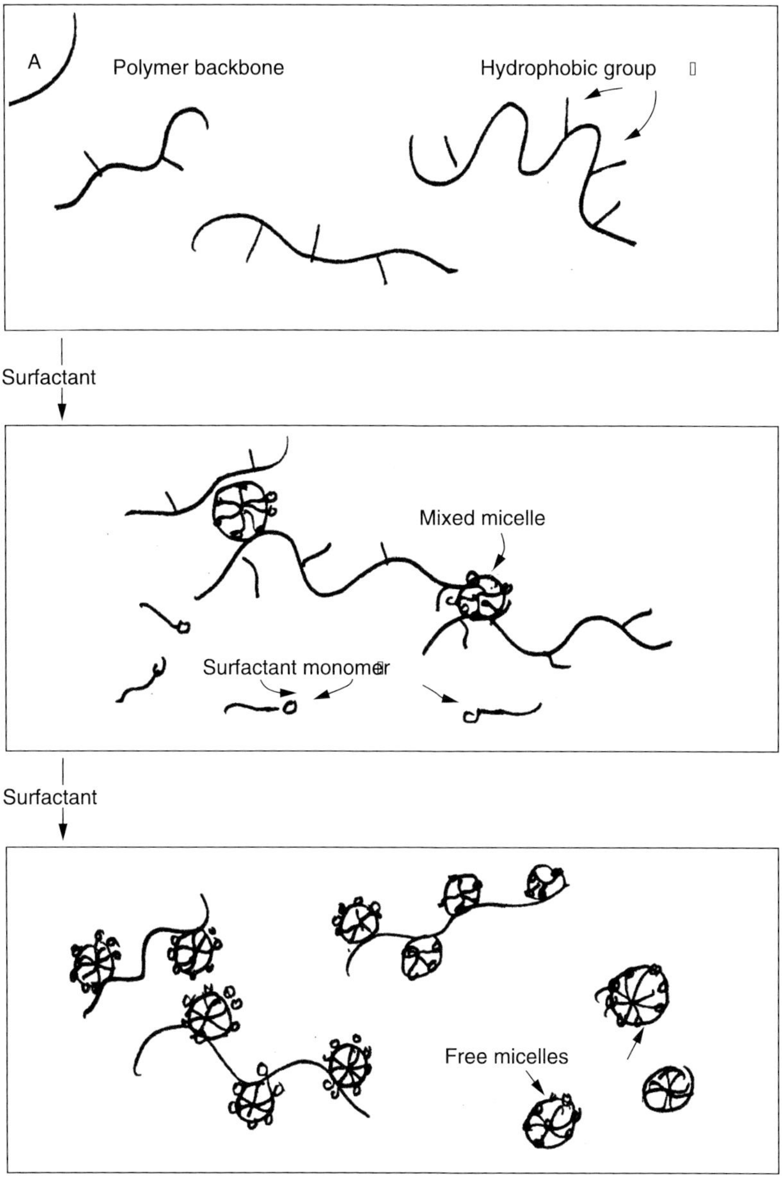

Fig. 15.4. Scheme of interaction between hydrophobically modified polymer chains and surfactant micelles.

mation of cross-links and the subsequent decrease indicates destruction of these cross-links (Figure 15.4).

15.6
Surfactant Association Structures, Microemulsions and Emulsions in Food

A typical phase diagram of a ternary system of water, ionic surfactant and long-chain alcohol (co-surfactant) is shown in Figure 15.5. The aqueous micellar solution A solubilizes some alcohol (spherical normal micelles), whereas the alcohol solution dissolves huge amounts of water, forming inverse micelles, B. These two phases are not in equilibrium, but are separated by a third region, namely the lamellar liquid crystalline phase. These lamellar structures and their equilibrium with the aqueous micellar solution (A) and the inverse micellar solution (B) are the essential elements for both microemulsion and emulsion stability [3].

As discussed in Chapter 10, microemulsions are thermodynamically stable and they form spontaneously (primary droplets a few nms in size), whereas emulsions are not thermodynamically stable since the interfacial free energy is positive and dominates the total free energy. This difference can be related to the difference in bending energy between the two systems [3]. With microemulsions, containing very small droplets, the bending energy (negative contribution) is comparable to the stretching energy (positive contribution) and hence the total surface free

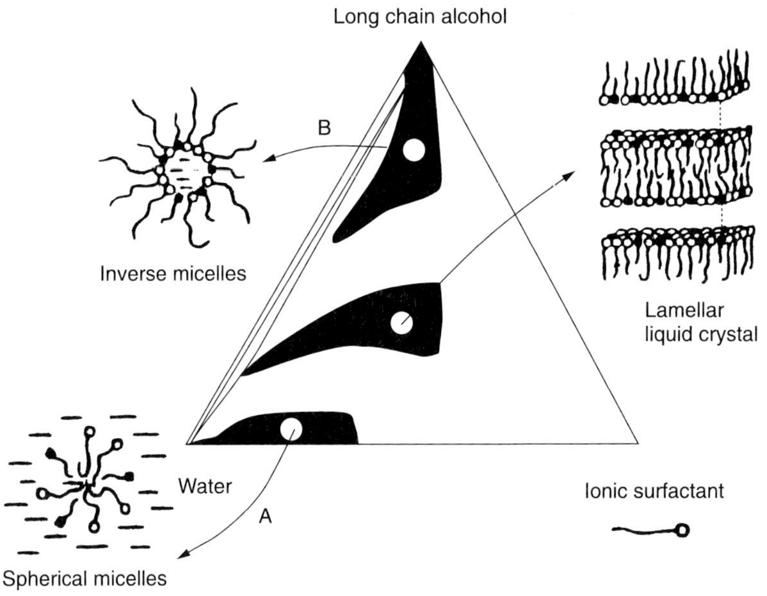

Fig. 15.5. Ternary phase diagram of water, an anionic surfactant and long-chain alcohol (co-surfactant).

energy is extremely small ($\sim 10^{-3}$ mN m^{-1}). With macroemulsions, however, the bending energy is negligible (small curvature of the large emulsion drops) and hence the stretching energy dominates the total surface free energy, which is now large and positive (few mN m^{-1}).

The microemulsion may be related to the micellar solutions A and B shown in Figure 15.4. A W/O microemulsion is obtained by adding a hydrocarbon to the inverse micellar solution B, whereas an O/W microemulsion emanates from the aqueous micellar solution A. These microemulsion regions are in equilibrium with the lamellar liquid crystalline structure. To maximize the microemulsion region, the lamellar phase has to be destabilized, as for example by the addition of a relatively short chain alcohol such as pentanol. In contrast, for a macroemulsion, with its large radius, the parallel packing of the surfactant/co-surfactant is optimal and hence the co-surfactant should be of chain length similar to that of the surfactant.

From the above discussion, a surfactant/co-surfactant combination for a microemulsion is clearly of little use to stabilize an emulsion. This is a disadvantage when a multiple emulsion of the W/O/W type is to be formulated, whereby the W/O system is a microemulsion. This problem has been resolved by Larsson et al. [13], who used a surfactant combination to stabilize the microemulsion and a polymer to stabilize the emulsion.

The formulation of food systems as microemulsions is not easy, since addition of triglycerides to inverse micellar systems results in a phase change to a lamellar liquid crystalline phase. The latter has to be destabilized by other means than adding co-surfactants, which are normally toxic. An alternative approach to destabilize the lamellar phase is to use a hydrotrope, a number of which are allowed in food products.

As discussed above, for emulsion stabilization in food systems lamellar liquid crystalline structures are ideal. At the interface, the liquid crystals serve as a viscous barrier to accept and dissipate the energy of flocculation [16]. Figure 15.6 illustrates this, showing the coalescence process of a droplet covered with a lamellar liquid crystal. It consists of two stages: at first the layers of the liquid crystals are removed two by two and the terminal step is the disruption of the final bilayer of the structure. Initiation of the flocculation process leads to very small energy changes and good stability is assumed as long as the liquid crystal remains adsorbed. This adsorption is the result of its structure. At the interface, the final layer towards the aqueous phase terminates with the polar group, while the layer towards the oil finishes with the methyl layer. In this manner, the interfacial free energy is a minimum.

15.7 Effect of Food Surfactants on the Rheology of Food Emulsions

Surfactants play a major role in the rheology of food emulsions. Both interfacial and bulk rheologies have to be considered and these will be summarized below.

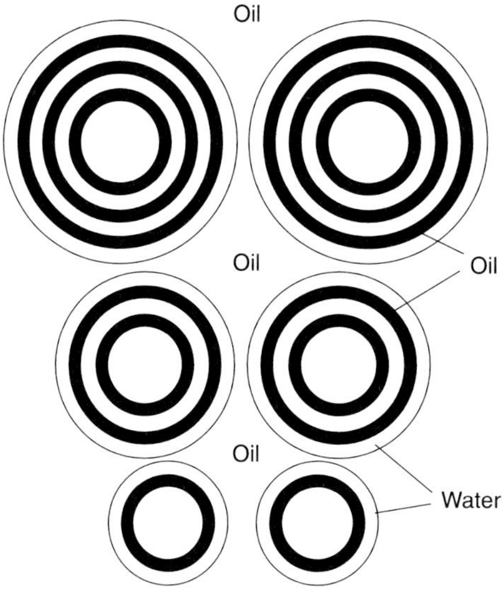

Fig. 15.6. Representation of emulsions containing liquid crystalline structures.

15.7.1
Interfacial Rheology

It has long been argued that interfacial rheology, namely interfacial viscosity and elasticity, play an important role in emulsion stability. This is particularly the case with mixed surfactant films (which may also form liquid crystalline phases) and polymers such as hydrocolloids and proteins that are commonly used in food emulsions. The interfacial viscosity is the ratio between shear stress and shear rate in the plane of the interface, i.e. it is a two-dimensional viscosity (the unit for interfacial viscosity is surface Pa s or surface poise). A liquid/liquid interface has viscosity if the interface itself contributes to the resistance to shear in the plane of the interface [17]. Most surfactants (and mixtures) and macromolecules adsorbed at the interface are viscous, showing a high induced interfacial viscosity. Usually, the interfacial viscosity is higher than the bulk viscosity. The high viscosity of the adsorbed film can be accounted for in terms of the orientation of the molecules at the interface. For example, surfactants orient at the oil/water interface with the hydrophobic portion pointing to (or dissolved in) the oil and the polar group pointing to the aqueous phase. Such films resist compression by a film pressure, π, given by the equation

$$\pi = \gamma_0 - \gamma \qquad (15.1)$$

where γ_0 is the interfacial film of the clean interface, i.e. before adsorption (of the order of 30–50 mN m^{-1}) and γ is the interfacial tension after adsorption, which be as low as a fraction of a mN m^{-1}. Thus, surface pressures of the order of 30–50 mN m^{-1} can be reached. As the film is compressed at the interface, the surfactant molecules become more closely packed and orient more nearly normal to the interface. Such vertically oriented layers resist further compression, producing a high interfacial viscosity. Macromolecular films also give high interfacial viscosity due to their orientation at the interface. Generally, the macromolecule adopts a train-loop-tail configuration (see Chapter 5) and the film resists compression as a result of the lateral repulsion between the loops and/or tails. With proteins, more rigid interfacial films are produced, particularly when these molecules adsorb unfolded and, in such cases, very high surface viscosities are produced.

Interfacial films show both viscosity and elasticity. Films are elastic if they resist deformation in the plane of the interface and if the surface tends to recover its natural shape where the deforming forces are removed [17]. Similar to bulk materials, interfacial elasticity can be measured by static and dynamic methods. Another important interfacial rheological parameter is the dilational elasticity, ε, that is given by

$$\varepsilon = A\left(\frac{d\gamma}{dA}\right) \tag{15.2}$$

where A is the area of the interfacial film and γ is the interfacial tension. The interfacial elasticity can be measured using, for example, a Langmuir trough with two movable barriers.

Several examples may be quoted to illustrate the relationship between interfacial rheology and emulsion stability. The first example is where mixed surfactant films were shown to give more stable films than the individual components. This is, for example, the case when a long-chain alcohol such as lauryl alcohol is mixed with an anionic surfactant such as sodium lauryl sulphate. Although the alcohol is not particularly surface active, its presence at the interface tends to lower the interfacial tension of the adsorbed film of sodium lauryl sulphate. Prins et al. [18] found that ε increased markedly in the presence of the alcohol and, therefore, they attributed the enhanced stability to such high interfacial elasticity. Other authors attributed the enhanced stability to a high interfacial viscosity, although Prins et al. [18] argued against this since they found that the film elasticity was not very sensitive to temperature changes (although of course η_s is) and to the concentration of the alcohol (which had a pronounced effect on η_s).

A second example in which surface rheology was applied to investigate emulsion stability is the work of Biswas and Haydon [19]. These authors systematically investigated the rheological characteristics of various proteins (which are relevant to food emulsions), namely albumin, poly(ϕ-L-lysine) and arabinic acid at the O/W interface, and correlated these measurements with the stability of the oil droplets at a planar O/W interface.

The viscoelastic properties of the adsorbed films were studied using two-dimensional creep and stress relaxation measurements in a specially designed

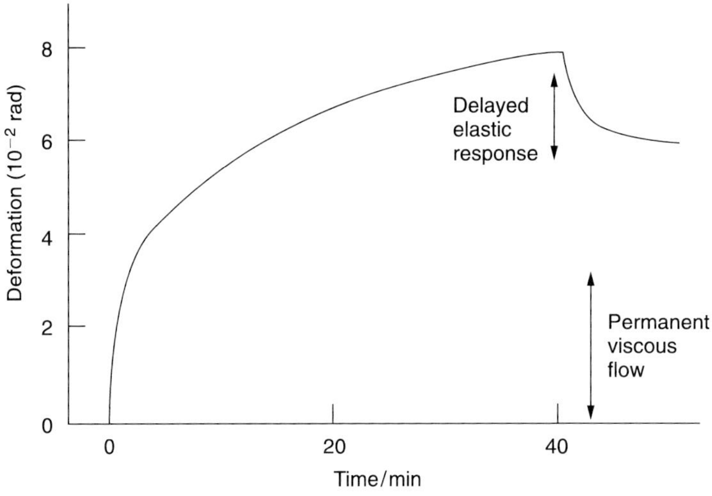

Fig. 15.7. Creep curve of an adsorbed Bovine Serum Albumin film (pH = 5.2) at a petroleum ether–water interface at a constant stress of 0.0116 N m^{-2}.

rheometer. In the creep experiments, a constant torque (expressed in mN m^{-1}) was applied and the resulting deformation γ (in radians) was recorded as function of time. The creep recovery was recorded by following the deformation when the torque was withdrawn. In the stress relaxation experiments, a certain deformation γ was produced in the film by applying an initial strain, and the deformation was kept constant by decreasing the stress.

Figure 15.7 gives a typical creep curve for bovine serum albumin films, showing an initial, instantaneous, deformation, characteristic of an elastic body, followed by a non-linear flow that gradually declines and approaches the steady flow behaviour of a viscous body.

After 30 minutes, when the external force was withdrawn, the film tended to revert to its original state, with an instantaneous recovery followed by a slow one. The original state, however, was not obtained even after 20 h and the film seemed to have undergone some flow. This behaviour illustrates the viscoelastic property of the bovine serum albumin film.

Biswas and Haydon [19] also found a striking effect of the pH on the rigidity of the protein film. Figure 15.8 illustrates this, where the shear modulus G and interfacial viscosity η_s are plotted as a function of pH. The elasticity is seen to have a maximum at the isoelectric point of the protein. Biswas and Haydon [19] then measured the rate of coalescence of petroleum ether drops at a planar O/W interface by measuring the lifetime of a droplet beneath the interface. The half-life of the droplets was plotted against pH, as shown in Figure 15.8, which clearly illustrates the correlation with G or η_s.

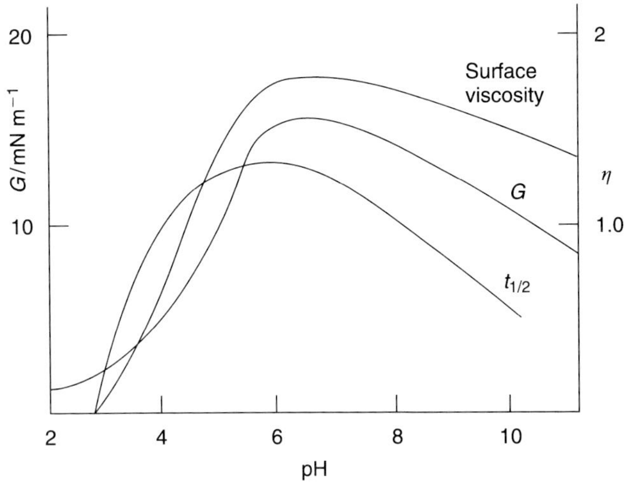

Fig. 15.8. Shear modulus, surface viscosity and half-life of petroleum ether drops beneath a plane petroleum ether–aqueous KCl (0.1 mol dm^{-3}) solution interface.

Biswas and Haydon's [19] results clearly indicate that no significant stabilization occurred with non-viscoelastic films. However, the presence of viscoelasticity was not sufficient to confer stability when drainage of the film was rapid. For example, highly viscoelastic films of bovine serum albumin or pepsin could not stabilize W/O emulsions; the same was found for pectin and gum arabic films. In these cases the drainage of the film was clearly too rapid, even from two rigid films. In fact, as expected, it was only after solvent drainage, and the disperse phases were still separated by a film of high viscosity, that enhanced stability occurred. These investigations concluded that experimental stability to coalescence requires a film of appreciable thickness. In addition, the main part of the film should be located on the continuous phase side of the interface.

15.7.2
Bulk Rheology

Several factors may be quoted that control the bulk rheology of food emulsions, such as the disperse phase volume fraction, the nature of the continuous phase and the presence of ingredients such as thickeners [20]. One important factor that affects the rheology of food emulsions is the presence of "networks" that are produced by the droplets or by the thickeners. These "networks" or "gels" control the consistency of the product and hence its acceptability by the customer. This can be illustrated from the work of van den Tempel [20] and Papenhuizen [21] who studied "gels" consisting of 25% glyceryl stearate in paraffin oil (a model system for margarine). Creep experiments at various stress values showed an increase in strain (shear), γ, under constant stress, τ, with time t. The data could be fitted empirically to an equation of the form

$$\gamma = \frac{\tau}{G_1} + \frac{\tau}{G_2} \log t \tag{15.3}$$

where G_1 and G_2 are the "rapid" and "retarded" elastic moduli, respectively. The results could be explained by postulating two types of bonds between the particles in a network. The primary bonds (crystal bridges) were assumed to remain unbroken, whereas the secondary bonds (assumed to be due to van der Waals bonds) were broken under the influence of a stress and will reform in another relaxed position. The latter process gives rise to a retarded elastic behaviour. Relaxation of the reversible bonds causes an increasing part of the stress to be carried out by the irreversible bonds. Steady-state stress–strain measurements, carried out at low shear rates, showed a rapid increase in stress, reaching a maximum that was followed by a decrease and, subsequently, an equilibrium value at large deformation (Figure 15.9). This behaviour was explained by assuming that the network structure was destroyed to such an extent that only non-interacting aggregates of particles remained. The only effect of the agglomerates was immobilization of the liquid.

The above behaviour at low and larger deformation has been analysed using a network model, in which the particles were assumed to be connected by van der Waals forces. The network was considered to consist of agglomerates of particles connected by chains. This is illustrated in Figure 15.10, in which the network structure is subdivided into small volume elements of characteristic size L, each consisting of one agglomerate. During the deformation process, stretching or tensile forces are applied to the network chain. Such forces will increase the distance between the rheological units (agglomerate or single particle). If this force reaches

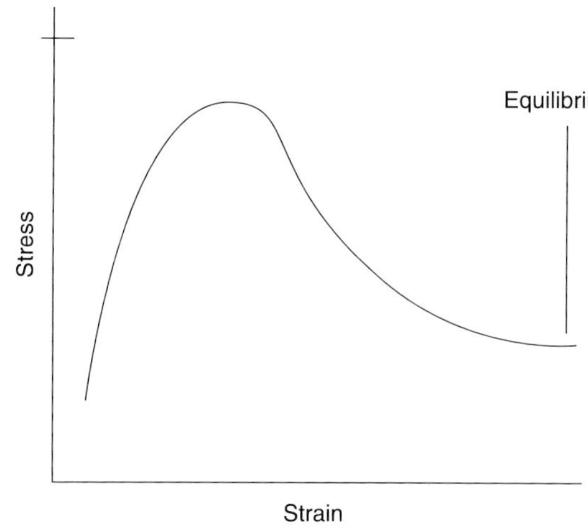

Fig. 15.9. Steady-state stress–strain relationship (at low shear rate).

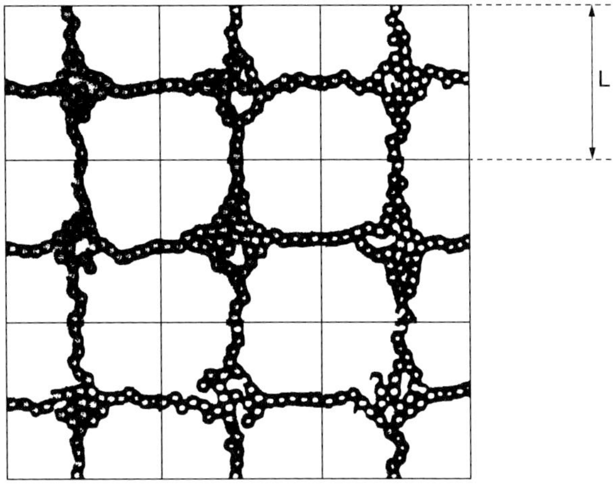

Fig. 15.10. Model of a network structure of a flocculated structure.

a critical value, the bond may break, depending on the time available. However, in large deformation, reformation of the bonds may also occur. This is due to compression, i.e. lateral deformation.

Using the above model, Papenhuizen [21] derived an expression for the viscosity coefficient, η_{I}, resulting from purely hydrodynamic effects, i.e.

$$\eta_{\mathrm{I}} = \frac{\eta_0 \phi a}{H} \tag{15.4}$$

where η_0 is the viscosity of the medium, ϕ is the volume fraction, a is the radius of the particles (assumed to be spherical) and H is the distance between two spheres.

Papenhuizen [21] derived an expression for the viscosity coefficient, η_{II}, resulting from the presence of an agglomerate. He considered the force required to move an agglomerate consisting of a large number of particles through a stationary viscous medium at a certain speed. Such a flow problem is similar to determining the velocity of a viscous liquid flowing through a stationary porous plug under the influence of a pressure gradient, e.g. using D'Arcy's law [20] and the Kozney–Carman equation [22, 23]. Proceeding in this manner, the following expression for η_{II} was derived,

$$\eta_{\mathrm{II}} = \frac{CS^2}{2^{1/2}} \left(\frac{\phi}{1 - \phi} \right)^2 \eta_0 L^2 \tag{15.5}$$

where C is a constant, that is equal to 5 for spheres, S is the surface area that is equal to $3/a$ for spheres. Equation (15.5) shows that η_{II} depends on S and hence on particle size. Large particles have small S resulting in a low η_{II}, whereas small particles give rise to a large η_{II}. The latter is also proportional to the square of the volume fraction of the disperse phase. This shows the importance of particle size and volume fraction in controlling the viscosity (consistency) of a food emulsion system.

15.7.3 Rheology of Microgel Dispersions

Many food colloids are thickened with elastic micro-networks of polymeric materials, e.g. gelatinised starch granules. The rheology of these systems is determined by particle swelling and deformability. Evans and Lips [24] developed a theory for the elasticity of microgel dispersions, which was tested using dispersions of Sephadex particles (spherical cross-linked dextran moieties). However, when using non-retrograded starch dispersions, deviation from theoretical predictions was obtained. This was attributed to the presence of solubilised amylose. The effect of addition of dextran on the elasticity of Sephadex dispersions was also investigated. The results could be explained by polymer particle bridging or depletion flocculation. However, it was concluded that bridging is unlikely since Sephadex and dextran are chemically similar. Thus, addition of dextran to the dispersion was assumed to cause depletion flocculation, which provides an attractive component to the pair potential.

15.7.4 Food Rheology and Mouthfeel

As mentioned above, food systems are complex multiphase products that may contain dispersed components such as solid particles, liquid droplets or gas bubbles. The continuous phase may also contain colloidally dispersed macromolecules such as polysaccharides, protein and lipids. These systems are non-Newtonian, showing complex rheology, usually plastic or pseudo-plastic (shear thinning). Complex structural units are produced as a result of the interaction between the particles of the disperse phase as well as by interaction with polymers that are added to control the properties of the system, such as its creaming or sedimentation as well as the flow characteristics. The control of rheology is important not only during processing but also for control of texture and sensory perception.

Well-defined rheological experiments are essential for adequate investigation of food rheology. These experiments fall into two main categories, namely steady-state shear stress–shear rate measurements, and the possible time effects (thixotropy), and low-deformation measurements of constant (creep) and dynamic (oscillatory) stress. During the flow process, both viscous (shear and normal) and inertial stresses act on the fluid matrix. Flow stresses tend to impede or influence the interactions of the structural components. Above a critical stress, flow-induced structuring may occur. The structural states may be reversible or irreversible. These structural changes influence the rheological behaviour of the fluid system and, consequently, the flow process itself is affected.

The above structural changes can have a significant effect on the technical performance of the food product. Problems of creaming or sedimentation and phase separation are directly related to the rheological characteristics. It is, therefore, crucial to control the rheology of the food product to avoid problems during manufacture, during storage and sensory perception of the product.

The sensory perception of food texture is significantly dependent on the structure of the system (e.g. the nature of the three-dimensional units produced and the nature of the "gel" produced in the system) as well as its rheological behaviour. In a multiphase food product, such as an oil-in-water emulsion that contains surfactants for emulsification and polysaccharides that are added to reduce creaming, it is essential to relate the structure of the system to its rheology. This allows one to define the quality of the product in terms of its sensorial function (texture and consistency) as well as its technical function such as flow, dosing and storage stability.

To achieve the above objectives, it is essential to understand the colloid-chemical properties of the system as well as its flow characteristics under various conditions. Many food products (e.g. yoghurt) can be compared with the microstructure of particulate gels. The structure is formed from a continuous colloidal network, which holds the product together and gives rise to its characteristic properties. A colloidal network can be formed from particles linked together forming strands, enveloping pores and/or droplets, inclusions, etc. The size and shape of the particles, strands and pores may vary, thus creating different product properties.

During mastication, the structure breaks down and the sensory perception of the texture reflects such breakdown processes. Various subjective tests for sensory evaluation are used, e.g. manual texture (touching) by a light pressure with forefinger, visual texture, and mouthfeel during manipulation of the sample in the mouth. To relate the rheological characteristics of the product to the above sensory evaluation, experiments must be carried out under various deformation conditions.

The basic principles of rheology and the various experimental methods that can be applied to investigate these complex systems of food colloids have been discussed in detail in Chapter 7. Only a brief summary is given here. Two main types of measurements are required: (1) Steady-state measurements of the shear stress versus shear rate relationship, to distinguish between the various responses: Newtonian, plastic, pseudo-plastic and dilatant. Particular attention should be given to time effects during flow (thixotropy and negative thixotropy). (2) Viscoelastic behaviour, stress relaxation, constant stress (creep) and oscillatory measurements.

In steady-state measurements, one applies a constant and increasing shear rate, $\dot{\gamma}$ (s^{-1}), on the sample (which may be placed in concentric cylinder, cone and plate or parallel plate platens) and the stress σ (Pa) is simultaneously measured. For Newtonian systems, the stress increases linearly with increasing shear rate and the slope of the shear stress–shear rate curve gives the Newtonian viscosity η (which is independent of the applied shear rate),

$$\sigma = \eta\dot{\gamma} \tag{15.6}$$

For a non-Newtonian system, as is the case with most food colloids, the stress–shear rate gives a pseudo-plastic curve (see Chapter 7) and the system is shear thinning, i.e. the viscosity decreases with increasing shear rate. In most cases the shear stress–shear rate curve can be fitted with the Herschel–Buckley equation (see Chapter 7),

$$\sigma = \sigma_\beta + k\dot{\gamma}^n \tag{15.7}$$

where σ_β is the yield stress (that gives a measure of the "structure" in the system, e.g. its gel strength), k is the consistency index and n is the shear thinning index.

By fitting the experimental data to the above equation, one can obtain σ_β, k and n. The viscosity at any shear rate can then be calculated,

$$\eta = \frac{\sigma}{\dot{\gamma}} = \frac{\sigma_\beta + k\dot{\gamma}^n}{\gamma} \tag{15.8}$$

Most food colloids show reversible time dependence of viscosity, i.e. thixotropy. If the system is sheared at any constant shear rate for a certain period of time, the viscosity shows a gradual decrease with increasing time. When the shear is removed, the viscosity returns to its initial value. This phenomenon can be understood by considering the structure of the multiphase food colloid that contains particles and/or droplets, surfactants, hydrocolloids, etc. On shearing the sample, this structure is "broken down". When the shear is removed, the structure recovers within a certain time scale that depends on the sample. Thixotropy is investigated by applying sequences of shear stress–shear rates within well-defined time periods. If the shear rate is applied within a short period, e.g. increasing from 0 to 500 s^{-1} in one minute, then, when reducing the shear rate from 500 to 0 s^{-1}, the structure of the sample cannot be recovered within this time scale. In this case, the shear stress–shear rate curves (the up and down curves) show large hysteresis, i.e. a large thixotropic loop is produced. By increasing the time of shear (say 5 minutes for the up curve and 5 minutes for the down curve), the loop closes. In this way one can investigate the thixotropy of the sample.

In constant stress (creep) measurements, one applies the stress (that is kept constant at each measurement) in small increasing increments. If the stress applied is below the yield stress, the system behaves as a viscoelastic solid. In this case, the strain shows a small increase at zero time and this strain remains virtually constant over the duration of the experiment (near zero shear rate). When the stress is removed, the strain returns back to zero. This behaviour will be the same at increasing stress values, provided the applied stress is still below the yield stress. Any increase in stress will be accompanied by an increase in strain at zero time. However, when the stress exceeds the yield stress, the system behaves as a viscoelastic liquid. In this case, the strain rapidly increases at zero time, giving a rapid elastic response characterised by an instantaneous compliance J_0 (the compliance is simply the ratio between the strain and applied stress, Pa^{-1}). At time larger than zero, the strain shows a gradual and slow increase with time. This is the region of retarded response (bonds are broken and reformed at different rates). Ultimately, the system shows a steady state (with constant shear rate), whereby the compliance increases linearly with time. The slope of this linear portion gives the reciprocal viscosity at the applied shear stress (slope $= J/t = \text{Pa}^{-1}\ \text{s}^{-1} = 1/\text{Pa s} = 1/\eta_s$). After the steady state is reached, the stress is then removed and the system shows partial recovery, i.e. the strain changes sign and decreases with time, reaching an equilibrium value. Creep curves are analysed to obtain the residual (zero shear) viscosity, i.e. the plateau value at low stresses (below the yield stress) and the critical stress σ_{cr} above which the viscosity shows a rapid decrease with further decrease in stress. This critical stress may be denoted as the "true yield value". In addition, by fitting the compliance–time curves to models, one can also obtain the relaxation time of the sample (see Chapter 7).

In dynamic (oscillatory) measurements, one applies a sinusoidal strain or stress (with amplitudes γ_0 or σ_0 and frequency ω in rad s^{-1}) and the stress or strain is measured simultaneously. For a viscoelastic system, the stress oscillates with the same frequency as the strain, but out of phase. From the time shift of stress and strain, one can calculate the phase angle shift δ. This allows one to obtain the various viscoelastic parameters: G^* (the complex modulus), G' (the storage modulus, i.e. the elastic component of the complex modulus) and G'' (the loss modulus or the viscous component of the complex modulus). These viscoelastic parameters are measured as a function of strain amplitude (at constant frequency) to obtain the linear viscoelastic region, whereby G^*, G' and G'' are independent of the applied strain until a critical strain γ_{cr}, above which G^* and G' begin to decrease with further increase of strain, whereas G'' shows an increase. Below γ_{cr} the structure of the system is not broken down, whereas above γ_{cr} the structure begins to break. From G' and γ_{cr} one can obtain the cohesive energy density of the structure E_c (Chapter 7). The viscoelastic parameters are then measured as a function of frequency at constant strain (that is kept within the linear viscoelastic region). For a viscoelastic liquid, G^* and G' increase with increasing frequency and, ultimately, both values reach a plateau that becomes independent of frequency. G'' shows an increase with increasing frequency, reaching a maximum at a characteristic frequency ω^* and then it decreases with further increase of frequency, reaching almost zero at high frequency (in the region of the plateau region of G'). From ω^* one can calculate the relaxation time of the sample ($t_{relaxation} = 1/\omega^*$).

The above measurements are essential before one can relate in detail the rheology to sensory evaluation, e.g. mouth feel, which is discussed below.

15.7.5 Mouth Feel of Foods – Role of Rheology

Food products are generally designed with an optimum "consistency" for application in cutting, slicing, spreading or mixing. During eating and mastication the food loses its initial "consistency", at least partially. The mouthfeel of food products may be related to the loss of this initial "consistency". During the first stage of mastication, the food is comminuted by the action of the teeth into particles (few mm in size). At this stage, the food is close to its initial "consistency".

Thus, in the first stages of mastication, the mouthfeel may be related to rheological characteristics. It is, therefore, possible to relate the mouthfeel during the first stages of mastication to the rheological parameters such as "yield value", "creep compliance", "storage modulus", etc. After the initial stages of comminution, the food particles "soften" as a result of temperature rise and moisture uptake in the oral cavity. This significantly reduces the "consistency", which may reach values of stresses comparable to the level encountered by the saliva flow in the oral cavity. When these stresses are reached, the food particles will be broken down into a much smaller size that is determined by the hydrodynamics of the "flowing" saliva. The flow in the saliva is rather complex and calculation of shear stresses is not straightforward.

When the above stage is reached, the food product will form a "homogeneous" mix with the saliva, and the mouthfeel will appear smooth. Clearly, if the "consistency" of the product does not decrease to a sufficient degree (such that the stresses are comparable to those encountered by the saliva flow), the masticated food will remain "thicker" and the mouthfeel becomes unacceptable to the consumer (feel of "graininess", "stickiness" or "waxiness"). Control of the "consistency" (rheological characteristics) of food products is essential for consumer acceptability and this may require sophisticated measurements and interpretation of the results obtained.

The reduction in size of food products during mastication controls the flavour release. Assuming the particles produced are spherical, the time required for release is directly proportional to the square of the radius of the particles R (which is a measure of the surface area),

$$t \approx \frac{R^2}{D} \tag{15.9}$$

where D is the diffusion coefficient of the flavour molecule, which is inversely proportional to the viscosity of the medium (D is of the order of 10^{-9} m^2 s^{-1} in dilute aqueous foods and can be as low as 10^{-11} m^2 s^{-1} in fat foods).

To achieve adequate release of food flavours, R has to be reduced to ~70 μm for aqueous foods and to much smaller sizes for fat continuous foods. The breakup of food products in the saliva is determined by the balance of two forces: (1) Hydrodynamic forces exerted by the saliva flow, which will deform the food produce. (2) Interfacial forces and rheological properties of the food product that resist the deformation.

To investigate the breakup of food products during mastication one needs to know (1) The stress exerted by the saliva flow. (2) The interfacial tension between the food material and saliva, relevant to both non-aqueous and fat continuous products. (3) The rheological properties of the food products.

The relationship between the above forces and the droplets size of the product is known exactly for Newtonian liquids (e.g. oils). The breakup of Newtonian fluids in purely elongational flow is the simplest to analyse. Each element of volume is being stretched without rotation of the direction of stretching. If the direction of stretching is not fixed, but rotates, then in simple "shear flow" the rate of rotation of the axis of stretching and the rate of stretching are equal.

Using the above assumptions it is possible to predict the droplet diameter of Newtonian oils during breakup by the flow in the saliva. In elongational flow, the stress σ_c acting on each drop is approximately equal to the stress in the continuous phase ($\eta_c\dot{\gamma}$, where η_c is the fluid viscosity and $\dot{\gamma}$ is the shear rate),

$$\sigma_c = \eta_c\dot{\gamma} \tag{15.10}$$

The interfacial tension γ resists the deformation (i.e. it tries to keep the spherical symmetry of the drops) and this effect can be accounted for by means of a "Young's modulus", E, equivalent to the Laplace pressure,

$$E = \frac{2\gamma}{R} \tag{15.11}$$

The degree of deformation of the drop, ε_d, is the ratio between σ_c and E, i.e.

$$\varepsilon_d = \frac{\sigma_c}{E} = \frac{\eta_c\dot{\gamma}R}{2\gamma} \tag{15.12}$$

When drop elongation exceeds a certain value, the drop breaks up into smaller drops; ε_d is related to the capillary number Ω,

$$\Omega = \frac{\eta_c\dot{\gamma}d}{\gamma} \tag{15.13}$$

where d is the droplet diameter. (Note that $\Omega = 4\varepsilon_d$).

Using Eqs. (15.12) and (15.13) one can obtain the droplet diameter from a knowledge of the stress acting on each drop (in elongational flow) and the interfacial tension at the oil/saliva interface. Alternatively, one can measure the droplet diameter of the oil drops produced in the saliva and, from a knowledge of the viscosity of the saliva and the interfacial tension of the oil/saliva interface, estimate the stress in the flowing saliva. This is illustrated below.

15.7.6
Break-up of Newtonian Liquids

The break-up of Newtonian liquids with various viscosities η_d can be investigated by mastication of small oil samples and measuring the resulting droplet size distribution, using a Coulter Counter or a Master sizer. Samples are expectorated into a suitable surfactant solution, e.g. Tween (to prevent coalescence during measurements). η_d can be measured at 37 °C (body temperature) using a suitable rheometer (e.g. Haake-Rotovisco). The interfacial tension γ at the oil/saliva interface can be measured using the Wilhelmy plate method. A typical result for an oil/saliva interface is ~15 mN m^{-1}. Interfacial tension between oil and saliva can be systematically reduced by dissolving various amounts of lecithin in the oil phase. To calculate the capillary number one needs to know σ_c. Initially, σ_c may be given an assumed value, say 1 Pa. The viscosity of saliva can be measured using the Haake and this is about 50 mPa s.

Experimental results using the above assumed σ_c can be compared with the literature value for elongational flow. The measured *d*s are ~50 times lower than the literature value, meaning that the actual saliva stress in the mastication process is ~50 Pa. Under shear flow, there is a rapid increase in capillary number when $\eta_d/\eta_c > 1$.

15.7.7
Break-up of Non-Newtonian Liquids

Food products are usually non-Newtonian and they may be approximated by Bingham fluids,

$$\sigma = 2\sigma_\beta + \eta_b\dot{\gamma} \tag{15.14}$$

$2\sigma_B$ = yield stress in elongation (assumed to twice the yield stress in shear flow); η_{pl} = Bingham plastic viscosity.

"Soft" foods, e.g. salad dressing and yoghurts, show a Bingham-like consistency at room temperature. More "solid" foods, e.g. fat spreads, cheese and puddings, become more liquid-like during mastication (melting and moisture uptake) – the "yield stress" may decrease by several orders of magnitude during mastication. A "Bingham fluid" will only break-up when the stress exerted in the saliva (~50 Pa) exceeds the yield stress of the food product. This means that the break-up of food products with a "yield stress" greater than ~50 Pa is difficult in the oral cavity.

An example of a "model" food product with varying "yield stress" is W/O emulsions that can be prepared by emulsification of water in an oil such as ricinoleic acid or soya oil using an emulsifier with low HLB number such as polyglycerol ester. The yield stress of the resulting W/O emulsions can be systematically increased by increasing the water phase volume fraction, ϕ. The ratio of water to emulsifier should be kept constant in the above system. When $\phi = 0.6$, the emulsion is nearly Newtonian ($\sigma_B = 0$) and it becomes gradually more non-Newtonian as the water volume fraction increases, i.e. σ_B increases with increase in ϕ and may exceed 50 Pa when $\phi > 0.6$. During mastication, all emulsions show large drops, but "Newtonian" emulsions with $\phi < 0.6$ showed a much larger number of small drops than did non-Newtonian emulsions.

The above investigations, using droplet size analysis and microscopy investigations can be used to study the effect of rheology on the "break-up" of non-Newtonian food products. It also allows one to study the mouthfeel, using panels, and some correlations between rheology and mouth feel may be obtained.

15.7.8
Complexity of Flow in the Oral Cavity

Flow in the oral cavity is not a "steady" flow and hence the break-up process is not simple. Break-up in the oral cavity can only occur when this flow is maintained long enough, longer than the relaxation time of the drops. For most viscous oils ($\eta_d \sim 6$ Pa s and $\eta \sim 15$ mN m^{-1}), the drop relaxation time is $\sim 5 \times 10^{-3}$ s, giving an ultimate drop size of ~20 μm. A range of 200–2000 μm is initially produced with relaxation times of 5×10^{-2}–5×10^{-1} s, respectively. Since these large drops break-up, the elongational flow remains steady for such periods of time. When one considers how the jaws and the tongue drive the saliva flow, one must conclude that the flow cannot be kept steady for much longer times. The limited duration of elongational flow in the oral cavity is more important for food products showing visco-elastic behaviour at large degrees of deformation, e.g. for products containing thickeners such as hydrocolloids. Many food products contain hydrocolloids such as xanthan gum, which is added for physical stability reasons and also to control the consistency of the product. In the presence of other food materials that increase the hydrodynamic stresses on the material of interest (e.g. bread) the drops produced could be much smaller.

15.7.9
Rheology–Texture Relationship

During any flow process, whether during manufacture or during mastication of the food product, the flow stress influences the "structure" of the system, which in turn affects its rheological characteristics. Sensory perception and the mouthfeel depend to a large extent on the structure of the system (e.g. its "gel" behaviour) as well as its response to the stresses exerted by flowing saliva in the oral cavity. Using colloid and interfacial methods to study the "structure" and various rheological methods to assess the response of the food material to various shear regimes allows one to obtain a "texture"–rheology relationship.

A good example to consider is oil-in-water (O/W) emulsions such as mayonnaise or sauces, which can be prepared using an industrial dispersing process. By controlling the energy input one can control the droplet size of the emulsion. These emulsions are usually "structured" by addition of emulsifier/"thickener" combinations such as proteins/polysaccharides. In laminar flow, the stresses acting in the gap of a dispersing process device are dominated by the viscous shear stress σ (viscosity × shear rate). For turbulent flow (which is the case for most dispersing devices) the so-called Reynold stress σ_R is the dominant factor. A critical shear stress σ_{crit} has to be exceeded for droplet break-up, i.e.

$$\sigma_{cr} = \frac{W_e \gamma}{d} \tag{15.15}$$

where W_e is the critical Weber number that is a function of the ratio of the viscosity of the disperse phase and that of the continuous medium,

$$W_e = f\left(\frac{\eta_d}{\eta_c}\right) \tag{15.16}$$

where η_d is the viscosity of the disperse phase, η_c is the viscosity of the continuous medium; γ is the interfacial tension, and d is the droplet diameter.

An O/W emulsion of mayonnaise (using, for example, sunflower oil) can be prepared at various oil weight fractions, e.g. 0.14, 0.65 and 0.85, using an emulsifier such as modified starch. The droplet size distribution of the resulting emulsions can be measured using a Coulter counter or Malvern Master sizer (based on measurement of the light diffraction by the droplets). The texture of the mayonnaise can be assessed according to "spoonability" and mouth feel (using panels). Various rheological methods may be applied as discussed above.

Using the above emulsion systems, in many cases the mean droplet size was found to decrease with increasing volume energy input E_v (J m^{-3}). In some cases, the mean droplet size showed an increase, after the initial increase, with increasing E_v. This could be due to emulsion droplet coalescence when E_v exceeded a critical value. Comparison of the various rheological results showed that the "structural" changes produced are determined by the elastic modulus G'. G' was measured at low strains (in the linear viscoelastic region) and at a frequency of 1 Hz. G' is an elastic parameter and hence it reflects the interdroplet interaction as well as any interaction with the thickener. Since G' is measured at low deformation, it causes a "minimum" change in the structure of the system during measurement. An increase in G' reflects an increase in interaction. For example, for O/W emulsions without any thickener, a decrease in droplet size increases the number of "contact" points between the emulsion droplets and this leads to an increase in G'. Any reduction in G' with increasing E_v (which leads to a decrease in droplet size) implies a reduction in the "networking" properties (produced, for example, by the emulsifier). In cases where G' increases with increasing E_v (particularly for high oil phase volume fraction) an increase in "network" stability is implied.

There seems to be a correlation between the sensorial texture parameter ("thickness" as measured by the spoon test) and the rheological parameters G' (the storage modulus, the elastic component) and G'' (the loss modulus, the viscous component). One of the most useful parameters to measure is tan δ,

$$\tan\delta = \frac{G''}{G'} \tag{15.17}$$

The reciprocal of tan δ is referred to as the dynamic Weisenberg number W_i',

$$W_i' = \frac{1}{\tan\delta} = \frac{G'}{G''} \tag{15.18}$$

W_i' is a measure of the relative magnitudes of the elastic to the viscous moduli. Many food products such as yoghurt, egg products, etc. can be compared with the microstructure of particulate gels. The structure is formed from a continuous colloidal network, which holds the product together and gives rise to its characteristic properties. A gel network structure can be produced from particles linked together, forming strands, enveloping pores and/or droplets. During mastication the gel structure breaks down and the new "structure" formed is perceived as "texture". An example of gel networks is protein gels formed, for example, from lactoglobulin. Several physical methods may be applied to characterise the gel produced. Image analysis and transmission electron microscopy could be applied to obtain the average pore size and particle size of the gel formed. Several rheological methods may be applied to study the properties of these gels: (1) Large deformation measurements, e.g. tensile tested by fracturing the sample using an Instron. (2) Viscoelastic measurements (low deformation measurements) to obtain the storage and the loss modulus as well as the phase angle shift δ. Low deformation measurements can be used to obtain quantitative information on the structure of the gel formed, for example the number of "cross-links", the gel rigidity and its behaviour under low deformation.

Sensory tests carried out by panels (subjective tests) include manual texture measurement using light pressure with a forefinger, visual evaluation of the texture produced in a newly cut surface and oral texture (mouthfeel):

- Manual texture, soft – resistance to light pressure by finger; springy – recovery of shape after light pressure.
- Visual texture, surface moisture – water released from a newly cut surface; grainy – of a newly cut surface.
- Oral texture, gritty – during chewing; sticky – adherence to teeth after chewing; falling apart – during chewing.

The perceived texture shows a non-linear dependence on the "microstructure". Gels formed at faster heating rates (12 °C min^{-1}) were more difficult to fracture than gels formed at slower heating rates (1 °C min^{-1}). Gels formed at high heating rates have smaller pores; higher resistance to falling apart. The perception of "soft" and "springy" is related to the strand characteristic of the gel. Gels formed at slower heating rates (1 °C min^{-1}) have higher G' than those produced at higher heating rates (12 °C min^{-1}). Gels formed at 1 °C min^{-1} have stiff strands formed of many particles joined together (resulting in higher G'). Gels formed of flexible strands have lower G'. The strand characteristics can explain the gel texture as assessed by viscoelastic measurements.

To analyse the texture of gels one can perform two tests: (1) Destructive (Instron test). This gives a measure of the overall network dimensions. (2) Non-destructive (viscoelastic measurements). The measured G's are sensitive to the strand characteristics, which can be evaluated using microscopy. These measurements are carried out on gels produced under various conditions, such as different heating rates, to arrive at the desired properties.

In conclusion, a combination of microscopy, sensory analysis and rheological properties (obtained under high and low deformation) using statistical evaluation methods can provide a correlation between sensory perception (as evaluated by expert panels) and the various characteristics of the gel. The relationship between microstructure and texture is important in optimizing the properties of food products as well as in the development of new products with desirable properties. Modern techniques of microscopy (such as freeze–fracture) can be applied to study the microstructure of gels. The viscoelastic properties of gels, which can be studied using oscillatory techniques (under various conditions of applied strain and frequency), can be correlated with the microstructure.

15.8
Practical Applications of Food Colloids

Processed foods are often colloidal systems such as suspensions, emulsions and foams [1]. Examples of food emulsions, which are the most commonly used products, are milk, cream, butter, ice cream, margarine, mayonnaise and salad dressings. Emulsions are also prepared as an intermediate step in many food processing items, e.g. powdered toppings, coffee whiteners and cake mixes. These systems are dried emulsions that are re-formed into the emulsion state by the consumer.

Milk and cream are oil-in-water (O/W) emulsions consisting of fat droplets (triglycerides partially crystalline and liquid oils) typically in the size range 1–10 μm. The fat content of milk is 3–4% by volume, and that of cream is 10–30% by volume. The aqueous disperse medium contains milk proteins, salts and minerals. The fat droplets are stabilized by lipoprotein, phospholipids and adsorbed casein. This produces a very stable system against coalescence, as a result of steric stabilisation and the presence of a viscoelastic film at the O/W interface. The only instability process in milk is creaming, since the gravity force exerted by the droplets exceeds the Brownian diffusion (see Chapter 6). This problem of creaming is eliminated by homogenisation of the milk using a high-pressure homogeniser. This reduces the droplet size to the submicron range and the gravity force becomes smaller than the Brownian diffusion.

Ice-cream is an O/W emulsion that is aerated to form a foam. The disperse phase consists of butterfat (cream) or vegetable fat, partially crystallised fat. The volume fraction of air in the foam is approximately 50%. The continuous phase consists of water and ice crystals, milk protein and carbohydrates, e.g. sucrose or corn syrup. Approximately 85% of the water content is frozen at −20 °C. The foam structure is stabilized by agglomerated fat globules that form the surface of air cells in the foam. Added surfactants act as "destabilizers", controlling the agglomeration of the fat globules. The continuous phase is semi-solid and its structure is complex.

Both butter and margarine are W/O emulsions with the water droplets dispersed in a semi-solid fat phase containing fat crystals and liquid oil. With butter, the fat is partially crystallised triglycerides and liquid oil. Genuine milk fat globules are also present. The water droplets are distributed in a semi-solid plastic continuous fat phase. With margarine the continuous phase consists of edible fats and oils, partially hydrogenated, of animal or vegetable origin. Dispersed water droplets are fixed in a semi-solid matrix of fat crystals. Surfactants are added to reduce the interfacial tension, so as to promote emulsification during processing. Considerable energy is needed to reduce the size of the dispersed phase droplets during preparation of the W/O emulsion. Once the emulsion is produced, the whole system is chilled to enable the final emulsification and crystallisation of the fat phase. The initial emulsion need not to be very stable since, by cooling, the water droplets become fixed in a semi-solid fat phase.

In the early development of margarine, egg yolk was first used as the emulsifier, since this contains lecithin and other phospholipids. Later, lipophilic emulsifiers such as mono-diglycerides of long-chain fatty acids (C_{16}–C_{18}) were used in combination with soybean lecithin. The emulsifiers produce water droplets in the size range 2–4 μm. The consistency of margarine is strongly related to the amount of crystalline fat (solid fat content, SFC), which can be determined using dilatometry or low-resolution NMR spectroscopy. The solid fat content of margarine is in the range 5–25% at 20 °C. It is desirable to use fat blends that form small needle-shaped β' crystals (about 1 μm long), which impart good plasticity. One should avoid transformation of these small needle shaped β' crystals into the large β crystals during storage. This results in undesirable grainy consistency ("sandiness"). Crystal morphology may be controlled by using sorbitan esters and their ethoxylates, ethoxylated fatty alcohols, citric acid esters of monoglycerides, diacetyl tartaric acid esters of monoglycerides, sucrose monostearate, sodium stearoyl lactylate and polyglycerol esters of fatty acids. Sorbitan monostearate and citric acid esters of monoglycerides are most effective in preventing the crystallisation of tristearin from the α to the β form. However, when used in emulsions, the surfactants become adsorbed at the O/W interface and only lipophilic surfactants with high oil solubility can act as crystal growth inhibitors.

Low calorie margarine contains at least 50% water, 40% fat and the balance being protein milks, salts, flavour, vitamins and emulsifiers (mainly monoglycerides and soybean lecithin). Some products are based on milk fat or a combination with vegetable fats. With such a high water content, a stable interfacial film is required. Saturated monoglycerides are superior to unsaturated monoglycerides in stabilising the water droplets due to the formation of a liquid crystalline films at the W/O interface.

An important class of O/W emulsions in the food industry is mayonnaise and salad dressings. Mayonnaise is a semi-solid O/W emulsion made from a minimum of 65% edible vegetable oil, with acidfying ingredients, e.g. vinegar and egg yolk phosphatides, as the emulsifying agent. The high volume fraction of oil does not favour the formation of O/W emulsion and it is necessary to disperse the egg yolk in the water phase before addition of the oil phase. Colloid mills and other homogenisers must be used with care in order not to produce too small oil droplets (with high surface area), whereby the emulsifier content is not sufficient to cover the whole interface.

The main difference between mayonnaise and salad dressing is the oil content, which is lower in the latter. Thickening agents such as starch, cereal flour or hydrocolloids may be used. Egg yolk is the main emulsifying agent, but other food grade surfactants may also be used, e.g. polysorbates or esters of monoglycerides. Addition of salt can enhance the emulsion stability as a result of its effect on the protein conformation.

Several other food emulsions can be quoted, such as coffee whiteners and cake emulsions. Coffee whiteners are O/W emulsions containing vegetable oils and fats covering the size range 1–5 μm and an oil volume fraction of 10–15%. The aqueous continuous phase contains proteins, e.g. sodium casinate, carbohydrates, e.g. maltodextrin, salts and hydrocolloids. The emulsifying system consists of blends of nonionic and anionic surfactants with adsorbed protein.

Cake emulsions are very complex systems of fats or oil in an aqueous phase containing flour, sugar, eggs and micro ingredients. The mix is aerated during the mixing process and then further processed by baking. In many cake emulsions the air bubbles formed during mixing are located in the fat phase instead of the water phase. This is the case with high-ratio cakes that may contain 15–25% plastic shortenings or margarine, based on total batter weight. Fat-free cakes or high-ratio cakes made with liquid vegetable oils are aerated in the aqueous phase and the foam stability is provided by egg yolk and added surfactants.

To obtain a satisfactory appearance, volume and texture, the shortening or margarine must have special properties with regard to the solid fat content and plasticity. Shortening containing fat crystals in the β' form are ideal for entrapping and stabilising the air cells. Unless egg yolk is present in the batter, the air cells in a fat particle tend to coalesce within the fat particles rather than be transferred as individual air cells in the aqueous phase. By heating during the baking process, the air cells are greatly enlarged by thermal expansion and by uptake of carbon dioxide from leavening agents and generated water vapour. At this point, the surface elasticity properties of the layers surrounding the air cells are very important. At the end of the baking process, the air cells become connected in an open network and the liquid fat droplets coalesce into a film that covers the inner surface of the air channels.

Surfactants play a major role in both fatless and fat-containing cakes. The types of surfactants commonly used are monoglycerides, polyglycerol esters, propylene glycol esters of fatty acids and polysorbates. These surfactants act as emulsifiers for the fat by reducing the interfacial tension, thus aiding the dispersion of the fat phase. Plastic shortenings may contain 6–10% lipophilic surfactants such as monoglycerides, or propylene glycol esters of fatty acids. These surfactants have no influence on the air/fat surface tension. Fat-based aeration is, therefore, highly dependent on the plasticity of the fat phase, which is controlled by the type of fats and surfactants used.

Surfactants such as monoglycerides may also interact with the starch fraction of the batter and form an insoluble amylose complex. This reduces gelatinisation in the cakes, resulting in a better cake structure with improved tenderness. In fat-free cakes, special surfactant preparations in gel form or α-crystalline powder forms are often used as aerating agents. Monoglycerides of palmitic and stearic acids form liquid crystalline mesophases in cakes containing corn oil. These monoglycerides encapsulate oil droplets at 94 °C by multilayer sheets. At higher temperatures, transition of monoglycerides from lamellar to cubic phases enhances the viscosity, and this plays an important role in stabilising the sponge cake batter during baking.

References

1 N. J. Krog, T. H. Riisom: *Encyclopedia of Emulsion Technology*, P. Becher (ed.): Marcel Dekker, New York, 1985, 321–365, Volume 2.
2 E. N. Jaynes: *Encyclopedia of Emulsion Technology*, P. Becher (ed.): Marcel Dekker, New York, 1985, 367–384, Volume 2.
3 S. E. Friberg, I. Kayali: *Microemulsions and Emulsions in Food*, M. El-Nokaly, D. Cornell (ed.): 1991, 448, ACS Symp. Ser., no. 448.
4 N. Krog, A. P. Borup, *J. Sci. Food Agric.*, **1973**, *24*, 691.
5 V. Luzzati: *Biological Membranes*, D. Chapman (ed.): Academic Press, New York, 1968, 71.
6 G. Lindblom, K. Larsson, L. Johansson, K. Fontell, S. Forsen, *J. Am. Chem. Soc.*, **1979**, *101*, 5465.
7 K. Larsson, K. Fontell, N. Krog, *Chem. Phys. Lipids*, **1980**, *27*, 321.
8 E. Pilman, E. Tonberg, K. Lartsson, *J. Dispersion Sci. Technol.*, **1982**, *3*, 335.
9 H. Mierovitch, H. A. Scheraga, *Macromolecules*, **1980**, *13*, 1406.
10 C. Tanford, *Adv. Protein Chem.*, **1970**, *24*, 1.
11 *Proteins at Liquid Interfaces*, D. Mobius, R. Miller (eds.): Elsevier Science, Amsterdam, 1998.
12 P. G. de Gennes: *Scaling Concepts in Polymer Physics*, Corenell University Press, Ithaca, New York, 1979.
13 K. Larsson, *J. Dispersion Sci. Technol.*, **1980**, *1*, 267.
14 E. Dickinson, P. Walstra: *Food Colloids and Polymers: Stability and Mechanical Properties*, E. Dickinson, P. Walstra (eds.): Royal Society of Chemistry, Cambridge, 1993.
15 *Polymer-Surfactant Interaction*, E. D. Goddard, K. P. Ananthapadmanqabhan (eds.): CRC Press, Boca Raton, FL, 1992.
16 P. O. Jansson, S. E. Friberg, *Mol. Cryst. Liq. Cryst.*, **1976**, *34*, 75.
17 D. W. Criddle: *Rheology, Theory and Applications*, F. R. Eirich (ed.): 1960, 429, Volume 3, Chapter 11, Academic Press, New York.
18 A. Prins, C. Arcuri, M. Tempel van den, *J. Colloid Interface Sci.*, **1967**, *24*, 811.
19 B. Biswas, D. A. Haydon, *Proc. R. Soc.*, **1963**, *A271*, 296.
20 M. Tempel van den, *Rheol. Acta*, **1958**, *1*, 115; *J. Colloid Sci.*, **1961**, *16*, 284.
21 J. M. P. Papenhuizen, *Rheol. Acta*, **1972**, *11*, 73.
22 H. D'Archy, *Les Fantaines Publique de la Vill de Dijon*, Paris, 1961.
23 P. C. Carmen, *Trans. Inst. Chem. Eng.*, **1937**, *15*, 150.
24 I. D. Evans, A. Lips, *Food Colloids and Polymers: Stability and Mechanical Properties*, E. Dickinson and P. Walstra (eds.), 1993, p. 214, Royal Society of Chemistry, Cambridge.

Subject Index

Applied Surfactants: Principles and Applications. Tharwat F. Tadros

ISBN: 3-527-30629-3